Ajayi, J F Ade

History of West Africa

DATE DUE			

History of West Africa

VOLUME TWO

Edited by
J. F. A. Ajayi and
Michael Crowder

Columbia University Press
New York
1973

Published in 1974 in Great Britain by Longman Group Limited

Library of Congress Catalog Card Number: 79–173985

ISBN: 0–231–03737–6 *Cloth*
0–231–03738–4 *Paperbound*

Printed in Great Britain

Contents

Acknowledgements

The publishers are grateful to the following for permission to reproduce photographs:
The Trustees of the British Museum, pp. 32, 37, 51; The London Library, p. 41; Establissement Cinematographique et Photographique des Armées, p. 494; Keystone Press, Paris, p. 602; Documentation Françaises, pp. 619, 674; Ghana Information Services, p. 631; Paul Popper, p. 691(a); Camera Press (photographer Ritchie), p. 691(b); Camera Press (Photographer David Lomax), p. 691(c); p. 691(d).

The publishers are also grateful to:
Stanford University Press for permission to adapt Map 10.1 from *The Staple Food Economies of Western Tropical Africa* by Bruce F. Johnston; The R.W.A.F.F. Officers' Association for permission to adapt Maps 13.1 and 13.2 from *The History of the Royal West Africa Frontier Force* by Haywood and Clark.

Notes on contributors

R. A. Adeleye, PH.D. (Ibadan), is Senior Lecturer in History at the University of Ibadan and Commissioner for Economic Development in the Kwara State Government of Nigeria. He specialises in Islamic movements in the Western Sudan, and in addition to articles in various journals he has published *Power and Diplomacy in Northern Nigeria 1800–1906* (1971).

A. E. Afigbo, B.A. (London), PH.D. (Ibadan), is Senior Lecturer in the History Department at the University of Nigeria. Previously he was a lecturer at the University of Ibadan. His publications include *The Warrant Chiefs* (1972) and contributions to many learned journals and books on West African History.

I. O. Aluko, PH.D. (London), is a Research Fellow in the Institute of Administration, University of Ife and lectures in International Affairs. He has recently completed a manuscript on *Ghana and Nigeria 1957–70: A Study in Inter-African Discord* and is at present editing a book on *Foreign Policies of the African States.*

A. A. Boahen, B.A., PH.D. (London), is Professor of History at the University of Ghana. His publications include *Britain, the Sahara and the Western Sudan* (1964), *Topics in West African History* (1966) and joint authorship (with J. B. Webster) of *The Growth of African Civilisation: The Revolutionary Years. West Africa since 1800* (1967).

L. Brenner, B.S. (Wisconsin), M.A., PH.D. (Columbia), is Assistant Professor of History and Research Associate at the African Studies Center, Boston University. He has published *The Shehus of Kukawa: A History of the Al Kanemi Dynasty of Bornu* (1973). His primary interest is the history of Islamic West Africa. He has done research on the history of Bornu and is now doing research on the history of Sufism in West Africa.

R. Cohen, B.A. (Toronto), Ph.D. (Wisconsin), is Visiting Professor at the Department of Sociology, Ahmadu Bello University. He first carried out research among the Kanuri of Bornu from 1955 to 1957; this has formed the basis for his interest in African Studies and has resulted in a number of published papers and books on African social change, marriage, history and local politics. His publications include *The Kanuri of Bornu* and *Dominance and Defiance* (1971). He has edited *Power in Complex Societies in Africa* and, with John Middleton, *Comparative Political Systems: Studies in the Politics of Pre-industrial Societies* and *From Tribe to Nation in Africa.* He is at present working on a manuscript entitled *The Origins of Polity.*

J. E. Flint, Ph.D. (London), is Professor of History at Dalhousie University. His publications include *Sir George Goldie and the Making of Nigeria* (1960), *Nigeria and Ghana* (1966) and contributions to the Oxford *History of East Africa,* Volumes 1 and 2. He is at present editing Volume V of the *Cambridge History of Africa.*

C. Fyfe, M.A. (Oxford), is Reader in African History at the University of Edinburgh. His publications include *A History of Sierra Leone* (1962), *A Short History of Sierra Leone* (1962), *Sierra Leone Inheritance* (1964) and *Africanus Horton* (1972).

J. D. Hargreaves, M.A. (Manchester), is Professor of History at the University of Aberdeen. His publications include *A Life of Sir Samuel Lewis* (1958), *Prelude to the Partition of West Africa* (1963), *West Africa: the Former French States* (1967) and *France and West Africa* (1969).

A. B. Jones, B.A. (Cuttington College) M.A. Ph.D. (Northwestern), is a member of the UNESCO Committee for the drafting of a General History of Africa and Director of Research with the Department of Education, Monrovia. She specialises in Liberian history and was a Professor of History at the University of Liberia. Her publications include *Grand Cape Mount County* (1964), (co-editor) *The Official Papers of William V. S. Tubman* (1968).

M. Last, B.A. (Cantab), M.A. (Yale), Ph.D. (Ibadan), is engaged on a study, under the direction of the Anthropology Department of University College, London, of the anthropology of sickness and its treatment among rural Hausa, both pagan and Muslim. He was formerly Research Fellow of the Northern Nigerian History Research Scheme at Ahmadu Bello University, concentrating on Arabic and Fulani manuscripts. His publications include *The Sokoto Caliphate* (1967) and articles in *Africa* (1970) and in the *Journal of the Historical Society of Nigeria.*

B. O. Ọlọruntimẹhin, B.A. (London), Ph.D. (Ibadan), is a Senior Lecturer in the Department of History, University of Ibadan. Before this he taught at the University of Lagos for two years. He is the author of *The Segu Tukulor Empire* (1972) and has contributed chapters to several symposia and many articles to scholarly journals.

D. Cruise O'Brien, B.A. (Cantab), M.A. (Berkeley), Ph.D. (L.S.E.), is Lecturer in Politics at the School of Oriental and African Studies, University of London. He specialises in the politics of West Africa and his publications include *The Mourides of Senegal: the Political and Economic Organization of an Islamic Brotherhood* (1971) and numerous articles on French West Africa.

Y. Person, Docteur ès lettres, is at present at the University of Paris. From 1948–63 he was a District Commissioner in the French colonial administration (Dahomey, Guinea, Ivory Coast) and then spent three years at the University of Dakar. He is the author of *Samori: Une Révolution Dyula*, 2 volumes (1968).

J. B. Webster, B.A. M.A. (Brit. Col), Ph.D. (London), is Professor of African History at Dalhousie University. He was formerly Head of the Department of History and Dean of the Faculty of Arts at Makerere University. His major publications include *The African Churches among the Yoruba 1888–1922* (1964), joint authorship (with A. A. Boahen) of *The Growth of African Civilisation: The Revolutionary Years. West Africa since 1800* (1967) and (with others) *The Iteso during the Asonya* (1973).

G. Wesley Johnson, A.B. (Harvard), M.A. Ph.D. (Columbia), is Assistant Professor of African History at Stanford University and also serves as Chairman of the Committee on Comparative Studies of Africa and the Americas at Stanford. His publications include *The Emergence of Black Politics in Senegal; The Struggle for Power in the Four Communes, 1900–1920* (1971) and (co-editor) *Perspectives on the African Past* (1971). He is at present co-editing a collection of original essays, *The Senegalese Political Tradition* and is also working on a biography of the Senegalese politician, Blaise Diagne.

The Editors:

J. F. A. Ajayi, B.A., Ph.D., is Vice Chancellor of the University of Lagos. Previously he was Professor of History at the University of Ibadan and was a Fellow of the Center for Advanced Study in the Behavioral Sciences at Stanford 1970–1971. His publications include *Milestones in Nigerian History*

(1962), *Yoruba Warfare in the nineteenth century* (with R. S. Smith), *Christian Missions in Nigeria 1841–1891* (1965) and articles in *The Journal of the Historical Society of Nigeria, Tarikh* and others. He has also edited (with I. Espie) *A Thousand Years of West African History* (1965) and is general editor of the Ibadan History Series. In recent years he has become the principal spokesman for the Ibadan approach to African history, which he himself did much to pioneer.

Michael Crowder, M.A., is Professor of History at Abdullahi Bayero College of Ahmadu Bello University and former Director of the Institute of African Studies at the University of Ife. He is also Director of the newly established Ahmadu Bello University's Centre for Nigerian Cultural Studies. His historical writing includes *The Story of Nigeria* (1962), *Senegal: A Study in French Assimilation Policy* (1962), *West Africa under Colonial Rule* (1969) and *Revolt in Bussa* (1973). He has also edited a number of symposia including *West Africa Resistance* (1971) and *West African Chiefs* (with O. Ikime) (1970). In addition Michael Crowder has a wide interest in African studies and in the development of African arts and has organised many conferences, symposia and festivals, including the First International Congress of Africanists, the proceedings of which he edited jointly with Professor Lalage Bown.

Publisher's note

Collation of a symposium work such as this involves accepting the contributions from authors at different times relating to their own writing programmes at the time of commission. As a result the work of some contributors was submitted at a much earlier date than that of others.

List of maps

Introduction

J. F. A. AJAYI AND MICHAEL CROWDER

Volume I of the *History of West Africa* covered the period from the earliest times to 1800. Its emphasis was on the theme of state formation and the development of social and political institutions. Outside influences such as the penetration of Islam into the savannah region and of Europeans from the coastal areas played an increasingly important but never predominant role. They were accordingly treated as subsidiary themes.

By contrast, in the nineteenth and twentieth centuries covered by this volume, both Islamic and European influences became increasingly dominant in the lives of the peoples of West Africa. By 1800, Islam had become effectively an indigenous religion in West Africa. In the nineteenth century, it became the dynamic force which created a series of major empires spreading in geographical extent over more than half of West Africa. Islamic influences spread even beyond the boundaries of those states and Islam has become the religion of the majority of West Africans.

At the same time, European missionaries, explorers and traders began to penetrate into the West African interior, gathering information, seeking to make converts to Christianity, and establishing outposts of European trade. Soon, there followed European soldiers and administrators who conquered and established colonial empires for Britain, France, Germany and Portugal. Europeans thus became dominant over the whole of West Africa, including the Islamic areas. Many West Africans became Christians. Both the traditional West African and Islamic religions, as well as the cultural and economic systems, came under pressure from European power, ideas and values. Resistance to this pressure has since led to a resurgence and revival of Islamic and traditional cultural and political life in West Africa.

This leading role that Islam and European Christian influences have played in the history of West Africa in the nineteenth and twentieth centuries has been the major theme of this volume. Two preliminary chapters discuss the ideologies of the two movements in the early nineteenth century. These are followed by seven regional chapters reviewing developments in the nineteenth century generally and examining the impact of these factors on internal institutions. With the colonial conquest, the traditional historical and even geographical boundaries became less important. In their place, colonial boundaries assumed greater prominence, particularly in the dichotomy between British and French territories. For this reason, the chapters discussing the history of the colonial conquest, patterns of rule, African resistance and resurgence have either attempted to survey the whole of West Africa or have followed the logic of the Anglo-French partition.

This approach is dictated also by the prevailing state of our knowledge, but it is not entirely satisfactory. For example, by taking this approach, continuity in the history of the development of internal institutions traced to the nineteenth century has not been sustained to the present day. Politics in independent West African States have shown the continued vitality and dynamic nature of many of these institutions. Yet the basic research on their developments in the colonial period has only just begun. This neglect makes early twentieth century West African history a less well-known period than the nineteenth. It is our hope that this volume will draw attention to such gaps in our knowledge, and that future editions will be able to cover the internal history of West Africa in the first half of the twentieth century more adequately.

As with Volume I, we have to point out that the chapters in this volume were commissioned in 1965. While some authors were able to respond in good time with their chapters, there have been delays on the part of others, as well as interruptions caused by the Nigerian Civil War, and the need to find alternative contributors rather late in the day. One result is that the chapters by authors who met our initial deadline may not have taken account of the more recent publications in the field. Another is that plans for a chapter on the Niger Delta States and their hinterland in the nineteenth century fell through and we have to await a new edition to fill in this gap. Nevertheless, we believe that these nineteen chapters together give a balanced and coherent picture of the main outlines of West African history in the nineteenth and twentieth centuries.

As in Volume I also, no attempt has been made to standardise the orthography of African names, and we must again express the hope that steps will soon be taken to establish a standard orthography for African personal and place names. Such a project would be of immense benefit to historians as well as to scholars in disciplines other than history.

Chapter 7 of this volume was originally written in French and we owe the excellent translation to Dr Joan White. We also wish to thank various colleagues for editorial assistance, particularly Professor R. J. Gavin.

Finally, this history could not have been published without the general support of colleagues and the secretarial staffs of the History Departments of the University of Ibadan and of Abdullahi Bayero College of Ahmadu

Bello University; the Institutes of African Studies of Fourah Bay College of the University of Sierra Leone and of the University of Ife; and the Centre for Advanced Study in the Behavioural Sciences, Stanford, California. To them, and to the Vice-Chancellors of these Universities and the Director of the Centre, we wish to express our sincere thanks.

J. F. Ade Ajayi,
University of Lagos
October 1972

Michael Crowder,
Abdullahi Bayero College,
Ahmadu Bello University

CHAPTER I

Reform in West Africa: the Jihād movements of the nineteenth century★

MURRAY LAST

Too often the Muslim states of nineteenth-century West Africa are described secularly as empires, while the warfare by which they were established is labelled jihād. But jihād should rather be considered not merely as warfare *against*, but as a struggle *for*, as a constructive reform movement and not simply as the destruction of pagan peoples. For the struggle was to continue throughout the century, whereas the period of intensive warfare was short: in Hausaland and Bornu only four years, in Masina less than a year. Only in the later period, in the very different struggles of al-ḥājj ʿUmar and Samory, was warfare chronic and, by comparison with Sokoto and Masina, destructive. It is true that in Sokoto and Masina expeditions were sent out every year, and these, in general terms, were jihād: but such campaigns, especially in the central emirates, were police operations designed more to check invasion than to conquer fresh lands. Jihād in the postwar period was concerned rather with maintaining politically and socially the ideals for which the war was fought. The war itself was an extension of intensive teaching and preaching; once war was over, the teaching had to continue as strongly as before, not least since ideals are apt to be among the casualties of victory.

Though the fighting may be seen as only a phase in the jihād it is nonetheless necessary to consider how fighting, as distinct from preaching, achieved its results, and what consequences sprang from it. To some extent the attitudes current in each jihād are reflected in the career of its leader, partly owing to the fact that more detail is available about him than about his followers. For example, the Shaikh ʿUthmān's primary role of teacher, scholar and reluctant fighter seems to have set the trend in Sokoto for tolerance: those in authority were too learned not to know the range of opinions embraced by the *Sharīʿa*. In Masina, where learning was less, there was a greater rigidity in the practice of Islam than the Law warrants.

★written in 1967

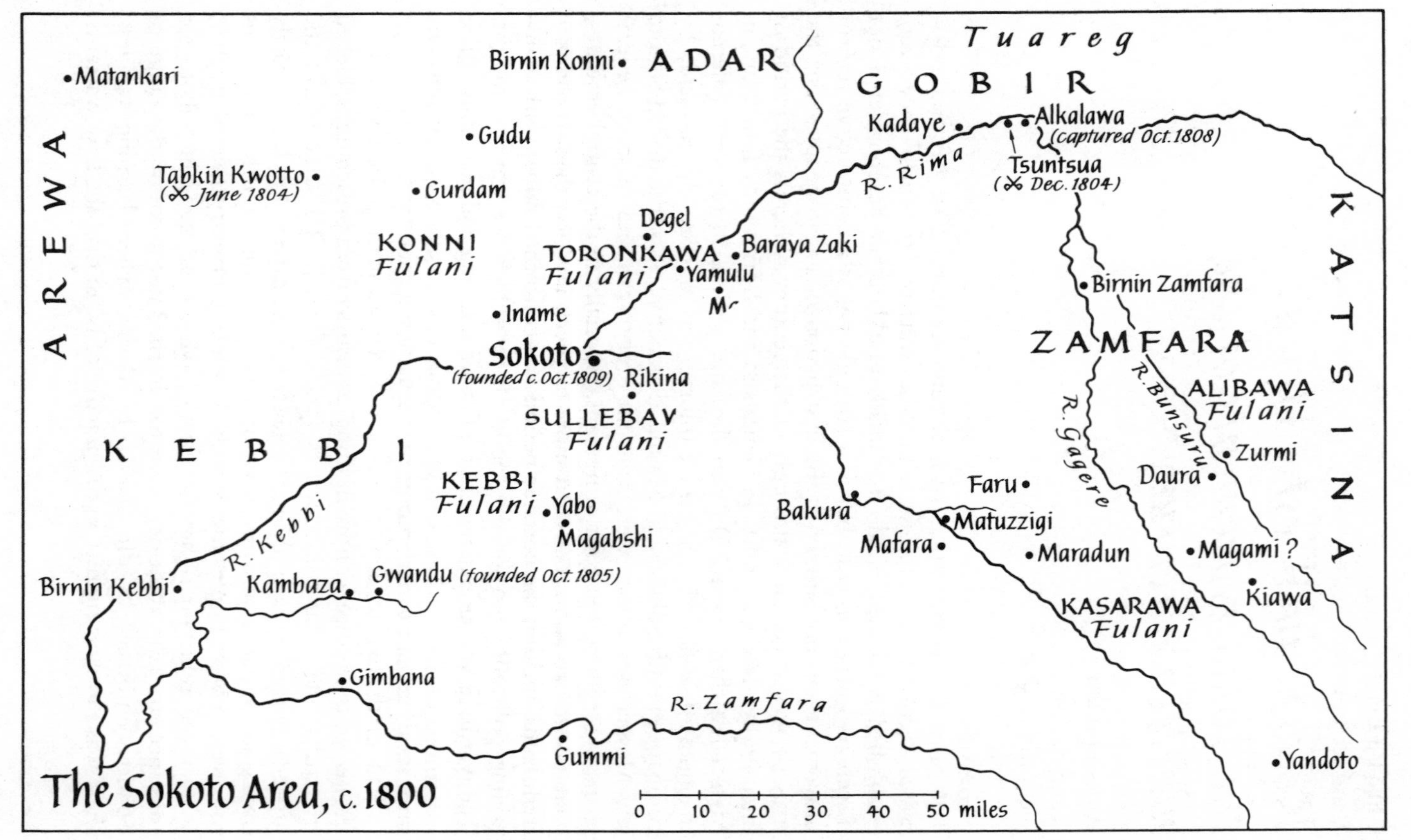

The Sokoto Area, c. 1800

1:1

Thus it will be useful to try to understand the men who led the movements and see how they personify the reasons for which they fought. The aftermath of the fighting can then be more easily understood.

Particular attention will be paid to Sokoto and Masina, as in both areas the jihād aimed at building a new society in which religion and prosperity were possible. The later movements largely failed to achieve this, since the forces involved were radically different in their composition and needs. No attempt will be made to give all the details here: published works have already tried to establish the basic facts and readers will find more complete accounts there.[1] This is, instead, an attempt to extract the significant trends and facets that differentiate the various movements or show up their common foundation.

It is important at the outset, however, to stress that the jihād movements to be discussed were not the only attempts at Islamic reform. Apart from the Fulani imamates of the eighteenth century in Futa Toro and Futa Jallon there were innumerable minor revolutions in the great states of West Africa in which, for example, the accession of a Muslim prince to power might take on the nature of a jihād. Thus the majority of jihād movements, and especially the failures among them, being within peasant societies, have tended to escape notice. By contrast, the dramatic coups that mark the nineteenth-century movements derive part of their drama from the fact that they were led by men from predominantly pastoral peoples who had never previously created a state.

Some of the interest, therefore, in the jihād movements lies in the reasons why the Fulani in the nineteenth century were able to take power and create states. One factor was the growth in the number of Muslims, both Fulani and non-Fulani, and their increasing awareness that the states in which they lived were by Islamic standards corrupt. When in one area a Fulani scholar of the calibre of the Shaikh ʿUthmān was able by his teaching to intensify both the awareness and the solidarity of the Muslims, the unplanned clash with, and the overthrow of, authority created a precedent having repercussions throughout the Muslim–Fulani world, from Adamawa to the Senegal valley, from the desert to the forests. In their diversity the many scattered Muslim communities inspired by the Shaikh defy simple analysis: though these semi-independent communities were the component parts of the wider jihād movement, it is possible now to discuss only the main source of their inspiration.

1 The main published work on Masina is Hampate Ba and J. Daget, *L'Empire Peul du Macina*, Paris, 1955, reprinted 1962. On Sokoto there are H. A. S. Johnston, *The Fulani Empire of Sokoto*, Oxford, 1967, and D. M. Last, *The Sokoto Caliphate*, London, 1967. An excellent essay on the jihād movements as a whole is H. F. C. Smith, 'The Islamic Revolutions of the Nineteenth Century', *Journal of the Historical Society of Nigeria (J.H.S.N.)*, ii, 2, 1961.

THE JIHĀD IN HAUSALAND

The centre of the reform movement led by the Shaikh 'Uthmān b. Fūdī lay on the northwestern edge of Hausaland, among the Fulani and Tuareg communities between Agades in the North and Kebbi in the South. The Fulani reforming scholars, though classified as *Fulanin Gida* ('house'—that is, settled, Fulani), seldom lived in the Hausa towns, preferring instead camps and hamlets outside the main areas of peasant farming. Initially much of the preaching was aimed at the neighbouring Fulani pastoral groups, but the partially Muslim states of Hausaland, in so far as they impinged on local Fulani life as well as on the freedom of Muslim scholars and students generally, soon became the more important target. Although the extent of the exploitation and luxurious living practised by the courts of Hausa kings is not known, it was enough to cause bitter resentment not only among austere Muslims but also among peasants and pastoralists.

Pre-jihād society in Hausaland and elsewhere has been described in the previous volume, but one aspect is important here: none of the city-states that came under attack by the reformers was plainly pagan. Islam, in varying degrees, had been established for several centuries in the trading quarters of the cities, while royal households in many cases had been at least nominal Muslims for generations. The concepts of Islamic justice were therefore known, though traditional beliefs continued to be widely held. The local repute of Muslims, many of whom came from the more sophisticated emirates of North Africa, was considerable even in wholly pagan societies: no insuperable barrier lay between neighbours of different faiths or peoples. Within such an open society the beginnings of a reform movement were possible.

There were certain factors in the early life of the Shaikh 'Uthmān b. Fūdī which favoured a career of teaching and, later, of political reform. As a young man he had been taught by the radical Jibrīl b. 'Umar al-Aqdasī (from Agades) but he had chosen, unlike most of the young Muslim intellectuals of his time, to teach the fundamentals of Islam instead of perplexing men with sophisticated arguments on the oneness of Allah in order to prove his hearers' ignorance.[2] Nor was the Shaikh content, like many of the scholars of his clan, merely to teach the Islamic sciences to a circle of students. His family consisted mainly of distinguished, if quietist, scholars, and among his neighbours were mystics devoted to a life of

2 For the education of 'Uthmān b. Fūdī, see F. H. El-Masri: 'The Life of Shehu Usuman dan Fodio before the jihad', *J.H.S.N.*, ii, 4, 1963. The main Arabic texts are 'Abdullāh b. Fūdī's *Tazyīn al-waraqāt*, ed. and trans. M. Hiskett, Ibadan, 1963; his *'Īdā' al-nusūkh* in M. Hiskett, 'Material relating to the state of learning among the Fulani before their jihad', *Bulletin of the School of Oriental and African Studies (B.S.O.A.S.)*, xix, 1957; and 'Uthmān b. Fūdī's *Asānīd al-faqīr*. Some relevant passages are to be found in Muḥammad Bello's *Infāq al-maisūr* (translated or paraphrased in E. J. Arnett, *The Rise of the Sokoto Fulani*, Kano, 1922; a new edition and translation is being prepared by M. A. Al-Hajj of Abdullahi Bayero College, Kano).

seclusion and recitation. The Shaikh's son says that Jibrīl was the greatest influence on his father. As Jibrīl is known to have been forced to leave his country because of subversive preaching (subversive, that is, to the scarcely Muslim aristocracy of the Tuareg), the activist element in the Shaikh's outlook is probably due more to Jibrīl's inspiration than to his actual teaching.[3] For Jibrīl taught him for only a year before leaving for the pilgrimage; and the Shaikh expressly wrote a book disagreeing with Jibrīl's extreme definition of the sins for which a man ceases to be considered a Muslim. To the influence of Jibrīl may be added the Shaikh's evident annoyance at the clever ploys of his fellow students in dismaying illiterate Muslims. Like most Fulani scholars he had close ties with the pastoral Fulani in their camps and their herdings; and it is probable that his sympathy with them and their ill-informed practice of Islam prompted him to write the instructional poems in Fulfulde which are so marked a feature of his early work.[4]

In addition to his work with his fellow Fulani, he preached in Hausa and wrote pamphlets in classical Arabic. He therefore appealed to all groups, the pastoral Fulani with his verse, the peasant Hausa with his preaching and his fellow students with his Arabic books. His concern for correcting the local practice of Islam extended to women, whom he was prepared to teach in his circle of students and to welcome at his meetings (where they kept to one side).[5] Furthermore, once the Shaikh's community at Degel was established and largely out of reach of Gobir's authority, it became, it seems, a place of refuge: it is likely that slaves escaped from pagan masters to become free Muslims. Similarly, during the Gobir conquest of Zamfara the Shaikh's preaching tours took him to areas that were oppressed; and once he had left, his students continued his work in their home areas.

Politically the decline of Gobir coincided with the growing influence of the Shaikh. While there was no armed confrontation—the often disastrous wars with Katsina and the continuing border struggles with the northern Tuareg and the southern Zamfarawa were the main military concern of the Gobirawa—the Shaikh had challenged the moral authority of Bawa, the King of Gobir, and had won from him concessions for freedom of religion.[6] Although the concessions were rescinded some years later, it was then too

3 The main sources of Jibrīl's career are those listed in n. 2.

4 Forty-two of the Shaikh's poems in Fulfulde are due to be published with Hausa translations by the Gaskiya Corporation, Zaria. They represent only a fraction of the jihād poetry still extant.

5 'Abdullāh b. Fūdī' *Tazyīn al-waraqāt* (trans. Hiskett), pp. 85–6.

6 Some data on the pre-jihād wars of Gobir are given in 'Abd al-Qādir b. al-Muṣṭafā, *Rauḍāt al-afkār* (trans. H. R. Palmer, *Journal of the African Society (J.A.S.)*, xv, 1916). The concessions granted were freedom of preaching and conversion to Islam; freedom to wear a turban (or a veil) and not to pay tax; release of prisoners (who had been illegally gaoled?). Source: Gidado b. Laima, *Rauḍ al-jinān* (National Archives, Kaduna). Some of the political grievances in Gobir are given in the Shaikh's *Kitāb al-farq*, ed. and trans. M. Hiskett, *B.S.O.A.S.*, xxiii, 1960.

late to reassert effective control over the widely dispersed Muslim community. Once the Shaikh had called on his people to arm themselves in self-defence, the threat of repression was no longer so meaningful.

The attraction of Islam, and in particular Islamic justice, must have been considerable if the abuses to which the Shaikh refers were common. Though the appeal had existed over centuries, the Shaikh, once his community was established, seemed to offer a viable alternative to the oppressive government of Gobir. Lying on the borders of Gobir, comparatively free from raiding and with no immediate danger threatened from neighbouring states, the community at Degel served as a focus for a para-political network of students and followers. Although it did not have its own authority, it was able, through the universal call of Islam, to wield influence far beyond its formal military capacity.

A factor that lent urgency to the Shaikh's teaching and his call for reform was the widespread belief in the imminent end of the world. Prophecies were known to predict the coming of the *Mahdī* in the years 1200 or 1204 of the *Hijra*; if those dates proved inaccurate, evidence that the thirteenth century after the emigration of the Prophet was to be the last seemed overwhelming.[7] The Shaikh denied being the Mahdī himself, but he acknowledged he was a forerunner, and in the course of the jihād promised his newly appointed emirs that their administration would last till the appearance of the Mahdī. The excitement such expectations aroused should not be underestimated: it gave martyrdom an extra value; it made risks seem less great with the end so near. Even the marginally Muslim must have been stirred to join in the final jihād; the Shaikh's preaching dwelt heavily on the joys of paradise and the terrors of hell. The call was urgent and unequivocal.

The jihād in Sokoto falls into four phases. The first was when the Shaikh was a young man, learning, teaching and preaching: still an individual scholar, despite the growing number of his students.[8] His arguments were more with other scholars and students than with kings and the politically powerful; he appealed to men as individuals rather than as citizens. During this phase, when the Shaikh was just thirty years old, he was summoned to a meeting with the King Bawa of Gobir, at Magami. He was reluctantly recognised there as a political force not merely in defeated Zamfara but also in Gobir itself. The Shaikh does not seem to have sought

7 The thirteenth Islamic century covered the period November 1785 to November 1882. Attention was drawn to the importance of Mahdism in the jihād by Muḥammad Al-Hajj in his 'The Sudanese Mahdiyya and the Niger–Chad Region', in *Islam in Tropical Africa*, ed. I. M. Lewis, Oxford, 1966. A further essay, 'The 13th Islamic century in Muslim eschatology', is in the *Research Bulletin, Centre of Arabic Documentation (R.B.C.A.D.)*, Ibadan, iii, 1, 1967.

8 The Shaikh was born in 1754 (1168 A.H.) and started preaching in 1774/5; he was in Zamfara (Faru) *c.* 1787–92, during which time he attended the meeting with the King of Gobir, Bawa. From 1792 to 1804 he remained at Degel.

such recognition, or at least he did not use it. He wanted to be free to teach and for people to be free to learn. But after this meeting, he moved away from the sensitive Zamfara area to Degel on the western borders of Gobir, a place of retreat where he had been brought up. At Degel he continued to teach and travel; but students often came to him, scholars settled nearby and thus a community grew up. This was virtually independent, according to the Shaikh's brother, of the Gobiri government at Alkalawa.[9]

The second period opened when the new King of Gobir, Nafata, realised that the community was becoming too great a threat to his authority to be overlooked. He therefore rescinded the rights of teaching and conversion and tried to disrupt the cohesion of the Muslims by banning their distinctive turbans and veils.[10] In addition, sufficient harassment started for the Shaikh to sanction the carrying of arms by his followers. Raids on the pastoral Fulani continued and free men were kidnapped.[11] The next King, Yunfa, finding himself powerless to control either the Muslims or the neighbouring King of Katsina, resorted to extremes. He summoned the Shaikh to his court, and drew a pistol, but it backfired, burning his own face.[12] Soon afterwards, a small refugee Muslim community, Hausa in origin but beyond the borders of Gobir, was destroyed by a king dependent on Gobir. The shaikh of the community had refused to respect the authority of the king, who was passing with an army on his way to a campaign.[13] When the Muslim captives from this community were being led past the settlements at Degel a skirmish ensued in which the prisoners were forcibly released. As a proscribed party the Muslims were forced to draw closer together and in their harassment it became clear that the only solution was to overthrow the Gobir establishment. Persecution bred revolution, a revolution linking Muslims, Hausa, Tuareg and Fulani, with the pastoral Fulani, who alone were to provide the base from which to launch the war.

After the blatant challenge to his power implicit in the release of the Muslim prisoners, Yunfa was forced to act, not least through fear of a coup by rival claimants to the throne of Gobir. Yunfa therefore ordered the Shaikh to leave Degel and allow the community to be dispersed. The

9 'Abdullāh b. Fūdī, *op. cit.*, p. 107.

10 The year was either 1795–6 or 1797–8.

11 One of the questions asked in 'Uthmān b. Fūdī, *Masā'il muhimma* (National Archives, Kaduna) concerns Fulanis held as slaves. Cf. C. E. Hopen, *The Pastoral Fulbe Family in Gwandu*, Oxford, 1958, p. 10.

12 The word used for 'pistol' in this incident is the same as that for powder—naptha: the exact type of weapon is therefore not known, see 'Abdullāh b. Fūdī, *op. cit.*, p. 108.

13 The community was at Gimbana, on the northern edge of the Zamfara river valley, under an Arewa scholar 'Abd al-Salām (his earlier name was Mika'ila). He had left his settlement near Degel owing, it seems, to harassment or threats from the Gobir authorities. He was a focus of Hausa support for the Shaikh and many of his followers may have been of low status, possibly ex-slaves: this was perhaps a further reason for the King of Gummi to sack the place.

Shaikh responded by bidding all Muslims to emigrate, in the manner of the Prophet, and from beyond the borders of Gobir to begin the jihād against the Unbelievers, for so the Gobirawa had shown themselves to be by their acts of persecution.[14] The emigrants formally elected the Shaikh as their leader and, to defend their camp at Gudu, dug a ditch such as the Prophet had helped to dig at Madina. After preliminary skirmishes and an attempt by the main force under Yunfa to cut off their retreat, the Muslims left their defensive positions for formal battle. The Muslims charged first, and although the wings fell back, the centre held, and the Gobir forces were eventually routed. Yunfa escaped.

Warfare in the jihād was largely improvised. The Muslims' chief weapon was the Fulani bow; they had little body armour, and none of the panoply of a Hausa state army. Their horses were too few to counter the cataphracts of the Gobir heavy cavalry designed to ride down lightly armed, undisciplined foot-soldiery. But with their poisoned arrows (poisoned despite Islamic Law) the Muslims were able to fight in rough country, keeping out of range of the cavalry.[15] After the initial struggle the Muslims were able to obtain horses, but they did not use the heavy padded armour until after the jihād.

Strategy was dictated by the need for food; for on food depended not only the health and strength of the army, but also its numbers. On many occasions there were no volunteers for a campaign and once the core of students had been killed in a disaster at Tsuntsua much depended on the freelance fighters. In the warfare many villages, their grain-stores and their crops had been destroyed, and the inhabitants swelled the number of those in search of food. The constant movement allowed no time for planting, hoeing and harvesting. According to the most active of the Shaikh's commanders, his son Muḥammad Bello, famine and sickness, not the enemy, were the major causes of death in the community.[16]

In such conditions, with banditry rife, the only strongholds were the towns. Political authority lay there, and their defences were almost impregnable. The long walls, which enclosed farm land as well as houses, were enough to keep horses out. If marauders did scale the walls, they were easily rounded up by mounted defenders within. Frontal assaults were seldom successful; the enemy had first to fight outside the town and the victors might then force entry through the gates. But even if the invading army was unsuccessful, its very presence in the area ensured hardship; crops destroyed and the inhabitants penned in. Therefore after the initial

14 The date of the emigration from Degel to Gudu is 21 February 1804; the decisive battle, at Tabkin Kwotto, took place on 21 June 1804.

15 Since the penetrating power of an arrow from a Fulani bow is low (the pull of a bow is 40–50 lbs compared with over 100 lbs for a Welsh bow), the effectiveness of Fulani archery lay largely in the poison and to obey Islamic Law in this regard would have been suicidal.

16 Muḥammad Bello, *al-Dhikrā* (also in his *Nuṣḥ kāfi*) (Shahuci School collection).

victory at Tabkin Kwotto the Shaikh moved for the next dry season up to the hinterland of the Gobir capital.[17] Alkalawa stood in a river valley, which was fertile enough to supply the Muslims with food while they attacked the only major town in the state. However, while the Muslims were encamped close to the city, they were caught unawares; instead of withdrawing, they offered battle and were severely defeated, with some two thousand killed. The disaster caused in the composition of the community a shift in emphasis which seriously jeopardised the ideals for which the jihād was being fought.

The fighting had already seen the rise of war-leaders, Fulani nobles, who commanded a following of their own; added to them were the free-lance soldiers whose homes had been destroyed. Neither group was particularly concerned with the niceties of the Law concerning booty or with the requirements of the Faith. The loss of some two hundred scholars at Tsuntsua made it still more difficult to restrain the soldiery. Although the Shaikh wrote a detailed handbook on the distinctions to be made in dealing with the property and families of an enemy who theoretically included 'neutral' Muslims, hypocrites, apostates, rebels, bandits and 'robber-barons', clearly the Law could not be applied in the heat of campaign.[18] Definitions of a Muslim became simpler: in effect, those against the Shaikh and his followers were counted as non-Muslims while the pagan groups who remained well-disposed to the Muslims might receive the *amāna* and become protected peoples (*ahl al-dhimma*). The Shaikh at the start of his campaign against Alkalawa had offered terms to the Hausa rulers, but with this offer rejected he was free to attack.[19] Neutrality was difficult to maintain. It was scarcely tolerable to either side, least of all was it tolerated in scholars; thus the intellectual centre of Yandoto was destroyed by Bello after it had driven out those scholars who favoured the Shaikh. Similarly, peasant villages in other parts of Zamfara gradually gave up neutrality or even withdrew their support for the community; although the hostility sometimes resulted in open warfare, more often, however, it meant that the villages were abandoned, with a consequent further shortage of purchasable food—traders ceased to come to markets and crops ceased to be grown.

The breakdown of the economy and general administration in Gobir

17 Dry season, 1804. The disaster at Tsuntsua took place in December 1804 during the fast of Ramadān.

18 'Uthmān b. Fūdī, *Bayān wujūb al-hijra 'alā 'l-'ibād.* An edition and translation is being prepared by F. H. El-Masri of the University of Ibadan. See also A. D. H. Bivar, 'The Wathiqat ahl al-Sudan', *Journal of African History (J.A.H.)*, ii, 2, 1961 and M. A. Al-Hajj and D. M. Last, 'Attempts at defining a Muslim in 19th century Hausaland and Bornu', *J.H.S.N.*, iii, 2, 1965.

19 Letters to all the Hausa rulers calling on them to reform were sent out in about July 1804. Only the ruler of Zaria accepted. For an attempt at collating data on the jihād in the other Hausa states, see D. M. Last, 'A Solution to the Problems of Dynastic Chronology in 19th century Zaria and Kano', *J.H.S.N.*, iii, 3, 1966, pp. 462–5.

and its provinces isolated the capital. Campaigns therefore could be directed solely at Alkalawa, the Muslims drawing their combined forces from all quarters. On the fourth attempt, with assistance from the Zamfara and the Katsina Muslims, the Shaikh's community succeeded in taking the town; Yunfa was killed, and the town broke up.[20] The major Hausa towns of Katsina and Kano were already in Muslim hands, the ancient Bornu capital of Birnin Gazargamu was temporarily occupied in 1808 by a Muslim force, while by the end of that year Zaria, the last of the important Hausa towns, had been captured. The jihād outside Gobir had been conducted by coalitions of basically Fulani groups fighting the Hausa or Kanuri rulers and their retinue. It is not known for certain whether peasants and traders participated in the fighting but from the accounts of such eye-witnesses as Ali Eisami of Bornu it seems that they were little involved.[21]

There was no popular resistance to the jihād as such. There was little support for the kings on ethnic or political grounds among the peasantry; they were only concerned about possible loss of property. On the Fulani side, there was growing up a cadre of professional fighters, men who had fought in several campaigns outside their home areas. The problem for the leaders was how to maintain their forces in the towns they had captured. The recapture of Birnin Gazargamu by Bornu forces and the death of the Fulani leader Gẇoni Mukhtār was due to the departure of most of his followers, leaving him to hold the town against the returning Bornu army. Since many of the rank and file had herds or families to care for, the towns held little attraction for them once capture was effected. Until a permanent camp could be established (such as that set up by the Shaikh at Gwandu), occupation could only be transitory; for again the problem was one of food. It is likely that there was an optimum size for a military group operating off the land. In Bornu two groups operated independently, one from the north-west, the other from the south-west. In Kano, despite occasional plans to unite, the eastern and western groups remained separate although the western group sometimes joined up with the combined Katsina and Daura Muslims. The Zaria group, mustered in southern Kano only a few weeks before the campaign, probably did not exceed about a hundred men. All these groups were usually under local leadership; men who were connected with the pastoral Fulani clans in the area and were generally known to people as scholars. Their knowledge of the neighbourhood and their established reputations made the struggle a revolution rather than an invasion. The attack was aimed at the Hausa aristocracy, and in these circumstances the peasant stayed aloof. Although the courtiers might flee, the towns remained; the only exceptions were Alkalawa where

20 October 1808.
21 Ali Eisami's account of his journeys is given in S. W. Koelle, *African Native Literature*, London, 1854, and has been reproduced with notes in *Africa Remembered*, ed. Curtin, Wisconsin, 1967.

an administrator appointed by the Shaikh resettled a Gobirawa community nearby, and Birnin Gazargamu which may have been largely a palace town with the bulk of the population living in villages in the hinterland. After a conquest there were immediate attempts to restore an equilibrium to life in the area, and after the prolonged hardship and dislocation any government that promised peace must have been welcome. In Bornu, mallams went round telling men to return to their homes. Nearly thirty years later, after pacifying middle Zamfara, Muḥammad Bello wrote to his brother stationed at Bakura and told him to give high priority to land distribution and the re-establishment of farms, even at the risk of contravening the Law.[22] Similarly after the war the courts had to settle claims and counterclaims concerning items of booty.[23] Until the dislocation of the war was cleared up, the Muslim state for which it had been fought could hardly exist.

The reorganisation of the administration and the creation of Muslim types of government form the final phase of the jihād.[24] Although modifications were made subsequently, as the campaigns against non-Muslims continued, their history lies outside this chapter. The Shaikh's administration was a model for the whole Caliphate. Under the Caliph in Sokoto were his agents the emirs of the provinces, and with them were the judges who advised on and administered the full Islamic Law. In addition there were the tax-collectors, who were responsible for certain specific taxes, inspectors to make sure the public welfare was not infringed and imams to lead public prayers and to teach. Although similar officials had undoubtedly existed before the jihād these offices were now formally defined; precedents were available (and quoted) from the histories of the Islamic world.

As early as 1805–6 the Shaikh had appointed his representatives for the rest of north-western Hausaland. In later years leaders from each of the other areas received their war-flags as accredited representatives of the jihād. The men appointed were usually from minor Fulani groups; the Shaikh took care to avoid identifying the new emir with any one clan or class, choosing learned men who would not only know the Law on which their administration was to be based, but would also attract the respect due

22 Bornu: Ali Eisami's account. Bakura: letter in ʿAbd al-Qādir b. Gidado, *Majmūʿ* (Ahmadu Bello University collection, Zaria).

23 ʿAbd al-Qādir b. al-Muṣṭafā: [*Baʿḍ tanbīhāt*] (Ahmadu Bello University collection, Zaria). I am indebted to F. H. El-Masri for drawing my attention to this point.

24 Published texts on the Shaikh's reforms are: ʿUthmān b. Fūdī, *Kitāb al-farq*, ed. and trans. M. Hiskett, *B.S.O.A.S.*, xxiii, 1960; ʿUthmān b. Fūdī, *Nūr al-albāb*, French trans. I. Hamet, *Revue Africaine (R.A.)*, Algiers, 41/227; 42/228, 1897–8; and passages from a number of texts in M. Hiskett: 'An Islamic Tradition of Reform in the Western Sudan from the 16th to the 18th century', *B.S.O.A.S.*, xxv, 1962. The Shaikh handed over day-to-day administration to his brother ʿAbdullāh and his son Muḥammad Bello in 1812. He died five years later, in 1817.

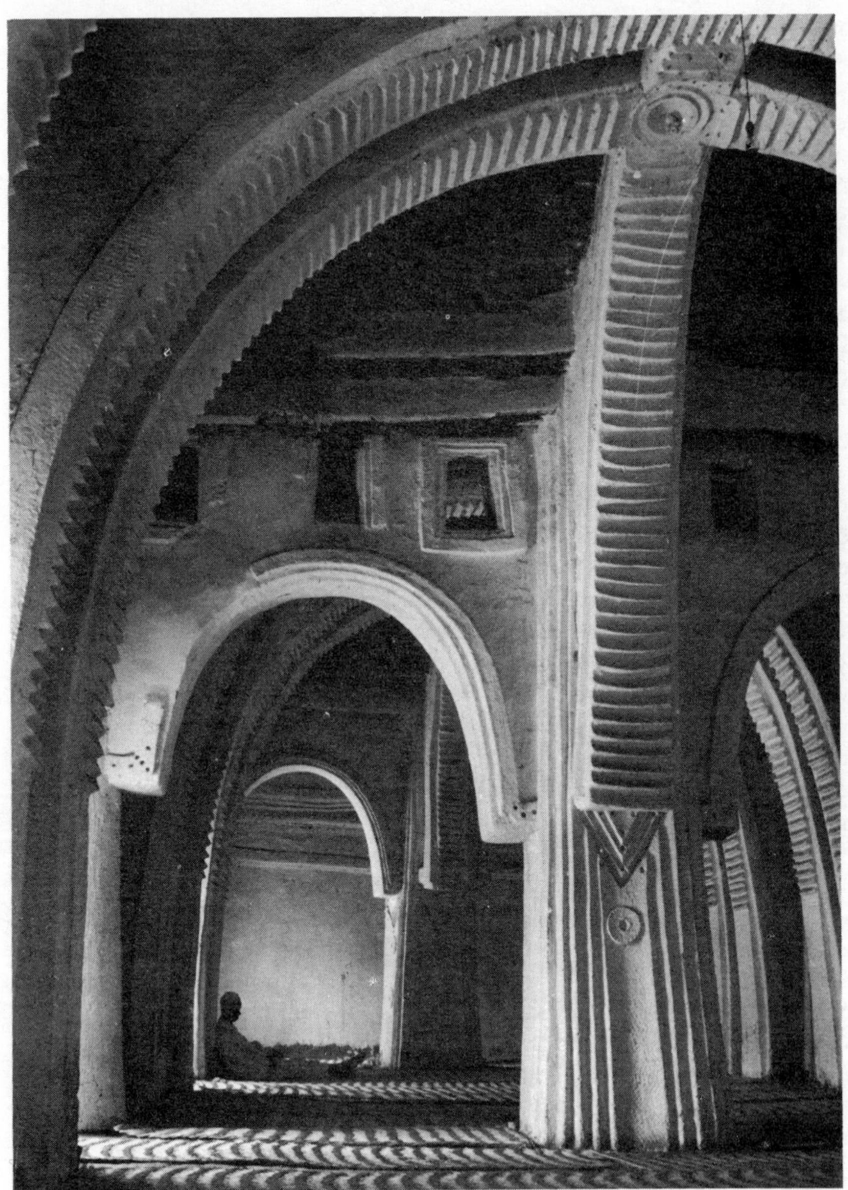

The Zaria Mosque

to their scholarship. As some were those who had left their own areas to join the Shaikh or another community, the new emirs tended to form an interregional élite with contacts that extended far beyond their home groups. In towns like Zaria and Kano, or communities like Adamawa, their former lowly position in local society earned them the scorn if not the hostility of the local Fulani aristocracy. But their external ties, the sanction of the Shaikh's appointment and their reputation for learning enabled them to overcome some of the envy and to avoid the kinds of malpractice through which 'a Muslim in the morning becomes a pagan by the evening'.[25] Nonetheless, the existence of local war-lords, Fulanis who with their clan following had often been decisive in giving the scholars their victory, necessitated a political system which could include them but limit their powers. In some cases, as in Zaria, the emirship with its retinue of officials rotated between the major groups; in Kano, one family (after the original appointee died) held the emirship while the other families retained their positions of power in the council. In Katsina or Adamawa the local leaders were almost independent of the emir, though they owed a theoretical subordination to him as the representative of the Caliph in Sokoto. Without professional standing armies the emirs relied on the co-operation of their subordinates for the application of their authority; equality of equipment denied anyone an overwhelming superiority. Horses were the only possible keys to power but against the long town walls the advantages of cavalry were lost. The balance was upset with the introduction of firearms, the effective use of which in no way depended on prerogatives of wealth and birth. These weapons, though known, were never used in the early jihād.[26]

The reforming movement in Hausaland underlines the difficulty of transforming a community of scholars and students into an Islamic government responsible for vast areas and diverse peoples. The Shaikh had to confront first, the demands of warfare—the need for armed men and skilled commanders, for food supplies and allies—and, second, the demands of a reformed administration in which Islamic ideals had to be reconciled with the power of newly established interests. In Masina, where both the extent and diversity of peoples and the duration and intensity of the war were less great, the problems were to lie in creating a centrally organised Muslim state out of a diffuse pastoral society.

25 Muḥammad Bello, *al-Dhikrā*.

26 The beginning of the revolution in power due to firearms is to be seen in appointments such as those in Zaria in 1897 and in Sokoto in 1902 when both Yero and Attahiru were chosen by electors under duress. An important aspect of the use of firearms in the Caliphate—and one which explains why the vast quantities of guns and ammunition found by the British in Kano were ineffective in stopping British-led forces—is that little or no arms practice was done: to use powder and shot in target practice evidently seemed a waste.

THE MASINA JIHĀD OF SHAIKH AḤMAD

Although outside the immediate sphere of the Shaikh's jihād and later in its inception, the jihād of Aḥmad Lobbo in Masina belongs to the same reform movement in a way that the still later jihād of al-ḥājj 'Umar does not. Aḥmad Lobbo looked to Sokoto, not only for the blessing of the Shaikh to start his campaign, but also for books on which to base his administration. Though he later resisted all persuasion to acknowledge the Caliph in Sokoto, he did recognise the primacy of the Shaikh.[27]

Aḥmad Lobbo was born of a minor scholar family from one of the less important Fulani clans.[28] While still a child he lost his father and was brought up by his mother's family. He was taught, with his cousins, by his maternal relations and their friends in an area far from his birthplace. While his education progressed he continued to herd the cattle of others. He was aggressive both towards his contemporaries (who teased him) and to the traditional Fulani authorities. Serious and solitary, he was a proud student of Islam, outspokenly opposed to the non-Muslim practices of the *Ardo'en* and their young men.[29] In a community of herders he was an exception. Yet he dreamed of the liberation of his people and in a vision he was the leader who threw off the Bambara yoke.

In his studies, Aḥmad Lobbo's closest friends and mentors were the scholars of his own clan and some learned men from a local dispossessed minority group, the Marka. One man in particular, 'Uthmān Bukhārī, Aḥmad's fellow clansman and some twenty years his senior, appears to have masterminded at least the start of his campaign for support among the young Muslims and their teachers outside the city. His other friends gave what help they could to win him acceptance among their own people. His initial following came primarily from four groups: men from his own area,

27 There is some divergence in tradition about Aḥmad Lobbo's links with Sokoto. He never travelled to Sokoto (or indeed anywhere outside Masina), but at the beginning of his jihād campaign he did send men to establish contact with Sokoto, to receive the Shaikh's blessing if not an actual flag. In the later correspondence of the 1830s Lobbo rejected repeated requests to pay allegiance and taxes to Sokoto. Lobbo depended heavily on Sokoto books for his textual authorities, though he later accused Sokoto of falling away from the austerity of the Faith they professed. cf. Ba and Daget, *L'Empire Peul du Macina*, pp. 36, 40, 62; letters of Abū Bakr Atiku and Aḥmad Lobbo in Institut de France library, Paris; H. Barth, *Travels and Discoveries in North and Central Africa*, Centenary edition, London, 1965, iii, p. 183.

28 The main dates of Aḥmad Lobbo's career are: born 1775/6 (1189 A.H.), starts teaching about 1797, meets 'Uthmān Bukhārī (then aged about fifty) in about 1807, has a vision and starts to prepare for the jihād in about 1816, wins the crucial battle of Noukouma 21 March 1818. Hamdallahi was founded in 1821, and Aḥmad died in 1845, two years after the death of 'Uthmān Bukhārī. Almost all the details, if not their interpretation, in this section are based on Ba and Daget, *L'Empire Peul du Macina*. I am greatly indebted to W. A. Brown for the discussions we had in Mali, when he was preparing his Ph.D. thesis, *The Caliphate of Hamdullahi*, University of Wisconsin, 1969.

29 Ardo (plural, Ardo'en) is one of the titles used by heads of particular Fulani clans (and approximately, therefore, places). It is frequently extended to cover any kind of Fulani chief.

his own clansmen from other areas, men from Muslim minorities in the rural areas and his own students (who may have been largely drawn from the previous three groups).

Although Aḥmad Lobbo was for a while taught by an elderly, relatively undistinguished scholar in Jenne, as far as possible he kept out of the town, then the only important centre of learning west of Timbuktu. He preferred to live in a small community close to Jenne at Roundé Sirou, and from here his students carried his teaching beyond his own circle. For twenty years he continued to teach but, in contrast to Sokoto, little of what he taught in Masina has survived. Aḥmad Lobbo and his disciples seem to have been distinguished more for their piety and zeal than for their scholarship.[30]

Apart from his preaching, Aḥmad Lobbo achieved a certain fame by antagonising the Ardo'en and their followers. He accused them not merely of non-Muslim habits (like drinking alcohol) but also of being tools of their Bambara overlords. He had also openly insulted the group of notable foreign scholars in Jenne, and attacked them for their laxity in the Faith and for their fine living. He appealed strongly to the more austere and nationalistic among the Fulani who on either religious or political grounds resented alien domination. Finally, in a society where castes were entrenched he seemed to oppose both tribal and academic aristocracies and their extortions. He offered, as the Shaikh ʿUthmān had offered in Sokoto, hope for the oppressed in the form of a Muslim society based solely on the rule of Law.

The clash which provoked the jihād was an incident in which an Ardo's son was killed after he had insulted some of Aḥmad's students. The Ardo'en closed ranks and called in their Bambara suzerains to help them suppress the movement; but in the crucial battles that decided the fate of the jihād the Fulani of the Ardo'en all discreetly held aloof, although they came within sight of the battlefield. Once Aḥmad had won despite all odds, he offered generous terms to the undecided and in the course of the week following the decisive battle he succeeded in creating a Fulani mass movement against the Bambara and those Ardo'en who remained with them.

Whereas the initial core of the community was composed of ardent Muslims the newcomers were, as in Sokoto, less concerned with Islam. It is Aḥmad's great achievement that he was able with such scanty resources to establish a clearly Muslim regime. By identifying Islam with Fulani independence he was able to make 'Fulani' and 'Muslim' synonymous. The discredited Ardo'en were forced to come in and repent. The dangers of the first few years made a closely-knit state imperative. The Bambara, who were the best fighting force in the area, had not been conquered, only dislodged from control over the province of Masina. Owing to the need for pasture, the Masina herds were exposed to raids every year. Thus

30 Aḥmad Lobbo's short book, *al-Iḍṭirār* has survived, and Hampate Ba refers to a book (possibly the same one) by Aḥmad which was distributed to Masina judges. Books generally were in short supply and were kept in a state library (Ba and Daget, p. 62.)

the state administration to a great extent concentrated on making the seasonal migration safe by establishing an elaborate defence system for the herds. The interest of the cattle owners, whatever their piety, lay therefore with the new state. The fact that the tenor of the state was dictated by the scholars on the Council may have borne heavily on the lighthearted (as later events tend to show) but it was offset by the genuine efficiency and zeal of the new system.

The main organ of administration was a Council of Forty, presided over by a noted scholar.[31] Aḥmad, although he had assumed the title of Amīr al-mu'minīn, was merely a member of the Council, delegating much of his authority to his fellow-councillors and frequently leaving important decisions entirely to them.[32] Nonetheless he retained a dominant influence over the Council; its president was one of his two perpetual attendants and always a possible spokesman of his wishes.[33] The emirs appointed over the five provinces, though members of Aḥmad's family or entourage, were responsible to the Council and were expected to render account to it. In the same way any extraordinary command, military or diplomatic, was answerable to the forty. In each province there was a local administrative structure reaching down to village level, with officials appointed as judges, tax-collectors, inspectors and army officers. To check on the honesty of officials and private citizens there was a state spy system,[34] regular weights and measures were introduced to bring standardisation throughout the state and welfare legislation dealt with both education for the young (with public examinations, and stipends for teachers) and pensions for the aged and widowed. Military service covered the whole male population, although a tax was levied on those who were normally exempt (for example, traders or smiths). A system of payment for armed service and a network of state granaries enabled levies to avoid living off the land and alienating local inhabitants. Such an elaborate structure of welfare and military commit-

31 The Council was appointed out of a larger assembly of scholars. Although there are no precedents for a formal council such as this in the early period of Islam, several of the states in North Africa had such councils; and the principle of a ruler-and-his-advisers became embodied in theoretical texts (for a discussion of this, see E. Tyan, *Institutions du Droit Public Musulman*, Paris, 1954–6). But a democratic council of this kind perhaps reflects also the continuing egalitarian tradition of a fundamentally pastoral people.

32 The formal use of the title 'Amīr al-mu'minīn' has external rather than internal implications: it emphasises Masina's independence and legal separation from Sokoto to whose ruler, also called Amīr al-mu'minīn, allegiance would otherwise be due. The difference in the roles of the two Amīr al-mu'minīn's is reflected in the terms used for the new states: Masina was impersonally called the Dina—*al-dīn* means the Religion (of Islam)—whereas the terms for Sokoto (e.g. *jamā'at al-Shaikh*; *al-'Uthmāniyya*; *Kadirawa*) refer more to its founder and the sufic-academic origins of the community.

33 The president was Nouhoun Tayrou (Nūḥ al-Ṭāhir), who had been to Timbuktu and Sokoto as a student. He was formally a disciple of the Kunta Shaikh at Timbuktu.

34 A spy was called *sa'i*, a term used in Sokoto for a collector of alms among pastoralists. In Arabic usage, it is simply an official.

ments was costly; in addition to the normal Islamic taxes, a special produce tax had to be levied.[35]

A new capital was built for the state and laid out according to a prepared plan with land allocated for certain groups. All the councillors were obliged to reside in the city, and an attempt was made to maintain there the ideal Islamic community. Outside, in the Masina and other provinces, the policy was to settle the Fulani into permanent villages which would be easy to defend and amenable to administrative surveillance. Jenne itself was left, but when the national and religious capital was moved away it lost not only much of its importance but also, once its ostentatious mosque was unroofed, its pride. Sumptuary legislation, while never specifically enforced, curtailed some of the luxury trade and Jenne was beginning to lose its foreign merchants at the time of Caillié's visit in 1825.[36]

The success with which the state supplied both the worldly and spiritual needs of its people seems to have been largely due to Aḥmad who managed to combine a minimum of overt personal authority with an ability to counteract the puritanical excesses of some of his councillors. Occasionally, as in his refusal to abolish castes (in which he differed from the Shaikh 'Uthmān), he preferred not to press idealism to its conclusion. Similarly, when dealing leniently with the aristocratic Galadio, he showed himself understanding and conciliatory.[37] The most notable characteristic that comes out of the stories of the later period of his life is his extreme humility; by rarely losing control of himself he was able to disarm his opponents. Though often defeated in verbal exchanges, he seems never to have lost respect. This, and the efficient security system, made his position unassailable; Masina remained united, despite the rigour of the regime.

THE BAMBARA JIHĀD OF AL-ḤĀJJ 'UMAR

The contrast between the reform movements of Shaikh 'Uthmān and Aḥmad Lobbo on one hand and the jihād of al-ḥājj 'Umar on the other is important not only for helping to define what is the essence of the West African jihād, but also for showing some of the alternative courses of action which the other leaders might have followed.

35 This tax, imposed on peasants, pastoralists, craftsmen and traders alike, was the object of criticism later by the Kunta Shaikh of Timbuktu, al-Bakkā'ī. For the other criticisms and the rebuttal by Aḥmad Lobbo's son, see Ba and Daget, pp. 277ff.

36 R. Caillié, *Travels to Timbuktu*, English edition, London, 1830, i, p. 466.

37 Castes: Ba and Daget, p. 67. Galadio, the Fulani lord of the area in which Hamdallahi had been built, had been equivocal in his support of Aḥmad Lobbo; while clearly despising the mallams, he did not dare break with the increasingly powerful state. Aḥmad Lobbo tolerated him despite pressure from the more fiery councillors, until Galadio openly opposed the state. He was with difficulty defeated and forced to flee East into the western territories of the Sokoto Caliphate, where Barth met him in 1853 (*Travels*, iii, p. 182). I am indebted to W. A. Brown for this reference.

Al-ḥājj ʿUmar was of a different calibre from his two predecessors.[38] He was born the youngest son of a proud scholar; who refused to pray alongside his fellow villagers and instead held his own Friday prayers within his compound.[39] ʿUmar imbibed this proud tradition; a precociously clever student, he distinguished himself early, and on his own initiative went to Saint Louis (then under French administration) and raised funds from local Muslims to enable himself to travel to Mecca.[40] It seems probable that before his pilgrimage he was inducted into the Tijāniyya order, then a new sufic brotherhood which claimed to be the culmination of all previous brotherhoods, just as the Prophet was the culmination of all previous prophets. Once in the Hijaz, he attached himself to the Tijānī leader, excelling both in humility and in his sufic studies. There and during his travels in Palestine and Syria he was able, by his learning and wit, to turn the tables on all who belittled him for his colour. He so impressed the Tijānī authorities in the Hijaz that they appointed him the Tijānī leader for West Africa. Armed with this honour he started home. In each state he passed through he antagonised the established ruler, by inducting men into the Tijāniyya. Where these men were influential, the new brotherhood then posed a threat to the ruler.[41] In Bornu he antagonised Muḥammad al-Amīn al-Kānamī, who, ʿUmar later claimed, tried to have him assassinated.[42] In Sokoto, though he was protected by the ageing Muḥammad Bello, he angered enough men (and converted enough) to

38 The main sources used are H. Gaden, *La Vie d'El Hadj Omar*, Paris, 1935 (this is a translation of a Fulfulde poem by Mohammadou Aliou Tyam); chapter xi of Ba and Daget, *L'Empire Peul du Macina* (this represents traditions among ʿUmar's descendants and the Tijani community in Mali); al-ḥājj ʿUmar's own *al-Rimāḥ*, Cairo, 1345 H. (particularly chs 28 and 29); a history of al-ḥājj ʿUmar in C. A. L. Reichardt, *Grammar of the Fulde Language*, London, 1876, pp. 287–319. Other studies available are B. O. Oloruntimehin, *The Segu Tukolor Empire*, London, 1972; J. M. Abun-Nasr, *The Tijaniyya*, Oxford, 1965; P. Salenc, 'La Vie d'el-hadj Omar', *Bulletin du Comité d'Études de l'Afrique Occidentale Française (B.C.E.A.O.F.)*, 1918; M. Sissoko, 'Chroniques d'el hadj Oumar et Cheikh Tijani', in *L'Education Africaine, Bulletin de l'Enseignment de l'A.O.F. (B.E.A.O.F.)*, pp. 95–6, 1936–7. I am indebted to Omar El-Nager for discussions on ʿUmar's pilgrimage, and to John Willis for discussions on ʿUmar's thought.

39 The main dates of ʿUmar's career are: 1794, born in Futa Toro; 1825–8, journey to Mecca; 1829–40, return to Futa Toro, via Bornu, Sokoto (*c.* 1831–8), Masina and Segu; 1840–8, spent touring, teaching, trading in Futa Toro, Futa Bondu, Futa Jallon, and writing *al-Rimāḥ* (1845); 1848–57, moves slowly North from Dinguiray and captures Kaarta; 1857–60, attempts to keep the French out of his main recruiting areas on the Senegal river, enforcing emigration of Muslims from Futa Toro and enlistment for Kaarta garrisons; 1860, moves on Segu; 1862, moves on Hamdallahi; 1864, dies in the Masina revolt against him.

40 There are some doubts about the accuracy of this tradition. It is likely that his contacts with the French at this time were slight.

41 An illuminating account of his travels, and what he thought of his hosts, is given by ʿUmar himself in *al-Rimāḥ*, ch. 29. For a recent critical study of the Tijāniyya, see J. M. Abun-Nasr, *The Tijaniyya*, Oxford, 1965.

42 Omar El-Nager has pointed out that ʿUmar won over members of the deposed royal family of Bornu, a move unlikely to please his host, the new ruler.

make it advisable for him to leave soon after Bello died. In Masina he again won adherents, but soon left for Segu where he was imprisoned, presumably for subversion aimed at the non-Muslim regime. Finally he returned to his home area of Futa Toro, but even here he was regarded with suspicion by the established Muslim authorities. His journey home was that of a highly sophisticated scholar who made a great impression in the provincial Muslim courts of the Western Sudan. He was consequently regarded as a subversive influence promoting a revolutionary if primarily sufic brotherhood and was unacceptable to any of the existing states. Sokoto alone under the caliphate of Bello was sufficiently cosmopolitan and tolerant to harbour him for some five years, but with his growing following of Tijānīs he seemed to have had political ambitions which alarmed the remaining members of Shaikh 'Uthmān's family. Owing to the close ties necessary between a shaikh and his students in sufic brotherhoods, any following al-ḥājj 'Umar would win was essentially a personal one.[43] As the Tijāniyya elevated its members into a spiritual class above all others, with claims to favoured treatment on the Day of Judgement, it inevitably attracted wide support in the excitement of the approaching end of the world and in the lull following the success of the jihād in Sokoto and Masina. As no regime could afford the disruption which his arrival and proselytising implied, al-ḥājj 'Umar was forced to find a permanent place for his community outside the existing Muslim states. From such a base he could then start a jihād and build his own state.

At this time, in the 1840s, the only remaining major pagan states in the western Sudan were the twin kingdoms of the Bambara at Segu and Kaarta. As Segu was within the nominal sphere of Masina, al-ḥājj 'Umar directed his forces against Kaarta and the buffer principalities on the upper Senegal which lay between Kaarta and the edge of Futa Jallon where he had established his Tijānī community.[44] Being without any local support, he had to rely on his students and any recruits he could attract from the Fulani of his home area of Futa Toro and neighbouring Futa Jallon.[45] To make his small numbers effective against the Bambara armies, he imported guns and powder with the profits of the trade he had built up. This is the first occasion therefore in the history of jihād in the Western Sudan that a professional army, trained to use guns, launched an invasion on a state with which it had no previous connection. The invasion was

43 In *al-Rimāḥ* 'Umar emphasises the utter pliability of students in the hands of their Shaikh (chs 18, 19, for example).

44 The main town associated with this period of 'Umar's life was Dinguiray, but he moved his headquarters several times. Dinguiray became well known later as the capital of the Futa Jallon province of the 'empire' after 'Umar had died. The relationship between Segu and Masina is not clear; it seems there was a truce, or at least a cease-fire, between them at this time.

45 In so far as all recruits might become members of the Tijāniyya it is proper to call them 'students' (or Tālibés). This does not necessarily imply a strict practice of Islam.

A letter from Muḥammad Bello (Caliph 1817–1837) to 'Umar Dadi (of Kanoma) giving the date and place of rendezvous for a campaign. The numerological square is a later addition.

intended as a permanent conquest with subsequent colonisation or garrisoning. Unlike other jihād movements it was neither a revolution nor an independence movement. Al-ḥājj 'Umar was chronically short of men; as his conquests increased, more garrisons were required. At the same time, his armies had to live off the land. On occasions shortages were so desperate that the women and children who followed the camps had to be driven off.

To keep the homelands of many of his students free from French control he had to divert some of his forces on to the upper Senegal. When he failed to stop the French militarily, spheres of influence were negotiated which left him free to hold Kaarta.[46] From Kaarta he turned towards the last remaining kingdom, Segu. The Masina caliphate under Aḥmad, grandson of Aḥmad Lobbo, grew alarmed at the prospect of a Tijānī-dominated Segu on their borders; so Segu was hastily claimed as a pagan protectorate.[47] Attempts at mediation by scholars on both sides failed to restrain the two leaders. The two jihād armies clashed and though 'Umar's smaller numbers were nearly overrun their superior technology overcame the traditional cavalry of Masina. 'Umar's forces were able to replace their ammunition in a few days, and with their guns they swept through Masina, to settle as the permanent rulers of the new caliphate.

For the first time one jihād movement had annihilated another. Though there had been incidents between them, none of the Sokoto emirates had conquered another within the caliphate; similarly, in the smaller state of Masina civil war had been restrained. But this was in the nature of those movements; the early jihād consisted of internal revolts, led and manned by local men. 'Umar's forces, his students of the Tijāniyya brotherhood, were drawn from all over West Africa and from different strata of society. Some of his leading commanders were Hausa and Bornu men, possibly ex-slaves, who were outside local Fulani politics and clan rivalries. Their loyalties, and their future, depended on one man in a way the position of commanders in the other movements did not.[48]

'Umar was perhaps the cleverest, certainly the most widely travelled, of the three jihād leaders; his major book on the Tijāniyya, *al-Rimāḥ*, was written in the margins of the one crucial Tijānī text, and it achieved a wide circulation. In it he writes condescendingly of the other states, he

46 Al-ḥājj 'Umar's conflict with the French, though it is the preoccupation of many students of his career, is not central to his jihād. It was important for him and his followers to keep their homeland free from Christians, but he had no desire, it seems, to make Futa Toro a part of his 'empire'.

47 When 'Umar threatened Segu, Aḥmad III (son of Aḥmad, the son of Aḥmad Lobbo) negotiated a protectorate over it, by which it would cease to be pagan in return for military aid from Masina. 'Umar, when he had captured Segu, made much of the idols he found in the town, and claimed that Masina was thus proved to be hypocritical.

48 For example C. A. L. Reichardt, p. 312.

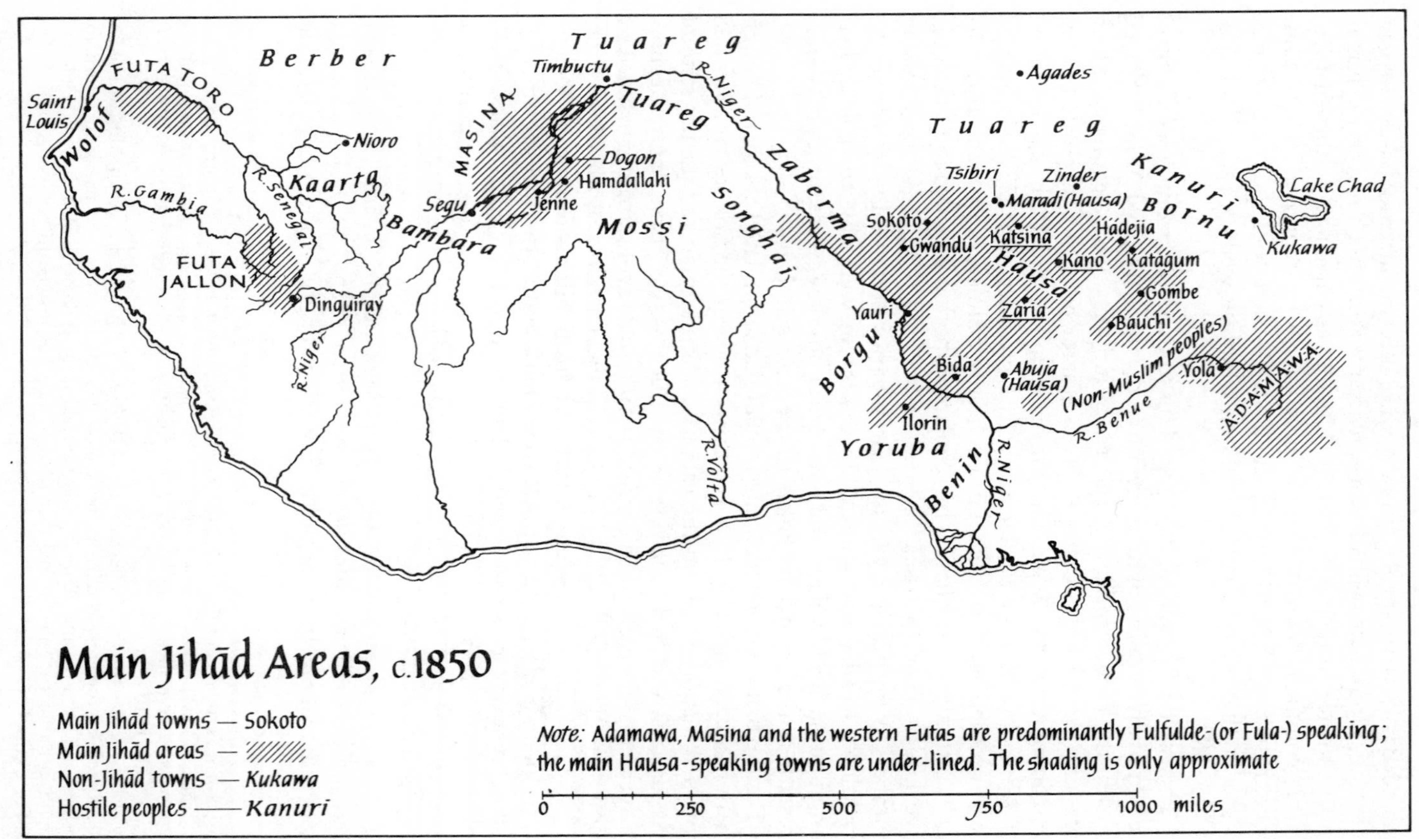
Main Jihād Areas, c.1850
Main Jihād towns — Sokoto
Main Jihād areas —
Non-Jihād towns — Kukawa
Hostile peoples —— Kanuri
Note: Adamawa, Masina and the western Futas are predominantly Fulfulde-(or Fula-) speaking;
the main Hausa-speaking towns are under-lined. The shading is only approximate
0 250 500 750 1000 miles
Saint Louis
Wolof
FUTA TORO
Berber
R. Gambia
R. Senegal
FUTA JALLON
Kaarta
Nioro
Dinguiray
R. Niger
Bambara
Segu
MASINA
Jenne
Hamdallahi
Dogon
Timbuctu
Tuareg
Tuareg
Mossi
R. Niger
Zaberma
Songhai
R. Volta
Borgu
Yoruba
Ilorin
Benin
Yauri
Sokoto
Gwandu
Bida
Tsibiri
Maradi (Hausa)
Katsina
Hausa
Zaria
Abuja (Hausa)
R. Niger
Agades
Tuareg
Zinder
Kano
Hadejia
Katagum
Gombe
Bauchi
Kanuri
Bornu
(Non-Muslim peoples)
R. Benue
Yola
ADAMAWA
Lake Chad
Kukawa

I : 2

describes Sokoto as 'a land in ruins', and emphasises his own supremacy.[49] Neither he nor the last of Masina's rulers was ready to compromise; although there is nothing to show that 'Umar had intended to conquer Masina after Segu (there were the lands south of Segu which he was planning to attack at the time of his death) the provocation was more than his pride forgave. The arguments, preserved in letters between the two states, echo the dispute between Sokoto and Muḥammad al-Amīn al-Kānamī; they merely justify the positions already taken. But even more clearly than in the Bornu–Sokoto clash, both 'Umar's students and the Masina forces were Muslims beyond doubt. Their dispute, therefore, ceased to be a matter of jihād.

The tragedy involved not only the destruction of the Masina caliphate but also the early death of 'Umar. Yet, though he died, his sons maintained his conquests and developed an administration which gradually transformed itself from military rule to a more broadly based regime using local rulers and advisory councils. For the most part, however, the Bambara remained unconvinced Muslims, if not pagans, and there could not be an intensive Islamisation of society. The rulers and the ruled remained distinct in a way they did not in the more homogeneous societies of Masina and, to a lesser extent, Sokoto. Furthermore al-ḥājj 'Umar died before he was able to create any plan of administration for the postwar period: his movement did not have the time to develop policies of social and political reform for the societies it governed.

SOME CONSEQUENCES OF THE JIHĀD

The explicit purpose of the jihād in Sokoto and Masina was to establish a society governed by the *Sharī'a*, a society where men could be Muslims under a Muslim administration. In both areas Muslims had been so harassed by nominally Muslim governments which had been in existence in one traditional form or another for several hundred years that by the nineteenth century there seemed to reforming Muslims to be no alternative but to replace the old regimes entirely. The organs of the new administration were to be based on classical Islamic models described by Arab theorists such as al-Māwardī, and the Law to be enforced was that developed in the Middle East.[50]

In order to accommodate the Law the old culture had to be transformed, especially with regard to cults and beliefs which implied the existence of

49 Al-ḥājj 'Umar, *al-Rimāḥ* (1345 H. edition), i, 210–11.

50 See, for example, 'Abdullāh b. Fūdī, *Ḍiyā'al-ḥukkām*, Zaria, *c.* 1965, written in 1807 to guide the Kano community on setting up their state. This book is quoted as an authority by Aḥmad Lobbo in his argument with Sokoto over paying tax. The experience of early Islam was not, of course, irrelevant to West Africa; Muslim administrators from the beginning had faced the problem of integrating the nomad Beduin, the urbanised merchants of Mecca and the peasant farmers of conquered lands such as Syria and Iraq into the new Islamic community.

more than the One Deity. It was argued that many of the pagan practices were social traditions rather than religious rites, in the same way as customs from the *Jāhiliyya* were still to be found in Arab lands.[51] But syncretic practices which seem innocuous in established Islamic societies appear too dangerous a precedent to overlook in partially Muslim areas like Bilād al-Sūdān. Thus, for example, the Fulani games in the forests were finally banned, although the arguments for their legality were well known from earlier disputes among scholars.[52] Similarly the laws against dances and ceremonies included a ban on the accompanying musical instruments. The use of molo, kalangu and goge was specifically mentioned by the Shaikh as contrary to the Laws of Islam.[53]

Women had been important participants not only in specific fertility cults but also generally in social gatherings. In pastoral Fulani society, as in the pagan peasant societies of the Hausa and the Bambara, women were free to work and play outside the house. In contrast, women of the Muslim communities before the jihād had been distinguished by wearing veils while some Fulani scholars went further and insisted on keeping their womenfolk at home.[54] With the introduction of Muslim government, an attempt was made to restrict women to activities within the compounds of their houses and to allow only young children and servants to circulate freely. At the same time households became larger; the number of wives was limited to four at any one time, but the number of concubines was not limited (the pastoral Fulani, due to cultural and economic conditions, were seldom able to have more than two wives at most). As children were crucial for the growth and strength of the Muslim communities and their maintenance was now no problem, large families were encouraged. The Shaikh ʿUthmān, for example, had thirty-seven children, his son Bello seventy-three, his son-in-law Gidado forty-eight. Although many children died young, genealogies record a vast growth in numbers in the nineteenth century.[55] But marriages were more than child-producing. The Shaikh encouraged selecting wives in order to break up the clannish patterns of the

51 This was the line of argument taken by al-Kānamī in his defence of Bornu against the crticisms of the Sokoto leaders. The argument (with a refutation) is quoted in Muḥammad Bello, *Infāq al-maisūr*.

52 Gloss on Zād b. Muḥammad Sa'd, *Khulāṣat al-qarā'iḥ*; cf. Muḥammad Bello, *Infāq al-maisūr*, ed. Whitting, London, 1951, p. 23.

53 The Shaikh's attitude to music, early expressed in his Fulfulde poetry, was later modified in such books as *Najm al-ikhwān,* (University of Ibadan). *Molo* is a kind of guitar, *goge* a fiddle, and *kalangu* a small drum. The war drum (Arabic, *ṭabl;* Hausa, *tambari*) is usually permitted in jihād.

54 The role of women is discussed, for example, in ʿUthmān b. Fūdī, *Nūr al-albāb;* ʿAbdullah b. Fūdī, *Tazyīn al-waraqāt,* pp. 86–7; *Irshād al-ikhwān ilā aḥkām khurūj al-niswān* (National Archives, Kaduna), attributed to the Shaikh.

55 Pre-jihād genealogies for the family of the Shaikh and their relations are given in Muḥammad Bello, *Ishāra wa-iʿlām* (collection of Wazirin Sokoto). Nineteenth-century genealogies are well remembered, especially by the women of the house.

old society; although the Fulani custom of marrying a close cousin as the senior wife persisted, the other wives were often taken from widely differing areas. With the increase in the number of children, and as the children of wives and concubines had the same legal status, the opportunity for marriage ties grew. In this way the community remained close-knit while extending its links throughout the Caliphate.

One consequence of the use of concubines soon affected the pattern of Muslim society. Concubines were mainly drawn from captive women who were thus never Fulani and, later, seldom even Hausa. In consequence, perhaps a majority of the children born in the new community in Sokoto were brought up by non-Fulani mothers. Furthermore, owing to the practice of giving a young man a concubine before he assumed the responsibility of marrying his cousin, the eldest son was often the child of a concubine. Thus ʿAlī Babba, the senior son of Muḥammad Bello and the Caliph for seventeen years, was the son of a concubine and his was not an isolated case. While undoubtedly the concubines, usually young girls, imbibed the Fulani culture of their new homes they in turn imparted something of their own upbringing, so creating a hybrid society united in its Islamic basis.

While a large proportion of captives were women and girls, considerable numbers of men were brought as slaves into the society of the Caliphate, and again these were often non-Hausa in origin. Apart from their influence on their owners and their children, certain economic changes resulted. Women, at least those of rank, were released from labour outside the house thus making purdah possible (and later causing a social distinction between those whose wives were cloistered and those whose wives had to work). Of greater importance was the fact that it enabled some men to be free from labour, thus creating a class who could devote themselves to administration, scholarship or, less idealistically, to 'being gentlemen'.[56] The problem of a leisured class losing its ideals was recognised early; Muḥammad Bello describes in a short pamphlet the various forms 'being a gentleman' might take, and much of the post-jihād literature in Sokoto and elsewhere was aimed at rekindling the ideals that had sparked off the jihād.[57]

Crucial to the maintenance of ideals was the attention paid to Islamic learning. The influx of slaves which allowed a life of leisure for the administrators also helped to support an economically unproductive class of students and scholars. Whereas earlier much of the scholarship of Hausaland had depended on either royal patronage or trade for a living, the new

56 The Hausa states were also great importers of labour in the form of slaves, but the scale of raiding and the size of the catchment area was nothing compared to that of the Sokoto Caliphate, where captives were a common form of present or tax-substitute for subordinate emirates.

57 Bello's pamphlet is paraphrased in Last, *The Sokoto Caliphate*, p. 59.

jihād states were, at least in part, governments by scholars for scholars. It is symbolic of this that Aḥmad Lobbo, during a crisis in Masina over the insolence and outrageous demands of students, should unequivocally take the side of the students.[58] Education was not given selectively; all children received elementary instruction and women were also expected to be educated. In some Fulani houses primary education was in the hands of the elderly women of the house and women scholars and poets were by no means unknown. Captives were also taught the principles of Islam, as were their children (although the status of a slave did not change on his conversion to Islam).[59]

One aspect of the general rise in Islamic knowledge was the spread of sufism. The Shaikh's community before the jihād had practised the Qādirī way, and the assemblies for recitation continued in the wider society of the Caliphate. The sufic brotherhoods seem to have provided the exclusive sense of community and religious enthusiasm which the Caliphate had otherwise ceased to satisfy. The popular orthodox definition of *taṣawwuf*, however, is 'being good' and this lacks the mystical element for which the early sufis were famous. It was perhaps because of this lack, particularly evident in the official Qādiriyya brotherhood, that adherents of the new and more esoteric Tijāniyya increased so rapidly in the mid-nineteenth century.

To initiate the necessary social and political reforms, state intervention in daily life was inevitable. In Masina, where among the pastoral Fulani a centrally organised system of government was lacking, the entire framework of Islamic administration had to be introduced. In order to establish this administration and the changes in culture which it implied, the state was bound to be deeply involved in creating and enforcing the new regulations. Thus in some aspects the Masina government was closer to the classical Islamic model than was the Sokoto Caliphate. In the long-established city-states of Hausaland the existing tradition of urban life had been adopted with little disruption. However, Masina did adopt certain traditional practices—the caste system is an example—and it is difficult to judge the extent to which the official regulations were modified in practice. Only the rules and extreme exceptions are recorded; the daily compromises have to be assumed.

Some account has already been given of the way in which the state in Masina controlled the life of the new society. The high degree of centralisation which made possible such a precise application of the Islamic model was due to a number of factors. The most important one of these was the

58 Ba and Daget, pp. 55–6.

59 Though conversion did not necessarily bring emancipation, the sale was restricted both of concubines who had borne their owners children and of the children themselves. Emancipation of slaves, however, was at all times regarded as a virtuous act.

small area and population of Masina (equal to about one major emirate in Hausaland), the unusually good communications afforded by the numerous waterways, and the ethnic homogeneity of the people. By contrast, the Sokoto Caliphate was vast and diverse. A special messenger, for example, might reach Yola from Sokoto in about seven weeks, and although most of the ruling groups spoke Fulfulde or Hausa, there were Nupe and Yoruba, and many smaller groups, within the Caliphate. While Masina could be centralised, Sokoto administration was necessarily a centrally supervised federation bound together by the common practice of Islam and by respect for the legitimacy which Islam, and the jihād, gave to the Caliph and his representatives.[60] Within the Caliphate, despite the universal validity and application of the Sharī'a, divergences were possible. The emirates, for example, show differences in the status and location of officials and their succession to office. The use and rates of tax also seem to have varied widely.[61] Due to the strong pressure against luxurious living in both Masina and Sokoto, wealth was seldom used for personal ostentation. A man's position, instead, was enhanced by his generosity, since those who received his gifts thus came under some obligation. As titles, rather than individuals, traditionally held the key to wealth, followers or clients tended to be linked to offices. In general every person whether corporately in his village or individually, was a client to some office or person. Furthermore, as the Islamic law of inheritance divides property between relatives and all the children, personal wealth was seldom maintained over generations. To some extent, therefore, a class structure by wealth was avoided.

Birth, however, was often a factor in official appointments. The family of Shaikh 'Uthmān, for example, in later years tended to hold the majority of important posts in the state. An office was usually given to the senior member of the family eligible for the title, the original titles having been awarded by the Caliph or Emir. But the large number of junior relations dependent on an office-holder made the office and its subsidiary titles almost a corporate appointment. Thus the family network, with its marriage links extending to other similarly titled families, fostered its own form of centralisation.

The degree of centralisation and class structure is reflected in the armies, which in both caliphates were bodies of non-professionals levied from the citizenry. In Sokoto by the 1820s, retainers followed their lords, titled officials or emirs on major expeditions, in the manner of the Hausa states. The followers depended largely on their leader to obtain, from the state or on

60 In its degree of decentralisation, Sokoto was similar to the early Caliphate of Islam with its distant provinces of Egypt and Iraq. The only virtue of centralisation lies in the extent, if any, to which the *Sharī'a* can thus be more closely enforced.

61 The emirate of Zaria, for example, as described by M. G. Smith, *Government in Zazzau*, Oxford, 1956, appears to be very different in organisation from Sokoto or the Fulani emirates of the Bornu/Sokoto frontier studied by Dr V. N. Low (*Three Nigerian Emirates*, Evanston, 1972).

his own account, sufficient arms and horses for them. In Masina where, despite castes, a more egalitarian social structure survived, conscripts were drafted from every family and the state supervised payments for arms and supplies of rations at granaries in strategic towns.[62] But nonetheless the key appointments of governors and military commanders seem to have been reserved for members of Aḥmad Lobbo's family or for his close associates and their sons. Later an élite cavalry corps grew up composed of young 'noble' volunteers. At the local level in Sokoto, an integral part of the defence system was the use of ribāts, village or town strongholds along the frontiers. These were manned by titleless sons of important families under the command of a resident emir to whose family (usually a branch of the 'royal' house) the ribāt had originally been assigned. The emir and the various young 'noblemen' constituted the core of the local forces with peasants and herdsmen from the neighbourhood making up the infantry levies alongside the household retainers.

Security was thus maintained at two levels. The local commanders were responsible for the immediate defence of their area and for giving advance warning to the central administration which was in overall charge of punitive campaigns and strategy. In the jihād states, defence involved not merely protecting towns and farming communities but also guarding cattle during the annual movements in search of pasture. In Masina, where cattle had the predominant role in the economy, military strategy centred on keeping the grazing grounds safe from Bambara raids or Tuareg competition. In particular Tuareg pressure southwards, especially in the area south of the Niger at Timbuktu, was fiercely resisted. The control held over the Songhai by the Tuareg and their allies, the scholastic Arab family of the Kunta, was broken in the course of several campaigns. In Sokoto where the process of settling the pastoral Fulani was more intensively applied, the early escorts for the herds gave way entirely to annual harvest-time raids against the refugee pagan states which, often in alliance with the Tuareg, used to raid across the frontiers.

It is easy to exaggerate the insecurity which such pagan raids might seem to have engendered, and which struck Barth so forcibly in the 1850s.[63] Local defence systems, whether of walls or warnings, were effective against most attacks and there is little evidence that farming was seriously curtailed or that men were forced to migrate to avoid warfare. Instead, areas were gradually opened to settlement and new towns built far from the metropolitan centres of power. Similarly, trade continued despite wars; traders were usually protected persons exempt from military duties (in Masina, in return for a tax) and were able to pass through the frontiers of states at war. The early fighting was largely a war between élites with

62 Ba and Daget, pp. 67ff.
63 For example, Barth, iii, pp. 117, 146.

peasants remaining neutral, as far as possible, paying dues to those who had the mastery, and sometimes therefore paying twice.

In general within the metropolitan areas of the Sokoto Caliphate increasing settlement and security brought prosperity. Not only did the forcibly imported labour contribute to agricultural and craft production but it also stimulated trade. The influx of so much new 'currency' contributed to the wealth of the community and with it imports from North Africa grew. In the trade boom there was a greater diversification of manufactures. Goods were exchanged with ease within the Sokoto Caliphate, credit was available on distant towns and the Benue–Niger traffic competed with the Saharan routes for supplying goods and shipping out the exports of a trade area of several million inhabitants.

Within the prosperity and internal unity of the two Caliphates of Sokoto and Masina, Islamic learning became not only more widespread but also more sophisticated. The books of Sokoto circulated widely; Arabic was the medium of official correspondence and the scholastic learning of the early leaders ensured at least an ideal that was other-worldly. The all-pervasive influence of faith affected practices such as medicine and fortune-telling, the use of eunuchs and the rights of prostitutes, the smoking and chewing of tobacco. Tradition gave way to Islamic law and the tenor of society at all levels became gradually, although never completely, Muslim–Arab. But undoubtedly the basic attitudes of Islam were accepted in the new society as they had never been before and in this way the jihād was successful. The cost was the denial of certain aspects of traditional culture among the Fulani, Hausa, Nupe and Yoruba, and, in the Sokoto Caliphate, the disruption of the lives of several thousand families of non-Muslim peoples every year. In Masina the jihād was more nationalist than expansionist; but it too had to fight the pagan peoples to the south and east. The Dogon were subdued and incorporated as protected peoples and the Bambara and the Mossi were pushed back. As non-Muslims they had been offered the choice of becoming Muslim or of becoming protected peoples with a right to their own faith guaranteed in return for a payment. The final choice was to fight; but peace was preferred to war, and in these jihād societies military prowess was rated lower than scholarship. In this way the early ideals of the Shaikh's community, the peaceful aspects of jihād, were perpetuated. In inspiration, the West African jihād for the most part was a war more against impiety than against peoples; and its success, as exemplified in Masina or Sokoto, lies more in the nature of the society established than in the extent of its conquests.

CHAPTER 2

Reform in West Africa: the abolition of the slave trade

CHRISTOPHER FYFE

From at least the sixteenth to the eighteenth centuries the slave trade was the basis of the import–export economy in coastal West Africa. It was a well organised trade. Usages grew up over the centuries which enabled it to be conducted in an orderly fashion. In many areas credit was given, an indication that trading partners, African and European, trusted one another. Nor is this surprising. Each needed the commodity the other provided. The Europeans needed slaves; the Africans needed manufactured imports. Even when trading relations were broken off by a quarrel (as they often were) hostilities were usually short-lived for neither party was ready to renounce for long the benefits conferred by the slave trade.

Both sides obtained the best bargain they could. African vendors resisted, usually with success, any attempt to alter the terms of trade to their disadvantage.[1] Slave prices rose steadily from the late seventeenth to the late eighteenth century. Nevertheless the European purchasers had the ultimate advantage. In return for expendable consumer goods they received slaves, who were sent across the Atlantic to work as a labour force to create wealth. The slave trade and the plantation economy it fed provided profits and employment on a large scale in Europe and America as did the allied shipping and manufacturing industries. In this way the West African market supported and extended the industry of Europe, and the West African economy was tied to the European capitalist economy and its transatlantic extensions.

The slave trade plainly affected the social structures of the coastal states. The existing distinctions between privileged rulers and the unprivileged masses became more marked. The former benefited by the trade, receiving the goods given in exchange for slaves. The latter were liable to be sold. Though some historians have maintained that slavery is

1 K. Polyani, *Dahomey and the Slave Trade,* Seattle, 1966; M. Johnson, 'The Ounce in 18th Century West African Trade', *Journal of African History (J.A.H.),* vii, 2, 1966, pp. 197–214.

'age-old' in coastal West Africa, it has been convincingly argued that before Europeans arrived on the coast to buy human beings there was no special category of 'slaves', in the sense of people liable to be sold, only subjects, obliged to obey.[2] If this is so, the slave trade brought a radical change to West African society, divesting a large section of the population of their human rights, and turning them into potentially saleable goods.

The slave trade also affected West African legal systems. Those found guilty of crimes could be sold, instead of being punished in other ways. Hence it could be in a ruler's interest for his subjects to commit crimes. He was tempted to exploit his subjects instead of protecting them—tempted to use the mechanisms of justice for the sake of profit rather than to maintain law and order. Warfare too became more profitable, and therefore presumably more prevalent, once the victors could sell their captives.

Nevertheless the basic structures of society seem to have survived. There was no breakdown of authority. The polarity of rulers and ruled continued. No middle class emerged. Those who benefited by the trade either were already themselves rulers, or became rulers, gaining political but not economic power. They did not generate wealth for themselves like the capitalists of Europe. In any case social pressures prevented them from accumulating private wealth. A ruler's obligation to redistribute his profits to his people generally prevented individual capital formation. Thus the balance of society, based on a principle of reciprocity which ensured a narrow spread of wealth differentials among the privileged, remained unimpaired. Coastal West African society proved flexible and resilient enough to adapt to the slave trade.

Indeed, far from destroying African society by innovatory contact, Europeans were obliged to conform to some of its usages. Those who settled down to trade put themselves in the power of their African 'landlord', and, as his 'strangers', paid him customary dues.[3] Though European governments (and subsequently many European historians) may have considered that the trading posts their citizens occupied in West Africa were colonies, they did not in fact obtain sovereignty over them. European traders had to pay rent or other customary dues for their trading establishments, whether hut or castle. Sovereignty remained vested in an African authority. Europeans had no rights over the soil and could be dispossessed if they refused to pay.

2 Walter Rodney, 'African Slavery and other Forms of Social Oppression on the Upper Guinea Coast', *J.A.H.*, viii, 3, 1966, pp. 431–44; Walter Rodney, *A History of the Upper Guinea Coast, 1545–1800*, London, 1970. I am deeply indebted to Dr Rodney for his illuminating analysis of the slave trade which has enabled me to reformulate my own views.

3 V. R. Dorjahn and C. Fyfe, 'Landlord and Stranger', *J.A.H.*, iii, 3, 1962, pp. 391–7; C. Fyfe, 'Landlords and Brokers', *Markets and Marketing in West Africa*, Centre of African Studies, University of Edinburgh, Edinburgh, 1966.

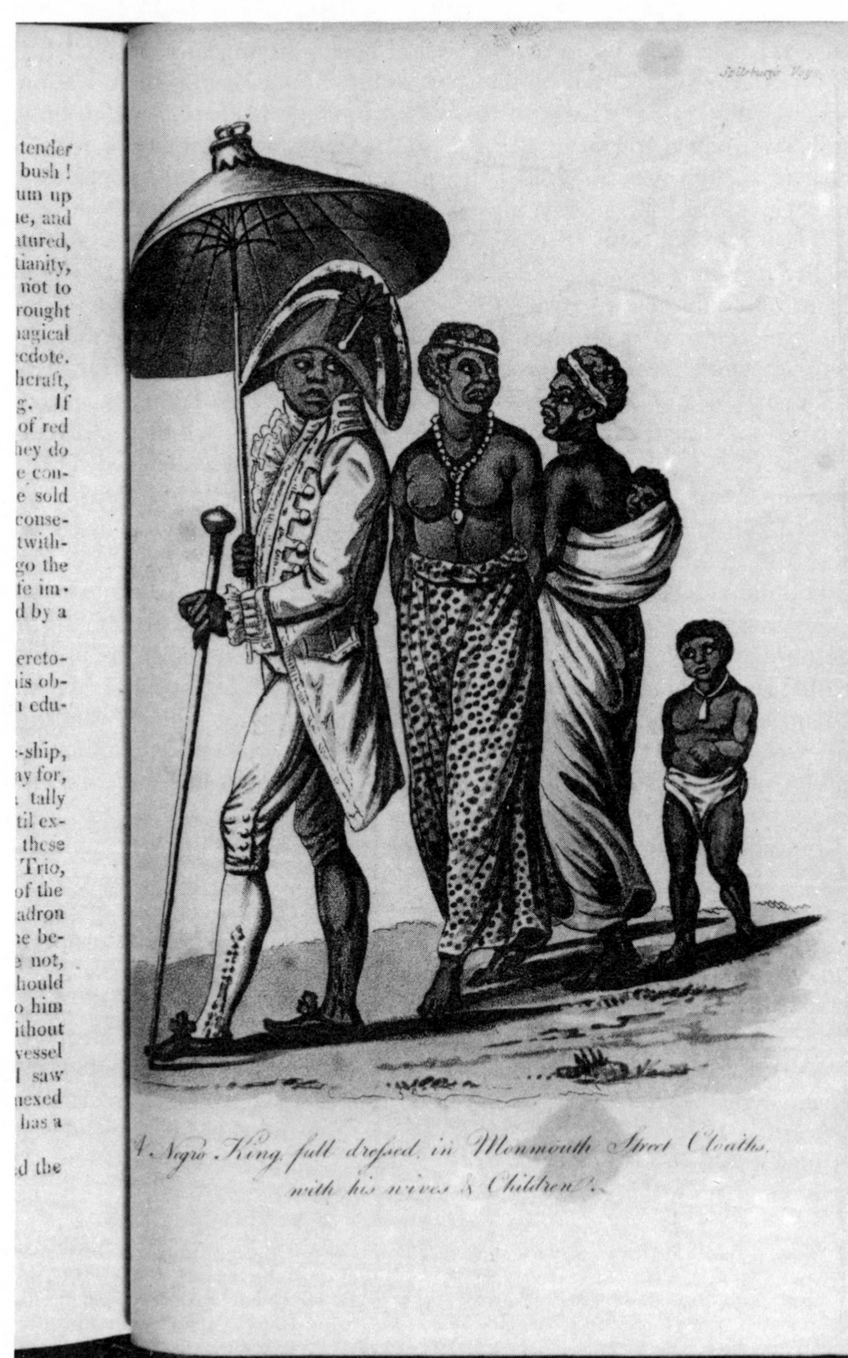

A Negro King, full dressed, in Monmouth Street Cloaths,
with his wives & Children.

An African Beneficiary of the slave trade. (The ticket round the child's neck indicates that he is in pawn for a debt incurred by his father)

Hence European slave traders did not appear in West Africa as invaders, rulers or masters. They acknowledged the sovereignty of the African authorities and paid them the respect and customary dues that recognised usage prescribed. If they claimed superiority, it could only be for their technical skills, not for any political pre-eminence. Nor, whatever some might believe, could they claim any valid moral authority. The trade they had introduced was evil and immoral. Occasional attempts that slave traders made to explain the Christian religion were received inattentively, if not with derision. The slave trade made buyers and sellers moral equals.

In coastal West Africa, then, during the slave trading era, black vendors and white buyers were on equal terms. Each traded willingly with the other. They lived on terms of mutual respect, sometimes friendship. To use a phrase still not invented in the eighteenth century, it was an era of 'good race relations'.

OPPOSITION TO THE SLAVE TRADE

There were, however, Africans who opposed the slave trade. The enslaved victims obviously—their unrecorded protests often acted out as mutiny or suicide. At least one African king, Agaja of Dahomey, tried to suppress it in his own country.[4] In Benin the kings prohibited the export of male slaves for the greater part of the eighteenth century.

Outside Africa, too, voices were raised against it almost from its inception. Even Bartolomé de las Casas, who originally proposed the introduction of African slave-labour to the Spanish-American colonies, lived to condemn his proposal. The beginning of the campaign that brought about its illegalisation and ultimate suppression is usually credited to some American Quakers who met together in 1688 in Germantown, Pennsylvania, to protest against it. Other Quakers joined the opposition. So did members of newly articulate groups within the other Protestant churches in Britain and America who were beginning to see their religion as a potentially ameliorative rather than a submissive force. By 1787 it was possible for an English clergyman, the Reverend Thomas Clarkson, to organise a group of influential people in London into a Committee for the Abolition of the Slave Trade and for William Wilberforce to speak on its behalf in the House of Commons.

These abolitionists (as they were called) did not oppose the slave trade merely as an evil in itself. By the end of the eighteenth century European technology was developing rapidly, far outstripping technology elsewhere. Increasingly Europeans came to assume that the manners, customs and morality of Europe were also self-evidently superior and to believe that their spread over the rest of the world could only be for the advantage of its inhabitants. European ways could therefore be regarded as 'civilisa-

4 *The History of West Africa,* eds J. F. A. Ajayi and M. Crowder, London, 1971, i, ch. 9.

tion' and be contrasted with the 'barbarism' of Africa. Hegel, perhaps the most influential European philosopher of the early nineteenth century, adopted this basic assumption unquestioningly.

Eurocentrism reinforced the exclusive claims of the Christian religion. The Protestant churches, hitherto scarcely conscious of any mission towards non-Christians, began from the late eighteenth century to found missionary societies with the specific aim of preaching Christianity to all mankind. Catholics too were inspired to reinvigorate their missions. The belief that the Christian religion (not in origin European, but annexed by Europeans) and European ways must be made to prevail over the earth was further strengthened by theories which assumed white people to be superior to non-white—theories that were to be scientifically formulated in the mid-nineteenth century as the ideology of race.[5]

The slave trade stood as an obstacle to the realisation of such dreams. While Europeans were still trading in slaves they could not in good conscience assert their moral superiority over Africans. Hence those who had visions of spreading European civilisation in Africa had to begin by destroying the slave trade. Only then could they claim to be morally superior. With the slave trade once abolished they could advance into Africa not merely unashamed, but with the self-righteous zeal of liberators who demand gratitude from the liberated. Ending the slave trade was therefore an essential stage in the European conquest of Africa.

The British opponents of the slave trade did not rely entirely on humanitarian arguments. Though Clarkson organised his movement on a nationwide basis with all the devices that contemporary propaganda methods could then command, the response would have been far smaller had he been able to use only the language of humanitarianism. Indeed it may be questioned whether humanitarianism alone could possibly have defeated the slave trade. For generations the slave trade had brought great wealth to the entrepreneurs who financed it, the plantation owners who depended on it for labour, and the manufacturers of the trade goods used as an exchange medium. Bristol and Liverpool had grown rich on it; the West Indian sugar industry relied on it; it gave employment to shipbuilders and sailors. Massive entrenched commercial interests were implicated.

Yet it could be argued that the slave trade was uneconomic. Adam Smith, the Scottish economist, the apostle of free trade, demonstrated in his famous book *The Wealth of Nations*, published in 1776, that slave labour was wasteful. Free trade principles, at that time a new and exciting technique for advanced intellectuals, also demanded that monopolies like the British West Indian sugar industry, which alone was allowed to import sugar into Britain, must be swept away. It followed that those who attacked

5 See pp. 55–6 below.

the slave trade were also undermining the sugar monopoly and striking a blow for free trade.

Also, the increasing volume of manufactured goods now being turned out by the factories of industrialising Britain demanded new, wider markets. It was argued that vast untapped markets lay open in Africa—could the slave trade only be abolished, and Africans be treated as customers instead of as merchandise.

Such economic arguments appealed to hard-headed businessmen who might have remained deaf to purely humanitarian appeals. Factory owners in new British industrial towns remote from slave-exporting centres were readier to sympathise with the sorrows of far-off slaves than with those of their own workers—particularly if they were given reasons to make them believe it might enrich them in the long run. So Clarkson was not content to publicise the cruelties of the slave trade. He also investigated what alternative African commodities could be discovered to replace its profits after it had been abolished.

Two Africans living in England also wrote books against the slave trade.[6] Ottobah Cugoano, a Fante, published *Thoughts and Sentiments on Slavery* in 1787; Olaudah Equiano (also known as Gustavus Vassa) published his *Interesting Narrative* in 1789. Both had made their way from West Indian slavery to freedom. Equiano's book is an autobiography and contains valuable accounts of his Ibo homeland.[7] To appeal to their British audience both used the economic as well as the humanitarian arguments. Cugoano called on the British Government to send out a fleet to suppress the slave trade. Most of his first readers must, in 1787, have thought his proposal fantasy. Yet only twenty-one years later the British anti-slave-trade squadron was to begin its long campaign in West Africa.

THE SIERRA LEONE SETTLEMENT[8]

Wilberforce and Clarkson receive most of the credit for the success of the British anti-slave-trade campaign. Yet they built on the labours of a less obtrusive collaborator, Granville Sharp. Unlike them, Sharp really got to know Africans in England personally—indeed he was befriending Africans twenty years before they took up the cause. It was he who brought before the law courts James Somerset, a former slave from the West Indies, who was being claimed by his old master, and succeeded in extracting from the Lord Chief Justice, Lord Mansfield, in 1772, the decision that he must be left at liberty.

Somerset's case is often misrepresented by historians. It did not establish unequivocally that slavery could not exist in England (whatever poets might imagine on the subject), only that a former slave could not be sent

6 For Africans in England see James Walvin, *Black and White*, London, 1973.
7 Both books have been reprinted with introductions by Paul Edwards (London, 1969).
8 For a detailed account see Christopher Fyfe, *A History of Sierra Leone*, London, 1962.

back from England to slavery in the colonies.[9] Nor did it impel black servants to leave their masters *en masse* (as some historians have asserted) to take to a life of freedom on the streets. Servants in eighteenth-century England, black or white, needed masters; those who left them knew they were heading for trouble. There is no evidence of any concerted movement by the African population of England in the years immediately following Somerset's case.

Not for another dozen years was there any large-scale vagrancy among them. Then, at the end of the War of American Independence, a big influx of people of African descent came in from across the Atlantic. Some were discharged soldiers and sailors who had served in the British forces. Some were former slaves who had escaped from their American masters. Many congregated in London, destitute and unemployed. Sharp made friends with them, formed a committee to raise money to support them, and proposed that they be sent to found a new home of their own in Africa. The British Government, only too anxious to be rid of them, agreed to ship them to Sierra Leone.

The Sierra Leone estuary had been an important trading centre since the first period of contact with Europeans in the fifteenth century. The wide river mouth with deep water along the southern shore offered one of the few safe anchorages on the surf-bound coast. A fresh stream supplied water for passing ships, and gave the ruler who controlled it regular opportunities for trade, as well as revenue from the watering dues paid by the ships' captains who took on water. By the eighteenth century it was controlled by a Temne ruler, the King of Koya. This was the site selected for the immigrants.

Sharp did not regard the settlement merely as a dumping ground for unwanted black vagrants as the British Government did. Unquestioningly dedicated to introducing into Africa the Christian religion and the moral principles he inferred from it, he was inspired with the utopian dream that these freed slaves should found an egalitarian African community of liberty-loving Christian people in the heart of the slave-trading coast.[10] He named it 'The Province of Freedom' and drew up a constitution to enable its citizens to govern themselves.

The settlers reached Sierra Leone in May 1787. The naval captain in charge of transporting them negotiated a treaty with King Tom, the Temne chief at the watering-place, who made over to them a strip of land to the east of the peninsula that still bears his name. The settlers landed and elected one of themselves, James Weaver, governor, according to Sharp's

9 E. Fiddes, 'Lord Mansfield and the Sommerset Case', *Law Quarterly Review*, l, 1934, pp. 499–511. The spelling 'Sommersett' was used in T. B. Howell, *A Complete Collection of State Trials, 1771–77*, London, 1814, xx, 1, p. 72. Granville Sharp's spelling was 'Somerset'.

10 John Peterson, 'The Enlightenment and the Founding of Freetown', *Freetown: A Symposium*, eds E. Jones and C. Fyfe, London, 1968.

The Province of Freedom. (View taken from near the site of the present State House, Freetown)

constitution. Then the naval ships sailed away and left them on their own.

From the first the settlement met with disaster. The heavy rains began as they landed before they had time to build adequate houses. Many fell sick and died. Those that survived into the dry season found the country was unsuited to the kind of farming they were used to. Gradually they began to drift away.

King Tom was subordinate to Naimbana, the Regent of the Koya Temne. When he died Naimbana declared that a new treaty must be negotiated. This was the accepted practice in relations between African rulers and their European 'strangers'. The tenancy was made with the individual ruler and at his death it had to be renewed (with payment of customary dues) with his successor. Naimbana therefore was following established precedent and assumed that the settlers were doing the same. Yet the new treaty he put his mark to (for he could neither read nor write English) in 1788 declared that he was giving up all claims over the land he had granted in return for a consignment of consumer goods. Without realising it he renounced sovereignty over part of his country to aliens—which no West African ruler had done before. A completely new principle of sovereignty was now established in West Africa.

The Province of Freedom was short-lived. In 1789 the settlers took sides with some neighbouring European slave traders against King Tom's successor King Jimmy. In retaliation he burnt their town and they dispersed over the countryside.

As Sharp had no money left to refound the settlement he sought help from richer supporters of the abolition movement. In 1791 the Sierra Leone Company was formed in London to take over and develop the land Naimbana had granted. Financed by shareholders' money, the Company was administered by a board of directors. The settlement was governed on a colonial pattern; Sharp's envisioned 'Province of Freedom' was forgotten. Important decisions were made in London and the senior officials were responsible to the directors, not to the settlers.

The Sierra Leone Company combined philanthropy with business. Like Clarkson, Cugoano and Equiano, its sponsors realised they could attract financial support for their humanitarian aims more easily if they offered sympathisers some return for their money. So they founded it as a trading company, hoping to make a profit for their shareholders by providing in Sierra Leone an alternative to trading in slaves. For, anxious as they were to abolish the slave trade, they believed that it would automatically come to an end once some other form of trading was introduced.

Their policy was set out in the formal instructions given to their first employees.[11] It was based on what was described as 'the true principles of Commerce'—the export of British manufactures in return for African

11 *Sierra Leone Studies,* old series, xviii, 1932, pp. 42–77.

produce. These were the principles slave traders had always followed. Thus the Sierra Leone economy remained tied to European industry, slave trade or no slave trade.

The instructions also declared that the Company's employees must not limit themselves to trade but must advance 'the introduction of Christianity and civilisation'. 'On you', they were warned, 'is devolved the Honourable Office of introducing to a Vast Country long detained in Barbarism the Blessings of Industry and Civilisation.'

Here a new note was sounded in West Africa. Africans were being condemned on ethical as well as on technological grounds. It was made clear that Europeans who renounced the slave trade were no longer equal partners in evil. Instead, their renunciation was held to confer a self-evident moral superiority. Europeans could appear on the coast as citizens of a better land, bringing moral benefits to a degraded population. They could claim to be benefactors, entitled to expect gratitude, and be resentful if it were withheld.

The gulf between Africans and Europeans was now bridged in a new way. Trade on 'the true principles of Commerce' could go on crossing it both ways (manufactures one way, produce the other) as it always had. Religion, moral standards, ideology, each until now autonomous in its own sphere, were henceforth to cross (but one way only) from the European to the African side.

Settlers for the new colony appeared across the Atlantic. Some former slaves who had escaped from their masters during the War of American Independence had been settled by the British in Nova Scotia. Planted in cold remote districts, denied the grants of land promised them, fearful that they might be pushed down into slavery again, they sent a delegate, Thomas Peters, to London to protest. He made contact with the directors who promised them a home in Sierra Leone.

About a thousand Nova Scotian settlers landed in 1792 and went to work building a town, which was given the name Freetown. Self-reliant pioneers, they had been ready to take the initiative of escaping from slavery and of making new lives for themselves in a hostile world. In Sierra Leone, after an initial period of sickness and mortality (for, like the 1787 settlers, they arrived at the onset of the rains), they settled down easily as a coherent, self-respecting community.

They too found the land unsuited for an agricultural settlement. Most of them soon abandoned farming and took to trading for produce with the neighbouring Temne and Bulom. They put their profits into building houses to live in, or to let to the Company's European employees, investing in house property as the safest (indeed almost the only available) way of obtaining a regular income from capital. Thus they inaugurated the house-owning *rentier* tradition of urban coastal West Africa.

Already in Nova Scotia they had organised themselves into churches—Methodist, Baptist and Huntingdonian (a Methodist offshoot)—small congregations grouped round a pastor. Their pastors were their leaders and led them ashore singing hymns when they landed on the continent from which they or their ancestors had been cruelly sold. They had therefore no need for Europeans to preach to them. Nevertheless they shared the belief held by their former masters, by the sponsors of the Sierra Leone Company and by the Christian missionaries who worked in Sierra Leone that European ways were superior to African, and therefore organised their churches on European models.

They were brought over to Sierra Leone by John Clarkson (a brother of Thomas Clarkson) who remained as governor until the end of 1792. Before they left Nova Scotia he swore to them that they would be granted land in their new home free of rent. Here he misrepresented completely the explicit rules laid down by the Company which stipulated that land was to be granted subject to a small annual quit rent.[12] So when, after his departure, the directors instructed the Company's officials to collect quit rents, the settlers felt they had been betrayed. Cheated of land in Nova Scotia, they feared they were being cheated in Africa too. Bred in slavery as most of them were, they were inevitably suspicious of anything that might deprive them of status; to them their land was a guarantee of freedom.

This the directors could not understand. They felt the settlers were not only unreasonable in refusing to pay a small rent, but ungrateful too, when so much had been done for them. Both sides entrenched themselves in emotional hostility—the settlers with the ardour of the persecuted, the directors with the outraged resentment of an unappreciated benefactor. Under a strong governor, Zachary Macaulay (who was later to play a large part in Britain in the campaigns against the slave trade and slavery), open conflict was averted. Deeply religious, like the settlers, he could speak to them in a language they could not ignore. He was also tough-minded enough to stop open rebellion. But after he left, the Company's government lost control. A group of extremists among the settlers began demanding that they be allowed to make their own laws and carry them out themselves without instructions from London. In September 1800 armed rebellion broke out. The government seemed to be facing defeat when a British transport unexpectedly anchored with troops and 550 new settlers on board. Together they put down the rebels.

The new settlers were Maroons from the Jamaican mountains, members of a free community of African descent who had been deported from Jamaica for making war on the government. In return for taking them into Sierra Leone the Company was supplied with a garrison of soldiers and

12 For the quit rent controversy see N. A. Cox-George, *Finance and Development in West Africa*, London, 1961, pp. 39–52.

Freetown in 1798

an annual grant. From now on the directors depended increasingly on official British support. Their hopes of making a profit had been disappointed when Britain and France went to war in 1793, for their trade was irretrievably damaged by war losses. Disillusioned by the failure of their dreams, social and financial, they became chiefly concerned with ridding themselves of responsibility for the settlement.

The Sierra Leone Company officials took Naimbana's renunciation of sovereignty over the land they occupied at its face value. Ignorant of the customs that had so long regulated relations between European 'strangers' and their landlords, they refused to admit any claim for rent from a Temne ruler. Pa Kokelly, who succeeded King Jimmy and took the former title of King Tom, demanded vainly that the Company make a new treaty with him. Neither he nor Farama, King of the Koya Temne after Naimbana's death, received the payments they felt they were entitled to. Nor were they allowed to mediate in disputes that concerned the settlers as established usage prescribed.

So relations between the Colony and its Temne neighbours deteriorated. At first, when it was assumed that the colonists were 'strangers' of the familiar type, bringing the benefits of trade, they were welcome. The dispute between King Jimmy and the first settlers did not imply any lasting hostility; such incidents were common enough in the slave-trading period and were soon forgotten. But once it was clear that the newcomers were not 'strangers' but settlers, who were taking over the land as if it was theirs, they could only be regarded as enemies.

The menace appeared all the greater when the garrison of British soldiers arrived and it got about that the rough wooden fort in the middle of the town was to be replaced with stone. King Tom realised that any delay would only enable the enemy to entrench themselves irresistibly. In November 1801 he took the offensive and attacked the fort. The defenders were taken by surprise but soon rallied and drove out the invaders. Then they counter-attacked, invaded the territory west of Freetown, burnt King Tom's towns, and drove out all the inhabitants.

King Tom took refuge to the north, gathered allies and five months later again attacked Freetown. Again he was defeated. From now on it was clear that no African force could hope to expel the settlers. The dispossessed Temne rulers had therefore to accept what they could not remedy. In 1807 a final treaty was negotiated which confirmed the Colony's right to the land that had been conquered in 1801. As a perfunctory concession to the old landlord and stranger system, the King of the Koya Temne was allowed a small annual payment in return for the customs dues formerly collected at the watering-place.

Hence European sovereignty over the Colony, based originally on a treaty, was maintained by force. Whatever legalists might assert to the contrary, the settlers' right to the land was henceforth derived from conquest, not from consent. Yet this aggression could still be justified in moral terms as the diffusion of ultimate benefits 'to a Vast Country long detained in Barbarism'. In this way the feelings of moral superiority introduced into West Africa by the opponents of the slave trade were acted out politically as the inauguration of an era of European colonial conquest.

THE SUPPRESSION OF THE SLAVE TRADE

In the early years of the nineteenth century, governments outside Africa began renouncing the slave trade. The Danish Government was the first in 1804; in 1807 a British Act of Parliament declared the trade illegal; in 1808 it was illegalised in the United States. Gradually other governments followed, some only under pressure from those who had already given it up. Nevertheless, despite all attempts at suppression, slaves were still shipped regularly across the Atlantic until the mid-1860s. Possibly (though there are inadequate statistics to prove it) as many as two million slaves were shipped during the nineteenth century.[13]

The British campaign to suppress the slave trade began during a period of warfare in Europe when British ships were in a position to capture any ship with slaves on board no matter what its nationality. A British enforcement squadron, based on Freetown, was stationed off the west coast of Africa. In 1808 the British Government took over the settlement from the bankrupt Sierra Leone Company and it became a British colony. A Court of Vice-Admiralty was constituted in Freetown where cases of slave trading could be tried. Slave ships captured by the naval squadron were brought in for adjudication. The captain and crew were prosecuted and if the prosecution was successful the slaves were freed. As freed people they became known officially as 'Liberated Africans', or 'recaptives' (captured once into slavery, then a second time into freedom).

After the European war ended in 1815 the British naval squadron was no longer empowered to intercept non-British ships. The Vice-Admiralty Court, similarly, could hear only British cases—and British subjects were deterred by severe penalties from trading directly in slaves (though some still participated indirectly, selling to Spanish or Portuguese traders the manufactured goods for which slaves were obtained). So international enforcement procedures were introduced. The British Government signed treaties with other European governments, by which each party agreed to let its own ships be intercepted by the navy of the other. In practice this meant leaving enforcement to the British navy with its large, full-time West African squadron. Only the governments of France and the United States (both recently at war with Britain) refused, preferring to use their own navies.

13 Philip D. Curtin, *The Atlantic Slave Trade,* Madison, Wisconsin, 1969, pp. 231–64.

The treaties also provided for international courts to be set up to hear cases of illegal slave trading. Courts of Mixed Commission (as they were called) were established in Freetown in 1819. Two judges heard the cases, one British, one representing the country whose nationals were accused. If they disagreed, the decision was referred to an arbitrator chosen by lot from two commissioners of the two nations concerned.[14]

The Mixed Commission system extended internationally (and also institutionalised) the moral assumptions which had prompted the foundation of the Sierra Leone Colony. The participant white governments formed an implied moral entity in contrast with, and superior to, an Africa 'long detained in Barbarism'. This international cooperation in an avowedly moral task may be seen as foreshadowing the international conferences of Berlin (1884–85) and Brussels (1890), where the participants were also to avow moral aims, justifying their actions by contrasting themselves with a morally inferior Africa.

Despite the apparatus of enforcement the transatlantic slave trade went on undeterred. Large expanding markets for slaves opened in Brazil and Cuba, which were developed intensively for sugar production during the early nineteenth century. Supply met demand. Though dozens of slave ships were captured every year the shippers were not discouraged. Profits were large enough to offset losses by capture. As marine technology improved faster ships were sent to outstrip the naval cruisers. By the late 1830s it was estimated that the slave trade, far from being 'abolished', had doubled since 1807.

A remedy was suggested by a British Member of Parliament, Sir T. F. Buxton. He proposed the rulers along the West African coast be induced to make treaties with British naval officers. Under these treaties they would renounce the slave trade, and cede land where planters and produce traders could settle and provide alternative commodities for export to Europe. Like the promoters of the Sierra Leone Company, Buxton saw his scheme as a way to raise a morally benighted Africa to a higher level by harnessing it more tightly to the European economy and bringing it under European political influence.

His project included a grandiose naval expedition up the Niger.[15] Its misfortunes discredited his plans. Nevertheless officers of the naval squadron adopted his treaty-making proposal. During the 1840s they took to cajoling or threatening the rulers in the slave-trading areas into renouncing the trade and promising to expel any European slavers who settled in their countries. If a European then remained or returned, a naval party would land, burn

14 Leslie Bethell, 'The Mixed Commissions for the Suppression of the Transatlantic Slave Trade', *J.A.H.*, vii, 1, 1966, pp. 79–93.
15 See ch. 10.

down his premises and free his slaves. This policy was held up for several years by a protracted legal action brought in the British lawcourts by dispossessed slave traders who claimed damages against the naval officers who had destroyed their property. Until it was settled the policy had to be suspended. Only in 1848 was it finally declared that the officers could not be held personally responsible for the damages. Then they could take the offensive again, and slowly clear out one pocket of slave traders after another along the coast. Yet, driven from one place, slave traders would reappear in another. Their last hideout was the swampy interlocking waterways of the Rio Nunez and Rio Pongas (on the coast of the modern Republic of Guinea).

It was action across the Atlantic that really ended the slave trade. During the early 1850s the Brazilian Government (under threat from the British navy) effectively stopped the import of slaves.[16] In the early 1860s similar measures were taken in Cuba. The United States Government also tightened up enforcement; in 1862 the captain of an American slave ship was hanged as a pirate. Once the transatlantic market was closed the slave trade had to stop. The last slave ship was condemned by the Freetown courts in 1864; in 1866 the last shipment of slaves was freed in Havana.

Many historians have depicted the 'abolition' of the transatlantic slave trade as a glorious chapter in the history of mankind. Certainly those who were ready to sacrifice time, money and their lives (for mortality from disease was high in the naval squadron) to suppress this evil traffic deserve applause. But the change of policy towards the slave trade, which, as much as anything else, enabled Europeans to present themselves as moral benefactors and liberators in Africa, was no sudden crisis of collective conscience. For over half a century policy-makers in Europe and America mouthed disapproval of the slave trade, and tried half-heartedly to suppress it. Yet so long as it brought profits to the developing plantation economies across the Atlantic it still continued. Only in an era of fast-growing machine technology, when the transatlantic slave plantation system and the human trade that fed it were no longer economically necessary, was it effectively put down.

With the Atlantic slave trade suppressed, the abolitionists switched their protracted campaign to Eastern and Southern Africa, fed by the publicity David Livingstone and his fellow adventurers provided. Here too suppression was the necessary prelude to European intervention. As its opponents, white men could advance into Africa as liberators. Thus by the period of the European partition of the 1880s the European partitioners could justify their aggressions throughout the continent in a spirit of moral rectitude.

16 Leslie Bethell, *The Abolition of the Brazilian Slave Trade*, London, 1970.

THE EXPORT ECONOMY AFTER THE END OF THE SLAVE TRADE

In places where the slave trade was suppressed people were deprived of the mainstay of their export economy. Some countries had long produced alternative commodities—gold from the Gold Coast, gum (for glazing textiles) from Senegal.[17] Elsewhere alternatives had to be found to provide an exchange medium against the imported European goods the inhabitants had come to rely on.

Coastal Sierra Leone was covered with fine forest trees. In 1816 an Irish trader, John McCormack, shipped a cargo of timber to England and found it saleable. A large timber trade grew up; most of the logs were sold for shipbuilding in the British naval dockyards. The Temne and Bulom rulers welcomed timber traders who supplied them with the rents and customs they lost when they had to give up the slave trade. The traders employed Temne or immigrant Kru labour to fell trees, tie them into rafts and float them to Freetown for export to Britain. When all accessible trees along one riverbank were cut they moved on to the next. Thus the country was gradually despoiled; the secondary bush that grew up was useless for export. By the 1860s most of the available timber had been cut. The market for it was declining anyway, as naval ships were now being built of iron. Hence in Sierra Leone, as in many parts of Europe and America during the entrepreneurial period of nineteenth- and twentieth-century capitalism, long-term natural resources were sacrificed for immediate gain.

Here, as with the slave trade, the purchasers had the best of the bargain. They acquired saleable timber while the vendors received expendable consumer goods and lost permanent assets.

Industrialising Europe opened an expanding market for lubricants. Palm oil, and oil expressed from palm kernels, ground nuts and benniseed were therefore in growing demand and could supply an export product to supersede slaves. In the Niger Delta the elaborate trading mechanisms worked out in terms of the slave trade switched to palm oil. King Gezo of Dahomey, instead of just selling slaves, took to employing them on oil palm plantations, exporting the oil through a French firm which opened in Whydah in 1841.[18] In Sierra Leone an enterprising Afro-European merchant, Charles Heddle, perceived the economic value of the fine colourless oil inside palm kernels. He was also a pioneer of the groundnut trade from the Gambia, which spread over the northern part of the coast. Groundnut oil was mostly exported to France where it was used as a lubricant or for cooking.

African rulers or entrepreneurs who could organise a large supply of groundnuts or palm produce could therefore obtain a good market for

17 John D. Hargreaves, *West Africa: The Former French States*, Englewood Cliffs, 1967, pp. 64–5.
18 John D. Hargreaves, *Prelude to the Partition of West Africa*, London, 1963, p. 17.

them. Slaves provided the necessary labour force. The European philanthropists, who hoped that trade in vegetable produce would drive out the slave trade, were in fact inaugurating a new kind of demand for slave labour with a slave trade to feed it. Driven from the Atlantic, the slave trade went on within Africa to recruit labour to grow produce for new European markets.[19]

Nor did substituting vegetable for human produce alter the pattern of export trade. The West African economies remained as closely tied to Europe as ever. Produce was exported in a raw state. No processing industries grew up; a few factories which were opened in Freetown in the 1840s for expressing groundnut oil soon closed. The French prohibited oil-processing in Senegal.

Indeed the colonial economic relationship ('the true principles of commerce')[20] was intensified. The volume of manufactured imports from the expanding factories of Europe increased steadily, pushing the frontier of import–export trading inland and bringing more peoples within the European trading sphere. Also African vendors tended to have less bargaining power when selling easily storable commodities like palm oil than they had with a perishable commodity like slaves who had to be shipped away as quickly as possible.

THE SIERRA LEONE CREOLES

The Sierra Leone Colony was explicitly Christian. Its settlers were Christians and its sponsors saw it as a potential centre of mission activity. A few British missionaries came out during the 1790s to preach to the neighbouring peoples but achieved nothing. In 1804 missionaries from the Church Missionary Society and in 1811 from the Wesleyan Missionary Society arrived in Freetown. Both societies regarded Sierra Leone as an important base from which the Christian Gospel could be spread over West Africa.[21]

The changeover from Company to Colonial rule in 1808 brought about a demographic revolution. Shiploads of recaptives were constantly being landed. By 1811 they outnumbered the Nova Scotian and Maroon settlers combined—indeed the number of recaptives freed during the year 1816 alone exceeded the total settler population.

Once freed, the recaptives had to be provided for in the Colony. Their homelands extended from the Senegal to the Congo; a few even came from East Africa. It was therefore impossible to repatriate them—and risk in any case their being immediately enslaved again. Some were employed as servants in Freetown households; some were enlisted in the British army; some were sent into the surrounding country to form villages.

19 Walter Rodney, 'Slavery on the Upper Guinea Coast', *J.A.H.*, vii, 3, 1966, pp. 431–43; A. G. Hopkins, 'The Lagos Strike of 1897', *Past and Present*, xxxv, 1966, p. 138.
20 See p. 38 above.
21 Stiv Jakobsson, *Am I not a Man and a Brother?* Uppsala, 1972, pp. 62–227.

Many officials saw the recaptives as a problem. Sir Charles MacCarthy, who governed Sierra Leone from 1814–24, saw them as an opportunity. Inspired by the prevalent dreams of introducing European civilisation to Africa, he proposed that government and missions should cooperate to transform them into a Christian population who would spread Christianity and European ways throughout West Africa. Recaptives would be settled in villages, carefully laid out in English style round a nucleus of parish church, parsonage and school house, where they would learn from missionaries a new religion, new skills and new behaviour patterns. Though his policy entailed a large outlay of British taxpayers' money (always grudged on any colonial project) he persuaded the government in London to adopt it.

The recaptives responded. Without a common language of communication they were ready to learn English. Cut off geographically from the community-based religions of their homelands, unable (unless they chanced to have a religious specialist among them) to perform their own rites, they listened with interest to Christian preaching. They took new names and began wearing European-style clothes. Realising what practical advantages education and technical skills could confer, they were ready to learn or at least to see that their children learned.

The missionaries' task was made easier by their being backed with the authority of the government and by their personal prestige as the dominant figures in the villages. If they chose they could exert strong pressures over their people to force them to conform. Yet unofficial agencies were equally effective. The Nova Scotian and Maroon settlers formed a reference group for recaptives to copy—a black community who were nevertheless Christians living in a European style.[22] Recaptives who worked in their households adopted their religion and ways. Settler pastors made many converts. Neither pastors nor missionaries questioned that they were doing God's work in bringing recaptives out of what both considered to be heathenism and barbarism.

As there were never enough missionaries to staff the villages (missionary disease and mortality rates were high), lay officials were often substituted—sometimes Afro-Americans or Afro-West-Indians who had immigrated to Sierra Leone to seek their fortunes. They, too, spread a Europeanising Christian influence. Recaptives also helped to convert and re-educate one another. Those already settled would take new arrivals (usually their own countrymen) into their own houses and prepare them for a new kind of life.

Within a few decades the recaptives were apparently transformed; Governor Temple in 1834 could describe his subjects as 'a nation of free

22 For 'awareness of alternatives' as a prerequisite for transition from traditional to scientifically-oriented societies see Robin Horton, 'African Traditional Thought and Western Science', *Africa*, xxxvii, 1967, pp. 155–86.

Black Christians (thirty-two thousand)'.[23] Here he certainly exaggerated. Some, perhaps 5 per cent, were Muslims. Others clung to the religions of their homelands and went on practising familiar rites. Yet by mid-century only a waning minority would openly admit that they were not Christians.

Despite the efforts of missionaries and pastors to enforce a European version of the Christian religion (which in origin is neither European nor African), a synthesis of old and new faiths emerged. Many Christians, particularly among the Yoruba, the largest recaptive group, remained members of secret societies they had brought from their homelands, without feeling that membership was necessarily incompatible with Christianity. Old customs such as protracted burial rites and communication with the family dead were grafted on to Christian forms.

Similarly the recaptive village communities evolved their own system of local government, based on the secret societies or other authority structures they had known in their homelands. Though the Colony government normally refused to recognise such institutions openly, they preserved law and order in the villages. Hence the recaptives did not turn into the 'Black Englishmen' of contemptuous stereotype. They retained much from their own past, adapting it unobtrusively to their new situation.[24]

Their speech may serve as an example. The Krio language that evolved during the nineteenth century grew out of the trade pidgins (based on European languages) long current in West Africa. But it developed under the influence of the recaptives' own African languages and that of the neighbouring Temne and Mende. It also incorporated many words and phrases only comprehensible in terms of their own distinctive culture.

The children born to recaptives became known as 'Creoles' (nowadays often 'Krio'), a name which has been generally applied to their descendants and to those who adopted their way of life.

Strong economic motivations underlay the rapid, to Christians almost miraculous, transformation of the recaptives. Freetown and the villages afforded only a meagre reward for subsistence farming. Those who wanted to prosper (and the Christian Gospel of the nineteenth century stressed the virtues of hard work, enterprise and economic success) had to trade. The growing produce trade with the interior opened opportunities as agents for Europeans or on their own. As Freetown expanded, so did the opportunities for retail trade. Street traders would save their penny profits, put up a permanent stall and eventually build a shop. By the 1840s recaptives were entering the import–export trade. Manufacturers in Europe, turning out more goods than ever before, were ready to give them

23 Temple's letter and other material about the recaptives reprinted in Christopher Fyfe, *Sierra Leone Inheritance*, London, 1964, ch. 7.
24 John Peterson, *Province of Freedom: A History of Sierra Leone 1787–1870*, London, 1969; E. A. Ayandele, *Holy Johnson*, London, 1970, p. 43.

credit to open up new markets. Thus they helped to extend the expanding European economy, as well as the expanding European way of life.

Those who made money found (as the Nova Scotian settlers had done) that Freetown house-property was a remunerative outlet for capital investment. Trading profits invested in land or houses could bring in a secure income and enable a *rentier* class to emerge. Hence, by mid-century, enterprising recaptives, who had entered the Colony naked and penniless from slave-ships, were forming a prosperous bourgeoisie.[25]

Ambitious themselves, they were even more ambitious for their children. Here they were helped by missionary education. In 1845 the C.M.S. opened a Grammar School in Freetown, offering a good secondary education for boys, and in 1849 a secondary school for girls. Other missions opened similar establishments. From its early days the C.M.S. had supported an institution to train teachers and missionaries; after various vicissitudes it was refounded in 1827 at Fourah Bay, east of Freetown. In 1876 Fourah Bay College, as it had become, was affiliated to Durham University in England, to enable students to take an external degree.

Education opened the way to the professions. By mid-century Creoles were officiating as pastors and teachers. In 1859 two qualified in Britain as doctors, the first of many. Others qualified there as lawyers. Many entered government service where by the last decades of the century they were holding senior posts. The transformation Governor MacCarthy envisioned had taken place, and a new distinctive community had emerged.

For though the Creoles assumed unquestioningly that the Christian religion and European customs and morality were superior to indigenous African religion, customs and morality, they still retained an African identity. As officials and missionaries they practised and preached the alien doctrines they had grown up to believe. Yet they remained Africans, proud of the community they had created and ready to transplant its idiosyncratic ways wherever they went.[26]

THE CREOLE DIASPORA

Yoruba predominated numerically among the recaptives—victims of the civil wars of the 1820s.[27] Some still yearned for their home. In 1839 a group of them bought a ship and sailed to Badagry. On their return to Freetown they asked the government to found a colony there, to impart to their countrymen the Christian religion and European ways they had learnt themselves.[28] Though the government refused them help, it did not stop them going privately. Within a couple of years several hundred had

25 Arthur T. Porter, *Creoledom*, London, 1961.
26 As witnessed in a book published by a Creole in 1868, James Africanus B. Horton, *West African Countries and Peoples*, 2nd ed., Edinburgh, 1968, pp. 23–4, 82–3.
27 See ch. 5.
28 Fyfe, *Sierra Leone Inheritance*, pp. 147–9.

The Gate of the Liberated African Yard, Freetown. (Photograph taken in 1870)

emigrated, returning to their astonished families like Joseph in the Bible returning from Egypt to his brethren.

At their invitation missionaries of the C.M.S. and W.M.S. established themselves among them, preaching to the Yoruba. Recaptives and Creoles cooperated with them, working in the missions as pastors, catechists and teachers. Those who traded at a distance from a mission station would often hold services in their own houses, bearing Christian witness among the unconverted. They believed, in the words of their petition to the government, that 'their poor Country people' were 'now living in darkness'—an echo of the 'long detained in Barbarism' of the Sierra Leone Company's instructions.[29] Thus they showed that they shared their white coreligionists' disapproval of the ways of life they had discarded. Assuming like Europeans that Africans who adopted Christianity and European norms had everything to gain, they were eager, as European missionaries were, to spread them wherever they went.

The return to Badagry and its hinterland was only part of a wider movement, the *diaspora* of the recaptives and Creoles over West Africa. Gathered in from the coast, they scattered over it again. The small Sierra Leone Colony offered only limited scope for an enterprising, ambitious population. But all along the coast their skills were in demand. As the produce-exporting economy extended, they found jobs as clerks and agents for European exporters or set up as exporters on their own. As missions

29 See p. 39 above.

spread they found jobs as pastors and teachers. There were openings in government service in the British colonies. Those who were skilled tradesmen could build and repair houses in the growing coastal towns. Enterprising women would often leave home to trade in produce in the Gambia or along the rivers north and south of Sierra Leone. By mid-century Creoles were scattered in communities from the Gambia to Fernando Po.

As well as new skills, they brought new ideas to the peoples they came in contact with. Bred in a British colony, protected by English law, they had a strong sense of their own rights. As British subjects, they had grown up conscious of their common citizenship with their British benefactors. Indeed many almost felt that, whatever the difference of skin colour, it was not just common citizenship but a kind of kinship that united them. Yet this sense of kinship and a deep loyalty and gratitude to the British Crown were not incompatible with the feeling (which they shared with loyal contemporaries in Canada and Australia) that they had also inherited the Englishman's liberties and were entitled to have a say in their own government. If therefore their *diaspora* can be seen as part of the Europeanising process that was to subject West Africa to aliens, it also sowed the seeds of the protest movements that eventually were to sweep alien rule away.[30]

THE FOUNDATION OF LIBERIA

The foundation of Liberia was in many ways like that of the Colony of Sierra Leone. The founders of both settlements were anxious to ship away undesirables of another skin colour; both settlements were intended as weapons to fight the slave trade; both served as instruments through which an alien morality could be introduced in a self-justifying spirit; both purported to be grounded on principles of peace but established themselves by war.

By the early nineteenth century there were in the United States about two hundred thousand free Negroes—to give them the descriptive label white Americans have traditionally attached to most, though not all, individuals of African descent.[31] In the slave-owning states they seemed a standing menace to society. In the northern states they were an anomaly—a community of allegedly free people who were nevertheless denied the opportunities and aspirations which American society was dedicated to extending to all, a living reproach to American idealism.

To give them real equality seemed to most white Americans unthinkable. Thomas Jefferson voiced their views when he declared in his *Notes on the*

30 See ch. 16.

31 Puerto Ricans, Cape Verdians and other Spanish or Portuguese speaking groups in the United States are not normally considered 'Negroes', however black their skin. 'The only valid definition of a Negro is a person who, for one reason or another, is known to be a Negro', Chester Hymes, *Pinktoes*, New York, 1966, p. 18.

State of Virginia that black and white could never live happily together, divided as they were by ineradicable memories of past subordination and superordination. He proposed instead that the black population be moved away to a distant home. Over the years his proposal found increasing favour and in 1816 the American Colonization Society was founded with the specific aim of sending them to settle in Africa.

The promoters emphasised in their public utterances that the project was designed to help Negroes find a happier home, where they could realise the opportunities denied them in the United States. But whatever was publicly professed, many (in time most) of the Society's supporters saw it primarily as a device to get rid of the black population. This most Negroes perceived. Despite all the pressures and propaganda the Society used, less than seventeen thousand were ever induced to emigrate. The vast majority no more wanted to settle in Africa than the white American population wanted to settle in Europe.[32]

The Society's aims resembled those of the Sierra Leone Company. Both opposed the slave trade and could thus take up a position of moral superiority. Both assumed they were bringing light 'to a Vast Country long detained in Barbarism'. Both heralded a new dawn for Africa.

The government of the United States, so recently risen from colonial status, could not directly participate in any process called 'colonisation'. But a means was found to give the Society indirect federal help. Its settlement was designated as the home where slaves captured in transit across the Atlantic by the U.S. Navy could be freed and settled, as British-liberated recaptives were in Sierra Leone. In this way government money was available to establish and maintain an agency for recaptives, and the prospectors for a site were accompanied by government agents. But the settlement was not an American colony. Nor did the aid given ever amount to much. The U.S. Navy seldom cruised off West Africa. Over a period of forty years only 5,722 recaptives were brought in. Called 'Congoes', because the first shipload came from the Congo, they formed a small, uninfluential minority in the population. Thus the Liberian demographic pattern differed from that of Sierra Leone. In Liberia American immigrants predominated —in Sierra Leone recaptives.[33]

Like the sponsors of the Sierra Leone Company, the Society and its protégés misconceived the existing relations of black and white in West Africa. Ignorant of African ways, they knew nothing of the many centuries of peaceful trading contact and the well-established usages it was based on. Africa was to them a *tabula rasa* for them to embellish. Convinced of their

32 P. J. Staudenraus, *The African Colonization Movement 1816–65*, New York, 1961.

33 For a comparison of Sierra Leone and Liberia at this period see John D. Hargreaves, 'African Colonisation in the Nineteenth Century: Liberia and Sierra Leone', *Sierra Leone Studies,* new series, xvii, 1962, pp. 189–203; the standard history of Liberia is still C. H. Huberich, *The Political and Legislative History of Liberia,* New York, 1947.

own moral superiority, contemptuous of African social organisation, they assumed that they must be welcomed as benefactors. Yet the coastal Temne could tell another story. For them the promised humanitarian dawn over Africa glowed like the warning flare of an approaching bush fire.

So it was not easy for the prospecting American agents to find land for a settlement. The rulers of the Sherbro country south of Sierra Leone, selected as a suitable site, were unwilling to receive them; nor was the Sierra Leone Colony Government anxious to have Americans so near. Another site was found further south at Cape Mesurado. Here too the ruler, King Peter, was reluctant. Ready to welcome traders, as his forbears had for centuries, he had a warning in the Temne experience against parting with land. Eventually Lieutenant Stockton, the U.S. Government agent, cut short the prolonged negotiations in a summary way by pointing a pistol at the king's head. The king then agreed—and just at that moment (according to Stockton's account) rays of sunshine burst symbolically through the clouds.

At Cape Mesurado, as formerly at Sierra Leone, a king put his mark to a treaty by which, in return for a consignment of miscellaneous consumer goods, he ceded an ill-defined tract of land in perpetuity. On 25 April 1822 the colonists took formal possession of what they considered, against all West African usage, to be their land. To mark its new alien status the settlement was named Liberia, and its capital Monrovia (after James Monroe, President of the United States).

Despite their tradition of friendship with Europeans (they told the Americans that never within living memory had there been a war in their country between black and white) the displaced inhabitants saw they would have to fight if they wanted to keep their land. But, like the Temne, they attacked in vain. The stockaded defenders fought with the tenacity their former white compatriots displayed against the Red Indians. Like them they beat off the attackers, and won their African 'frontier'. Like the Sierra Leone Colony (and most states in the world) Liberia was maintained by force of arms.

In succeeding years other settlements were established on the coast north and south of Monrovia.[34] Wherever the colonists settled the slave trade came to an end. Sometimes European slave traders and their African allies had first to be driven out. Thus by making suppression a pretext to extend their own territory, they used it as Europeans used it, as a device to justify expansion.

Their indigenous neighbours, unable to get rid of them, came to accept them. Those who lived inland, away from their aggressions, in any case saw them less as a threat than as the familiar means, long presented to them

34 Svend E. Holsoe, 'A Study of Relations Between Settlers and Indigenous Peoples in Western Liberia, 1821–1847', *African Historical Studies,* iv, 2, 1971, pp. 331–62.

by aliens, of obtaining imported manufactures. So apart from occasional petty wars, relations between the immigrants and the indigenous peoples tended to be amicable at this period. Some settlers felt their missionary duties strongly; the Reverend Lott Carey, a Baptist pastor, opened a school in the interior. But the relationship, however amicable, remained that of the would-be benefactor to those he conceives to be his cultural inferiors.

The Board of the American Colonization Society ruled Liberia as a colony, sending out a governor. But it lacked the means to administer it adequately from across the Atlantic. In 1839 most of the settlements were constituted a virtually self-governing commonwealth, which in 1847 became an independent republic with an American-type constitution drawn up by a white Harvard law professor. But, dependent or independent, Liberia remained, like Sierra Leone, an alien community, buoyed up by humanitarian pretensions into assuming unquestioning superiority over the neighbouring peoples.

THE IDEOLOGY OF RACE

By the nineteenth century Europeans were unquestionably superior to Africans technologically. By renouncing the slave trade they could also feel they were superior morally. These feelings of Eurocentric superiority were reinforced by theories of race.

Ever since the Swedish botanist Linnaeus published his classification of plants in 1752, scholars had been inspired to try to classify everything, including mankind. Tentative theories were advanced that the varieties of man belonged to different species, like plants—and that Negroes were an inferior species.[35]

Such theories had their greatest appeal in the slave-owning societies across the Atlantic, where they could be used to justify keeping Negroes in slavery or at least subjection. But they were first formulated as a fully developed theory of race in Scotland by Dr Robert Knox. In his *Races of Men*, published in 1850, Knox declared that mankind was divided irrevocably into races, each with its own fixed unalterable characteristics, physical and psychological. He explained all human history and behaviour in terms of the interaction of these races. He asserted not only that the white race was, and must always remain, superior to the non-white races but that as the superior white race advanced across the earth the inferior non-white races would be conquered and must ultimately perish.

Knox's theory had many loopholes. He could not, for instance, exhibit any examples of an unmistakably pure race, as his theory obviously demanded. Nor could he demonstrate his alleged correlations between physical and psychological attributes. But in an era when whites were

35 For the development of racial theories see Philip D. Curtin, *The Image of Africa*, Madison, Wisconsin, 1964.

plainly spreading over the world almost unchecked, it appeared immensely plausible. Charles Darwin's theory of the survival of the fittest seemed to confirm it; the fittest must be the whitest (an interpretation Darwin himself repudiated). It was therefore widely and increasingly believed, and assumed to be scientifically valid.

Armed with such a tool, the European impact could be devastating. European technology conferred the power, though not the right, to conquer. Belief in European moral superiority conferred the right to dominate spiritually and intellectually. Belief in European racial superiority conferred the right to dominate, conquer and even destroy, and declared domination, conquest and destruction not only desirable but inevitable.

SUMMARY

During this period the transatlantic slave trade and its horrors were very slowly brought to an end. Produce trade succeeded slave trade and tied the West African economy even more firmly to Europe. The internal slave trade expanded to supply labour to grow exportable produce. European and American settlements were founded, each based on a principle of alien sovereignty hitherto unknown. In them social structures emerged which were similar to those of contemporary Europe and North America but unfamiliar in West Africa. Moral values of a new type were introduced and preached militantly by Europeans, Afro-Americans and by most Africans who adopted them—particularly by the Sierra Leone Creoles who dispersed all along the coast carrying new skills and ideas. Belief in European superiority was reinforced by the ideology of race.

In this short span of two generations, then, coastal West Africa was changed far more radically than it ever was in the long period of the slave trade. These changes softened up the country for the coming European partition.

CHAPTER 3

*The Sokoto Caliphate in the nineteenth century**

R. A. ADELẸYẸ

The most striking consequence of the jihād of the Shaikh 'Uthmān dan Fodio was the creation of the Sokoto Caliphate. The jihād itself arose from the Islamic reformist movement which the Shaikh invigorated, canalised and led. Thus the history of the Caliphate is essentially a history of the legacy of the jihād. It is well known that the jihād affected not only the states out of which the Caliphate was created but also their neighbours. Bornu and Yorubaland provide apt illustrations of this point. However, the history of the jihād itself is outside the scope of this chapter.[1] For the congeries of states and principalities affected by the jihād, it is enough to note that the reformers achieved their immediate aim of regenerating society within the context of the Islamic faith and law and of making the countries safe for Islam. The jihād also gave rise to a profound political, social, demographic, cultural, religious, economic and intellectual revolution which was totally unprecedented in its scope and intensity in the history of the West-Central Sudan. The complexity and depth of these changes can be seen in the legacy of new socio-political problems inherited by the rulers of the Caliphate.

In spite of the spectacular fall of ancient dynasties, the obliteration of their state boundaries and the emergence of an almost exclusively Fulani class of rulers with new political frontiers, probably the most significant result of the creation of the Sokoto Caliphate was the supra-ethnic ethos which the Islamic ideology of the *Umma* brought to the fore as the strongest agency for large-scale political integration. The power politics which had characterised the imperial struggles of the heterogeneous states and peoples of the area since about the fifteenth century gave way before this new philosophy of state. Large-scale political integration was an

* Written in 1968.

1 See above, ch. 1, also D. M. Last, *The Sokoto Caliphate in the 19th Century with special reference to the Vizierate,* Ph.D. thesis, Ibadan, 1964; R. A. Adeleye, *The Overthrow of the Sokoto Caliphate 1879–1903,* Ph.D. thesis, Ibadan, 1967, ch. 1, and H. A. S. Johnston, *The Fulani Empire of Sokoto,* Oxford, 1967, p. 35 ff. and passim.

incidental result of the jihād—unforeseen and therefore not specifically prepared for by the jihadists. Yet it was a central factor in the shaping of the history of the Caliphate.

This integrating ideology won no universal acceptance despite the military sanctions of the jihadists. There were states which not only successfully rejected integration with the Caliphate but even went on to challenge and threaten its very existence. Moreover the degree of acquiescence of the several peoples within the emirates varied. The problem for the Caliphate was therefore firstly one of survival and thereafter of security against its enemies. In the circumstances the ferment of the jihād remained a permanent feature of the Caliphate. A second aspect of the Caliphate's history is that, the jihād having created an Islamic *Umma*—the embryo of a nation state, sharing identical ideals and aspirations—the central authority had to translate into practical reality of administration the principles of primordial Islam which the reformers had held up to society as the only worthy alternative to contemporary governments. The success or otherwise with which this problem was tackled will become manifest in a consideration of the relations which subsisted between the Caliphs and their Emirs, and the relations between individual emirates as well as their internal administration.

ESTABLISHMENT AND DEFENCE OF DAR AL-ISLAM

The foundation of the Sokoto Caliphate dates from the swearing of the Oath of Allegiance *(bay'a)* to the Shaikh 'Uthmān dan Fodio as Amīr al-Mu'minīn by the leaders and the generality of the Muslim community at Gudu.[2] By constituting the shaikh's followers into an independent polity, the *hijra* from Degel proved the final phase of the separatist movement which Gobir opposition had forced on the reformers. Similarly, the opposition of other governments beyond Gobir was pushing the *jamā'as* into rebellion and separation.[3]

The Caliphate expanded territorially, not only through the success of the Gobir jihād from 1804 to 1808, but notably through that of the jihād of scattered Muslim jamā'as throughout Hausaland and beyond. The leaders of these jamā'as had received both authority and investiture from the Shaikh 'Uthmān to prosecute the jihād in their localities. Each leader, known as a *nā'ib* (deputy), received a flag as a symbol of the Shaikh's authority.[4]

At the height of its territorial expansion, the Sokoto Caliphate, in addition to the Sokoto and Gwandu hinterlands carved out of Gobir, Zamfara and Kebbi, comprised the rump of Kebbi, Arewa, Zaberma,

2 Muḥammad Bello, *Infāq al-Maisūr,* ed. Whitting, London, 1957, pp. 70–1.

3 *Ibid.*, pp. 83, 105 ff., also Adeleye, pp. 18–38.

4 Adeleye, pp. 104–5, 190 for a short list. A fuller list of the *nuwwāb* who came to be popularly known as 'Flag Bearers' is given by Last, p. 111.

Dendi, Gurma, while some influence was exercised by the Caliph over Adar and Ahir. All of these territories were in the west or to the immediate north of Sokoto. Also within the Caliphate were Katsina, Daura, Kano and Zazzau, the last having vassal states and sub-emirates extending as far south as the Benue river. The states of the eastern part of the Caliphate were Hadejia, Katagum, Jama'are, Missau, Gombe, Bauchi, Adamawa and Muri. In the south and south-west were Yauri, a semi-independent emirate enjoying the status of a protected people, that is *Dhimmi*, Kontagora, a tributary to Sokoto, and Nupe, Lafiagi, Agaie and Ilọrin, all tributaries to Gwandu.

It is of great significance to the history of the Caliphate to note that the founding of the emirates took place over several decades. Whereas the oldest emirates such as Katsina, Daura, Kano, Zaria, Bauchi, Gombe, Adamawa, Hadejia and Katagum had been founded by 1810, Ilọrin and Nupe were not founded until late in the 1820s, Missau about 1830, Muri about 1833 and Kontagora in about 1859. This means that the jihād specifically directed towards the founding of *Dār al-Islām* continued for more than half a century. Another significant observation is that the emirates, as finally founded, were not contiguous throughout the Caliphate. The founding of the later emirates reduced the lack of contiguity but regions of *Dār al-harb* persisted within the boundaries of the Caliphate throughout the century. Such were: Zuru (between Zamfara and Yauri); part of Gwari-land—bordering on the Kontagora, Zaria and Nupe emirates; Ningi Sultanate (between Kano and Bauchi); and a huge area of the Bauchi plateau bordering on Bauchi Emirate in the north and stretching as far as Shendam and the Benue to the south. In addition to these clearly non-Muslim territories, there were also the states of Gobir, Kebbi and Abuja which, though Muslim, remained persistently hostile to the Caliphate. It is thus clear that in terms of territorial consolidation, the jihād was an uncompleted revolution. This meant that the jihād became a permanent war sustained by hostility between the Caliphate and the enclave enemy states within it.

Another level of opposition existed in the rejection of integration with the Dar-al-Islām by ethnic groups within the confines of certain emirates notably, Gombe, Bauchi and Adamawa against whom the jihād became permanent throughout the century.[5] In Nupe the process of internal consolidation was delayed by the opposition of indigenous rival dynasties and their supporters as well as fission which led to active hostilities between rival factions of the descendants of Mallam Dendo, the founder of the

5 See Adeleye, pp. 73–5. See also Muḥammad Bello b. Aḥmad b. Idris al-Sūdānī, *Tarīkh 'Umarā' Bauchi*, University Library Ibadan, 82/377, passim; H. K. Strumpell, *A History of Adamawa Compiled from Verbal Information*, mimeographed, Hamburg, 1912, passim, and C. Vicars-Boyle, 'Historical Notes on the Yola Fulani', *Journal of the African Society (J.A.S.)*, 1910, ix, 27.

emirate.[6] The Nupe example raises the question of a vital distinction which must be made between emirates created from an at least quasi-Islamic cultural and political context (i.e. the Hausa states) and the others which were carved out of states of predominantly non-Islamic background. The process of internal consolidation in the former was swift while consolidation in the latter followed a more chequered course.

In addition to these sources of opposition there was challenge from states on the perimeter that became enemies of the Caliphate as a result of the jihād. Among these, on the western frontiers were Borgu, scorched but not conquered by the jihād, Gurma, Kebbi, Dendi, Zaberma, Arewa, Gobir. All of these were initially conquered regions which either could not be permanently held or persisted as undigested conquests and as centres of independence movements as well as sources of unrelenting opposition to the Caliphate. To the north of Katsina and Daura were the states of the Tessawa and Maradi region founded by the fugitive pre-jihād ruling classes of Katsina and their supporters. These, with the Bornu province of Damagaram (whose capital was Zinder), were, throughout the nineteenth century, inveterate enemies of the Caliphate. The greatest source of anxiety to the Caliphate was, however, probably the Bornu empire which, while it narrowly escaped being engulfed by the Caliphate, had lost territories like Gudiri (Hadejia, Katagum, Missau) as well as parts of Gombe to the Caliphate. The enmity persisted throughout the century and the Bornu frontiers with the Caliphate remained unstable and highly inflammable. Furthermore, between Adamawa and Bornu, the country of the Mandara constituted a hotly contested no-man's-land, not to mention the constant frictions with the pagan tribes on the banks of the Benue, most notable of which were the Jukun and the Tiv. Indeed, throughout the nineteenth century it was only to the south of the Nupe and Ilorin emirates that the Caliphate may be said to have had expanding frontiers. Even in this direction expansion was violently resisted. The result of all this was that the Caliphate was throughout the century preoccupied with problems of military security which can only be briefly assessed here.

THE NORTHERN AND WESTERN FRONTIERS

The victories of the jihadists in battles against Gobir, culminating in the fall of Alkalawa, the capital, in 1808,[7] did not result in the state's final subjugation. Although a good deal of territory was permanently lost to the Muslims, Gobir remained sufficiently strong to pursue an active policy

6 E. G. M. Dupigny, *Gazetteer of Nupe Province*, London, 1920, p. 10 ff.; A. Burdon, *Northern Nigeria: Historical Notes on Certain Emirates and Tribes*, London, 1909, pp. 52–4, and 'Umar b. Muḥammad b. Al-Ḥasan b. Adam, *Tarīkh min bilād Bida wa Tarīkh al-bilād Gbara*, National Archives, Kaduna (N.A.K.), Kadcaptory, box 2, item 22.

7 Muḥammad Bello, pp. 115–16, and Wazir Junaidu, *Dabt al-Multaqatāt*, University Library Ibadan, 82/31, f. 30.

of belligerency against the Caliphate. The same was true of Kebbi and, to a lesser degree, Zamfara, which in spite of being conquered in 1805 remained restive.[8]

The 'Abd al-Salām revolt, immediately after the death of the Shaikh 'Uthmān, was the signal for a general rising of these territories. 'Abd al-Salām, whose flight from Gobir territory and subsequent defiance of Sarkin Gobir's authority had led directly to the initial outbreak of the jihād, had been dissatisfied in 1812 when the Shaikh divided supervision over all the emirates between Muḥammad Bello and 'Abdullāh dan Fodio[9] and assigned him to rule only Sabiyel. He had been prevented from open revolt some two years later only by the Shaikh 'Uthmān's authority. When Muḥammad Bello became Caliph in 1817, 'Abd al-Salām supported by people of his tribe—the Arewa—and abetted by Sarkin Banaga of Zamfara, embarked on the path of revolt.[10] A sharp paper warfare between Muḥammad Bello and 'Abd al-Salām preceded military confrontation. 'Abd al-Salām was finally defeated in January 1818 and died shortly after from wounds sustained in battle.[11]

Bello's difficulties were further exacerbated by disaffection between him and his uncle 'Abdullāh since the latter felt that he rather than Bello should have succeeded to the Caliphate. The rising of the Kebbawa, allied with the migrant remnant of 'Abd al-Salām's forces against the Caliphate early in 1818, was a threat to both uncle and nephew. This revolt immediately threatened Gwandu. Muḥammad Bello gave 'Abdullāh military aid to crush the uprising, and as a result the new Caliph and his uncle were reconciled.[12] Sokoto faced hostility from Zamfara and Gobir, which persisted after 1818, in addition to that of the Katsina Habe principalities of Tessawa, Gazawa and Maradi as well as the unpredictable Tamesgida Tuareg.

In spite of annual dry season expeditions by Sokoto against its enemies, with contingents from the eastern emirates, Zamfara was not decisively beaten until 1820–21[13] and Gobir continued to be a perennial preoccupation for Sokoto. Under their new Sarki, Ali b. Ya'qūb, the Gobirawa soon reorganised their army and during Clapperton's visit to Sokoto in 1826 Muḥammad Bello was busy investing Konya and Magariya—new Gobir strongholds.[14] The Gobirawa suffered a serious reverse at the hands

8 Muḥammad Bello, pp. 90–4, 97, 104.
9 Muḥammad Bello, *Sard al-Kalām*, University Library Ibadan, 82/212, *Majmū'* Haliru Binji, p. 58.
10 See correspondence between 'Abd al-Salām and Muḥammad Bello, *ibid.*
11 *Majmū'*, pp. 62–3.
12 See Last, pp. 131–2.
13 Wazir Junaidu, ff. 35–8.
14 R. Lander, *Record of Captain Clapperton's last expedition to Africa*, London, 1830, ii, pp. 37–48.

of Sokoto, its eastern emirates' contingents and its Kelgere and Ullimiden Tuareg allies, at the celebrated battle of Gawakuke[15] in 1836 in which the Gobirawa with their Maradi and the Tamesgida Tuareg allies were routed.

Meanwhile, Sokoto hostilities with Kebbi had been continuous until a truce was imposed on the latter in about 1831 following a severe defeat in which Karari, the Sarkin Kebbi, lost his life. Karari's son, Yaʿqūb Nabame, was captured and thereafter kept under surveillance, a virtual prisoner, at Sokoto. The truce with Kebbi lasted till 1849. On the Gobir and Zamfara frontiers Muḥammad Bello instituted a new military and social policy through the building of stronghold frontier towns—*ribāts*—and the resettlement of the population in them.[16] However, in spite of this policy and the victory at Gawakuke Sokoto knew only a short respite from war. The next Sarkin Gobir, Mayaki, founded a new capital at Tsibiri in 1842.

From 1819, the Katsina successor-principality of Maradi, under its Sarki, Dan Kassawa (1819–31),[17] had persistently attacked Katsina and the aid provided to Katsina by Sokoto and its emirates had proved ineffectual. Like Gobir, the Habe Katsinawa successor towns resumed their attacks on Sokoto a few years after their defeat at Gawakuke.

Muḥammad Bello, by constant warfare (forty-seven regular campaigns are ascribed to him from 1817 to his death in 1837), consolidated in great measure the early conquests of the jihād. But both his successors, Abubakar Atiku and Ali b. Muḥammad Bello, had to face the combined forces of Kebbawa, Gobirawa, Tuareg and the Habe-Katsina. Atiku, aided by the Emir of Gwandu and the eastern emirs led annual expeditions against Gobir and its allies until his death in 1842.[18] The alliance continued its hostilities with Sokoto and Katsina for another twenty years.

From the 1840s Maradawa attacks were directed not only against Sokoto and Katsina but also against Daura, Kano and even Zaria. In 1843 Sokoto and Kano forces confronted Sarkin Maradi's (Dan Mari, 1835–44) forces in Kano territory. Under successive Sarakuna of Maradi—Binoni (1844), Dan Mahedi (1844–50) and Dan Baura (1850–51)—Maradawa raids into Katsina, Sokoto, Daura and Kano territories were intensified. From 1851 to 1877, under Dan Baskore and his son Barmu, Maradawa raids as far as Kano had become a permanent feature of the life of the northern frontier emirates. One of Baskore's expeditions is reported to have penetrated as far south as Fika in Zaria territory, while Barmu (1873–77) overran Kano territory and for some time settled at Yan Gwarzo near Bichi.[19] However, under Dan Mazoge (1877–80), respected for his religious devotion and

15 Y. Urvoy, *Histoire des populations du Soudan Central*, Paris, 1936, p. 282; Wazir Junaidu, p. 41.
16 Last, pp. 139, 141–5; Wazir Junaidu, f. 42.
17 Urvoy, *Histoire des populations,* pp. 281–2.
18 See Wazir Junaidu, ff. 43–5 and Last, pp. 148–9 for Atiku's expeditions.
19 See Urvoy, pp. 283–4, for an account of the Maradawa attacks.

piety, a policy of rapprochement with the Caliphate was vigorously pursued. This change in Maradawa policy had the incidental result of provoking Gobir enmity against Maradi,[20] a situation which was advantageously exploited by Sokoto during the rest of the century. Indeed by 1880 the hostility of Maradi and Gobir to the Caliphate was largely spent. It ceased to be a threat but remained a nuisance. The situation worsened during the 1890s when Dan Barahiya of Maradi raided as far as Zaria territory.

Still on the northern frontier, Zinder, the capital of Damagaram, a vassal state of Bornu, was the centre of virulent and unabated hostility to the Caliphate throughout the century. The targets of Zinder attacks were Daura, Katsina, Kano and Hadejia. Sarkin Zinder Ibrahim b. Sulaimān (*c.* 1822–41) penetrated south of Kazaure in the early years of his reign and in 1829 he began systematic incursions into Daura.[21] The greatest threat to Kano came from Damagaram under Tenimu b. Suleimān (*c.* 1851–84).

The rise of Damagaram as a strong militarist state was in a sense a direct result of the jihād. The attempted conquest of Bornu by the jihadists ushered in a period of internal instability for Bornu. Not only were there independence movements in the vassal states, there was also the conflict between the Seifawa on the one hand and Muḥammad al-Amīn al-Kānemī on the other. It was in the general confusion arising from this disturbed situation that Bornu's erstwhile power declined and Damagaram asserted its *de facto* independence. Under Tenimu it developed into a fully fledged militarist state with a strong army of several thousand men, armed with imported firearms and locally manufactured gunpowder.[22] Tenimu is said to have had as many as six thousand guns and forty cannons.[23] Although he was unable to secure a permanent hold on either Kano or Hadejia, the Damagaram threat remained an awesome preoccupation for both emirates up to the European occupation. Kano–Zinder hostilities were particularly frequent in the 1890s. Aḥmad of Zinder in about 1892 raided as far as Malikawa and rampaged into neighbouring Kano towns.[24] Early in the 1890s the Emir of Hadejia wrote to Sokoto to report his inability to send a messenger because he had learned that Zinder was resolved on raiding his territory.[25] Late in 1898, Aḥmad of Zinder raided Gazawa in Kano and although Aliyu of Kano successfully retaliated, the raid proved a painful experience for Kano.[26] The northern emirates, particularly Daura, Kano

20 *Ibid.*, p. 291.
21 Tilho, *Documents Scientifiques de la Mission Tilho*, Paris, 1911, ii, p. 441.
22 *Ibid.*, p. 445.
23 *Ibid.*, p. 446.
24 *Ibid.*, p. 450; also N. A. K. Sokprof 25, file vii, letter 36, Muḥammad Bello (Emir of Kano, d. 1893) to Wazir (of Sokoto) Muḥammad Bukhari (1886–1910).
25 N.A.K. Sokprof 25, file ii, letter 2, Hadejia to Sokoto; N.A.K. Kadcaptory, box 44, letter 86, Hadejia to Sokoto.
26 N. A. K. Kadcaptory, Box 38, letter 12, Emir of Kano, Aliyu, to Emir of Gombe, 'Umar.

and Hadejia, were compelled to maintain constant vigilance against Zinder right up to the last critical years of the nineteenth century.

On the Gobir and Kebbi frontiers the Sokoto Caliphate faced determined enemies particularly during the Caliphate of Ali b. Muḥammad Bello (1842–59). In the west Gwandu had a loosely defined frontier to defend since Liptako, Gurma, Dendi, Zaberma, Arewa and Kebbi were only vaguely under its suzerainty. In spite of numerous expeditions made by ʿAbdullāh into these territories, he had failed to subjugate them effectively. Vigorous efforts by Muḥammad b. ʿAbdullāh (1829–32) and Khalīl (1832–58) likewise failed to consolidate these territories under Gwandu. Chronic enmity to the Caliphate was particularly marked in Kebbi, Dendi, Arewa and Zaberma. The early jihād in these regions, especially in Dendi, Arewa and Zaberma, eventually succeeded in welding under one authority a congeries of petty principalities in a politically fragmented region. Gwandu authority had been exercised through exploitation of traditional divisions and the building up of a royal clientele by recognition of particular lines of the pre-jihād ruling families against their rivals.[27] In the end Gwandu rule unwittingly created a coalition of the aggrieved parties who naturally found common cause in opposing the power that had deprived them of their claims to power. Rankling discontent was given overt expression in revolt in 1849 when an able leader was found in Daud, a prince of the Zaberma ruling house.[28]

Meanwhile with a strong base at Tsibiri and a determined and able enemy of the Caliphate in the person of Mayaki the Sarki, the Gobirawa had launched and successfully maintained an offensive against Sokoto by constant attacks on the Zamfara frontier from 1842 to 1848.[29] During 1848 the forces of Sokoto and the eastern emirates suffered a heavy reverse at the hands of Gobir. From 1844, Gobir's hand had been strengthened by an alliance with Siddiq, Emir of Katsina, who had been deposed by the Caliph in the same year. In 1848, Zamfara and the surrounding countries apostasized while Kebbi followed in the general wake of a state of endemic revolt in 1849. Yaʿqūb Nabame, prince of the Kebbi ruling house, lately released from eighteen years of captivity in Sokoto and allowed to settle on the Kebbi frontier, led the revolt. The revolting countries to the west placed themselves under his leadership. For the next twenty years a chronic state of war reigned with varying fortunes on the western frontiers of the Caliphate. The independence of Kebbi, Zaberma, Dendi and Arewa became for the rest of the century an irreversible fact.

27 Jean Périe and Michel Sellier, 'Histoire de la population du cercle de Desso', *Bulletin de l'IFAN*, xii, 1950, p. 1046.
28 *Ibid.*, pp. 1046 ff.
29 See D. M. Last, pp. 151 ff. and Adeleye, pp. 81 ff. for the Yaqub Nabame and allied revolts.

At the death of the Caliph Ali (1859) this revolt had ceased to be a threat to the rest of the Caliphate. The territories that were lost to the Caliphate had never been strongly held. In 1867 the Caliph Aḥmad al-Rifa'ī (1867–73) concluded a truce, *Lafiyar Toga*, with the Sarkin Kebbi ʿAbdullāh (1863–80).[30] Though the truce was broken in 1875, Kebbi was not to cause undue apprehension in Sokoto until the first year of the Caliphate of ʿAbd al-Raḥmān (1891) when the forces of Sokoto and the eastern emirates suffered a severe defeat.[31]

The hostility of Gobir after 1859 was generally contained. The Caliph, Aḥmad al-Rufa'ī, exploited divisions within the Gobir state by encouraging Dan Halima, a Gobir prince, to found Sabon Birni in opposition to Tsibiri in the valley where Alkalawa had once stood.[32] It served as a buffer between Gobir and Sokoto and although Mayaki's successor, Bawa, succeeded in capturing the town, thereby necessitating expeditions against it by the Caliphs Muʿadh (1877–81) and ʿUmar (1881–91), Gobir was thenceforth no more than a nuisance, albeit one that could not be ignored.[33]

THE EASTERN AND SOUTHERN FRONTIERS

Following the early conquests of the jihād which had resulted in the loss of the western marches of Bornu out of which the Katagum, Hadejia, Missau and Gombe emirates had been carved, relations between Bornu and Sokoto continued to be hostile. The attempts of Muḥammad al-Amīn al-Kānemī to overrun the Caliphate in 1827 after successful exploits in Hadejia and Katagum between 1824 and 1826 are well known.[34] So also is the account of the dramatic reverse suffered by Bornu forces near Kano at the hands of a Caliphal army led by Yaʿqūb, Emir of Bauchi.[35] This latter event ended large-scale offensives aimed at territorial expansion by both Sokoto and Bornu. But thereafter the frontier between them was rendered highly unstable and explosive by mutual border raids.

During the Bukhari revolt[36] Bornu's potential threat to the Caliphate was demonstrated by the surreptitious aid which the Shaikh ʿUmar gave to the rebel, thus strengthening the latter's ability to defy Sokoto and its emirates. The extent of Bornu's threat is attested to by the constant fear of attack from Bornu felt by Kano during the revolt.[37] Jema'are Emirate, a

30 E. J. Arnett, *Gazetteer of Sokoto Province*, London, 1920, p. 12.
31 F. Edgar, *Litafi na Tatsuniyoyi na Hausa*, Belfast, 1911, ii, pp. 337 f.
32 Tilho, pp. 477–8; Last, pp. 198–9.
33 Last, pp. 199–200 for an account of internal strife and consequent weakness in Gobir, see J. Périe, 'Notes historiques sur la région Maradi', *Bulletin de l'IFAN*, i, 1939.
34 Y. Urvoy, *Histoire de l'Empire du Bornou*, Mémoire IFAN no. 7, Paris, 1949, p. 107.
35 R. Lander, ii, pp. 37–48; Muḥammad Bello, *Tarīkh 'Umarā' Bauchi*, pp. 121–31.
36 See below, p. 80.
37 H. Barth, *Travels and discoveries in North and Central Africa*, London, 1857 and 1858, iii, p. 14, iv, p. 496.

traditional enemy of Bornu, under its Emir Sambolei (1825–54) and the Shaikh 'Umar of Bornu, was a constant target of Bukhari's attacks.[38]

Border hostilities with Bornu on the Caliphate's eastern frontiers were chronic. Skirmishes with Bedde abounded. On at least one occasion, in 1869, a joint Hadejia–Katagum expedition to Bedde suffered defeat. In about 1866 a Bornu army led by Yerima Bakr made a successful incursion into Adamawa territory. Bakr was always a menace to the more northerly border emirates. In about 1870 he invaded Missau and turned back only on the orders of the Shaikh 'Umar who was anxious to avert complications with Sokoto.[39] At some time between 1877 and 1880 the frequent border clashes between Bakr's forces and Katagum were composed through negotiations with the Caliph initiated by Shaikh 'Umar. A truce was concluded and Bornu promised that 'there shall not be anything from us but good and reliance on the treaty'.[40] Traditional hostilities were, following the truce, reduced to a minimum. The short and weak reigns of the Shaikhs of Bornu after 'Umar, culminating in Rabīh's conquest of Bornu in 1893, appear to have promoted peace on the borders.

On the Bornu–Adamawa frontier war was endemic, with Mandara and Marghi territories the major battle grounds. In 1824, for example, there was a major confrontation between Adamawa and Bornu forces.[41] In turn Kilba, Song, Holma, Batta, Koncha, Banyo, Tibati, Ngaundere, Chamba and numerous other peoples were integrated into the Adamawa emirate under Modibo Adam b. Al-Ḥasan.[42] But these conflicts belong more to the internal history of Adamawa.

Muri, Adamawa's neighbour emirate, was founded by Buba Yero and his brother Hamman Ruwa as a result of expansion against frontier foes from Gombe. The expansion of Muri was continued in the 1860s with the addition to it of Gassol, Jibu, Bantaji and the foundation of Bakundi sub-emirate.[43] Under the Emir, Muḥammad Nya (*c.* 1874–95), further expansion was undertaken against the Tiv with raids southwards as far as the River Katsina Ala without any permanent success. Tiv hostility contained Muri expansion southwards while similar raids on Jukun territory also failed to secure a safe Muslim foothold.[44]

For the southern sub-emirates of Zaria, the river Benue was an effective

38 J. M. Fremantle, 'History of the region comprising the Katagum division of Kano Province, *J.A.S.*, x, 1910–11.

39 *Ibid.*, pp. 411–14 and Tilho, ii, p. 371, for accounts of these border hostilities.

40 From 'Umar b. Muḥammad Al-Amīn El-Kānemī to Amīr Al-Mu'minīn Mu'adh in A. D. H. Bivar, 'Arabic Documents of Northern Nigeria', *Bulletin of the School of Oriental and African Studies (B.S.O.A.S.)*, xxii, 1959, pp. 332–3.

41 Strumpell, pp. 22–4.

42 *Ibid.*, p. 21; A. H. M. Kirk-Greene, *Adamawa past and present*, Oxford, 1960, pp. 133–6; C. Vicars-Boyle, pp. 77–9.

43 J. M. Fremantle, *Gazetteer of Muri Province*, London, 1922, p. 17.

44 Kirk-Greene, p. 157.

boundary beyond which there was no peace and where dreams of expansion were better not indulged. In Nupe, however, the Niger was no deterrent to expansion. With its internal conflicts reasonably composed by 1859 Nupe not only had no external enemies to fear but it rapidly extended its suzerainty over the politically fragmented regions to the south of the Niger in the Bunu, Yagba, Owe, Ijumu, Akoko, Igbirra and Afenmai districts. The peoples of this region, victims as they were of annual dry season raids by the Nupe, became reluctant subjects, terrorised by Nupe mounted hordes and perfunctorily ruled through Residents *(Ajeles)* whose sole concern seems to have been the collection of tribute, which often meant taking free-born people as slaves. Nupe had to contend with the frequent revolts of these subject peoples particularly in what are now Kabba and Akoko divisions. A permanent camp was maintained at Arigidi, three miles from Ikarẹ, where Nupe forces once confronted the Ibadan. Nupe tyranny over the Kabba and Akoko peoples was broken only by the 1897 Ogidi War (eight miles from Kabba) in which the Royal Niger Company's intervention proved decisive.

In the south-west of the Caliphate, Ilọrin was carved out of territory of the old Ọyọ kingdom and was not internally consolidated until the reign of ʿAbd al-Salām (*c.* 1830–42). Before and during these years the petty Yoruba chiefdoms into which the old Ọyọ empire had disintegrated were constant targets of Ilọrin attacks. In the course of numerous campaigns Ilọrin expanded to Igbomina country and took Ikirun.[45] The power of Ẹdun of Gbogun, considered the most powerful Yoruba general of his day, was completely broken. The Kanla War saw the fall of Ikoyi. This was finally accomplished in about 1837 during the Eleduwe War in which Ilọrin received a Gwandu contingent while Ọyọ secured the alliance of Borgu. The Ilọrin thereafter remained the major threat to the rising power of Ibadan and other Yoruba principalities. After the severe defeat of the Ilọrin forces by the Ibadan at Oshogbo (*c.* 1841), the Ilọrin jihadists under Emir Shitta directed their expansionist forays eastwards. Ilọrin suzerainty was for a while imposed on several western Yagba and Ekiti towns until checked as from 1851 when the Ibadan scored a victory against the Ilọrin in the Awtun war and subsequently wrested most of Ekiti from them.[46] Under Emir Aliyu, the forays of Ilorin expeditionary forces are said to have extended as far as Igbirraland.[47] There was never any danger, throughout the century, that the initial jihād conquests in Ilọrin would be reversed. Rather, during the last quarter of the century the Ilọrin threat to Yoruba country became more serious than ever with the involvement of the

45 Aḥmad b. Abubakar, *Ta'lif Akhbār al-qurūn mīn 'Umarā'bilād Ilorin*, Centre of Arabic Documentation, University of Ibadan, uncatalogued.
46 S. A. Akintoye, *The Ekitiparapo and the Kirija War*, Ph.D. thesis, Ibadan, 1966.
47 Aḥmad b. Abubakar.

emirate first on the side of the 'Ekiti Parapo' and subsequently on its own behalf in the sixteen years war into which Yorubaland was plunged with the Ekiti revolt of 1877.[48]

The task of consolidation and maintenance of security against enemies on the perimeter of the Caliphate was but one aspect of the total military commitments of the Caliphate. Within the borders of the emirates or immediately outside them there were enemy peoples against whom the jihād had to be persistently waged. The military preoccupation of Sokoto with Kebbi, Gobir and Zamfara is a case in point. However, within the northern Hausa emirates, founded within a largely Islamic cultural and political context, the process of consolidation had proceeded faster than elsewhere. Even though pockets of *Maguzawa* (pagan) communities persisted in these emirates they constituted neither a threat nor a serious nuisance. In Missau, Katagum and Hadejia, although they had been carved out of non-Hausa and largely non-Islamic peoples,[49] internal consolidation appears to have been relatively easy to achieve. The large Fulani immigration into these emirates seems to have been a strong factor in the efforts at consolidation.[50]

The area south of Kano was markedly different. Emirates in this region, Zaria excepted, had been carved out of almost wholly, and in most cases completely, non-Hausa and non-Islamic states. The early jihād conquests had succeeded in founding Islamic outposts consisting mainly of the capitals of the emirates. The task of political integration was made more difficult than ever, not only by the weakness of Islamisation before the jihād but also by the fact that for centuries most of the region had been an area of political fragmentation among a legion of disparate ethnic groups.

Starting from the west, it is noteworthy that as late as 1830 the Lander brothers described Yauri as a flourishing kingdom disposing of a strong military force which had successfully resisted the power of the jihādists. Yauri was subsequently granted dhimmi status.[51] To the east of Yauri were vast expanses of Gwariland which, in spite of the conquest of Birnin Gwari early in the jihād,[52] had remained unsubjugated. It was not until the early 1860s that the military exploits of 'Umar Nagwamatse, a scion of the Sokoto ruling family, who had seen service among the Kotonkarfi 'pagans' and in Nupe Muslim armies against numerous foes including Abuja, resulted

48 S. A. Akintoye, *The Ekitiparapo*; B. A. Awe, *The Rise of Ibadan as a Yoruba Power in the Nineteenth Century*, Ph.D. thesis, Oxford, 1964.

49 Low, 'The Border Emirates: A Political History of Three North East Nigerian Emirates, *c.* 1800–1902, unpublished Ph.D. thesis, U.C.L.A. 1968, pp. 2–3.

50 *Ibid.*

51 Lander, pp. 131–2, 240–1; Barth, iv, Appendix II, p. 545.

52 Muḥammad Bello, *Rauḍāt al-Afkār*, trans. H. K. Palmer, *J.A.S.*, xv, 1916, p. 271; al-Waraqāt Muḥammad Bello, pp. 111, 120–1; Abdullah b. Fodio, *Tazyīn*, trans. M. Hiskett, Ibadan, 1963, pp. 130–1.

in the creation of Kontagora Emirate out of Gwariland.[53] 'Umar was appointed Emir of Kontagora by Aḥmad b. Atiku in about 1859, and given the title of Sarkin Sudan. In addition to perennial military expeditions against recalcitrant Gwari towns, 'Umar had to war persistently against the Zaria successor state of Abuja which disputed lordship over the Gwari with him. It was not until the reign of his son, Modibo, that the conquest was sufficiently consolidated to permit the establishment of a permanent headquarters at Kontagora. Even so the process of consolidation was not complete for there continued to be unrest within the emirate until the European conquest. Indeed, the detestation with which the British viewed Kontagora and its Emir, Ibrāhim, on the eve of their occupation confirms the incompleteness of this process of consolidation of the emirate. Perennial wars of consolidation were mistaken by the Europeans for indiscriminate slave raids.

The Bauchi plateau tribes constituted a constant source of anxiety. Secure in their mountain fastnesses, they could launch offensives against the emirates without undue risk to themselves. As they were not only numerous and scattered but also lacked political cohesion, they offered no common head which these emirates could smash. Separate expeditions against one after another of the tribes meant endemic though desultory warfare. Thus the Angas of Pankshin division, the Pyem of Gindiri district, the tribes of Kanam district, the Yergam, Montol, Ankwe, Dass and numerous others were constant but irreducible targets of attacks by Bauchi under Ya'qūb b. Datī, first Emir of Bauchi, and his successors who, according to recorded oral traditions,[54] spent all their lives fighting these tribes. The recorded oral history of Gombe is also in the main a catalogue of the wars of the emirs against dissident pagan tribes.[55] The expedient of conferring dhimmi status on these tribes, subjecting them to payment of tribute, which was adopted by both Gombe and Bauchi (and indeed by all the emirates of the Caliphate), paid off only in the sense that it provided short respites from war against particular tribes at different times. However, these tribes provided rich sources of slaves for which Bauchi was particularly noted. At best the pagan enclaves were barriers to desired expansion rather than threats to survival. The impediment they must have constituted to smooth administration can be imagined.

More potent an enemy than the splinter tribes was the Ningi Sultanate situated between the Bauchi and Kano emirates. Ningi's power had its origin in about 1844 in the emigration of a certain Mallam Hamza from

53 For the history of Kontagora see *Tarihim Kontagora* in Kontagora Divisional Office and F. C. Duff, *Gazetteer of Kontagora Province*, London, 1920; the latter is apparently based on the former.

54 Muḥammad Bello, *Tarīkh Umarā' Bauchi, passim.*

55 *Tabyīn Amr Buba Yero*, anon but provisionally ascribed to Alkali Garga, copy in possession of Professor H. F. C. Smith, Ahmadu Bello University, Zaria, Nigeria.

Kano following a dispute with the Emir Ibrahim Dabo over the weight of taxation.[56] Initially guests of Emir Ibrahim of Bauchi, Hamza and his followers soon became a law to themselves and rose in a rebellion which was quelled by Bauchi at the battle of *Kanghar* (*sic* Arabic) in which Hamza was killed. Under Mallam Aḥmad, Hamza's community settled first at Tabulam and finally in Ningi. From this time the rulers of Ningi referred to themselves, and were known, as Sultans of Ningi. From the late 1870s, under Abubakar dan Maji, the Ningawa not only abetted revolts of Bauchi's pagan tribes but they also began raiding into the Kano and Zaria emirates. As Galadima, Yūsuf b. ʿAbdullāh (later to win fame in the 1893 revolt) had inflicted serious defeats on the Ningi.[57] Kano on at least two different occasions between 1886 and 1891 overwhelmingly defeated Ningi during the reign of Dan Maji's son Harūna and his successor Dan Yahaya.[58] It is understandable, therefore, that during Harūna's Sultanate Ningi raids on Bauchi abated. Zaria however suffered heavily from attacks by Ningi which continued until nearly the end of the century. In the 1880s Galadima Suleimān of Zaria allied with Ningi in his bid for the Emirship, while Sambo, Emir of Zaria (1878–88), was deposed by the Caliph partly for inability to defend his territory against raids by Ningi.[59] The pagan challenge to Muslim authority was probably at its most threatening in Adamawa. The wars against pagan tribes mentioned above, particularly those with the Batta, persisted throughout the century. The internal conflicts among the Fulani in the emirate rendered the successful challenge to Muslim authority easier than would otherwise have been the case.[60]

On the whole pagan and other enemy threats to the Caliphate, dictated persistent vigilance on the part of the emirates in the defence of their territories. The survival of pockets of enemy peoples shows that the emirates had more on their hands than they could cope with, and is explained by parity in military equipment of the emirates and their enemies. The very limited use of firearms and inadequate training seem to have made total victory the pursuit of a mirage. The pagan and other enemy peoples had been unable to jeopardise the survival of the Caliphate because they had smaller armies and lacked the degree of cooperation and common purpose which neighbouring emirates often had.

Besides their external enemies, certain emirates, notably Adamawa, Nupe and, to a lesser degree, Ilọrin, were handicapped in their task of consolidation by internal dissensions between the Muslims resulting in

56 For Ningi history see Muḥammad Bello, *Tarīkh 'Umarā' Bauchi*, pp. 161–91 and R. M. East, *Labarun Hausawa da Makwabtansu*, Zaria, 1932, i, pp. 51 f.

57 *Tarikh Kano*, i.e. Kano Chronicle, University Library, Ibadan, 82/165, p. 105.

58 N.A.K. Sokprof 25, file 7, letter 15, and file 8, letter 5.

59 M. G. Smith, *Government in Zazzau,* Oxford, 1960, p. 187. See p. 101 for details of Ningi attacks on Zaria.

60 Strumpell, pp. 26–7, 29–31, 64–5, 73–4; and Barth, pp. 508–9.

armed hostilities. The founding of Adamawa Emirate was accompanied by rivalry among leading Fulani clans. The leadership of Adam Modibo b. Al-Ḥasan, a member of the small Ba clan, was disputed by clan chiefs of the larger Wollarbe and Jillaga clans. Disintegration was averted by the intervention of the Shaikh who gave the leadership to Adam. Forces of dissidence, largely latent under Adam, broke out in the form of independence movements under his successors. No sooner did he die than the Fulani district of Rei asserted its independence and defied or obeyed the Emir Lawal (1847–72) and Sokoto as it found convenient.[61] Tibati followed suit, and proceeded to attack Ngaundere and Banyo, two other districts of the emirate. When, therefore, Lawal told Dr Barth in 1852 that Adamawa was 'a fresh and as yet unconsolidated conquest',[62] he had aptly summed up Adamawa's political predicament. These and other similar revolutionary movements were inherited by the Emirs, Sanda (1872–91) and Zubeir (1891–1901).

In Nupe Emirate a tradition of acrimonious dispute had started with rivalry between Dendo's sons following the latter's death.[63] The pre-jihād dynasties had by now been reduced to mere ciphers but were used by Masaba (Muhamman Saba) and 'Uthmān Zaki to feed rivalry between them. Soon, open war supervened. The successors of the rival Etsus, Majigi and Majia, whose rivalry had been exploited by Dendo to seize power in Nupe, changed sides with baffling unpredictability between the parties of Masaba and 'Uthmān Zaki, thereby prolonging, intensifying and varying the fortunes of the conflict. After a general Nupe rising against the Fulani *Ajele* system and successive defeats of Zaki in battle the latter was exiled to Kebbi by Khalīl, Emir of Gwandu, in about 1842, and Masaba was installed in his place. But old Nupe resentment against the Fulani denied Masaba a long tenure of office as it led to a revolt which culminated in his defeat. A period of indigenous Nupe rule succeeded lasting from *c.* 1855–57 when Zaki was released from exile and reinstated as Emir as dissensions in the alliance against Masaba dissolved in civil war. Nupe could not be described as properly consolidated until Masaba's second accession in 1859.

Nonetheless indigenous Nupe resistance, now forced to go underground, reared its head first in the Kwenti war—a serious revolt of Nupe tribes in the Kaduna river district—which was with great difficulty suppressed in *c.* 1872 at a battle near Leaba.[64] Again in 1881–2 the Nupe, led by a scion of the pre-jihād ruling family, rose in a rebellion which might have proved fatal for the Emir, 'Umar, but for the timely assistance he

61 Strumpell, p. 31.
62 Barth, p. 497.
63 In about 1832, see 'Umar b. Muḥammad b. Al-Ḥasan b. Adam, and Dupigny for full accounts.
64 Dupigny, p. 17.

received from European trading firms.[65] Finally the continuation, though in a much-attenuated form, of Nupe hostility to Fulani authority was shown by the collaboration of the Ganagana with the invading Royal Niger Company's forces in 1897.[66] In Ilọrin, following the death of Alimi, internal consolidation was delayed until the 1830s by the struggle for succession between 'Abd al-Salām, son of Alimi (Sālih) the founder of the Emirate, and Ṣọlagbẹru of Oke-Sunna (i.e. Muslim quarters), the Yoruba Muslim collaborator of Alimi.[67]

The foregoing summary shows the military preoccupation of the Sokoto Caliphate and provides a framework and a background for its history in the nineteenth century which must be borne in mind, and against which other aspects of the Caliphate's history often have to be appraised. It is clear that, at least until about 1880, militarily speaking, the Caliphate continued to show elements of a state in formation. In this connection, the evidence shows that it is not easy to draw a sharp distinction between a period of conquest, completed early in the century, and a succeeding period in which the main problems were purely administrative. In most emirates, the two processes went on hand in hand far into the century. It is also clear that the Caliphate was not a militarily powerful polity even when religious enthusiasm was at its height. Largely for this reason, and partly because of lack of a centrally directed permanent or regular military force, the presence of hostile enclaves within the Caliphate persisted throughout the century. Yet the course of the history of the emirates outlined above shows that they had grown stronger in relation to their enemies by the last quarter of the century, to the extent that the latter's ability to harm them had been either destroyed or curtailed to managable proportions.

> What was going on [during the century] was not merely a series of military conquests which were not completely successful, but a far-reaching religious, political and social revolution, the necessarily incomplete nature of which was reflected in many different kinds of opposition.[68]

THE CENTRAL GOVERNMENT

The central authority of the Sokoto Caliphate was the Caliph. From a viewpoint of purely Islamic constitutional theory, he embodied the Caliphate, and his authority was subject only to conformity with the *Sharī'a,* the enforcement of which was his *raison d'être*. In the theory of the Caliphate—an Islamic theocratic nomocracy—the discharge of the duties

65 Adelẹyẹ, pp. 174–6.
66 *Ibid.*, p. 269.
67 *Ibid.*, p. 35; also Rev. S. Johnson, *History of the Yorubas,* 5th ed., London, 1960, pp. 197–201, 203–5.
68 Adelẹyẹ, p. 93.

of Caliph involved effective control over determination of policy not only in the general administration of the emirates and their relations with the Caliph but also in their relations with one another. The theory is organic and its implementation implies the establishment of a closely integrated polity pursuing a common ideal through a common policy. Centralism is concerned not only with vertical ties between individual emirates and the Caliph but also with lateral links between them. Thus the theory of state as put forward by the Sokoto reformers from classical and other earlier sources was conceived within the framework of one *Umma*. Hence the physical separation of many of the emirates had few disruptive effects in practice. Above the organic theory of centralism, the circumstances of the component emirates of the Caliphate as well as the historical development of the central machinery of government to a large extent determined the effectiveness and even the actual nature of the central authority.

Perhaps the first point of note is that outside the Sokoto heartland the Caliphate's expansion was the direct outcome of successful jihād revolts which were carried out by Muslim communities scattered throughout what was (until recently) the Northern Region of Nigeria. These jihāds which were sympathetic to the Shaikh's and inspired by it, were followed by voluntary submission and subordination to the Shaikh ʿUthmān dan Fodio and his successors. This spontaneous demonstration of loyalty to the Shaikh ʿUthmān followed naturally from the stature which a charismatic Muslim leader had acquired among the scattered *jamāʿas* before the jihād. A large number of the leaders of the various jihād movements which founded the emirates had either studied under the Shaikh or were known to him personally and they had all received authority (symbolised by a flag) from him to begin their jihād. The personal relations between the early leaders and the Shaikh became a traditional bond of cohesion of the emirates with the Caliph. Spontaneous, though independent, voluntary submission to the Shaikh as Imām was the strongest basis of the Caliphate's corporate existence as well as of its vast area. The scattered jamāʿas, as required by law, gave their allegiance to an Imām as soon as one became 'visible'.[69] The *bayʿa* paid to the Imām (Amīr al-Mu'minīn) is a symbol of complete surrender to his direction.[70]

The uniformity with which the scattered jamāʿas submitted to the Shaikh may be seen as a manifestation of a pre-jihād Fulani ethnic solidarity based on common identity as well as a common political and social experi-

69 On the necessity of appointing a Caliph and swearing allegiance to him, see Ibn Khaldūn, *Muqaddimah*, trans. F. Rosenthal, London, 1958, i, pp. 388–9; Al-Mawardi, *Al-Ahkām al-Sultaniyyā*, French trans. Alger, 1915, pp. 5–7 and M. Khadduri and H. J. Liebesny, *Law in the Middle East*, Washington, 1855, i, pp. 7–8. The obligation is stressed in Sokoto sources, e.g. the Shaikh ʿUthmān was chosen as Caliph at Gudu because it was considered anathema that Muslims should be left to their own devices without a ruler. See Muḥammad Bello, p. 70.

70 Al-Mawardi, pp. 96, 428; Qur'an IV, 62.

ence of oppression and injustice by non-Fulani governments. Such a 'Fulani-family' feeling may help explain the cohesion later enjoyed by the Caliphate. That this feeling should not be lightly dismissed is evidenced by the fact that the Cattle Fulani, who joined in the jihād, were generally pagan at the inception of hostilities while Fulani conflicts with traditional governments outside the religious reference were not unknown before the jihād.[71] Yet the concept of the jihād as principally a 'Fulani rising' *per se* is hardly tenable in view of the patent numerical inferiority of the Fulani *vis-à-vis* their 'host' peoples. This factor made military success conditional on massive non-Fulani support. Besides, it is now a commonplace of the jihād's history that Fulani and others fought on both sides. Ignoring Fulani participation against the jihādists, non-Fulani support for the war need prove no more than that the latter's grievances against contemporary governments were similar to those of the Fulani. But this deduction gives the motivation for participation in the jihād a supra-ethnic character of protest against unjust governments generally. This is the key point. Once granted that such protest was the basis of the success of the jihād the argument that the philosophy of the revolution was the substitution of the *Islamic Way* for unjust rule assumes overriding significance. From whatever motive, embracing the jihād meant acceptance, be it ever so unwittingly, of its reform programme; hence its philosophy. It is well known that the reformers' call was first to reform and only later, out of necessity, to war. The leaders at the centre and in the other jamā'as owed their positions to their piety and learning.

Being the most deeply Islamised, and with large numbers of their learned men generally excluded from government and therefore without a vested interest in defence of the status quo, the Fulani as a group not surprisingly led the reform movement, the jihād into which it developed and the state that emerged from it. The important point is first to grasp the great significance of the Islamic framework of the revolution both with regard to the cause as well as the territorial scope of the jihād. In the latter connection it should be noted that an Islamic ideological ferment of reform which conditioned the mental outlook of peoples over a wide area and their world-view was the prerequisite of successful jihāds. Secondly it should be noted that the totally comprehensive world view implicit in an Islamic philosophy of revolution concerns all aspects of the life of the society and the individual and thus renders partial any interpretation which depends on isolating certain aspects out of context of all the others. This argument is valid with the proviso that it is not intended by it to make Islamic revolutions immune to criticism and doubts as to their genuineness. Thirdly, the special position of the Fulani at all stages of the processes which culminated in the founding of the Sokoto Caliphate must have been an element of

71 Adeleye, p. 14.

cohesion in the Caliphate, particularly during the early years.

In comparison with imperial expansions generally and particularly with that of the early Arab–Muslim Caliphate, the expansion of the Sokoto Caliphate was marked by the peculiar feature that it was not achieved through expeditions organised and sent out from one centre. It was a coming together (voluntarily) rather than forcible integration by Sokoto. There was therefore no necessity to create a standing or regular central army. Also just as the emirates had been independently founded without material help from Sokoto, so they continued to defend themselves without direct aid from Sokoto through the century of war described above. It followed that the leaders and subsequent rulers of the emirates were local leaders and not resident agents from Sokoto. The emirates, in consequence, enjoyed a kind of home rule and, indeed, with the important qualification of the overthrow of pre-jihād dynasties, a kind of indirect rule. Since the jihād became permanent throughout the century, the factors just referred to emphasised localism in government above common policies and organic political integration from the centre.

The Caliph remained the symbol of unity and the guide and mentor of the emirates. Held together at the top by their loyalty to him, the emirates were separately controlled from the centre, thus making the nature of the central machinery of government imperial. Bound in obedience to Sokoto, the emirates did not participate in formulation of central policy. Bonds between the emirates were expressed by common membership of one Umma dedicated to the upholding of Islam and of Dār al-Islām. Lateral relations, though not absent, were expressed by a feeling of common brotherhood evinced in occasional cooperative efforts which, however, did not rule out occasional friction. Relations between emirates were underdeveloped and were not indispensable to the corporate existence of the Caliphate.

The picture above was modified in 1812 by the division of the Caliphate into East and West by the Shaikh ʿUthmān dan Fodio. The western sector, after the Shaikh's death, was under the control of the Emir of Gwandu who did not come under regular or close Sokoto control. Indeed the Emir of Gwandu's relationship to the Caliph was that of an unusually independent Wazir.[72] In this respect, though the Caliphate remained a single political entity in theory, and was constitutionally one in practice, the more striking feature was the diarchy existing within it. Though Gwandu owed allegiance to Sokoto and its Emir was, at least formally, dependent on Sokoto for his appointment, Sokoto–Gwandu ties, in terms of practical control, were tenuous.

The road to close political integration of the Caliphate was fraught with other difficulties. The chronic state of warfare already described made

72 *Ibid.*, pp. 96–9, for more details on Sokoto–Gwandu relations.

coordination of the forces of defence by means of a central standing or regular army impossible. Even if a central army had existed, it is doubtful if it could have been an adequate or a deployable instrument to operate concurrently against the enemies of individual emirates. Built-in localism in defence matters was only an aspect of the built-in instability occasioned by the permanence of the jihād. Not only in the military but also in the administrative sphere, the great distances between Sokoto and most emirates coupled with the slow means of communication ruled out any central control which required close scrutiny of emirate administrations.

Furthermore, it must be borne in mind that the protraction of the era of conquest and of the founding of emirates necessarily involved a slow development of the central as well as the local machinery. Considering how Islamic or un-Islamic some of the administrative practices were, it must also be noted that the creation of the Caliphate, with the reformers and jihādist leaders at its head, was an incidental result of the reformist movement. Largely inexperienced in government, the early rulers whatever theories they relied on for guidance, could not afford a clean break with traditional usages of government. In addition, it was not always necessary or indeed possible to discard such usages. Rulers had the traditional susceptibilities of their subjects to reckon with.

Even when all these reservations have been made, it remains indisputable that the polity born of the jihād was Islamic and that it is in terms of its Islamic bonds that the high degree of control exercised by the Caliph, without the backing of immediate military sanctions, is explicable. The administrative autonomy of the emirates was valid only so long as emirs and their councils discharged their Islamic obligations to the ruled and to their overlord—the Caliph—and did not overstep their constitutional bounds.

Thus, each emir exercised full delegated authority from the Caliph *(imarat al-tafwīd 'ala'l-'Umūm)* in the day-to-day running of his emirate. To maintain the ties with the Caliph as well as to ensure equitable administration of the emirates, each of them came under the supervision of a high Sokoto or Gwandu official as the case might be. In the eastern emirates (i.e. those directly under Sokoto) the Wazir himself supervised the emirates of Kano, Zaria, Hadejia, Katagum, Missau, Gombe and Adamawa.[73] Owing to the vastness of the Caliphate the other emirates were supervised by several high ranking officials of Sokoto. In fact, it was not unusual for the Wazir, even though he spent a great part of the year touring his emirates, to delegate his supervisory authority to lesser officials. It seems, for instance, that the Wali had close supervisory connections with Gombe. The supervisors acted as direct links between the Caliph and the emirates and as the instruments for the exercise of the Caliph's authority. However, in

73 Barth, p. 155, footnote; D. M. Last, pp. 258, 259.

doubtful situations or in cases of dispute within the emirates, they could act only after the Caliph had given specific instructions. On two different occasions during the century, Emirs of Gwandu visited Nupe Emirate personally to settle disputes.[74] Direct appeal from the emirates lay to the Caliph's Court through the supervisor of the emirate concerned.

To take the example of the eastern emirates, the overriding authority of the Caliph was symbolised by the fact that the emirs depended on him for their appointment and remained in office at his pleasure. His pleasure, however, could not be his whim. At the inception of the Caliphate disputes about leadership of the nascent emirates had been settled by ʿUthmān dan Fodio's personal intervention to impose his choice. Thus Modibo Adam b. Al-Ḥasan, first Emir of Adamawa, owed his appointment to such intervention.[75] Similarly a dispute among the chiefs of five leading Fulani clans in Kano was settled by the Shaikh's appointment of Sulaimān as Emir.

In straightforward appointments to the emirship, the power to recommend a candidate lay with the local electoral council. M. G. Smith's study of Zaria Emirate, in his *Government in Zazzau*, demonstrates clearly that the power of the Caliph in matters of appointment and deposition was only too often final. Through intervention in succession disputes Sokoto's power over appointments and distribution of offices in Zaria grew throughout the century. In 1855 Sidi Abdulkadir of Zaria was deposed by the Caliph for misgovernment and acts of insubordination.[76] In 1870 Emir ʿAbdullāh of Zaria was deposed for disobeying an order from the Caliph to withdraw from his attack on Keffi. Abubakar, the successor of ʿAbdullāh, was appointed without any reference to Sokoto. At the appointment of Muḥammad Sambo in 1878 the local electors of Zaria were required to nominate three candidates indicating their order of preference. Sokoto's appointee was thereafter specifically told whom to appoint to the three top-ranking offices of Madaki, Wombai and Dan Galadima.[77] Sambo, the successful candidate, was deposed by the Caliph in 1888 for inability to defend his emirate against its enemies, namely Ningi, Maradi and Abuja.

There had been earlier instances of deposition. In 1816 Liman Adandaya of Katagum was deposed and Dan Kawa appointed in his place.[78] The Emir of Gombe, Sulaimān, came close to being deposed by the Caliph in 1843–4, a letter of deposition having actually been sent. He received pardon only after he had visited Sokoto to sue for it and had shown his loyalty by distinguishing himself in a battle against Gobir. While his fate trembled in the balance he could not count on the support of his officials and the majority of his subjects openly demonstrated their loyalty to the Caliph

74 Dupigny, pp. 11–12 and 17.
75 Strumpell, p. 18, and see above, p. 71.
76 Smith, pp. 163–4.
77 *Ibid*, pp. 178–9.
78 Low, p. 163.

against him.[79] In 1844, Emir Siddiq of Katsina was deposed for charges of oppression brought against him by his subjects.[80]

The practice of sending to the Caliph the names of several nominees for an emirship was not limited to Zaria, nor was it begun there. At the death of ʿAbdullāh, Emir of Kano (1855–83) the local electors sent the names of four candidates to Sokoto.[81] Similarly in 1896, before the appointment of Abdulkadiri as Emir of Katagum, the names of four candidates had been forwarded for the Caliph's consideration. This tradition was started in about 1851 on Qadiri's appointment.[82] Further study may well show that recommendation of a number of candidates to guide the Caliph in the choice of an emir was a widespread practice throughout the emirates. The Caliph's decisive power over appointments and depositions is beyond dispute and it would be monotonous to multiply instances to prove this role, not only with regard to the emirship but often also lesser offices. The turbanning ceremony, marking the formal installation of an emir by a Sokoto official, was no mere formality but the symbol of an authority which the emir could defy only at his own peril.

The most generally known way by which the emirates demonstrated their vassalage to the Caliph was through payment of tribute. However, it is erroneous to claim that such tribute was a voluntary gift. In extant correspondence of the Caliphate during the late nineteenth century, such payments of tribute are described as 'well-known covenanted gifts' *(al-Hadīyat al-maʿ lūmat al-ma ʿhūdat)*. If gifts they were, they had become so established by tradition that non-payment would tantamount to a serious breach of obligation. Between 1886 and 1893, Muḥammad Bello of Kano had to explain to the Wazir of Sokoto his inability to meet an increased demand of tribute from Sokoto.[83] On one occasion between 1891 and 1900, Emir Abdulkadir of Katsina apologised for a delay in sending his tribute to Sokoto.[84] In a letter from Emir ʿUmar of Bauchi, the latter apologised not only for lateness in sending his tribute but also for his inability to attend personally at the Caliph's Court owing to illness.[85] The Wazir of Sokoto saw to the collection of tribute from Zaria in person. As Sokoto's demands were probably rising during the century, taxes were increased in Zaria during the Caliphate of Aliyu Baba (1842–59).[86] The government of Lugard had evidence between 1901 and 1902 that the already conquered emirates of the south continued surreptitiously to pay

79 *Ibid.*, p. 215.
80 Al-Ḥajj Saʿid, *Tarikh Sokoto*, p. 25.
81 Adeleye, pp. 129–30.
82 Low, pp. 263–83.
83 N.A.K. Kadcaptory, file 7, letter 51.
84 N.A.K. Kadcaptory, box 44, letter 38.
85 *Ibid.*, letter 15.
86 Smith, p. 74.

tribute to the Caliph.[87] In addition to regular payments it appears that the emirates often paid *khums,* the fifth part of booty captured in war as fixed by the sharīʿa to the Caliph.[88] Emirs also paid *Kurdin Sarauta* (money for accession to office) as well as inheritance money—*irth.*

Subservience of the emirates to Sokoto manifested itself in various other ways. Each emir was under obligation to pay a visit to Sokoto soon after appointment. Further, it was usual for emirs to attend an annual gathering near Sokoto during the dry season every year. This meeting was, as it were, the annual 'parliament' known as the meeting of the '*Manya Sarakuna*' of the Caliphate's eastern emirates. It was usual on these occasions for the emirs to come with military contingents to aid Sokoto in its annual dry season raids against its enemies. While representatives could in certain circumstances be sent to this gathering by the emirs, non-attendance by emirs was normally viewed by the Caliph as tantamount to rebellion. Refusal to obey the Caliph's summons to Sokoto could earn an emir severe punishment or even deposition. The narrow escape from deposition of Sulaiman of Gombe in 1843–4, mentioned above, arose from disobedience to honour a summons to appear in Sokoto. Sidi Abdulkadir of Zaria was summoned to Sokoto in 1855 before he was deposed. For sending a representative to Sokoto instead of appearing in person when summoned Caliph ʿUmar deprived Muḥammad Manga of Missau of his extra-territorial rights over Bornu Fulani settled in other emirates between 1881 and 1891.[89] The Emir's rights were restored to him only after reconciliation effected by the intervention and plea of the Emir ʿUmar of Bauchi. In a letter from Emir Muḥammad of Muri to ʿUmar of Bauchi, the latter explained his cancellation of a proposed visit to Bauchi on the grounds that he was going to Sokoto—'a journey' as he described it, 'which is a duty for all who are set over a people'.[90]

The Caliph's obligations to his emirates did not end with appointments and depositions. General supervision involved ensuring that direct obligations to the Caliph were met and that equitable administration was maintained in the emirates. His expectations are exemplified in a letter of appointment (which may have been typical) which ʿUthmān dan Fodio sent to Yaʿqūb,[91] first Emir of Bauchi. The latter was enjoined to ensure (1) that his community should be one in speech and in action and with no dissension amongst them, (2) that he should be zealous in the repair of mosques and in praying in them, (3) that he should ensure that the Qur'ān

87 P.R.O., CO 446/25, Lugard to Colonial Office, No. 597 of 21 November 1902.
88 Adeleye, pp. 111–12.
89 *Missau Chronicle.* English translation obtained from V. N. Low. See also uncatalogued letter in University of Ibadan (trans. A. N. Skinner) from Emir of Bauchi's collection; from Amīr al-Mu' minin ʿUmar to Emir of Bauchi ʿUthmān.
90 *Ibid.*, untranslated.
91 Reproduced in Muḥammad Bello, *Tarīkh Umarā' Bauchi*, pp. 26–7.

and the Islamic sciences are learned and taught, (4) that the markets should be kept in good repair and illegalities in them prohibited and (5) that the Emir should wage the jihād as a duty imposed on Muslims. The Caliph's role in respect of appointments and deposition provided him with a large scope for intervention in the 'internal' affairs of the emirates, the more so as succession and distribution of offices were a fruitful source of internecine disputes.

The intervention of the Caliph in resolving incipient disputes over leadership at the foundation of certain emirates has been noted. In subsequent disputes the Caliph traditionally played the role of final arbiter. However, on a number of occasions, the Caliph did get his fingers burnt by unpopular intervention in succession disputes. For instance the succession of 'Umar Sanda of Adamawa had been challenged by a certain Hamidu who sought and received the Caliph's initial support. As Hamidu was not a popular candidate with the electors the latter had politely set the Caliph's instructions aside while the affair was settled with an undoubted finality by the failure of Hamidu's armed revolt which ended in his death.[92] In 1873 and 1897 open rebellion against the Caliph's support of unpopular candidates to Zaria Emirship was only narrowly averted.[93]

The two contrasting episodes and the Bukhāri and Yūsuf revolts in Hadejia and Kano Emirates in *c*. 1848–63 and 1893–5 respectively provide apt illustrations of the limitations on the Caliph's power of intervention in legal and practical terms. In 1848, Bukhāri b. Muḥammad Sambo succeeded to the throne of Hadejia through means of doubtful legality.[94] When, however, he not only disobeyed the Wazir of Sokoto but also lost the confidence and loyalty of his subjects he was invited to Sokoto for questioning. Following his disobedience, a protracted war broke out between Bukhāri and Sokoto. Owing to assistance received from Bornu and the difficulty (for the attackers) of the terrain in which they fought, he defied the Caliph's authority till his death, in about 1863. The significant point is that the Caliph was able to command the loyalty and the military support of Hadejia's neighbouring emirates while his course won the moral support of the whole Caliphate. The Caliph could not have enjoyed such widespread support in any of the cases cited in the preceding paragraph because by ignoring patent popular opposition he was pitting himself against the public welfare—*maslaha*; an important consideration in the Maliki code. In the Bukhāri affair, the Caliph's cause enjoyed widespread sympathy and support because its justice was as patent as the rebel's cause was reprehensible and illegal. The Bukhāri incident illustrates the military weakness of the Caliphate as it shows that in practice the Caliph could

92 East, ii, p. 140, 'Labarum Sarakuna Adamawa'.

93 Smith, pp. 174–6, 193–4.

94 On the Bukhārī revolt, see Barth, ii, p. 175 ff.; J. M. Fremantle, *J.A.S.*, x, pp. 404, 412; Al-Ḥajj Sa 'id, pp. 27–8 and Wazir Junaidu, ff. 49–50.

prove incapable of enforcing his decision even when the justice of it was beyond cavil.

The Yūsuf revolt in Kano (1893–5) demonstrates the fact that the Caliph was in a still weaker position when it came to enforcing a decision considered unjust and that his power of control over the emirates depended ultimately on his keeping within the bounds set by the sharīʿa, which demanded that he be just in all his dealings. Since Caliph ʿAbd al-Raḥmān's appointment in 1893 of Tukur, the son of the late Emir Muḥammad Bello, rather than Yūsuf, the son of the preceding Emir Muḥammad ʿAbdullāh, was unpopular and was considered unjust, the Caliph's attempt to sustain his decision failed primarily because, although it did not arouse overt hostility in the emirates, it enjoyed no sympathy.[95] In the end the attempt of the Tukur faction in Kano to cling to office foundered owing to the lukewarmness of the support temporarily and half-heartedly given to it by the Emirs of Katsina and Hadejia and the apathy of the other emirs.

Yet in all these actions the Caliphs were only trying to play their role as supreme authority in the Caliphate. On balance, their role as mediator was productive of more positive achievements for the Caliphate than failures. A few examples from disputes between emirates illustrate this. Muri emirate was originally part of Gombe emirate, having been founded by Buba Yero's brother Hamman Ruwa. In about 1833 Buba Yero, anxious to maintain Muri succession in his own line, plotted Hamman's death but the issue was decided by the Caliph, Muḥammad Bello, who reorganised Muri as a separate emirate and settled the succession on Hamman Ruwa's line.[96] Thus continuous conflict between Gombe and Muri was averted for the future. In Kano, the rising of Dan Tunku, an early leader of the jihād in Kano, against Emir Ibrahim Dabo resulted in war. When after about five years of fighting, characterised by reversals of fortune for both sides, a stalemate resulted, both the Emir of Kano and Dan Tunku submitted their dispute to the Caliph for arbitration.[97] Dan Tunku was recognised by Caliph Muḥammad Bello as Emir over an independent emirate of Kazaure. Such appeal from disputants was indeed the way that the Caliph's arbitration was usually brought to bear on disputes both within and between emirates. Sokoto's arbitration, which was usually decisive in restoring order, was usually accomplished by the sending of high-ranking Sokoto officials to investigate the dispute on the spot. Ruling thereafter was binding on the parties to the dispute. Thus during the years 1891–1900 a dispute between Kazaure and Daura was settled through the arbitration of Galadima ʿUmar, delegated with the Caliph's authority for the purpose.[98]

95 Adeleye, pp. 129 ff. for further details.
96 Fremantle, *Gazetteer of Muri Province*, pp. 17–18.
97 Johnston, pp. 178–9.
98 Adeleye, p. 117.

In the emirates of the eastern marches there were numerous instances of inter-emirate disputes settled by Sokoto arbitration. Throughout the Bukhārī rebellion there were frequent hostilities between Hadejia and Katagum. Muḥammad Manga of Missau was involved in a protracted conflict with Katagum over the Fulanin Bornu town of Isawa in about 1896.[99] The town revolted against Missau's authority with the connivance of Sarkin Shira of Katagum. In the ensuing hostilities, Missau was aided by the Emir of Kano, Aliyu. After much desultory fighting and mutual raiding the disputants submitted to Sokoto arbitration. Through the good offices of the Caliph, an episode that nearly plunged all the border emirates of the East into civil war was effectively brought to an end. Isawa was awarded to Missau. Muḥammad Manga was also involved in a revolt by the Galadiman Ako against Emir 'Umar of Gombe between 1899 and 1901. Largely because of Manga's support for the Galadima, the rebellion proved insoluble until it was submitted to the Caliph's arbitration. The Caliph's agent, Wali Shehu, made an on-the-spot investigation which was followed by instructions to all parties to the dispute to withdraw from it and the Galadima to seek the pardon of his Emir.[100]

These instances of the Caliph's intervention in inter-emirate disputes prove Sokoto's authority, influence and acceptability, and the fact that the emirates looked up to the Caliph in their difficulties. Inter-emirate disputes show that the feeling of brotherhood among the emirates did not always rise above serious differences of a cold politico-economic nature. Just as the emirates had their disputes, which were often settled by Sokoto, so they had their occasions of cooperation against a common enemy and these were also often inspired or even organised by Sokoto.

The summoning of contingents to help Sokoto in its annual dry season raids on enemy territories or to assist in major wars was the closest approximation to a regular central army ever attained by the Caliphate. While it was usual for Sokoto to receive contingents from the emirates, the reverse was hardly known. Only during 1805 was the Caliph known to have sent an army to Katsina.[101] But after that the jihād in Katsina, Daura and Kano was fought by the local people. Cooperation between those three emirates was, however, the result of inspiration from the Caliph. From the inception of the jihād there was marked cooperation between the emirates of the eastern marches. The advance of Bornu forces towards Kano early in 1827 was halted at Garko by a joint army of neighbouring emirates under the leadership of Yaqūb of Bauchi acting on the Caliph's orders.[102] Similarly, the raising of a force against the rebel Bukhārī of Hadejia from several emirates was in obedience to Sokoto's instructions. When from the late

99 *Ibid.*, pp. 119–21.
100 *Ibid.*, pp. 121 ff.
101 Muḥammad Bello, p. 95.
102 See above, p. 65, n. 35.

1880s the power of Mallam Jibril Gaini, a Mahdist, threatened the emirates of Bauchi, Gombe and their neighbours, it was the Caliph who instructed the emirates affected to unite under the leadership of ʿUmar of Bauchi against their common enemy.[103] Arrangements for a truce were also referred to Sokoto. Throughout the century, emirates were in the habit of reporting their difficulties to Sokoto and Sokoto also kept them informed about its own wars and truces. In times of general danger, such as the expected invasion of Rabīḥ Faḍlallāh between 1894 and 1897, after he had overrun Bornu, the Caliph issued necessary instructions to all emirs to close their roads to Rabīh's people. Throughout these anxious years the emirates of the eastern marches kept the Caliph well informed about the threat of Rabīh's impending approach.[104] Indeed, after the fall of Bida and Ilọrin to the Royal Niger Company in 1897 the emirates seem to have received a general circular from the Caliph to close their roads against European infiltration.[105]

As custodians of general welfare, the Caliphs were able to use this position and their authority to strengthen the defence and security of the Caliphate by coordinating the efforts of neighbouring emirates on a regional basis. It is clear that within the practical limitations to the full discharge of their constitutional obligations as the supreme authority, the Caliphs maintained the cohesion of the Caliphate at a surprisingly high level. In the final analysis the control of Sokoto, which was never remote, was no less real than the local autonomy of the emirs.

It is clear that the Caliphate had a coherent machinery of government which was kept in strikingly good repair right up to its fall. Failures that may be detected in the actual functioning of this machinery cannot be attributed to intrinsic weakness in the system itself or indeed to degeneracy in its working. Extrinsic factors, such as the historical circumstances of the Caliphate and the geographical distribution of the emirates, must claim the attention of the student who wishes to see the central government in focus.

The interpretation of Sokoto history which has gained very wide acceptance is that there was a golden age ending with the Caliphate of Muḥammad Bello, during which the ideals of the Caliphate conceptualised in the jihād literature were realised, followed by a period of progressive departure from the 'Islamic Way' and a consequent degeneration in government and society at all levels, which, accompanied by disruptive centrifugal forces, dragged the Caliphate to the brink of inevitable destruction at the end of the century. This could not be farther from the truth. Such an interpretation

103 Adelẹyẹ, p. 149.
104 During these years, the eastern emirates lived in constant fear of Rabīh's invasion. They sent frantic letters to the Emir of Kano and to Sokoto giving information about Rabīh's movements. In early 1897 Rabīh's forces on the eastern borders of the Caliphate had to return eastwards to repel a French expedition under Captain Gentil.
105 See P.R.O., CO 537/15 Secret, No. 318, enclosure, Goldie to Bertie, 2 March 1898.

is based on inadequate understanding of the evolutionary history of the Caliphate. Indeed the lasting foundations of administrative cohesion, which in practice were to sustain the Caliphate to its end, were only being worked out and consolidated in their details under Muḥammad Bello. The student is liable to search for some easily discernible physical sanction on which, he believes, the Caliph's position as supreme authority must ultimately depend. Failure to find it may lead the student to conclude that the Caliph's authority must necessarily have been ineffective in practice. But in the light of the nature and the working of the Caliphate, it is evident that the extra-physical obligations of the emirs to the Caliph, as leader of the Islamic Umma, and the implications of acceptance and rejection of that leadership, so long as it was not abused by the incumbent, proved more valuable than an army would have been. As 'Uthman dan Fodio wrote in his book *Bayan Wujūb al-hijra*:

> Seeing to the welfare of subjects is more effective than a large number of soldiers. It has been said that the Crown of a King is his integrity, his stronghold is his impartiality and his Capital is his subjects. There can be no triumph with transgression, no rule without learning *(fiqh)* of the law and no chieftaincy with vengeance.

It may be said that the degree of success with which the Caliphate faced problems of survival, expansion and cohesion is, in the main, largely a reflection of the degree of the rulers' adherence to the foregoing political testament.

GOVERNMENT AND SOCIETY IN THE EMIRATES

The jihād was waged to pave the way for the establishment of Islamic government. The preceding section shows that relations between the Caliph and his emirs were largely determined by the bond of Islam. However, there is no better vantage point from which an appraisal can be made of the degree of success with which the aims of the Sokoto reformers were realised than through an examination of the local government of the emirates. These emirates were charged with the day to day administration of the practical details of the Islamic way.

It is useful to first take a brief look at the mother of the emirates—the Sokoto heartland. Here, zeal in the cause of the re-creation of the conceptualised ideal Caliphate *(al-khilāfā' al-Faḍīla)* was shown not only by the detailed attention paid to the ideal in the jihād literature but also by the appointments actually made.[106] The office of the Caliph (referred to in early Sokoto literature as Amīr al-Muminīn or Imām), the Wazīr, Qāḍīs with a Qādī al-Quḍāt (chief justice), the Muhtasib (a general welfare officer charged with commanding Good and prohibiting Evil), the Imāms (i.e. of

106 My account of Sokoto offices is based on D. M. Last, pp. 101–8.

the mosques), the Sa'i (collector of *jangali* which was designated as *Zakāt* or *jizya* respectively according to whether it was paid by a Muslim or *Dhimmi*), the Walī al-shurta (chief of police) and the Amīr al-Jaish (commander of the army). Commanders of the army were not only appointed for the Sokoto metropolitan province but also for all emirates. Initially, each emir was primarily an Amīr al-jaish for his area. The difference between the emirs and the Sokoto Amīr al-jaish was that whereas the latter was a specialised office *(Imarat 'alā 'l-husūs)* the former, as *nuwwāb* (deputies) of the Shaikh, exercised jurisdiction of general import *(Imarat 'alā-'l-'Umūm)*. The Sokoto offices follow closely those specified by the Shaikh in *Kitāb al-Farq* and *Bayān Wujūb al-hijra*, as essential offices of state of long standing in Muslim tradition.

D. M. Last has demonstrated the survival, even in Sokoto, of such pre-jihād Habe titles as Galadima, Magajin Gari, Magajin Rafi among many others.[107] Although in the beginning these titles were honorific rather than functional, it seems that owing to the influential position of their bearers and their contribution to the jihād, there later came to be attached to them territorial appanages or some other official position in the administrative hierarchy. This feature of administrative development reflects the compromise which had to be struck between pure Islamic theory and the pre-jihād culture of the society. Contrary to the inference which may be drawn from the *Kitāb al-Farq*, the survival of non-Islamic nomenclature of offices did not imply functions repugnant to Islam.[108] This is one generalisation which is valid for all the emirates. The pure Islamic culture, which has not been known in history to operate anywhere, had, as it were, to find its own level within the traditional culture, the spirit of which it was transforming as distinct from entirely replacing its form.

Another feature of the administrative arrangements in the Sokoto metropolitan province, as demonstrated by Last, is that the role of the dominant Fulani families and local leaders in the jihād was reflected in the distribution of jurisdiction over land after the jihād.[109] Generally, land was enfeoffed either to officials resident at Sokoto, who acted as supervisors over their fiefs and owed direct vassalage to the Caliph, or to local leaders, who lived on their fiefs and administered them virtually as sub-emirates. The latter class of officials also derived their authority directly from the Caliph.

The emirates were, by and large, replicas of the Sokoto model in all essentials. However, neither for Sokoto nor for the other emirates as a whole is it possible, with available data, to present an adequate historical survey which will reflect the details of the changes that took place within

107 *Ibid.*, pp. 160–2.
108 'Uthmān dan Fodio, *Kitāb al-Farq,* trans. M. Hiskett, *B.S.O.A.S.*, xxiii, 1960, p. 563. The Shaikh specifically condemns the use of non-Islamic names as titles.
109 Last, pp. 107–8, 160 ff., particularly pp. 167–9.

the administrative framework or with regard to particular offices through the century. With the possible exception of M. G. Smith's *Government in Zazzau* (and the analysis in this work is more sociological than historical) there has been no detailed evolutionary study of the administration of any of the emirates. Available data cannot do much more than provide a framework of functional relations, of a rather static character, within the administrative hierarchy. It is enough here to draw attention to some of the major factors which lay behind change within the administrative organisation of, and the socio-political relations in, the emirates.

The statement that the emirates were replicas of the Sokoto model does not amount to a claim of uniformity of structure and practice except in general terms. In reality, emirates tried to mirror the Sokoto model but their differing historical antecedents meant a wide range of diversity in the practical problems of administration. It will be observed that the structure of the administrative cadre and relations in Sokoto itself was contingent on the antecedent culture of the Gobir–Zamfara–Kebbi region, the distribution of the pre-jihād population among the Fulani clans and Habe elements, and the role of these in the jihād. It is a demonstrable inference that religious fervour is often accompanied by religious mysticism and this, though a testimony to piety, narrows the scope of flexibility in practice because it concedes so much to the unknown and to taboos. Nonetheless, although Islam is popularly regarded, not without reason, as a fatalistic religion, mysticism is not an essential of orthodox purist Islam. As a way of life Islam is as much concerned with life here as with the hereafter, or what some Western medieval philosophers would term the two ends of Man. For this life, Islam, in its origin and development, is a religion of the practicable. It is this factor which enables it to adapt to widely divergent cultures and find a genuine and far-reaching accommodation within them. This is not to deny that accommodation, unguided by close attention to the *sharī'a* (and this implies no contradiction), can easily drift to syncretism.

It is with this in mind, that an appraisal of the factors which supplied the dynamics of administrative practice and development in the emirates must be sought. Firstly, the jihād had brought about a demographic revolution—a factor which is often overlooked. In the migrational aspect of this revolution, it will be recalled that the Habe rulers of Gobir, dislodged from their capital, Alkalawa, in 1808, did not found another permanent capital (Tsibiri) until after the battle of Gawakuke. The migration of the Katsinawa Habe to the Maradi and Tessawa region and that of the Zazzau Habe to Abuja are well known. It is not unlikely that with the flight of Sarkin Daura and Sarkin Kano during the jihād, a substantial proportion of the ruling Habe were dispersed.

In the area of Adamawa—a vaguely defined and politically fragmented region known as Fumbina—it is known that the Fulani, though present in large numbers, were a subject population before the jihād. When this occurred they came into prominence while their erstwhile masters sank into obscurity. We do not hear of the emigration of the pre-jihād rulers. What in fact happened was a repatterning of society with the core of the pre-jihād rulers' resistance concentrated in those centres in Adamawa which have already been described as the stumbling block to political integration in Adamawa Emirate. Here, the regularity and intensity of the permanent jihād must have profoundly influenced the structure of the administrative hierarchy as well as movements and functions within it. The political fragmentation of the Adamawa region and the heterogeneity of its tribes were paralleled by the number of scattered Fulani clans. The problem of political integration was consequently in the forefront throughout the century. The jihād itself had been waged by locally appointed Fulani leaders. Twenty-four of them are said to have received flags from Modibo Adam b. al-Ḥasan.[110] The number may have increased as the century progressed. The main implication of this was that the Emir of Yola had a large number of powerful vassals enjoying a high degree of independence as rulers of compact fiefs.[111] Hence the revolt of some of them and their *de facto* assertion of independence in the course of the century.

In the emirates of the eastern frontiers, what is to be remarked is the stream of Fulani immigration from Bornu at the beginning of the jihād. The Mai of Bornu complained to the Shaikh 'Uthman about this emigration from his territory,[112] which indicated that its volume was considerable. This process of Fulani immigration and proliferation through intermarriage with local groups may account for a steady rise in the percentage of Fulani in the population of the emirates of Katagum, Hadejia and Gombe. The settled Fulani population may also have drawn upon their nomadic counterparts as the latter became converted to Islam or became involved in the economic or even the political nexus of the population of new urban areas resulting from resettlement and administrative requirements consequent on the jihād. These observations on population must have been largely applicable to all the emirates and to the non-Fulani as well.

In his study of the emirates of Katagum, Hadejia and Gombe Dr V. N. Low observes that they were created out of 'an almost wholly non-Muslim cultural context—by leaders of pastoral or semi-sedentary Fulani bands lately resident in Bornu rather than Hausaland'.[113] It is thus obvious that, added to the divergent political, social, religious and economic antecedents of the heterogenous communities of these emirates, the alien or quasi-alien character of the Fulani conquerors must have made the problems of

110 C. Vicars-Boyle, pp. 82–4.

111 Barth, iv, pp. 508–9. Barth described many Yola vassals as, in effect, overmighty subjects with a tendency to assert their independence.

112 Muḥammad Bello, p. 122.

113 Low, p. 2.

political integration even more complex than in areas of relatively homogeneous culture like the Hausa states. But the dominance of the Islamised Fulani in the course of conquest and in government made Islam the common denominator of intensive and rapid acculturation.

Control of political authority remained in Fulani hands. The structure of land distribution was akin to the Habe feudalistic system. In Gombe and Katagum, lands outside the capital were distributed in compact fiefs among high-ranking officials.[114] In Katagum the major fief-holders were men in the lines of succession to the emirship and they lived in their fiefs.[115] The distribution of offices was, no doubt, closely related to the part played by various families or groups in the jihād conquests. Hadejia was like Zaria, where, as shown by M. G. Smith, the fiefs of individuals were usually scattered. The complexity of administration grew as the century wore on. Increasing patronage became a strong factor in contests for succession to emirship and the creation of ever-expanding clienteles by rival groups became a prerequisite for the pursuit of power. In these circumstances offices and perquisites necessarily proliferated. It is claimed, for instance that in Katagum, Ibrahim Zaki had appointed only ten officers before his death. No doubt the number must have multiplied after him. Under Muḥammad Kwairanga in Gombe, successor of Buba Yero, twenty-five new titled offices were created.[116]

A general inference which can be drawn from the migrations induced by the jihād is that migrations by both Fulani and non-Fulani encouraged new and varying degrees of departures from traditional usages as well as new developments. Habe emigrations from their main centres of government, particularly from Zazzau and Katsina, removed substantial groups of the old order who would otherwise have been capable of putting up a virulent opposition to the new establishment, which could have resulted in political and social fissures or even ruled out political integration within the emirates. As it happened, in Zaria for instance, the remaining Habe population left behind was more amenable to subjection to the new aristocracy. By contrast, in Nupe, where the pre-jihād dynasties were overthrown but yet contained within the society and where the Fulani ruling group was relatively small, not only was conquest protracted but the subsequent rivalries between the sons of Mallam Dendo were rendered chronic for about three decades, owing to the shifting support which the old Nupe dynasty, itself divided against itself, afforded to the warring Fulani factions.

Also, the presence of diverse Fulani clans within emirates, all of which had participated actively in the jihād conquests, meant that their leaders were transformed into a ruling aristocracy with clienteles vital to success in political rivalries. At the inception of the jihād, dispute over the leadership

114 *Ibid.*, p. 23.
115 *Ibid.*
116 *Ibid.*, p. 223.

(i.e. emirship) had broken into the open among the Fulani clans of Kano and Adamawa. Zaria did not start with such a dispute but the influence exercised by succession disputes over the emirship among the Mallawa, Bornawa, Katsinawa and Suleibawa Fulani on the structuring of the administrative hierarchy and changes attendant on the position of many office holders has been demonstrated by M. G. Smith.[117] Changes in the personal fortunes of individuals in the disputing member-groups produced corresponding changes among their clients down the rungs of the social ladder. It seems that important changes in the political, social and economic life of Zaria must have turned on the abilities of each member-group to tilt the political balance in its favour.

It is noteworthy that even in these emirates where succession was in one line but actual choice could be made within various branches of that same line, rivalry over succession, its party-forming implications, the political manoeuvres of the rival branches of the line of succession and attendant changes similar to those in Zaria, formed the basis of political and social changes. In this respect the succession disputes in Kano on the appointment of Muḥammad Bello in 1886 and the more acrimonious dispute between the latter's son and the sons of the Emir ʿAbdullāh which gave rise to the Yūsuf revolt will be readily recalled.[118] All the disputants descended from Emir Ibrahim Dabo. Dispute in other emirates was the rule rather than the exception. The practice of nominating several candidates for Sokoto's final decision, which has been specifically mentioned with regard to Zaria, Kano and Katagum, was evidently a device to resolve local succession disputes. In Gombe in about 1886 there was dispute over the succession between Zaillani and Tukur, both children of Kwairanga, and ʿUmar's succession in 1898 led to a serious constitutional crisis occasioned by the challenge of the group who supported his brother against him.[119]

Indeed it seems that succession disputes became as endemic as the jihād. One point which must be remembered is that the period of conquest and consolidation in most emirates was very long. This preoccupation was necessarily reflected in the setting up of the administrative machinery. It delayed the full evolution of the process and profoundly influenced the structure and nature of it. It was often the exigencies caused by the foregoing factors— the jihād, the ethnic composition of Fulani and non-Fulani, intra-lineage disputes—that determined the peculiar structure of authority (including the offices to be found in the emirate) and the nature of political and social relations in different emirates. Whatever the specific offices in any emirate, the general pattern was that the emir administered his territory through further delegation to subordinates of the power delegated to him

117 Smith, pp. 74–88.
118 Adeleye, pp. 129 ff.
119 Low, p. 255.

by the Caliph. On the officers and vassals of the emir rested the actual translation of Islamic policy into the practical reality of administration. To a large extent, the offices which existed in the various emirates derived from pre-jihād offices. In no emirate were the offices of the local administration of the Sokoto hinterland reproduced with their traditional Arabic names. Thus in Gombe only the offices of Wazir and Qādī were at first appointed out of those mentioned in the *Kitāb al-Farq.*[120] In Nupe, probably an extreme case, the traditional names like Etsu (for the Emir), Ndegi (for the Wazīr) and others like them were retained.

The inhabitants of the emirates were divisible into free and slave. The first group was further divided into the aristocracy—divided, in turn, into those of royal blood and others—the clients of the aristocrats, the *'Ulamā'* (learned men) and the *talakawa* (the common people). In general, varying degrees of mobility were possible from one class to the other, although strong elements of heredity established by tradition in the families of the first holders of an office tended to make the top aristocratic cadres a closed group. Otherwise, the slave could become a free man by manumission. Also, even as a slave, he was not denied opportunities. Slaves could, and often did, own slaves. Slaves, as is well known, held offices of state. The talakawa would move high up the scale through acquisition of learning or sponsorship from the higher strata. It is important to note also that the stratification of society did not follow a strict Fulani and non-Fulani division into high and low. To the socially immobile talakawa, offices like occupational heads, e.g. Sarkin Makera (chief of the Blacksmiths), Sarkin Pawa (chief of the Butchers), etc., were open. They enjoyed the privileges of occupational control over members of their groups and also collected taxes from them.

The emirs collected taxes through the agency of subordinate officers and vassals who had jurisdiction over fiefs or districts. The rising burden of taxes not only for running the day to day administration but also for consolidating the power of the emirs against that of rival factions opposed to them was ultimately borne by the talakawa whose taxes might pass through as many as four pairs of hands before they, or a fraction of them, ultimately reached the emir.

The enforcement of the sharī'a was the object of all government. For this purpose grades of courts existed ranging from the village level to the Alkali (Qāḍī) and the emir's court. The deposing of various emirs, for oppression, inefficiency or disobedience, indicates that even the emirs did not always keep the sharī'a. Cases of offences against the law, or perversions of it, which did not merit deposition or similarly condign punishment may have been numerous. They are such, however, as escape documentation. The incidence of non-adherence to the sharī'a would be expected

120 *Ibid.*, p. 144.

to have been higher among the lesser officials, particularly those who came in direct contact with the ordinary people.

As observed for the central government, it is misleading to postulate a 'golden period' of rule based on pious and punctilious regard for the sharīʿa succeeded by one of laxity. The confusion which leads to interpretations of this kind arises from inadequate distinction between at least two periods. One was a period of conquest and consolidation when cooperation between groups was not only at its maximum, but was also keeping alive the enthusiasm born of the success of the jihād. The other was the period when, with the successful establishment of the emirates, the urgent task was to face the practical complexities of the details of administration involving a re-organisation and definition of group relations and the demands of seeing to the day-to-day working of the machinery.

It is often claimed that the reduced interest in literary and intellectual activities in the Caliphate after the death of Muḥammad Bello proves the decline of religious zeal. The argument cannot be sustained that the zeal for the realisation of the ideology of a revolution is ever as strong as during the period of struggle for the triumph of that ideal. With triumph, the outward edge of the passion, in evidence during the climax of the crisis, tends to be dulled or rather consumed by efforts at administering as distinct from proclaiming and fighting for the ideology. Such was the case with the Sokoto Caliphate. In fact, as early as 1806–7, ʿAbdullāh dan Fodio had noted disturbing departures from the original ideals of the jihād even among the Shaikh's community. He charged many of these followers with vain-glorious worldliness.[121] Both the Shaikh and Muḥammad Bello agreed with ʿAbdullāh's assessment.[122] When ʿAbd al-Salām accused the Community of Muslims of aberrations in devotion to the ideals of the reformist movement, Muḥammad Bello accepted the charge.[123] In common with all Muslim history the ideal was never realised in the Sokoto Caliphate.

On the fall in literary productivity, it should be noted that Muḥammad Bello, with his father, the Shaikh ʿUthmān and his uncle ʿAbdullāh, were the real Founder-Reformers *(Mujaddidūn)* and it was they who laid down the ideological–intellectual basis of the revolution. Other emirs, even during their lifetimes, were not noted for literary production. Muḥammad Bello was the last of the triumvirate of mujaddidūn to die. The problem of the contemporaries and successors of the mujaddidūn was to execute the tasks laid down by the founding fathers. The tradition of continued spread of

121 ʿAbdullāh dan Fodio, *Tazyīn al-Waraqāt,* pp. 121–2.
122 Last, p. 118, refers to a work by Muḥammad Bello, *Fī Aqwām al-Muhajirīn,* in which the author classifies the Muslim community into ten divisions, only one of which stood by the original ideals of the reformist movement.
123 *Sard al-Kalām*, pp. 56–8, from ʿAbd al-Salām to Muḥammad Bello; p. 59, from the latter to the former.

learning cannot be denied; neither can the distinct existence of an 'Ulamā' group up to the fall of the Caliphate and even after it. The correspondence of the Caliphate during the late nineteenth century bears testimony to the development of the chancery not only in the Caliph's court but also in the courts of the emirs.

The object of this chapter has been to present a perspective for the detailed study of the Sokoto Caliphate. It is evident that a meaningful study cannot ignore the jihād, its nature and its duration, and the fact that these, in turn, created problems which had not been finally resolved when the Caliphate fell to the British. The fall is, however, outside the scope of this chapter.

CHAPTER 4

Bornu in the nineteenth century

RONALD COHEN AND LOUIS BRENNER

The Kanuri of Bornu have one of the oldest known political histories in Africa, and a very brief review of its scope and origins will serve as a fitting introduction to nineteenth century Bornu. The exact historical beginnings of Bornu are buried in the sands of the Sahara but the archaeological factors that entered into its origins can be enumerated. Neolithic men came to the central Sahara at about 3 500 B.C.; the area seems to have been wet and green then and it probably remained fairly wet, or wet and dry cyclically, until at least 500 B.C. Some time after that period the two great regions of 'desert capture', or the desert lowlands north of the Niger Bend and northeast of Lake Chad, began drying up. As they dried southwards people moved with them and drew the ancient trade routes southwards into the Sudan. It is not, therefore, by accident that the two earliest areas of state building in the western Sudan occur in the upper Niger Bend and the Chad Basin. In the latter area, as desiccation proceeded, the people probably placed increasing reliance on nomadic and semi-nomadic ways of life. In this part of the world (Africa and the Near East) such a way of life is highly correlated with patrilineal descent groups and clans, sometimes all putatively related to one another, and sometimes with stories of heterogeneous origins for each clan. As Chapelle has shown, even today such clans are constantly forming and reforming and often differentiate ethnically in the desert regions to the northeast of Chad.[2]

As desiccation took place the increasing population density of the Kanem area must have led to competition for territorial rights, the control of trade both east–west and north–south, and for eventual dominance of the area. Out of this (still hypothetical) situation there emerged one dominant clan,

1 Field work for this chapter has been carried out under the auspices of a number of agencies. Cohen and Brenner have both received support from the Foreign Area Fellowship Training Program (Cohen, 1955–7; Brenner, 1965–7). Cohen has also been in Bornu with the assistance of the National Institutes of Mental Health (summer, 1964) and the National Science Foundation (summers of 1965 and 1966).

2 'Le clan naît, vit, et meurt', Jean Chapelle, *Nomades Noirs du Sahara,* Paris, 1957, p. 349.

the Magumi, and one lineage within that clan, the Sefuwa, who gained superiority over the others. The heads of clans and major lineages *(Mais)* formed a council in the Kanem area under the headship of the leading man of the Sefuwa. This group and their fellow clansmen formed the basis of the great Kanem state of the eleventh and twelfth centuries. Traditions have told that competition among the clans in the thirteenth century, especially between the Bulala and the Magumi, as well as succession disputes within the Sefuwa dynasty, led to the loss of Sefuwa dominance in Kanem and to their migration with their followers into what is today known as Bornu.[3]

In Bornu the Magumi Sefuwa and their followers found congeries of pagan tribes whom they called 'So', and who seem to have been much less well organised politically and less technologically skilled in warfare. Out of this amalgam of Magumi leaders, subjugated pagan tribes and captured slaves from elsewhere came an identifiable nation, Bornu, and a people—the Kanuri—whose language, brought in by the Magumi, was that of Kanem and of the central Sahara where it had originally evolved. They set up a capital city on the Yo River and, under a series of able monarchs or Mais, they were able to establish a strongly centralised conquest state that became one of the great powers of the Sudan during the fifteenth and sixteenth centuries.

With the death of Mai Idris Alooma, however, the kingdom of Bornu entered a period of steady decline. Although the capital, Birni Gazargamu, maintained its reputation as one of the greatest cities of the Sudan and was indeed one of the foremost centres of Islamic learning in the eighteenth century, the absence of strong leadership allowed the kingdom to disintegrate into its component parts. By 1800, only the loosest of bonds united Bornu and its former vassals in Hausaland, Baghirmi and Kanem. Even in Bornu proper (present-day northeastern Nigeria and southern Niger), which was inhabited primarily by the Kanuri people, the subjects of the Mai were growing more defiant of his authority. It should be noted, however, that most of the evidence for this decline comes to us from nineteenth century reports made after the founding of the second Bornu dynasty. It is only natural that such accounts would denigrate the last days of the first dynasty and thus speak of it in terms of 'decline', 'weakness', and so on. Only when detailed research on seventeenth and eighteenth century Bornu history is carried out will we be able objectively to assess this period now referred to in oral tradition as one of failing strength for the kingdom. Whether this decline was real or not, it was at this time that the Sefuwa dynasty of Bornu was to face the most serious threat to its existence since the Bulala wars of the fourteenth century.

3 Ronald Cohen, 'The Bornu King Lists' and 'The Dynamics of Feudalism in Bornu', *African History*, ed. J. Butler, Boston, 1966, ii, pp. 40–83, 86–105; Ronald Cohen, *The Kanuri of Bornu*, New York, 1967, pp. 12–16; Abdullahi Smith, 'The Early States of the Central Sudan' in *History of West Africa*, i, J. F. A. Ajayi and M. Crowder eds., London, 1972.

THE RISE OF AL-KANEMI

The jihād of Shehu Usuman dan Fodio and his adherents in Hausaland was not without far-reaching consequences in Bornu. There, as in Hausaland, the forces which opposed established authority were primarily Fulani who had been driven towards rebellious activity by a combination of religious motivation and political ambition. Although the Fulani had long been a source of political concern to the Mai's administration, there can be little doubt that Shehu Usuman's successes to the west encouraged unrest in Bornu. The first Fulani hostilities which seem related to the jihād erupted in the territory of Galadima Dunama under the leadership of three men: Umar ibn Abdur, Ardo Lerlima and Ibrahim Zaki. Of these, only Zaki was a highly trained *mallam*. Ibn Abdur had become chief of the Hadejia Fulani in about 1800.[4] His maternal cousin, Ardo Lerlima, was a functionary in the administration of the Galadima as well as his son-in-law.[5] Both men commanded sizable followings and were subject to the Galadima, although Ibn Abdur nursed considerable animosity towards his superior who had been responsible for his father's death.[6] It was Ibn Abdur who first attempted to ally himself with Shehu Usuman by sending his brother, Sambo, to Gobir for a flag, perhaps as early as 1805. The two brothers began a series of conquests first taking Hadejia and then attacking Auyo.[7] These activities were a direct affront to the authority of the Galadima, whose initial attempts to contain the Fulani resulted in the desertion of Lerlima who joined forces with his cousin to drive Dunama from his capital in Nguru.

Local political conditions also played a significant part in creating Fulani discontent in Deya, the second most important centre of rebellion in Bornu. At the end of the eighteenth century Mai Ahmad deposed the ruler of Deya for recalcitrance. This chief had been favourably disposed towards the Fulani residents in his district, but his successor was not. Declaring that Bornu was a land of unbelievers and claiming Shehu Usuman as their leader, two local Fulani mallams, al-Bukhari and Goni Mukhtar, launched hostilities against the ruler of Deya.[8] As the fighting increased anti-Fulani feeling in Bornu reached a high pitch, and the Fulani, fearing for their own safety, began to flee the country. Mai Ahmad wrote from his capital at Birni Gazargamu to the jihādi leaders to inquire into the causes of these incursions and claimed that his people were Muslims and that he himself was

4 W. F. Gowers, *Gazetteer of Kano Province*, London, 1921, p. 21.

5 J. M. Fremantle, 'A History of the Region Comprising the Katagum Division of Kano Province', *Journal of the African Society (J.A.S.)*, x, 1910–11, p. 312.

6 Gowers, p. 21.

7 *Ibid.* The dates here are estimates only as no hard evidence exists as to exactly when these events took place.

8 S. L. Koelle, *African Native Literature*, London, 1854, pp. 212 ff. Information in this account was given to Koelle in Sierra Leone by Ali Eisami, a liberated Kanuri slave. His story, though perhaps not entirely accurate, indicates that the Deya Fulbe initiated hostilities after the beginning of the jihād, but before Mai Ahmad had attacked them.

Commander of the Faithful in Bornu. Muhammad Bello, son of Shehu Usuman, answered this letter for his father and agreed with the Mai's position. He said he would order the Fulani to desist from their attacks in Bornu and he invited the Mai to join in the jihād. But the reply came too late or else was ineffective. Mai Ahmad had already set out with Galadima Dunama against Lerlima who was killed in battle, although later in the campaign the Bornu army suffered a serious reverse at the hands of Umar ibn Abdur in which the Galadima was killed.[9] Meanwhile the hostilities in the south were increasing. Al-Bukhari and Goni Mukhtar were successfully ravaging the area around Deya and Damaturu. The Mai sent several expeditions against them, all of which failed, so that by 1808 the Mai's position appeared desperate. He had unsuccessfully sought to contain the southern Fulani and had suffered a serious defeat in the west. In that year Ibrahim Zaki took the leadership of the western forces and advanced against Birni Gazargamu. He carried his attack almost to within reach of the capital but was forced to retreat. This was followed by a southern onslaught led by Goni Mukhtar which succeeded.[10] Gazargamu was abandoned to the Fulani, and the Mai fled eastward. Mai Ahmad, now advanced in age and blind, abdicated in favour of his son, Dunama.

These Fulani attacks should not be interpreted as a monolithic, well-organised and united attempt to overthrow Bornu as the initial rebellions of the Bornu Fulani were planned and executed by the various leaders independently of one another. Furthermore, although the Bornu Fulani were fired with the spirit of jihād and some of them had submitted to the authority of Shehu Usuman, the Sokoto leadership was only vaguely aware of events in Bornu prior to 1808 and was not directing activities there. Muhammad Bello's response to a letter from Mai Ahmad makes it quite clear that Sokoto had no quarrel with Bornu, for the Bornu Fulani were commanded to cease fighting and Mai Ahmad was asked to join in the jihād. But the Mai could not ignore the activities of the Bornu Fulani which he interpreted as rebellion. The southern group had taken over the entire Gujba–Damaturu area and the western group was fast absorbing the former fiefs of the Galadima: Hadejia, Techena, Shira and Auyo. The struggle continued as the Mai fought to defend his territory and the Bornu Fulani tried to increase their own conquests and draw closer to the Sokoto leadership in an effort to legitimise their gains.

In the years immediately following the capture of Birni Gazargamu, when al-Kanemi had begun corresponding with Shehu Usuman and Bello,

9 This occurred perhaps late in 1807, but again there is no evidence as to the exact date.
10 There is no consensus on the date of the first defeat of Birni Gazargamu. The conclusion offered here is that the attack took place in the late summer or autumn of 1808, since al-Kanemi does not refer to it in his letter to the Bornu Fulani dated May 1808 (see below, p. 100), although, in his first letter to Sokoto, he mentions the attack coming after his receipt of a reply to the May letter. See Muhammad Bello, *Infāq al-Maisūr*, ed. by C. E. J. Whitting under the title *Infaku'l Maisuri*, London, 1957, p. 125.

these leaders were obliged to put their arguments in writing, and the Sokoto leadership was hard pressed legally to justify the attacks against Bornu. In Bello's first letter to al-Kanemi, he stated specifically that he did not know how warfare broke out there, nor was he aware of the state of Islam in that country. The entire letter is an explanation and justification of the jihād in Hausaland. The only knowledge of Bornu which Bello displayed is that he knew the Mai 'rose up in order to harm [his] Fulani neighbours who were emulating the Shaikh [Usuman]'.[11] This, Bello concluded, was equivalent to aiding the Hausa unbelievers and rendered an attack on Bornu justifiable. This theme was subsequently elaborated as he accused Bornu of having attempted to aid Kano, Daura and Katsina. It is possible that the Mai sought to lend assistance to his vassals but any such aid must have been negligible; the Mai was much more concerned to crush the conquerors of his own territory in western Bornu. In subsequent letters, Bello developed a second theme in his argument; that the people of Bornu were unbelievers because they performed various pre-Islamic rituals. This theme was apparently introduced after consultation with the southern Bornu Fulani leadership, who could hardly claim they were opposing the Hausa kings, since the Deya–Damaturu area was inhabited by Kanuri and related groups. Al-Bukhari and Goni Mukhtar had justified their aggression in this way in their letter to al-Kanemi just before the defeat of Birni Gazargamu. Bello had, therefore, constructed a two-pronged argument which was designed to justify the attacks of each Bornu Fulani group.

This correspondence continued for several years after the victory of Goni Mukhtar, so the story has been anticipated, to some extent. The developments in Bornu just after the fall of Birni must be considered. The new Mai and his advisers searched feverishly for a way to rejuvenate the spirits of the army and recapture the vast portions of Bornu that had fallen to the Fulani. It was at this time that their attention was directed to al-Hajj Muhammad al-Amin ibn Muhammad al-Kanemi, then living in Ngala. Al-Kanemi[12] had only recently settled in Ngala, where he was studying and teaching, but he had already gained renown as a learned mallam as well as a capable military leader, who had repelled several attacks launched by local Fulani groups. He had also entered into correspondence with the southern Fulani leadership demanding that they justify their aggression. For these reasons Mai Dunama summoned him and enlisted his support against the enemy now occupying Birni Gazargamu. It was therefore a

11 *Infāq al-Maisūr*, p. 133.

12 This most prominent personality of nineteenth century Bornu is remembered by various names. Traditions recall that, during his early years in Bornu, he was known simply as Mallam al-Amin. After his rise to power in Bornu, he took the Arabic title *Shaikh* (or *Shehu* in Kanuri) and is even today remembered as Shaikh al-Amin (Arabic) or Shehu Laminu (Kanuri). In twentieth-century published works he is generally called by his *laqab*, al-Kanemi, and, although this name enjoys limited currency in modern Bornu, it will be used in this chapter.

reorganised and strengthened Bornu army which marched against Goni Mukhtar, whose Fulani forces had been seriously depleted by desertions after the sacking of Birni. Greatly outnumbered, the Fulani were defeated; Goni Mukhtar was killed and Mai Dunama was restored to his palace less than two months after his father's flight. Although this victory was the result of military superiority it is not without significance that popular accounts of these events attributed the success to the efficacy of al-Kanemi's prayers. Traditions state that for several days before the battle he had engaged in prayers and special recitations. It is also said that he prepared a charm or fetish in the form of a small calabash upon which verses of the *Qur'ān* were written. Just as the two armies were to make contact the sacred calabash was thrown to the ground and smashed to pieces, and the Fulani fled.[13]

This event and the popular explanation of it reveal much about the character of al-Kanemi and the role he was to play in Bornu during the next thirty years. He was possessed of a rare combination of personal qualities: a sound foundation in Islamic scholarship, a strong religious charisma and a natural ability for political leadership. The origin of the story of the calabash is not known; it probably contains a large proportion of truth and it quickly became widely known in Bornu. The most important fact is that al-Kanemi does not seem to have denied its veracity. On numerous subsequent occasions when political or military power had favourably settled an issue, al-Kanemi attributed his success to his ability to enlist the support of the Almighty. In fact, it would appear that he only gradually became aware of his own political talents as a result of his involvement in Bornu affairs. His earlier life was almost solely devoted to religion. Born in the Fezzan of an Arab mother and a Kanembu father, he had as a youth pursued his Islamic studies in Murzuk and Tripoli.[14] Father and son performed the pilgrimage together when the latter was in his teens. Traditions state that he remained in the east for several years to pursue his studies after his father died in 1790. He then returned westwards via Wadai, Kanem, Baghirmi and finally Bornu where he settled at Ngala having married a daughter of the local ruler of that area. Apparently his intentions were to pursue a life of study and teaching,[15] but he became involved in the growing

13 Primary sources for information on the Fulani wars in Bornu are: Muhammad Bello, *Infāq al-Maisūr*, pp. 121–74; Henry Barth, *Travels and Discoveries in North and Central Africa*, London, 1965, ii, 599–600; S. W. Koelle, *African Native Literature*, London, 1854, pp. 212–33. Information from oral sources has been gathered by the authors of this chapter and can also be found in J. M. Fremantle, 'A History of the Region Comprising the Katagum Division of Kano Province', *J.A.S.*, x, 1910–11, pp. 298–319, 328–421, xi, 1911–12, pp. 62–74, 187–200; and also in *Documents Scientifiques de la Mission Tilho 1906–1909*, Paris, 1911, ii, 358–9.

14 Ahmad Bey al-Nā'ib al-Ansārī, *Kitāb al-manhal al-'adhb fī tārīkh tarābulus al-gharb.* (The History of Tripoli), Cairo, 1318 A.H., p. 324.

15 Barth claimed al-Kanemi had wanted to return to the Fezzan, but had been dissuaded from this plan by his father-in-law. *Travels and Discoveries*, ii, p. 600.

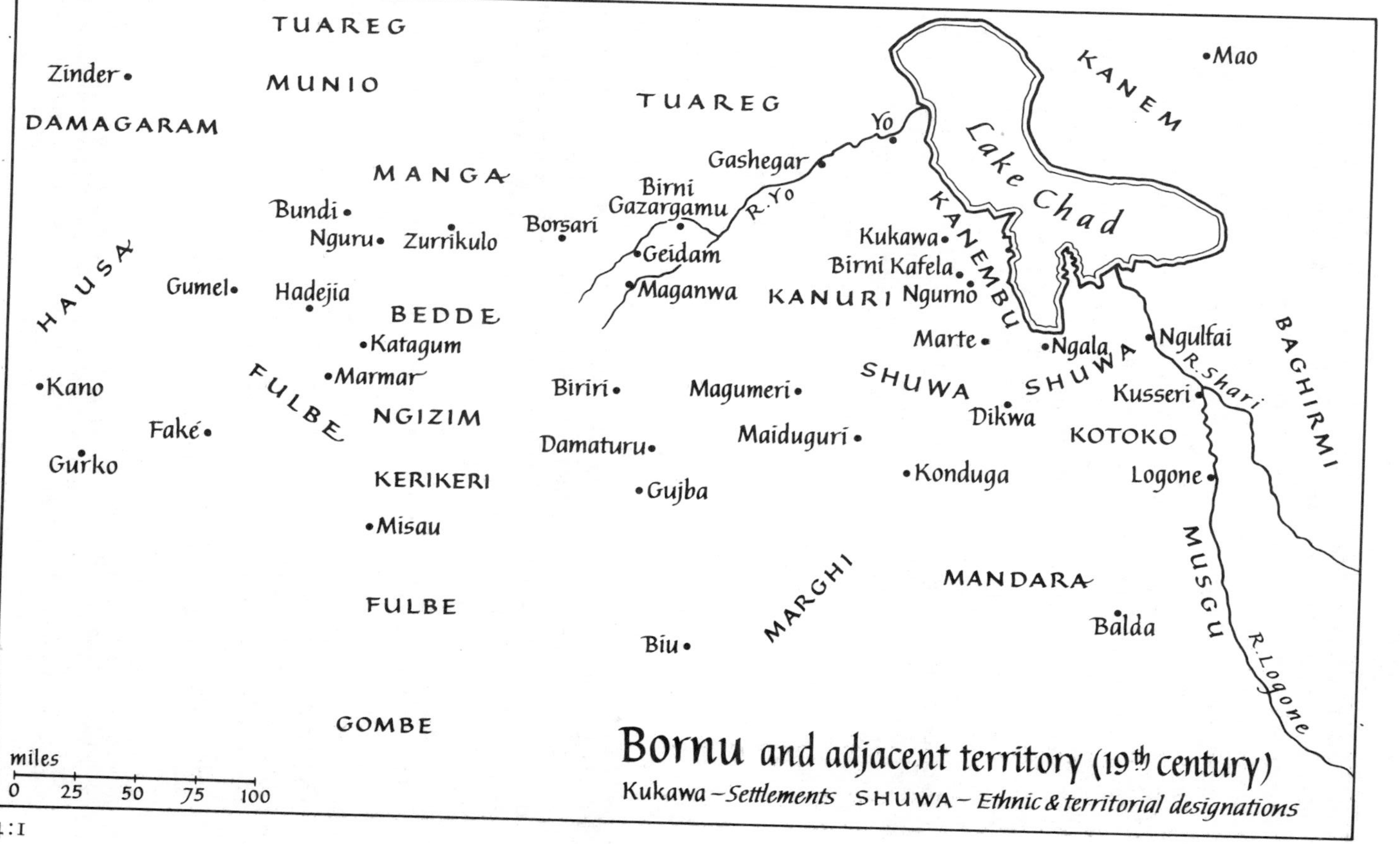
TUAREG
Zinder
MUNIO
DAMAGARAM
TUAREG
Yo
KANEM
Mao
Lake Chad
Gashegar
MANGA
Birni Gazargamu
R. Yo
Bundi
Nguru
Zurrikulo
Borsari
Kukawa
KANEMBU
HAUSA
Geidam
Birni Kafela
Gumel
Hadejia
Maganwa
KANURI
Ngurno
BEDDE
Katagum
Marte
Ngala
Ngulfai
BAGHIRMI
Marmar
FULBE
Kano
Biriri
Magumeri
SHUWA
SHUWA
R. Shari
Kusseri
NGIZIM
Faké
Dikwa
Damaturu
Maiduguri
KOTOKO
Gurko
KERIKERI
Konduga
Logone
Gujba
Misau
MARGHI
MANDARA
MUSGU
FULBE
Balda
R. Logone
Biu
GOMBE
Bornu and adjacent territory (19th century)
Kukawa – Settlements SHUWA – Ethnic & territorial designations
miles
0 25 50 75 100
4:1

strife caused by local Fulani incursions. Early in 1808 he wrote to al-Bukhari and Goni Mukhtar to inquire into the wisdom and legality of their aggressions.[16] His initial reaction to the Fulani threat was therefore that of a scholar.

Just before he marched against Birni Gazargamu, Goni Mukhtar had replied to al-Kanemi's letter of May 1808. The reply is not extant, but al-Kanemi later referred to it as 'a poor answer which could come neither from an intelligent man, nor a teacher, nor a reformer'.[17] This reference appears in al-Kanemi's first letter to the leaders of the Sokoto jihād, addressed to the 'ulamā' of the Fulani and their principal men. In this letter al-Kanemi reiterated the argument he had presented to the Bornu Fulani and thus began the lengthy correspondence between himself and the Fulani leaders of Sokoto, referred to above.[18] The inconsistency of the Fulani position during this correspondence has been discussed; on the other hand, al-Kanemi's argument was simple and remained unchanged. He contended that the inhabitants of Bornu, like those of any Islamic country, were a mixture of good and bad Muslims, apostates, and unbelievers, and he admitted that certain practices forbidden to Muslims occurred there. Nevertheless, the proper approach to these people was to teach them the correct way of Islam, not to make war against them, for the accidental killing of Muslims in the waging of a jihād was a far greater evil than the tolerance of practices of unbelievers. As for the Mai's attacks against the Bornu Fulani, these were justified since they were in defence of established authority. Neither their political activities nor their intellectual attainments qualified the Bornu Fulani as reformers of Islam, and al-Kanemi even went so far as to question the motives of the Sokoto leadership in defending them and suggested that the Sokoto quest for a 'worldly kingdom' might have adulterated their own standards of Islamic reform. This was an accusation which must have greatly shocked Shehu Usuman and his followers.[19]

This correspondence continued for years and, although the legal differences were never resolved, a political accommodation was later reached between the two parties. Nevertheless, this debate had immediate and significant consequences for both sides. The Fulani leaders recognised the opposition of al-Kanemi as the most serious they had yet encountered. It was to refute his arguments and justify the jihād that a large number of

16 University of Ibadan manuscript 82/237. This is actually the second letter written to the Fulani. The first never neached its destination and is not extant.

17 *Infāq al-Maisūr*, p. 124.

18 A good portion of the correspondence is reproduced in *Infāq al-Maisūr*, pp. 124–74. Al-Kanemi's first letter to Sokoto was translated into French in O. Houdas, 'Protestations des habitants de Kano', *Homenaje a D. Francisco Codera*, Zaragoza, 1904, pp. 121–31. Houdas mistakenly thought al-Kanemi was an inhabitant of Kano. The correspondence is partially translated into English in Sir Richmond Palmer, *The Bornu Sahara and Sudan*, London, 1936, pp. 260–7.

19 *Infāq al-Maisūr*, p. 127.

the Arabic works of Sokoto were written.[20] On the other hand, al-Kanemi's communications with the leaders of Sokoto served to enhance his prestige in Bornu. There is no evidence to indicate that he wrote these letters at the request of, or as the agent for, the Mai. It seems as though he initiated the correspondence independently because of his interest in the matter as an educated, and probably ambitious, man who undoubtedly believed that his judgements about this controversy were, or should be, of importance to those leaders directly involved.

After the defeat of Goni Mukhtar, al-Kanemi returned to Ngala, and Mai Dunama rewarded him handsomely with slaves and goods. He had probably urged the mallam to remain with him in Birni, but al-Kanemi preferred to live away from the pomp of the court and chose to remain among his learned following in Ngala. Already famous for his contribution to the defeat of Goni Mukhtar, al-Kanemi was now able to embellish his reputation by challenging the legality of the Sokoto jihād.

For about a year Bornu remained in relative peace although some western portions of the country had been seriously devastated and others occupied by the Fulani. Late in 1809 Ibrahim Zaki advanced with an army from Katagum and again captured Birni Gazargamu. The Mai was once more put to flight and al-Kanemi set out to join him but before a counter-attack could be prepared Ibrahim Zaki abandoned the capital and retreated. Nevertheless the noblemen had lost their faith in Dunama as a ruler. Having only grudgingly accepted him from the first, because he ascended the throne before his father's death, they now forced him to abdicate in favour of his uncle, Muhammad Ngileruma. Birni Gazargamu was not reoccupied, but a new capital, Birni Kafela, was built.

It is not known whether al-Kanemi nursed any political ambitions when he first involved himself in Bornu–Fulani hostilities but by the time he was called a second time to Dunama's assistance his intentions were unmistakable. Before he agreed to join in an attack against Mallam Zaki, he requested that he be allowed to take up residence in Ngurno as the leader of the local Kanembu and administrator of the Ngurno fief. The granting of this wish by Dunama placed al-Kanemi on the path of political ascendancy, for this fief provided him with an income from taxes as well as a political following among the Kanembu there. His learned companions from Ngala joined the mallam in his new home. With this in view, the deposition of Dunama may have also been an attempt by the Kanuri leadership to restrict the influence of al-Kanemi at court. If this was so, the plan failed, for Ngileruma came to rely on al-Kanemi just as his nephew had done and it was the mallam, with his Kanembu followers and some Shuwa Arab cavalry who had joined him, who led the attacks against the still aggressive Fulani.

20 D. M. Last and M. A. al-Hajj, 'Attempts at Defining a Muslim in 19th Century Hausaland and Bornu', *Journal of the Historical Society of Nigeria (J.H.S.N.)*, iii, 2, December 1965, p. 239.

Even with the aid of al-Kanemi, however, the Mai could do little more than hold the eastern portions of Bornu while the Fulani were free to ravage the western provinces. By 1813 the courtiers were beginning to grumble once more about their Mai, not only because of his failure to contain their enemies but also because Ngileruma's strong, religious leanings inclined him to rule Bornu by strict, legal standards, which resulted in the removal of many of the privileges and much of the freedom formerly enjoyed by the courtiers.

Al-Kanemi carefully followed these developments and realised the situation had progressed in a manner favourable to himself. He struck a deal with Dunama that, if he could persuade the noblemen to accept the former Mai back on the throne, then he, al-Kanemi, should receive as a fief all the land between Ngurno and Ngala. With this agreement confirmed, al-Kanemi encouraged the growing conspiracy and secured the throne for Dunama.[21] This reinstatement was another critical step in al-Kanemi's rise to power. He was now in a very favourable position as close adviser to a Mai, but this Mai was indebted to al-Kanemi for having obtained the opportunity to regain the throne of Bornu. The enlarged fief meant increased wealth with which he could attract new clients and develop a larger personal following and, thereby, increase his influence upon affairs of state. It was to this task that he now applied himself.

Al-Kanemi had already assembled a small group of followers to whom he could add others. His earliest companions in Bornu had been three Shuwa Arabs: Mallam Muhammad Tirab of Baghirmi, Mallam Ibrahim Wadaima of Wadai and Mallam Ahamed Gonimi, who had been living in the Ngala area when al-Kanemi came there. These four men had first been drawn together by their mututal interest in Islamic learning but, as al-Kanemi became increasingly involved in the Fulani wars and in Bornu affairs, the three Arabs became his advisers and comrades-in-arms. After 1813, when Dunama had been reappointed Mai and al-Kanemi had received his enlarged fief, it was decided to invite various Shuwa Arab clans to immigrate to the new lands. The invitation was a welcome one since it offered the opportunity for these cattle-raising peoples to move to favourable pasture lands as well as to become the followers of one of the most respected men in Bornu. Meanwhile al-Kanemi had encouraged his Kanembu brothers-in-law (brothers of the mother of al-Kanemi's eldest son) to join him in Bornu. The eldest brother was the leader of the Kanembu Kuburi in Kanem and, as such, possessed the title of *Shettima Kuburi*. The brothers came and, with them, a sizeable number of Kanembu followers, who also

21 Field notes. There are a number of versions of al-Kanemi's relations with the Mais Dunama and Ngileruma. Most extant traditions state that Dunama and al-Kanemi were friendly and that Ngileruma received the throne through the influence of other titled men at court. Barth blamed al-Kanemi for Dunama's deposition, *Travels and Discoveries*, ii, p. 601, and Palmer apparently accepted this interpretation, *Bornu Sahara and Sudan*, p. 259.

settled in al-Kanemi's fief. He also invited friends and relatives from the Fezzan to join him. Among the several men who came, two are worthy of special mention. Al-Hajj Sudani was a childhood friend from Murzuk; and al-Hajj Malia was a Tubu trader, who was a friend of the family and had been travelling to Birni Gazargamu for years before al-Kanemi's rise to fame. It was with these six men that al-Kanemi emerged as the single most most powerful person in Bornu.[22]

As al-Kanemi's following expanded with Arab and Kanembu immigrants, many Kanuri also began to enter his service. In about 1814 al-Kanemi moved from Ngurno and built his own town, Kukawa, about twenty miles north of Birni Kafela. It was at this time that he took the title *Shehu*. This move triggered off new intrigues as certain courtiers began to encourage the Mai's suspicions by suggesting that this so-called mallam was developing even greater political ambitions. Although he had shown no overt opposition to Dunama, al-Kanemi's detractors pointed to his expanding following and to the fact that, although for some years he had lived in nearby Ngurno, he now chose to remove himself to Kukawa. Dunama was not easily convinced by these arguments but it gradually became clear to him that al-Kanemi intended to establish himself as the dominant leader in Bornu. Since moving to Ngurno, al-Kanemi, with his Kanembu and Shuwa followers, had provided the Mai with his most reliable military force. After the construction of Kukawa, however, he began to act more as an agent for himself than for the Mai. His campaigns against the still hostile Fulani or against recalcitrant vassals resulted in tribute and allegiance being directed towards Kukawa rather than Birni Kafela. It was probably al-Kanemi's employment of Fezzani troops to further his ambitions which finally convinced Dunama that he must crush him.[23] To accomplish this the Mai sought an alliance with Baghirmi, a former vassal of Bornu which had received a good share of al-Kanemi's military attention. The alliance would be of mutual advantage. The Mai would gain unencumbered control of his kingdom and the Sultan recognition of his independent status.

However, the plot went awry. As the two armies approached one another, the Sultan of Baghirmi sent a letter to the Mai and instructed him to take up position to the north of al-Kanemi in the line of battle. Both armies would then fall upon the mallam and his followers. Before the letter could reach Dunama, it fell into the hands of al-Kanemi, who placed his contingent to the north of the Mai, and the latter was killed in the encounter.[24] Al-Kanemi then presided (about 1819–20) over the appoint-

22 Field notes.

23 Louis Brenner, *The Shehus of Kukawa: a History of the al-Kanemi Dynasty in Bornu*, London, 1973, ch. 3.

24 Field notes. This is a very popular story among the people of Bornu and is told almost without variation.

ment as Mai of Dunama's younger brother, Ibrahim, who was a young man in his late teens or early twenties.

With the death of Dunama, al-Kanemi became the virtual ruler of Bornu. Realising his inability to modify the situation, Mai Ibrahim remained quietly with his courtiers in Birni Kafela and waited patiently for the time when the good fortune of the mallam might be reversed. But al-Kanemi was not content merely to enjoy his newly gained power; he acted to consolidate his position and to extend his authority. Still beset by the Bornu Fulani and Baghirmi, he moved to drive out the former and to conquer decisively the latter. The gains of the western Fulani were never successfully challenged and even the southern Fulani, under the leadership of Muḥammad Manga, son of Goni Mukhtar, continued to reside in the Deya–Damaturu area until the 1820s. Al-Kanemi then launched a campaign to drive Manga from Bornu and, in 1826, actually pursued him as far as Gurko in Kano emirate. The Sokoto Caliphate was thrown into panic because of this invasion of its soil and armies from various emirates converged on al-Kanemi, who suffered a defeat at the hands of Yaqub of Bauchi, but only after al-Kanemi had routed that army on the first day of combat. The Bornu forces withdrew, and the nineteenth century saw no further serious military engagements between these two states. Nor did Muhammad Manga return to Bornu. The final defeat of Baghirmi occurred in 1824,[25] but that came only after more than ten years of vicious fighting.

The Fulani and Baghirmians were not, however, al-Kanemi's only enemies. Within Bornu itself his authority was not unanimously accepted. After Dunama's death (about 1819–20) Bornu consisted of two parts: that controlled by al-Kanemi and that by Mai Ibrahim. Al-Kanemi's portion consisted of the fief Dunama had given him plus a few minor fiefs which had been transferred to his control with the shifting allegiances of their owners. In 1820 this latter group was small. A major extension of al-Kanemi's authority to the southeast and west was accomplished by force of arms. In the southeast the struggles with Baghirmi enabled him to bring a number of the Kotoko city-states (Kusseri, Ngulfai, Logone[26]) west of the Shari River and south of Ngala under his control. Previously this area had been semi-autonomous, but it was now placed under the administrative direction of Barka Gana, a slave of al-Kanemi. In the west, when Muhammad Manga and the Fulani were forced out of the Deya–Damaturu area, al-Kanemi gave administrative control of the fief to another of his slaves, Ali Gana.[27] Obviously, during this time, some localities freely withdrew their allegiance from the Mai in favour of al-Kanemi but there is no record of this.

25 This battle was witnessed by Dixon Denham. See D. Denham, H. Clapperton and W. Oudney, *Narrative of Travels and Discoveries in Northern and Central Africa in the Years 1822, 1823, and 1824*, Boston, 1826, pp. 190–1.
26 Barth, ii, 446–7.
27 National Archives, Kaduna, *Gujba District Notebook* and *Gujba District Report*.

On the other hand there are numerous references to leaders and districts which were brought into line by the application, or threat, of force. There were Yedi, Marte and Yale Garua, communities to the southeast of Kukawa; the Kanembu Sugurti of Jemu, only ten miles from Kukawa; Muna, near Maiduguri; Tunbi, in the westernmost part of Bornu; the Shuwa Arabs of Shago, near Damaturu;[28] and the once powerful Galadima of Nguru whose power was almost completely eclipsed in the nineteenth century, largely as a result of his opposition to al-Kanemi. Thus, al-Kanemi gradually usurped administrative control of vast portions of Bornu from the Mai and his courtiers and, with it, the attendant income from taxes and tribute.

These developments left very little authority to the Mai although many of his courtiers remained loyal to him and much of the peasantry accepted only grudgingly the increasing power of al-Kanemi. Mai Ibrahim spent most of his time holding court in Birni Kafela. He and his noblemen occasionally went on military campaigns with the Shehu and he was kept advised of political and military developments but he had virtually no influence upon the course of political events. That al-Kanemi chose to allow the Mai to continue in this way offers an interesting insight into the political situation and the Shehu's character. Since al-Kanemi had predominant control in Bornu there was no necessity to depose the Mai, a move which might have brought upon him a great deal of ill will since his rule was not uniformly accepted. Although relatively little information is available on the subject it would appear that al-Kanemi was viewed by the Bornu populace not only as a successful military leader but also as a mallam possessing exceptional powers.[29] There is no question about the fact that he exerted great efforts to maintain and embellish this image at every available opportunity.[30] To depose the Mai and establish himself in his place might have destroyed this image.

Al-Kanemi has often been portrayed as a religious reformer, and there is some evidence to support this view. Oral traditions legitimise his usurpation of power by claiming that the practice of Islam in Bornu had deteriorated under the later Mais and he had received divine instructions to rectify the situation. His concern with religious matters is suggested by the fact that

28 This community of Shuwa Arabs is one of the many communities of so-called 'black', as opposed to 'red', Shuwas who were living in Bornu before the coming of al-Kanemi.

29 Current oral traditions about al-Kanemi are almost universally legendary in nature ascribing his accomplishments to his supernatural powers. An example of this is the story of his victory over Goni Mukhtar quoted above. Traditions concerning other nineteenth century political figures contain much less of this element. These legends originated during al-Kanemi's lifetime. See Denham's account of the subjugation of the Manga leader, Mallam Fannami in Denham, Clapperton and Oudney, p. 133.

30 See Denham's account of the 1824 battle against Baghirmi in which two recently acquired cannon caused panic among the enemy and brought victory to Bornu. Al-Kanemi, however, attributed the defeat to the effectiveness of his own prayers. *Ibid.*, p. 191.

the ruling group in Kukawa was composed largely of mallams and further by his puritanical attitude towards the morals of the women of Kukawa.[31] Furthermore, his correspondence (unfortunately, that which is extant is fragmentary) indicates that he at least occasionally implored his people to follow more strictly the demands of the *sharīʿa*. Nevertheless, al-Kanemi's impact on Bornu was political rather than religious. There is no evidence that religious reform extended beyond the walls of Kukawa itself. Indeed, some of the most serious opposition encountered by al-Kanemi was led by provincial mallams seeking to avoid his political impositions.[32] The history of the Kukawa leadership after the death of al-Kanemi reveals only a minimal concern for the religious standards of government.

On the other hand, the political changes brought about by al-Kanemi were far-reaching. He succeeded in driving the Fulani from Bornu, for which purpose he had been summoned by Mai Dunama, but he then proceeded to replace the administration of the Mai with that of his own. The final result was a new ruling class in Bornu composed of Fezzani, Kanembu and Shuwa Arabs. The political disruption resulting from the Fulani wars had enabled al-Kanemi to accomplish this. In the process the territory of Bornu proper was somewhat reduced in size from that which it had been in 1800. The Fulani had incorporated into the Sokoto Caliphate most of the former possessions of the Galadima, who subsequently became a vassal of al-Kanemi. The Deya–Damaturu region and the Kotoko area south of Lake Chad, both of which had enjoyed a sort of semi-autonomous status under Birni, were brought under the direct control of Kukawa, and Baghirmi, although defeated, had escaped from her former tributary status. An alliance had been established with Yusuf Karamanli of Tripoli, and this it was hoped would encourage increased trade along the Bilma–Fezzan route. Such trade would benefit the Shehus of Kukawa, not the Mais of Birni. Thus, the heartland of Bornu was secure and so was the Kukawa leadership. In spite of all this, however, Mai Ibrahim still hoped that authority would revert to his own court when al-Kanemi died.

THE REIGN OF THE AL-KANEMI DYNASTY

At the death of al-Kanemi in 1837–8 the changes for which Ibrahim had been hoping did not take place. Before his death the Shehu had agreed with his six companions that his eldest living son, Umar, should succeed him. When this became known to Ibrahim he realised that more positive action would be necessary if he were to rid himself of these usurpers. He immediately began to search for allies willing to invade Bornu and drive away Shehu Umar and his people. It was not until 1846, however, that the Sultan of Wadai, Muhammad ash-Sharif, agreed to undertake this mission. In the interim, Umar had been ruling Bornu much as his father had done

31 *Ibid.*, pp. 160, 212.
32 Mallam Fannami of the Manga and Shehu Abdullahi of Yale Garua in Konduga District.

before him. He was, however, neither as intelligent nor as strong a personality as al-Kanemi and he was very dependent upon the advice of the elder councillors who were still alive. Only al-Hajj Malia and al-Hajj Sudani of his father's councillors had died and these had been replaced on the council by their eldest sons. These men remained firmly in control of Bornu affairs; and indeed, their hegemony was slowly expanding in the west and south. This expansion was due not to Umar but to the good advice he received from the councillors and the excellent military leadership of Kachella Bilal, successor to Barka Gana, and Abba Abdurrahman,[33] his own younger brother.

It was when Abdurrahman was in Damagaram on a military expedition that Muhammad ash-Sharif wisely chose to invade Bornu. With a large portion of the Bornu army occupied elsewhere the Wadai Sultan could be reasonably certain of victory. The approach of the Wadai was a surprise to Umar although he was soon informed that the attack was at the instigation of Mai Ibrahim, who despite his denials was placed in chains to await the outcome. The armies met at the Shari River, but were not able to fight until the people of Kusseri, ever anxious to slip from the authority of Kukawa, collaborated with the Wadai people and directed them to a ford across the river. The Wadai army was thus able to surprise the flank of the Bornu force, most of whom fled. Mallam Tirab, Ahamed Gonimi and Shettima Kuburi were among the numerous Bornu dead. Shehu Umar fled to Kukawa where he killed Mai Ibrahim and then marched to the northwest to join Abdurrahman. The Wadai army, under Muhammad ash-Sharif, penetrated into Bornu as far as Ngurno where Muhammad presided over the appointment of Ibrahim's son, Ali, as Mai. He remained for about forty days, but then learned that Umar and Abdurrahman were approaching and withdrew from Bornu.[34] Mai Ali and his supporters met the Bornu army and were soundly defeated. The Mai was killed and his family decimated. His courtiers were hunted down and executed; large numbers of them fled from Bornu. It was decided to destroy Birni Kafela and disperse its inhabitants. Many of these people moved to Kukawa, while others settled in the provinces and began life again as farmers. It was in this way that in 1846 the one-thousand-year-old Kanem–Bornu dynasty was finally replaced by that of the al-Kanemi Shehus.

With the fall of the Sefuwa dynasty and the dispersal of its supporters, Shehu Umar set about consolidating his authority over all Bornu. Fiefs abandoned by the Mai's courtiers were reassigned to Umar's followers, and those who pledged allegiance to Umar were allowed to retain their

33 *Abba* is a title given to the royal princes of Bornu under the Shehus. The Kanuri word was *maina* and is still used although not as a title.

34 Some accounts state that Umar gave Muhammad ash-Sharif a large sum of money to retreat but it is not possible to confirm this. Barth does not mention it in his narrative. Barth, *Travels and Discoveries*, ii, pp. 603–4.

fiefs, or portions of them, although they enjoyed little influence at court. Several localities, like Kusseri, attempted to take advantage of this period of unrest to break away from the control of Kukawa but such attempts were crushed by force of arms. The Shehu's court was expanded by additions from that of the Mai and Umar responded to his sole claim to Bornu royal status by increasing courtly formality and adopting titles previously held only by the ancient Bornu court. Shehu Umar's mother became *Ya Magira* (the Queen Mother) and his senior wife, *Ya Gumsu* (the royal senior wife). These women discharged important duties in the palace, acted as *chimas* (fief-holders) of numerous fiefs and participated in certain state ceremonies. The palace slaves were given the ancient Bornu titles of *Mustrema, Yuroma, Fuguma, Digma* and *Zaifata*. All but the last two were eunuchs, who managed the royal household and also acted as chimas. A Chief *Qādī* was appointed to take some of the judicial burden from the Shehu's court. The title of *Shettima Kanuri* was created and given to a man who had been an early supporter of al-Kanemi. Umar made him chief of court protocol, chima of the noble Kanuri clans, supervisor of the Kukawa market, and the holder of an extensive territorial fief.[35] The original council of six still existed although by now the seats were occupied by the descendants of al-Kanemi's companions. The names of the six companions were made into titles so that, for example, al-Hajj Bashir, son of Mallam Tirab, bore the title of *Mallam Tirab*. But the council of six had, by this time, become more of a ceremonial than a political body. Umar sought advice from those in whom he had the greatest confidence: the Chief Qādī, the Shettima Kanuri and al-Hajj Bashir. He had also relied heavily upon his brother, Abdurrahman, but the latter's jealousy and unfettered ambition led Umar to fear rather than confide in him.

Ironically, at the moment that the Kukawa leadership acquired full authority in Bornu it was split by internal power struggles among the various royal and noble lineages. Such rivalry was to be expected in a political system in which loyalty and service to the ruler were the principle prerequisites for political success. This competition did enable the ruler to control his courtiers for they were constantly seeking to please him. He also had at his disposal the physical force necessary to enforce any decisions. Nachtigal estimated that the number of royal slaves in 1870 was about three thousand.[36] This force plus a number of loyal nobility should have enabled any Shehu to demand and receive obedience from dissatisfied courtiers or family members. In addition to this, the monarchy was surrounded by certain juridical and religious sanctions which encouraged general respect for that office.

35 See below; the Kanuri had a fief system that could apply both to landed, or territorially based, fiefs and to organised groups who were unified not by territory but through common descent or ethnicity.
36 Gustav Nachtigal, *Sahara und Sudan*, Berlin, 1879, i, pp. 725–6.

Nevertheless, the political situation in Bornu deteriorated largely because of Umar's failure to rule with decisiveness. His natural propensities were more towards scholarship than administration and he avoided direct involvement in government matters. He relied upon his favourite slaves and courtiers to govern the kingdom. Al-Hajj Bashir emerged as the person closest to Shehu Umar, and he was given the title *Waziri*, or first minister. But Umar placed too much authority upon the Waziri, for at the height of his power, so traditions record, not even the royal princes could gain access to the Shehu without first visiting al-Hajj Bashir. This earned him the increasing enmity of his fellow courtiers who felt he had appropriated the kind of power which should be reserved for the Shehu alone. Abba Abdurrahman was particularly enraged at the growth of Bashir's influence, for he felt himself responsible for saving his brother's throne in 1846 and thus entitled to a greater reward and recognition.

Abdurrahman finally determined to depose his brother, and he found few opponents among the courtiers, for the prevailing mood was to condone any action that might result in the removal of al-Hajj Bashir. Abdurrahman and his followers left Kukawa in late 1853 and established themselves west of Damaturu in defiance of Umar's authority. The Shehu and his Waziri pursued the rebellious prince but when it became evident that Abdurrahman would be victorious al-Hajj Bashir fled. He set out for Baghirmi but before he had crossed the River Shari he was persuaded by emissaries from the newly appointed Shehu Abdurrahman to return to Kukawa where he was promised that his life, if not his wealth and influence, would be secure. The former Waziri returned but the guarantee proved valueless and he was quickly put to death. On the other hand, Umar was not harmed but allowed to take up private residence in Kukawa.

The success of Abdurrahman was of short duration. He was a quarrelsome and arrogant man, who soon managed to alienate many of his courtiers, as well as to erode the loyalty of certain of his leading slaves. Within a year of his deposition Umar was able to recoup enough support to challenge his younger brother in the streets of Kukawa. This resulted in a victory for Umar who once again took over the throne as Shehu of Bornu. Abdurrahman fled but later returned to Kukawa and was allowed by Shehu Umar to live unpunished in his own house. He was, however, unable to bear his reduced circumstances and when he again began to complain and plot against Umar he was put to death.

This brief episode and its aftermath are illustrative of the relationship between clientship and politics in Bornu. Theoretically, every citizen of Bornu was (and is today) the follower of the Shehu. But the ruler's position was not so secure that he could expect the automatic loyalty of his subjects. The Shehu, through his justness and generosity, was expected to earn the loyalty of his followers, who, in turn, should act as his obedient servants.

This hierarchical arrangement of earned loyalty and obedient service was repeated at every level of authority in Bornu, even in the household.[37] The autocratic freedom of any leader, in particular a ruler, is limited to the amount of support he can muster for his actions. He cannot afford to alienate the loyalties of all men close to him but must be expected, if he is to maintain his status, to balance the ambitions of his people in order to be in control of the political forces of the state. This is no simple matter and success necessitates a clever, diplomatic, and constantly alert individual. Umar may have been clever, but he was not an astute politician. He apparently recognised this shortcoming and chose al-Hajj Bashir to conduct the internal affairs of state for him. Al-Hajj Bashir failed in his task, and Abdurrahman enjoyed no greater degree of success. After his return to the throne, Umar came to rely upon a man who was eminently successful in this sphere, Laminu Njitiya. From 1854 until his death in 1871 Laminu Njitiya was the most powerful man in Bornu; he succeeded to the wealth, position and power of al-Hajj Bashir but was much better able to fulfil the role of political leadership.

This is the same Laminu whom Barth described as the 'evil left hand of the Vizier'.[38] In 1867, Rohlfs claimed that Laminu was 'the wealthiest and most powerful man after the Shaikh'.[39] And Nachtigal, who was present in Kukawa when Laminu died in 1871, recorded the following reaction in the Bornu capital:

> The grief extended no less to the Shaikh and many of the citizens of the capital. What would he do without this honest adviser, this pathfinder through every difficulty, this one true man among a court of sycophants, this one asset among a collection of debts? In the streets of the city one perceived the general calamity brought about by the death of this man. All business stopped; the cheerful character of the people of the capital seemed for a day to have died; people spoke and thought of nothing save the great grief.[40]

The story of Laminu illustrates another aspect of Bornu's social and political life. He rose to political prominence from the most simple background. Although he became Umar's closest adviser during the latter's second reign, Laminu Njitiya did not possess a title and nor was he a descendant of a noble family. Barth claimed that he began his career as a highway robber.[41] This may well be true, but by the 1850s Laminu was the *Wakil*, or chief assistant, to al-Hajj Bashir. Like the Waziri, Laminu was a Shuwa

37 See Ronald Cohen, 'Some Aspects of Institutionalized Exchange: A Kanuri Example', *Cahiers d'Etudes Africaines (C.E.A.)*, v, 1965, pp. 353–60; Ronald Cohen, *Kanuri of Bornu*.
38 Barth, *Travels and Discoveries*, ii, p. 43.
39 Gerhard Rohlfs, *Reise durch Nord-Afrika*, Gotha, 1868, p. 35.
40 Nachtigal, ii, pp. 10–12.
41 Barth, ii, p. 319.

Arab and traditions state that the two men were related by marriage, but neither of these connections was the cause of Laminu's rise to prominence. Rather it was his administrative ability and his willingness to do al-Hajj Bashir's dirty work.[42] During Abdurrahman's brief reign Laminu remained out of Kukawa and when Umar resumed the throne he returned to the capital and offered his services. His experience under al-Hajj Bashir first influenced Umar to seek his advice but it was soon apparent that Laminu's abilities exceeded those of his former master. Like al-Hajj Bashir, Laminu had powerful enemies, but unlike the former Waziri, he kept the balance of power well in his favour.

During the period of Laminu Njitiya's ascendancy (1854–71), internal political strife was controlled but it was nonetheless festering. The most significant rivalry was between Umar's two eldest sons, Bukar and Ibrahim. On several occasions their mutual animosities flared in public and finally Umar forbade their participation in the same military campaigns for fear that one might attempt to murder the other.[43] From about 1860 it was clear that Umar had decided to prepare Bukar for the succession. As a proven military leader he was often sent on campaigns against Bornu's neighbours to the west and south. By 1870 Nachtigal observed that Bukar was surpassed in wealth and clientage only by the Shehu and Laminu Njitiya.[44] After Laminu's death in 1871 Bukar became his father's closest adviser, almost his regent, and when Umar died in 1881 there was little dispute as to who would succeed to the throne. Bukar already commanded the loyalty of the majority of the royal slaves and courtiers; indeed, many of them owed their favoured positions at court to his influence with his father.[45]

Bukar ruled for only three years, dying when he was about forty-five years of age. His brief reign served only to isolate and frustrate Ibrahim who now saw all his supporters purged from the Kukawa court. One of his fathers-in-law, Ahmad ibn Ibrahim Wadaima, had briefly and unsuccessfully attempted to oppose Bukar's succession,[46] but it does not appear that Ibrahim seriously sought to become Shehu on this occasion. Bukar's premature death in 1884, however, plunged Kukawa into a crisis. There was no prince who had an unchallenged claim to the throne and Bukar had not groomed a successor as his father and grandfather had done. Few of the titled nobility favoured Ibrahim, whose regime they feared would be inimical to established interests. They, therefore, agreed that Abba Masta, a younger brother of Umar, should become Shehu. When he heard of the situation, Ibrahim set about gaining the throne by force. He bribed with

42 *Ibid.*
43 Field notes.
44 Nachtigal, i, p. 727.
45 Field notes.
46 Field notes. Ahmad died very soon after Bukar's accession, some say because of his hatred of the new Shehu.

money the royal slave riflemen, who were leaderless during the interregnum. With this formidable force supporting his cause he demanded that the people install him on the throne. Resistance melted away and all came to give allegiance to the new ruler except Abba Masta who, with his large family and following, left Kukawa. Shehu Ibrahim and many noblemen followed and several unsuccessful attempts were made to reconcile Abba Masta to the situation. Even though Masta was killed in a subsequent skirmish Ibrahim did not rest securely on the throne. He and his followers were malcontents, junior sons of courtiers or minor title-holders, formerly excluded from what they considered to be their rightful share of wealth and power. They were seeking a greater share of wealth and could acquire it only from the incumbent title-holders. Dispossession was the obvious method, but that was not easily accomplished. The title-holders would unite against any far-reaching rearrangements among their ranks and Ibrahim knew this. So the new Shehu settled on a plan of wholesale re-appointments amongst the slave officers and assassinations of free title-holders who resisted him. Ibrahim ruled for only eleven months; he died, so the traditions state, just before he was about to assassinate a dozen of his more serious opponents. The slave reappointments were already well under way and one of Shehu Bukar's closest advisers had died of mysterious causes, his vast wealth confiscated.[47]

The primary importance of Ibrahim's reign was that it convinced the titled nobility of the foolishness of appointing a Shehu who might seek to reorganise affairs at court. They preferred a passive ruler, whom they found in Umar's fourth son, Hashimi, a religious recluse with a reputation for laziness.[48] Now perhaps more than at any other time Bornu was, however, in need of a strong and decisive ruler. The political strife in Kukawa engendered by Ibrahim did not abate, and his closest adviser, Moman Tahr ibn Ahamed Gonimi, continued to act as a free agent stirring up discontent. Hashimi was afraid to move against him for fear of failure. Furthermore, since about 1850 Bornu influence over her neighbours and vassals had been in a gradual state of decline. This reached serious proportions during Hashimi's reign. In addition, during the 1880s Bornu fell into the grip of an extended economic crisis.

The decline in Bornu's influence abroad and the economic crisis were not unrelated and the origins of both can be found in Umar's lengthy reign. Zinder was made a tributary of Kukawa by Abdurrahman; an alliance with Baghirmi was reached largely through the influence of Umar's mother, a former Baghirmi princess. These were achievements which furthered the policies formulated earlier by al-Kanemi but Umar did not supervise these affairs himself. It was a client of al-Hajj Bashir who was Kukawa's chief

47 Field notes.
48 P.-L. Monteil, *De Saint-Louis à Tripoli par le Lac Tchad*, Paris, n.d., p. 342. Monteil is supported in his evaluation by current oral traditions.

agent in Zinder. It was also al-Hajj Bashir who enlisted the assistance of the Aulad Sulaiman Arabs in controlling Kanem, a region which after 1846 increasingly became a devastated no-man's-land as a result of internal struggles for power and successive attempts by Wadai and Bornu at imposing their suzerainty. Numerous military campaigns were undertaken in the south and west but these were usually in quest of slaves rather than territory. It would appear that Bornu had for a long period not expanded to any significant degree into the non-Islamic regions to the south, retaining these under her sphere of influence as slave-producing territories.[49] Income from military campaigns and more especially from the sale of captives was crucial to the maintenance of power of the ruling class. This income was redistributed by the leadership to their clients as a reward for loyalty and services rendered. Bornu's attention was therefore always directed towards the north to protect the caravan routes by which the North African traders imported their luxury goods and exported Bornu's products, primarily slaves. It was along these routes that political change was most rapid by the middle of the nineteenth century.

The alliance established by al-Kanemi with Tripoli dissolved on the return of Turkish power there in the 1830s. The Turks extended their authority to the Fezzan, but Bornu always regarded the Ottomans with ambivalence, in fear that they might also attempt to subject Bornu. The Turkish governors of the Fezzan alienated the Aulad Sulaiman who began to threaten the caravan routes. It was then that this Arab tribe entered the service of al-Hajj Bashir, an arrangement which brought little benefit to Bornu.[50] The Tuareg also threatened from the north and by 1850 large sections of northwestern Bornu had been ravaged by them. Little concerted effort was made to regularise this situation. Rather the Kukawa leadership dabbled in adventures of questionable benefit, such as supporting the revolt of Bukhari in Hadejia and considering a plan to invade Kano.[51] This reflects the failure of Umar and his advisers to understand the long-range values to be derived from secure trade routes. An alliance with Sokoto to crush Bukhari would have secured east–west communications. There was no reason why this could not have been done; the mutual animosities between Sokoto and Kukawa had been largely resolved by the 1830s. Further cooperation might have succeeded in containing the Tuareg and Aulad Sulaiman and opening the desert routes to increased trading activity. Such a policy would not have been easy to implement but it would have been in the best interests of Bornu before 1880.

By that year, however, it is doubtful whether any policy could have preserved the Kukawa ruling class without significant internal re-

49 See Ronald Cohen, 'Incorporation in Bornu', *From Tribe to Nation in Africa*, eds. R. Cohen and J. Middleton, Scranton, Pa., 1971.
50 Barth, ii, p. 265.
51 *Ibid.*, ii, p. 236.

adjustments. European intervention in North Africa had begun to decrease the level of the slave trade and North African traders in the Sudan were beginning to purchase ivory and ostrich feathers, destined for the luxury markets of Europe. Just as the Bornu leadership was becoming accustomed to the new demands of the traders a depression in Europe destroyed its market for luxury items.[52] The impact of this depression reached Bornu during the reign of Shehu Bukar and the Kukawa leadership found itself without the source of income upon which it had come to rely. In order to raise funds Bukar issued an extraordinary decree known as *Kumoreti* (splitting or halving the calabash) according to which the peasantry was to surrender one half its wealth to the Shehu.[53] This kind of extortion could not offer a long range solution to the problem and the economic situation did not improve. The crisis continued through the reign of Shehu Hashimi, who is remembered as the least generous of the Shehus.[54]

Although the reign of Umar cannot be praised for its foreign policy, the three Shehus who succeeded him in the 1880s virtually ignored it. Bukar might have adopted a more forward policy; his accomplishments as a soldier were laudable, but he died young. Ibrahim was preoccupied with politics in Kukawa and Hashimi, who became Shehu in 1885, was immobile. By that year Zinder and Baghirmi were virtually independent of Kukawa. It was in this state of disunity and confusion that the al-Kanemi dynasty faced its most serious threat since 1846, the invasion of Rabeh Zubair.

Rabeh, a former slave commander of Zubair Pasha in the Nilotic Sudan, had left that area after the defeat of his leader and made his way westwards, leaving a path of destruction in his wake. When Rabeh threatened Baghirmi, Hashimi foolishly refused to assist the Sultan of Baghirmi against him. When Rabeh threatened Bornu in 1892 Hashimi still failed to grasp the importance of the crisis. He remained in Kukawa and sent Moman Tahr against the invader. His hope was that Moman Tahr might be killed, which would thus solve a serious problem for himself, and then the Shehu and his forces could fall upon Rabeh and drive the invader away. Moman Tahr, on the other hand, accepted this as an opportunity to gain great prestige and then install a Shehu of his own choosing in Kukawa. Rabeh, however, foiled both plans. Moman Tahr was killed in the first encounter and Hashimi was defeated in a second one. Hashimi, and all those who were able to, fled Kukawa, which was looted and razed to the ground by Rabeh and his followers.

Hashimi finally settled his followers near Geidam in western Bornu,

52 C. W. Newbury, 'North African Trade and Western Sudan Trade in the Nineteenth Century: A Re-evaluation', *J.A.H.*, vii, 2, 1966, pp. 233–46.

53 Field notes. See also Tilho, ii, p. 373. Although *kumoreti* has been attributed to Bukar's anger with the peasantry in not supporting one of his military campaigns as well as to his need for funds to prepare for war against Wadai, it undoubtedly was precipitated by a need for income.

54 Field notes.

but he no longer showed any desire to rule and few of the titled men showed any desire to have him as the Shehu. Not all the nobility had accompanied Hashimi; many were now scattered in small clusters throughout the western provinces of Bornu. One such group decided to appoint a new Shehu in an attempt to regroup Bornu's army and advance against Rabeh. Their candidate was Kiyari, eldest son of Shehu Bukar. Kiyari had been fully grown when his father died and had been considered a likely contender for the throne at that time. He had never advanced his candidature with any determination as he felt that his chances of succeeding Hashimi in due course were good. Now that Hashimi had all but abdicated many leaders urged Kiyari to fill the breach. The offer was accepted and Kiyari moved to Geidam to accept the allegiance of the people. Among those who attended was Hashimi himself who came to indicate that he did not intend to contest Kiyari's appointment. Support for Kiyari, however, was not unanimous; there were those who felt Kiyari was excessively militant and that he would lead them against Rabeh and into another disaster.

It was not long before the fears of these men were justified. Preparations were soon under way to march against Rabeh and talk began to spread among the timorous that perhaps a more reasonable course would be to rejoin Hashimi and withdraw to Damagaram out of reach of the invader. Kiyari realised that this move would seriously weaken his army and that, as long as Hashimi was alive, there would be a threat of divided loyalties and desertion. To avoid this Hashimi was put to death by a group of Kiyari's slaves. The Bornu army met Rabeh northwest of Kukawa near the River Yo, and the first encounter saw Rabeh turn in flight as the Bornu forces overran his positions and entered his camp. Although it was dusk some people felt Kiyari should pursue and destroy the enemy, but the majority advised waiting for daylight. The latter group prevailed and Kiyari's people turned to looting and rejoicing. This was a fatal misjudgement. Rabeh regrouped his army and counterattacked that night. Kiyari was wounded and his army took to flight. Attempts at dawn to reorganise the troops were unsuccessful and forced Kiyari to meet Rabeh in disarray. It was not long before most of those who had not fled during the night were in retreat. Kiyari and a few loyal and courageous supporters stood their ground but were finally overwhelmed and Kiyari was captured and subsequently executed by Rabeh. The Bornu leaders who were not killed in battle or drowned attempting to swim the river scattered in all directions. Bornu resistance crumbled and, with it, the al-Kanemi dynasty, which was destined to be restored only by the advancing Europeans.

THE ORGANISATION OF THE NINETEENTH CENTURY STATE

Bornu society in the nineteenth century was that of an organised state among other major and minor, or vassal states to the north, east and west, as well as small acephalous societies to the south and south-east of the emirate. The population lived in different sized communities, varying from small hamlets of only a few hundred to large towns of several thousand people. Social distinctions were based on a fundamental two-class system which was complicated through subdivision and the addition of modifying status criteria.[55] These two classes were known as commoners *(tala'a)* and nobles *(kəntuoma)*. The nobles will be discussed below as part of the Bornu state organisation. Commoners differed by occupation, wealth and rural residence as opposed to living in the capital city. Differences of speech, dress, household furnishings and architecture all separated the classes as well as subdivisions within them.

Marriage and residence rules produced patrilocal, extended families whose male agnates lived close together and whose common occupation often produced residential specialisation so that districts of towns and entire hamlets were often homogeneous or semi-homogeneous as to their socio-economic status in the society. This was more common in smaller settlements because as towns increased in size they became more heterogeneous both in terms of kinship associations and socio-economic specialisation within districts. Economically speaking, most people farmed and kept some cattle as well as doing a job during the dry season, such as craftwork or trading.

Cutting across all these distinctions were those of ethnic heterogeneity and that of slave versus free status. Slavery was a widespread institution throughout the society. Slaves were obtained as part of the booty in raids or by purchase in the markets. They worked for their masters in the household and on the farms, and female slaves served as concubines in the larger households. They could be freed by decree from the master or, if female, manumission was possible through marriage to a free man. Furthermore, the child of a freeman and a slave woman had non-slave status. The actual status of slaves varied with that of the master to such an extent that many slaves were in a position to have their own large households and their own slaves. As has been mentioned some of the highest offices of the state were reserved for people of slave status.

Ethnic heterogeneity was a constant feature of Bornu society. It stemmed first from clear-cut and widely recognised distinctions between the Kanuri, or people of Bornu, and those of different cultures, such as Fulani or Shuwa nomads who had wandered into the Bornu area. Such people were considered to be different and they maintained their cultural distinctions as alien subject peoples within the Bornu state. Secondly and much less intensely, ethnic or, more correctly, sub-ethnic distinctions within Kanuri society itself were recognised, such as that of the Manga or Budduwai peoples. These had adopted Kanuri culture many years before but were still remembered as groups who were originally separate and who

55 Ronald Cohen, 'Stratification in Bornu', *Social Stratification in Africa*, eds. A. Tuden and L. Plotnicov, New York, 1970, pp. 225–268.

still practised a few of their own customs. In addition such distinctions were often applied to older Kanuri clan groupings such as Magumi or Ngala's or Tera. These people were also seen as having a few cultural distinctions that made them somewhat different as groups even though their overall identification was as citizens of the Bornu state.

The society was fundamentally an Islamic one, although just how profound was the religious attachment it is difficult to tell. Certainly, as a state religion, it goes back for many centuries. On the other hand, the people of Bornu were in constant contact with non-Islamic peoples to the south and with so-called Islamic states, especially to the east, who practised a number of non-Islamic rites and rituals as part of their state religion. Furthermore, they were separated from Islamic North Africa by people such as the Tubu who, even today, are much less dominated by the religion than they are themselves. This can be seen in the Tubu practice of clan exogamy as opposed to cousin marriage, especially preferred unions with the father's brother's daughter, which is a widespread Islamic idea, often mentioned in Bornu but not among the Tubu.[56] In their state religion peoples to the east of Bornu practised annual sacrifices with prayers by the monarch at the tombs of the royal ancestors, or at the tops of mountains. There was a special retreat by the monarch at the time of his investiture, and the sacrifice of a pre-pubescent boy and girl at the same ceremony.[57] Of these practices the Kanuri did (and still do) carry out annual ceremonies at the royal tombs and have a period of royal retreat immediately after the official investiture ceremony. There are, however, no traditions or records of human sacrifices that have any proven validity. As far as can be ascertained, Islamic law was practised and its basic elements with respect to crime, property, inheritance and marriage were well-known throughout the society.

In terms of the political organisation, the beginning of the nineteenth century was a difficult period for the inhabitants of the Bornu state because the ancient Mai held a titular claim to the royal throne until 1846, although the Shehus held effective power in the state from at least the 1820s onward, even though they may not have been universally accepted as the ruling family. In structural terms, however, the organisation of the state changed very little with the advent of al-Kanemi's power. Whether authority emanated from Birni or Kukawa, at the centre of the state was still the monarch who held court with the titled nobility. This nobility was divided into princes *(maina)* and several female members of the monarch's family, especially the monarch's own mother and senior wife. There were, as well, free and slave nobles; the latter were divided into eunuchs and non-eunuchs with some of the eunuchs being close to the household and person of the ruler. The chief adviser to the monarch was the Waziri, or vizier, who, as we have seen, could obtain great power in the realm. All these

56 Chapelle, *Nomades Noirs du Sahara.*
57 Marie-José Tubiana, *Survivances Pré-Islamique en Pays Zaghawa,* Paris, 1965.

nobles, with the exception of the Galadima, lived traditionally in the capital, attending court and forming the central administrative body of the state. Traditionally the Galadima was allowed to maintain a separate residence in the Nguru area and was supposed to defend the western border and approaches to Gazargamu. As we have seen, however, the position and power of this office declined during the nineteenth century. It has, however, been resuscitated in the twentieth century—the present (1972) Waziri of Bornu is the holder of this title.

The central figure of the Bornu court, the monarch, was the heart of the Kanuri state and the symbol of its unity and continuity. It was not by accident that the Kanuri ruler was called 'Commander of the Faithful', for, in theory, Islam blends religion and government inextricably together. This has resulted in the Kanuri's considering their royal leader to be the nominal head of all Muslims in the state. He was said to have the potentialities for supernatural power and, in Bornu, the monarch and his court have always been accepted as the last court of appeal in the state. His authority was so clearly defined that there was no danger of anyone upsetting a successful appeal from any of the citizens. On the other hand, as we have seen, the nobles of the court could protest against the monarch's power and even plot to have him removed in favour of another royal heir more amenable to their interests. Such actions were clearly proscribed by Islamic law. As leaders of the various segments of the Bornu administration, the nobles had some power to influence their leader. As we have seen in the case of a weak monarch and a strong, politically astute noble, the latter, in the position of chief adviser, could come to dominate Bornu.

Succession to the throne was by first generation, patrilineal inheritance; the rule was that any man whose father had been appointed to the monarchy was eligible for the office. As lineages usually expand over time while dynasties generally do not this means that a constant tension tends to exist over the possibility that an individual, or a royal lineage segment, might become deroyalised when no one in the segment succeeded in having his father take the throne. Thus, although Abdurrahman held the Bornu throne for only about a year in the midst of Umar's reign, Abdurrahman's sons were made *maina* (princes), rather than remaining as *maidugu* (grandsons of a monarch). In effect, this means that the potential competition among heirs to the throne was increased every time a collateral lineage segment obtained the throne and, conversely, competition among the heirs decreased when royal lineage segments were deroyalised so that royal grandsons did not become royal princes.[58]

These strict rules of succession and the central religious and symbolic position of the Bornu ruler help to explain some of the problems of Bornu politics at the beginning of the nineteenth century. Al-Kanemi, as we have

58 Cohen, 'The Dynamics of Feudalism in Bornu'.

seen, tried to emphasise his position as a religious leader rather than a political one in an attempt to avoid any accusation that he was pursuing the 'worldly' kingdom, for which he had condemned the Sokoto leadership. To have challenged the Mai directly or to have usurped the throne, thus arbitrarily creating a new dynasty, would have given rise to grave problems of legitimacy for al-Kanemi and his heirs. A situation in which an adviser could wield great power was not a new one in Bornu. The replacement of a monarch, however, meant the replacement of a dynasty whose traditions took their lineage back for nearly an entire millennium. Thus, al-Kanemi wisely chose not to do this. The feeling of the people of Bornu for their ancient ruling family was such that an outright usurpation of the throne might have meant that he would be faced with having to subdue all Bornu by force of arms. Instead of a clear-cut usurpation of the throne al-Kanemi was able to capitalise on the events of his own times and take over the kingdom without ever confronting the problem of a dynastic change. In other words, the period of al-Kanemi's rule in Bornu and the way in which he came to power are indications of the great significance of the monarchy in traditional Kanuri culture. The Emir or Mai was not a simple tyrant and the state was not a simple dictatorship by a royal person and his court, in which local communities felt themselves to be separate from, but ruled by, a central despotic government. Citizenship in the Bornu state meant something beyond local loyalty. Each person was in fact a subject of his ruler and felt this to be a personal bond. Even today the traditional practice of singing to young babies about their kin members, their local leaders and their Shehu (Emir or monarch) is still carried on. Thus a child learns almost before he can talk, about his relationship to the monarch within the same context as his relationship to his own kin group.

The organisation of the Bornu administration was a simple, yet flexible system that allowed for change, catastrophe, patronage and the incorporation of ethnic or occupational groups as wholes or parts. This flexibility also contributed to the relatively smooth shift of authority from the Sefuwa to the al-Kanemi dynasty. All titled nobles of the royal court held the office of *Chima Kura*, or senior fiefholder. Some of these (the *Chima Kura Chidebe*) held rights in settled, and usually dispersed, villages and were represented in these fiefs by a trusted and obedient client, the *Chima Gana*, or junior fiefholder, who was a free or slave client or a relative of the senior fiefholder. He acted as the representative of the interests of the senior fiefholder in the local area and served as a link between the locally based leadership and the central administration of the state. The local leaders were of several ranks which depended upon the size of the settlement and the number of extended, or satellite, communities. Small settlements started out under a local leader *(Mbarma)* who was either the founder or a patrilineal descendant of the founder. Patrilineal criteria for recruitment could be changed by superiors in the political hierarchy, and this was done on occasions where there was no male heir or where political superiors in the fief hierarchy wished to appoint a more loyal local leader who was not himself a member of the founding lineage of the settlement. When a small settlement increased in size so that it had wards and perhaps a few hamlets in the immediate vicinity, founded by people from the original settlement, then the leader was referred to as a *Bullama* who had *Mbarma* under him. If this process continued, a Bullama could be taken to the royal court of Bornu in the capital and receive a turban and the title of *Lawan*. He was thus the leader of a large town and its satellite communities, many of whose headmen would in their turn be Bullamas. In the far north of Bornu, the fiefholder representative (Chima Gana) was empowered by the state to 'turban'—i.e. to ennoble local leaders with the title of Lawan so that they could remain in their settlements and protect the area against raiding Tuaregs. Ultimately, in such a system, the Chima Gana held the final authority in the local government, at least on a day-to-day basis and could, if necessary, depose local leaders and replace them with more cooperative ones. Thus local communities were effectively tied to the central government since the senior authority in a local settlement was the client or slave follower of a noble fiefholder in the royal capital.

Other groups in the state, nomads, such as the Shuwa Arabs, and some Kanuri and Kanembu clan groupings, were represented at court by a fiefholder (Chima Kura), who was in charge, not of a territory, but of the group itself. In most of these cases there was no junior fiefholder as there was in the case of the sedentary fiefs. Into this category also came a number of pariah occupations, such as those of blacksmith and drummer as well as the alien trading community, primarily from North Africa. In contrast to the fiefholders of sedentary, agricultural populations, these fiefholders were drawn from the leadership of the group itself, and were called *Chima Kura Jilibe,* or senior fiefholder of the variety.

Local leadership was the court of first appeal for the settlement of disputes and adjudication and most cases seem to have been settled in this manner. However, the political hierarchy from the local leadership to the royal court was an appeal system and in general capital cases were taken direct to the highest court as the monarch was supposed to have had power of life and death over his subjects. This was supposedly true for both types of fief systems and there were special interpreters attached to major courts in order to translate for the non-Kanuri speakers. Although there were some changes in the law brought about by al-Kanemi, especially with respect to adultery—whose punishment he made more severe, including death if the act was committed on a Muslim holy day—in general civil law and the rules of courtroom procedure remained unchanged. In contrast to present-day Bornu women were not allowed in court and if they had to give evidence they did so from a special room attached to the side of the courtroom itself.

The two types of fief system were in a sense interchangeable. If a group of settled people from one fief, many of whom formed part of a patrilineal lineage core, its clients and closely related lineage groups, was dispersed because of disease or the incursions of warring groups, then connection to the state would still be operative. The group would have become a *jili*, or variety, and would still owe some allegiance to their former fief-holder even if they had singly, or as a group, joined other settlements and thus become included in another fief. In other words it was possible for people to belong to a settled fief but through the exigencies of local history, migration or other factors to belong as well to an ethnic unit that was also organised as a 'fief of variety'. This was especially true of nomadic or semi-nomadic groups who, for some reason or other, settled down. They would owe allegiance to the political leadership of the area but would maintain a connection with their former ethnic or sub-ethnic group and its political integration into the state.

In theory the monarch owned all the land by virtue of being the ruler of all the people of Bornu, or, as the Kanuri say, 'We are his [the ruler's] people—Bornu is his land'. As we have seen, however, the land and its people were, in effect, divided among the fiefholders. These were distributed by the monarch among his clients as rewards for services rendered although in order to maintain themselves all courtiers had to have some fiefs. The monarch had the power to revoke fiefholdings, or to divide and/or redistribute them as he saw fit. Differences among the royal courtiers were expressed in the size and wealth of their holdings. Politically, a fiefholder exercised almost complete control over his fiefs, as he was responsible for the protection of the inhabitants, law enforcement, representation at the royal court and tax collection. Since the fiefholder kept large portions of the tax revenue for himself, it was to his advantage to encourage the wealth and prosperity of his fief. The method by which this was accomplished touches on some of the basic features of political life, and this point will be considered after the revenue collection system.

In all likelihood the revenue system of nineteenth-century Bornu began before that century in the earlier kingdom of the Magumi Mais. Although there were numerous minor court fees, market fees, validation fees for taking up new land and other minor taxes, there were only two major, annual taxes. *Sadaga*, due at harvest time, was officially considered to be 10 per cent of the crop and belonged, formally, to the ruler although small portions of it went to the various officials responsible for its collection. At harvest time courtiers were assigned various sections of the state, not necessarily their own fiefs, in order to collect the monarch's sadaga. People close to the capital brought in the tax themselves, while those at a greater distance were supposed to store it in readiness for the collectors when they came around. Since the fief system was not used for the royal tithe, this tax served to produce a flexible reward and patronage system which the monarch could utilise to his own advantage on an annual basis without going to the more serious step of redistributing the fiefs.

Fiefholders, through their clients in the fief, collected the *binəmram*, a winter tax, which was collected during the harmattan months from January to March. The amount varied with the needs of the fiefholders and the assessed wealth for that particular year in the fief. Nomad groups were usually assessed in terms of herd size for any particular year. Local headmen, the local junior fiefholder, and messengers from the senior fiefholder in the capital all took a small portion of the tax for themselves as part of their rightful claim for tax collection services performed for their superior. The senior fiefholder in the capital, however, kept the major portion for himself after he had passed a small tribute to the monarch.

Several important exceptions to the fief and the taxation system should, however, be noted. Fiefholders could settle slaves, either on their own or on other fiefholders' territories (with the latter's permission). These settlements *(kaliyari)* were generally tax-exempt and almost all surplus produce from this group of slaves was the property of the owner. Girls from such settlements could be used as concubines by the owner or given as gifts to his friends, superiors or dependants.

Some groups, by virtue of service to the throne at an earlier date, might be wholly or partially tax-exempt and would generally possess a document *(mahram)* to prove this special privilege. For the most part, these groups were made up of mallams who offered up prayers for the success of the kingdom and the throne. As has been noted elsewhere,[59] these exemptions were subject to renegotiation through time, and many were pre-nineteenth-century in origin, although they were still honoured by the nineteenth-century administration, in one form or another. Al-Kanemi also used such tax exemptions or reductions to attract Kanembu settlers into the Ngurno area when he first set up his centre of activities near Lake Chad. Thus tax exemptions could also be used as patronage to increase or maintain a political following. There are also traditions suggesting that they were temporarily applicable during times of drought or disaster so that a local community could direct all its efforts to the resuscitation of its productive capacities and then later return to full participation in the taxation system. The system itself may well have been uniquely Kanuri, for there are no records of such exemptions in other West African areas.[60]

One final method of revenue collection has not been recorded but is remembered in oral traditions and is rather difficult to evaluate. It is referred to in Kanuri as 'rolling up the mats'. According to informants,

59 Ronald Cohen, 'From Empire to Colony: Bornu in the Nineteenth and Twentieth Centuries', *Colonialism in Africa 1870–1960*, iii: *Profiles of Change*, ed. V. W. Turner, Cambridge, 1971, pp. 74–126.

60 Such a system does turn up in southern Arabia (Abdullahi Smith, personal communication).

the ruler could, on occasion, grant what amounted to special rights of taxation to a favourite, often one of his own sons who wished to start off in life with a certain amount of capital. The person receiving such a privilege could then go to stipulated villages and forcibly take revenue for himself, 'rolling' the people's possessions in their mats for transport to his own household in the capital. Presumably such an action could be used as a punishment inflicted on an area by the central government; however, this point is unclear in the traditions.

Although the administrative structure and the revenue collection system sound oppressive—and there are many tales to corroborate such a judgement—the system was not simple exploitation. It was despotic, but there were concepts of responsible leadership, justice, and a sense of protection from which the people were supposed to benefit by their participation in this hierarchical and centralised state. Leaders redistributed their wealth openly, generously and ostentatiously. At certain ceremonies, al-Kanemi is said to have given away a thousand robes at a time. Loyal subordination was blessed by gifts, privileges and marriage alliances with the family of one's superiors. The kingdom was made up of many factions of nobles, who were allied to various segments of the royal lineage and to various groups within the state. To balance such a system and manipulate it for one's advancement took a clever, diplomatic politician. Thus the contradiction between Barth's view of Njitiya as an evil schemer and that of Nachtigal, who saw the same man as generous, just and beloved by his people, is interesting.[61] Obviously, to be successful, Njitiya had to be both. He had to be generous and just to his followers and the people at large in order to maintain his following and his popularity, but he had to maintain spies and plot against his enemies constantly in order to maintain his superior position in the political system.

When the people of Bornu grieve for a great political leader who is either going away, or who has just died, they quite consciously grieve for their own loss. Thus, when a local judge was seen to be leaving his district, his friends held a going-away party, not to celebrate his promotion, but to grieve loudly over their loss. This was a public proclamation of his worth to the community. In Bornu society, an important man is a generous one, by definition. He distributes wealth freely, takes on supporters easily and rewards obedient and loyal ones well. If he is also a political leader he dispenses justice and is one to whom people can take their troubles and always be assured of obtaining a sympathic hearing. His power enables him to help redress a wrong even if it is with a person's own political superior. This is the traditional view of leadership and its relationship with the common man. Thus, the widespread grief over Njitiya's death is an indication that he was, in fact, an extremely powerful Bornu leader. Indeed,

61 Barth, ii, p. 43; Nachtigal, ii, pp. 10–12.

recollections available about him in contemporary Bornu confirm this impression.

Although the virtual autonomy of a chima in his fief offered the opportunity for extortion the resulting disadvantages discouraged many of the courtiers from such activities. Furthermore, fiefs, however vast, could provide only food, a few local manufactures, such as cloth, and some money, whereas a successful raid against a non-Muslim neighbour might produce hundreds of slaves, not to mention many head of cattle, sheep and goats. It was, therefore, by predatory excursions that the surplus wealth which maintained the nobility was obtained. The loot would be distributed among all the participants in such a raid, peasant as well as noble, although the latter group would normally take the larger portion. Slaves were the most valuable commodity obtained from these raids. Some of them would be retained by the captors, some given away as gifts to loyal followers or impressionable superiors, but the majority would be sold in the slave market to north African traders. The wealth of the nobility, foreign trade, and slave-raiding forays were therefore inextricably linked in nineteenth-century Bornu and fluctuations in the fortunes of any one of these factors affected the other two.

Finally, a word should be said about Bornu's relations with surrounding peoples.[62] To the north, east and west, Bornu had tributary relationships with a number of smaller and less powerful principalities and attempted to rule these through the declared subordination, obtained by right of arms, of their leaders. In some cases, such as in Zinder in the 1850s, this technique was insufficient and a Bornu consul had to be appointed to reside in the tributary state to ensure that Bornu policies were carried out. Tributary status involved the payment of large annual fees to Bornu through a noble of the Bornu court, often the Waziri, who was responsible for the administration of a particular tributary. It also included freedom of passage for Bornu trade and to Bornu military forces when such freedom of access was necessary. Some pressure was also put on the tributary to direct its own foreign policy in relation to the wishes of the Bornu state. Thus Bornu refused to allow states like Zinder and Muniyo to raid for slaves among southern Daura (Hausa) under Bornu hegemony but nothing was said if such raids were carried out among Daura villagers who came under the hegemony of Sokoto. In the south, Bornu raided for slaves and sent a few colonists into the area north of the Benue River in order to maintain some influence there after the founding of the Fulani state of Adamawa, south of the river. Some pagan peoples, however, could obtain through their leaders a protected or *amana* status with Bornu and for a time at least they would not be molested by slaving expeditions. It should also be noted that for the tributary states subordinate status to Bornu was often related

62 Cohen, 'Incorporation in Bornu'.

to their own internal politics. Usually such states found themselves *between* rather than *within* great spheres of power influence. They could, of course, pay tribute to both, and this did occur, or they could try to seek protection from one of the great powers. In the latter case it was often possible for competitive factions within the tributary state to seek help from different states. Thus competition between great powers, for example, between Wadai and Bornu, was reflected by the competing contenders to a local throne. An example was the case of Baghirmi where each faction promised its loyalties in a different direction should it gain power locally.

THE DYNASTIC CHANGEOVER

The changeover from the dynasty of the Magumi Sefuwa Mais to that of the Kanembu Shehus involved fundamentally a change in the person of the monarch, or central ruler of Bornu. Whether or not it also resulted in changes in the basic structure of the Bornu state is a much more difficult question to answer and one that requires more intensive research into the nature of the pre-nineteenth-century state.

In brief the historical events from the point of view of the take-over are as follows. Al-Kanemi commented on the Fulani–Bornu conflict as an educated person and defended the Bornu position. His military and religious aid to Bornu were not so important as his subsequent move to Ngurno where he was able to collect taxes and assemble a large and expanding Kanembu following. This put him in a special position in Bornu. If he were a close adviser to the Mai he should have established residence in the same place as the monarch. On the other hand if he was the head of a local ethnic group or 'jili', he should have been connected to the throne by way of a court noble who would have been his *chima jilibe*. There is, however, no record of his having done either. Whatever the reasons for this anomaly, al-Kanemi's fear for his safety in the Mai's court, or a desire to remain separate for religious or ethnic reasons, or political ambition, or a combination of all these and other psychological factors which we cannot at this juncture understand, the situation created two distinct centres of power and authority in the kingdom and set the stage for the first dynastic change in Bornu's very long history.

But al-Kanemi's move to Ngurno did not give him a kingdom. Along with his success as an adviser and military–religious aide to the Bornu throne it merely put him into a position of prominence in company with many others in the state. The establishment of his dominance came to fruition when he was able to put Dunama back on to the throne of the Mais of Bornu. At this point not only was al-Kanemi powerful and semi-independent in the state but the holder of the ancient throne was now indebted to him for the office itself against the wishes of many of the powerful nobles of the realm. Finally, his power was consolidated and made permanent with the building of Kukawa which in effect became the capital city of al-Kanemi's Bornu. Henceforth, successful military campaigns against Fulani or recalcitrant fiefs and vassals produced revenues and loyalties not to the ancient Bornu throne but to Kukawa, the capital of al-Kanemi's political organisation of the Bornu state.

These are the basic historical events which describe al-Kanemi's rise to power. They are *historical* because they are unique events that occurred but once in the entire history of the kingdom. Usurpers are mentioned in the royal genealogies of Bornu but until the time of al-Kanemi they were unable (so far as we know) to found new dynasties. It must be asked, however, whether this dynastic changeover created any basic or fundamental change in the nature of the state organisation itself.

At first, al-Kanemi did not preside over a formal court. Decisions concerning the administration of Bornu and the adjudication of whatever cases came before him were handled by the Shehu and his six close companions, as mentioned above. In addition, he had around him his sons, other male relatives and a number of title-holders who had deserted the Mai's court. Any courtly procedure introduced into the setting was probably brought by this group of deserters,[63] who retained their titles, although in general al-Kanemi did not appoint, create, or repeat titles already present in the Mai's court. As he needed a military organisation capable of defending Bornu and enabling him to develop and expand his power within it, he set up a group of his own slave followers to administer his fiefs and command groups of soldiers. These slaves were called *Kachella*, a title which had been used in the court of the Mai, as well as having been a title of clan leadership among the Kanembu (similar to the Arabic, *Shaikh*). The first two appointed were Barka Gana and Ali Gana who headed large contingents of Kanembu spearmen. Others were later appointed to serve as resident chimas with the primary purpose of resisting foreign incursions on the boundaries of the state. By mid-century, the Kachella organisation was already akin to that of a standing army. There were two separate divisions: the so-called 'house', or 'palace', Kachella who were attached to the royal person, and the 'field', or 'bush', Kachella whose followers lived outside the capital. Although there were many others among the free nobles who could aid the Shehu and raise armed forces for him the slave status of the Kachellas made them a strong and dependable support for their leader.[64]

With this armed force at his command and the ready income from the fief ceded to him by Dunama, al-Kanemi was able to use force and threats to subdue other areas of the kingdom. By building a city at Kukawa, giving

63 G. J. F. Tomlinson, *Report on Nganzei District*, National Archives, Kaduna, 1918. This report and current oral traditions relate that the Mai sent some of these men to al-Kanemi to assist him in organising court protocol.

64 For a list of the Kachellas and other nobles and the size of their respective mounted followers on a war campaign in the early 1850s, see Barth, ii, Appendix IV, pp. 638–40; for a similar list in the early 1870s, see Nachtigal, i, 725–8.

fiefholdings to subordinates who, in turn, gave him their loyalty, and by taking a controlling interest in the appointment and deposition of Mais he was, in fact, usurping the kingdom but not the dynasty or its royal court. He kept his own court as a high court of appeal for the country and dealt with major crimes and all disputes involving the nobility. These actions showed him to be, at least from the time of the building of Kukawa onwards, a strong, political leader interested in religion and the law but with major political ambitions and goals. Certainly by the 1820s, when Denham and Clapperton arrived, the Shehu was in full administrative and political control of the Bornu state.

The fact that he did not reproduce the formality and leading title-holders of the court was of little moment to the structure itself. Bornu culture is such that powerful households, or men, can, when necessary, take on new titles which are derived either from the household founder or the name of the town from which the political leader comes. Thus the local head of Titiwa could use the title *Titima* and though this did not become a title of the royal court its potential for becoming so is available in the culture. Similarly, the six friends of al-Kanemi founded powerful households in Kukawa, and the successors to the headship of these households took on the names of their founders and thus turned these names into titles of the Shehu's court.

The pre-nineteenth-century division of the state into quartiles under four great noble leaders, two of whom were eunuchs and two free nobles, was not taken up and practised by the Kanembu Shehus. Traditionally, this quartile system had been superimposed over the fief system of administration much as it was in traditional Baghirmi, Wadai and Darfur to the east of Bornu. Although exact data are not available it is likely that this quartile administration was used to collect a royal tax. It probably functioned, as well, as an alternative administrative hierarchy so that local people had two paths or appeal systems upwards to the royal court which must have served as something of a check on the excesses of fiefholders. However, from a functional point of view, the fact that in the nineteenth century courtiers who were not necessarily the fiefholders could be appointed to collect royal taxes in other people's fiefs meant that the local people had the same type of opportunity to contact members of the central administration who were not their direct superiors in the fief-holding hierarchy of the state. In other words, how great a change it was actually to abandon the quartile system, which was not directly pertinent to the everyday administration of the countryside, is difficult to assess at this time.

Finally, after 1846, Umar was free to organise the new royal court in any way he chose. The Kachella, or slave, military title remained important throughout the century and is remembered today as one of the central elements of the nineteenth-century political organisation. Some older, pre-nineteenth-century titles were taken over by Umar after the final demise of the Sefuwa Mais while other titles died with the ancient dynasty. Again it is questionable whether this change meant that a significant structural change had occurred in the organisation of the Kanuri state. Certainly increased use of slave followers all classed as Kachella must have enlarged, at least on the face of it, the power of the monarch over his own court. However there was still intrigue, and there were still individuals like al-Hajj Bashir whose fortunes rose and fell because of success and failure in Bornu politics. It is also important to note that the Kachella themselves were not standardised military leaders each with exactly the same number of platoons or regiments. Instead they varied in the size of their own followings and their fortunes waxed and waned through time. Nor were they the only nobles of the nineteenth-century state. There were other titled officials at the court, both slave and free, all of whom competed for favour and power at the centre of the state. It is in this sense then that the system remained the same. The fief system as an administrative structure also remained the same. True, the Shehu never appeared in a *fanadir* or cage, as had the Mai, and the quartile system of nobles over segments of the kingdom superimposed over a fief system did not continue. But the nature of Bornu politics, its basic administrative procedures, including a royal court, a set of fragmented fiefs and a means of linking local groups to the central administration, changed but little throughout the entire century.

Given the antiquity of the pre-nineteenth-century ruling family and the relative rapidity with which al-Kanemi and his son, Umar, were able to assert the ascendancy of their own dynastic claims to the throne of Bornu, it is surprising that the country was not rocked by serious civil wars between supporters of the old and the new rulers throughout the century. Indeed the very peacefulness of the changeover is an indication that Bornu was in no way revolutionised by the dynastic change in the first part of the century. The change was seemingly primarily historical and can be traced to historical events while basic structural features remained, with some minor changes, essentially the same.

CHAPTER 5

The aftermath of the fall of Old Oyo[1]

J. F. A. AJAYI

The collapse of the Old Ọyọ empire with the resulting movement of populations, wars and political rivalries among successor states, dominated the history of the Yoruba people in the nineteenth century before the establishment of British rule. It had important consequences for the Aja-speaking peoples to the west as well as for the Benin empire to the east.

At the beginning of the century the Yoruba people shared a common language and basically the same culture as reflected in religious and social values, a common pantheon, the same monarchical system in which all paramount rulers claimed to have originated at Ile-Ifẹ, as well as other similar social and political institutions. But they were not united politically. They did not even have a common name for themselves and it was only in the process of evolving a written form of the language that the name Yoruba, formerly applied only to the Ọyọ, came to be applied to all the Yoruba people. Besides the Ọyọ, there were the Ketu, Sabẹ, Dassa, Ẹgbado and Ẹgba to the west. To the south were the Awori, Ijẹbu, Ikalẹ and Ilajẹ. At the centre were the Ifẹ and their neighbours the Ijeṣa and Owu. To the east were the Ekiti, Ondo, Ọwọ and Akoko. Each of these spoke a different dialect of Yoruba and had distinguishable subcultural traits.

These subcultural groups were organised in kingdoms of varying size and degree of autonomy.[2] Some of the groups were small and compact and the paramount authority of a single ruler was acknowledged. This was true of the Ifẹ and Ijeṣa. Among the Ijẹbu, the Awujalẹ was paramount, but his

1 The reinterpretation of early nineteenth century Yoruba history attempted in this chapter has benefited from the work of Mr S. O. Babayẹmi, Junior Research Fellow, Institute of African Studies, University of Ibadan, who has been working with me on the study of ancient ruins in Ọyọ division. His two unpublished seminar papers: 'Upper Ogun Historical Sketch' and 'The Ọyọ communities and the Fulani Jihad' have been especially helpful. I also wish to thank Dr J. A. Atanda and others of the History Department, University of Ibadan, who read the initial draft and made valuable suggestions.

2 J. F. A. Ajayi and R. S. Smith, *Yoruba Warfare in the Nineteenth Century*, Cambridge and Ibadan, 1964, p. 2; W. Bascom, *The Yoruba of Southwestern Nigeria*, New York, 1969, p. 5.

power was challenged by the Ṛẹmọ who claimed a separate identity. Later events, even within the rest of Ijẹbu, were to demonstrate that the tendency for subordinate towns to grow powerful and claim autonomous status was always present. In Ondo, three principal towns, Ondo, Ilẹ-Oluji and Idanre, each claimed to be autonomous. In the hilly country of the Ekiti no fewer than sixteen towns, often more, claimed the status of independent monarchies.[3] The Ẹgba comprised three subgroups—Gbagura, Oke Ọna and Ake—each under an ọba but the councils of the *Ogboni* (secret society), the warriors and the trade guilds *(parakoyi)* in each town and village were so powerful that it has been disputed whether these can properly be called kingdoms. On the periphery of the Yoruba country, among the Ẹgbado, Awori, Akoko and Ilajẹ, though there were market towns, the typical monarchical system based on the town and its surrounding district was lacking and political authority tended to be vested in the village head and a council. Each independent village community claimed its own basis for cultural identity and political autonomy.

Of all the subgroups, the Ọyọ were the largest and they achieved the greatest measure of political and cultural homogeneity. The kingdom, outside the capital city and its immediate neighbourhood, was divided into two main provinces, the western *Ẹkun Ọtun* to the right and the eastern *Ẹkun Osi* to the left of the river Ogun. The western province was also called the Oke Ogun, Upper Ogun. Johnson called the Ẹkun Osi the Metropolitan province since it was the larger and more politically powerful. Each had two subordinate districts, the Onko and Ibarapa in the western province and the Ibọlọ and Epo in the eastern province.[4] Substantial towns had their own provincial rulers, some with the status of ọba. The titles of some, their oral traditions and political structures indicate non-Ọyọ origins,[5] but they all acknowledged the supreme authority of the Alafin. The rulers gathered at Ọyọ every dry season for the king's festival, bringing tribute and gifts of local produce and manufactures for the maintenance of the royal household and grass for the symbolic act of repairing the palace roof.[6] They were obliged to supply military contingents whenever required to do so. They did not merely pay homage and annual tribute; they acknowledged the right of the Alafin and his chiefs and officials to interfere in their day-to-day administration. In theory, only the Alafin claimed origin at Ifẹ with the right to wear beaded crowns with

3 While the Ekiti kingdoms usually consisted of only a few square miles and no more than ten to fifteen towns and villages, Ọyọ kingdom was over ten thousand square miles and according to tradition, contained six thousand six hundred towns and villages. See Samuel Johnson, *The History of the Yorubas from the Earliest Times to the Beginning of the British Protectorate*, London, 1921, reprinted 1969, p. 183; A. L. Mabogunjẹ, *Urbanization in Nigeria*, London and New York, 1963, ch. iv.

4 Johnson, pp. 75–7; he does not give a breakdown of the *Ẹkun Ọtun* though he mentions Ibarapa in other connections (e.g. p. xxii). See also Bascom, pp. 5–6.

5 Johnson, p. 77.

6 Johnson, pp. 49–50; G. J. A. Ojo, *Yoruba Palaces*, London, 1966, pp. 63–6.

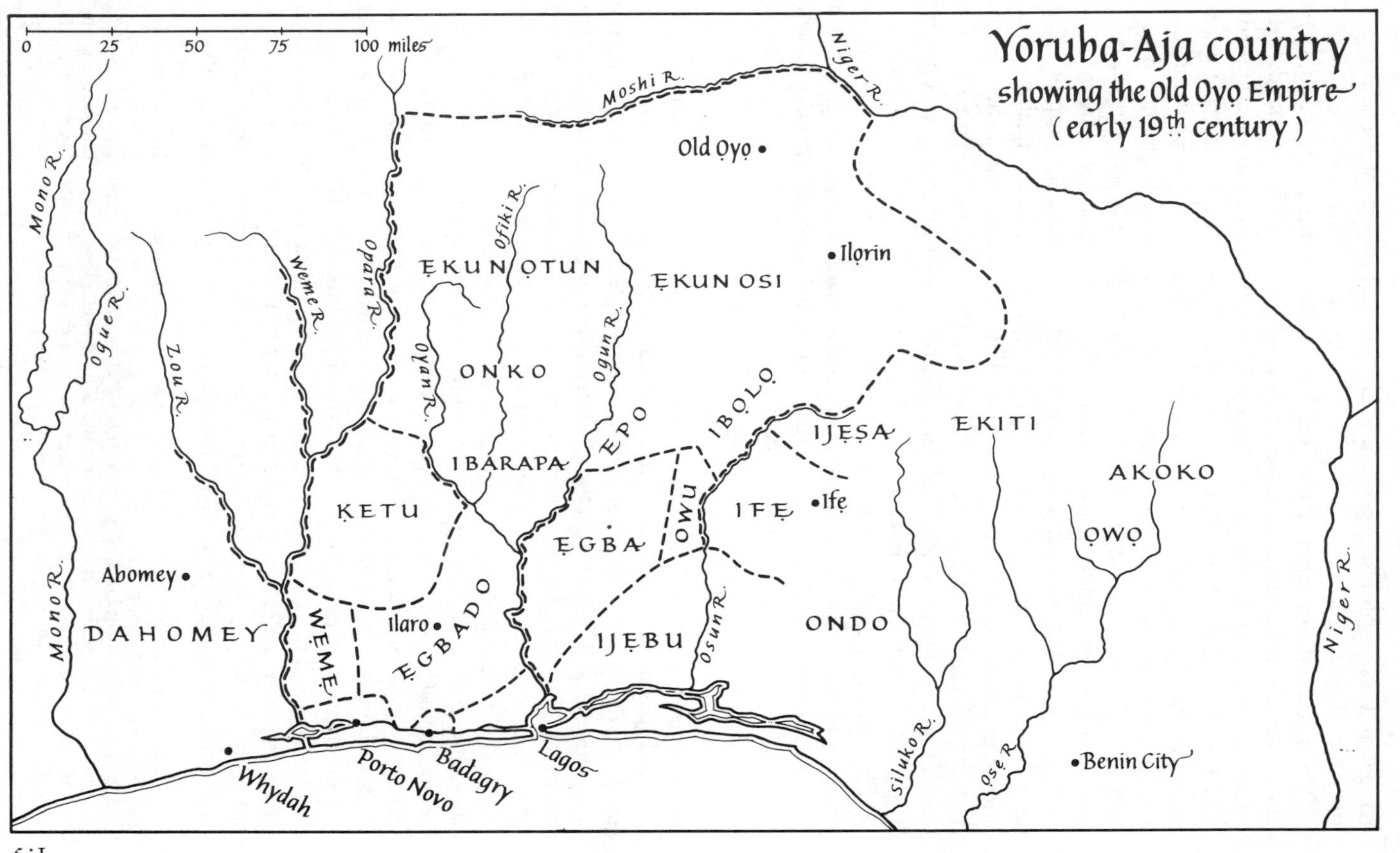
Yoruba-Aja country
showing the Old Ọyọ Empire
(early 19th century)
0
25
50
75
100 miles
Niger R.
Moshi R.
Old Ọyọ
Ilọrin
ẸKUN ỌTUN
ẸKUN OSI
ONKO
IBARAPA
ẸPO
IBỌLỌ
IJẸSA
EKITI
AKOKO
ỌWỌ
IFẸ
Ifẹ
ẸGBA
OWU
KETU
ẸGBADO
Ilaro
WẸMẸ
DAHOMEY
Abomey
IJẸBU
ONDO
Whydah
Porto Novo
Badagry
Lagos
Benin City
Mono R.
Ogue R.
Zou R.
Weme R.
Opara R.
Oyan R.
Ofiki R.
Ogun R.
Osun R.
Siluko R.
Ọse R.

5:1

fringes and to exercise sovereignty in the judicial power of life and death over the Ọyọ people.

The degree of homogeneity was due to several factors. First of all, Ọyọ lay in the sub-savannah area, more open than the rest of Yorubaland, with easier communication and better facilities for close administrative control. Given such an area, the Alafin was able to establish a more effective hegemony in the Ọyọ area than was possible in other parts of Yorubaland. The provincial Ọbas, who in other parts would probably have claimed origin from Ifẹ, claimed descent from Ọyọ, usually from the Ọyọ ruling family. Above all, Ọyọ exploited its position effectively, built up a formidable cavalry force which conquered and helped to control an empire. This empire provided opportunities for Ọyọ people to expand as traders, royal messengers and priests of religious cults. This in turn increased the wealth, power and prestige of the Alafin far above that of other Ọyọ rulers and facilitated his supremacy as well as a degree of homogeneity among the Ọyọ not possible elsewhere.

The Yoruba people had different levels of political allegiance. For day-to-day political life and administration, maintenance of law and order and defence, the effective unit was the autonomous kingdom or independent community. Then came the subcultural group for immediate needs of alliance, common religious festivals, cycle of market days and other social and economic relationships. Over and above that there was no Pan-Yoruba 'nationalism' or 'tribalism', but there was a significant consciousness of cultural identity. Lander was surprised that in the face of Fulani pressure on Ọyọ the Yoruba did not rise as one people.[7] Political and military allegiance did not follow ethnic lines: there were many non-Yoruba peoples in the Ọyọ empire as there were several Yoruba peoples who owed allegiance to Benin. But in spite of the subcultural differences, the Yoruba emphasised the importance of speaking the same language, worshipping the same gods and sharing similar political institutions. Throughout the whole country sovereign power, however achieved, was validated by reference to an origin at Ifẹ from where the divine rights of monarchs were believed to have been derived. This ideology acted as a bond of union. At the coronation of many of the rulers the ritual link with Ifẹ was celebrated. The regalia and other symbols used at the ceremonies were often borrowed from Ifẹ.[8] The rulers exchanged customary gifts and often intermarried. The conventions governing the friendly and orderly relationship between the rulers and between their kingdoms were by no means always

7 R. L. and J. Lander, *Journal of an Expedition to explore the Course and Termination of the Niger*, London, 1832, i, pp. 178, 188.

8 At Ọyọ the regalia from Ifẹ included the *Ida Ajase* (sword of authority) and the *Igba Iwa* used for forecasting the nature of the king's reign. See I. A. Akinjogbin, 'The Prelude to the Yoruba Civil Wars of the Nineteenth Century', *Odu: University of Ife Journal of African Studies (Odu)*, i, 2, 1965; I. A. Akinjogbin, *Dahomey and Its Neighbours, 1708–1818*, Cambridge 1967, pp. 14–17.

observed but they did act as a norm. Traditional accounts of Ọyọ intervention in Ijẹbu, Ondo, Ifẹ, Ileṣa or Ekiti, recorded by Johnson,[9] did not imply that all Yorubaland was ever one kingdom ruled by the Alafin but, at the same time, their significance should not be completely overlooked. At the peak of its power Ọyọ intervention, directly or through an ally or vassal state, was always a possibility. In this way Ọyọ could act as a factor of stability which helps to explain the comparative infrequency of major intra-Yoruba wars until the 1820s.[10]

With the Ọyọ kingdom at its core, the Old Ọyọ empire consisted of a number of kingdoms and states over which the Alafin was able to claim suzerainty at various times, for longer or shorter periods. This power seems to have been most effective in neighbouring Ẹgba and Ẹgbado, and perhaps in Igbomina, Ketu and Sabẹ as well. The separate identity of the people in these kingdoms was acknowledged and their rulers were left virtually autonomous. The rulers were, however, subjected to close supervision from Ọyọ through the *ajẹlẹ*, resident officials, who saw to it that the annual tribute, the Alafin's share of the wealth of deceased chiefs, and other similar levies were forwarded on time. These officials intervened in disputed successions and ensured that no conspiracies were hatched against the interest of the Alafin. The political sanction and administrative supervision of the *ajẹlẹ* were supplemented by the religious oversight of the priests of Ṣango, the god of thunder. The god was a deified Alafin, the object of a royal cult whose worship symbolised the Alafin's power and spread with the expansion of the Ọyọ empire. Among the Ẹgbado, on the main route to the coast, there were in addition colonies of Ọyọ officials to safeguard the route and collect duties at important market towns and toll posts.[11] The ultimate sanction lay with the cavalry force and the punitive expedition that could be mounted either from Ọyọ or a neighbouring loyal part of the empire, during the dry season when the level of the rivers was low.

Beyond the Ọyọ kingdom and its immediate Yoruba neighbours were the tributary states, defeated on the battlefield from time to time, but too distant or too independent to be ruled. These included the eastern parts of Borgu (the Ibariba), the riverine parts of Nupe, especially south of the Niger, and the Aja-speaking peoples.[12] These were areas of military dominance rather than close administrative supervision. Ọyọ imperial control rested

9 Johnson, *passim*, especially pp. 17–25 on 'the Origin of the Tribes'. Also Bada of Ṣaki in files of the Yoruba Historical Research Scheme, cited in R. S. Smith: *Kingdoms of the Yoruba*, London, 1969, p. 79.

10 S. O. Biobaku, *The Egba and their Neighbours 1842–72*, Oxford, 1957, pp. 12–13.

11 P. Morton-Williams, 'The Oyo Yoruba and the Atlantic Trade, 1670–1830', *Journal of the Historical Society of Nigeria (J.H.S.N.)*, iii, 1, 1964.

12 See ch. 9 of vol. 1, *History of West Africa*, eds. Ajayi and Crowder. Nupe once succeeded in driving the Alafin into exile in Borgu and Borgu may in fact have originally laid claims to the land on which Old Ọyọ was established. However, since the seventeenth century the Alafin had claimed suzerainty over parts of Nupe and Borgu.

largely on the ability to compel these autonomous states and peoples to pay tribute, to tolerate the occasional envoys, and to supply military assistance when required. The intention was not to subject them to Ọyọ rule but rather to create satellites or involuntary allies who were left alone as long as they did not stand in the way of Ọyọ's economic and external policies. However, if, like Dahomey in the second quarter of the eighteenth century, they tried to be more assertive, the degree of control could be tightened through annual raids or the prolonged visits of envoys.

For the rest, the Aja-speaking people were, like the other Yoruba, left autonomous. The main subcultural groups were the Gun, the Fon and the Ewe. Each of these could be further subdivided, the Ewe being distinguished from the Anlo to the west and the Watyi to the east.[13] In the same way, the Gun of Porto Novo could be distinguished from the neighbouring Ahori and Ifonyin. The Aja claimed a remote origin from Ifẹ and Ketu and, west of the Aja, pockets of Yoruba-speaking people still exist around Atakpeme. Nevertheless the Aja languages were no longer mutually intelligible with Yoruba. Tado and Nuatja were regarded as the more immediate centres from which the Aja ruling dynasties dispersed.[14] The rise of Dahomey at the end of the seventeenth century and its successful absorption of the older states of Allada and Whydah on the coast created the only large-scale kingdom in the area, a centralised conquest state that thoroughly assimilated all the autochthonous people on the Abomey plateau in a manner similar to that of Old Ọyọ. However, its effort to challenge Ọyọ's hegemony in the area only provoked a tightening up of Ọyọ control. But as long as Dahomey paid tribute regularly and played the role of loyal satellite, it was left free to expand northwards and westwards and to build up its economic ties with Europeans on the coast. Ọyọ controlled Porto Novo more closely as its main access to the European trade.

To the east of Yorubaland was the Benin empire.[15] As in the Ọyọ empire the capital exercised varying degrees of control over the constituent parts. There was the Benin kingdom, the heartland of the empire, closely ruled from Benin through chiefs who held fiefs in different parts of the kingdom. Then there were other Edo-speaking peoples—the Ishan, Etsako, Ivbiosakon, Northern Edo, and Urhobo-Isoko—who had accepted Edo language and culture but whose rulers, often descended from the Benin monarchs, maintained a good deal of internal autonomy. In fact, the more inaccessible parts of Etsako, Urhobo and Isoko were never fully subdued by Benin. There were also the outlying areas on Benin's

13 M. Manoukian, *The Ewe-speaking People of Togoland and the Gold Coast*, London, 1952.
14 Aja myths of origin are summarised in C. W. Newbury, *The Western Slave Coast and Its Rulers*, Oxford, 1961, pp. 6–16.
15 R. E. Bradbury, 'The Kingdom of Benin', *West African Kingdoms in the Nineteenth Century*, eds. Daryll Forde and P. M. Kaberry, Oxford, 1967, pp. 1–35; A. F. C. Ryder, *Benin and the Europeans, 1485–1897*, London, 1969, ch. 1.

trade routes, to the north through eastern Yorubaland and to the east through Ibo villages west of the Niger. These were maintained as satellites paying annual tribute and offering facilities for Benin traders and military assistance when required. Sometimes people banished from Benin settled at such places and resident officials were posted there to protect Benin trading interests. As Ryder has said,

> Some of these bonds between Benin and its vassals were in truth light as air demanding little more than a formal acknowledgement of the oba at the beginning of his reign and a friendly attitude thereafter. Within such a system disruption at the centre led to a loosening of ties, a lapse in the performance of customary obligations, rather than to sudden rebellion, or to a wholesale repudiation of links with Benin. The loose-knit, heterogeneous character of the 'empire' helped also to ensure that revolts in one part did not flare into a general conflagration.[16]

Benin's similarities with Ọyọ went further.[17] The ruling dynasty in both places claimed not merely a common origin at Ifẹ, but in fact a common founder in the mythical hero, Ọranmiyan. While the language and culture of the Edo people remained distinct and the monarchy developed along separate lines, the ritual link with the Ifẹ was maintained. Perhaps partly because of this but even more because of the differences in terrain there was little conflict between the Ọyọ and Benin spheres of influence. Old Ọyọ with its cavalry force dominated the more open country to the north and west; Benin, fighting with swords, cutlasses and European muskets, dominated the more forested areas of eastern Yorubaland, especially Ọwọ, Ekiti and Ondo, without much challenge from Ọyọ. The trade of the two areas was complementary. The most important southern extension of the trans-Saharan network of routes in this area crossed the Niger to Old Ọyọ. From there one route went south through Upper Ogun to Porto Novo. Another, from Nupe, went westwards through Kaiama to Borgu and Gonja. A third went south-east through Ekiti and Ọwọ to Benin.[18] In the pre-Portuguese period this last route was probably the most important channel by which copper and other luxury goods from North Africa and the Sudan reached Benin. In the era of the European slave trade Benin opened up other contacts towards Idah and the Lower Niger, just as Old Ọyọ traded to the coast westwards through Porto Novo. Nevertheless Benin remained down to the nineteenth century one of the most important sources of European-manufactured goods reaching Ọyọ. To protect this trade Benin sent periodic expeditions and maintained colonies of officials

16 Ryder, p. 21.

17 R. E. Bradbury, 'The Historical Uses of Comparative Ethnography with special reference to Benin and the Yoruba', *The Historian in Tropical Africa*, eds. J. Vansina *et al.*, Oxford, 1964.

18 Mabogunjẹ, *Urbanization in Nigeria*, pp. 85–90.

at a few centres like Akurẹ, Ikẹrrẹ and Isẹ. In this way Benin exercised political and cultural influence over Ọwọ, southern Ekiti and Ondo. The effects of this are still evident today in the regalia of the rulers, the titles and sometimes structure of the chieftaincy itself, as well as in the traditions about the route of migration taken by the founding rulers which often included some reference to early contact with Benin.[19] Special markets were created, as at Ọttun in north-western Ekiti, where Ọyọ and Benin traders met to exchange goods. Such markets were regarded as the boundary between the two empires.[20]

Thus Old Ọyọ was at the centre of an interlocking chain of connections that gives some unity to the history of the whole area between the Volta and the Niger. It was the capital of a formidable kingdom, surrounded by vassals and related Yoruba kingdoms. As an emporium for the collection of European manufactured goods from the coast, as well as kola, indigo, gourds, calabashes, parrots and other products of the forest belt, and local manufactures which were then exchanged for horses, leather goods, rock salt, and other items from the north, Ọyọ had wide-ranging economic connections. The peace and prosperity of the whole area depended to a large extent on the existence and stability of the Old Ọyọ empire. It is not surprising, therefore, that its sudden collapse at the beginning of the nineteenth century had far-reaching consequences.

CAUSES OF THE COLLAPSE

Among the probable causes of the collapse of the Old Ọyọ empire, the European trade has hitherto received disproportionate attention. It is sometimes suggested that the dispute between the Alafin and his chiefs was about the control of the European trade. At other times the blame is put not on quarrels over control of the trade but on the depression in the slave trade in the closing decades of the eighteenth century. This depression coincided with the first overt signs of crisis in the empire and it has been suggested that the prosperity of the empire had probably come to depend on the trade and that the depression therefore caused a fatal economic weakness.[21] Yet again it is suggested that it was not the slave trade as such but its abolition that contributed the catastrophic factor.[22] All these conflicting suggestions exaggerate the importance of the European trade in the total economy of the Old Ọyọ empire and the surrounding states. Their economic life depended far more on agriculture and on the cycle of periodic markets for the exchange of agricultural produce and local manufactures such as pots, mats, baskets, iron implements, cloths, etc., which were produced region-

19 S. A. Akintoye, 'The Benin Kingdom and the north-eastern districts of the Yoruba country', paper read to the 14th Annual Congress of the Historical Society of Nigeria, 1969; F. Willett, 'On the Funeral Effigies of Ọwọ and Benin . . .', *Man* (n.s.), i, 1.
20 J. U. Egharevba, *A Short History of Benin*, 3rd edn., Ibadan, 1960, p. 32.
21 I. A. Akinjogbin, 'The Ọyọ Empire in the Eighteenth Century', *J.H.S.N.*, iii, 3, 1966.
22 Smith, *Kingdoms of the Yoruba*, pp. 137–8.

ally. There is little basis for the assertion that 'the Ọyọ derived their wealth largely from the export of slaves'.[23] Participation in the European trade brought wealth to individuals, especially along the coast. The Kings of Dahomey and Benin took an active interest in trying to regulate the trade as an important source of royal revenue and a possible source of ammunition. But the emphasis was on regulating the trade and the markets, not on excluding individual chiefs or traders. For Old Ọyọ, the European trade was not important for ammunition, which was largely home-made or imported from the north, nor for royal revenue which came largely from tribute. The coastal trade supplied European-manufactured goods which, with the exception of salt, were luxuries and its main importance was in generating further trade. Some of these goods must have been valuable in the trade for horses which was of supreme military importance. However, the trade in horses did not really depend on European goods, as several products of the forest zone and some local manufactures were at least as valuable in the northern trade.

It is reasonable to suggest, therefore, that the European trade did not dominate the Ọyọ empire to the extent of controlling its destiny. It was a stimulant providing opportunities for expanding a trade which existed before and was independent of the coming of the Europeans. This included trade along the main axes leading from Porto Novo and Benin to Ọyọ, as well as along the lagoons and in the entire series of local markets where European and Sudanese goods were retailed.

The slave trade no doubt had insidious adverse effects such as the general coarsening of social life, the discouragement of local manufactures and inventions, especially in metallurgy, textiles and household utensils, because European imports appeared much cheaper since the actual cost by payment in slaves could never be assessed in purely economic terms. There was also some potential danger to a northward-looking empire in a trade which enriched coastal peoples and gave them access to European markets and gunpowder. But it should be noticed that Dahomey's possession of muskets did not save it from Ọyọ cavalry. A line of horsemen keeping outside the range of muskets could, before the muskets were reloaded, bombard the attackers with poisoned arrows with decisive effect. Until the beginning of the nineteenth century, Ọyọ remained militarily in control of developments in southern Dahomey and Porto Novo. It was the Benin empire, not Ọyọ, that suffered from the revolt of coastal peoples enriched by the slave trade and this was because control of the lagoon-borne trade was vital to trade in the Benin area. Benin lost control of the lagoons and creeks to Itṣẹkiri, Ijọ and Ijẹbu traders who had the canoemen and the naval power which Benin failed to develop.[24] But even in Benin, decline of the coastal trade meant a shrinking of Benin influence not a collapse of the

23 Akinjogbin, 'Prelude to the Yoruba Civil Wars', pp. 27–8.
24 Ryder, pp. 212–14.

kingdom. The Benin kingdom survived intact until the British conquest of 1897 because, in spite of various tensions, the monarchical structure in the heartland of the empire held together. What has to be explained is the collapse of the monarchy and the internal disintegration at Old Ọyọ. This can hardly be explained in terms of the European trade.

It seems more useful to explore internal factors within the empire. Here we are handicapped by our ignorance of the eighteenth-century background. Johnson recorded traditions about the working of the Ọyọ monarchy after Atiba had deliberately remoulded it in the middle of the nineteenth century. Social anthropologists, bearing Johnson in mind, have studied the system as it existed in the middle of the twentieth century after the impact of the Colonial period.[25] There is therefore considerable doubt as to the extent to which the model of the Ọyọ Constitution which we now have is a useful guide to the politics of Old Ọyọ at the beginning of the nineteenth century or in the middle of the eighteenth.

In the traditions recorded by Johnson not a single Alafin of the eighteenth century died a natural death. Every one was either compelled to commit suicide or was poisoned. None of them led his troops in war, as they did in the earlier period or in the nineteenth century. Only two, Ojigi early in the century and Abiọdun Adegolu towards the end, can be regarded as significant. The former was the conqueror of Dahomey and the latter tried to bring back order and monarchical control after a turbulent period of rule by Baṣọrun Gaha. The others are depicted as minor rulers or worthless individuals. By contrast, the Baṣọruns like Yau Yamba, Jambu and Gaha, and the Kakanfos, especially Ọyabi and Afọnja, emerge in the traditions as the principal figures.[26] When it is remembered that it was in the eighteenth century that the Ọyọ empire reached its greatest expansion and height of power, the importance of the new role of the Baṣọrun, the war chiefs and the provincial rulers can be better appreciated.

Peter Morton-Williams stresses the ritual importance of the Baṣọrun as head of the cult of the king's spiritual double *(ọrun)* with the power of signifying to the Alafin that he had been rejected by the gods. We should also emphasise his political power as head of the most powerful lineage in Ọyọ after the royal lineage, leader of the Ọyọ Misi (the State Council of Senior Chiefs) and administrative head of a ward of the capital city and part of the surrounding district. He was also head of the levy of troops raised in the capital city, as the Onikoyi was head of the troops raised in Ẹkun Osi and the Onibode of Igboho head of the troops in Ẹkun Ọtun and the Kakanfo head of the cavalry force. It would appear that a good deal of the military

25 P. Morton-Williams, 'The Yoruba Kingdom of Ọyọ', *West African Kingdoms*, pp. 36–69.

26 Johnson, pp. 174–87. He says that Basorun Gaha 'had great influence with the people and a great many followers who considered themselves safe under his protection. . . . Gaha was not without friends, even among the chiefs' (pp. 178, 186). Yau Yamba, Jamba and Gaha were presumably onamatopoeiac cognomens.

expansion and imperial administration of the Ọyọ empire in the eighteenth century depended more on the initiative of these chiefs than on that of the king. It seems that, at the same time, the sacerdotal nature of the king's office was emphasised. This is reflected in the regularity with which he was asked to commit suicide. Johnson says that 'the practice had gained ground by this time that kings should not be allowed to die a natural death' and also that 'the custom began to prevail for the Aremos to die with the father as they enjoy unrestrained liberty with the father'.[27] It would seem, therefore, that the position of Baṣọrun Gaha can be seen as an extraordinary aberration of the political system in the light of the mid-nineteenth-century model of the constitution, but that his career makes more sense within the context of what we know about the eighteenth-century system as a whole. The Baṣọrun, in alliance with provincial rulers, became the real power in the land and the Alafin had become virtually only a ritual figure.

The military system helped to reinforce the position of the chiefs rather than that of the king. The expansion of the Ọyọ empire and its administration, as we have seen, depended more on the cavalry force than on the traditional levies. A cavalry force depended on regular training in horsemanship, perfected over the years. This tended to encourage professionalism in the army and a career of warfare among the chiefs.[28] Furthermore, horses were expensive. They could not be bred in Ọyọ because of the tsetse fly. They had to be imported regularly from the north and cared for by skilled veterinarians, usually experienced Hausa slaves. The king had no monopoly over horses and owning a horse, like owning a car today, became a status symbol among the chiefs. It was therefore the lineages who controlled the cavalry force as they could afford the luxury of horses, train their young men in horsemanship, and engage in constant military expeditions. Initially the Alafin tried to control the cavalry by placing them under, not hereditary lineage chiefs, but a group of seventy captains called the Ẹṣọ, appointed by the king himself, usually from warriors of slave origins with no powerful lineage connections. Their administrative head was a eunuch, the Osi Ẹfa, who deputised for the Alafin on political and military matters as the Ọna Ẹfa deputised for him on judicial and the Ọtun Ẹfa on religious matters.[29] Their professional head was the Arẹ-ọna-Kakanfo who was also expected to be of slave origin and was forbidden to live in the capital. In the course of the eighteenth century, however, these checks ceased to be operative and the king gradually lost control of the cavalry force. The slave origin of the Ẹṣọ became obscured; they developed an esprit de corps of their own and gained recognition as a highly respected nobility, next in rank to the Ọyọ Misi, whom they now assisted,

27 Johnson, pp. 117, 174.
28 J. F. A. Ajayi, 'Professional Warriors in Nineteenth Century Yoruba Politics', *Tarikh*, i, 1, 1965.
29 Morton-Williams, 'Yoruba Kingdom of Ọyọ', pp. 62–3.

and sometimes rivalled, in the administration of the wards of the capital city.[30] Though in theory still appointed by the Alafin, Ẹṣọ titles tended to run in lineages which had acquired and were in a position to retain a reputation for excellent horsemanship and valour. The Kakanfo became an office to which an ambitious prince like Afọnja, who lost the contest for the throne, chose to be appointed.[31] The Kakanfo became as powerful, sometimes more powerful, than the Baṣọrun and the Alafin receded more into the background.

It was not that the Alafin readily accepted such relegation. Yau Yamba, the Baṣọrun who in the reign of Ojigi led three expeditions to conquer Dahomey and parts of Borgu and Nupe, comes down in tradition as 'one of the most famous men in Yoruba history'. Yet Alafin Gberu, who succeeded Ojigi, is reported to have decided to interfere with the selection of the Baṣọrun because the lineage that had formerly filled the office 'became as wealthy and great as, or even greater than the sovereign himself'.[32] To deprive a powerful lineage of a chieftaincy they regarded as their possession could not have been done without conflict. The new lineage, or perhaps segment of the former lineage, was even less agreeable to the monarchy. Baṣọrun Jambu turned against Gberu. Gaha, who became the notorious tyrant who dominated the Alafins of the mid-eighteenth century and sent four to their death within twenty years, was presumably Jambu's grandson. It was also reported that Gaha's sons and relations began to occupy strategic positions throughout the empire and to divert tribute meant for the king into their own pockets. Alafin Abiọdun patiently organised a revolt against the Baṣọrun. The precautions he was reported to have taken, of travelling in disguise to intermediaries to help win over the Kakanfo, indicate not merely the power of the Baṣọrun, but also the extent to which the king had been divested of active participation in political and military organisation. With the support of Kakanfo Ọyabi and Ojubu, a brother of the Baṣọrun who was promised the title, Alafin Abiọdun overthrew Gaha. Gaha's sons and relations, including Ojubu, as well as his leading supporters among the chiefs were systematically eliminated.

It seems that Abiọdun, an energetic ruler and former long distance trader who knew the country well, then tried to set up a more effective and personal control of the administration. He was barely successful, in spite of the purge. The situation internally appeared stable enough but signs of military decline and shrinking in the extent of the empire began to show. The successful overthrow of Ọyọ rule by the Ẹgba and the defeat of the Ọyọ army by Borgu in 1783 are ascribed to this reign.[33] Not only did Aolẹ who reigned after Abiọdun fail to maintain this position of personal

30 Johnson, pp. 73–4.
31 Johnson, p. 189.
32 Johnson, p. 71.
33 I. A. Akinjogbin, 'A Chronology of Yoruba History', *Odu*, ii, 2, 1966.

leadership but his relatively brief reign of seven years was followed by a complete collapse of both the power and the authority of the Alafin.

The extent of the collapse needs to be set out more fully. The military decline had continued as the military leaders concentrated attention on the internal struggle for power. Nupe was reported to have defeated an Ọyọ army in 1791. What is more, in about 1797, the Ọyọ army mutinied and Kakanfo Afọnja joined with the Baṣọrun, provincial rulers like the Onikoyi, war chiefs like the Owota (an Ẹṣọ) and Chief Opele of Gbogun, to force Alafin Aolẹ to commit suicide. Alafin Adebọ who succeeded Aolẹ was 'without the authority and power of the king'. He accepted the offer of Ojo Agunbambaru, a surviving son of Baṣọrun Gaha, who tried to rally the chiefs against Afọnja, but his own personal vendetta, the ravages of his Borgu soldiers and the divisions among the Yoruba leaders, many of whom supported Afọnja, were enough to ensure defeat outside Ilọrin. Maku, the next Alafin, enjoyed even less support. He led his troops himself and when he was defeated, he could not return to the capital and had to commit suicide. Rebellion was so widespread and the central administration had broken down to such an extent, that for several years no successor to the Alafin was appointed.[34] There was a prolonged interregnum of perhaps up to twenty years when no one was really in charge at Ọyọ. The war chiefs and provincial rulers began to act as independent sovereigns.[35] Afọnja's power grew, especially in Ẹkun Osi and Ibọlọ provinces. The rising power of the Muslims in Ilọrin, which eventually destroyed Afọnja himself, must have aided the revival of the monarchy. Gradually, support for the monarchical cause was built up and the monarchy was again established. But even then several chiefs who resisted subjection to Ilọrin chose to collaborate with it in their effort to frustrate the rebirth of the Old Ọyọ monarchy.

The argument here is that this rather sudden collapse of the Ọyọ monarchy and widespread rebellion were the culmination of a deep-rooted and long-standing struggle for power between the Alafin and various groups of supporters on the one hand and lineage chiefs, provincial rulers and military leaders on the other; and that the struggle concerned the effective control of the kingdom and empire, an issue that was wider and much more fundamental than the control of the slave trade.

There is an attractive suggestion that Aolẹ precipitated the final collapse by ordering an attack on the market town of Apomu in Ifẹ territory in

34 Johnson, pp. 193–222.

35 R. C. Law ('The Constitutional Troubles of Ọyọ in the Eighteenth Century', *Journal of African History (J.A.H.)*, xii, 1, 1971, p. 43, published after this chapter was written), in addition to the collapse in the capital emphasises the tension between the capital and the outlying provinces. He argues that the conflict with the Ọyọ Misi broke the monarchy, but that it was Alafin Abiọdun who, in bringing in the provincial rulers to subdue Baṣọrun Gaha, made them aware of their strength and unwilling to return to earlier subordination to the capital.

violation of his coronation oath.[36] This caused disaffection among his leading chiefs and the officials who played important roles in the coronation ceremony and their refusal might explain the interregnum. We have already drawn attention to the importance of the cultural ties that bound the Yoruba kingdoms together and anything that weakened those ties undermined the stability of the Yoruba people, including the Ọyọ empire. But the order to attack Apomu—never carried out since the Balẹ of Apomu took his own life—was not enough to explain such fundamental cleavages at Ọyọ. Cultural relationships were important as an ideology, but individual acts of rebellion against convention were more likely to be the result than the cause of the weakening of traditional ideologies. The position of arbiter that Ọyọ held in Yorubaland was based initially more on its power than on tradition but convention had come to accept it and tradition was modified to keep pace with change. The tensions and acute divisions in the body politic were indications that the existing constitutional arrangements were no longer adequate to control the empire at the centre, or on the periphery. Neither could they maintain the relationship of Ọyọ with the rest of Yorubaland. Traditional ideologies were unable to make up for the inadequacy. The Benin empire was in similar difficulties.[37] Faced by the kind of challenge posed by the rise of smaller, more compact, more efficiently organised states like Dahomey and Asante, the Old Ọyọ empire was becoming something of an anachronism. It could either have become more like the empires of the Western Sudan with an effective and not a ritual head, with literate officials, a standing army and a firmer control over tributaries, or else have been broken up into smaller, more efficient and better organised states.

There is no evidence that this state of affairs caused Aolẹ to turn towards Islam.[38] There were already a few Muslims in major towns like Ọyọ itself, Ikoyi, Ogbomọṣọ, Isẹyin, Iwo and Kuwọ. There were itinerant Muslim preachers ministering to these little Muslim communities, making charms for and trying to influence leading rulers and warriors. The early history of Islam in Yorubaland is still unknown. Johnson refers to a Nupe Muslim priest admonishing an Alafin as early as the Igboho period[39] (early sixteenth century). The trade to the north must have encouraged contact with Muslims. It is this that has led to the *parakoyi* (traditional leader of the guild of long-distance traders) becoming associated or confused in various places with leadership in the Muslim communities.[40] By the nineteenth century there was also an Ifa divination chant which, if the omens pointed

36 Akinjogbin, 'Prelude to the Yoruba Civil Wars'.
37 Bradbury, 'The Benin Kingdom', pp. 5–6.
38 Akinjogbin, 'Prelude to the Yoruba Civil Wars', p. 36.
39 Johnson, p. 164.
40 P. G. Gbadamosi, *The Growth of Islam among the Yoruba*, Ph.D. thesis, Ibadan, 1969; B. A. Agiri, *The Development of Local Government in Ogbomoso, 1850–1950,* M.A. thesis, Ibadan, 1966.

to it, enjoined the inquirer to educate his child in the Muslim way and several persons thus became attached to Islam.[41] As the career of the Muslim preacher, Al-Salih Alimi, shows, Islam was already well established at a few centres even before Afọnja's revolt.[42] As Crowther's narrative also shows, Ọyọ Muslims from Isẹyin and Dada played a prominent part in the destruction of his town Oṣogun in 1821.[43] Several Yoruba recaptives were Muslims on arrival in Sierra Leone in the 1820s.[44] Nevertheless, Islam had not yet become a major factor in Yoruba affairs, and was not responsible for the collapse of the kingdom. But it was important in Afọnja's military revolt because it was allied to three other factors, each with military implications. The leading Yoruba warriors believed in the potency of Muslim charms and consequently frequently patronised Muslim priests and charm-makers. There were prominent wealthy traders among the Yoruba Muslims and they were of importance in the northern trade in horses. Most significant of all were the Muslim Hausa slaves recruited for their skills in the care of horses and without whom the size of the cavalry force could not be sustained. Afọnja was able to recruit all three groups to aid his rebellion. Alimi was the most respected and most feared Muslim priest. Afọnja's own well-known military valour allied with Alimi's famous magical powers made the rebellious army virtually invincible, even before it took to the field. Ṣọlagberu, a wealthy Yoruba Muslim trader, joined Afọnja probably for personal rather than any religious reasons. But the leading Yoruba Muslims, as a minority group, stuck together and when Ṣọlagberu settled at Oke Suna near Ilọrin, he was able to attract the support of fellow Yoruba Muslims. Furthermore, Muslim slaves were encouraged to revolt and join the rebellion as free men in a special task force called the *Jama'a*.[45]

Afọnja soon discovered that he was no longer in control of the situation at Ilọrin. The loyalty of the Jama'a became increasingly directed towards Alimi. They behaved with indiscipline and committed excesses which alienated the non-Muslim friends of Afọnja who were seeking power within the traditional Ọyọ system but were confronted by an Islamic community at Ilọrin with a new ideology challenging the very foundations of that system. Afọnja refused to become a Muslim or even to put away symbols of the traditional religion.[46] When he tried to discipline the Jama'a, they turned and destroyed him. Toyejẹ of Ogbomọṣọ, who succeeded as

41 Gbadamosi, *The Growth of Islam*,

42 *Ibid.*; also Ahmed b. Abu Bakre (alias Ọmọ Kokoro) in his history of Ilorin (cited in Gbadamosi, *The Growth of Islam*, p. 12) states that al-Salih Alimi had stayed three months at Ogbomọṣọ, a year at Ikoyi and three years at Kuwọ where he built a stone mosque.

43 J. F. A. Ajayi, 'Samuel Crowther of Ọyọ', *Africa Remembered: Narratives by West Africans from the Era of the Slave Trade*, ed. P. D. Curtin, Madison, 1967.

44 John Peterson, *Province of Freedom: A History of Sierra Leone, 1787–1870*, Evanston, 1969, p. 164.

45 Johnson, pp. 193–4, 197–9.

46 Ahmed b. Abu Bakre.

Kakanfo, twice attempted to rally Yoruba rulers to avenge Afọnja's death and reconquer Ilọrin. He failed and his own quarrels with the Onikoyi frustrated further attempts. Alimi died soon afterwards and his son Abdul Salaam took the initiative and won the contest for power at Ilọrin from Ṣolagberu who had tried to assert Yoruba Muslim control. It was Abdul Salaam who succeeded in keeping Ilọrin independent and with the support of his brother attempted to expand Ilọrin power and Muslim influence in Yorubaland. He eventually won the blessing of the Sokoto Caliphate and became the first emir. Ilọrin thus developed from a rebellious province of Ọyọ into a frontier post of the Fulani jihād.

The restored monarchy at Old Ọyọ failed to regain the loyalty of the rebellious chiefs even in the face of the Fulani menace from Ilọrin.[47] Many must have feared the consequences of giving up a rebellion which had gone so far. There was much suspicion, rivalry and internal quarrelling. The Alafins tried to secure allies abroad, notably from Borgu, but, instead of helping to rally the Ọyọ leaders, this tended to alienate some of them and cause further divisions. On their part the Fulani, by playing up one leader against another, made general reconciliation impossible. They presented Islam, not as a radical force subverting the Yoruba way of life, but as a new enlightenment that the leaders could use in their new state-building activities. Several not only became allied with Ilọrin but were also attracted to Islam. Notable among these was Prince Atiba who spent some time at Ilọrin, thinking, it is said, of learning wisdom from the Muslims before he realised that Fulani wisdom was useless to him.[48] Meanwhile, large numbers of people were deserting the capital and the surrounding ambitious to revive the former glories of the monarchy, also went south to to seek refuge in Ilọrin or else to flee southwards. Several of the war leaders went to find new homes and new ambitions to the south. Atiba himself, ambitious to revive the former glories of the monarchy, also went south to a village called Ago-Ọja, near his mother's home.[49] There, initially as a

47 Johnson, p. 217–19, 258–68; H. B. Herman-Hodge, *Gazetteer of Ilorin Province*, London 1929, especially for the Eleduwe war in which Oluewu and leading members of both the Ọyọ and the Ibariba army perished.

48 *Oriki*, collected by S. O. Babayẹmi from Sule Ajeniyi, an *arokin* from Ọyọ now living at Isẹyin, can be translated as:

Atiba whose folly was so great
That he went to Ilọrin to learn wisdom
Aladeleye! It was the wisdom of others
That Olukuewu relied upon
Atiba did not know that Ilọrin wisdom was unbecoming
He did not know that Fulani wisdom was useless.

49 Atiba's mother was born at Akeitan although she probably grew up at Gudugbu. During the interregnum at Ọyọ, Atiba was brought up by his maternal uncle at Akeitan. By the late 1820s Agọ-Ọja was the nearest urban settlement and Atiba chose to make it his capital. Presumably he spent some time, perhaps as a hostage, in Ilọrin in the mid-1820s. (Johnson, pp. 274–83).

friend and ally of Ilọrin, he organised a task force like the Jama'a and established some authority over the Epo district. When Alafin Oluewu died in war in about 1835, Atiba sought the support of the rising generation of war chiefs and the Ọyọ Misi to get himself appointed Alafin so as to make a new start at Ago-Ọja.

POPULATION MOVEMENTS

The movement of people southward was one of the most far-reaching consequences of the collapse of Old Ọyọ. The semi-circle sixty to seventy miles in radius south of Ọyọ was the most thickly populated part of Yorubaland before 1800. By 1830 two-thirds of this area, south and south-west of Ọyọ, had become the least populated. Famous towns such as Igboho, Igbẹtti, Ikoyi, Kiṣi, Igbọn, Irẹsa, Oje and several others were destroyed and evacuated. There may have been over half a million people involved. Some took refuge in Ilọrin or in towns and villages that came under Ilọrin hegemony. Others began to live in hilltop settlements in the Upper Ogun area or in the major towns of Ibarapa and Epo, southern provinces of the Old Ọyọ kingdom, which could be defended against Ilọrin. These included Ogbomọṣọ, Isẹyin, Okeiho and Iwo. Some people found a new life among neighbouring Yoruba people like the Ẹgbado, Ẹgba, Ifẹ, Ijẹbu, Ijẹṣa and Ekiti. Some of the problems arising from this enforced Ọyọ colonisation will be discussed later.

Meanwhile, even before Old Ọyọ was completely evacuated, bands of immigrants, particularly of the warrior class, began to roam the countryside restlessly, living off the land, intervening in local disputes, acting as mercenaries and sometimes initiating quarrels of their own. One such group went to intervene in the Owu war which had been going on inconclusively for quite a few years.

The Owu war[50] began as a contest for the market town of Apomu which Ọyọ, Ifẹ, Ijẹbu and Owu traders used to frequent. Though it was on Ifẹ territory, Ifẹ was not in a position to maintain control over the diverse traders there. It was the kind of situation in which Ọyọ's authority had succeeded in keeping the peace. With the decline of Ọyọ, however, kidnapping and riots became rampant. Owu, probably the most powerful of the states in the area, tried to take over control. Ifẹ challenged this but was defeated. After some time Ifẹ secured the alliance of the Ijẹbu and laid siege to Owu. Bands of Ọyọ warriors living off Egba land came to join in the siege which lasted roughly from 1820 to 1825. This Owu war marked a turning point in Yoruba warfare in two ways. Firstly, the Ijẹbu were armed with muskets. While this was probably not the first time that muskets were used in Yorubaland it was the first time that a whole army was equipped not only with clubs or cutlasses like the Owu or with bows and arrows like the

50 A. L. Mabogunje and J. D. Omer-Cooper, *A History of Owu*, Ibadan, 1971.

Ọyọ but substantially with European-manufactured muskets and gunpowder.[51] In the event, the efficacy of the muskets was quite limited. The Owu retreated behind their city walls ready to face a long siege. They could not be defeated and had to be starved into surrender. Secondly, the Owu war marked the beginning of a new era of total warfare with Owu being levelled to the ground and put under an interdict prohibiting rebuilding.[52] Even the Fulani of Ilọrin dealt more gently with Old Ọyọ.

The Ifẹ, Ijẹbu and Ọyọ allies then proceeded to deal in the same spirit of total warfare with practically every Ẹgba town and village, except for those such as Ibadan, Ijaye, Ọjọ̃ and others in the Gbagura area, which the allies used as camps or which had already been occupied by Ọyọ migrants. Scattered Ẹgba and Owu refugees joined Ọyọ migrants in these places, or began in their own turn to press southwards and southwestwards across the Ogun river. Ibadan, because of its defensible position, became the main camp of the allied warriors and a location for uprooted Ọyọ, Ẹgba and Owu. The allied warriors were soon struggling for leadership in the growing town. The Ọyọ gradually became dominant as they replaced the Ifẹ who were the initiators of the alliance, but the town itself remained cosmopolitan. Around 1830, most of the Ẹgba and Owu withdrew under the leadership of Ṣodẹkẹ to found a new capital at Abẹokuta. There they tried to re-establish their people, and sought access to the sea through Badagri.[53] This increased Ẹgba pressure on the Ẹgbado, Ọtta, Egun and other coastal peoples.

The result of this series of shifts of population was a complete reorganisation of the demographic situation in Yorubaland. As a result of the Fulani pressure and the collapse of Old Ọyọ the Yoruba people became increasingly pressed into the forest area. The Ọyọ abandoned the open savannah, except around Ilọrin, and poured into the forest zone, pushing the Ẹgba and Owu further into the forest and the coastal swamps. The forest belt became more thickly populated and heavily cultivated and the coastal areas attracted more settlements than ever before.[54] Through this enforced colonisation, Ọyọ cultural influences on the rest of Yorubaland became more intense. Cultural traits, such as the Ọyọ narrow male-operated loom as opposed to the women's broad loom, clothed *egungun* as opposed to those covered with grass or palm fronds, drums slung from the shoulder as opposed to standing drums, and the royal cult of Ṣango, all

51 Johnson, pp. 207–10. There are traditions of Benin soldiers in Ekiti using firearms, perhaps as early as the sixteenth century; also of Ọyọ cavalry in Dahomey encountering firearms early in the eighteenth century but no record yet of a Yoruba force armed with guns before the Owu war. See Ajayi and Smith, *Yoruba Warfare*, pp. 17–18.

52 Johnson, p. 210. Owu was in fact resettled in the colonial period.

53 Biobaku, *The Egba and their Neighbours*, pp. 13–15.

54 P. C. Lloyd, *Yoruba Land Law*, Oxford, 1962, pp. 51–3.

spread more rapidly in the days of Ọyọ decline than in the days of Ọyọ hegemony.[55]

Islam, which began to be associated more with the Ọyọ than with any other Yoruba subcultural group, also spread widely as a result of this Ọyọ diaspora. Though in a few places the establishment of Islam can be traced to non-Ọyọ sources—former Muslim slaves and a distinguished Nupe warrior in Lagos, Badagri and Ẹpẹ, Ilọrin and Nupe penetration of northern Ekiti, Akoko and Afẹnmai—the most important centres of diffusion were in the Ọyọ area. Although the numbers of Muslims were few, they were usually influential members of society, largely warriors and long-distance traders who soon began to acquire high political office. By 1858 the Timi of Ẹdẹ was a Muslim. Iwo and Isẹyin were regarded as Muslim strongholds. In 1871 Ibadan had its first Muslim ruler. By 1878 the new Ọyọ was reported to have at least twelve mosques. Koranic schools were spreading. The occasional Islamic scholar from Ilọrin or further north spent a few years with different Islamic communities, teaching the imams, improving their knowledge of Islamic theology, law and traditions, and generally helping to organise the communities along Islamic lines. Ọyọ migrants from these centres helped to spread some of this Islamic knowledge and influence to other parts of the Yoruba country.[56] But while Islam and the northern factor in Yoruba society remained important, the southward push had turned the attention of most Yoruba people to the possibilities of trade and other opportunities along the coast, hitherto a monopoly of the coastal peoples and the few inland traders.

One important reason for this new and active interest in the coastal trade was the changing nature of Yoruba military strategy. With the defection of Muslim slaves and with the Ilọrin attempt to cut off supplies from the north, the number of horses available to Ọyọ warriors was quickly reduced and cavalry soon ceased to be regarded as the élite corps of the army, except among the Ilọrin. The military engagements in the forests of Owu and Ẹgbaland dictated fresh tactics. Contact with the Ijẹbu showed the possibilities of musketry. In the 1830s muskets began to be introduced more generally. Around 1832 the Ẹgba who had begun to acquire muskets were reported to have been still 'more afraid of poisoned arrows than of bullets'. Soon after that the Ibadan had to defend their new settlement against the Ifẹ leaders they had expelled. The Ifẹ had a superiority of musketry but the Ibadan won the war largely, it is said, by charging bravely at the men with guns and forcing the muskets away from them; for this reason the battle was nicknamed *Gbanamu* (grasping fire).[57] In 1840, when Ibadan confronted Ilọrin at the battle of Ọṣogbo, muskets

55 For a discussion of the consequences of this in the field of drama, see J. A. Adedeji, *A History of the Yoruba Theatre*, Ph.D. thesis, Ibadan, 1969.
56 Gbadamosi, *The Growth of Islam*; Mabogunjẹ, *Urbanization in Nigeria*, pp. 94–5.
57 Johnson, pp. 241, 251.

had become widespread and Yoruba cavalry warfare was already a thing of the past.[58] Horses continued to be used principally for transporting the more wealthy war chiefs and for sending urgent messages.

The most important legacy of the period of cavalry warfare was the professionalism which it had bred in the old warrior classes, who now became masters of the new art of warfare. Like the Ẹṣọ, the warrior classes continued to devote themselves constantly to warfare. The frequent wars of the century and the general instability of the period enhanced the status of warriors and further encouraged the development of warfare as a profession. Instead of horsemanship, the warriors now trained their young men through a system of apprenticeship in marksmanship, leadership and military valour. They learnt tactical manoeuvres of deception and surprise attacks, and scouted for information about enemy locations and intentions, and also organised regular supplies of muskets and gunpowder. Such developments were most advanced at Ibadan, and warriors from different parts of the country came there to learn the art of warfare. Each state had its own military tradition and style of fighting; the Ẹgba, for example, specialised in siege warfare.[59] The older weapons, such as spears and bows and arrows, continued to be used but musketeers became the élite corps of the new armies. A few cannon were imported later, but the great difficulty of transporting them and the large amount of powder they consumed prevented artillery from being important.[60] More sophisticated guns came in and by the late 1870s breech loaders and repeater rifles, ex-Prussian war, became features of Yoruba warfare.

NEW URBANISATION

Even before the nineteenth century the nucleated settlement was a typical feature of Yoruba society.[61] The unit of administration was a town and its surrounding villages, inhabited largely by people of the same ethnic or sub-ethnic group. The town was the centre of government, religious festivals, litigation, markets and crafts. Usually the head of the town was an ọba, a

58 Smith and Ajayi, *Yoruba Warfare*, pp. 33–6. This did not rule out their occasional use as on the occasion mentioned by Johnson (p. 451) when in 1881 a group of horsemen from Ibadan went to disperse Ijẹbu troops raiding the outskirts of the city.

59 A British officer reported in 1861 that the Ẹgba followed the scriptural example of 'going out to war, sitting down and encamping against a Town'. Ajayi and Smith, *Yoruba Warfare*, pp. 90, 136.

60 *Ibid.*, pp. 19–21. Britain at missionary insistence gave the Ẹgba six field pieces in 1851 and sent a naval officer to instruct the Ẹgba in gunnery. By 1861 the guns were reported to have been neglected and one of them in the missionary compound was 'a roost for cocks and hens'.

61 For studies of Yoruba urbanism, see Bascom, 'Urbanization among the Yoruba', *American Journal of Sociology (A.J.S.)*, lx, 5, 1955; and 'Some Aspects of Yoruba Urbanism', *American Anthropologist (A.A.)*, lxiv, 4, 1962; G. J. A. Ojo, *Yoruba Culture*, London 1966, ch. 5; Mabogunjẹ, *Urbanization in Nigeria*, ch. 4; Eva Krapf-Askari, *Yoruba Towns and Cities*, Oxford, 1970.

divine king; that of the village, a balẹ. Each was advised by a council of senior chiefs representing lineages or age grade associations. Some towns had succeeded in dominating others and therefore not every ọba was a supreme authority. As mentioned earlier, the Yoruba method of validating autonomous power was to establish claim to an Ifẹ origin with the consequent right to wear a fringed, beaded crown.

The collapse of Ọyọ left its mark on the development of the Yoruba town in many ways. For, in the trials and tribulations of the time, the Yoruba made full use of this, their most characteristic social institution. They built new and larger towns and cities to meet the need for security and defence, the organisation of food supply, the succour of refugees, and the establishment of a stable political and economic life. As new towns sprang up the older ones were expanded. Each town tried to become as large and powerful as possible, often destroying the surrounding villages so as to gather larger populations together at one defensible spot. The estimates of Clapperton and Lander in their journals suggest that pre-nineteenth-century Yoruba towns would have had populations of between five and twenty thousand, though a few like Old Ọyọ might have been larger.[62] By the 1850s, when missionaries were visiting and describing

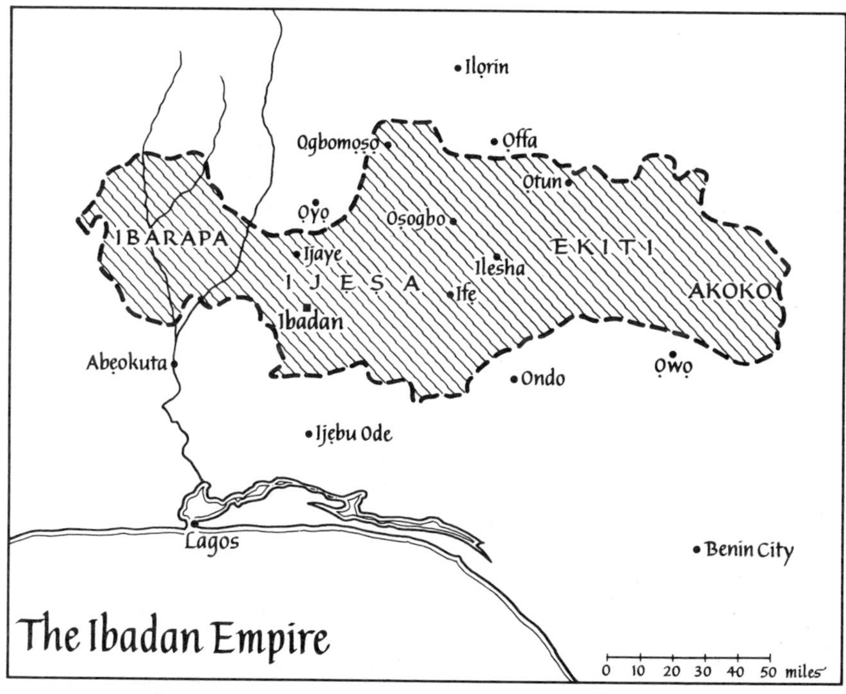

5:2

62 H. Clapperton, *Journal of a Second Expedition into the Interior of Africa from the Bight of Benin to Soccattoo*, London, 1829, pp. 1–59.

Yoruba towns, the populations of such towns as the new Ọyọ, Ijaye, Ibadan, Abẹokuta, Iwo, Ogbomọṣọ, Isẹyin and others were usually reported as ranging from twenty to over sixty thousand.[63]

The character of many of these towns also changed. Ọyọ migrants moved into non-Ọyọ areas. Sometimes they were few in number and became acculturated in varying degrees, merging with the local populations without much trouble as long-term visitors in the past had tended to be assimilated. At other times, the migrants became the dominant group, assimilating the existing population and, where the territory was contiguous with or near to others settled by the Ọyọ, the ethnic character of the town, the political system and political affiliation were changed. Thus Oṣogbo, formerly an Ijẹṣa town, became an Ọyọ town. This also happened to some Ifẹ towns, notably Ikire, Gbọngan and the disputed market town of Apomu.[64] Other Ifẹ towns, such as Ipetumodu and Ẹdunabọn, dominated by Ọyọ migrants, were so far inside Ifẹ territory that while the ethnic character, political system and political loyalties changed, the political affiliation could not be changed. Ile-Ifẹ itself barely escaped being swamped by creating a separate town of Modakẹke just outside its own walls for Ọyọ migrants. The Ifẹ people in their period of alliance with Ọyọ warriors during the attack on Owu and the Ẹgba towns had welcomed many Ọyọ migrants in the hope that they were acquiring valuable tenants and cheap labour for their farms. However, when the alliance broke down and Ifẹ leaders were expelled from Ibadan, hostility developed between the Ọyọ migrants and their Ifẹ hosts. The Ọyọ complained that they were treated like slaves, being made to build Ifẹ houses, clean their farms, perform all sorts of menial tasks and above all pay heavy rent for the land they tilled. It also appeared that some of the Ọni of the period were trying to keep their turbulent chiefs in check by building up the Ọyọ migrants with their Ibadan connections as a counterweight. This only helped to alienate the Ifẹ people from both their ruler and the Ọyọ migrants. Ifẹ politics became very turbulent, with the Ọni trying either to use the Ọyọ to stay in power and risk the hostility of the Ifẹ, or to please the Ifẹ by attempting to keep the Ọyọ in their place and risk rebellion from the Ọyọ and hostility from Ibadan. Twice, in 1850 and 1881, Modakẹkẹ sacked Ile-Ifẹ and drove the Ọni and his people into exile.[65]

Even when the bulk of the migrants were from the same ethnic group, as happened in the remaining Ọyọ territories or that part of Ẹgbaland completely settled, in fact annexed, by the Ọyọ, there were problems. When a town or village was destroyed its inhabitants tended to stick

63 Missionary evidence summarised in Mabogunjẹ, *Urbanization in Nigeria*, p. 91.
64 Johnson, p. 247. For an account of Oṣogbo, see W. B. Schwab, *Political and Social Organisation of an Urban African Community*, Ph.D. thesis, Pennsylvania, 1952.
65 Johnson, pp. 230, 239, 525; Bascom, *The Yoruba of South-western Nigeria*, p. 14.

together and preserve their identity. They would name their new settlement after their previous home. Whether they were on their own or formed part of a larger community, they wished to retain their own rulers and the structure of administration they were used to. When such groups found themselves as refugees in an existing town they might begin, while a pressing emergency lasted, by accepting the overall control of the existing ruler. However, as soon as the emergency passed, tension usually arose between the different competing political systems. Usually, representatives of the migrant groups were absorbed into the existing council or new 'federal' councils were created. The situation was even more complicated when, as at Ogbomọṣọ, a subordinate town ruled by a balẹ found itself host to several groups which included ọbas like the Arẹsa and Olugbọn previously senior to the balẹ in rank. The problem of working out a new system to satisfy both the hosts and the refugees has plagued such towns as Okeiho and Ogbomọṣọ down to this day.[66]

The situation was probably easier where the migrants were founding a new town and, depending on the astuteness of the initial leaders, could work out an efficient, new system, representative of the various groups or acceptable to them. For example, Atiba, the new Alafin, took over the small settlement called Agọ-Ọja and renamed it Ọyọ. He invited the old Ọyọ aristocracy and surrounding villagers to swell the population, and he tried to recreate the monarchical system and structure of administration of Old Ọyọ.[67] In other places, however, the leader was often not a traditional ruler attached to any particular political system but a single warrior or perhaps a group of warriors, and the field of experimentation with constitutional ideas was wider. The warriors as a class began by accepting the importance of an ọba as a stabilising factor in the political system. But increasingly in day-to-day administration they showed a marked awareness of their enhanced status and an increasing impatience with traditional protocol. The warriors at Ijaye and Ibadan, the two most powerful new Ọyọ towns, solved the problem by recognising the overall authority of Alafin Atiba at Ọyọ but ignoring the monarchical principle in the administration of their own towns. Kurunmi established a personal ascendancy at Ijaye and his appointed generals were the leading chiefs. He was traditionalist enough to organise the various cults and Ogboni society, but sceptic enough to subordinate all of them to his own personal control.[68] At Ibadan, no one single warrior was dominant, not even Oluyọle who died before he was able to achieve any personal hegemony.

66 B. A. Agiri, *The Development of Local Government in Ogbomọṣọ*; J. A. Atanda, *The New Ọyọ 'Empire': A Study of British Indirect Rule in Ọyọ Province, 1894–1934*, Ph.D. thesis, Ibadan, 1967, ch. 2.
67 Johnson, pp. 274–84.
68 Ajayi and Smith, *Yoruba Warfare*, pp. 67–8.

After some experimentation, therefore, the leaders evolved a non-hereditary promotional system of chiefs in which civil and military chiefs were supposed to be parallel but the military were usually the more powerful. The idea was to create an open society where administrative and military talent, no matter from where, could rise to the top. The system attracted many able men who would have had no chance at Ijaye. The result was a military aristocracy, highly competitive at least to start with and republican in outlook.[69]

Abẹokuta, the new Ẹgba capital to which the Owu had also been invited, had the most acute administrative problems. Before the crisis there were some one hundred and sixty-seven Ẹgba towns and villages, each with its own ruler, its own Ogboni council and group of war chiefs, the *Ologun*. These towns were grouped into three districts each under an ọba—the Alake, the Agura and the Osilẹ. There were representative meetings of the Ogboni and the Ologun at the district level, but none for all Ẹgbaland. The remnants of all these towns and villages, as well as the Owu, each preserving its own identity and institutions, were resettled at Abẹokuta. Ṣodẹkẹ, who led them, was more concerned with the task of getting the warrior class of each town to cooperate in defending the new settlement against the Ọyọ migrants to the north, the Ijẹbu to the south-east who saw the Ẹgba as potential rivals along the coast, and the Ọtta and Ẹgbado on whose land the Ẹgba were encroaching. In the process, some organisation of a federal Ẹgba army emerged and ensured that the new town survived, but no satisfactory system of internal administration was evolved. Was it to be a military autocracy in which the leading warriors held rule, or a federal council representative of all the existing Ogboni councils, or the supremacy of one of the ọbas, in particular the Alake, or some other new system representing these interests and groups? Various expedients and experiments continued to be tried throughout the century in search of the right constitution.[70]

RIVALRIES AMONG THE NEW TOWNS: 1830–55

The new towns' fight for survival and the general search for a new political system to replace the broken Ọyọ kingdom bred rivalry and war. Ilọrin remained powerful and ambitious. Once Abdul Salaam and Shitta, his brother and successor, were firmly established in control, they treated the lineages and supporters of both Afọnja and Ṣọlagberu with politic generosity and admitted their representatives to the central council.[71] However, effective power lay with the emir and the community of Muslim

69 *Ibid.*, pp. 68–9; B. A. Awẹ, *The Rise of Ibadan as a Yoruba Power in the Nineteenth Century*, D. Phil. thesis, Oxford, 1964.

70 Biobaku, *The Egba and Their Neighbours*, pp. 21–2, 85–7, I. S. O. Biobaku, 'An Historical Sketch of Ẹgba Traditional Authorities', *Africa*, xxii, 1, 1952, pp. 35–9.

71 Johnson, pp. 199–200; Herman-Hodge, *Gazetteer of Ilorin Province*. There is need for a good history of Ilọrin.

scholars, priests and judges. Their ambition was to extend Fulani power down to the coast and to convert the Yoruba into Muslim allies or subject or tributary states. Failing all that they were determined, through calculated military intervention and adroit diplomacy, to ensure that the Yoruba were never allowed to form a general alliance threatening the survival of Ilọrin.

Atiba, now based at the new Ọyọ, did try to reorganise the Ọyọ kingdom, even if his promises to lead the people back to Old Ọyọ and revive the glories of the former empire were no more than a manifesto.[72] He filled the old offices, built a palace on the model of that at Old Ọyọ and patiently worked to attract to himself the accustomed homage and allegiance of all Ọyọ people. His greatest problem came from the warriors, few of whom agreed to live at the new Ọyọ. Most of them were at Ibadan or Ijaye where they were virtually a law unto themselves. To win them over and secure their allegiance, he offered their leaders the highest titles. He appointed Kurunmi of Ijaye the Arẹ-ọna-Kakanfo, with a charge to guard the western provinces particularly against Dahomey. Even though he had no intention of living at Ọyọ to lead the Ọyọ Misi, Oluyọle at Ibadan accepted the title of Baṣọrun with the responsibility of guarding the eastern provinces against continued encroachment from Ilọrin. Thereafter, rivalry developed between Ijaye and Ibadan for leadership within the Ọyọ area, each protesting continually that it was upholding the prestige and authority of the Alafin.

Initially, Ijaye appeared the more important champion of the Alafin and the new system he was trying to create. The soldiers who settled at Ijaye included a significant part of the old cavalry warchiefs, the Ẹṣọ, who fled from Ikoyi and other places rather than submit to Ilọrin. Kurunmi was an experienced general and administrator and the personal rule he established at Ijaye had created stability while Ibadan was still experiencing frequent internal upheavals. But while upholding the Alafin's authority vis-à-vis Ibadan, Kurunmi was quietly establishing control over the Upper Ogun, the western half of the Ọyọ area, forcing the towns and villages there to pay homage and tribute to Ijaye and not to Ọyọ. As Ibadan achieved greater internal stability, it too began to do the same in the eastern half of the kingdom. It achieved a notable landmark in 1840 when at a battle near Oṣogbo it broke the myth of Ilọrin invincibility and established the fact that well-drilled soldiers armed with muskets could defeat the Ilọrin cavalry force.[73] This effectively checked the expansion of Ilọrin southwards into the Ọyọ area and allowed Ibadan to establish its sway over the old Epo and Ibọlọ districts. It also created in the inhabitants of Ibadan the attitude that they were the saviours of Yorubaland. Heady with success, Baṣọrun

72 Johnson, pp. 279–84.
73 Ajayi and Smith, *Yoruba Warfare*, pp. 33–6.

Oluyọle[74] challenged the authority of Atiba and decided to test his strength against Ijaye in the Batedo war. It was an unpopular war and after two years of desultory fighting, Atiba was able to mediate. In this way he demonstrated how much everyone needed the stability of the monarchical system. His authority thus enhanced, it could be said that Ọyọ again functioned as an effective monarchy. However, the authority of the new Alafin remained largely moral. Ibadan and Ijaye allowed very little territory to pay tribute directly to Ọyọ. The Alafin, having no adequate military force of his own, had to rely at any particular time on either Ibadan or Ijaye. With this balance established in the Oyọ area, Ibadan gradually shifted the scene of its ambition eastwards into the Ijẹṣa, Ekiti and Akoko country where Benin authority was crumbling and Ilọrin and Nupe were beginning to seek new expansion.[75]

Meanwhile, Dahomey had asserted its independence of Ọyọ control since the early 1820s. Local Ọyọ officials in Dahomey and around Ilaro had got an army together but they were defeated and annual celebrations of this great victory were instituted at Abomey. This was a period of revival in Dahomey, following the depression in the slave trade of the closing years of the eighteenth century, the assassination of the king in 1797, and the failure of an attempted *coup d'état*. Gezo initiated the revival when, through the support of leading chiefs and Brazilian traders, especially Felix da Souza, he seized power and offered a more dynamic leadership.[76] Freed of Ọyọ control, he was able to expand northwards into the Mahin country and westwards into Eweland, where Dahomey was eventually to confront Asante power. Dahomey's influence also spread to Porto Novo, seeking to control Porto Novo's policies without taking over the government. Its main attention, however, was turned eastwards to Ẹgbado.

Gezo found in the cataclysm in the Yoruba country not merely a chance to gain national independence, but also an opportunity to avenge years of subjection and humiliation by expanding into neighbouring parts of Yorubaland. Dahomey was a compact, effectively organised kingdom. It had no intention of weakening that effectiveness by building up an empire. What the Dahomey rulers sought in their wars of expansion was manpọwer to be sold as slaves or used as cultivators. With Sabẹ and Ketu as sizable monarchies and Ijaye becoming a formidable power, Dahomey

74 Johnson, p. 305. He says that Oluyọle was the great grandson of Baṣọrun Yamba and that his mother was a daughter of Alafin Abiọdun. Current traditions (e.g. collected from Chief Aderibigbe, the Agunpopo of Ọyọ, in 1969) claim that Oluyọle was a posthumous son of Baṣọrun Gaha. This could indicate that Gaha's lineage was closely related to that of Yamba, as one would expect, and that in his lifetime Oluyọle preferred to stress the connections with Yamba rather than with Gaha. Because of the links with the Ọyọ royal family, he emerged in the 1840s as a personal rival of Alafin Atiba.

75 Awẹ, *The rise of Ibadan*, pp.

76 D. A. Ross, *The Autonomous Kingdom of Dahomey, 1818–94*, Ph.D. thesis, London, 1967; D. A. Ross, 'The First Chacha of Whydah: Felix Francisco De Souza', *Odu*, n.s., 2, 1969, pp. 19–28.

turned its attention to begin with, not north-eastwards into Ọyọ territory but south-eastwards into Ẹgbado which had dissolved into several factious towns, villages and communities when Ọyọ control was removed.[77] The divisions in Ẹgbado had attracted others: Lagos and Ijẹbu adventurers were already active in the area. The Ẹgba, recently resettled at Abẹokuta, looked to the Awori and Ẹgbado countries for farming land and direct access to the sea. At the battle of Owiwi, in 1832, the Ẹgba were able, with the support of a faction of the Lagos ruling dynasty, to establish their predominant interest vis-à-vis the Ijẹbu. The main confrontation in Ẹgbado was, therefore, to be between Dahomey and Abẹokuta. The Ẹgba were invited to intervene in Ilaro against Dahomey and they took the opportunity to establish a base there for themselves. Soon, they also conquered Ọtta. Then in 1842 they besieged Ado, the main obstacle to their power in southern Ẹgbado and on their route to the sea through Badagri. Dahomey came to Ado's help and frustrated the attempt. The war was to last until 1853, when missionaries persuaded the Ẹgba to give up the siege. Meanwhile in 1845 some Ẹgba troops ambushed a Dahomey army on the march and captured the royal drum, state umbrella and other regalia. Gezo vowed to avenge the humiliation by destroying the new Ẹgba town. After years of preparation, Gezo mounted a major expedition to attack Abẹokuta in March 1851. Abẹokuta, with some help from European arms and European advisers, routed the Dahomey army.[78] Ẹgba-Dahomey rivalry was to continue and it eventually made the Ẹgbado one of the few African peoples genuinely to welcome European rule as a form of deliverance.[79]

The town had succeeded the kingdom as the effective unit of political organisation. Though loyalty to the subcultural group was still of considerable importance it was often subordinate to loyalty to the town which afforded immediate security and a chance of survival. As we have seen, although Ibadan became essentially an Ọyọ town it welcomed warriors who came from different parts to be apprenticed to leading war chiefs to learn the art of war. Ibadan's loyalty to Ọyọ was more from political calculation than from ethnic solidarity. Its rivalry with Ijaye was just as intense as its rivalry with Ilọrin or Abẹokuta. As the scope of the rivalry of the towns expanded or shifted to other areas, the pattern noted above was repeated: the destruction of smaller towns and villages, the movement of peoples, regrouping, the rise of new towns and the expansion of old ones. In the Ẹgbado area, Ilaro and Okeọdan became major centres of refuge and commerce until Okeọdan was sacked in 1848 and its population scattered once again to Ilaro, Ado, Badagri and other places. In this way, the Ọyọ experience was extended not only to the Ifẹ and Ẹgba who felt

77 K. Fọlayan, *Ẹgbado and Yoruba-Aja Power Politics*, M.A. thesis, Ibadan, 1965.
78 Ajayi and Smith, *Yoruba Warfare*, pp. 37–43.
79 Fọlayan.

the immediate impact but also to the Ẹgbado and eventually the Ijẹṣa, Ekiti, Ondo, Akoko and Ijẹbu, until there was no part of Yorubaland that escaped the effects of the collapse of Old Ọyọ.

THE IMPACT OF THE ABOLITIONISTS

By 1855, the European factor had become of some importance in the Yoruba country, especially in Badagri, Abẹokuta and Lagos. However, it should be clear that the effect of the actions of Europeans on the chain of events described above was very limited both in the days of the slave trade and those of abolition. In fact, such was the force of local factors that they usually diverted the external European factor into producing unexpected results. It was in the days of abolition that Yoruba participation in the Atlantic slave trade expanded at a dramatic pace. While the trade from the ports of Benin declined sharply[80] slaves from the Yoruba wars boosted the trade of Lagos, Porto Novo and Whydah.[81] Lagos in particular attracted Brazilian traders and reached a new peak of prosperity. What the abolitionists did was to restrict the areas to which Yoruba slaves were being carried and thus to ensure the survival of Yoruba culture in the New World. The British naval squadron failed to stop the flow of slaves, but its success in contributing to the increase of liberated Yoruba slaves in Sierra Leone was significant.

Hardly any Yoruba were listed in the registers of the liberated Africans in Sierra Leone in 1816. Soon after that, Akus (Acoos, Ackoos), as they were called, began to feature in large numbers.[82] At the beginning, they were mostly from the Ọyọ area. By 1830, the Yoruba had become the dominant group in Sierra Leone and the Ẹgba were the largest single subcultural group among them. Ethnic groups in Sierra Leone were encouraged to stay together, succour one another and provide, through mutual assistance and traditional organisations, those services in which the colonial administration was deficient. Some who were trading in condemned slave ships down the coast realised that they were not too far from home and that they

80 Ryder, pp. 231–2.

81 The figures for British, French and Portuguese trade suggest that between 1741 and 1811, except for the decade 1781–90, the volume of slave trade from the Bight of Biafra increased steadily and was always more than double that from the Bight of Benin. The unusual flow from the Bight of Benin in the decade 1781–90 when its trade slightly exceeded that of the Bight of Biafra was probably due to the disturbances associated with the period of Baṣọrun Gaha. One of the best indicators for the relative volume of trade after 1811 is the Sierra Leone Census of 1848 which indicated that 63.7 per cent of liberated Africans in Sierra Leone had come from the Bight of Benin, and 20.2 per cent had come from the Bight of Biafra. P. D. Curtin, *The Atlantic Slave Trade*, Madison, 1970, Tables 66 and 71, pp. 221 and 245.

82 Horton to C.M.S. Secretary, 15 March 1817, 'List of Boys and Girls supported by the C.M.S. at Leicester Mountain as at last quarter of 1816' indicated that there were out of 337 children, 150 Ibos, 49 'Calabars', 3 Hausa, 11 'Popo', but no recognisable Yoruba. By contrast, the List in Renner to Secretaries, 29 July 1818, shows out of 140 children, 27 Accoo and 20 Ibo. (C.M.S. CA1/E7.)

could make the journey back. The self-repatriation began in 1838, and the stream swelled with people longing for home and for opportunities that were lacking in Sierra Leone. Many had come under missionary influence, embraced Christianity and had usually picked up some English, while others remained attached to the traditional religions or Islam. Some had learned carpentry, tile- or brick-making, masonry, sawyering, printing or similar trades. When they were not well received at Lagos, they moved to Badagri and several Ẹgba found their way to Abẹokuta where friends and lost relatives welcomed them. There was a similar movement of emancipated slaves from the New World, especially from Bahia. In 1851 there were in Abẹokuta an estimated three thousand of such returned 'emigrants', as they were called (Saros from Sierra Leone and Amaros from Brazil or Cuba) and there were others in Badagri, Lagos, Ijaye, Ibadan and other places.

This movement of largely emancipated Yoruba from Sierra Leone and the New World was important in that it attracted the attention of the abolitionists to the Yoruba country. European missionaries and traders followed them and later invited a British Consul and warships after them. The 'emigrants' became the agents of the missionaries, traders and Consuls, their liaison with the rulers of the towns in the interior and the excuse for several actions taken to promote and protect European penetration of Yorubaland.[83]

Christian missionaries followed the emigrants to Badagri, the Methodists in 1842 and the Church Missionary Society, after an exploratory visit, in 1845. They moved in 1846 to Abẹokuta which they tried to help build up as the most powerful of the new towns. They built schools and encouraged trade with Europeans. They taught new skills and advised on town planning. They supplied ammunition, including some cannon. Above all they tried to secure for Abẹokuta access to the sea and to this end persuaded the British government to intervene in a disputed succession in Lagos at the end of 1851 and install a pro-Ẹgba puppet. The missionaries expanded inland and located their agents at Ibadan (C.M.S.), Ọyọ (C.M.S.), Ijaye (C.M.S. and Baptists) and Ogbomọṣọ (Baptists). They established schools and mission stations. This further encouraged the return of the emigrants whom they tried to group together near the mission stations and encouraged to practise their new skills. The 1850s were thus a decade of expansion of missionary activities. There was little success in evangelisation outside the ranks of the emigrants but it became for the new towns something of a status symbol to have a resident European missionary. In this way, some contacts were made which later proved useful for the spread of Christianity.[84]

83 J. F. A. Ajayi, *Christian Missions in Nigeria, 1841–91: the Making of a new Elite*, London, 1965, ch. II.
84 *Ibid.*, ch. III.

Even before the arrival of missionaries British traders based at Cape Coast and Accra were pioneering trade in palm oil at Badagri. The Brazilians while continuing to deal in slaves began also to collect palm oil if only because the trade goods imported by the European legitimate traders—printed cloths, metal goods and, above all, guns and gunpowder—were in greater demand than rum and tobacco, the staples from Brazil and Cuba. Since slaves were more readily available than palm oil the Brazilians acted as brokers, buying cloths and guns for money and selling these for slaves.[85] The presence of the missionaries and emigrants helped to boost the palm oil trade so that in time the leaders of the new towns saw the wisdom of keeping their slaves on the farms to manufacture palm oil and to carry it to the coast to exchange for cloths and guns. Kosọkọ, the Ọba of Lagos, like Gezo of Dahomey, refused to compromise his sovereignty by accepting dictation on what to sell or to whom but actively encouraged the palm oil trade as soon as there was a market for it. The formal prohibition of the importation of slaves into Brazil in 1850 reduced the demand for slaves in the coastal markets. It was these factors, rather than the activities of the naval squadron or the British occupation of Lagos, that caused a sharp decline in the overseas slave trade from the Yoruba country after 1850. The traders who supported the British intervention in Lagos did so in the knowledge that the palm oil trade was already flourishing there. Soon, it became obvious that the increase of legitimate trade, far from stopping internal slavery by encouraging the emergence of a sturdy European-type peasant class, was expanding the class of domestic slaves.[86]

The occupation of Lagos demonstrated the superiority of European warships, cannon and rockets available along the coast over Yoruba armaments. In Lagos, therefore, the Europeans began to dictate the pace of events and, within ten years, economic and political power had passed to the British and their agents. But there were limits to the initiative open to the Europeans. Their power was still limited to the coast, and Dahomey and Abẹokuta could not be dealt with like Lagos. Even on the coast, since Lagos depended on trade and the towns in the interior could control that trade, British authorities were obliged to take account of the wishes and intentions of people in the interior until the time when the British had superiority in the weapons which could be easily transported inland.

Meanwhile, in the hinterland of Lagos, the initiative remained with the Yoruba, and the European intervention made little change in the power politics of the Yoruba towns. European support helped Abẹokuta but was not a decisive factor in either its survival or its defeat of Dahomey. Other towns knew how to obtain enough ammunition without missionary

85 Newbury, chs. II and III.

86 See for example the evidence of R. F. Burton to the 1865 British Parliamentary Committee, Questions 2165–6, *Report of the Select Committee on State of British Settlements*, Parliamentary Papers, v, 1, 1865.

assistance. And when the missionaries themselves saw that Abẹokuta had no chance of becoming the dominant power in the Yoruba country or of assisting missionary expansion outside the Ẹgba area, they abandoned their pro-Ẹgba policy,[87] and the British administration in Lagos allied with Ibadan instead. Missionary activities were to be of great importance later in their influence on the rising class of westernised Yoruba. In particular, studies in the Yoruba language helped to foster a new awareness of the common nationality of all Yoruba people, and a school of historians, typified by Samuel Johnson, who saw the Yoruba people as formerly belonging to one single united kingdom.

IBADAN HEGEMONY (1855–75)

Missionaries arrived in Ibadan early enough to witness the gradual build-up of its power. After the death of Oluyọle in 1847 Ibadan had stabilised its internal political system. There was to be collective, not personal, leadership. Office was to be not hereditary but achieved through promotion as a reward for outstanding abilities shown in war or administration, particularly in war.

The base of Ibadan's power was in the Ibọlọ and Ẹpo districts, secured through the effective control and the loyalty of Iwo and Ẹdẹ. Ogbomọṣọ was a major town in its own right, and it continued to behave independently until the fear of Ilọrin was strong enough to induce it to seek the protective oversight of Ibadan. Since the battle of Oṣogbo, Ilọrin itself had accepted the limits to its cavalry force when it came to fighting in the forest areas and its infantry was never a match for that of Ibadan.

Thus, Ibadan's power was well established in the eastern half of the Ọyọ areas. Except for occasional challenges to Ijaye in Ibarapa, Ibadan from about 1845 onwards turned its attention eastwards. Some Ijẹṣa had been challenging Ibadan's hold on Oṣogbo. Ibadan invaded their territory and terrorised them. Ileṣa, their capital, was as yet unsubdued but in spite of dissentient groups the Ijẹṣa acknowledged Ibadan's overlordship. Then Ibadan moved into Ekiti. The internal problems in Benin had tempted Ilọrin troops to seek expansion in the more open parts of northern Ekiti. Ibadan followed and harassed them. In a series of systematic campaigns between 1851 and 1855 most of the Ekiti towns were brought under Ibadan's control. Soon Ibadan troops were going further into Akoko and Afẹnmai where they confronted not only Benin but also Nupe troops.[88]

In this way the core of a new Ibadan empire was created. The local rulers were left in charge of the day-to-day administration but they were obliged to pay tribute and give military service to Ibadan. They also had to support an Ibadan ajẹle and his often expensive entourage. The ajẹle had to confirm in office newly appointed rulers and, in the event of

87 Ajayi, *Christian Missions*, pp. 99, 170–7.
88 Awẹ, *The Rise of Ibadan*.

dispute, he tilted the balance in favour of whichever candidate Ibadan supported. In this and other ways he supervised the local administration more closely than had been the case even within the metropolitan provinces of Ọyọ or Benin. But at Ibadan itself there was no unified control of the tributary peoples. Groups of towns and their ajẹlẹs were responsible to leading chiefs in Ibadan who supervised their administration, nominated the ajẹlẹs, received their own tribute and acted as spokesmen for the towns in Ibadan's councils.[89]

Tribute from the subjected towns and slaves from the wars who could be used as troops, labourers or carriers constituted the economic base of Ibadan's military power. The successful military chiefs built large households comprising their kin, hosts of retainers who attached themselves to the warriors and different categories of slaves. Some of the slaves tilled the farms to feed the large households or the soldiers at the front. Others were the soldiers, the war boys who composed the private armies of the chiefs who organised them and led them in war. Others helped in the administration of the chief's fief, bringing tribute in slaves, cowries or palm oil which was taken down to Porto Novo, Ikorodu or Benin to exchange for the necessary guns and gunpowder. The more successful the war chief became, the more retainers and slaves he acquired to further increase his resources, influence and fighting capabilities. A daring act of bravery at war, recognised by rapid promotion at home, could set off an ambitious young warrior on the path to building up power and wealth. In the same way, a major military disaster or political disgrace resulting in heavy fines or loss of property could just as quickly send him on the downward path. Wealth meant slaves, horses, a rich arsenal, rare and expensive finery and cowries. An aristocratic style of life and a system of patronage and clientage similar to that associated with life in the capital city of Old Ọyọ emerged. There was intense rivalry among the war chiefs and in the early days, feuding, vendettas and civil war were rife. Gradually, the position became more stable as second and third generations of war chiefs began to achieve office. It was a system geared to war which needed more wars to nourish it. Yet it attracted men of talent from different parts of the country who came to learn the art of war as retainers. And as they, or emancipated slaves, returned home, the Ibadan system began to be copied in other places.[90]

By 1860 Ibadan felt strong enough to challenge Ijaye once again for supremacy in Ọyọ territory. Ibadan made sure this time that it went to

89 B. A. Awẹ, 'The Ajẹlẹ System: A Study of Ibadan Imperialism in the Nineteenth Century', *J.H.S.N.*, iii, 2, 1965, pp. 47–60.

90 Awẹ, *The Rise of Ibadan*, especially ch. III; S. A. Akintoye, 'The Economic Background of the Ekitiparapọ, 1878–93', *Odu*, iv, 2, 1968; R. G. Gavin, 'The Building of Ibadan', Ibadan, History postgraduate seminar paper, December, 1968; *The City of Ibadan*, eds. P. C. Lloyd *et al.*, Cambridge and Ibadan, 1967.

war, not on its own, but as the loyal supporter of Alafin Adelu, Atiba's eldest son who, contrary to usual practice, had succeeded his father. Kurunmi of Ijaye refused to recognise Adelu as king. When Adelu, with obvious encouragement from Ibadan, went on to challenge Ijaye's control of the Upper Ogun districts, war broke out in 1860. In spite of Abẹokuta's support, Ijaye was closely besieged and starved till its resistance collapsed in 1862 after Kurunmi's death. The troops and the rest of the population fled, a substantial part of them to Abẹokuta.[91] With Ijaye destroyed, only the tactful acknowledgement of Ọyọ's suzerainty qualified Ibadan's dominance among the Ọyọ subcultural group. Of the area hitherto subject to Ijaye, Ibadan took Ibarapa and conceded the Upper Ogun to Ọyọ.

Ibadan's greatest need thereafter was a means of access to the sea, more direct than Porto Novo or Benin, and more closely under its own control. Ibadan had been working towards this during the Ijaye war by dealing with and encouraging a secessionist movement among the Rẹmọ in western Ijẹbu. The Ẹgba and Ijẹbu therefore came together after the fall of Ijaye to prevent Ibadan from breaking up the Ijẹbu kingdom or acquiring any such direct access to the coast. Ibadan allied with the British in Lagos and in the name of free trade scattered the allied army that was besieging Ikorodu, the main trading port of Rẹmọ.[92] Although this did not succeed in establishing Ibadan's control over Rẹmọ, it did encourage the divisive forces which continued to plague the Ijẹbu for the rest of the century. It also helps to explain the great suspicion and hostility with which the Ijẹbu treated all foreign travellers.

Nevertheless Ibadan's hegemony was real enough and it had altered the balance of power throughout the Yoruba country and beyond. The destruction of Ijaye had provided fresh opportunities for Dahomey. While Abẹokuta was busy trying to prevent Ibadan from getting a stranglehold on Rẹmọ, Dahomey sent various expeditions into Ẹgbado which, among other places, destroyed Iṣaga in 1862. Then in 1864 Dahomey launched a major attack against Abẹokuta itself but as in 1851 Abẹokuta was ably defended and the Dahomey army was routed with great losses.[93] Thereafter, although Dahomey continued to raid into Ẹgbado and occasionally into Ẹgba territory, its attention was concentrated on the north-east. Eventually it succeeded in bringing both Sabẹ and Ketu under its control and forced more towns in the Upper Ogun to seek the safety of hilltop settlements.

On the other hand Ibadan's expansion eastwards helped to weaken the Benin empire and accelerated the process of pushing Benin back into its heartlands. As Professor Ryder has shown, the stoppage of the slave trade did little damage to Benin since slaves no longer formed a major item in

91 Ajayi and Smith, *Yoruba Warfare*.
92 Biobaku, *The Egba and their Neighbours*, p. 74.
93 *Ibid.*, pp. 75–6.

Benin trade.[94] The shift to palm oil favoured the Itṣẹkiri, who could trade directly in the major oil-producing country of the Urhobo-Isoko. But most of these areas still acknowledged Benin suzerainty, and participation in trade there enabled Benin to obtain the ammunition it sold to the Yoruba country. The more significant and irreplaceable losses to Benin came from the Ibadan occupation of Ekiti and their interference in Afẹnmai at a time when the Benin empire was preoccupied with a disputed succession and prolonged civil war (1854–80) in which Afẹnmai was the base for the party opposed to the Ọba.

Nearer home, Ibadan had helped to break the previous alliance between the Ẹgba and the British and in 1867 Abẹokuta had expelled all European missionaries and traders.[95] This increased the understanding between the Ẹgba and the dominant groups in Ijẹbu, strengthening their determination to prevent further British penetration of the Yoruba hinterland. They banned Lagos and Ibadan officials or traders from passing through Abẹokuta or Ijẹbu without specific permission. It was to break this blockade that Lagos sought for an alternative route to Ibadan.

By the early 1870s Ibadan was at the height of its power. Its hold on its empire was consolidated and the occasional revolt, such as that at Ileṣa in 1869–70, was severely put down. Some recalcitrant warriors refused to surrender, notably Ogedemgbe of Ileṣa, who preferred to live a life of adventure in exile on the borders of Ekiti and Akoko where he occasionally clashed with Ibadan troops. Nevertheless Ibadan had established a sizable area of peaceful and orderly administration. It was the most successful of the new towns and its history illustrates clearly the mechanisms by which the Yoruba people faced the tribulations caused by the collapse of Old Ọyọ.

There were war, destruction and oppression but there were also reconstruction, regrouping, adaptation and reorganisation. Ibadan emerged as the real successor of Old Ọyọ. By the 1870s it was exchanging correspondence with Nupe and sending troops to intervene in the disputes going on there over the succession. Ibadan was said to be receiving troops and envoys from Dahomey and Asante. Locally, Ibadan was also playing the arbiter. When Modakẹkẹ sacked Ifẹ in 1850 it was Chief Ogunmọla of Ibadan who negotiated the return of the Ifẹ in 1854. In 1878 Ibadan was able to insist that its candidate should become the Ọni.[96] Before then, when Ifẹ refugees seized Okeigbo from Ondo and encouraged a civil war which destroyed Ondo town and sent the Oṣemawe into exile, Ibadan also took part in negotiating for their return. In this the British came to their assistance since the new route the British were interested in was to pass through Ondo to Okeigbo, Ifẹ and Ibadan. By 1872 the route was opened to traffic. The C.M.S. was asked to send missionaries to Ondo and to

94 Ryder, pp. 225, 233.
95 Ajayi, *Christian Missions*, pp. 200–4.
96 Johnson, pp. 230, 525.

encourage 'emigrants' and others in Lagos to use the new route and settle at important market centres on the way. Through this route, European influences hitherto confined to western Yorubaland began to penetrate eastern Yorubaland.[97] The new route may have contributed to the new confidence of Ibadan, but it was to prove as useful, if not more so, to the opponents of the Ibadan hegemony as to Ibadan itself.

A WAR TO END ALL WARS

From dominance in the Ọyọ areas, some rulers of Ibadan began to nurse greater ambitions, perhaps to bring all Yorubaland together. Only the Ẹgba and Ijẹbu appeared formidable. The Ijẹbu were divided and experience at Ijaye had shown that, although the Ẹgba were good at defending the walls of Abẹokuta, they were no great tacticians in open battle. It began to appear that the seemingly endless chain of Yoruba wars could be broken by a war to end all wars if only Ibadan's supremacy could thereby be extended to the Ẹgba and Ijẹbu. Such were the dreams that have been ascribed to Chief Momoh Latosisa, who bore the title of Arẹ-ọna-Kakanfo and was Ibadan's ruler from 1871 to 1885.[98] He was a fervent Muslim, the first to rule in Ibadan, and the example of the Muslim rulers of the Sudan may have encouraged him. Such dreams were soon to prove unrealistic.

The Ẹgba and the Ijẹbu, of course, would not hear of Ibadan supremacy and though they could be bypassed in finding other routes to the sea their position near the coast had its advantages. Ifẹ was not powerful and it had the Ọyọ settlement at Modakẹkẹ at its throat but Ibadan's role of king-maker had created such resentment that the Ifẹ joined the ranks of those determined to stop Ibadan's ambitions. An Ifẹ prince, Derin, controlled Okeigbo which occupied a highly strategic position for protecting or disrupting traffic on the new Ondo route. Even Adeyẹmi, the new Alafin who succeeded Adelu in 1874, soon realised that he was Ibadan's suzerain in name only and that Ibadan's power was no longer an extension of the powers of the Alafin or the collectivity of the Ọyọ people. Ibadan was harbouring Lawani, the former Arẹmọ (Crown Prince), who had contested Adeyemi's claims and remained a pretender to the throne.[99] Though Alafin Adeyẹmi hesitated openly to challenge the ambitions of the Ibadan he turned to the old Ọyọ art of diplomacy to frustrate them behind their backs.

Thus, when Arẹ Latosisa attacked Ẹgba troops in 1877 he discovered that the rest of Yorubaland was united against Ibadan. His efforts to detach the Ijẹbu from the Ẹgba alliance were rebuffed. The Ekiti rose up against Ibadan's ajẹlẹs, massacred them, declared themselves independent, and became the core of the anti-Ibadan alliance. Ilọrin, coming as usual

97 S. A. Akintoye, 'The Ondo Road, Eastwards of Lagos, *c.* 1870–95', *J.A.H.*, x, 4, 1969,
98 B. A. Awe, 'The End of an Experiment: The Collapse of the Ibadan Empire 1877–1893', *J.H.S.N.*, iii, 2, 1965, pp. 221–30.
99 Johnson, p. 459.

to fish in troubled waters, joined the allies. Ibadan responded with vigour and at the Jalumi battle near Ikirun in 1878 imposed heavy losses on the allies. This daunted but failed to crush them. Ogedemgbe brought his warriors to join them and to lead the Ekiti and Ijẹṣa in what became a formal confederacy—the Ekitiparapọ. The allies dug themselves in at Imesi-Ipole and the Ibadan troops confronted them from their base at Igbajọ. After some sharp initial fighting the Kiriji war, as it came to be known, dragged on. By 1883 everyone was tired of the fighting. Yet even when peace was made in 1886 through British intervention and both sides broke up camp several issues remained unresolved. The Ekitiparapọ secured their independence. Ibadan gave up its ambitious scheme for dominating all Yorubaland. But the demand of the Ifẹ that Modakẹkẹ be destroyed or removed was difficult to implement. Similarly, there were disputes about the control of border towns such as Igbajọ and Ọffa. When the British came to dictate peace in 1893 Ilọrin and Ibadan were still fighting over Ọffa.[100]

There were many reasons for the eventual failure of Ibadan. There was the military factor. The alliance against Ibadan was unwieldy. It was the negative desire to stop Ibadan's tyrannical ambitions that held it together and the allies rarely agreed on any positive goals. Yet, unwieldy as it was, it was a formidable alliance compelling Ibadan to fight on many fronts and making the organisation of military supplies extremely difficult. If Ibadan had been able to win in a few short battles such as at Jalumi there might have been a chance of success. But the beginning of the war coincided with the introduction into the coastal markets of the more sophisticated weapons from Europe. The Ekitiparapọ National Association in Lagos first got hold of the new rifles—Mausers, Schneiders, Martini Henris, Remingtons, and Manchester repeaters—and began to transport them and some trained marksmen along the Ondo route to the allies at the front. For a while Ibadan was in great difficulties. Early in 1883, as a result of a *coup d'état*, the Ijẹbu declared themselves neutral in the war and allowed Ibadan to trade at the border town of Oru. The supply situation improved but the time for a quick victory was long past. The siege war that had developed became a stalemate.[101]

Ibadan's problems were not, however, merely military. It is necessary to explain not why Ibadan was unable to defeat such an alliance but why there was such unanimity against Ibadan. This is connected with some of the issues we raised earlier about the nature of Yoruba society, especially in the transitional period when much was changing but the past was being invoked as a factor of stability. For example, there is no doubt that Ibadan was a sovereign state exercising all the attributes of

100 S. A. Akintoye, *The Ekitiparapọ and the Kiriji War*, Ph.D. thesis, Ibadan, 1966.
101 *Ibid.*; S. A. Akintoye, 'The Economic Background of the Ekitiparapọ, 1878–93', *Odu*, iv, 2, 1968.

sovereignty including the judicial power over life and death. Everywhere the warriors of the new towns had taken over prescriptive rights which used to belong to the divine monarchs alone. Yet not even in Ibadan was the monarchical principle completely forgotten. Ibadan kept up for as long as possible the façade that the Ibadan empire was only part of the Alafin's kingdom with its real capital at Ọyọ. This was because the support of the Alafin was necessary to give legitimacy to the Ibadan regime. Arẹ Latosisa himself was installed in office by an *ilari* from Ọyọ.[102] At the height of the crisis following the collapse of Ọyọ people were willing to take refuge and protection wherever it existed. When the worst of the emergency had passed, Ibadan discovered that while its military power could compel obedience it had failed—or had not yet succeeded—in creating legitimacy and instinctive loyalty. All the crowned monarchs of the Ekitiparapọ bitterly resented subjection to a parvenu power, and the Ẹgba and Ijẹbu determined to curb its ambition.

The resentment was no doubt reinforced by the oppressive nature of Ibadan's rule. Large empires such as Old Ọyọ and Benin were able to survive among different cultural groups because usually they demanded only homage and imposed a minimum amount of interference. Ibadan began as a smaller, more efficient state faced with a crisis of survival. This dictated the necessity to build up a powerful war machine, demanding heavy tributes and regular military service from those owing allegiance to Ibadan. From this resulted an elaborate, and expensive, organisation to supervise the collection of tribute and the supply of military equipment. This demand for increased taxes was not accompanied by any new notions of improving the lot of the governed beyond the guarantee of defence in case of attack. It is not surprising that the people regarded such rule as unwarranted exploitation. Many Ekiti who had tolerated an undemanding overlordship from Benin bitterly resented the heavy tributes and the domineering excesses of the Ibadan ajẹlẹs.

There was also an ethnic factor involved, but this should not be exaggerated. Rivalry among the Yoruba subcultural groups was not always the decisive factor in the power politics of the new towns. Ethnic sentiment was useful when the Ijẹṣa wanted to throw off Ibadan rule but it does not explain why the Alafin secretly supported Ibadan's antagonists. When Ibadan fought against Ijaye it was the Ẹgbas who—for their own reasons—went to defend Ijaye and many of those who deserted Ijaye found refuge in Abẹokuta. It would appear that appeals to the subcultural group, pan-Yoruba feeling, and the monarchical tradition, were all ideologies in a grim political struggle for survival and power. Like other ideologies, each had its devotees, but ideologies were not sacrosanct. They were invoked to support the necessary line of action as occasion demanded. When Ibadan

102 Johnson, p. 503.

stood for checking Fulani incursions and creating order and stability, it received wide support in spite of its having broken many traditions. When it tried to acquire too large an empire and was creating a leisured wealthy class at Ibadan it lost its claim to indispensability and was opposed on the grounds that it did not conform to the traditional political structure.

The failure of Ibadan meant that no satisfactory solution was found to the problem of a new political system to replace the one underpinned by Old Ọyọ and this created opportunities for the British administration in Lagos. It was not that the wars provoked British intervention, the reasons for and even the timing of which were determined largely by external factors. The wars, however, do explain the method of British expansion in Yorubaland.[103] The military intelligence and the war maps of Yorubaland which the British War Office had compiled with missionary cooperation did not have to be used as the Yoruba wars and the divisions among the Yoruba made large-scale conquest unnecessary. The British emerged in the role of peacemakers sending missionaries as envoys to negotiate peace terms. In 1886 they succeeded in getting both camps at Imesi Ile and Igbajọ to break up, but the war dragged on, especially between Ibadan and Ilọrin, Modakẹkẹ and Ifẹ. The British then decided to demonstrate what new capabilities they had acquired for military exploits on land by making an example of the Ijẹbu in 1892. Only the Ilọrin failed to heed the warning and the Royal Niger Company dealt with them in 1897. The British thus emerged as the power most able to fill the vacuum left by Old Ọyọ. With this threat in the background, in 1893 they got Ọyọ, Abẹokuta and Ibadan to accept 'treaties of friendship' which extolled the powers of the Alafin, guaranteed the 'independence' of the Ẹgba and permitted the posting of a 'friendly' Consul at Ibadan. It was only gradually that the British revealed through individual punitive acts that they were no longer friends or allies but rulers far more effective and domineering than Old Ọyọ ever was.

103 E. A. Ayandele, 'The Mode of British Expansion in Yorubaland in the Second Half of the Nineteenth Century', *Odu*, iii, 2, 1967.

CHAPTER 6

Politics in Ghana, 1800–1874

ADU BOAHEN

In 1807, two absolutely unrelated events occurred, one in Ghana and the other in Britain. These were jointly to determine the course of the history of Ghana in the nineteenth century. The first was the successful invasion of the Fante states by the Asante and the second was the abolition of the slave trade.

ASANTE AND THE FANTE STATES

The Asante conquest of Fante brought to a conclusion the expansion of the small Oyoko principality of Kumasi which had begun during the last two decades of the seventeenth century. From 1807, the main political preoccupations of the Asante were the consolidation of their grip on Fante and the maintenance of the integrity of their empire. The abolition of the slave trade generated a host of social and economic problems. It was the solutions to these problems together with the emergence of imperialism in Europe during the last quarter of the nineteenth century which were to lead first to the overthrow of the Asante empire, then to the annexation of southern Ghana and finally to the formation of the British colony of the Gold Coast, the colony that was to attain independence about half a century later under the name of Ghana.

The conquest of the Fante states in 1807 was more or less an historical inevitability. The relationship of Fante's position to the Asante empire by the last two decades of the eighteenth century made its conquest only a matter of time. The Asante empire had extended over an area including virtually the whole of modern Ghana, parts of modern Ivory Coast and Togo and included Gonja and Dagomba to the north, Gyaman to the north-east, Sefwi, Dunkwa, Twifo, Wassa and Nzema to the south-west, Akyem, Assin, Akuapem, Kwahu and Accra to the east and south-east and even Akwamu and Anlo across the Volta.[1] The only states to the south of

1 T. E. Bowdich, *Mission from Cape Coast Castle to Ashantee*, 3rd edn., 1966, pp. 232–7; J. Dupuis, *Journal of a Resident in Ashantee*, 2nd edn., 1966, pp. xxxiii–xl; G. A. Robertson, *Notes on Africa*, 1819, p. 177.

Kumasi, the capital of this sprawling empire, still free from Asante domination by the beginning of the nineteenth century were the Fante states stretching in an elliptical shape from the mouth of the Pra river to the western borders of the Ga kingdom. As Cruickshank correctly pointed out, the Asantehene must have found this most irritating and must have 'looked with ambitious eyes at the narrow strip which interposed between him and the sea and longed for some opportunity of adding it to his territories'.[2]

The independent existence of Fante was not merely an irritant but it also constituted a real danger to Asante's economic and political interests. Despite appeals to the Dutch and English, the Fante refused the Asante direct access to European trading forts and insisted on trading with the Asante in the inland market towns such as Fosu and Manso. Had their dealings with Asante traders been honest all would probably have been well. But as Osei Bonsu told Bowdich in 1817, the Fante received pure gold from the Asante and mixed it with base metal before selling it to the Europeans, 'ten handkerchiefs are cut into eight; water is put into rum, and charcoal to powder, even for the king'.[3] Moreover, the Fante did not hesitate to plunder Asante traders or close the trade paths and the inland markets on the slightest provocation. Nor would the Fante allow Europeans to visit the Asantehene's court. As Osei Bonsu wrote to S. H. Smith, 'the old king wished to see some of them (white men) but the Fante stopped it'.[4]

The Asante never abandoned their determination to gain access to the ports and it is quite evident from the records that most of the disputes between Asante and Fante in the eighteenth and early nineteenth centuries centred on the question of trade and trading routes. The nineteenth century opened on such a conflict. In October 1800, the Governor of Cape Coast Castle reported: 'The trading paths have for many months been shut up, by a misunderstanding between the Fantees and the Ashantees. This has occasioned a great stagnation of trade at Annamboe where there is much competition.'[5] The Asante must have found their position really exasperating and it was clear that they would break through to the castles sooner or later. Moreover, the Fante did not hesitate throughout the eighteenth century whenever it suited them to offer assistance or grant protection to vassal states such as Assin, Wassa and Akyem that rebelled against Asante. The immediate occasion for the invasion and conquest of Fante in 1806 was, as we shall see presently, the refuge given by the Fante to two rebel Assin chiefs, Otibu and Aputae.

Elmina presented yet another bone of contention between Fante and Asante. Since the conquest of Denkyira in 1700 and the acquisition of the

2 B. Cruickshank, *Eighteen Years on the Gold Coast of Africa*, 2nd edn., 1966, i, p. 61.
3 Bowdich, p. 72.
4 *Ibid.*, p. 72.
5 P.R.O. T70/34, Dalzel to Committee, 13 October 1800.

Red Volta R.
Boundary of modern Ghana
MAMPRUSSI
White Volta R.
Yendi
Black Volta R.
GONJA
DAGOMBA
Salaga
Begho
Kintampo
R. Volta
Wenkyi
Nkoransa
GYAMAN
BONO
AHAFO
Mampon
Dwaben
KRAKYE
KWAHU
Kumasi
SEFWI
Bekwai
AKUAPEM
AKYEM ABUAKWA
R. Pra
PEKI
AKWAMU
ADANSI
AOWIN
AKYEM KOTOKU
KROBO
Dodowa
ANLO
Keta
DENKYIRA
ASSIN
ADANGBE
TWIFU
WASSA
FANTE
GA
Accra
NZEMA
Winneba
Apam
Mankessim
Saltpond
Anomabo
Moure
Cape Coast Castle
Elmina
Komenda
Sekondi
Dixcove
AHANTA
Apollonia
Ghana, 1800-1874
The Asante Empire by 1824
The British 'Protectorate' 1830-1874
The Fante Confederation
0 20 40 60 miles 100

6:1

'Notes' for the Elmina Castle,[6] the Asante had formed an alliance with the state of Elmina giving them direct access to the castle from where they obtained urgent supplies of guns and ammunition. Elmina divided the Fante allied states of Fetu and Komenda from Fante proper and also gave Asante a coastal trading port. Obviously, this position was intolerable to Fante since it constituted a real threat to her political and economic interests. Throughout this period, therefore, Fante endeavoured to win over Elmina and end her association with Asante. As we shall see, the Asante invaded the coast in 1811, broke off peace negotiations with Fante and the British in the 1820s and finally invaded the coast between 1869 and 1873, all with the aim of maintaining their ancient hold on Elmina.

Contrary to what such historians as Claridge and Ward have maintained,[7] the attempts of the Asante to conquer Fante dated not from the first half of the nineteenth century but rather from the third decade of the eighteenth century. In 1726, 1765 and 1777, the Asante invaded Fante and throughout the second half of the century, especially in the 1760s and 1770s, there were constant rumours and threats.[8] Indeed, by October 1767, the Council in Cape Coast thought war so probable that it resolved to rebuild the castle and to ask the committee in London, 'to make application to the Lords of the Admiralty for ships of war to be sent and remain here while the present Disturbances continue'.[9]

Effect of the European presence

Asante attempts during the eighteenth century to defeat and annex the Fante states failed for a number of reasons. During this period, the Asantehenes were fully absorbed in their constitutional reforms at home.[10] The Fante were also more united for most of that time and therefore more able to resist Asante pressure. Also, the rebel vassal states, Wassa, Assin and Akyem, which lay between Asante and Fante, had to be effectively subdued and this was not accomplished until the last three decades of the century. But most important of all, the British adopted a policy of preventing a direct conflict and, in the event of war, of assisting those 'whom it is in our interests should be the conquerors, and these are certainly the Fantes'. Giving reasons for this policy and condemning the Dutch for supporting the Asante, the then Governor, Hippisley, wrote:

The Dutch avowedly espouse the cause of the Shantees and this from a

6 D. Coombs, *The Gold Coast, Britain and the Netherlands, 1850–1874*, 1963, pp. 7–11.
7. W. W. Claridge, *A History of Gold Coast and Ashanti*, 2nd edn., 1964, v, p. 238; W. E. F. Ward, *A History of the Gold Coast*, 1948, pp. 140–2.
8 J. Ade Ajayi and I. Espie, *A Thousand Years of West African History*, 1965, pp. 182–5; Valckenier to Haes, Chama, 10 June 1726, Furley Papers.
9 T70/31, Mutter to Committee, 20 July 1765; T70/31, Mill to Committee, 27 June 1772 and 12 August 1772; T70/31, Act of Council, 25 October 1767.
10 I. G. Wilks, *Ashante Government in the Nineteenth Century*, 1964.

principle the most erroneous in African Politics that ever was adopted. They urge the insolence of the Fantees ever since Establishments were made, and the frequent outrages they are guilty of to white men, their confused Government and the consequent difficulty of appeals to them. Whereas, say they, if the whole Gold Coast was under one powerful prince, there would be only him to satisfy, and this accomplished as it easily might by presents, we should then be in a condition to bid Defiance to the rest; but ought they not to consider that we derive the supremest Advantage from the mixt government of the Fantees? At variance with one Town or District we are Friends with the rest, and no very great harm ensues. Besieged in one Fort (a most common case) we still have communication by Water to the next, every Town in the way being Independent of the other, and never willing to enter into quarrels with White Men but for their own particular concerns.... Master of the Coast, Master to the Water's edge, what communication could we have between one Fort and another if the Black Monarch put an Interdict upon it? Deprived of all necessaries from the Country, what supplies could we receive from Shipping at Landing Places where the Sea is so boisterous that no boat dares approach, fit only for canoes and those canoes in the hands of the King?... Cut off by the Sea and Land from every assistance, from all Advice and from every kind of Food, what must become of us?[11]

Pleading for the continuation of that policy in the 1770s, Mill also wrote:

I think it very clear that our trade would be totally ruined should that event take place [Asante victory over Fante], for the whole trade being in the hands of a powerful absolute Monarch, he certainly would put what price he thought proper on Slaves, and also on our commodities; this will never be the case as long as the Fantees (who are divided into so many petty Republicks, and always at war amongst themselves) exist.[12]

Fear of the British prevented Osei Kwadwo (1764–77), whom they described as 'the young enterprising king', from attacking the Fante. However, by the first decade of the nineteenth century, most of these impediments had been removed. Wilks has shown in his brilliant study that Osei Bonsu was able to continue with outstanding success the constitutional reforms initiated by Opoku Ware and continued by Osei Kwadwo and Osei Kwame (1771–1801). These reforms resulted in the conversion of most of the leading stools or offices from hereditary to appointive ones.[13] Indeed, by the 1810s Bowdich observed that there were only four direct descendants then living of the noble families 'who assisted the enterprise of Sai Tootoo, the founder of the kingdom, and all of whom were beggars'.[14]

11 T70/31, Hippisley to Committee, 13 July 1766.
12 T70/31, Mill to Committee, 22 June 1772.
13 For details of these reforms, see Wilks, *op cit.*
14 Bowdich, p. 255.

Asante's policy of centralisation

Osei Bonsu also continued reforms in provincial administration begun by Osei Kwadwo. He imposed on the system in which members of the hereditary aristocracy represented the conquered states in Kumasi, a network of Asante resident commissioners hierarchically organised at regional and district levels. In 1776, for example, Osei Kwadwo posted Boakye, Ankra and Nkansa to Dutch, English and Danish Accra respectively and all of them were placed under the Resident Commissioner in Akuapem.[15] As we shall see, as soon as he gained complete mastery over Fante, Osei Bonsu posted three commissioners there. All these reforms meant that by the first decade of the nineteenth century, a new bureaucracy had emerged in Kumasi whose membership depended not on birthright but on ability and competence. And since appointments were, to use Robertson's words 'in the king's gift' there occurred a tremendous concentration of power in the king's hands. It must be pointed out, however, that the king wielded this absolute power only in Kumasi and the provincial states but not in other Asante divisional or metropolitan states such as Dwaben and Mampong. Secondly, since retention of office depended on efficiency, and it is absolutely clear from accounts by Bowdich and Dupuis that the king did not hesitate to dismiss anybody who proved incompetent or insubordinate, the Kumasi state in particular and provincial Asante in general were exceedingly effectively administered by the first decade of the nineteenth century. Daendels, the Dutch Director-General of Elmina, wrote in 1816: 'Law and Order is just as great in the Ashantee Kingdom as with the Asiatic Eastern peoples. There thus exists no palavers between one town and another and Panyarring finds no place.'[16] Dupuis was also greatly impressed with the internal peace and security in Asante adding that its military resources were 'great indeed'. He was assured by his Muslim friends that the king could raise an army of eighty thousand men and could arm fifty thousand of them with muskets and blunderbusses.[17]

As a result, partly of the wars of the eighteenth century and partly of the creation of provincial commissioners, the impediment that was placed in the way of Asante by Wassa, Assin and Akyem had been removed and Asante's hold over these vassal states greatly strengthened.

Conquest of Fante

While the history of Asante during this period is one of centralisation, and increased stability and order, that of Fante is one of increasing disunity, anarchy, rivalry and internecine warfare. We have seen reports from Hippisley and Mill describing this political disintegration. In February 1789 there were reports of a war between Cape Coast and Komenda, between

15 Dupuis, p. xxvi.
16 H. W. Daendels, *Journal and Correspondence*, November 1815 to January 1817, p. 256.
17 Dupuis, p. xxviii.

Anomabo and the inland Fante in August 1791, between Cape Coast and Anomabo in April 1802 and between Shama and Komenda in July 1805.[18] Panyarring, that is kidnapping debtors and selling them as slaves, was common while the activities of the 'Braffoes', a group of priests, had rendered life in many Fante towns exceedingly insecure. Colonel Torrane, Governor of Cape Coast castle, wrote in 1805, 'the Braffoes . . . are a com-common nuisance, they lay with traders, take a part of the goods from every bundle they purchase . . . come to people's houses, they obtain everything

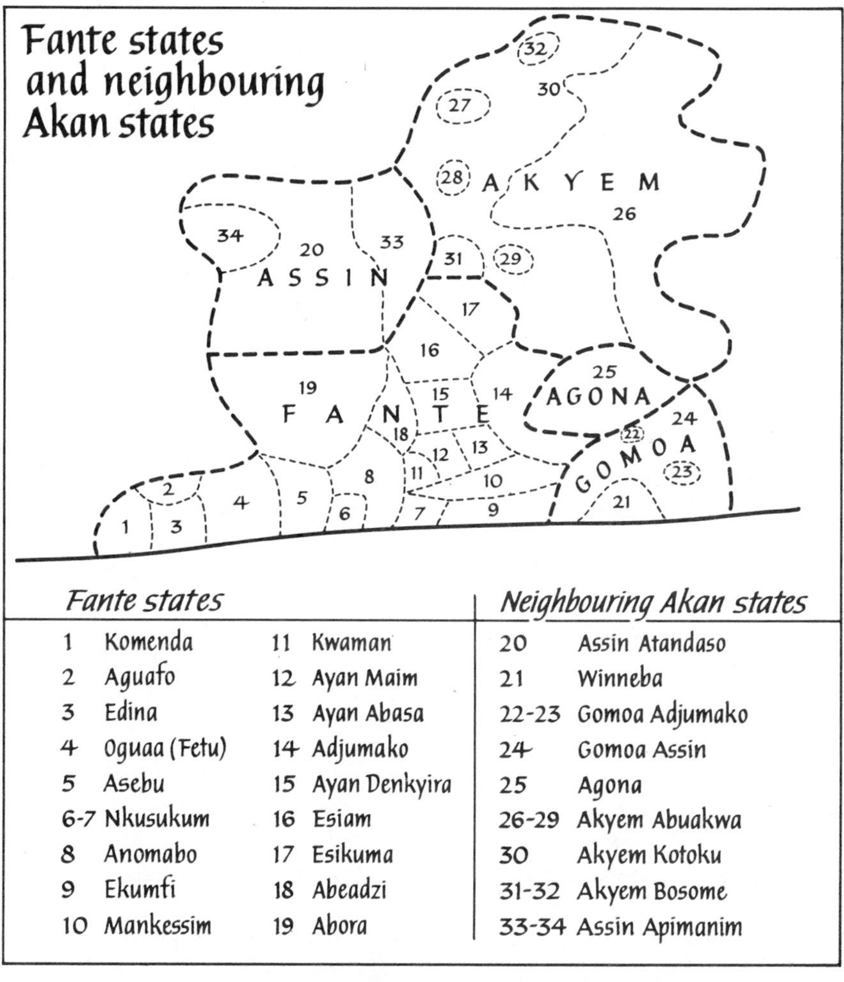

6:2

18 T70/1081, Cape Coast Day Books, Entry 21 April 1801.

they ask for gratis, and in reward of their hospitable reception, they plunder every individual . . .'.[19]

Finally, though the British had not abandoned their policy of Fante support, it is clear that their own position was so weakened by the turn of the century, due mainly to the Seven Years' War, the American War of Independence and the Napoleonic Wars,[20] that this policy could not be effectively enforced. It is significant that not until after 1821, when the British Government had assumed direct control of the affairs of the forts from the African Company of Merchants, that the British raised an army against Asante, at first with disastrous results as we shall see below.

With stability and order prevailing in Asante, the Fante disunited and troubled with internecine warfare and the British in no position to resist, it was obvious that the Asante would at last pluck this century-old Fante nettle. What finally made it all the more obvious was the personal ambition of Osei Bonsu. The two of his predecessors whom he admired most and sought to emulate, Osei Tutu and Opoku Ware, had triumphed in war, the former having conquered Denkyira and won the note for Elmina Castle, the latter having subdued Akyem and Accra and won the 'Notes' for all the castles in Accra.[21] As he constantly urged on both Bowdich and Dupuis, he wanted to be considered the greatest king in Africa just as 'the British king was the greatest in Europe'. The acquisition of the Notes for Cape Coast and Anomabo was a temptation too difficult to resist. It is significant that in all his negotiations with the British, Osei Bonsu insisted upon these Notes until he got them. As he wrote to Hope Smith,

> When the English made Apollonia fort he [Osei Tutu] fought with the Aowins, the masters of that country and killed them; then he said to the Caboceer, I have killed all your people, your book is mine; the Caboceer said, true! so long as you take my town, the book belongs to you. He went to Dankara and fought, and killed the people, then he said; give me the book you got from Elmina, so they did, and now Elmina belongs to him.[22]

On conquering the Fante, Osei Bonsu made his way to the sea and waded in it and thereby acquired the appellation of 'Bonsu' or whale because not even in the sea had he found an enemy to withstand him.[23] As we have seen, Osei Bonsu was also enraged by the obstacles being placed in the way of Asante trade with the coast by the Fante and he hated the dishonesty of Fante traders. Only an excuse was needed to touch off an invasion, and this spark was provided by the Assin episode.

This episode is well documented by at least two contemporary accounts

19 T70/158, Torrane to Committee, 11 April 1805.
20 Ward, pp. 138–9.
21 Bowdich, p. 71.
22 Osei Tutu Kwamena to Gov. Hope Smith, 28 May 1817, quoted in Coombs, p. 10.
23 Ward, p. 149.

and two others based on traditions collected within about ten and thirty years of the event.[24] Assin, an Asante tributary state, was divided into two, one part under Tsibu assisted by Kwadwo Aputae and the other under Amo. One of Tsibu's people rifled the grave in which a wealthy subject of Amo's had been buried with a considerable treasure. Amo appealed to the Asantehene for redress. Summoned to Kumasi, Aputae, who stood in for the sick Tsibu, was asked to remain behind until full restitution had been made. Aputae fled refusing to pay the fine and Amo mounted an attack on him. After further unsuccessful appeals to Tsibu and Aputae, who not only seized the Asantehene's two gold swords and his golden axe but even killed some of his messengers, the Asantehene declared war on them. Both fled to Fante where they won support mainly through heavy bribes including the gold swords and axe. The Asantehene appealed to the Fante chiefs for permission to pass through their territories to apprehend the two Assin rebel chiefs. The Fante agreed to this request, but nevertheless joined the Assin and attacked the Asante army. However, they were defeated and among the captives was Atie, King of Abora. He was to be exchanged for the swords and axe but the King of Asikuma allowed him to escape. At the same time, Aputae offered to accept the Asantehene's meditation on condition that he agreed to pay all the debts he had incurred. The Asantehene accepted this offer and sent presents of cloth and gold to Tsibu and Aputae. But the two rebels beheaded the king's messengers. The Asantehene now had his excuse and declared a full-scale war against the Assin and Fante.

The Asante army defeated all the armies sent against it and Tsibu and Aputae were chased from town to town until Asante attacked Anomabo, where the Assin had sought refuge, and besieged the castle in June 1807. Tsibu and Aputae slipped away to Cape Coast but Tsibu was captured by Governor Torrane and handed over to the Asantehene. This act left an indelible impression on Osei Bonsu of the friendly attitude of the British towards him. Having executed Tsibu, the Asantehene marched eastward and attacked and burnt Winneba in October 1807. After an outbreak of smallpox among his forces, the Asantehene broke camp and returned to Kumasi in about January 1808 after, as Torrane wrote, 'having unquestionably conquered the [Fante] country and leaving thousands of Fante captives to be sold as slaves'.[25]

24 The eye witness accounts are those by H. Meredith, who took part in the action himself, in his book, *An Account of the Gold Coast of Africa*, 1812, and the despatch by Col. Torrane, Governor of the Cape Coast Castle, in T70/35, Torrane to Committee, 20 July 1807, printed in *Great Britain and Ghana: Documents of Ghana History 1807–1957*, ed. G. E. Metcalfe, 1964, doc. 4, pp. 7–12. The other sources are Dupuis, pp. 250–64 who says he got his information from the 'King himself and which was confirmed by the Moslems who participated in the victories over the Fantees, and at Anomaboe', and Cruickshank, i, pp. 62–82. Secondary sources are Claridge, i, pp. 237–57 and Ward, pp. 140–50.

25 T70/35, Torrane to Committee, 27 October 1807; T70/35, White to Committee, 10 February 1808; T70/35, Hamilton to Committee, 27 October 1807.

It is perfectly clear that in spite of Osei Bonsu's own grievances towards the Fante, he showed extreme forbearance in his attitude towards them. Only after three appeals to the Assin, a number of missions to Fante and after his messengers had been killed on at least two occasions, did he resort to arms. Either he was being cautious or he was being true to his maxim which he impressed on Dupuis that he would 'never appeal to the sword while the path lay open for negotiation'.[26] However, since he insisted on having the Notes for the castles in Fante and retained sovereignty over the conquered states, it may not be wrong to conclude, as Cruickshank did in the 1850s, that the Assin episode was a pretext and that his ulterior objective was 'the subjection of the Fante'.[27]

The Asante withdrawal from Fanteland did not bring peace. The Fante remained defiant and in December 1809 and March 1810, in alliance with Cape Coast and Wassa, they attacked Elmina and Accra for assisting the Asante during the invasion. Akyem and Akuapem joined the Fante in revolt. However, after assuring the Dutch and English of his friendship, the Asantehene, in a series of campaigns in 1811 and between 1814 and 1816, succeeded in crushing all the revolts and consolidating his hold on the Fante who were forced to pay a heavy indemnity. The Fante remained submissive to the Asante until 1823 when they were instigated to join a revolt of the Wassa by the British.

Asante administration of Fante

It has generally been believed by historians that though the Asante 'could conquer, they could not govern: in fact, they never made any serious attempt to do so'.[28] If this is true of the Asante kings of the first half of the eighteenth century, it is not true of those of the second half, and is certainly not true of Osei Bonsu. Having withdrawn his army from Fanteland towards the end of 1816, he appointed Kwadwo Akyeampon Boakye, Addo Bradie and Kwame Butuakwa to Elmina, Cape Coast and Abora Dunkwa respectively as Resident Commissioners or, as they were referred to in the Cape Coast 'Daybooks', as 'Ambassadors'.[29] A principal Asante man was also dispatched to Anomabo as Viceroy. He was recalled shortly afterwards because the Governor of the British fort there, Swanzy, refused to allow him any authority. However, the other three remained at their posts, each with a large retinue.

26 Dupuis, p. 263 footnote.
27 Cruickshank, p. 226.
28 Claridge, i, p. 228; Ward, p. 134.
29 T70/1603 (1), Zay Tootoo Quamine to J. H. Smith, 1 October 1817. Smith to Hutchison, 2 November 1817. T70/111, Cape Coast Day Books, entries on 4 February, 20 March, 6 April, 30 December, 1817; T. C.O.267/52, Collier to Crocker, 14 April 1820; Bowdich, pp. 79, 83, 87–9, 119, 121, 390, 418; Wilks, pp. 42–3; Cruickshank, i, p. 106.

On the eve of his war against Gyaman, Osei Bonsu recalled his provincial officials and all the Asante from Fanteland, closed all the paths and asked the British Resident, Hutchison, to leave Kumasi so that he might 'unobserved prosecute his wars in the interior'.[30] Addo Bradie and his retinue left Cape Coast on 14 February 1818 and returned to Kumasi in accordance with the order.[31] However, it seems he left a deputy described by Dupuis as a 'young athletic man'. This man died suddenly in March 1819, not without a strong suspicion of poison, and no successor was immediately appointed.[32] After the Gyaman war, the king's nephew, Owusu Adom with his Captain, Amacon, were appointed to Cape Coast. Amacon entered Cape Coast on 14 January 1820 with a retinue of about twelve hundred people, of whom about five hundred were armed.[33] Adom and Amacon remained until January 1821. However, Osei Bonsu's exasperation at his failure to hear from Dupuis five months after his return from Britain caused him to recall Adom and place Cape Coast in a state of blockade.[34]

What purpose did these Resident Commissioners fulfil? They were regarded in their districts as 'the organ of his [the Asantehene's] government', and their main duties were 'to reside with the fallen chiefs, to notify them the royal will, to exercise a general superintendence over them, and especially to guard against and to spy out conspiracies that might be formed to recover their independence'.[35] They were also in charge of negotiations between the Asantehene and the British. They were to collect tribute and protect Asante traders and to see to the administration of justice, especially in disputes involving the oath of the Asantehene. For example, Kwame Butuakwa, the Commissioner in Abora, intervened on behalf of Samuel Brew, a notorious slave trader and agent of the Asantehene. Expelled by the Governor from Cape Coast in 1816, Brew lived in Amanfo, near Winneba, where he pursued his illegal slave trade in co-operation with Osei Bonsu. To ensure his safety, Butuakwa asked the people of Amanfo to swear the Asantehene's oath that they would not molest Brew. The people refused and Butuakwa threatened to destroy the town and the English fort there. The Governor made a strong complaint and though the Asantehene denied knowledge of this, he instructed his Captain in Cape Coast to plead with Hope Smith for Brew to be allowed to return, a request which was firmly turned down.[36]

30 Cruickshank, i, p. 126; T70/1603 (1), Hutchison to Smith, 23 November 1817.
31 T70/113, Cape Coast Day Books, entry 14 February 1818.
32 Dupuis, pp. xiii–xiv.
33 T70/117, Cape Coast Day Books, entry 14 January 1820; Dupuis, p. xxviii; Claridge, i, pp. 310–13.
34 T70/119, Cape Coast Day Book, entry January 1821; Dupuis, pp. 199–201; Cruickshank, i, pp. 143–4.
35 Cruickshank, i, pp. 340–1.
36 T70/1603 (1), Smith to Hutchison, 7 November 1817.

It seems that the Commissioners were also instructed to intercept any letters between the coast and Kumasi entrusted to ordinary Asante traders and not sent by special messengers. In fact, Butuakwa intercepted a letter from the Governor of Cape Coast Castle to Bowdich and is also reported to have detained in Abora a trader who had received a letter and to have had the letter returned.[37] When the British Governor complained, Osei Bonsu again denied any knowledge and promised to stop the action as well as to punish Butuakwa for exceeding his powers. It would appear that in his anxiety to win the friendship and alliance of the British, Osei Bonsu always repudiated any action of his Commissioners which gave offence.

Asante's effective control over Fanteland through the Commissioner is evident from the fact that, apart from Cape Coast and Komenda, Fante remained submissive until the 1820s. However, there is no doubt that the Asante rule was oppressive and extortionate and the Fante naturally looked for an opportunity to revolt. They saw their opportunity towards the end of 1823 when the Denkyira and Wassa revolted. On this occasion, the British not only encouraged the Fante to join the revolt but for the first time dramatically entered the conflict. By command of the Governor, Sir Charles MacCarthy, British troops marched into the field in 1824 and the fourth Asante–Fante clash turned out to be the first of the Anglo-Asante wars of the century. These were eventually to lead to the defeat and disintegration of the Asante empire by 1874.

BRITAIN AND ASANTE

British involvement was prompted partly by the conquest of Fante and partly by the abolition of the slave trade in 1807. The final conquest of Fante produced a situation the British had done everything to prevent during the eighteenth century, the domination of the entire coastline by Asante. Had they reconciled themselves to this and allied with Asante, most of the Anglo-Asante wars of the nineteenth century would have been avoided and the Asante empire would probably have endured until the last two decades of the nineteenth century. But after Torrane's death in 1808, the British, far from recognising Asante authority, challenged it, though without any legal justification. In 1817 when the Asantehene claimed the Fante as his 'slaves' by conquest, Hope Smith instructed Hutchison in Kumasi to inform the king

> and in the most decided terms that the people of Cape Coast are not his slaves, nor have they ever been acknowledged as such; neither can they nor any of the natives residing under the British protection be included in that most degrading title. I have in a former letter mentioned that any

37 Bowdich, p. 79.

interference on the part of the king in matters concerning the people residing under the protection of the Forts, cannot possibly be allowed.[38]

The British were also exasperated by the overbearing and hostile attitude of the people of Elmina towards Cape Coast. They correctly attributed this hostility to 'their connections with the Ashantees' and they were anxious to see either an effective Fante control over Elmina or the King of Asante exercising an effective check on Elmina and exerting 'his influence and prevent the Elminas from provoking a war between the two peoples'.[39]

Abolition of the slave trade

British intervention was made even more likely by events arising from the abolition of the slave trade in 1807. The abolition, necessitated not only by humanitarian considerations but also by the economic and industrial needs of Britain at the time, rendered illegal as much as nine-tenths of European trade with the coast of West Africa.[40] A huge economic vacuum was thus created and the British hoped that it might be filled by the encouragement of the cultivation of exportable commodities such as white rice, indigo, cotton, coffee and palm oil.[41] Humanitarians were not only concerned with economic considerations but wanted to atone for the slave trade by the introduction of Christianity and European education and civilisation into West Africa. Between 1790 and 1804 a number of missionary societies were formed in Britain—including the Church Missionary Society, the Wesleyan Missionary Society and the Baptist Missionary Society—to undertake the planting of Christianity and education in Africa.[42] However, it became clear that neither legitimate trade nor missionary activity could progress without peace, order and stability. Thus the local wars, which had ensured good supplies for slave traders, were now a liability and had to be halted once and for all.

African traders, particularly the Asantehene, like their European counterparts, suffered from the abolition of the slave trade. They at first found abolition incomprehensible. Like other powerful rulers, the Asantehene now had the problem of the disposal of war captives. Though he strongly denied going to war simply to acquire slaves, he pointed out that one of the readiest ways of disposing of war captives who could not be absorbed locally was to sell them. At the time of abolition he had twenty thousand captives on his hands whom, he frankly confessed, he could not feed, adding that 'unless I kill or sell them, they will grow strong and kill my people'.[43]

38 T70/1603 (1), Smith to Hutchison, 21 November 1817.
39 Bowdich, p. 12.
40 Metcalfe, *Great Britain and Ghana*, p. 3.
41 C.O. 267/24, Zachary Macaulay to Lord Castlereagh, 6 May 1807; Metcalfe, p. 5.
42 F. L. Bartels, *The Roots of Ghana Methodism*, 1965, pp. 11–12.
43 Dupuis, p. 164.

Also, he had always relied on using slaves to pay for goods needed from the British and since abolition he could only use gold dust which he would prefer to keep. He urged both Dupuis and Bowdich to revive the slave trade. 'I think', he told Dupuis, 'that the great king will do me much good, if he likes to make a proper trade for slaves as before. You must not forget that palaver.'[44] The British, of course, flatly refused and in the 1810s and 1820s stepped up their naval action for the suppression of the illegal trans-atlantic slave trade.[45]

British missions to Asante

With the British refusal to recognise Asante authority over Fanteland or even to hand over the Notes for the Fante forts, and because of disputes over the abolition of the slave trade, relationships between Britain and Asante were strained. It was decided to send a British mission to Kumasi in order to settle these outstanding differences, to ensure the regular exchange of views through a British resident in Kumasi, to bring peace to the area and allow the development of legitimate trade and missionary activities, in order to counteract the Dutch mission to Kumasi led by the mulatto W. Huydecooper, and finally to explore the hinterland of the coastal area.[46] The Asantehene was most anxious to welcome the mission. He was aware of the practical benefits that might accrue from a smooth relationship between Asante and the British: the possibility of the revival of the slave trade, expensive gifts, formal recognition of his claim to Fante and, above all, the fame and honour that would be his as equal of the British king. The mission, originally led by Frederick James, Governor of James Fort, but later taken over by Bowdich, left Cape Coast on 22 April 1816 and entered Kumasi on 19 May 1816.

Unfortunately, instead of bridging the gap between the British and Asante this mission was to widen it even further. The discussions opened amicably. The first palaver raised by the Asantehene was that of Komenda whose people had assisted the Fante and Cape Coast against Elmina. He insisted that settlement by payment of two thousand ounces of gold was one of the essential prerequisites of lasting peace between himself and 'the people of the forts'.[47] Hope Smith succeeded in persuading Komenda to agree to acknowledge their fealty to the king and to pay one hundred and twenty ounces in gold. Osei Bonsu accepted this offer and a treaty was signed.[48] Realising the king's uncompromising attitude, Bowdich also agreed to hand over the Notes for the Cape Coast Castle and Anomabo Fort to the

44 *Ibid.*, p. 171.
45 For details, see C. Lloyd, *The Navy and the Slave Trade.*
46 P. G. James, *British Policy in Relation to the Gold Coast, 1815–1850*, M.A. thesis, London, 1935, pp. 8–10.
47 Bowdich, pp. 43–9, 87–8.
48 Bowdich, pp. 101–3, for the text of the treaty see p. 113.

Asantehene, and even agreed to increase the value of the former from twelve ackies or three-quarters of an ounce a month to two ounces a month.[49] However, no agreement was reached on the question of reviving the slave trade or on lifting the ban on Sam Brew.[50] Failure here did not ruin negotiations and a treaty of perpetual peace and harmony was signed.

According to the version published by Bowdich,[51] the first two clauses of the treaty gave mutual assurances of peace and harmony between 'the subjects of the Kings of Ashantes and Dwaben' and 'the nations of Africa' under the protection of the Forts. In the third clause, the king guaranteed security to Cape Coast from hostilities threatened by the people of Elmina. In case of aggression by those under British protection, in the fourth clause Asante agreed to seek redress from the Governor in order to avert the horrors of war and not to resort to hostilities 'even against the other towns of the Fantee territory' without endeavouring to effect an amicable settlement. In the fifth clause, the King of Asante agreed to allow a British Resident to be stationed in Kumasi and in the sixth Asante and Dwaben were to countenance, promote and encourage trade with Cape Coast Castle and its dependencies. The next two clauses enjoined the Governors of the respective forts to protect the persons and property of the Asante at the coast and allowed them the right to punish people of Asante and Dwaben for secondary offences though 'any crime of magnitude' was to be referred to the king 'to be dealt with according to the laws of his country'. The final two clauses committed the kings to sending their children to be educated in Cape Coast Castle and to making diligent enquiries about a British expedition into the interior under Major John Peddie and Captain Thomas Campbell.

Had this treaty been implemented fully and in the right spirit by both parties, peace and harmony would have existed and most of the misunderstandings and wars would not have occurred. However, the treaty failed for several reasons. Some of the clauses were quite vague and therefore subject to differing interpretations as in clause two where 'the nations of Africa' were not qualified and no mention was made of British power over those living outside the Forts. In clause eight, both 'secondary offences' and 'any crime of magnitude' were without definitions and since both the British and Asante had different conceptions of these categories, deadlock was bound to ensue. Also, the copy of the treaty retained by the Asantehene differed in significant respects from that submitted by Bowdich to the British Government and subsequently published.[52] No

49 T70/40, The King of Ashantee to Gov. Hope Smith, 26 May 1817; Bowdich, pp. 68–72, 77–81. T70/1111, Cape Coast Day Books, entry 8 August 1817.
50 Bowdich, pp. 81, 105, 149, 390, 392, 413.
51 Bowdich, pp. 126–8; see also Metcalfe, *Great Britain and Ghana*, pp. 46–7, doc. 33.
52 Dupuis, pp. cxix–cxx, 135, 138–9; see also Metcalfe, *Great Britain and Ghana*, pp. 46–7, doc. 33.

mention of the King of Dwaben was made in the Asantehene's copy and his outburst on learning about this is significant. Dupuis reports:

> 'The king (of Juabin),' replied the monarch, 'Who is he? Am I not the King? Is there another King then besides me? Does the book [i.e. treaty] say that too? If so, it spreads a shameful lie in the white country.' 'Ah!', said he, with increased irritation, 'I see, I see white men can tell lies and put them in books too. Ask my Captains,' added the monarch, throwing himself ostentatiously back in the chair, 'ask all the Fantees, if they know any other King besides me. Shame, great shame!'[53]

It should be noted that the British did indeed consider the King of Dwaben to be just as powerful and encouraged him to establish direct contracts with them. The Dwabenhene sent messengers to Cape Coast in September 1817 and again in February 1818 and the British Governor had a uniform made for the king at a cost of £14 16s.[54] Again, in clause four, 'even against the other towns of the Fantee country' was omitted from the king's copy as well as the important clause 'affording the Governor the opportunity of propitiating it [amicable settlement] as far as he may with discretion'. Thus a categorical assurance stated in Bowdich's version was never in fact given by the Asantehene.

But the most important reason for the failure of this treaty was the British attitude towards implementing it. Osei Bonsu definitely attached great importance to it and was eager for it to be effective. Indeed, he requested Hutchison, left in Kumasi as Resident, to write to all the Governors of English forts to order a representative from each to gather in Cape Coast and then to come to Kumasi to swear by oath in the king's presence 'that none may plead ignorance of the treaty concluded between his Majesty and the British'.[55] The British on their part refused to honour the treaty. Though they had, by agreeing to hand over the Notes of the Fante forts to the king, acknowledged Asante's sovereignty over Fanteland, they refused to do so formally. Moreover, the Asantehene was ignored when he sent messengers on three separate occasions, in March, June and September 1819, to report insolent behaviour in Cape Coast and Komenda where false rumours had been spread of Asante defeat in the Gyaman campaign.[56] The king's nephew was then sent with a large retinue to demand a fine of sixteen hundred ounces of gold from the Governor-in-Chief and the people of Cape Coast for breaking clauses four and seven of the treaty.[57] This claim was rejected with indignation when there is no doubt that the Asantehene was justified in making it. As Cruickshank admitted, 'It is humiliating

53 Dupuis, pp. 138–9.
54 T70/1111, Cape Coast Day Books, entries 16 September, 9 October 1817; T70/1113, Cape Coast Day Books, entries 16 September, 24 October 1817, 16 February 1818.
55 T70/1603 (1), Hutchison to Smith, 23 September 1817.
56 Dupuis, pp. xii–xx, 130–1.
57 Dupuis, p. xxxi.

to be compelled to make the admission and to confess that a King of Ashanti had greater regard for his written engagements than an English Governor'.[58]

Unaware of these strained relationships, the British Government at home decided to post Joseph Dupuis, in accordance with clause five, as full Consul to Kumasi to maintain peace and cordial relations and to encourage trade and commerce. Dupuis arrived on the coast in January 1819 but, due to ill-health and a misunderstanding between himself and the Council, he did not arrive in Kumasi until a year later. He returned to Cape Coast within five weeks.

It was not the Asantehene's fault that Dupuis's mission failed to heal the rift. He had accorded Dupuis a royal welcome, clear evidence of his anxiety to win the friendship of the British, and concluded a second treaty with him on 23 March 1820.[59] The first clause recognised Dupuis as Consul and allowed him complete freedom to live in either Cape Coast or Kumasi and to travel unhindered between these towns. In the second and eleventh clauses, the Asantehene and his Captains, unaware of its implications, made an 'oath of allegiance and fidelity to the Crown of Great Britain' and in the fourth and eighth clauses, the king undertook to support, protect and encourage commerce with Britain and to keep all the trade paths clear. He renounced his claim to the fine from Cape Coast and promised to consult the Consul before resorting to arms against the people at the coast. On the British side, the persistent claim by Asante to sovereignty over Fanteland was conceded in clauses five and ten and action promised on the exorbitant prices demanded for the Notes to forts in Fanteland. In supplementary clauses, the people of Gyaman were allowed to trade directly with the British forts, missionaries were given protection to operate in Asante and schools were to be established in Abora Dunkwa (or Payntree). Both parties agreed to exclude Cape Coast from the benefits of the treaty 'as the King is resolved to eradicate from his dominions, the seeds of disobedience and insubordination'.

It seems that this treaty should have settled nearly all the major outstanding differences between the British and Asante. Although the slave trade was not revived and the implications of clauses two and five not fully understood, by having his claim to Fanteland officially recognised, his long outstanding complaints over the prices of British goods noted and, above all, his objection to the British plea to trade direct with Salaga and Yendi upheld,[60] the Asantehene felt extremely pleased with the outcome of the negotiations. So anxious was he to see harmony at last prevail that he sent a high-powered embassy including his treasurer, Owusu Adom, two royal bearers of the gold-gilted sword, two counsellors, a priest and a court

58 Cruickshank, i, pp. 140–1.
59 Dupuis, pp. 158–72, for the text of the treaty see Appendix no. III, pp. cxx–xxiii. See also Metcalfe, *Great Britain and Ghana*, pp. 58–61, doc. 43.
60 Dupuis, p. 167.

crier to accompany Dupuis to Britain with valuable presents including two tigers, 'and then my master [i.e. the King of England] will know that I have a good heart for him and his people and that only bad men tell lies about me'.[61]

Had this second treaty been ratified by the British, the Ambassador and his retinue allowed to proceed to Britain and had Dupuis or even a deputy remained in Kumasi as Consul, all might have been well. But the Governor and his Council as well as King Aggrey of Cape Coast rejected the treaty because of the recognition of Asante sovereignty over the coast and the exclusion of Cape Coast from the treaty.[62] Yet, after the British had agreed to hand over the Notes, these objections were groundless. According to Dupuis, the Council went even further and forbade the payment of any tribute by Cape Coast to the Asantehene and, what was worse and in spite of pleas from Dupuis, refused the embassy permission to leave for Britain and even for Owusu Adom to enter Cape Coast Castle. Having reminded Osei Bonsu of his oath to the British king and asked him not to proceed with hostilities against Cape Coast until he heard from him, Dupuis left for Britain on 15 April 1820.[63]

Naturally, Osei Bonsu was furious with British behaviour at the coast but he nevertheless strictly observed the terms of the treaty assuming that good sense would prevail in Britain if not in West Africa. It was not until ten months later, in February 1821, two clear months beyond the period stipulated by Dupuis, when he failed to hear from the British, that he ordered his Ambassador, Owusu Adom, to leave Cape Coast and place it under blockade.[64]

By a remarkable coincidence, the House of Commons were at this time debating an Act, passed in May 1821, to transfer the powers of the African Company of Merchants together with their forts and possessions to the British Government.[65] The British settlements were placed under the Governor of Sierra Leone, Sir Charles MacCarthy, who arrived in Cape Coast on his first visit on 28 March 1821.

MacCarthy

This change of administration afforded an excellent opportunity to settle numerous pending problems, particularly the strained relations between the British and their allies and the British and Asante. MacCarthy arrived to find Cape Coast still under blockade. As he informed Earl Bathurst, the then Colonial Secretary, 'He [the Asantehene] has strictly forbidden them to deal with our people at Cape Coast, and though they frequently take

61 *Ibid*, p. 168; C.O. 267/52, Collier to Cooker, 16 April 1820.

62 C.O. 267/52, Collier to Cooker, 16 April 1820; C.O. 267/52, Smith to Collier, 12 April 1820.

63 Dupuis, pp. 182–93.

64 *Ibid.*, pp. 197–203.

65 Metcalfe, *Great Britain and Ghana*, pp. 164–5, doc. 149; C.O. 267/9, Bathurst to MacCarthy, 19 September 1821.

this path as the shortest cut to and from Elmina, our merchants have not been able to remove the injunction against them.'[66] Dupuis assures us that he had many discussions with MacCarthy in London in which he impressed upon him 'a just conception of the impolicy and injustice of late proceedings' before he departed for the west coast and convinced him that peace with Asante would pay the best dividends.[67]

MacCarthy was not, therefore, completely unaware of the Asante problem but he seems to have abandoned his convictions soon after his arrival in Cape Coast. Indeed, in his first despatch to the Colonial Secretary he told him of the formation of the 'native troops into a Colonial corps of three companies' under his command and of the appointment of Captain Chisholm of the Second West Indian Regiment as Brigadier Major.[68] But, worse still, he failed to inform the Asantehene of the change in administration or to send the customary presents. Major Chisholm attributed the Asantehene's hostilities solely to this last breach of custom.[69] This was an oversimplification, but the sending of gifts would undoubtedly have convinced the king of the new Governor's peaceful disposition and might have led to renewed negotiations. MacCarthy's references to the Asantehene in letters and despatches were always derogatory. In his second despatch from Cape Coast dated 16 May 1822 he referred to the Asante as 'true barbarians'[70] and in a letter written from Sierra Leone in November he again referred to the Asantehene as 'this Barbarian Chief' and expressed the hope that 'matters may be arranged . . . without a war on *an extensive scale* [my italics] with the Chief of the Ashantees'.[71] On no occasion did he send any envoy or mission to Kumasi and, according to Claridge, when a Captain Laing volunteered to go to Abora or Kumasi permission was refused.[72]

Clearly, if MacCarthy had arrived in Cape Coast with intentions of peace, by the end of 1822 he was resolved on war 'on a limited scale' with Asante. He told Bathurst in a letter that the Asante situation could not be resolved by peaceful means but only by war.[73] This change of attitude was due mainly to pressure from local British officials and merchants whom he met during his tour soon after his arrival.[74] For example, in a long memorandum submitted in September 1822, the British merchants resident in Cape Coast pointed out the monopoly enjoyed by the Asante of trade with the north and advocated that they be destroyed or humbled 'before

66 C.O. 267/56, MacCarthy to Bathurst, 18 May 1822.
67 Dupuis, p. 209.
68 C.O. 267/56, MacCarthy to Bathurst, 24 April 1822.
69 C.O. 267/56, Chisholm to MacCarthy, 30 September 1822.
70 C.O. 267/56, MacCarthy to Bathurst, 16 May 1822.
71 C.O. 267/56, MacCarthy to Bathurst, 11 November 1822.
72 Claridge, i, p. 339.
73 C.O. 267/58, MacCarthy to Bathurst, 11 November 1822, 7 April 1823.
74 C.O. 267/56, Report on the Forts and Settlements on the Gold Coast, enclosed in MacCarthy to Bathurst, 16 May 1822.

our footing in this country can be considered safe and respectable' and British education and Christianity established.[75] MacCarthy was also told during his tour of 'the extortions of the Asante' in Dixcove, of the destruction of plantations in Accra and the 'greatly overrated power' of Asante in Anomabo.[76] Mends, Commodore of the west coast, advised against the dismantling of the Anomabo fort on the ground that the Asantehene would 'greedily seize' the town and 'incalculable mischief would arise' in the area with 'the probable entire extirpation of the remainder of the Fantees'.[77] Chisholm, another local official, reported Asante intentions to injure British establishments, instancing a number of cases of Asante 'imposture'. He demanded 'immediate and effective efforts to check the proceedings of the ambitious author of our disgrace and their [British-protected persons'] sufferings'.[78]

In view of this unanimous conviction of the need to humble or destroy Asante, MacCarthy's attitude can be understood. He became convinced that the 'misconduct' of the Asantehene was due to the mistaken policy and weakness of measures pursued by former British governments of the forts.[79] But if his attitude can be understood, it cannot be justified. At least he thought he could have and should have sent an envoy to Kumasi to find out for himself without swallowing the prejudiced opinion of the coastal people and snubbing the Asantehene completely. In spite of what both Claridge and Ward maintain, he is to blame for rejecting Dupuis's advice and concluding from prejudiced opinion that it was 'useless to negotiate with the Ashanti'.[80]

Only an excuse was necessary for hostilities to commence and this was provided on 16 August 1822 when a sergeant of the British Royal Colonial Light Infantry in Anomabo was arrested for abusing the Asantehene, sent to Abora Dunkwa and held there until 1 February 1823 when he was executed.[81] This was seized upon as evidence of the king's brutality and anti-British attitude and immediate action was called for. The incident was described as 'a crisis that was fast approaching' and as 'fortunate than otherwise as it affords an opportunity of destroying or humbling the Asantehene'.[82] Yet the sergeant had committed a grave offence, punishable by death according to Asante law and custom, and British officials were told

75 C.O. 267/56, Memorial of the Cape Coast Merchants, 30 September 1822, enclosed in MacCarthy to Bathurst, 11 November 1822. This memorial was signed by thirteen merchants.

76 C.O. 267/56, Report on the Forts.

77 C.O. 267/56, Mends to MacCarthy, 4 April 1822.

78 C.O. 267/56, Chisholm to MacCarthy, 30 September 1822.

79 C.O. 267/56, MacCarthy to Bathurst, 11 November 1822.

80 Ward, p. 168; Claridge, i, p. 337.

81 C.O. 267/56, Chisholm to MacCarthy, 30 September 1822; C.O. 267/58, MacCarthy to Bathurst, 10 February 1823.

82 C.O. 267/56, Memorial of the Cape Coast Merchants; C.O. 267/56, Chisholm to MacCarthy, 30 September 1822.

that the Asantehene regarded the sergeant as his subject 'having ascertained that he was born in the Fantee country'.[83] It should be noted that the sergeant was not taken to Kumasi but to Abora Dunkwa only eighteen miles from Cape Coast and that he was not executed until four months later. One cannot help but conclude that the Asantehene hoped for a British envoy to be sent to negotiate with him. In fact, Chisholm, who sent some Abora people to plead with the Asante official in Abora only, wrote to inform MacCarthy that the official 'hinted that he thought it possible that I would wish to send a person to intercede with the King'.[84] No such envoy was sent and it must have been out of sheer frustration that the execution was finally ordered.

MacCarthy was in Sierra Leone when he received the news of the sergeant's arrest. He cancelled his intended visit to the Gambia and returned to Cape Coast bent, as we have seen, on dealing with the Asantehene.[85] He left Freetown with as many troops as he could safely take from the garrison there, one captain, two sergeants and thirty other ranks of the Second West Indian Regiment.[86] He landed in Cape Coast on 9 December 1822, yet he sent no envoy to Abora or Kumasi but sat in Cape Coast entertaining the hope that the 'chief of the Ashantees would deliver to him the sergeant whom he had so unjustly detained' and the grounds for this hope were 'reports emanating from principal merchants . . . that messengers were actually on the path for that purpose'.[87] MacCarthy's immediate reaction to news of the execution was to send a contingent to Dunkwa to arrest the Asante chiefs who had executed the sergeant to 'give an example as would prevent a similar occurrence'. Though this expedition failed to effect the arrests and indeed a contemporary observer described it as a reverse,[88] MacCarthy deemed it a success and reported that it had 'the happiest effects since it dispelled the terror of the Fante and other native tribes, who had, for several years, been held under the most abject state of oppression by the Ashantees'.[89]

From then on, until his departure for Sierra Leone in May 1823, MacCarthy and his officials used persuasion and force to build up anti-Asante alliances with all the tributary states to the south, organising militias whom they supplied with arms, strengthening the fort at Cape Coast and asking for a man-of-war.[90] So successful was the anti-Asante campaign that by the end of July virtually all the southern tributary states were in open rebellion. Indeed, according to Claridge, shortly after MacCarthy's visit to Accra the

83 C.O. 267/56, Chisholm to MacCarthy, 30 September 1822.
84 *Ibid.*
85 C.O. 267/56, MacCarthy to Bathurst, 11 November 1822.
86 *Ibid.*
87 C.O. 267/58, MacCarthy to Bathurst, 10 February 1823.
88 J. Ricketts, *A Narrative of the Ashantee War*, 1831.
89 C.O. 267/58, MacCarthy to Bathurst, 7 April 1823.
90 C.O. 267/58, MacCarthy to Bathurst, 7 April, 8 April, 26 April 1823.

militia he had formed there attacked two parties of Asante traders, killing fifty-four of them and taking twenty-four prisoners together with a booty valued at about £500.[91] His overtures to the Dutch Governor having failed,[92] the Asantehene had no choice but to move to crush these rebellions and in June an advance party of his army crossed the Pra followed at the end of July by a second contingent. Both were withdrawn on the approach of troops led by Captain Laing.

Hearing of these moves, MacCarthy returned to Cape Coast in November with a third company of the Royal African Colonial Corps and visited the allies at Yankumasi where he promised not to make peace with the Asante without first consulting them.[93] Shortly afterwards news arrived that the main Asante army had crossed the Pra. This had been precipitated by the escape of Tsibu, King of Denkyira, from Kumasi and the army's objective was to recapture Tsibu and to establish contact with their ancient ally, Elmina, not to fight the British.[94] But MacCarthy now abandoned any hope 'that the barbarian chief would be anxious to send messengers for peace' and decided to put an end to a 'war which is attended with considerable trouble and fatigue and has put a stop to all my endeavours of improvement'.[95] He ordered the troops in Accra and Anomabo to move forward. He himself marched with no more than four hundred and forty untrained men first to Abora Dunkwa and then across the Pra to Nsamanko where he found the Denkyira and Wassa retreating before the Asante army. On 21 January 1824, about twenty-eight miles from Nsamanko, a direct clash occurred between the Asante army of ten thousand and MacCarthy's small contingent. After a brief encounter, MacCarthy's tiny army was cut to pieces and among the slain was Sir Charles MacCarthy himself.[96] The victorious Asante army crossed the Pra and entered Komenda in April 1824. By the end of May, it had moved into Fetu, only five or six miles from Cape Coast, sixteen thousand strong. The officer administering the country admitted that the Asante army was 'in undisputed possession of the country'.[97]

Many historians, especially British ones, have persistently maintained that MacCarthy is to be pitied rather than blamed.[98] Claridge explains

91 Claridge, i, pp. 340–1.
92 C.O. 267/58, Poolman to MacCarthy, 2 May 1823; MacCarthy to Poolman, 6 May 1823; Poolman to MacCarthy, 11 May 1823; MacCarthy to Poolman, 12 May 1823.
93 Ricketts, pp. 40–1.
94 C. C. Reindorf, *History of the Gold Coast and Asante*, 1895, pp. 181–3; Claridge, i, pp. 364–5.
95 C.O. 267/58, MacCarthy to Bathurst, 12 December 1823; Adm 1/1815, MacCarthy to Filmore, 17 January 1824.
96 C.O. 267/61, Chisholm to Horton, 23 February 1824.
97 C.O. 267/61, Sutherland to Bathurst, 31 May 1824; C.O. 267/61, Sutherland to Hamilton, 1 June 1824.
98 Claridge, i, p. 355; Ward, p. 168; G. E. Metcalfe, *Maclean of the Gold Coast*, 1962, pp. 39–42.

that 'as a stranger to the country, he had no idea of the strength and fighting qualities of his enemy nor of the unreliable character of some of his allies with many of whom the mere approach of an Ashanti army was sufficient to cause panic and flight'. Yet MacCarthy could and should have discovered this, either through secret missions to Asante or through an attempt at negotiations. But he imbibed the contempt of the local officials and traders and therefore refused to seek first-hand information on Asante. On the contrary, he sat in Cape Coast expecting the 'ruffian' and 'barbarian chief' to come before him in supplication so that he could impose his own idea of 'honourable' terms or crush him in a war 'on a limited scale'.[99] The small contingent he took to confront the Asante forces is evidence of his contempt for the enemy and not, as historians like Claridge, Ward and Metcalfe maintain, of his foolhardiness. In a despatch to Bathurst of 8 April 1823, he stated that he 'did not entertain a very exalted idea of the skill or bravery of the Ashantees' and later wrote that the size of the Asante army was 'exaggerated by the dread they have so long caused the natives around'.[100] Sutherland, who arrived in Cape Coast on 18 May 1824 to assume command from MacCarthy, described the war as 'ill-fated, ill-advised and unprofitable'.[101] No one can deny this assessment.

ASANTE RULE IN THE NORTH

The Asantehene could now consolidate his hold over the southern tributary states but the states to the north were equally refractory during the first three decades of the century. However, as with the south, the Asantehene was successful in suppressing their revolts. The century had opened with Gyaman or Abron in open rebellion and Dupuis, Beecham and, more recently, Wilks are of the opinion that it aimed at restoring Osei Kwame who was deposed in about 1801 allegedly because of his pro-Islam sympathies. Dupuis categorically states that this was because the Asante elders feared the Muslim idea of levelling all ranks and 'placing them at the arbitrary discretion of the sovereign'.[102] However, it is now generally agreed that the rebellion 'was one of a series of wars of independence that occurred at the beginning of each reign, to keep the Ashanti off the north-west territories where the trade system was dominated by Kong and Gyaman'.[103] A year after Osei Bonsu's accession, Gonja also revolted but the rebellion was crushed. Dupuis wrote: 'this defeat led to the occupation

99 Adm. 1/1815, MacCarthy to Filmore, 17 January 1824; C.O. 267/58, MacCarthy to Bathurst, 12 December 1823.

100 C.O. 267/58, MacCarthy to Bathurst, 8 April 1823; C.O. 267/58, MacCarthy to Bathurst, 27 August 1823.

101 C.O. 267/61, Sutherland to Bathurst, 28 May 1824.

102 Dupuis, p. 245; I. Wilks, *Northern Factor*, p. 22; J. Beecham, *Ashantee and the Gold Coast*, 1841, p. 23.

103 J. R. Goody, *Ashanti and the North-West*, 1965, p. 30; N. Levtzion, *The Spread and Development of Islam in the Middle Volta Basin in the Pre-Colonial Period*, Ph.D. thesis, London, 1965, p. 45.

of a considerable territory bordering the desert which, heretofore, belonged by a sort of feudal, yet limited title, to the Ashantee monarchy'.[104]

It appears that Gyaman (or Abron) took advantage of Asante's preoccupation with Gonja to transfer her allegiance from the Asantehene to 'the nephew of their late monarch' but this partial revolt was also suppressed and Asante power in the north-west was not challenged again until 1817 when Abron once more rebelled. After elaborate preparations,[105] the Asante inflicted a decisive defeat on the Abron and their king, Adinkra, was killed. This victory, says Dupuis, led 'to the annexation of that kingdom to Ashantee as a province in lieu of the tributary rank it enjoyed before'.[106] Asante's power was not challenged again in the north-west until towards the end of Kwadu Dua's reign (1834–67) when Nsawkaw revolted in about 1864.[107] Throughout the period, and especially after the Asante defeat in the south, they tightened their grip on the northern territories, stationing Resident Commissioners there and corresponding with the official Imams of Gonja, Mamprusi and Dagomba to whom they regularly sent gifts.[108] Thus peace and order was maintained in those areas.

Two scholars, Levtzion and Braimah, have recently discussed Asante rule in northern Ghana and both of them have pointed out that the disastrous civil war in Gonja in 1892, between two segments of the royal family, would not have occurred if the Asante hegemony in Gonja had not been overthrown in 1874. Braimah, himself a member of the Gonja ruling dynasty, wrote:

> The Kanyase people could not fight for their rights in the early stages because of the Ashanti, who had resident ambassadors posted to supervise the collection and payment of the annual tribute in slaves *(ayibuadie)*; a civil war would have involved the Ashanti, who would have taken sides with those in authority. Although plans for war were secretly made, they had to be passed on from generation to generation until the Ashanti, during the reign of King Karikari, were forced to withdraw because of their defeat by the British.[109]

Levtzion, citing an amicable settlement in 1817 at Kumasi of a dynastic dispute in Gonja, also concluded that it is likely that such a civil war [Gonja 1892] could not have developed a quarter of a century earlier under Asante rule, and even if it had, interference from Nanumba and Dagomba was unlikely. He goes on to say that Asante rule eased, or at least held in

104 Dupuis, pp. 248–9.
105 A detailed description of these preparations in which both the Muslims and the traditional fetish priests took part is given in Hutchison's diary in Bowdich, pp. 381–421.
106 Dupuis, p. 264.
107 Goody, p. 36.
108 N. Levtzion, 'Early 19th Century Arabic Manuscripts', *Transactions of the Historical Society of Ghana (T.H.S.G.)*, viii, 1966.
109 J. A. Braimah, *The Two Isanwurfos*, 1967, p. ix.

check, the tension between Gonja and Dagomba.[110] In Dagomba, too, serious disputes over succession are recorded only in the second half of the century. In Mamprusi, however, which never came under effective Asante rule, succession disputes often led to civil wars.[111] Asante rule in the north was certainly oppressive and led to the enslavement of many northerners but it did ensure peace and order in the area.

This steady expansion and consolidation of Asante political influence in the north had two interesting effects, ignored by historians until recently. One was the steady penetration to the south of Muslims, Islamic faith and aspects of Islamic culture, the other the phenomenal increase of trade between Asante and the savannah regions, particularly to the north-east.

Muslim influences on Asante

Islam had penetrated the trade routes from Jenne and Timbuctu in Mali under the agency of the Dyula or Mande traders in northern Ghana from the fourteenth century when Mali was at the peak of her power under the rule of her two renowned rulers, Mansa Kankan Musa and Mansa Sulayman.[112] Until the eighteenth century and the beginning of Asante expansion to the north, Islam was essentially the religion of these traders. However, during the first half of the century, it became the official religion of Dagomba and Mamprusi and in the second half of the century the first Imam of Mossi was appointed and the first mosque built at Wagadugu, both acts marking the conversion of the Mossi King to Islam.[113] Islam thus took firm root in northern Ghana and the Muslim communities became firmly integrated in the social and political systems of Gonja, Dagomba, Mamprusi and Sansanne-Mango and to some extent in Mossi, Gyaman and Banda. Numerous Koranic schools were established in these areas. The Imams wielded considerable influence in politics as well as religion and Islamic culture was steadily absorbed, particularly in Dagomba and Gonja.[114]

With Asante expansion northwards, Muslim traders began to visit and live in Kumasi where they soon began to influence the court. It seems that this influence became particularly strong during the reign of Osei Kwame (1777–1801), who is said to have been a believer at heart. Dupuis thinks he was overthrown because of his attachment to Islam.

Whether Osei Kwame was overthrown for religious or more mundane reasons as Bowdich and, more recently, Levtzion have maintained, it is absolutely clear from accounts of all those who visited Kumasi during the first half of the nineteenth century, Huydecooper, Bowdich, Dupuis,

110 Levtzion, *The Spread and Development of Islam*, pp. 103–6.
111 J. S. Rattray, *Tribes of the Ashantee Hinterland*, 1938, p. 548.
112 J. S. Trimingham, *A History of Islam in West Africa*, 1962, pp. 60–83; Levtzion, *The Spread and Development of Islam*, pp. 10–12, 29–40.
113 Levtzion, *The Spread and Development of Islam*, pp. 118–42, 174–85, 190, 207–20.
114 I. G. Wilks, 'Note on the Spread of Islam in Dagomba', *T.H.S.G.*, viii, 1966.

Hutton and Freeman, as well as from Arabic manuscripts recently discovered in the archives of Copenhagen and which consist of correspondence between the Asantehene and the resident Muslims on the one hand, and the Muslims in the northern states on the other; that Muslim influence in the court grew rather than declined after Osei Kwame's fall and that the Muslims played a particularly important role during the reign of Osei Bonsu. By the time of the visits of Bowdich and Dupuis, there were about a thousand Muslims in Kumasi, headed by the son of the Imam of Mamprusi, the learned Muhammad al-Ghamba, referred to as Baba in European records,[115] together with other influential leaders. Among the community were visitors and merchants from Tripoli, Tunisia, Egypt and even from Hijaz as well as regular traders from Jenne, Timbuctu, Katsina and Kano.

The Muslims not only monopolised the cattle industry but, with deliberate encouragement from the Asantehene, controlled the trade of primary products of Asante, gold, kola, slaves and salt along the trade routes to the north.[116] Muslims were always found in the company of the king at all important negotiations with foreign visitors and ambassadors and played an active part. Bowdich referred to 'the Moorish chiefs and dignitaries by whom the King is surrounded, whose influence is powerful not only from their rank but their repute'.[117] Dupuis also reported that 'the character of true believers . . . stood very high with the King, for he consulted them upon many important occasions, where the interests of the nation were concerned' and he mentioned as among those surrounding the king 'Mohammad al Ghamba or Baba . . .' and many other Muslim names, pointing out that 'many of these people enjoyed rank at court or were invested with administrative powers, entitling them even to a voice in the senate'.[118] Baba told Dupuis that he not only ruled over the Muslims but was also 'a member of the King's Council on affairs relating to the believers of Sarem and Dagomba', a claim confirmed by a study of the Copenhagen Arabic manuscripts.[119] As Wilks has shown, Muslims served as secretary to the Gyaasehene, head of the Kumasi bureaucracy, recording political events and noting casualties in war, while others served Asante Resident Commissioners in the provinces. As the Copenhagen manuscripts show, they were also responsible for relations between Asante and the northern Muslim states.

Finally, Muslims played an active part in the Asante campaigns and the head of the Kumasi Muslims led the Muslim contingent. As Dupuis pointed out, the king 'never engaged in any warlike enterprise without their society'; during the invasion of the Fante in 1807 the Asantehene was, as

115 I. G. Wilks, 'Position of Muslims in Metropolitan Ashanti in the early 19th Century', *Islam in Tropical Africa*, ed. I. M. Lewis, 1966.
116 Wilks, 'Position of Muslims'.
117 Bowdich, p. 53.
118 Dupuis, pp. 94–5, 97.
119 Wilks, 'Position of Muslims'; Levtzion, 'Early 19th Century Arabic Manuscripts'.

Torrane noted, 'attended with many moors' and their leader was graphically described in Meredith's eye-witness report as 'tall, athletic and rather corpulent . . . of a complexion resembling an Arab or an Egyprian . . . a native of Kassina, a country . . . to the South of East from Timbuctou. He said he had been at Tunis and at Mecca . . . and described the method of travelling over the great desert'.[120] During the invasion of Gyaman, Baba, then head of the Muslims, commanded seven thousand Muslims drawn, as Wilks suggests, from Kumasi and other large towns in central Asante.[121]

In addition to wielding political and economic influence, the Muslims were permitted to practise and even proselytise their faith. Though they won few Asante converts there is no doubt that their religion had an impact on the society, particularly in Kumasi court circles, and Osei Bonsu seems to have become a virtual convert though he never formally abandoned his traditional state cults. He told Dupuis that 'the book [Koran] is strong and I like it because it is the book of the great God; it does good for me, and therefore I love all the people that read it', and Dupuis added that 'the Moslems instantly prostrated themselves, and prayed aloud. The King too extended his arms looking upwards as if to receive a blessing'.[122] Osei Bonsu always supplicated the Muslims for their prayers, especially for success in war. He never killed Muslim war captives and regarded the God of the Muslims as 'the God of all gods and men'.[123]

Particularly widespread in Asante among all ranks of society was the belief in magic as a result of Muslim activities. As Bowdich and Dupuis observed, 'The most surprising superstition of the Ashantees, is their confidence in the fetishes or saphies they purchase so extravagantly from the Moors'.[124] According to Wilks, a six-line amulet might cost almost half an ounce of gold (about £2) and for a complete amulet-covered war coat, between thirty and a hundred and twenty slaves could be paid. This certainly became the main production of Kumasi Muslims as is evident from the fact that over 90 per cent of the Copenhagen manuscripts are 'magical formulae or prescriptions for preparing amulets'.[125] It was a very lucrative occupation; Bowdich observed, 'a sheet of paper would support an inferior Moor in Coomassie for a month'.[126] Belief in these amulets is still common in Asante and among Ghanaians in general. Islamic education was also introduced into Kumasi. At the time of Bowdich's and Dupuis's visits Baba had established an Islamic school with seventy students, some of whom 'were Ashantees given to him by the King'.[127]

120 H. Meredith, *An Account of the Gold Coast of Africa*, 1812, pp. 157–8.
121 Wilks, 'Position of Muslims', n. 73; Dupuis, p. xxxviii.
122 Dupuis, p. 161.
123 *Ibid.*, pp. 97, 98, 99, 243.
124 Bowdich, p. 271; Dupuis, p. xi.
125 Levtzion, 'Early 19th Century Arabic Manuscripts'.
126 Bowdich, p. 272.
127 Dupuis, pp. 97, 107–8.

Wilks points out that at the turn of the century Asante seemed to be considering the official adoption of Islam[128] but, contrary to this happening, Islamic influence declined in the nineteenth century. Wilks points to two main factors; the defection of Baba's Muslim contingent from the campaigns of 1818 and the 'general decline that overtook the Ashanti nation as a result of the abolition of the maritime slave trade . . . that left the Ashanti shorn of one of its main supports'. However, Wilks seems to have grossly exaggerated the nature of Islamic influence in Asante for there were hardly any converts among members of the Asantehene's court or even among the ordinary people. The religion was that of immigrant northerners, as it is today, and the syncretic tendencies of Osei Bonsu did not spread to many of his advisers or subjects. There was thus little chance of Islam becoming the official cult of Asante. Wilks is also wrong to think that Islamic influence declined because of Baba's defection. T. B. Freeman, who visited Kumasi in 1839, observed, as Bowdich and Dupuis did twenty years earlier, that 'the King is surrounded by moors who poison his mind against missionaries'.[129] Many instances are cited of Muslims being in the presence of the king and of the extensive use of Muslim amulets. Moreover, the descendants of Kramo Tia, head of the Kumasi Muslims during the reign of Kwaku Dua I (1834–67), still live in Kumasi, and his grandson, until his death in 1964, was Imam to the late Asantehene.[130] It seems, therefore, that Islamic influence did not decline. This influence, interesting as it was, need not be exaggerated. It remained throughout the nineteenth century as it was at the beginning—marginal: Islamic elements were not incorporated into the Asante society and conversions remained very limited if not negligible.

Trade and Asante expansion

The second main effect of the Asante expansion northwards was the strengthening of the traditional trade links with the north and especially with the Hausa states to the north-east. This trade, mainly in gold and kola nuts, dates from the fourteenth and fifteenth centuries as is shown by the *Kano Chronicle*.[131] Until the collapse of the Songhai Empire at the end of the sixteenth century, trade with the north-west had been more important, centring on Jenne, Begho and Timbuctu.[132] With the Kong and Gyaman determination to control trade with the Mande in the north-

128 Wilks, *Northern Factor*, pp. 26–7.
129 T. B. Freeman, *Journal of Various Visits to the Kingdoms of Ashanti, Aku and Dahomey In Western Africa*, 1844, p. 48.
130 Levtzion, *The Spread and Development of Islam*, pp. 344–5.
131 H. R. Palmer, 'The Kano Chronicle', *Journal of the Anthropological Institute*, xxxviii, 1908.
132 A. A. Boahen, *Britain, the Sahara and the Western Sudan, 1788–1861*, London, 1964, pp. 106–8, 118; Levtzion, *The Spread and Development of Islam*, pp. 45–6, 56.

west during the seventeenth and eighteenth centuries, Asante turned more and more towards the north-east. By the second decade of the nineteenth century, as Bowdich observed, the Kumasi–Jenne route was 'much less frequented by Moors than that from Dagwumba through Haussa'.[133] The importance of this trade was greatly increased by the abolition of the transatlantic slave trade and the strained relations between Asante and the British and the coastal peoples. Also, there was political stability in Hausaland after the jihād of Usuman dan Fodio. With increased attacks on Asante traders travelling southwards, trade naturally turned northwards. Moreover, with abolition of slave trading, the Asante had to trade with the coast in gold which they preferred to keep. In the north, they could still barter with kola nuts for the same European manufactured goods, except guns and gunpowder. These goods had begun to reach Salaga and Kumasi in increasing quantities from North Africa, Fezzan and Hausaland and, later in the nineteenth century, through Badagri and other ports in Dahomey and Nigeria by way of Nupe and Bornu. Imported commodities included white cotton cloth and shea butter from Dagomba and Gonja, woven and unwoven silk and carpets from Turkey, Tripoli and Fezzan, natron from Bornu and cloth from Hausaland. Asante's principal export continued to be kola nuts. Bowdich, who investigated the trade between Asante and the north, reported:

> The preference of the Ashantees for the Dagwumba and Inta [i.e. Gonja] markets, for silk and cloth results not merely from their having been so long accustomed to them, but because they admit of a barter trade. Boosee or Gooroo nut, salt (which is easily procured and affords an extravagant profit), and small quantities of the European commodities, rum, and iron, yield them those articles of comfort and luxury, which they can only purchase with gold and ivory from the settlements on the coast. Gold, they are all desirous of hoarding.[134]

Both trading routes from the north-east converged on Salaga and ultimately on Kumasi.

Trade with the north was rigidly controlled by the Asante authorities. As an Asante *Batahene* (chief's trader) told Rattray,

> The *Omanhene* sent his heralds with us, carrying an *afona* (state sword) as an insignia of office to show we had been sent by the Chief. As soon as we passed, these heralds closed the path until we had disposed of our kola, i.e. about twenty days. . . . It was a law in olden times that all kola must first be exchanged for cowries (sidie) [with which] you bought what you had been commissioned to purchase. . . . The heralds . . . exacted a toll of twenty-five nuts on each load (of 1,500–2,000 nuts) from all other

133 Bowdich, p. 181.
134 Bowdich, pp. 334–5.

> traders. . . . There was not any tax on imports from the north. . . . Northern Territory men were not permitted to trade south of Salaga.[135]

Though the *Qissat Salagha* states that Hausa traders were not allowed to cross the river Yeji (Volta),[136] the resident Muslim population in Kumasi seems to show this rule was not rigidly enforced.

The growth of Salaga best illustrates increased trade with the north-east. Until the very end of the eighteenth century Europeans had not heard of Salaga since Buipe, Mpaha and Kafaba had been the main trading centres of the Volta basin. Salaga is not shown on any of the European maps made in the eighteenth century of this area. Thus, though oral tradition and the *Qissat Salagha*, written in the 1890s,[137] tell us that the town was of some importance before the end of the seventeenth century, its true growth dates from the late eighteenth century, a period when Gonja was under Asante domination and when, as Levtzion suggested, Salaga was a more attractive market for the Asante than the remote Kafaba.[138] By the end of the visit of Bowdich and Dupuis, Salaga had certainly become a large town. Dupuis described it as 'twice the size of Coomassy and its population of whom nearly the sixth part of Moslems to be about four hundred thousand souls'.[139] Salaga continued to prosper until 1874 when, with the overthrow of Asante and the consequent diversion of trade to Kete-Krakye and Kintampo, the town began to decline.

DECLINE OF ASANTE

By January 1824, with the defeat of the British and their allies of the southern states, with Asante commissioners firmly entrenched in the provinces, Asante dominance of the northern states, thriving trade with Hausaland and an Asante army estimated by Dupuis at eighty thousand men of whom forty to fifty thousand could be armed with 'muskets and blunder busses',[140] together with a king ruling 'with unrivalled sway' and in 'undisputed possession' of his claimed territories, it seemed that the Asante Empire was in for a long run. However, this was not to be.

Osei Bonsu

At the moment of Asante's highest peak of fame and power, the defeat and death of MacCarthy, indeed, according to many authorities, on the same day that MacCarthy was killed, Osei Bonsu died. Architect of Asante's prosperity in the nineteenth century, he was undoubtedly one of the greatest of the Asantehene. Yet his reign was one of many paradoxes. Pacific in

135 R. S. Rattray, *Ashanti Law and Constitution*, 1929, pp. 110–11.
136 *Qissat Salagha*, G.N.Q., no. 3, 1961.
137 *Ghana Notes and Queries,* part 1, no 3; part 2, no 4, 1961–1962.
138 Levtzion, 'Early 19th Century Arabic Manuscripts'.
139 Dupuis, p. xi. This figure is obviously an exaggerated one.
140 Dupuis, p. xxxvii.

intention and a firm believer in the maxim 'never to appeal to the sword while a path lay open for negotiation', his entire reign was a series of wars. Anxious for the friendship of the British, he died hated by them. Honourable in his dealings and regarding his word as sacred, he was mistrusted by all those with whom he entered into treaties. And yet he brought his empire to its widest possible extent as well as preserving its integrity.

Had the British supported him and not, as Austin Freeman rather ungraciously put it, 'consistently endeavoured, after the fashion too common among philanthropists, to secure in the most perfect manner the survival of the unfittest',[141] there might never have been the series of Anglo-Asante conflicts. But, as Fuller stated in his endearing tribute to Osei Bonsu, 'Honourable in all his dealings with white men, he was much misunderstood by them, especially by the English, who unwittingly, did all they could to turn a would-be friend into a bitter enemy.'[142] Judged in his relation to his work for Asante, Osei Bonsu deserves the title 'the Great', but in relation to his dealings with the British 'the Misunderstood'.

The period from 1824 to 1874 saw a complete reversal of fortunes; Asante's history is one of decline, disintegration and defeat, Britain's one of steady expansion and growth of power. Indeed, by 1874 Asante was little more than metropolitan Asante while the British had emerged as the only European power on the coast having taken over the Dutch and Danish possessions and, in spite of opposition, converted the former states south of the river Pra into a Protectorate and their settlements into the British Crown Colony of the Gold Coast.

Asante's decline began with the defeat of the hitherto invincible Asante army near Cape Coast in May and July of 1824 followed by a dispirited retreat into Kumasi a month later.[143] Before it had recuperated, the army was marched out again early in 1826 and was able to overrun the Fante states without too much resistance. However, in July it moved across to Accra to punish the Ga for deserting the Asante to join the anti-Asante coalition and, again, suffered a heavy defeat at the famous battle near Dodowa on 7 August 1826 where an estimated seventy commanders were killed, the Asantehene wounded, hundreds of captives taken including members of the Asante royal family and even the Golden Stool captured, though this was retrieved by the Dwaben.[144] The Asante defeat was due partly to the spirited fight put up by the allied forces, partly to the openness of the battlefield which prevented the Asante from adopting their usual effective method of bush-fighting but, above all, to the use of the Congreve

141 Austin Freeman, *Travels and Life in Ashanti and Jaman*, 1898, p. 471.

142 F. Fuller, *A Vanished Dynasty*, 1920, p. 70.

143 C.O. 267/61, Sutherland to Bathurst, 30 June 1824; C.O. 267/61, Grant to Bathurst, 22 July 1824; C.O. 267/61, Grant to Bathurst, 31 July 1824; C.O. 267/61, Grant to Bathurst, 8 December 1824 and encl.; Claridge, i, pp, 373–80.

144 Fuller, p. 76; Metcalfe, *Great Britain and Ghana*, pp. 103–4, doc. 75; Reindorf, pp. 276–84.

Rockets at a critical moment when, according to Claridge, 'the issue of the battle trembled in the balance and there seemed every likelihood that the Ashantis would after all gain the victory'. Cruickshank, too, claims the introduction of these rockets as a main factor in the Asante defeat.[145] The British now refused to pay for the Notes for the forts on the coast and, in the Maclean treaty of 1831, the King of Asante 'renounced all right or title of any tribute or homage from the kings of Denkera, Assin, and others formerly his subjects'.[146] From that time onwards, Asante was confined to the area north of the Pra. Thus the defeats of 1824 and 1826 and the Maclean treaty of 1831 marked the first stage of the decline and disintegration of the empire.

Dwaben

The second stage was the civil war that broke out in 1832 between Kumasi and Dwaben. The full causes of this war need not concern us;[147] it seems to have been the product of two main events, the tremendous increase in wealth and power of Dwaben during the first two decades of the century and the personal enmity of the Asantehene, Osei Yaw Akoto, towards the Dwabenhene, Kofi Boaten. Bowdich, though not Dupuis, reports that Dwaben was at this time as powerful and wealthy as Kumasi. It is obvious that a decision would have to be made as to who was in real command of Asante. Osei Bonsu had certainly been aware of this impending problem as is evidenced by his outburst at finding the Dwabenhene a party to the Bowdich treaty and, according to Bowdich, he contemplated 'the reduction of the king of Dwaben from an independent ally to a tributary'.[148] The clash between these two proud rulers was precipitated by the two main events of the late 1820s; the ill-fated Dodowa war and the Nsuta enstoolment episode.[149] As we have seen, the Asante defeat at Dodowa included the capture of the Golden Stool. The Dwabenhene swore to retrieve it and succeeded in doing so but the result was only to increase his own prestige whilst intensifying Osei Yaw Akoto's envy. Tradition reports that the Asantehene, far from praising Kofi Boaten, accused him of having stolen the treasure kept with the Golden Stool when he retrieved it,[150] an accusation calculated to excite the whole Asante nation against the Dwabenhene.

The Nsutahene had been one of the victims of the Dodowa war. The Asantehene not only put his candidate on the throne without consulting Dwaben as he was constitutionally bound to do, but also arranged for the

145 Claridge, i, p. 387; Cruickshank, i, p. 162.
146 For the full text, see Claridge, i, pp. 409–11.
147 I have discussed this in my article, 'Dwaben and Kumasi relations in the nineteenth century', 1964. See also Reindorf, pp. 276–84; Claridge, i, pp. 422–4 and G. E. Metcalfe, *Maclean of the Gold Coast*, 1962, pp. 124–44.
148 Bowdich, p. 245.
149 Boahen, 'Dwaben and Kumasi'; Reindorf, pp. 276–7.
150 Boahen, 'Dwaben and Kumasi'; Metcalfe, *Life and Times of George Maclean*, p. 125.

murder of the Dwabenhene's candidate and supporters.[151] So enraged was the Dwabenhene that he refused to go to Kumasi for a whole year and had the Asantehene's Ambassador pelted with stones. Osei Yaw Akoto took this opportunity to launch an attack on Dwaben in 1832. The Asantehene inflicted a decisive defeat and the Dwaben people migrated to Kibi and did not return until 1841.

This absolutely unnecessary war advanced the decline of Asante even further. Dwaben had been the most powerful of Asante's metropolitan states and she never regained her power. She was therefore never able to play her eighteenth-century role as the most courageous and most martial of the Asante states and controller of the vassal states to the east and north-east. Furthermore, Asante greatness had depended on the cooperation between Dwaben and Kumasi. Tradition says that even Okomfo Anokye, architect of the Asante Confederacy, realised this and had it embodied in the unwritten constitution of the Confederacy that under no circumstances should Kumasi ever take up arms against Dwaben. Bowdich had also realised the importance of this cooperation and was greatly impressed by it.[152] Though, as we shall see, the Dwaben were prevailed upon to return home and they did so in 1841, active, fruitful and brotherly co-operation was never restored and bitter rivalry and ill-feeling ensued. It is significant that in the wars of 1874 and 1900 between Asante and the British, Dwaben took no part. Clearly, this civil war had accelerated Asante's collapse.

Kwaku Dua I

Two years after this civil war, that is, in 1834, Osei Yaw Akoto died and was succeeded by Kwaku Dua I. Though the latter reigned for thirty-three years, he was unable to reverse the trend of affairs. Advisers and officials were anxious to regain the southern territories and revive the empire but, until the last three years of his reign, the Asantehene refused to resort to force in order to accomplish this. His policy was to stay on friendly terms with the British and to use diplomacy and negotiation to realise this aim. Thus, largely with the help of the British, especially Maclean, and the Dutch, he was able to persuade the Dwaben to return to their homeland and receive assistance to rebuild their homes and farms.[153] Kwaku Dua I strengthened the integrity of his northern possessions by settling disputes between Kumasi and Nkoranza and Gyaman as well as encouraging trade with the north. From evidence so far available, no revolts broke out in the north until the last three years of his reign.[154]

151 Reindorf, p. 277; Boahen, *op. cit.*
152 Bowdich, p. 232.
153 Reindorf, p. 296; Boahen, 'Dwaben and Kumasi'; Metcalfe, *Maclean of the Gold Coast*, pp. 124–30.
154 Goody, pp. 36–7.

Salaga continued to enjoy her position as the commercial metropolis of the Volta Basin during this period. In 1835 the Asantehene successfully appealed to the Europeans to halt the development of inland carrying trade.[155] In 1844 Kwaku Dua I reported the murder of an Asante woman trader to the British and the murderer was tried and executed in the presence of the Asantehene's messengers.[156] He also welcomed missionaries into Kumasi in the 1840s giving them land on which to build their mission. He further prevented war with the British by recalling a contingent he had despatched into the Protectorate to assist the refugee Assin under Tsibu to return to Asante. This episode ended with Tsibu's execution by the British for conspiracy[157] and afterwards Kwaku Dua I declared his readiness to affirm the Treaty of 1831 and even asked for a British resident to be stationed in Kumasi. Since they could not spare a British officer, George Musgrave, an educated African was sent.[158]

In 1863 Kwaku Dua I at last yielded to the persistent clamour for war by his chiefs and sent an army across the river Pra for the first time since 1826. In 1864 he sent another north against Nsɔkɔ and in 1867 a third across the Volta into the Krepi country. The move to the south was precipitated by the refusal of the British to hand over two fugitives and was, as usual, a three-pronged attack into Wassa and Denkyera, into Assin and through Akyem. Largely through blunders by Major Cochraine, in charge of the allies, the south was defeated at Asikuma and Bobikuma and the Asante army once more swept all before them. However, they returned to Asante at the commencement of the rains without any achievement apart from proving that they were still a force to be reckoned with. The invasion to the north succeeded in crushing the rebellion in the towns of Nsɔkɔ and Asono Menum.[159] The third campaign was to assist the Anlo, Asante's traditional allies, in the war they had been waging with the British since 1866 and also to safeguard important trade routes to the coast. This campaign was still in progress when Kwaku Dua I died in April 1867. Thus, Asante was still confined to the area north of the Pra though its military prestige was slightly redeemed.

Defeat and disintegration

It was during the reign of the next Asantehene, Kofi Karikari (1867–74), that the fate of the Asante empire was finally sealed. Like his predecessor, he was determined to regain Asante's traditional hold on Elmina and control of the southern states. He wrote to the British in December 1870

155 Metcalfe, pp. 132–3.
156 Claridge, i, pp. 453–5.
157 For details, see Claridge, i, pp. 486–92, and Hill to Newcastle, 8 April 1853 in Metcalfe, *Great Britain and Ghana*, pp. 240–2, doc. 188.
158 Claridge, i, p. 494.
159 Goody, pp. 36–7.

stating that Elmina was his 'by right'. In 1873, when his armies were everywhere victorious, he again wrote to Governor Harley:

> His Majesty further states that your Honour's restoring him these tribes, viz Denkerahs, Akims, and Assims back to their former position as his subjects and also restoring the Elmina fort and people back in the same manner as they were before, will be the only thing or way to appease him, for he has no quarrel with white men.[160]

The Asantehene had sworn on his succession that his 'business should be war' and he remained true to his word. As soon as he completed the burial and enstoolment ceremonies he mounted the usual triple attack against the south from Kumasi at the end of 1868 to relieve Elmina, then under blockade by the Fante Confederation, and to conclude the unfinished war against the British begun in 1863. After a circuitous route through Assini and Axim, an Asante column under Akyeampon entered Elmina in 1869 and ably supported the resistance of the Elmina to the Fante Confederation. Akyeampon remained there until 1872 when the Dutch banished him to Half Assini.[161] A second Asante column won what Ward has accurately described as a 'series of barren and extremely costly victories' capturing two missionaries in Anum and a French trader, M. Bonnat, in Ho in 1869 whom they sent to Kumasi. This column finally returned to Kumasi at the end of 1870 victorious but without much to show for it and with the loss of about a hundred and thirty-six chiefs and nearly half the troops.[162] Because the king wished to extract a ransom of at least £2,000 from the British for the three European prisoners and these negotiations did not reach a deadlock until November, the third and central Asante column did not move until December 1872. The occasion for this movement was clearly the cession of the Dutch forts in April 1872 which Ramseyer and Kühne, then in Kumasi, observed 'occasioned much vexation in the palace'. From that time earnest but secret preparations were made for war. The Asante army defeated the Assin army at Nyankumase on 9 February 1873 and, on 10 March, routed the allied forces consisting of Denkyira, Assin and Cape Coast troops as well as a hundred Hausas from Lagos under the command of a British officer.[163] It was after these victories that the Asantehene wrote to Harley demanding the return of Elmina and the southern tributary states. Though repulsed in two desperate battles at Dunkwa, the Asante virtually annihilated a large army of

160 King Kofi Kalkaree to Governor Harley, 20 March 1823, in Metcalfe, *Great Britain and Ghana*, p. 349, doc. 290.
161 D. Coombs, *The Gold Coast, Britain and the Netherlands, 1850–1874*, 1963, p. 53.
162 F. A. Ramseyer and J. Kühne, *Four Years in Ashantee*, 1875, pp. 201–5; Claridge, i, pp. 576–83, 590–3.
163 For fuller details of the wars between the Asante and the British and their allies between 1869 and 1874, see Claridge, i, pp. 567–649 and ii, p. 1–193 on which the following sections are mainly based.

Fante and Denkyira at Jukwa in June 1873. This victory convinced the local and British governments of the need for more effective measures against the Asante.

By this time it seemed that Asante had at long last won revenge for their losses of 1826 and that they would reoccupy their former southern provinces. But this did not happen. Having realised the significance of recent Asante victories, the British Government, in the words of none other than the Leader of the Conservative Party, Disraeli, believed that 'the honour of the country now requires that we should prosecute the war with vigour necessary to ensure success' and the more severe and effective the victory 'the better chance of avoiding future inroads'.[164] A very experienced and able military commander, Major-General Sir Garnet Wolseley, was therefore sent with instructions to rally the allied forces and to demand that Asante withdraw from the Protectorate, pay compensation and give assurances of good behaviour for the maintenance of future peace. If the Asantehene refused to accept these demands, he was to 'inflict in default of such [honourable] peace an effectual chastisement on the Ashantee force'.[165]

Wolseley arrived in Cape Coast on 2 October 1873 and soon decided that the one method of freeing the settlements from the continued Asante menace was

> to defeat the Ashantee army in the field, to drive it from the protected territories and, if necessary, to pursue it into its own lands, and to march victorious on the Ashantee capital, and to show not only to the King, but to those chiefs who urge him on to constant war, that the arm of Her Majesty is powerful to punish, and can reach even to the very heart of their kingdom.[166]

He later wrote that he hoped 'the flag of England from this moment will be received throughout Western Africa with respectful awe'.[167] Ultimatums were sent and at the same time a successful attack was made on Elmina.

Amankwa Tia, commander of this column of the Asante army, sent his reply which is interesting in both style and content:

> . . . Those four nations [Assin, Dankra, Akyim, Wassaw] belong to the king of Ashantee, and they refused to serve the king, and they escaped away unto you. . . . And those nations ordered the King of Ashantee that he may come and fight with them. Therefore I said they are not a

164 Minute by Sir E. A. Knatchbull-Hugesson, 27 February 1873, in Metcalfe, *Great Britain and Ghana*, pp. 347–8, doc. 288.
165 Kimberley to Wolseley, 10 September 1873 and 6 October 1873, in Metcalfe, *Great Britain and Ghana*, pp. 351–2, doc. 293 and 294; Coombs, p. 134.
166 Wolseley to Cardwell, 13 October 1873, in Metcalfe, *Great Britain and Ghana*, p. 352, doc. 295.
167 Wolseley to Kimberley, 28 November 1873, C.O. 96/101, C921 of 1874.

> friends with the King. On account of that I shall come down here to catch those four thieves who ordered the King of Ashantee to come and to fight with them. . . .
>
> And they fought with me six times, and I drove them away, and they escaped to be under you . . . and when you deliver Assin, Dankra, Akyim and Wassaw unto me, I shall bring unto the King there is no any quarrel with you.[168]

But four days before this letter had been written, plagued with smallpox, dysentery and fever and aware of the rallying of allied forces under Wolseley, the Asante column began its orderly retreat from the south. They entered Kumasi on 22 December, their numbers reduced from forty to twenty thousand and none of their objectives achieved.

Wolseley began his operations towards the end of October. After receiving reinforcements of British troops of battalions of the 2nd Rifle Brigade, 23rd Royal Welch Fusiliers and 42nd Highlanders (Black Watch), together with the 1st West India Regiment from Jamaica and a contingent under Captain Glover from Nigeria, which was to march into Kumasi from the east through Anlo and Dwaben, the main British force crossed the Pra in January 1874.[169] At Amoafo the Asante put up a magnificent fight to beat back this invasion but were then defeated with very heavy losses of between eight and twelve hundred men including Amankwa Tia. Victorious a second time at Odaso, the British army entered Kumasi on 4 February 1874 only to find it deserted. After waiting to hear from the Asantehene, the British army looted the palace, blew it up, set fire to the town and began the return journey to the coast on 6 February.

This crushing defeat of the Asante was caused by three major factors; the refusal of the Dwaben to participate in the war, the superior weapons used by the British and the experience and ability of the British commander. Before the battle of Amoafo it had been arranged that the Dwaben should contain Glover's column moving in from the east, destroy the bridge over the Pra and cut off supply lines.[170] If Dwaben had acted, disaster might have been avoided and the British forced to retreat across the Pra without provisions. But contrary to what Ward says, the Dwaben refused to take part in the war.[171] Secondly, there is absolutely no doubt that militarily the Asante were at a great disadvantage. Not only were the guns they were using completely outmoded, but owing to the effective suppression of the slave trade and the departure of the Danes and the Dutch, they found it increasingly difficult to get adequate supplies of guns and powder from the coast. The British, on the other hand, were armed with seven-pounder guns

168 Quoted by Claridge, ii, p. 59.
169 Claridge, ii, pp. 44–119.
170 *Ibid.*, p. 138.
171 See Boahen, 'Dwaben and Kumasi'; Ward, p. 274.

and Enfield rifles, the most up-to-date guns in the field then. It was clearly the use of these guns that decided the battle of Amoafo in favour of the British. The British also relied on well-drilled detachments of soldiers brought from Britain, the West Indies and Lagos rather than on the Fante and allied troops. There can also be no doubt about the value to the campaign of the initiative and ability of Sir Garnet Wolseley, an experienced soldier who had served with distinction in four campaigns and who had been the brilliant leader of the Canadian Red River Expedition of 1870.

This decisive defeat of the Asante in February 1874 finally and permanently sealed the fate of their empire. First, the heavily one-sided treaty of Fomena was forced on the Asante on 13 February 1874[172]. By this the Asantehene was compelled to pay the huge sum of fifty thousand ounces of approved gold in war damages, and once more to renounce all claims by himself and his successors to the southern states and Elmina, to withdraw all troops from the coastal areas and to abolish human sacrifice. To this day, the southern allied states form part of the Eastern, Central and Western Regions of modern Ghana and not part of the Asante Region. Secondly, the defeat provided a splendid opportunity for the hitherto obedient northern states to reject Asante dominance. On hearing the news of the Asante defeat, Gonja and Dagomba rebelled, killing or imprisoning Asante traders and agents and declaring themselves independent.[173] The Brong states did the same and became the Brong Confederation under the spiritual leadership of Krakye Dente. Gyaman and Sefwi also threw off the Asante yoke and in 1876 despatched messengers to the coast to ask for flags from the British. In spite of several attempts the Asante never reconquered these northern states. As Kimberley had correctly predicted in August 1873, a decisive victory over the Asante in the south also disrupted the integrity of the north.[174]

Thus, within a few months of the defeat, the Asante empire was confined to the area within about thirty miles of Kumasi. Nor was all well even within this traditional core. Adansi asserted her independence and this was recognised in the treaty of Fomena. The British also instigated a revolt by Dwaben immediately after the withdrawal of British forces and in a treaty virtually dictated by the British, Dwaben's independence was conceded in August 1874.[175] According to Fuller, even Mampon, Agona, Nsuta and Bekwai openly defied the Asantehene. The final conclusion to this humiliating disintegration was that in August or September 1874, Kofi Karikari, the Great Spendthrift, was deposed for rifling the mausoleum and stealing

172 For full text see Metcalfe, *Great Britain and Ghana*, pp. 356–7, doc. 299.
173 Goody, pp. 38–9; F. Agbodeka, *The African Protest Movement and its effects on British policy in the Gold Coast*, Ph.D. thesis, Legon, 1968.
174 Kimberley to Glover, 18 August 1873; Metcalfe, 1964, *op cit.*, p. 350, dos. 292.
175 Adm. 1/465, Strahan to Caernarvon, 31 August 1874; Adm. 1/465, Strahan to Caernarvon, 3 September 1874 and encl.; Claridge, ii, pp. 189–93; Boahen, 'Dwaben and Kumasi'.

other precious ornaments buried with some of his predecessors.[176] The fame and glory of the Asante empire were both destroyed when Kumasi burned in 1874.

Not only did the British destroy the Asante empire, they inherited parts of it. Indeed, five months after setting fire to Kumasi and forcing the Asantehene to renounce all claim to the southern states, the British declared this same area a formal British Protectorate and their settlements as the British Crown Colony of the Gold Coast. This declaration was the culmination of the growth of British power, jurisdiction and trade during the nineteenth century at the expense not only of Asante but also of other European powers who had traded on the coast of modern Ghana since the seventeenth century.

GROWTH OF BRITISH AUTHORITY

The British Government's growing interest and influence on the coast was not the outcome of any clearly defined policy nor was it guided by any well defined theory. Their approach was mere pragmatism and the growth was rather an accidental product of the uncoordinated and at times illegal activities of officials, traders and missionaries on the coast. It was thus marked by periods of advance followed by periods of stagnation or even retrogression.

At the beginning of the century the Dutch and Danes were also trading on the coast, the Danes in the eastern area in which their principal forts at Accra, Keta and Ada were situated, while the Dutch and British forts were situated alternately along the coast west of Accra. The principal Dutch forts were at Accra, Elmina, Shama, Butri and Axim and those of the British were at Accra, Anomabo, Cape Coast, Sekondi, Dixcove and Apollonia. Authority over the people of the coastal areas was never clearly defined and any jurisdiction enjoyed depended ultimately on the cooperation of the people. Cruickshank, a contemporary observer, wrote: 'Nominally masters, we yet exercised no authority or only such as the natives did not care to dispute. They willingly made up a present of their submission and deference in words, but belied their expressions by their acts.' He goes on to explain that local chiefs had to be given presents at regular intervals so that trade could be carried on and that by the turn of the century attempts at breaking through 'the trammels of native chicanery and imposition' were usually unsuccessful.[177] As one European visitor to the coast at the beginning of the nineteenth century admitted, 'Toutes les possessions des Européens là-bas ne sont que tolerées par les indigènes, pour autant qu'ils y ont des avantages' and therefore that the only true possessions of the Europeans were the forts.[178]

176 Fuller, p. 145; Ward, p. 280.
177 Cruickshank, i, pp. 32–3.
178 Quoted by Coombs, p. 6.

Each European power had some sort of alliance with one or more of the local states on the coast and in the interior. The British maintained fairly strong ties with the Fante, Wassa and Nzima, the Dutch with Ahanta, Elmina and above all Asante, and the Danes with Asante, Akyem, Krobo, Akwamu, Anlo and Krepi. By this time the Danes were even claiming some sovereignty rights over the latter four, a claim which the chiefs never recognised.[179] But the Europeans still had to pay the rulers of these states monthly stipends and send regular presents in order to retain their goodwill and because of the Notes they held for the castles.

British ascendancy at this time really dates from the crushing defeats they inflicted on the Asante in May and July 1824 and more convincingly in August 1826. These victories enabled them to revoke all the stipends and gifts paid to local rulers and restored the prestige and influence which they had lost during the first two decades of the century. It is significant that instead of taking full advantage of this very favourable turn of events, the British Government decided to hand over the administration of the forts to a committee of merchants trading there.

Reasons for this unexpected decision were threefold. The British Government found the administration of the forts too expensive both in terms of money and lives. Between September 1823 and December 1825, 1,554 soldiers were sent to West Africa of whom eight hundred died, forty-two were killed in action and twenty-two were invalided home.[180] As Turner also pointed out, though the British had no clear rights over the local people, the salaries of her local officials amounted to a large sum of money.[181] The British had failed to promote legitimate commerce and education in the country. The total value of British manufactures exported to the coast of modern Ghana between April 1822 and June 1826 was estimated at £108,560 7s. 10d. and the government naturally drew the conclusion that they 'would not be justified in purchasing such trade at a large sacrifice of human life and expense to the country'.[182] But probably the most important reason was the failure to conclude peace with Asante. This failure was due mainly to the intransigent attitude of the coastal allies especially towards Elmina which they besieged in 1828 and 1829. The Asantehene flatly refused to consider any treaty which did not include the Elmina whom he regarded as his subjects.[183] The third and final reason, and one closely related to the Asante issue, was the total disillusionment of the British with the policy of 'native alliances'. Faced as they were with a serious economic depression at home, increased expenditure on the Coast

179 Coombs, pp. 8–10; James, pp. 262–4.
180 *Ibid.*, p. 52.
181 C.O. 267/65, Turner to Bathurst, 24 March 1825.
182 Hay to Lack, 27 June 1827, in Metcalfe, *Great Britain and Ghana*, pp. 112–13, doc. 81.
183 Denham to Hay, 1 February 1828, in Metcalfe, *Great Britain and Ghana*, pp. 115–16, doc. 84; King Osei Yaw to the Governor of Cape Coast Castle, 12 April 1828, in Metcalfe, *Great Britain and Ghana*, p. 116, doc. 85.

and no immediate prospects of increased trade, they abandoned their allies, salving their conscience with the plea that it was their allies who had prevented the conclusion of peace with the Asante. Clearly, they considered Ghana a useless and irritating liability and as early as April 1826 the British Government appears to have arrived at the decision to withdraw, but delayed mainly because of the vehement opposition of the British merchants.[184] Not until October 1828 was a compromise reached and an agreement drawn up by which only the Accra and Cape Coast forts were to be maintained and their affairs regulated by a committee of three merchants in London chosen by the government and a council of five in Cape Coast, elected by the British merchants resident in Cape Coast and Accra.[185] The members of this council were also appointed Justices of the Peace and empowered to form among themselves a court to try all offences not amounting to felonies and misdemeanours which were to be tried in Sierra Leone. The council's president was to be elected by its members. It is important to note that jurisdiction was to be exercised only over the people living within the two British forts and a request for an extension of this power to those living outside was firmly rejected by the government.[186] The British Government agreed in turn to grant an annual maximum subsidy of £4,000 for the maintenance and defence of the forts. The British obviously expected the merchants to abstain from any interference with the affairs of the local people and to confine their activities entirely to trade and commerce.

George Maclean

The administration of the Committee of Merchants lasted from 1828 to 1843 and ended with the British so deeply immersed in local affairs that the government had to take over the administration once again. The growth of British influence and jurisdiction at the time was due almost entirely to the work of the second President of the Council, Lieutenant George Maclean of the 3rd West Indian Regiment, who had accompanied Colonel Lumley to Ghana in 1827 as his Military Secretary and who had greatly impressed the merchants.[187] Maclean arrived in Cape Coast to assume office on 19 February 1830 and remained in Ghana until his death in 1847. Conditions on his arrival were pretty grim, as Cruickshank observed: 'a rankling hatred still subsisting between the Fantees and the Elminas, with our intercourse with the interior cut off, our trade annihilated, our allies

184 *The Times*, 20 April 1826, 27 April 1826; J. D. Fage, 'The Administration of George Maclean on the Gold Coast, 1830–1844', *Transactions of Gold Coast and Togoland Historical Society*, i, part IV, 1955.

185 Metcalfe, *Great Britain and Ghana*, pp. 121–2, doc. 87.

186 Hay to Barnes, Brown and Forster, 4 November 1828, in Metcalfe, *Great Britain and Ghana*, pp. 122–3, doc. 88; C.O. 267/97, Hay to Committee, 29 November 1828; James, p. 64.

187 Metcalfe, *Maclean of the Gold Coast*, pp. 33–4, 45; Cruickshank, i, p. 165.

fighting and squabbling with each other, and our authority so limited as scarcely to be any protection to the oppressed.'[188] Maclean was undaunted and tackled each of these problems with tact, patience and dedication. He soon realised that unless peace and order reigned, trade could not thrive. He therefore deliberately ignored his instructions and actively interfered in local politics.

His first task was the war between the Asante and their traditional ally, Elmina, on the one hand and the Fante and the other southern states on the other. He began negotiations by sending selected coloured men whom he had carefully briefed. A sister of the Asantehene, held captive in Accra since the Dodowa War, was released and also played an active role in these negotiations.[189] The Fante were weary of war while the Asante were eager for the revival of trade with the coast and the restoration of the King's relatives still held in Cape Coast Castle. It is not surprising then that fourteen months after assuming office, a tripartite peace treaty was signed between the Asante, the allied chiefs and the British.[190] The Asante recognised the independence of their former vassal states south of the Pra river and agreed to refer all disputes with the coastal peoples to the British for peaceful settlement. They also agreed to deposit six hundred ounces of gold in Cape Coast and hand over two princes as security for six years. The allied states agreed to keep the paths free to all persons engaged in lawful traffic, not to force them to trade in any particular market and not to insult their former master. Finally, all agreed to stop panyarring. As James has pointed out, Maclean's treaty 'is at once more precise in terms and limited in scope than either of the abortive ones of 1817 or 1820. The former asserted jurisdiction over the Ashanti subjects at the coast, the latter made the king a British subject. No such claims were advanced in 1831.'[191]

Maclean certainly endeavoured to see this treaty strictly enforced. As he wrote to Hay, 'it is only by strictly keeping faith with the natives that we maintain our influence; and to break it even when otherwise justified in doing so is dangerous'.[192] He even returned the gold deposited with him in the very bags in which it had been given which greatly impressed the Asante. He also sent the two Asante princes, Owusu Ansa and Owusu Nkwantabisa, to school in Cape Coast and in 1836 he took them to England where their education was continued at a boarding school in Clapham, London. In 1840 when the Asantehene asked for their return this was promptly complied with and they were back in Kumasi by the end of 1841. It is interesting to note that these princes were each placed on a stipend of £100 paid by the Secret Service fund in the hope that they would further

188 Cruickshank, p. 170.
189 Metcalfe, *Maclean of the Gold Coast*, pp. 84–5.
190 For the full text of the trèaty, see Metcalfe, 1964, *Great Britain and Ghana*, pp. 133–4, doc. 98 or Claridge, i, pp. 400–9.
191 James, pp. 80–1.
192 Quoted by James, p. 81.

British interests.[193] Furthermore, Maclean protected Asante traders and either returned or redeemed at his own expense Asante fugitives at the coast.[194] It was Maclean who brought pressure to bear on the Dwaben to persuade them to return to their traditional area from Kibi in 1841.[195] As a result of all this, confidence developed steadily between Kumasi and Cape Coast and friendship continued between the British and Asante throughout Maclean's long tenure of office.

Through persuasion and negotiation, Maclean also tried to maintain peace among the chiefs of the southern states and to suppress human sacrifice, panyarring, attacks or raids on peaceful traders and slave trading. But he did not hesitate to use force when necessary. In 1833 he arbitrated in a dispute between Wassa and Denkyira imposing a fine on the latter's king whom he held responsible for the commencement of war.[196] He did the same later in a quarrel between Akuapem and Krobo and in 1836 and 1838 he tackled the long-standing dispute between Upper and Lower Wassa. In 1835 Maclean invaded Nzima, because the local king failed to abandon his slave-trading and raiding practices, and imposed a treaty which included an indemnity of three hundred ounces of gold on him. Other expeditions were sent to Wassa in 1836 and 1842 and to Asikuma in 1837 though only in Nzima did fighting actually take place.[197]

Maclean also realised that peace could not be maintained without the proper and impartial administration of justice. He told Nichols, 'the great means whereby the local government has succeeded has been the exercise of even-handed justice'.[198] He closely supervised the work of his magistrates stationed in Dixcove, Anomabo and Accra and paid them regular visits. From 1836 he constantly toured the courts of local chiefs or sent a member of his council to watch the process of trial and see that justice was done.[199] At these trials he often pointed out the inhumanities of some local laws and thus steadily disseminated ideas of British justice. Cruickshank describes how the courts became 'a species of lecture room, from which the principles of justice were disseminated far and wide throughout the country' and also 'a very successful, and not an invidious medium of attack upon their most objectionable laws and customs'.[200] Maclean allowed chiefs and subjects to appeal to his court imposing fines or imprisonment on the guilty. By 1841 there were ninety-one prisoners in Cape Coast

193 Ramseyer and Kühne, p. 75 n.; James, p. 81. For their return and activities in Kumasi see Metcalfe, *Maclean of the Gold Coast*, pp. 270–5; Ramseyer and Kühne, pp. 97–116.
194 Metcalfe, *Maclean of the Gold Coast*, pp. 134–7.
195 *Ibid.*, pp. 124–30.
196 *Ibid.*, pp. 152–3.
197 *Ibid.*, pp. 157–68, p. 150 n. 3.
198 C.O. 267/136, Maclean to Nichols, 3 November 1836.
199 J. D. Fage, 'The Administration of George Maclean on the Gold Coast, 1830–1844', *Transactions of the Gold Coast and Togoland Historical Society*, i, 4, 1955; Pine to Labouchere, 31 August 1857, in Metcalfe, *Great Britain in Ghana*, p. 141, doc. 109; *Ibid.*, doc. 267.
200 Cruickshank, ii, pp. 24, 26.

castle. Swanzy, a merchant and magistrate of Dixcove Fort, gave evidence before the Select Committee of 1842 of the scope and authority of Maclean's justice.[201] Maclean had not been the first British official to administer justice in Ghana. Indeed, between 1822 and 1826, one hundred and forty cases between Africans had been dealt with by magistrates including theft, assault, burglary, slaving, forging coins and obtaining money on false pretences.[202] But under Maclean judicial control was widely extended and firmly established. This extension of judicial control encroached on the chiefs' powers and the consequence will be discussed later.

In order to preserve law and order, Maclean also employed members of the local militia, numbering about sixty, as police. They were stationed singly or in small groups under non-commissioned officers at various posts throughout the southern states and especially at the residences of important chiefs. These men were usually people of the district and their intelligence, training and local knowledge enabled them to be keen observers and to make accurate reports of incipient troubles. Lieutenant-Governor Hill who replaced Maclean described this corps as an effective police force who were 'known and acknowledged by the tribes over whom they exercised control'. The local chiefs found the system agreeable and even clamoured for its revival in the 1850s as Pine reported to the Colonial Office.[203]

As a result of Maclean's administration there was peace at the coast throughout the 1830s and 1840s and, in the words of Cruickshank, 'the most complete protection and security were afforded to every one'. In this atmosphere Maclean succeeded to a considerable extent in promoting trade and missionary activities. He gave every encouragement to the manufacture of palm oil, begun about ten years before his tenure of office, and by 1844 it was Ghana's leading export.[204] He also introduced cowrie shells for the first time as a medium of exchange for small quantities of oil and articles of trifling value.[205] Taking advantage of the established peace, British traders returned to the abandoned forts and many adventurers were attracted by the security and increased profits to trade in the area and often used agents to sell their goods inland. Cruickshank observed:

> There was not a nook or corner of the land to which the enterprise of some sanguine trader had not led him. . . . In the principal towns on the main line of communication with Ashantee, extensive depots were formed, where every species of goods suited to the traffic might be got in abundance; and in Coomassie, the capital, many agents constantly

201 Swanzy, Evidence before the Select Committee of 1842.
202 James, pp. 90–2; Metcalfe, *Maclean of the Gold Coast*, p. 129.
203 Pine to Labouchere, 31 August 1857, in Metcalfe, *Great Britain and Ghana*, p. 263, doc. 207; James, p. 96.
204 Cruickshank, ii, pp. 41–2.
205 *Ibid.*, pp. 44–5.

> resided. . . . The land carriage of all this merchandise . . . gave employment to many thousands in transporting the goods.[206]

Significantly, neither the British nor the Dutch, nor indeed the Asantehene, welcomed the development of an inland carrying trade since it prevented the Asante from trading directly with the coast. In February 1834 the British and Dutch signed an agreement aimed at eliminating unnecessary competition between their traders which carried a fifth clause that '*the carrying trade into the interior* [my italics] shall be free and open to everyone and that it be allowed to remain so until, with the consent of all parties, some change can be made or a stop put to this system which is on all hands allowed to be most *pernicious and dangerous*'.[207] In the following year, the Asantehene also sent in a protest against 'the establishment by Fante traders of a stationary market in his dominions contrary to the understanding entered into when the treaty of peace and commerce was signed in 1831 and whereby his people were greatly injured'.[208] On 30 November 1835, the European merchants readily agreed that 'the conveyance of goods by merchants from the coast to the interior for the purpose of sale should be prohibited, leaving the inland traders free to bring down their gold and ivory direct to the European forts'.[209] Cruickshank points out that the collapse of this carrying trade, with the Asante becoming their own carriers, was caused by the breakdown of the credit system on which it was based and also by the attraction of reduced prices at the coast.[210] The deliberate moves by the Europeans and the Asantehene to curb this trade must also have contributed towards its collapse. From then on the Asante again began to go to the coast in very large numbers, boosting trade in the forts. Exports through the British forts increased in value from £90,000 in 1830 to £325,508 in 1840 while imports rose from £131,000 to £422,170.[211] This testifies to the phenomenal growth of trade under Maclean's administration.

Maclean was also interested in promoting missionary activities and education. The Wesleyan Missionary Society began its activities during his administration and Maclean worked closely with its missionaries, particularly the mulatto Thomas Birch Freeman, guaranteeing half the cost of the schools that Freeman established at Anomabo and Accra under Ghanaian teachers educated at Cape Coast.[212] His magistrates regularly visited and reported on these schools. When the parliamentary grant proved inadequate for the maintenance of the establishment on the coast,

206 *Ibid.*, pp. 32–3; Metcalfe, *Maclean of the Gold Coast*, pp. 131–2.
207 Quoted in Metcalfe, *Maclean of the Gold Coast*, p. 132.
208 *Ibid.*, p. 132.
209 *Ibid.*, p. 133.
210 Cruickshank, ii, pp. 35–9.
211 Fage, 'The Administration of Maclean'.
212 Metcalfe, *Maclean of the Gold Coast*, pp. 270–2.

Maclean wrote: 'Where the money for the support of these schools is to come from, I do not at present know, but they shall be supported from what quarter it may'.[213] Undoubtedly, it was due to government meanness that Maclean failed to accomplish much in the field of education. However, the Basle Missionary Society was also able, thanks to the peace and order of the Maclean era, to lay a firm foundation for the growth of Christianity and Western education during that period.

Maclean's other two achievements were the suppression of the external slave trade and of human sacrifice in the southern states. However, he refused to attack domestic slavery and confined his work to seeing the condition of these slaves improved. He left this institution alone since he felt it was delicately woven into the fabric of Ghanaian society and he might lose his authority by abolishing it. As we shall see, his stand on this issue led to the fall of his administration.

Even Maclean's bitter critic, Madden, admitted the success of his administration and the Select Committee of 1842 also confirmed its beneficial effects and popularity.[214] To all intents and purposes, a British Protectorate had been created in southern Ghana by 1840, the direct result of Maclean's own work. British prestige and power were increased at the expense of other European rivals and, above all, British merchants were handling the lion's share of the trade.

How was it, then, that Maclean's administration was brought to a premature end in 1843? First, whatever his success in the judicial and political fields, Maclean's activities were illegal and depended on his own popularity and moral influence. Sooner or later, it would have to be given a legal basis. Secondly, an over-sensitive Whig Government in Britain heard reports that Maclean and the merchants were conniving at the slave trade and slavery and two committees were set up, one under Dr R. R. Madden and the other a Parliamentary Select Committee under Viscount Sandon, in 1840 and 1842 respectively.[215] Though both committees disagreed in many respects on their conclusions, both agreed that the British Government should once more assume control of the affairs of the settlements from the Committee of Merchants. The 1842 Select Committee further proposed that judicial control be regularised and that all jurisdiction over the people and districts beyond the influence and protection of the forts should be considered optional and should be made 'the subject of distinct agreement, as to its nature, and limits with the native chiefs'.[216]

213 Quoted by James, p. 101.
214 Report of the Select Committee, 1842.
215 For details of the circumstances leading to the setting up of these committees and their proceedings and recommendations, see Metcalfe, *Great Britain and Ghana*, pp. 241–93.
216 Report of the Select Committee on West Africa, 1842, in Metcalfe, *Great Britain and Ghana*, doc. 135.

Bond of 1844

As a result, in March 1843 Hill was appointed Lieutenant-Governor and Maclean Judicial Assessor. Hill arrived on 13 February 1844 charged with three main functions;[217] to gain *de jure* recognition of the jurisdiction, to establish a system of government to replace the Council of Merchants and to promote trade and western civilisation. The first problem was tackled more by accident than design. Shortly after his arrival, several of the chiefs came to pay their respects to the new Governor and to enquire whether rumours that domestic slaves were to be freed were true or false. Having assured them that it was 'quite an idle report', Hill saw this opportunity to draw up an agreement which, he says, the chiefs 'readily signed'.[218] This document is the now famous Bond of 1844 with three main clauses; the chiefs in the southern areas where Maclean had exercised jurisdiction acknowledged formally the power and jurisdiction for the protection of individuals and property; human sacrifices, panyarring and other 'barbarian customs' were abolished and thirdly murders, robberies and other crimes and offences were to be tried by the Queen's judicial officers and local chiefs 'moulding the customs of the country to the general principles of British Law'.[219] The original document was signed on 6 March 1844 by eight chiefs and kings of Denkyira, Abora, Assin, Donadie (part of Dominase state), Dominase, Anomabo and Cape Coast. During the same year, the chiefs of Twifu, Ekumfi, Adwumako, Gomoa, Asikuma, James Town, Nsaba, Wassa, Amenfi, Dixcove and Wassa Fiasi also signed.[220] This document has been regarded as the 'Magna Carta' of Ghanaians by later politicians while some scholars consider that it constitutes the legal basis for British rule in Ghana.[221] Indeed, it was the date of this document that gave us the date for our Independence (6 March). But the importance of this bond has been greatly exaggerated.

In the first place, as Dr Danquah has correctly pointed out, the Bond was neither a treaty nor a charter but rather a mere unilateral declaration on the part of the Fante and other coastal rulers or, to use his own words, 'a one-sided obligation having a binding effect only on the person whose rights are handed over to another', and the British who drafted it never considered themselves bound by any article in it. It seems clear from the first clause that the signatories were only recognising jurisdiction exercised by Maclean; in other words, the Bond itself did not create this jurisdiction but only made formal acknowledgement of it. It is therefore not true to contend, as Danquah does, that 'the Bond is the only living document

217 Stanley to Hill, 16 December 1843, in Metcalfe, *Great Britain and Ghana*, pp. 192–3, doc. 142.

218 Hill to Stanley, 6 March 1844, in Metcalfe, *Great Britain and Ghana*, p. 195, doc. 144.

219 See the full text in Metcalfe, *Great Britain and Ghana*, doc. 145, p. 196.

220 *Ibid.*, p. 196, doc. 746; Metcalfe, *Maclean of the Gold Coast*, pp. 306–8.

221 J. B. Danquah, 'The Historical Significance of the Bond of 1844', *T.H.S.G.*, iii, part 1, 1957; Ward, pp. 187–8.

which bears permanent testimony to the fact that but for the express recognition, acknowledgement, and consent of the Chiefs of the Gold Coast, the British could have had no claim in law to administer the first essentials of a constituted community, namely, the protection of individuals and of property.'[222] Maclean had administered these first essentials and moreover, as we shall observe presently, the exercise of British jurisdiction after the Bond took a course utterly different from that envisaged when it was drawn up and signed. Indeed, it became a dead letter after 1847, and British judicial power grew despite, rather than because of, the Bond. Finally, the widely held belief that the Bond contained a clause stating that the British were to rule this country for exactly one hundred years and then leave is a myth.

Nor did the Bond introduce much that was new from the point of view of the signatories for no territorial or political sovereignty was surrendered to the British. As the Earl of Caernarvon pointed out, it 'is silent as to the Queen's right by her officers and delegates to collect customs, to administer civil justice, to legislate for the public health, to erect municipalities, to provide for education, to construct roads, and regulate the industrial and social economy of the Protectorate'.[223] But he even forgot to add that the limited rights which had been granted to the Queen's officers were to be exercised only with the concurrence of the chiefs. This point was underlined by the Secretary of State's letter of 22 November 1844 to Hill when it stated, 'the power of the Assessor in his judicial capacity . . . must be founded on the assent and concurrence of the sovereign power of the state within which it is exercised' and further that 'the system upon which Mr Maclean has proceeded in the exercise of judicial powers over the natives is to be taken as a guide for the exercise of the powers of the Assessor for the future'.[224] In other words, no new rights or liberties were conferred on the local rulers and the Bond cannot therefore be described as their 'Magna Carta'. In fact it was the very reverse since it did surrender formally part of their powers over criminal affairs. Had the terms of the Bond been adhered to, Ghana would never have become a 'British territory'[225] but, after Maclean's death in 1847, the Bond was completely ignored and had no effect on the subsequent history of Ghana. The importance of the Bond of 1844 lies more in the myth that later grew around it than in its terms and impact.

Administration of justice

Until his death, Maclean as Judicial Assessor continued to dispense justice

222 Danquah, 'Historical Significance'.
223 Caernarvon to Strahan, 20 August 1874, in Metcalfe, *Great Britain and Ghana*, pp. 369–70, doc. 304.
224 Stanley to Hill, 22 November 1844, in Metcalfe, *Great Britain and Ghana*, pp. 198–9, doc. 149.
225 Danquah, 'Historical Significance'.

on much the same lines as he had done before. After his death, however, the chiefs were gradually eliminated from the position accorded them by Maclean and the Bond and even by 1851 this had reached an advanced stage as is shown by James Bannerman's letter to the Colonial Office: 'Your Lordship is to understand what may already be known to you, that the Assessor instead of being assistant and adviser, has been elevated into the supreme judicial authority, even where purely native law is administered; and . . . he holds court more frequently without the presence of a single native ruler than otherwise'. To preclude the abuse of power by the Assessor, Bannerman advocated the establishment of an Appeal Court.[226]

The extension of British judicial control was further advanced by the establishment of the Supreme Court in the Supreme Court Ordinance of 1843 and subsequent developments. It was to be presided over by a Chief Justice and was given civil, criminal and Admiralty jurisdiction though not jurisdiction in equity until 1857.[227]

Jurisdiction over the Protectorate was still exercised by the Assessor until 1856 when an order-in-council extended it to what was now officially recognised as 'the protected territories' in certain cases such as bankruptcy which had hitherto been tried by the Assessor and a panel of chiefs.[228] Moreover, from the time of the Supreme Court Ordinance when the Assessor and Chief Justice became one individual, James Coleman Fitzpatrick being the first, the position became in practice 'an English judge administering English law' and, as the chiefs bitterly complained to Freeman in 1850, 'almost all cases in which chiefs were concerned were being referred to the police court of the coastal towns'. The Select Committee of 1865 reported: 'The Judicial Assessor does not fulfil the intentions of the office, assisting the chiefs in administering justice, but supersedes their authority by decisions according to his sole judgement'.[229]

The result was the strengthening and consolidation of British authority and the corresponding erosion of the authority and prestige of the chiefs. As early as 1850, Cruickshank and Bannerman drew the attention of the Colonial Office to this development. They saw the danger of this loss of authority by the chiefs and recommended the setting up of an 'Assembly of Native Chiefs' to meet twice yearly in Cape Coast to frame laws with the assistance of the Judicial Assessor and other magistrates'.[230] For this reason and also in order to raise revenue, as will be seen below, the famous assembly which passed the Poll Tax Ordinance of 1852 was convened. Had this

226 Bannerman to Grey, 6 May 1851, in Metcalfe, *Great Britain and Ghana*, pp. 229–30, doc. 179.

227 See the Supreme Court Ordinance in Metcalfe, *Great Britain and Ghana*, pp. 242–3, doc. 189; F. A. R. Bennion, *The Constitutional Law of Ghana*, London, 1962, pp. 12–13.

228 Ward, pp. 191–2.

229 Report of Select Committee of 1865.

230 Bannerman and Cruickshank to Winniet, 22 August 1850, in Metcalfe, *Great Britain and Ghana*, pp. 219–20, doc. 174.

body continued to meet, the chiefs might have regained their former stature. But it was never reconvened and the chiefs' power continued to be eroded.

Pine underlined this trend in his letter of 31 August 1857: 'there has been too much interference with the authority of the native chiefs, and this is one of the causes of the decline of our influence. This interference has been exerted by ignoring the native tribunals, and by allowing the chiefs to be summoned before our courts in comparatively trivial cases . . . what is this but at once to abrogate their authority.'[231]

By 1858 the authority of the chiefs had virtually collapsed in the main coastal towns and it was partly to check this and also to deal with 'the disorderly and filthy state of the towns under the walls of the forts' that Pine introduced the municipal system of government in Cape Coast and James Town, Accra, in May 1858.[232] Under the Municipal Ordinance, the towns were to be governed by a council elected from the chiefs as well as 'the large class of native merchants and traders' that had grown up. It was to have powers to levy rates, pass bye-laws for the suppression of nuisances, etc., and to establish a court to have jurisdiction over civil cases where the matter in dispute did not exceed the value of £50 and over criminal cases where the matter did not exceed six months' imprisonment with hard labour. The first municipal elections in Ghana took place in James Town shortly after the enactment of the Ordinance; seven chiefs and merchants were elected and James Bannerman, son of the late Commandant of the fort, was chosen as its Mayor or leader. These councillors began their work with every enthusiasm; for example, they passed several bye-laws to suppress nuisances and assessed the houses and fixed rates varying from 5s. to £3.[233]

Thus, by the 1860s, contrary to the clauses of the Bond of 1844, the post of Judicial Assessor had virtually disappeared, British law and law courts had in many cases replaced customary law and the courts of the chiefs of 'the protected territories' while judicial power in the two principal coastal towns was being exercised jointly by the chiefs and the merchants.

System of government

Hill and his successors also had to deal with the problem of establishing a new system of government to replace the Committee of Merchants in London and the Council of Merchants in Cape Coast. At first the Lieutenant-Governor was subordinate to the Governor of Sierra Leone and had to refer to him for advice and for approval of his decisions. Local merchants, loath to lose their direct rule in affairs, described this as 'too cumbrous for efficiency' since it appeared to 'involve a formula of reference and

231 Pine to Labouchere, 31 August 1857, in Metcalfe, *Great Britain and Ghana*, doc. 207.
232 Pine to Stanley, 7 May 1858, in Metcalfe, *Great Britain and Ghana*, pp. 274–5, doc. 216.
233 Kimble, *A Political History of Ghana, 1850–1928*, 1963, pp. 185–6.

reports utterly unsuited to the exigencies of the country where action was required'. They suggested a council presided over by the 'Lieutenant-Governor and including the Judicial Assessor and four merchants residing within the settlements'.[234] The Colonial Office rejected this idea on the grounds that a few merchants did not share the interests of the population as a whole.[235]

However, it was partly in response to persistent pressure from the merchants that in 1850 the British Government decided to separate the Gold Coast from Sierra Leone and to give the Gold Coast her own Governor, Legislative and Executive Councils from which, with varying vicissitudes, evolved the President, National Assembly and Cabinet of Ghana of the 1960s. A Royal Charter, dated 24 January 1850, stated that the Legislative Council was to consist of the Governor and 'not less than two persons appointed by the British Government' with powers 'to make and establish such laws, institutions and ordinances as may from time to time be necessary for the peace, order and good government of our subjects and others within our said present or future forts and settlements on the Gold Coast.' The Executive Council, appointed by the Governor, was to 'advise and assist' him. The Governor himself could 'make grants of land for public and private use, constitute and appoint judges, commissioners, justices of the peace' and other officials he felt necessary.[236] In April 1851 together with the Governor, Judicial Assessor and Collector of Customs, two merchants were given membership of the Legislative Council while the Judicial Assessor and Collector of Customs also sat on the Executive Council. In 1853 the Collector of Customs was replaced on both councils by the officer holding the post of Colonial Secretary and three years later the Chief Justice and commander of the new Gold Coast Corps joined the Legislative Council.[237]

The merchants' demand for a majority of seats on the Legislative Council had been rejected and because of this they vehemently opposed the local administration and refused to cooperate with the Governor. As Merivale minuted in February 1855, 'They have never been reconciled to the change. There has always been a contest between the officers representing the Government and this trading interest.'[238] Sir William Molesworth, five months later, said: 'We have reason to suspect that the merchants (as a body) are ill-disposed to the present regime, which has no doubt, interfered with

234 Bannerman, Cruickshank and Clouston to Forster, 12 February 1847; Forster to Grey, 22 March 1847, in Metcalfe, *Great Britain and Ghana,* docs. 156 and 157.

235 G. E. Metcalfe, 'After Maclean: Some Aspects of British Gold Coast Policy in the Mid-Nineteenth Century', *T.G.T.H.S.*, i, pt. V, 1955.

236 Metcalfe, *Great Britain and Ghana*, pp. 212–15.

237 Pine to Labouchere, 12 June 1857, in Metcalfe, *Great Britain and Ghana*, pp. 260–1, doc. 206; Bennion, pp. 1–10.

238 Minute by Merivale, 29 February 1855, in Metcalfe, *Great Britain and Ghana*, doc. 196. See also Connor to Herbert, 7 April 1855, in Metcalfe, *Great Britain and Ghana*, p. 250, doc. 198.

their influence, formerly almost supreme, at the establishments on the coast.'[239]

Not surprisingly, Ord, the Commissioner appointed to investigate the affairs of the country in 1855, recommended that two merchants from Cape Coast and one from Accra should be added to the Council.[240] Pine, the Governor, not only accepted this but added one more merchant and two civil commandants though for three years only and not for life. Pine said, 'The present Legislative Council . . . so constituted is, in my opinion, extremely unsatisfactory . . . it fails and must fail to command the confidence of the people, from the feeling that it does not possess within itself that intimate knowledge of the nature of the country . . . without which knowledge, no man can safely legislate for any country, more especially for one in so anomalous a state as that of these settlements and the protected territory.'[241] Pine's proposals were turned down however, because the Colonial Office deemed them too radical.[242]

By the late 1850s, then, British courts and law had replaced those of the chiefs and a Legislative Council was operating in southern Ghana.

Taxation

The third problem for Hill and his successors was, as Earl Grey, then Secretary for War and Colonies, described it, 'the extension of British commerce and the improvement and civilisation of the barbarous inhabitants of this portion of Africa'.[243] The British Government considered that this meant the establishment of schools and industrial institutes because of its belief that 'intellectual and industrial instruction should go hand in hand', the promotion of religion and legitimate trade, the abolition of the slave trade and slavery and 'the improvement and extension of the judicial system, in affording greater facilities of internal communication, increased medical aid, etc.' But the British Government insisted that these developments should be paid for locally[244] through direct and indirect taxation and Grey stated his preference for the former, partly because of 'the stage of society in which the tribes of our neighbourhood are at present placed' and partly because of his sincere belief that it would 'provide a motive for exertion among the otherwise indolent people of the tropics and thus ensure the attainment of a high degree of civilisation'.[245] During the 1850s, both

239 Minute by Molesworth, 31 July 1855, in Metcalfe, *Great Britain and Ghana*, pp. 251–2, doc. 200.
240 Ord to Labouchere, 16 May 1856, in Metcalfe, *Great Britain and Ghana*, pp. 255–6, doc. 203.
241 Pine to Labouchere, 12 June 1857, in Metcalfe, *Great Britain and Ghana*, p. 261, doc. 206.
242 Kimble, p. 407.
243 Grey to Winniet, 20 January 1849, in Metcalfe, *Great Britain and Ghana*, p. 207, doc. 164.
244 *Ibid.*
245 Earl Grey, *The Colonial Policy of Lord John Russell's Administration*, London, 1853, ii, pp. 280–2.

methods of taxation were in fact tried. Among the reasons for the British Government's purchase of the Danish forts for £10,000 and subsequent occupation of Anlo were to prevent part of the coast falling into French or Belgian hands, to settle the question of jurisdiction over Akyem, Akuapem, Krepi and other eastern states and to suppress the slave trade throughout the coast but the overriding reason was undoubtedly the opportunity the purchase provided for raising revenue through the imposition of increased customs duties.[246] And it was to raise revenue through direct taxation that at the assembly of chiefs convened by the Governor in Cape Coast on 19 April 1852, a Poll Tax 'Ordinance' was passed.[247] This ordinance stated that the assembly, to be entitled the Legislative Assembly of Native Chiefs upon the Gold Coast, agreed voluntarily to pay to the government the annual sum of '1*s.* sterling per head, for every man, woman and child residing in the districts under British protection'. The enactments of the assembly, sanctioned and approved by the Governor, would become law and binding on the whole population. The tax was to be collected by officers appointed by the Governor with the assistance of the chiefs. Hill confidently expected an annual income of £15,000 from the western area alone.[248] By October 1852, all the chiefs west of the Volta as far as and including Nzima had subscribed to the ordinance and agreed to pay the tax.[249] An even larger income was therefore expected. Metcalfe says that Grey hailed this assembly, which had been his idea, with 'extravagant optimism and paternal pride' as an important step in the direction of a civilised African form of self-government. Indeed, Grey called it, with grotesque exaggeration, 'a rude Negro Parliament' which had 'converted a number of barbarous tribes, possessing nothing which deserves the name of a government, into a nation, with a regularly organised authority'.[250]

246 Forster to Howes, 30 April 1849, in Metcalfe, *Great Britain and Ghana*, p. 209, doc. 166; House of Commons Committee of Supply, 19 July 1850, *ibid*, pp. 171–2, doc. 171; Convention . . . for the cession of the Danish possessions on the Coast of Africa to Great Britain, signed London, 17 August 1850, *ibid*, p. 218, doc. 173; James, pp. 217–27. The Danes had been anxious to sell their forts and leave the coast since the 1820s because they considered them a drain on their finances, and in the 1840s they began negotiations with the French. The local British merchants got alarmed and urged their government to purchase them. They pointed out that the presence of the French would lead to a revival of the slave trade and a decline of legitimate trade; they even went on to advocate the reoccupation of the fort at Whydah and the building of another fort at Badagry. By the end of 1848, Palmerston had become convinced of the need to purchase the Danish possessions. In December he pointed out the inexpediency of letting slip the opportunity offered 'of assisting in the suppression of the slave trade and at the same time avoiding the inconvenience of the acquisition of the Danish forts by a rival naval and commercial power'. However, it was not until the purchase price was brought down from £40,000 to £10,000 that both the Colonial Office and the Treasury backed Palmerston and that was why the agreement was not concluded until 1850.

247 The Poll Tax Ordinance, 19 April 1852, in Metcalfe, *Great Britain and Ghana*, pp. 230–2, doc. 181.

248 Hill to Grey, 24 April 1852, in Metcalfe, *Great Britain and Ghana*, p. 233, doc. 183.

249 Metcalfe, *Great Britain and Ghana*, p. 232, n. 1.

250 Grey, ii, pp. 284–7; Metcalfe, *Great Britain and Ghana*, pp. 233–4, doc. 184.

Both schemes failed. Despite the purchase of the Danish forts, the British soon realised that unless the Dutch agreed to adopt the same system of import duties as theirs, it would always be impossible to collect any duties worth mentioning. In 1850, 1856 and 1857, therefore, they negotiated with the Dutch for a common system but the Dutch broke off negotiations because of the project's unpopularity in Dutch commercial circles. Undaunted, the British then suggested an exchange of forts on either side of the Sweet River. Had the Dutch cooperated, as the Duke of Newcastle minuted, it would 'have been the first great step towards effecting the objects which brought us to this unhealthy region, and would before long so increase the pecuniary resources of the colony, by rendering moderate customs duties possible as to relieve the Parliamentary grant altogether'.[251] The main cause for disagreement was the British refusal to give up their jurisdiction over the inland territories of eastern and western Wassa and their insistence that the exchange should be confined only to their respective possessions on the coast.[252] The British consequently only increased their duties from a ½ per cent *ad valorem* imposed in 1839 to 2 per cent in 1855,[253] thereby failing to earn any increased revenue from indirect taxation. In view of this the British withdrew their garrisons from Keta and Dzelukofe in 1855 and abandoned the entire Anlo country in 1860.[254]

Attempts to raise revenue from direct taxation also failed. Only £7,567 6*s.* 1*d.* was collected in the first year, half Hill's estimate, and this was the highest figure ever actually received. From then on receipts fell rapidly until they petered out after 1861 when only £1,552 3*s.* 4½*d.* was received.[255] Failure was due to the system of collection, the manner in which the tax was expended and, as is explained below, the resistance of chiefs and people to the payment of the tax. Rejecting Grey's original suggestion,[256] Hill refused to allow the chiefs to collect the tax on the grounds that they would abuse their authority and demand oppressive sums in order to enrich themselves.[257] Hill appointed his own officials to

251 Minute by the Duke of Newcastle, 31 August 1859, in Metcalfe, *Great Britain and Ghana*, p. 277, doc. 218; Pine to Labouchere, 10 October 1857, *ibid*, pp. 265–6, doc. 209.
252 Bird to Newcastle, 12 November 1859, in Metcalfe, *Great Britain and Ghana*, pp. 282–3, doc. 222; Ord to Barrow, 16 May 1860, *ibid.*, pp. 283–4, doc. 223; Napier to Russell, 1 August 1860, *ibid.*, pp. 285–7, doc. 225; Lewis to Andrews, 22 September 1860, *ibid.*, pp. 287–8, doc. 226; Andrews to Newcastle, 8 November 1860, *ibid.*, pp. 288–90, doc. 227; Coombs, pp. 22–34.
253 Metcalfe, *Great Britain and Ghana*, p. 265, n. 1.
254 Andrews to Newcastle, 4 July 1860, in Metcalfe, *Great Britain and Ghana*, pp. 284–5, doc. 224.
255 Claridge, i, p. 495. The receipts were as follows (in round figures): £7,567 in 1853, £3,625 in 1854, £3,990 in 1855, £3,353 in 1856, £3,192 in 1857, £2,921 in 1858, £2,351 in 1859, £1,725 in 1860 and £1,552 in 1861. See also Kimble, p. 187.
256 Minute by Earl Grey, 16 November 1850, in Metcalfe, *Great Britain and Ghana*, p. 221, doc. 176.
257 Hill to Grey, 24 April 1852, in Metcalfe, *Great Britain and Ghana*, p. 233, doc. 183; Connor to Herbert, 7 April 1855, in Metcalfe, *Great Britain and Ghana*, doc. 198.

supervise the whole exercise and divided the Protectorate into twelve districts for the purpose. Thus the Assembly of Native Chiefs never met again. However, his successors found it exceedingly difficult to obtain reliable collectors, and bad communications and huge distances to be covered made collection extremely costly. Some collectors plundered the people and robbed the government. But the British did not change their minds and use the chiefs. A Commissioner, appointed in October 1855 to investigate the collection, defended the system and refused to recommend a measure that 'would tend to exalt the power of the chiefs which it is part of our policy to endeavour to lower and would give them opportunities for exaction and extortion'.[258]

It is important to note that there was no opposition to the collection of the tax when it began in October 1852. Indeed, Hill reported that the collectors noted 'the cheerful disposition evinced by the masses of the people in the payment of this voluntary revenue'.[259] The fact that the amount collected in the first year was the highest ever received bears this out. But resistance to the tax and a consequent sharp drop in receipts began from 1854 mainly because of the utter disgust of the chiefs and people with the manner in which the proceeds were expended. It is evident from the Poll Tax Income and Expenditure account of 1852–3[260] that it was used to pay the collectors, Governor and other officials and only 8 per cent of the total was spent on schools and roads. What is worse, though some allocation was made for stipends to the chiefs, in fact most of them were not paid their promised salaries as they informed Hill in Accra in January 1854.[261] It is evident from the minute by Merivale on the Governor's reports of 4 May and 16 May 1855 on the proceeds of the Poll Tax, and from Ord's report in May 1856, that there was not much change in the expenditure of the proceeds from the tax. Merivale minuted:

> Out of a receipt of scarcely £4,000—not much more than half what was at one time anticipated—we pay £850 for the salaries of commandants, who may be partially useful to the natives as magistrates; £190 for a surgeon for them, whose services they do not use; £200 for roads, buildings and 'superintendence' whatever that may mean. And besides these sums, I do not see any item, of the smallest kind, which can be set down as for the public good of the natives. In fact, almost the whole goes, in one shape or another, in what may be termed expenses of collection.[262]

258 Ord to Labouchere, 16 May 1856, in Metcalfe, *Great Britain and Ghana*, pp. 255–6, doc. 203.
259 Kimble, p. 175.
260 See Kimble, p. 177, for the Balance Statement of the Poll Tax Income and Expenditure, 1852–3..
261 Hill to Newcastle, 29 January 1854, in Metcalfe, *Great Britain and Ghana*, pp. 246–7, doc. 193; Kimble, p. 156.
262 Minute by Merivale, 28 July 1855, in Metcalfe, *Great Britain and Ghana*, pp. 250–1, doc. 199.

Ord also reported a year later that the coastal peoples complained that 'the promises made to them had not been fulfilled, in respect, to the improvement of their towns and roads, etc.'.[263]

Shortly after his assumption of office in March 1857, the energetic and progressive Benjamin Pine realised that the chief complaint was the expenditure of the tax and he therefore announced a number of changes towards the end of the year calculated to improve the administration of the proceeds.[264] There were now to be two common chests, in Cape Coast and Accra, into which proceeds from the western and eastern dis-districts were to be respectively placed and the funds were to be used in the districts in which they were collected. Two-thirds of the funds were to be spent 'in the extension of the magistracy and the establishment of schools and hospitals and the remaining third in physical improvements, such as roads, bridges and the like', this expenditure to be supervised by local councils appointed by the Governor. Each council was to consist of local merchants and other inhabitants and there was to be a total of three such councils in three districts or divisions which Pine created, namely, Cape Coast, Anomabo and the Eastern or Accra district. Pine reported that the councils in Accra and Cape Coast which he set up had held several meetings and passed some 'very useful bye-laws for cleaning and improving the towns, and establishing police regulations'.[265] These measures were adopted by the Legislative Council and embodied in the District Assemblies Ordinance passed in May 1858, the same day on which the Municipal Corporation Ordinance, which was part of Pine's schemes, was passed. However, opposition to the tax had become too crystallised for the new changes to affect the situation. Pine himself reported in March 1858: 'The Tax is very unpopular throughout the Country and the dislike of it increases as we go from the West to the East. . . . It is now utterly detested!'[266] In September 1859, the Colonial Office disallowed District Assemblies because they considered 'the state of society on the Gold Coast is not yet ripe for an enactment of the kind'.[267]

Opposition to the tax was also part of the general resistance to the growth of British power and jurisdiction to be discussed later and developed into open rebellion first in the eastern districts in January 1854 where the chiefs and people organised protest meetings in Accra and refused to pay the tax. When Hill rushed to Accra, he found several thousand armed men not from Accra only but from all the coastal villages between Accra and the Volta,

263 Ord to Labouchere, 16 May 1856, in Metcalfe, *Great Britain and Ghana*, doc. 203.
264 Pine to Labouchere, 30 April 1857, in Metcalfe, *Great Britain and Ghana*, pp. 259–60, doc. 205; Pine to Labouchere, 4 December, *ibid.*, pp. 268–9, doc. 211.
265 Metcalfe, *Great Britain and Ghana*, pp. 268–70, doc. 211.
266 Kimble, p. 184.
267 Newcastle to Bird, 17 September 1859, in Metcalfe, *Great Britain and Ghana*, p. 282, doc. 221; Kimble, pp. 176–8.

waving Danish flags and menacing the fort.[268] This riot was stopped and a fine of £250 imposed but an even more violent riot broke out eight months later and involved people from Akyem, Akuapem and Krobo as well.[269] The chiefs and elders of these states and those of Accra assembled together and swore 'never to pay the tax again'. Riots followed and soldiers attempting to collect the tax were attacked. The British ship *H.M.S. Scourge* bombarded Labadi, Teshi and Christiansborg but this only exacerbated the chiefs and people who retaliated by attacking the British castle with an army of over four thousand men. The rebellion was finally quelled after a second bombardment in which many houses were destroyed and much property damaged. Similar protests occurred in the western districts and the Governor confessed that he met 'the greatest opposition from the Cape Coast chiefs and headmen' who protested at the use of the tax to pay chiefs and magistrates. They agreed to pay the tax only after Hill had threatened to use force to punish them according to the law passed by the Native Assembly of Chiefs.[270]

Riots were renewed in 1855 and, as Ord reported in May 1856, he found opposition to the Poll Tax general. He attributed this more to 'a general feeling of soreness and irritation, an impression that they were despised and neglected' than 'to any specific evil in the Poll Tax itself'. He went on to suggest that infants, poor women having no husband or head of family and aged persons unable to work be exempted from the tax, adding that the coastal people complained about the use of the fund.[271] He thus either misunderstood or oversimplified the causes for the opposition to the tax. However, neither these changes nor those introduced by Pine reconciled the people and opposition grew until it finally exploded with even greater violence and, rather significantly, in the eastern districts again, this time in Krobo in September 1858. As Claridge has pointed out, this rebellion, led by Ologo Patu, Chief of South Western Krobo or Yilo Krobo, had its origin in rivalry between Patu and another local chief of Western Krobo but was due principally to the opposition to the tax in the area.[272] After defeat by a British force, a punitive fine of £8,125 was imposed which the Krobo refused to pay and staged the palm oil hold-up of 1858–60.[273] Resistance to the tax continued and became so widespread by 1860 that Pine decided to stop the collection of the tax in the Volta areas and withdraw British

268 Kimble, pp. 176–8.
269 *Ibid.*, pp. 179–81.
270 Hill to Newcastle, 31 May 1854, in Metcalfe, *Great Britain and Ghana*, pp. 247–8, doc. 195.
271 Ord to Labouchere, 16 May 1856, in Metcalfe, *Great Britain and Ghana*, pp. 255–6, doc. 203.
272 Bird to Bulwer, 10 November 1888, in Metcalfe, *Great Britain and Ghana*, pp. 279–81, doc. 219; Claridge, ii, p. 499.
273 Kimble, p. 188.

protection and jurisdiction.[274] Two years later no serious attempt was being made to collect the Poll Tax anywhere in the Protectorate. In 1864 Governor Richard Pine, younger brother of Benjamin Pine, gave up the idea of pressing for payment or imposing other taxes because, as he confessed, he was 'not sure that resistance will not take the place of cheerful obedience'.[275]

The collapse of the Poll Tax scheme was not, as Kimble contends, due so much to the failure to eradicate the numerous abuses in the system of collection, as to the methods of expenditure and a refusal to allow the chiefs a part in its collection and expenditure. Nonetheless, it was one of the important episodes in the history of Ghana during the nineteenth century. As Kimble says, its failure created the impression in Ghana that direct taxation was an unjust imposition to be resisted at all costs, an impression which endured until after independence.[276] It was also the first time that the British Government had given in to persistent pressure brought to bear on it by the local people, a lesson which was never lost sight of again. Finally, the anti-tax 'movement' was the first of its kind organised on a supra-ethnic basis by the chiefs and people against the Colonial Government and can therefore be regarded as the first nationalist movement in Ghana and, as Kimble concluded, 'it set the pattern for a century of strenuous, if intermittent, political opposition to taxation in any shape or form'.[277] The protest movement can therefore be regarded as an early straw in the wind of change that was to engulf the country in the following century.

Asante invasion

Despite the failure of the Poll Tax Ordinance and measures to increase indirect taxation, the growth of British power and jurisdiction in the southern states, especially in the area west of the Volta, would have continued in the 1860s as it had done in the 1850s but for a series of events begun by the resumption of hostilities between Asante and the British in 1863. There had been a welcome period of peace since Maclean's administration. The renewed hostilities are attributed by many historians to the refusal by the British to hand over two Asante fugitives[278] but this was more the occasion for rather than the cause of war. It is rather significant that the Asante moved first against the Assin and Akyem Kotoku in 1863. As long as the Asante refused to reconcile themselves to the loss of the southern states, and in particular to Assin, there could be no real lasting peace between them on the one hand and the British and their

274 Andrews to Newcastle, 4 July 1860, in Metcalfe, *Great Britain and Ghana*, pp. 284–5, doc. 224.
275 Pine to Cardwell, 13 July 1864, in Metcalfe, *Great Britain and Ghana*, pp. 299–301, doc. 242.
276 Kimble, p. 189.
277 *Ibid.*
278 Claridge, i, p. 502; Ward, pp. 205–7.

allies on the other. Nor did the British and their allies after Maclean develop any genuine respect for Asante nor feel really secure as long as Asante power was left unchecked. The reasons given by Pine, the Governor, for renewed attacks in 1863 are almost identical to those given by MacCarthy in the 1820s. Pine wrote:

> It is with deepest regret that I find myself involved, in spite of all my precautions, in a serious and, I fear, lingering war: but such being the case, I will not conceal from your Grace the earnest desire that I entertain that *a final blow shall be struck at Ashantee power, and the question set at rest for ever, as to whether an arbitrary, cruel and sanguinary monarch shall forever be permitted to insult the British flag and outrage the laws of civilisation* [my italics].[279]

Only Maclean's tact and the pacifism of Osei Kwaku Dua I had postponed this inevitable clash. What then was the extradition episode that Pine took as the reason for the renewal of hostilities?

In 1862, two Asante, one a slave, the other an old man named Kwesi Gyanin, fled to the coast and sought refuge with the British.[280] The slave first claimed that he had been maltreated by his master and the old man had escaped because he had failed to hand over to the Asantehene a piece of 'gold rock' that he had found, as he should have done in accordance with Asante law and custom. In December 1862 the Governor rejected the Asantehene's request for the extradition of the two fugitives and gave as his reasons that the old man solemnly denied his charge.

> He is a man of property (the Governor continued) and declares that the King desires only to entrap him, take his head and afterwards possession of his property. The King's messengers offer to swear that the accused will be fairly tried, and even if found guilty, will not lose a hair of his head. The old man imploringly cries to me, 'Kill me if you like; that will be better than giving my head to the King.' And no one can assure me that I may rely upon the King's word [of a fair trial]. . . . Gladly would I try an experiment and send back these subjects of Ashantee, for if confidence were once created between this Government and Ashantee, the greatest obstacle in the way of amicable relations between us, would be removed; and if against the old man there were the slightest shadow of prima facie case of criminality, my course would be clear; but as it is, I dare not deliver him up, much less the runaway boy. . . . And yet I feel that I am

279 Pine to Newcastle, 12 May 1863, in Metcalfe, *Great Britain and Ghana*, pp. 295–6, doc. 235.

280 For details of this episode see Claridge, i, pp. 503–30 and Pine to Newcastle, 10 December 1862, in Metcalfe, *Great Britain and Ghana*, pp. 291–2, doc. 230. This section is based on these two sources.

estranging, if not exasperating, the most powerful King on this coast. . . .[281]

Clearly, Pine saw himself to be on the horns of a dilemma. But was he? Why did he not return the fugitives to Kumasi with an embassy to see that justice was done? Why did he accept the words of the two accused against those of the King's messengers? The messengers had sworn an oath that they would not have been killed if found guilty. Cruickshank had stated in writing that there was no instance of this oath having been broken[282] and Pine could have found this out. Furthermore, Pine had precedents to guide him. The Asantehene wrote two months later drawing his attention to the agreement signed between him and Maclean over extradition.[283] Cruickshank, who himself acted once as Governor, confirms that an agreement between the British and Asante did exist.[284] Even if there had been no such agreement, Pine could have discovered that such a mutual extradition of fugutives had become common practice since Maclean's time. Claridge quotes two instances of the exchange of fugitives in this manner.[285] Had Pine bothered, he could also have looked up Lord John Russell's ruling in 1841 that British laws on slavery were 'binding only within the British dominions and that with regard to people living within the vicinity . . . the same rule does not apply'.[286] Pine therefore had no right to hold the runaway slave.

Pine, in his evidence before the Select Committee in answer to the question why he did not send Kwesi Gyanin away, admitted that he would have done so but that he was prevailed upon by the chiefs and headmen as well as an overwhelming majority of the merchants not to do so, and both Horton and Ellis maintain that the chiefs and elders of Cape Coast did this because they were bribed by Gyanin.[287] Significantly, the Asantehene did not immediately resort to hostilities but sent a second deputation with a letter to Pine himself and also an embassy to the Dutch Governor complaining, as Pine informed the Colonial Office, 'of my conduct nearly in the same terms as that he addressed to me'.[288] Pine rejected this second request as well as the Dutch Governor's suggestion 'to despatch an Officer to the King of Ashantee as mediator between us' and the Colonial Office fully endorsed Pine's conduct. Kwaku Dua I had no choice now but to give in to the demand from his advisers for war

281 Metcalfe, *Great Britain and Ghana*, doc. 230; Claridge, i, p. 504.
282 Cruickshank, ii, pp. 236–7.
283 Asantehene to Pine, 9 February 1863, in Metcalfe, *Great Britain and Ghana*, p. 293, doc. 232.
284 Cruickshank, ii, pp. 236–8.
285 Claridge, i, pp. 507–8.
286 Claridge, i, p. 570.
287 Claridge, i, pp. 511–12, 570; Metcalfe, *Great Britain and Ghana*, p. 293 n.
288 Asantehene to Pine, 9 February 1863, in Metcalfe, *Great Britain and Ghana*, p. 293, doc. 232; Pine to Newcastle, 10 March 1863, *ibid.*, pp. 293–4, doc. 233.

and the three-pronged attack on the Protectorate was launched. Clearly, because of sheer and deliberate ignorance of precedents, or out of lack of respect for Asante law and custom, or simply because he succumbed to the pressures of the chiefs and merchants of Cape Coast, Pine was to blame for the resumption of hostilities between the British and Asante.

As we have seen, the Asante army defeated the allied forces a number of times before retiring across the Pra in June 1863 and it was six months after this that the Colonial Office reluctantly granted permission for Pine to march to Kumasi 'if no opportunity be found of striking such a blow without entering the Ashantee territory'. After receiving this permission, Pine moved more soldiers to Praso and waited for further reinforcements which did not arrive until April 1864. But the rains had begun and by the end of the month a quarter of the soldiers were in an inadequately equipped hospital.[289] Barely a month after the arrival of the 1st West Indian Regiment, four of its seven officers and about half the other ranks were on the sick list. When news of this reached Britain, the government ordered the withdrawal from the Pra which was carried out in July. Claridge concluded that this ended 'the greatest failure in the history of the British occupation of the Gold Coast' and as the Asantehene is reported to have boasted, 'the white men bring many cannon to the bush but the bush is stronger than the cannon'.[290]

Debate over British withdrawal

This disastrous fiasco which became a scandal in Britain had far-reaching effects on the consequent activities of both the southern peoples and the British in Ghana. It caused the British to lose the prestige they had enjoyed since the battle of Dodowa and it led directly to the resumption of the debate, started in Britain by Benjamin Pine's pessimistic dispatches of 1857 and early 1858 about conditions in Ghana, as to whether Britain should withdraw from or maintain her connection with Ghana.[291] In order to help the British Government to reach a decision, Colonel Ord was sent to Ghana for the second time in 1864 to investigate the condition of the British forts and settlements and a Select Committee of the House of Commons was appointed in the following year under Sir Charles Bowyer Adderley, a violent opponent of colonial expansion, to consider Ord's findings and make recommendations on future British policy in West

289 Claridge, i, p. 526.
290 Claridge, i, p. 529.
291 See Metcalfe, *Great Britain and Ghana*, docs. 204–18, pp. 258–78, and especially the C.O. minutes on these between March and August 1859. The demand for complete withdrawal was turned down on the grounds that it would lead to a more complete anarchy and more bloody wars than existed and to the revival of the slave trade. As the then Secretary of State minuted on 31 August 1859, 'The attempts now making by forcing nations to revive the slave trade would, if no other consideration interfered, render abandonment of our position on the coast impossible.'

Africa.[292] These recommendations, as Metcalfe has pointed out, 'reflected the views of Adderley and Stanley rather than those of the witnesses examined by it'.[293] One recommendation was a central government over all the four British settlements in West Africa to be re-established in Sierra Leone. The second, often quoted, was

> That it is not possible to withdraw the British Government, wholly or immediately, from any settlements or engagements on the West African coast. . . . That all further extension of territory or assumption of Government, or new treaties offering any protection to native tribes would be inexpedient and that the object of our policy should be *to encourage in the natives the exercise of those qualities which may render it possible for us more and more* to transfer to them the administration of all the Governments, with a view to our ultimate withdrawal from all except, probably, Sierra Leone.[294]

It is important to note that these resolutions were immediately enforced. In November 1865 T. F. Elliot of the Colonial Office minuted that he was working on a scheme to withdraw all the outlying stations on the Gold Coast and concentrate on Cape Coast Castle. The Colonial Secretary asked the Administrator in Ghana to recall a notice declaring the territory within five miles of eight separate British forts as British territory on the grounds that 'the extension of British territory . . . cannot receive my sanction'.[295] In January 1866 it was officially announced in Cape Coast that in the event of any future invasion of the Protectorate, the people were to rely entirely on their own efforts to repel it unless the forts themselves were directly threatened.[296] A Commission issued a constitution, outlining instructions by Major S. W. Blackall on his appointment as Governor-in-Chief of British West Africa, stating that the Charter of 1850 be revoked and all the British settlements were once more united under one government.[297] The Legislative Council was retained but the Executive Council was abolished. The Permanent members of the Legislative Council were to be three principal officers of the settlement, namely, the Administrator of the Government, who replaced the Governor and the Colonial Secretary, the Collector of Revenue, who was also the Treasurer, and the Magistrate. The Supreme Court was also abolished and replaced by 'the Court of Civil and Criminal Justice' presided over by the Chief

292 Claridge, i, pp. 531–56.
293 Metcalfe, *Great Britain and Ghana*, p. 305.
294 Resolutions of the Select Committee, 26 June 1865, in Metcalfe, *Great Britain and Ghana*, pp. 311–12, doc. 248.
295 Minute by T. F. Elliot, 22 November 1865, in Metcalfe, *Great Britain and Ghana*, pp. 312–13, doc. 251; Cardwell to Conran, 23 November 1865, *ibid.*, p. 313, doc. 252.
296 Claridge, i, pp. 537–8; Cardwell to Blackall, 23 June 1866, in Metcalfe, *Great Britain and Ghana*, pp. 315–16, doc. 256.
297 Cardwell to Blackall, 23 February 1866, in Metcalfe, *Great Britain and Ghana*, pp. 314–15, doc. 255; Bennion, pp. 14–15.

Magistrate. The Judicial Assessor was retained and he and the magistrate were to continue to exercise jurisdiction outside the forts.

Thus, since the Municipal Councils had been abolished five years earlier,[298] except for the Legislative Council, the British position in Ghana had reverted to where it was at the time of the end of Maclean's administration but without the prestige and power upon which it had rested. It looked in 1866 as if the British were preparing to abandon Ghana.

The exchange of forts

However, this set-back turned out to be temporary, for within eight years of the promulgation of the 1866 constitution, the British had become even more deeply involved in the affairs of the country than at any time during the nineteenth century, and converted their forts and settlements into the full-fledged British Crown Colony of the Gold Coast. The first step in this direction was the resumption of the negotiations for the exchange of forts on either side of the Sweet river.

Initiative for this came from the British in Ghana who found the 1865 resolutions impracticable in view of the realities with which they were confronted. Conran, Acting-Governor from 19 August 1865, not only issued a proclamation extending the sphere of British jurisdiction in Ghana, which, as we have seen, was forbidden by the Colonial Office, but urged the exchange of forts. He felt that a division of the coast and protectorate would be the safest plan for the removal of the difficulties facing the British, 'the natives not having the slightest prospect of self-government', he stated, 'and even if they had, the total want of union which has from time immemorial existed . . . and will for ever exist, I fear, renders the prospect of their being able to govern so remote that we must either adopt some plan such as I have suggested, or content ourselves with the prospect of having a most difficult and unsatisfactory task to labour under in governing the Gold Coast.'[299] He also contended it would also compel the Dutch to participate in 'the common defence of the country which has hitherto devolved upon us', would help them to cut off supplies from Asante in the event of war and, finally, would enable each power to raise a revenue of

298 The reasons for the repeal of the Municipal Ordinance in 1861 were given in the preamble to the repealing Ordinance, which stated: 'It has been found by experience that in the existing state of the said towns and of those in Her Majesty's Settlements on the Gold Coast generally, the satisfactory working of such elective municipal institutions is impracticable, particularly in consequence of the coexistence of the courts of the native kings or chiefs, from the want of a sufficient proportion of educated residents able to understand and willing to assist in their operation, from the impossibility of raising a revenue. . . . The effect of the establishment of them has been to produce serious quarrels, disturbances, and ill-will between different classes of the people. . . . With a few individual exceptions, the people of both towns earnestly desire that such municipalities should be discontinued. . . .' Quoted by Bennion, p. 14.

299 Conran to Blackall, 6 September 1866, in Metcalfe, *Great Britain and Ghana*, p. 316, doc. 257.

not less than £12,000 a year by means of import duties. As we saw, the main reason for the failure of earlier negotiations was the refusal of the British to include Wassa in the deal. But the British were now ready to abandon the area partly because it was in line with the current policy of retrenchment but mainly because of Conran's timely report on the change of attitude of the Wassa, their people being 'so bad and disobedient to our rule that in place of its being a hardship in placing them under Dutch protection, owing to their great distance from here and their proximity to Axim, a Dutch settlement, I consider it would be a blessing'.[300]

The Dutch, too, were prepared for re-negotiations because their trade, declining throughout the century, had become very insignificant by the 1860s and their settlements were being maintained at an annual loss of about 120,000 florins, with complaints in the States General about 'retaining these burdensome and profitless settlements'.[301] It would appear that the main reason for their continued interest on the coast was that since the 1830s they had been buying slaves from the Asantehene to serve as soldiers in Java and their other East Indian colonies. But even this scheme had lost its importance by the late 1860s. In fact, from 1867 to 1872 only three shipments of recruits totalling two hundred and nineteen men were made from Elmina.[302]

Thus, with renewed willingness on both sides, a settlement was finally reached and the Convention for the exchange of forts was signed in London on 5 March 1867 and ratified on 5 July.[303] In the Convention[304] Britain conceded 'all British forts, possessions and rights of sovereignty or jurisdiction, which she possesses on the Gold Coast to the westward of the mouth of the Sweet river where their respective territories are coterminous'. These possessions were Apollonia, Dixcove, Sekondi and Komenda. In return, the Dutch conceded from east of the river Mori, Kormantin, Apam and Dutch Accra.[305] The boundary between the two possessions would be a line from the centre of the mouth of the Sweet river northwards 'as far as the boundary of the present Ashantee kingdom'. Thus Wassa and Denkyira now fell to the Dutch. Finally, both European powers agreed to impose uniform customs duties.

Had this exchange worked smoothly, the Dutch and British might have consolidated their positions in their respective spheres but from the

300 Coombs, p. 42.
301 Claridge, i, pp. 557–8; Coombs, p. 14.
302 Claridge, i, pp. 558–9.
303 For a detailed and authoritative discussion of the complicated and rather protracted negotations leading to the transfer of the Dutch forts to the British, see Coombs, pp. 57–120. The following section is based mainly on that study.
304 Convention between Her Majesty and the King of the Netherlands, 5 March 1867, in Metcalfe, *Great Britain and Ghana*, pp. 320–1, doc. 259.
305 The Dutch fort at Accra, Fort Crèvecoeur, which the British took over in January 1868, was repaired and renamed Ussher Fort, after the administrator of the British settlements at the time.

beginning it provoked violent reaction and rebellion among the peoples of the coast who were not consulted at any stage in these negotiations and were bartered, as the *African Times* put it in 1870, 'like so many bullocks'.[306] These revolts were particularly vehement in the Dutch areas and the Van Bosse Ministry which took office in June 1868 soon expressed their desire to abandon the coast altogether. A year later they were so anxious to leave that they were prepared not to charge for the forts themselves but only for 'the stores and movable articles'. In 1870 Granville, the British Colonial Secretary, said during negotiations that by agreeing to take over the forts 'England was doing the Dutch a favour—and the Dutch envoy in London agreed'.[307] Granville was, of course, bluffing and he knew it, for the British, especially the officials and merchants on the coast, had been pressing for the purchase of the Dutch and Danish forts since the 1820s if not earlier. MacCarthy and the merchants had strongly advocated this in 1822 and had successfully persuaded the Colonial Secretary of it; only MacCarthy's death had put an end to the scheme.[308] Turner, his successor, had felt the same and the Legislative Council expressed the same opinion in 1851 as did Ord and Pine about six years later. In March 1868, Ussher, the Administrator, reported that the departure of the Dutch 'would be hailed with joy, and our rule is the only one desired' and a month later pressed for the purchases 'as a remedy for present ills, but also with a view to future commercial benefits and forestalling France' which he described as 'a willing purchaser'.[309] By 1869 Ussher's superior, Kennedy, the Governor-in-Chief, had also become a convert to this cause. Granville, in October 1869, informed Clarendon, the Foreign Secretary, 'If the Colony can pay for it, which I believe Kennedy thinks it can, it would be well worth buying Elmina. Whether we remain on the Coast, or retire from it, it is equally desirable for Trade and Peace to get rid of the Dutch.'[310]

From this time Britain welcomed an initiative from Holland for the cession of their forts but two hurdles had still to be cleared; the King of Holland was reluctant to cede any Dutch territory and the British insisted that the Dutch obtain a renunciation of Asante's claim to Elmina. Knatchbull-Hugessen minuted that nothing should be settled 'until the Dutch have set matters right with the Ashantee King, and got rid of his savage relation [i.e. Akyeampon] from Elmina'.[311] By 1870 the Dutch king's scruples were overcome but the second hurdle proved more difficult to clear. As soon as news of the probable transfer reached

306 *African Times*, 24 October 1870.
307 Coombs, p. 8.
308 *Ibid.*, p. 25.
309 *Ibid.*, pp. 58–9; Ussher to Kennedy, 6 April 1868, in Metcalfe, *Great Britain and Ghana*, p. 323, doc. 262.
310 Coombs, p. 63.
311 Coombs, p. 83.

Kumasi, the Asantehene immediately reasserted his claim to Elmina maintaining, 'The fort of that place has from time immemorial paid annual tribute to my ancestors to the present time by right of arms when we conquered Intim Gackidi, King of Denkyira' and that his ancestors had paid £9,000 demanded by the Dutch for this right.[312] The Dutch Governor, Nagtglas, to whom the Asantehene had addressed this letter, expressed utter surprise pointing out that the annual payment of tribute to Asante was a 'present' and denied Asante's claim to Elmina. Though the Dutch case was clearly unconvincing, the British Colonial Office expressed satisfaction 'as to the invalidity of the King of Ashantee's title to Elmina'[313] and the treaty of cession was signed on 25 February 1871.[314] However, the treaty was not ratified until a year later.

Certificate of Apology

Many historians[315] have believed that the ratification was delayed because of British doubt of Asante's renunciation of Elmina and that it was not until after the Asantehene's celebrated 'Certificate of Apology' dated 19 August 1871 had been received that the treaty was ratified. Some historians go further and claim that the Apology was a forgery by the Dutch envoy and gave the Asantehene cause to invade the coast in 1873. For instance, Ward wrote: 'This document, if genuine, was the very thing that the two governments wanted, for it removed all scruples', but he added, 'a forgery it almost certainly must have been'. Yet, Ward maintains, both governments accepted it as genuine.[316] The Apology did seem to be, as Claridge calls it, 'an extraordinary document'[317] written by Plange, the Dutch envoy. In it the Asantehene claimed that he had been misinterpreted, that the money from the Dutch was 'wages or salaries' and not tribute and that the £9,000 was paid to 'ensure friendship and goodwill . . . *but not for purchasing the said Elmina Castle or Fort*'.[318] And in a covering letter in Dutch to Nagtglas he further declared 'if anyone says I consider Elmina my rightful property, it is a lie'.[319]

With painstaking and scholarly scrutiny, Coombs has shattered these former cherished beliefs about the significance of the Certificate of Apology.[320] He shows that the British had become convinced of Asante's

312 Claridge, i, p. 603.

313 Minute by Rogers, 20 March 1871, in Coombs, p. 101, see also pp. 64 and 84 of Coombs.

314 For the full text see Metcalfe, *Great Britain and Ghana*, pp. 341–2, doc. 284.

315 A. B. Ellis, *History of the Gold Coast*, 1893, i, pp. 270–3; Claridge, i, pp. 610–13; Ward, p. 242.

316 Ward, p. 242.

317 For the full text, see Claridge, i, pp. 609–10; Coombs, pp. 98–9.

318 The italicised phrase has been added by Coombs presumably from the original document.

319 Coombs, p. 99.

320 *Ibid.*, pp. 98–120. The following sections are based mainly on this chapter.

renunciation five months before the Apology and that, though the Apology was received at the end of August, it was not until six months later that the treaty was ratified. Coombs suggests that the lack of interest in the Apology on its receipt proves its practical insignificance, that not a single member of the Colonial Office 'thought it called for any comment whatsoever'. Thus the Apology was not as important in the signing or ratification of the Treaty as historians have claimed. I endorse Coombe's conclusion that 'this document, like the so called "Bond of 1844", would seem to have played a part in Gold Coast historiography that is out of all proportion to its actual importance'.

Coombs also claims that the Apology can be proved to be genuine. He says that it was written by the Asantehene with the deliberate aim of deceiving the Dutch and thereby collecting an outstanding tribute of £320 for the purchase of ammunition and arms desperately needed for the war that he was actively preparing. Coombs further quotes a letter written by Ussher to Crawford, another British envoy, reporting an interview with the Asantehene on 5 August during which he again claimed that he had been misinterpreted saying the Elmina and Asante were 'allies, friends and brethren' and Elmina had never paid tribute. He added that he disagreed with the cession of the fort to the English because of the 'present' to him by the Dutch and not because of 'tribute by right of conquest'. Moreover, neither Ansa, who had written the Asantehene's letter of 24 November 1870, nor Afrifa and Koliko, the Asantehene's ambassadors, denied the authenticity of the Apology. The Asantehene did not show any surprise or anger when in December 1871 Kühne read the Dutch Governor's letter acknowledging receipt of the Apology to him. It seems clear, therefore, from facts and circumstantial evidence, that the Apology was indeed genuine. It is true, as Ellis and Claridge point out, that the Asantehene still claimed his right to Elmina shortly afterwards in a letter written to the British Governor[321] but this was surely because, since he had no claims to make on the British, he had no reason to dissimulate his intentions. There is no doubt that the Dutch refusal to pay the outstanding £320 was one of the factors that made an Asante invasion of the coast an inevitability for the Asantehene's original plan when he sent the Apology had misfired.

Delay in the ratification of the Treaty was rather due to the delay of the approval of it by the two houses of the States-General and to the presence of Akyeampon in Elmina. Owing partly to the opposition of the Elmina and some commercial circles in Holland and partly to the negotiations over the Siak and Suriman treaties, the treaty of cession was approved by the Second Chamber of the States-General in July 1871 but the First Chamber did not pass it until 18 January 1872.[322]

321 Claridge, i, pp. 612–13.
322 Coombs, pp. 88–98, 108–11.

Coombs has shown that after March 1871 British anxiety about Akyeampon's continued presence in Elmina was the one remaining obstacle.[323] In April the Dutch Governor arrested and imprisoned him in the castle but released him in May only to have him rearrested in June for failing to leave the fort within thirty days of his release. Akyeampon was not released again until November, after receipt of the Apology, with a view to placating the Elmina who were becoming increasingly restive.[324] These continued arrests and rearrests convinced the British of Akyeampon's weakening position in Elmina and the ratification did at last take place on 17 February 1872 though the actual ceremony of transfer was not held until seven weeks later on 6 April. This last delay was undoubtedly due to the indecision of the Elmina because of Akyeampon's continued presence. However, by the end of February, five of the eight quarters had declared in favour of the transfer and a unanimous vote was taken the following month after which Akyeampon was tried on a charge of sedition and banished with his followers to Assini.[325] They were transferred to Cape Coast in October and in December they were allowed to return to Kumasi. The ceremony for the handing over of the Dutch forts was publicly carried out at Elmina without incident and thus ended a two-centuries-old connection of the Dutch with the coast of modern Ghana.

This withdrawal was a significant landmark in the conversion of the area into a British colony. The four-hundred-years-old inter-European rivalry was over and the British from 1872 were supreme over the entire stretch of territory from Apollonia to the Volta. Animosity and internecine warfare among the coastal states had also ended and, more important, there remained no obstacle to the extension of British authority and influence, the major one having been revenue from import duties and other local resources. The raising of revenue had been the main cause of British interest in the Dutch possessions. British trade boomed after 1872 and revenue increased from £28,609 in 1871, the year prior to the cession, to £93,347 in 1877 and exceeded £100,000 in 1878.[326] Local British officials and merchants now embarked on more ambitious projects while Colonial Office objections to expansion based on increased expenditure and uneasy trade lost their force.

ORIGINS OF THE FANTE CONFEDERATION MOVEMENT

The departure of the Dutch was the first clear landmark in the process

323 *Ibid.*, pp. 83–4.
324 *Ibid.*, pp. 83, 85–6, 94, 108–9.
325 *Ibid.*, pp. 116–17; Claridge, i, pp. 612–13.
326 Duties increased from £28,609 in 1871 to £40,165 in 1872, £65,706 in 1873, £74,868 in 1874 and, after dropping to £67,368 and £64,788 in 1875 and 1876 respectively, obviously because of the Sargrenti War, rose again to £93,347 in 1877 and exceeded £100,000 in 1878. For the trade and revenue figures from 1853 to 1956, see Metcalfe, *Great Britain and Ghana*, appendix D, pp. 750–2.

of the conversion of the isolated forts into the British Crown Colony of the Gold Coast in 1874 and the second landmark was the suppression of the Fante Confederation Movement and its eastern counterpart, the Accra Native Confederation. Many historians have concentrated on the former movement but its origins, true nature and significance have not been adequately explained. Many attributed its origins to the political events of the 1860s: Claridge reports that it owes its origins to 'the disgraceful fiasco of 1863'; Fuller states quite categorically that the Sweet River Convention 'gave rise to the Fante Confederation'; Ward maintains that 'it began with the meeting held at Mankessim [early in 1868] which decided to support Komenda in resisting the Dutch flag'.[327] Others talk of the Confederation in disparaging terms. In fact, Coombs has referred to it as 'a so-called Fante Confederation'.[328] The Fante Confederation, which in spite of its name, consisted of Wassa, Denkyira, Assin and Twifu as well, was a genuine and real nationalist movement whose origins go much further back than the 1860s. It was nothing more than the culmination of years of resistance to the extension of British power and jurisdiction in Ghana dating from the late 1830s, of the perennial fear of Asante domination of the coast and of the activities of the new educated élite generated by the phenomenal economic and social changes that occurred in the nineteenth century. The events of the 1860s did not give rise to the Fante Confederation Movement, they merely precipitated it.

Role of the chiefs in early protest movements

The movement was primarily a reaction against the growth of British power and jurisdiction, a growth which has already been traced and which brought about a corresponding decrease in the power and prestige of the traditional rulers and a steady encroachment on the independence won from Asante. Cruickshank, a contemporary witness, describes in terms typical of his age how the attempts at impartial laws and justice and elevation of the people which curtailed the power and sources of revenue of the chiefs 'made them sigh for a return of former lawlessness, and almost regret that they had thrown off the yoke of the king of Ashantee, who, provided he obtained his revenue, did not care how much his vassal chiefs oppressed their people'.[329] Naturally individual chiefs 'upon several occasions attempted to assert their independence, and to resist the authority of the government'.

These protests began as early as the 1830s. In 1834 the King of Denkyira, Kojo Tsibu, who had been fined £200 by Maclean for sacrificing human beings on his sister's death, sent a petition to the Secretary of State for the Colonies complaining of Maclean's action. The King of Nzima, the notor-

327 Claridge, i, p. 64; Fuller, p. 101; Ward, pp. 233, 246.
328 Coombs, p. 52.
329 Cruickshank, ii, pp. 11–12.

ious Kweku Aku, also openly defied Maclean.[330] Resistance towards the end of Maclean's service in Ghana was even stronger. For instance, in 1846, when Maclean sent a policeman to Tantum to arrest the chief who had refused to answer a summons, the people attacked him, rescued the chief and declared openly their determination to resist by force any attempt to enforce British law in the town.[331] In the same month the King of Gomoa also ignored a summons by Maclean. Maclean's reaction was sharp and he obtained permission to send a force of twenty-five regulars with some policemen to quell the rebellion in Tantum and impose a heavy fine. He similarly imposed a heavy fine on Gomoa.[332]

However, resistance intensified, especially after Maclean's death in 1847 when British judicial control was wider and more direct. When Tsibu, King of Assin, began his intrigues with the Asantehene in 1852 it was mainly because of the way he had been treated for, as Claridge has pointed out, 'rightly or wrongly . . . the Government was ignoring the power and dignity of the chiefs too much'.[333] In 1853 the educated Africans in Accra not only set up their own court under the presidency of J. R. Thompson, a schoolmaster, but four African merchants in Cape Coast, namely, Henry Barnes, Joseph Smith, Thomas Hughes and William de Graft, also challenged the authority of Fitzpatrick, then Judicial Assessor, in bringing the King of Cape Coast to trial in a civil action and this led to the virtual suspension of Fitzpatrick from office.[334]

This early phase of protest came to a head in the 1850s in the agitation over the Poll Tax scheme. Though, as we have seen, this protest was mainly against methods of collection, it was also against the growth of British jurisdiction. It is significant that opposition to the tax was strongest in the eastern districts, which the Danes had recently left, and where British control was relatively new. Hill correctly pointed out in 1854, while commenting on the Christianborg rebellion,

> our stringent rules . . . are strictly enforced, whereas under the Danish government, too weak to control them, these people did whatever they pleased, and committed many acts of barbarity, without notice or punishment. . . . The whole of the tribes transferred from the Danish Government are most uncivilised, excitable, and difficult to manage . . . and they would gladly shake off their sworn allegiance to the British Government, if it were not for the protection afforded them against more powerful tribes . . .'.[335]

330 Metcalfe, 1962, *Maclean of the Gold Coast*, pp. 151–8.
331 *Ibid.*, p. 323.
332 *Ibid.*, pp. 323–4.
333 Claridge, i, p. 486.
334 Kimble, p. 196.
335 Hill to Newcastle, 29 January 1854, in Metcalfe, *Great Britain and Ghana*, pp. 246–7, doc. 193.

Thus, when the Christiansborg people hoisted a Danish flag during the rebellion, they were performing more than a symbolic act.

The Krobo rising in 1858 was also partly attributable to resistance to British authority. One of the officials reported in November of that year:

> For years past, the Crobboes, in common with all the former Danish subjects, have shown a restless and turbulent disposition, and a strong desire to get rid of the controlling power of the British Government. . . . The chiefs have certainly felt what they never experienced under Danish rule—themselves being governed and their extortions being curtailed.

He continues significantly though unkindly,

> before the poll tax ordinance was passed, this irritation was confined to the kings and chiefs. They, however, by accepting the ordinance and complying, for the first year, with its provisions, have probably used it as an engine to excite dissatisfaction among the masses towards the new government.[336]

If the Poll Tax scheme reinvigorated anti-British feeling in the central and eastern districts, the abortive Anglo-Dutch exchange of forts in 1859 and 1860 generated strong suspicions of the intentions of the British in the western districts, especially in Eastern and Western Wassa and Denkyira. Until 1859 the Wassa, mainly because of their morbid fear of Asante, were exceedingly docile towards the British and, according to Bird, 'always paid our government a considerable amount of poll tax'. However, when news of the proposed exchange of the 1850s leaked out, Bird reported that the people were determined never to submit to Dutch authority. Though the scheme was dropped the Wassa never regained their former confidence in the British and grew steadily more disobedient so that by September 1866 Conran reported their conduct to be 'so bad and disobedient to our rule that in place of its being a hardship in placing them under Dutch protection . . . it would prove a blessing'.[337] Wassa's resistance paved the way for renewed exchange negotiations in 1867 and the eventual withdrawal of the Dutch from the coast.

Resistance to British authority reached its climax between 1864 and 1865 with the issuing of the 'Grievances of the Gold Coast Chiefs'[338] and the enstoolment of John Aggrey as the King of Cape Coast. In this document the chiefs outlined their grievances as being deprived of the power of holding slaves, which had 'greatly reduced their power and dignity', that 'they are for very trifling cause now and then cast into prison by the officials,

336 Bird to Lytton, 10 November 1858, in Metcalfe, *Great Britain and Ghana*, pp. 279–81, doc. 219.

337 Coombs, p. 42.

338 Grievances of the Gold Coast Chiefs, 9 August 1864, in Metcalfe, *Great Britain and Ghana*, pp. 300–1, doc. 243.

which insures to them great disgrace, and places them upon the same par with their subjects', and they urged that they should once more be involved in justice in the courts, that the abolition of household slavery has meant the country is 'totally deprived of its wealth. That the kings and chiefs are not to convert themselves to shepherds, farmers and such like' and finally they urged the British 'not to tender the method of peacemaking to the enemies until they [the enemies] asked for it, because it will add more shame to the British flag'. Two months later the chiefs again protested to the Governor for introducing an annual licence fee of £2 to be taken out by all traders in wines and spirits without first consulting them.

Aggrey of Cape Coast

However, it was Aggrey of Cape Coast who was the first to challenge British authority with open defiance. Shortly after his enstoolment in February 1865, and working in close collaboration with the educated class, the most important of whom was Charles Bannerman, described in British despatches as 'a native advocate possessing much talent but totally devoid of principle',[339] Aggrey objected to appeals against the decisions of his court being sent to the British courts, denied that his courts were irresponsible and even went on to question the whole basis of British authority. In his now famous letter to the Governor in March 1865[340] he vehemently censured Maclean who, he stated, 'in a peculiar, imperceptible, and unheard of manner, wrested from the hands of our kings and chiefs and headmen, their power to govern their own subjects'. He pointed out that Cape Coast, 'in the eyes of law, is not British territory' and he further challenged the Governor's request that the two of them should come to a mutual understanding about the establishment of a court to be called the 'King's Court' on the grounds that 'the King's court is not of yesterday' but from 'time immemorial it had existed, and it even existed before Cape Coast Castle itself was erected, and the ground on which the castle stands was originally taken from my ancestors at an annual rent. . . . We have already protested . . . against the inhabitants of Cape Coast and other places being regarded as British subjects.' He ended his letter by declaring his intention of referring the question to the British Government. It is not surprising that Hackett of the Colonial Office informed Pine that he read Aggrey's letter 'with pain because it is the first instance within my recollection in which the supremacy of British tribunals on the Gold Coast has been disputed' and that 'if the claims of the king of Cape Coast be conceded in this case, British influence in the protected territories is at an end and the office of Judicial Assessor become practically useless.'[341]

339 Kimble, pp. 200–1.

340 Aggrey to Pine, 16 March 1865, in Metcalfe, *Great Britain and Ghana*, pp. 308–9, doc. 245.

341 Hackett to Pine, 22 May 1865, in Metcalfe, *Great Britain and Ghana*, pp. 309–10, doc. 246.

Pine's reaction was not only to threaten to withdraw recognition of Aggrey as King of Cape Coast but, in the hopes that Aggrey had little support, he hastily convened a public meeting to discuss 'all matters connected with the Protectorate, and the king of Cape Coast's attempt to sever it from British rule'.[342] Chiefs and captains of the Cape Coast companies boycotted the meeting but it was attended by more than a hundred inhabitants of Cape Coast, Anomabo and others including the King of Winneba. They drew up a petition dated 14 April 1865 in which they dissociated themselves from Aggrey and insisted on the continuance of British protection.[343] Shortly afterwards Aggrey, undeterred, sent a deputation of two, Joseph Martin and D. L. Carr, to England to appear on his behalf before the Select Committee of 1865. These witnesses emphasised the lack of respect shown to Aggrey and the local rulers, the increasing cost of administering justice because of the use of attorneys and, above all, criticised the Judicial Assessor for virtually superseding the authority of the chiefs and recommended the abolition of that office.[344]

The recommendations of the Select Committee and the subsequent steps taken by the British in Ghana have been disussed already. They made the situation even worse for they suggested that Aggrey and other educated Africans were right, and not wrong as Claridge contends,[345] in thinking that the British would withdraw sooner rather than later and that, as Aggrey told Conran, 'the Government in England had expressed its desire that we the kings and chiefs of the Gold Coast are to prepare ourselves for self-government and no protection'. Their defiance therefore became sharper.[346] In March 1866, a few weeks after central government had been re-established for all the British settlements, Aggrey wrote to the newly appointed Governor-in-Chief, William Blackall, then visiting Ghana, asking him to define 'the relationship between the King's Court and the British Magistrates' court; between the King and the Governor, and between the King and his brother Kings on the coast'.[347] He also repeated many old grievances, asked why he did not receive a share of the customs and other revenues and declared his intention to set up his own military corps for the purpose of self-defence. Conran, the Administrator, condemned this letter as seditious and the Colonial Office supported the local officials' opposition to its contents; 'Aggrey's proposal to establish a military force is mischievous,' the Secretary of State ruled, 'and his claim to revenue inadmissible, and his pretensions should be effectually dis-

342 Kimble, p. 193.
343 *Ibid.*, pp. 204–5.
344 1865 Report, see n. 294.
345 Claridge, i, pp. 538–40.
346 *Ibid.*, pp. 547–52.
347 Kimble, pp. 214–15.

countenanced.'[348] He asked Blackall to inform Aggrey that in return for British protection he should defer to British authority.

Aggrey continued his defiance and when Conran refused to recognise Aggrey's appointee, Thomas Hughes, as 'Representative' of the King and people of Cape Coast to assist Aggrey 'as regards order, civilisation, improvement and welfare of the people', Conran's refusal was described as curtailing the civil liberties of the people of Cape Coast in a manner unheard of in a civilised country. In September 1866 Aggrey and his followers again objected to the indignities to which they were subjected and the whole arbitrary system of British rule.[349] This clash between Aggrey and Conran reached a climax in December when Conran released several of Aggrey's prisoners as a result of private petitions. Aggrey wrote to Conran:

> The time has now come for me to record a solemn protest against the perpetual annoyances and insults that you persistently and perseveringly continue to practise on me in my capacity as legally constituted king. . . . However much you may wish to have me and my people under martial law, you will never have that pleasure.

He went on to say that he would appeal to Carnarvon in London and

> if some tangible satisfaction is not accorded to me and those whose interests I am bound to protect, it will be time enough for me to adopt those measures which will ensure to me and my people something unlike the slavery that you are endeavouring to place us in. . . .[350]

Taking this letter as the last straw Conran promptly arrested Aggrey, declared him deposed and exiled him to Sierra Leone.[351] This action was fully endorsed by the Colonial Office. There is no doubt that it greatly exacerbated feeling on the coast and, as an eye-witness reported, throughout the Protectorate people were losing faith in the British Government and especially in Conran for punishing the king 'for giving notice that he will complain to the Earl of Caernarvon'.[352]

Not until 1869 was Aggrey allowed to return to Cape Coast and then only as a private citizen, a condition which was of course ignored by his loyal subjects, and he died later in the same year. It is a pity that this great ruler has not yet been accorded his deserved place in the history of the struggle for self-government and the challenge to British rule and

348 *Ibid.*, p. 215.
349 Petition of September 1866 from Aggrey and forty-three others, quoted by Kimble, p. 216.
350 Claridge, i, p. 553; Kimble, p. 217.
351 Kimble, p. 218.
352 *Ibid.*, p. 219.

institutions. His reign had been a brief but momentous one. Conran himself admitted in February 1866 that the changed attitude of the coastal peoples can be attributed to 'the efforts made by King Aggrey and his Councillors to spread reports amongst the natives'.[353] He was the first king to work in close collaboration with the new educated élite—indeed Claridge thinks he was their pawn, but he is wrong. He was also the first to present his case in reasoned terms and the first to suffer deposition and exile for championing the cause of self-government. Most important are the things he struggled for, namely, self-government and respect for traditional laws and custom. His methods, cooperation between the chiefs and the use of the educated élite, outlived him. His deposition and exile only a year before the launching of the Fante Confederation Movement was no coincidence.

Neither his death nor exile ended the explosive situation created by years of resistance to British jurisdiction. His successor, Kwesi Atta, chosen by the British because he was 'possessed of little or no influence', was outlawed and deprived of his house and property only a year later for defying the instructions of the Administrator.[354] This open defiance was also present in Anomabo, Abora, Gomoa, Agona and Wassa in the western districts and in Accra in the eastern districts.[355] The challenge to British authority in Accra was particularly strong. It was led by the three Kings of Accra, Cudjoe of James Town, Dowounah of Christianborg and, the most influential and powerful of them all, Tackie of Dutch Accra. Their defiance embraced the states of Akyem, Akuapem, Krobo, Ada and Peki.[356] The three Accra kings flatly refused to move their camp in Ada after their war with the Anlo in 1866 and in the following year the Ga refused to end the celebration of a custom which the Governor considered barbaric. The constables and officials sent to stop the celebration were pelted with stones. Ussher himself wondered 'whether to attribute this to any organised attempt at revolt, or a new wild desire to return to their former barbarous practices on the part of the Accras'.[357] There were other instances of defiance in this area at the time. As Agbodeka has pointed out,

> so widespread was the disobedience in the Eastern districts that Ussher was [by July 1867] appealing to the Duke of Buckingham, Head of the Colonial Department, to give favourable consideration to his suggestion that a resident officer be appointed in the distant portions of the

353 Conran to Cardwell, 5 February 1866, in Metcalfe, *Great Britain and Ghana*, p. 313, doc. 254.
354 Kimble, p. 221.
355 Claridge, i, p. 547; F. Agbodeka, 'The Fanti Confederacy, 1865–1969', *T.H.S.G.*, vii, 1964, pp. 82–123 (especially pp. 88–90).
356 Agbodeka, pp. 90–4.
357 Ussher to Young, 24 September 1867, quoted by Agbodeka, p. 93.

> 'Protectorate' to travel . . . and generally promote British interests in the interior.[358]

The widespread and deep-seated nature of this challenge to British authority should be enough to disprove the official attitude at the time, adopted by some recent historians, that it was simply the outcome of intrigues and ambitions of semi-educated or educated Africans. It was the culmination in the middle sixties of a long-term dissatisfaction with the gradual extension of the authority and jurisdiction of the British.

Fear of Asante and the Fante Confederation Movement

However, this was not the only factor leading to the launching of the Fante Confederation Movement and the Accra Native Confederation. Equally, if not more, important was the perennial fear of the domination of the coastal areas by Asante. This fear reached a new climax in the mid-sixties following the Asante invasion of 1863 and the British reaction to it. As we have seen, the victories won by the Asante, the absolutely ineffective assistance initially given to the Fante by the British and the disastrous end to their rather belated counter-invasion of 1864 not only damaged British prestige but convinced the coastal peoples of the need for a strong, united front against their inveterate enemy. The Colonial Office itself advocated unity among the coastal states maintaining in a letter to the Governor that

> If they are not united, and will not take upon themselves the principal part of the exertions necessary, it will not be possible to defend them without exposing the Queen's Forces to the risks of a deadly climate, and to the hazard of being virtually defeated by the disastrous consequences of that climate, before they have been able to bring the native enemy to the issue of arms. The proper course, therefore, is to take every possible means for bringing the chiefs to a united and decided system of defence. . . .

The letter went on to outline possible assistance that could be given for this purpose.[359]

At a meeting held on 5 July 1864 Pine conveyed these instructions to the chiefs who naturally received them with surprise and Pine added in his report that 'a feeling of discontent and dissatisfaction pervades the Protectorate'.[360] Instead of assuaging fears of Asante, the Colonial Office two years later asked Blackall to impress 'on the natives of the Gold Coast,

358 Agbodeka, p. 94.
359 Cardwell to Pine, 23 June 1864, in Metcalfe, *Great Britain and Ghana*, pp. 298–8, doc. 241.
360 Pine to Cardwell, 13 July 1864, in Metcalfe, *Great Britain and Ghana*, pp. 299–300, doc. 242.

the necessity of relying on their own exertions for their self defence'.[361] It is not surprising that some of the educated and wealthy men began to talk of forming private armies for their self-defence. It will be recalled that Aggrey announced his intention of forming his own army in March 1866 and, though the Colonial Office described this intention as mischievous, there is no doubt that the need for some form of self-defence in the face of the clearly declared policy of the British and the constant threats in 1866 and 1867 of Asante invasions must have been felt throughout the 'Protectorate'. Thus the coastal peoples were bound to come together in some form of defensive union or unions sooner or later.

What finally precipitated union of the coastal and inland states between the Pra river and the borders of the Ga kingdom into the Fante Confederation Movement were first the Anglo-Dutch exchange of forts of March 1867 and secondly the activities of the educated élite or, as they were described in reports, 'a small class of discontented and unprincipled mulattos and semi-educated blacks who appear to be an evil inseparable from all negro communities'.[362] News of the agreed exchange came in January 1868 and infuriated not only the Fante but also the Denkyira, Wassa and the Twifu, partly because they were not consulted but mainly because they saw it as a prelude to the occupation of the areas that would fall to the Dutch by Asante. As Ussher reported in March 1868, the loyalty of the Fante chiefs has been 'shaken to its foundations . . . by the provisions of the convention of March 5 by which, they assert, the English have sold them and their brothers to the Dutch, and subsequently, as they say, to the Ashantees'.[363] Those states to come under Dutch authority were genuinely alarmed. They looked upon the Dutch, as Ussher reported in April 1868, 'with contempt for their apparent want of power, and contrast their cautious policy with the vigorous and determined attitude so long maintained by Great Britain against Ashantee'.[364] It was the King of Denkyira who hurriedly despatched ambassadors to the Fante people to ask about their reaction to the news, and it was the rising of Komenda in a determined resistance to the exchange in January 1868 and the shock of the bombardment of their town by the Dutch that led to the meeting at Mankessim of the Fante Kings as well as delegations from Denkyira, Wassa, Twifu and Assin. The meeting passed 'a series of resolutions . . . to support and protect themselves, regardless of British interests or protection', to oppose the

361 Cardwell to Blackall, 23 June 1866, in Metcalfe, *Great Britain and Ghana*, pp. 315–16, doc. 256.

362 Ussher to Kennedy, 6 April 1868, in Metcalfe, *Great Britain and Ghana*, p. 323, doc. 262.

363 Ussher to Kennedy, 19 March 1868, in Metcalfe, *Great Britain and Ghana*, .p 322, doc. 261.

364 Ussher to Kennedy, 6 April 1868, in Metcalfe, *Great Britain and Ghana*, p. 323, doc. 262.

Anglo-Dutch exchange and assist Komenda, and to form a government which would 'be to ourselves a head, having no king under the British'.[365]

Role of the educated élite

This assembly of chiefs in Mankessim was also prompted by the activities of the 'scholars'. Ussher attributed at least part of the trouble to these people in the same report of April 1868, stating that this group

> is active in its endeavours to persuade the ignorant, impressionable, and childlike Fantees, that the time has come to govern themselves and to throw off our rule, retaining us here as advisers only . . . and I can trace their evil effects too plainly to the disrespect I have received lately from the central Fantee tribes at Mankessim. . . . They undoubtedly urged the chiefs to assemble a solemn council at Mankessim, avowedly to collect money and concert measures to repel an Ashantee invasion . . . but in reality to foster the great scheme of armed opposition to the provisions of the treaty of interchange. . . .[366]

The scholars' role is no longer denied. As early as the 1850s their activities had become sufficiently important to attract official comment. In February 1855 Merivale of the Colonial Office had minuted:

> But there is growing up an intermediate class—half caste and half civilised—men with a certain amount of English knowledge and ideas derived from the missionaries or elsewhere—and they seem to exert an influence among the Africans by insisting on their anomalous state of dependence—assuring them that they are not 'subjects' of Great Britain, and therefore owe no obedience—which results in movements like this [1854 protest] at Cape Coast, of which more are to be expected.[367]

It was three such men, Joseph Smith, Henry Barnes and T. Hughes, who took a leading part in the opposition to the Poll Tax demanding the right of the African people to a share in the government.[368]

The numbers of educated Africans, products of missionary and economic forces at work since the 1840s, increased with the years and it was they, particularly Carr and Martin, who spread the idea that the British would withdraw sooner rather than later in accordance with the recommendations of the Select Committee, while they, with Thomas Hughes, a West Indian, and Charles Bannerman, became the advisers of King Aggrey. By the middle sixties, these educated Africans had become the scapegoats for all

365 Ussher to Kennedy, 19 March 1868, in Metcalfe, *Great Britain and Ghana*, p. 322, doc. 261; Kimble, pp. 223–4.

366 Ussher to Kennedy, 6 April 1868, in Metcalfe, *Great Britain and Ghana*, p. 323, doc. 262.

367 Minute by Merivale, 29 February 1868, in Metcalfe, *Great Britain and Ghana*, p. 248, doc. 196.

368 F. L. Bartels, *The Roots of Ghana Methodism*, 1965, p. 87.

the troubles on the coast and the main butt for scathing and caustic attacks by British officials. They were described by the Administrator of the 1860s as people 'who cling like leeches to the skirts of their more ignorant kings and chiefs, for the sake of gain, for the mere writing of the commonest letters to the Government and especially the Chief Justice's department, giving the greatest trouble, and causing, what is worse, the greatest discontent'.[369] Indeed, there is no doubt that the meeting at Mankessim was instigated by people like Joseph Africanus Horton, the Sierra Leonean, R. J. Ghartey, James F. Amissah, James Hutton Brew and F. C. Grant. The Constitution of the Federation, drawn up in 1871 to be discussed later, equally bears out the important role of the educated Africans.

The most influential of these men was J. Africanus Horton, a Sierra Leonean then holding the post of Staff Surgeon to the Army in Ghana. In 1868, the year of the launching of the Confederation, he published his political ideas in a book entitled *West African Countries and Peoples*. Its main purpose was to dicuss 'the Requirements necessary for establishing that self-government recommended by the Committee of the House of Commons, 1865'[370] and his main recommendation was that the Gold Coast should be divided into two parts: a Fante kingdom under a king chosen for his sagacity and education and a Republic of Accra each of which was to remain under the auspices of the British Government.[371] Other educated men in the movement were active Methodist laymen or schoolmasters or clergymen in training. Of these the most important was F. C. Grant who was elected first Treasurer of the Confederation. He was described as 'a native gentleman who is certainly not the inferior of any European on the Gold Coast in character, ability or mercantile position'.[372] Educated in England, he became the first headmaster of Freeman's school in Accra and later entered business but remained throughout such a strong and zealous member of the Methodist Church that he was among the first group of laymen ever admitted to the Methodist Synod.[373] The second powerful spokesman was R. J. Ghartey, also 'a well-known businessman and Methodist layman' who was to be elected President of the Confederation in 1869 and to be enstooled as King Ghartey IV in June 1872. Others were J. de Graft Hayford, James F. Amissah, elected Secretary, and James Hutton Brew, elected Under-Secretary, three of the eight men arrested in November 1871 after a copy of the Constitution of the Confederation had been presented to the British. They were all active members of the Methodist Church. A Ghanaian

369 Conran to Cardwell, 5 February 1866, in Metcalfe, *Great Britain and Ghana*, pp. 313–14, doc. 254.
370 Kimble, p. 230.
371 For a detailed discussion of Horton's ideas and recommendations, see Kimble, pp. 230–2.
372 Quoted by Bartels, p. 88.
373 *Ibid*, p. 87.

historian, Bartels, has contended that the Confederation was 'in every way Methodist in its inspiration'[374] and, though the important roles played by Horton, kings such as Aggrey, Otu of Abora and Edu of Mankessim are not included, his conclusion is to a considerable extent justifiable.

The launching of the Fante Confederation was thus the product of political, economic and missionary forces which had been operating in Ghana as far back as the 1830s. From 1868 to 1871 it remained very active but from then on it began to lose the initiative and to decline so that, in Claridge's words, it 'ceased to exist' by about May 1873. What were its achievements during that brief spell of life and why did it cease to exist?

The Confederation's achievements

Claridge and Kimble give the completely wrong impression that the Confederation was so completely preoccupied with drawing up constitutions and establishing a *modus vivendi* with the British that it achieved nothing. But it did register some positive achievements and these, though they proved evanescent, were no mean feats, considering the circumstances of the time. It did succeed in declaring itself independent of the British in August 1868. As Ussher stated, the Fantes 'are bold enough to reject and deny our right of interference in peace and war, in other words they throw off our allegiance', and he went on to admit, with disengaging frankness, that 'our relations with the interior, for the present at least have ceased; and there remains now only the supervision of our trading posts along the seaboard and the collection of revenue'.[375] In other words, the western part of the 'Protectorate' had become the independent Fante Confederacy.

It was hoped to make this confederation a lasting one by giving it an agreed constitution and, for reasons to be discussed later, three were in fact successfully drawn up,[376] in January 1868, November 1871 and April 1872. According to that of January 1868, there was to be a Fante Council to consist of kings and chiefs and seven members elected by each member state, three Joint Presidents, a Magistrate and a Secretary. The leading Fante kings, Edu of Mankessim, Otu of Abora and Ortabil of Gomoa, were made the 'Joint Presidents of the Fantee Nation', Samuel Ferguson was appointed Magistrate and Joseph Dawson or G. Amissah Secretary. The Councillors were sworn in and all the chiefs took an oath to cooperate with each other. The following year, since the kings could

374 *Ibid.*, p. 89.
375 Ussher to Kennedy, 7 August 1869, in Metcalfe, *Great Britain and Ghana*, p. 324, doc. 264; Agbodeka, p. 107.
376 Kimble, pp. 224–5, 247–8, 256–8; Claridge, i, pp. 617–22; Metcalfe, *op. cit.*, 'Constitution of the New Fante Confederacy', 24 November 1874, in Metcalfe, *Great Britain and Ghana*, pp. 336–8, doc. 279; 'Proposals of the Leaders of the Fanti Confederation', 16 April 1872, *ibid.*, pp. 342–4, doc. 285.

not agree among themselves on which one was to become the single President, R. J. Ghartey was made President of the Confederacy Council.[377] This first constitution had been unwritten.

The second constitution was carefully written out and formally adopted at a meeting in Mankessim which lasted from 16 October 1871 to the last week of November. It was undoubtedly inspired and greatly influenced by the ideas of J. B. Horton as outlined in his earlier book and in his *Letters on the Political Condition* published in 1870. This constitution was meant to rouse the Confederation from what Ussher described as its 'apathetic slumber'. The now officially designated 'Fante Confederation' was to be headed by a King-President elected by a body of kings; he was to preside over the meetings of the National Assembly and to give assent to all laws passed by the Representative Assembly. There were to be three representative bodies: an Executive Council to consist of the Vice-President, Secretary, Under-Secretary and Treasurer, all of whom were to be 'men of education' and position, and 'such others as may be hereafter from time to time appointed', with responsibility for preparing and administering legislation, for examining 'carefully the financial condition of the Confederation' and to decide disputed cases of enstoolment and as the final court of appeal; a Representative Assembly of two representatives, 'one educated, the other a chief or headman of the district' appointed by each king or principal chief, with responsibility, when convened by the Secretary, for exercising all the functions of the legislative body including the levying of taxes (its meetings were to be presided over by the Vice-President to be appointed from the educated group by the National Assembly and its members were also to see to the execution of the laws of the Confederation in their constituencies); and, finally, a National Assembly composed of all the kings, principal chiefs and others within the Confederation which was to meet each October to hear reports on the work done by the Representative Assembly, to elect the King-President for each year, discuss the programme for the ensuing year and elect members of the ministry. It was also laid down that provincial assessors were to be appointed to preside over provincial courts that were to be established in each province or district. Article 8 of this constitution also defined the aims of the Confederation as the promotion of 'friendly intercourse between all the kings and chiefs and to unite them for offensive and defensive purposes against their common enemy' and 'the improvement of the country at large' by the construction of good roads, building of schools, promotion of agricultural and industrial pursuits and the introduction of 'such new plants as may hereafter become sources of profitable commerce to the country'. Articles 21–27 gave details about the school-building and road-building programmes. Finally, Article 43

377 Kimble, p. 237.

laid down that 'the officers of the Confederation shall render assistance as directed by the executive in carrying out the wishes of the British Government'.

Some historians have expressed doubts about the ability of the Confederation to accomplish its objectives, describing the November 1871 Constitution as 'utterly impracticable unless the whole country was united in the matter, which was not the case'.[378] This seems an unnecessarily pessimistic view of the situation. Since the states had formed a pact, there is no reason why the constitution could not have been implemented or taxes collected. Nor was there anything inherently impracticable in the constitution itself. It all depended on a smooth cooperation between the chiefs and the educated élite which, from all accounts, had prevailed since the formation of the union. Nor were the development programmes impracticable once funds were made available and it is certainly uncharitable to suggest, as Claridge does, that 'the whole Constitution seems to have been framed by a few educated and semi-educated men, primarily no doubt for the good of their country, but also for the benefit of themselves'.[379] It was agreed upon after weeks of discussions between the chiefs and the educated men, nor do I see what benefit the latter could have derived from it. Thirty-one kings and chiefs not only signed and sealed the constitution (I discount later allegations that some chiefs were not content or not present as propaganda by British officials), but went on to hold the first meeting of the National Assembly and to elect the officers.[380] The Executive was also empowered to appoint officers in the field 'for the purpose of carrying on the Government' and a law was renewed providing for the collection of a poll tax and another was passed for the collection of import and export duties.[381]

As we shall see presently, most of these decisions were enforced and there is no doubt that had the Confederation been allowed a free hand in the execution of its plans, much would have been achieved and, as Kimble says, a considerable fund of local initiative and enthusiasm could have been aroused and tapped.[382] But this was not so. In fact, only five months after the signing and sealing of the constitution, the leaders of the Confederation, for reasons to be discussed later, had to submit new constitutional proposals to the British which were a drastic amendment to those of 1871.[383] In the very first proposal it was blandly admitted that the

378 Claridge, i, pp. 618–19.
379 *Ibid.*, p. 619.
380 W. E. Dadson, Vice-President; J. F. Amissah, Secretary; J. H. Brew, Under-Secretary; F. C. Grant, Treasurer; J. M. Abadoo, Assistant Treasurer. R. J. Ghartey, G. Amissah, S. Ferguson, G. Blankson Jnr., J. M. Insaidoo and J. D. Hayford as members of the Fanti Executive Council.
381 Agbodeka, pp. 118–19; Kimble, p. 249.
382 Kimble, p. 248.
383 'Proposals of the Leaders of the Fanti Confederation', 16 April 1872, in Metcalfe, *Great Britain and Ghana*, pp. 342–4, doc. 285.

Confederation 'must have the recognition and support of Her Majesty's Government' in order to function properly; in other words, the independence of the Confederation had been renounced. In the second, two clear and distinct jurisdictions were agreed upon; confederacy and local government. The fourth proposal suggested that the courts of the Confederation be recognised as those for the first hearing of matters or disputes between its subjects but that there should be freedom to appeal to the British courts. Moreover, it was suggested that 'the Vice-President and four other gentlemen, natives or residents of the Gold Coast' were to be members of the Legislative Council by election in the eastern and western districts 'so that the interests of all might be represented' and not by nomination by the Administrator. Finally, a number of financial proposals were made; half the expenditure of the Federation estimated at £20,000 (half of which was to be used for education, roads and medical facilities) was to be 'placed at the disposal of the Confederation out of the revenue of the Gold Coast' and the other half was to be raised by the Confederation itself by way of court fees, fines, etc.; the government was to clear the Confederation's outstanding debt of £7,200 'since the government was opposed to their imposing any taxes or duties'. The proposals ended with the significant note that if the British Government would not grant the Confederation the revenue required or allow them to raise it through taxation, the British would have to take over government of 'the whole country and govern it as vigorously and on the same principles as it does her other colonies; but not permit us to be governed in the shameful and neglectful way in which we have been for years past. . . .' These proposals clearly implied a capitulation. But the note of frustration, disappointment and disgust is quite evident and the reasons for this and the reaction of the British Government to these proposals will be discussed later.

The Fante Confederation did much more than simply spend all its time making constitutions. First, after declaring itself independent in 1868, it adopted a national seal consisting of the now familiar elephant standing against an oil palm tree and encircled by the words, 'The Government [later changed to Council] of the Fante Confederacy, Mankessim'.[384] Secondly, the Confederation raised an army which was about fifteen thousand strong; its right wing had troops from Denkyira, Anomabo, Ekumfi, Twifu, Komenda and Wassa, its left wing had troops from Kormantin, Mouri, Western Gomoa and Cape Coast and the central body was composed of troops from Abora, Adjumako, Ayan and Mankessim.[385] This army saw continuous though uneventful action from its formation until about the end of 1870. A contingent was sent to the help of Komenda in February 1868 while the bulk was sent against Elmina, the headquarters

384 Kimble, pp. 242–3.
385 Agbodeka, p. 114.

of the Dutch and the traditional ally of Asante. The siege was lifted in June for negotiations with the Elmina but was resumed when they broke down. Fighting continued for the rest of that year and throughout 1870 as well. Early in 1870, because of urgent appeals from eastern coastal states against the Dutch and because of the continued fighting at Elmina, a despatch was sent to many of the other states in the Confederation for five hundred troops from each but these recruitments were very unsuccessful and only a few troops could be sent against the Dutch.[386] The Confederate army was therefore not able to give substantial help to any state and the military record of the Confederation remained rather poor. But there is absolutely no doubt that it was because of the actions of the army that the Anglo-Dutch exchange scheme failed and the Dutch felt obliged to abandon the coast in 1872.

Thirdly, the Confederation set up an administrative machinery. By 1869 officials and correspondents had been appointed who were responsible for the administrative and defence matters of the various districts and who were sending in regular reports.[387] It seems clear that the articles in the Constitution of 1871 about the administration merely embodied the existing machinery which in itself shows that it had proved its effectiveness.

Fourthly, the Fante National Supreme Court was established and R. J. Ghartey was appointed as Chief Magistrate in 1868.[388] Though Kimble does not say so, this court did function. Ghartey moved to Mankessim, the headquarters of the Confederation and a number of cases, as shown by his papers, were in fact referred to his court, including those by King Anfo Otu of Abora between 1 July and 5 July 1869. One of King Otu's cases was to the effect that 'these men came to me that they came from Amanfu stating to me that some one seized[?] the case to me and being the cases of such sorts is not my rule so am oblige[d] to direct them to you at Mankessim.'[389] This letter is of particular interest in that it shows a clear distinction prevailed between cases to be sent to the federal court and those to be sent to the court of the chiefs and kings. On 9 July 1869, Joseph Eshun from Kormantin brought a case of improper conduct towards a pregnant woman to the court, while in the following month another case was referred to that court from Saltpond. Though there is no record of the judgements delivered in each case, the fact that cases were brought

386 Agbodeka, p. 116.
387 *Ibid.*, pp. 115–16 n. 151. Amissah to Ghartey, 1 February 1870, in Ghartey Papers, Ghana National Archives; Gibbs and Johnson to Ghartey, 23 December 1869, Ghartey Papers; Dawson to Ghartey between May and December 1869, Ghartey Papers; Clement to Ghartey, 3 February 1870, Ghartey Papers.
388 Kimble, p. 236. Kimble adds that Ghartey carefully preserved his sealed letter of appointment, dated 5 December 1868.
389 Otoo to Ghartey, 1 and 5 July 1869, Ghartey Papers; Eshun to Ghartey, 9 July 1869, Ghartey Papers; letter to Ghartey, 28 August 1869, Ghartey Papers.

regularly shows that members did recognise the court and also that they were satisfied with its proceedings.

Fifthly, the Confederation imposed and collected poll tax and other duties. The states were divided into districts for this purpose each with a poll tax commissioner who was not always the administrative officer of the Confederate Government.[390] There are letters in the Ghartey papers to the effect that certain chiefs had been asked to persuade the people of some of the member states to pay the tax.[391] All the sums collected were sent to the Treasury at Mankessim. The Confederation decided to collect import and export tax in 1871 and as late as August of that year, the Confederation was collecting salt as a form of taxation, one third of which was retained by the local chief and the remaining two-thirds by the Confederation.[392] There are documents in the Ghartey papers to show that the Confederation was in a healthy financial state by July 1869, due mainly to the payment of taxes by the districts, but by 1871 it was in debt and financed mainly by the wealthy Ghartey brothers. By April 1872, it is clear from proposals made to the British that this outstanding debt amounted to 'some two thousand ounces of gold, equal to £7,200'.[393] But the Confederation had remained solvent during the first two or three years of its existence. Finally, the Fante Confederation did inspire a similar movement in the eastern districts in 1869, namely, the Accra Native Confederation.

We have seen that protest against the growth of British authority was as strong in Accra as it was in Cape Coast and Horton's books were as avidly read. At the inauguration of the Fante Confederation, the leaders extended invitations to the people of the eastern districts to attend their meetings and it appears that a delegation from Accra was present at the meeting in October 1868 and many subsequent ones. Certainly, when Simpson attended the meeting of April 1869 he saw a small delegation from Accra.[394] However, it was the Asante campaigns in the Krepi areas during that year that occasioned the launching of the Confederation on the initiative of the educated community.[395] The inaugural meeting was convened by William Lutterodt, 'the acknowledged head of the educated natives of Accra', in his own house in August 1869. About sixty people attended and, since they were not impressed by what they were told about their defence plans by the Kings of James Town, Dutch Accra and Christiansborg whom they invited to the meeting, they decided to set up a Managing Committee of six 'to act in conjunction with the Kings and to draw up rules for their joint guidance'. The members of this committee

390 Agbodeka, p. 118.
391 Dawson to Ghartey, 22 December 1869, Ghartey Papers.
392 Kimble, p. 251 n. 2.
393 'Proposals of the Leaders', 16 April 1872.
394 Kimble, p. 251.
395 *Ibid.*, p. 239; Claridge, i, pp. 615–16.

were Lutterodt, the President, and James Bannerman, L. Hesse, G. F. Cleland, W. Addo and J. E. Ritcher. Bannerman wrote to Horton: 'I doubt that, with God's help, we shall soon learn practically the art of self-government'[396] and they were reported to be 'in daily communication' with the Fante Confederation towards the end of that year. However, apart from evoking enthusiasm for war, the Committee achieved nothing and ceased to exist after 1869. Unlike its Fante counterpart, the Accra Native Confederation never gained the backing of the chiefs whose power and support was too strong to be easily set aside by the educated community.

Reasons for the Confederation's collapse

Historians have differed in their accounts of the collapse of the Fante Confederation. As we have seen, Claridge simply says that it ceased to exist. According to Kimble, 'by May 1873 the Fante Confederation was dead presumably of natural causes'. But the unkindest and fortunately the most inaccurate verdict is that of Metcalfe: 'In face of this new threat [the Asante threat of early 1873], the Fante Confederation disappeared without trace.'[397] Metcalfe's view can be readily dismissed since, if he was right, the Confederation should have disappeared in 1869 when threats of a large scale Asante invasion reached alarming heights. Yet in August 1869 it in fact met precisely to discuss what Kimble has called 'disturbing reports of Ashanti advances' and remained intact and active throughout the year and in the following year as well. In any case, when the Protectorate found itself faced with full-scale war, the Confederation had already virtually disappeared.

The first cause of the collapse was the origins and nature of the Confederation itself. We have already seen that, disgusted as they were with British authority, what finally precipitated the formation of the Confederation was the immediate need to prevent the take-over of the western forts by the Dutch. It was this that brought the non-Fante states, Twifu, Denkyira and Assin, into the union. Thus when the Dutch threat lost its force, the cause of this unity was gone and, when the treaty was ratified, links between the Fante and non-Fante states were snapped. As Pope-Hennessy reported to the Colonial Office in 1872, 'whatever may have been the case before the transfer, there is no doubt whatever now, that the majority of the chiefs in the whole Protectorate as it exists at present, would rather see British rule extended and made more of a reality.'[398] It is to the Confederation's credit that they realised this and sought to maintain unity by formal means, hence the promulgation of the Constitution

396 Bannerman to Horton, 21 September 1869, quoted by Kimble, p. 240.
397 Claridge, i, p. 622; Kimble, p. 262; Metcalfe, *Maclean of the Gold Coast*, p. 347.
398 Pope-Hennessey to Kimberley, 29 October 1872, in Metcalfe, *Great Britain and Ghana*, p. 344, doc. 286.

of 1871, three months before the ratification of the Convention Treaty. But, as we have seen, the Constitution was never effectively implemented and from that time the Confederation had become Fante in essence.

The second was the rivalry between the kings of Abora and Mankessim. This began from the first meeting of the union and it was mainly because of it that the kings and chiefs failed to elect a single king as head in 1868, 1869 and again in 1871. In 1869 an ordinary commoner was elected President (R. J. Ghartey), while in 1871 both kings were jointly elected 'pending the election of the said King-President'. Not until 1872 was Edu elected when Otu agreed to accept a position as 'General Chief Marshal of the Fantee Nation'.[399] But, by then it was too late since their rivalry had considerably weakened the movement at the top and much time must have been wasted by the educated men in attempting to settle it.

We have already seen that the military strength of the Confederation was never strong and its record was poor. This was the third cause of its collapse. Had the army succeeded in its siege of Elmina, its reputation would have been enhanced and it would have been able to give effective assistance to Wassa and Dadiase against the Dutch and also could have considerably minimised the Asante threat. This weakness must have disposed the non-Fante states to respond favourably to the active intrigues of the British. The army was finally dissolved by the end of 1871.

The fourth reason was financial. Because it failed in its efforts to collect revenue, the Confederation was never able to implement its development programmes. Without finances, the administrative machinery must also have suffered.

Britain's attitude to the Confederation

But in spite of all this, the Confederation could have lasted and achieved more but for the final factor, namely the hostility of the British towards it and their active interference. The British persistently attributed the whole Confederation to the machinations of the educated élite and were clearly determined never to implement the recommendation of the Select Committee of 1865 about their eventual withdrawal. On the contrary, the local British officials were in favour of entrenching themselves, hence, for example, their insistence on the Anglo-Dutch exchange of forts. Everything was done, therefore, to sabotage the Confederation from its inception. At the time of the inaugural meeting, Ussher sent Thomas Hughes, the former adviser to Aggrey who had since changed sides, to Mankessim with 'private and confidential instructions' to find out the purpose of the meeting and 'to endeavour to counteract some mischief which I fear these people in their ignorance may be preparing'.[400] Hughes was further

399 Kimble, p. 258.
400 Private and Confidential Instructions of 29 January 1868, to Hughes from Ussher, quoted by Kimble, p. 225.

instructed 'to warn the chiefs still under British protection that any treasonable practices against the supremacy of Her Majesty's Government would be punished as speedily as in the case of Aggrey, and that if they did not at once leave Mankessim and go home they would be arrested on suspicion alone.'[401] Hughes was even told to warn delegates from Wassa, Denkyira and Komenda that their presence in British territory could not be tolerated. Clearly, Hughes's mission failed but Ussher did not give up. As soon as the Confederate army attacked Elmina, he suspended the sale of arms and ammunition and asked the people of Cape Coast to observe strict neutrality in the war. But we have seen that Cape Coast joined the attack shortly after this warning. Ussher then persuaded the Fante to negotiate for peace and in June talks began in which Ussher himself took part but these broke down when the Fante rejected the Elmina's offer to suspend their alliance with Asante for six months but not to cancel it.[402]

Alarmed by the declaration of independence, Simpson, acting Administrator while Ussher went on leave, used even more subtle and active methods to substitute British authority for what he called 'the interested services and pernicious interference of the "scholars" or semi-educated in uncontrolled ascendancy'.[403] He summoned the chiefs of the Protectorate to a meeting in Cape Coast and offered to recognise and support 'the Council of the Kings assembled at Mankessim'. Of course, most chiefs saw the trap and only a few attended. However, the following month he began direct appeals to individual kings and wrote to the King of Wassa, for instance, warning him not to allow his power to be challenged by 'the persons styling themselves Councillors of Mankessim'. He added that if the Councillors attempted to tax his people, he should 'arrest them immediately and report them to me'.[404] When these moves had also failed, by a 'certain amount of diplomatising' such as the return of Aggrey to Cape Coast, he got himself invited to Mankessim. At this meeting in April 1869, Simpson admitted that the Fante were right in their anger about not being consulted over the exchange deal and in their quarrel with Elmina but said that they should confine their acts to defensive purposes and not interfere with the internal affairs of the country. He also offered to help them draft a constitution for the achievement of 'the unity of the Fante nation and its progress towards self-government' and to assist them improve the existing roads.[405] He, however, insisted that taxes on transit trade should stop, the British courts should be obeyed and the Judicial Assessor was to be allowed to go to Mankessim two or three times a year to assist them in settling their problems. The chiefs seem to have swallowed

401 Kimble, p. 225.
402 Kimble, p. 229.
403 Kimble, pp. 233–4.
404 Simpson to King Ennimil, 15 February 1869, quoted by Kimble, p. 234 n. 2.
405 Kimble, p. 235.

this since they declared their readiness to disclose everything and do nothing without consulting the Acting Administrator and gaining his concurrence. They even appointed him 'Field Marshal Commanding-in-Chief of the Fantee Forces', which Simpson graciously declined though he promised to lead them if they were invaded. Simpson returned fully satisfied that British authority had been re-established in the western districts and that he had secured 'reconciliation and union under the British flag which in their hearts the natives love so well'.

But Simpson assumed too much or was tricked because immediately after his departure the councillors went on to discuss how to establish the organisation on a more permanent footing and, according to Kimble, it was 'at this juncture that the term Confederation [or Confederacy] seems first to have been used, especially by some who were anxious to draw in more educated men "for the great business"'.[406] Ghartey was elected the first President at this meeting. Because of the Dutch attack on Dixcove and Simpson's vehement criticism of the Confederation's decision to take action against Asante, relations between the Confederation and the Administrator had become exceedingly strained by September 1869.

Ussher, on resumption of office, renewed his efforts to weaken the confederation by openly befriending it but secretly detaching the chiefs one by one and humiliating its leaders. In January 1870, for example, he called Ghartey and severely reprimanded him for using the term 'Government' in the seal and Ghartey was forced to apologise and change the term to 'Council'.[407] Indeed, Ussher was so successful that he talked of introducing a town council for Cape Coast which, he stated, would 'be more calculated to infuse a spirit of self-reliance as a commencement of self-government, than any hare-brained schemes of crafty and illiterate adventurers and drunken and barbarous "Bush Chiefs"'.[408]

Horton and Dawson and other leaders attempted to revive the Confederation during the second half of 1870 and throughout 1871 and met with much success. By the beginning of 1871 Ussher reported, in his usual alarmist manner, that the government was 'gravely menaced by the proceedings of the small knot of persons desirous of re-establishing the old grievances and mischievous scheme of a "Bush" Fantee Government or Confederacy' and, holding Horton mainly responsible for this, urged the Governor-in-Chief to expel him from Ghana. His renewed campaign against the Confederation had some success for in March 1871 he reported, 'Of late, I have brought down and punished chiefs from the interior, who have for many years set the Government at defiance.'[409]

406 *Ibid.*, p. 237.
407 *Ibid.*, p. 243.
408 Ussher to Kennedy, 11 October 1870, quoted by Kimble, p. 242.
409 Ussher to Kennedy, 6 March 1871, in Metcalfe, *Great Britain and Ghana*, p. 334, doc. 276.

The educated élite were very much aware of the hostile attitude and intrigues of the British officials and so they decided in 1871 to adopt new tactics in their efforts to resuscitate the movement. This is clearly brought out in a pencilled memorandum found later among their papers which included in an agenda in note form:

> Ghartey's issuing circulars creates distrust, first with Govt. & with natives, Confederation to be commenced by Kings—Confederation to be confined as clearly as possible to chiefs in interior leaving the coast for present. Ground work of constitution to be first formed—then others can join—by these means collision with Govt. will be avoided—& when things are properly worked others will join. Kings and Chiefs should avoid communicating with people at C.C. Confederation should be confined as closely as can be to the Chiefs of the interior for the present because the B. Govt. exercises supreme authority on the sea coast. The educated people may be sent for—& any interference of the people with Coast towns may bring the Confedrn into collision with the Government. Confederation for express purpose of improving the interior and developing resources of country.[410]

It was in this clandestine way that the discussions leading to the drafting and adoption of the detailed constitution as well as the election of officers took place in 1871. It is interesting to note that in this constitution the leaders made a rather tactical declaration to cooperate with the British Administrator and sent two men, Davidson and Brew, to Salmon, the then Acting Administrator, to tell him of recent developments and to give him, among other papers, a copy of the new constitution. Salmon, furious at not having previously been consulted, described their action as tantamount to a 'dangerous conspiracy' which must 'be destroyed for good, or the country will become altogether unmanageable'[411] and promptly arrested all the members of the Ministry except Grant and decided to try them in the Judicial Assessor's Court. There was no legal justification for Salmon's action since Britain did not enjoy sovereign rights in the country and it is not surprising therefore that, though the Colonial Office admitted that some of the articles of the constitution were inconsistent with the jurisdiction of the British government, the Colonial Secretary condemned the arrests and told Salmon to release the prisoners on bail and stop court proceedings.[412]

Salmon, undaunted, continued his interference by sending letters to individual chiefs asking if they approved of the proceedings at Mankessim

410 Kimble, p. 247.
411 Salmon to Kennedy, 4 December 1871, in Metcalfe, *Great Britain and Ghana*, p. 338, doc. 280; Kimble, pp. 249–50.
412 Kimberley to Kennedy, 16 January 1872, in Metcalfe, *Great Britain and Ghana*, p. 339, doc. 281.

and naturally 'favourable replies' were received from many of them. He also succeeded in stopping the tax imposed by the King of Mankessim on goods in transit through his territory. He then embarked on a detailed investigation that would expose the 'half-educated schemers' who he believed were responsible for drafting the constitution. His final report on the chiefs and leaders was so scandalous that the Secretary of State was forced to comment, 'If the Administrator's account is correct the chiefs of the "Confederacy" are a set of drunken scoundrels.'[413] Finally, in February 1872, he issued a warning against 'all loyal subjects . . . taking office under the said Constitution'. This proclamation further declared that almost all the kings and chiefs had repudiated the constitution and therefore that the government deemed it 'subversive of those relations which have for a long time past subsisted between Great Britain and this Country, and certainly leading to the discontinuance therefore'.[414] Ussher, again returning from leave, threatened to try anyone 'committing any overt acts on the part of the said Confederation, especially the levying of taxes, assumption of judicial power, and molestation of peaceful inhabitants' and at the same time hinted to the chiefs of Abora and Mankessim his willingness to discuss the formation of a chiefs' council.[415] In view of all this, Salmon's report that by January 1872, 'the Confederacy is at present confined to the small district of Mankessim and the larger one at Abrah' might well have been true.[416]

However, the leaders were still not prepared to capitulate and, encouraged by the friendlier disposition of the new Administrator, J. Pope-Hennessy, with whom they had had a series of interviews in the castle at Elmina, the leaders drew up their third scheme for a constitution which has already been discussed.[417] It is not surprising that it emphasised the need for the 'recognition and support of Her Majesty's Government', asked for the Confederation's outstanding debt of £7,200 to be cleared and ended on a strong note of disgust and frustration. And it is yet another evidence of the renewed determination of the leaders to meet the British challenge that, while waiting for the British reply to their April proposals, they elected a single King-President at their meeting in July 1872 and successfully persuaded the chiefs who had been driven out of the Confederation by Salmon's threats to return and take the oath of allegiance.

However, Pope-Hennessy, instead of saving the Confederation, achieved

413 Kimble, p. 252.
414 Proclamation of 12 February 1872, quoted by Kimble, p. 254.
415 Proclamation of 9 March 1872, quoted by Kimble, p. 256; Metcalfe, *Great Britain and Ghana*, p. 344 n. 1.
416 Salmon to Kendall, 20 January 1872, in Metcalfe, *Great Britain and Ghana*, p. 339, doc. 282.
417 See above, p.

its final collapse in his despatch of 29 October 1872[418] after having declared the latest proposals impracticable. He first repudiated the scandalous allegations about the leaders and pointed out that as far as he could observe 'every educated native of Cape Coast sympathised with the Confederation' and went on to agree that the judicial authority of the chiefs had been usurped and nothing put in its place. He attributed recent disturbances to this vacuum and suggested two alternatives: 'either a scheme of active government, with certain financial and judicial powers, should be recognised, or steps should be taken to gradually introduce throughout the Protectorate the same system of Government that exists in the other Crown Colonies.' He recommended the latter. However, he added that another educated African should sit on the Legislative Council and he suggested F. C. Grant, who, he stated, 'seems to have the confidence of the educated natives as well as of the chiefs'. In other words, Pope Hennessy wanted the liquidation of the Confederation in favour of what he called 'a firm extension of Her Majesty's authority'.

This despatch and its enclosures started a spate of minutes in the Colonial Office[419] about future administration on the Coast and the recommendations of the Select Committee of 1865 were called in and seriously discussed. The decision, however, rested with Kimberley, the Colonial Secretary, and it was to affect British policy for a long time to come. He stated: 'I would have nothing to do with the "educated natives" as a body. I would treat with the hereditary chiefs only, and endeavour as far as possible to govern through them'. He also rejected the idea of ultimate self-government because it would destroy all hopes of opening the interior to commerce, a result that 'would commend itself neither to philanthropists nor traders'. He further turned down Pope Hennessy's suggestion, strongly supported by some Under-Secretaries in the Colonial Office, for the establishment of a firm system of government on the grounds that it would be too costly in lives and money and would involve the recognition of domestic slavery. However, he did agree that 'the Powers and obligations of the Protecting state should be clearly defined and outlined'. These were included in a despatch to Pope Hennessy in March 1873 which were to be made the subject of agreement with the chiefs on the lines of the Bond of 1844. Britain was to defend the forts while the Africans were to defend themselves against Asante. Britain was to control finance and revenue, to enforce her orders in the Protectorate, maintain peace within it, protect trade and improve the country. Finally, Britain was to pay stipends to the chiefs and encourage education 'but in the vernacular

418 Pope Hennessey to Kimberley, 29 October 1872, in Metcalfe, *Great Britain and Ghana,* pp. 344–5, doc. 286.
419 W. D. McIntyre, *The Imperial Frontier in the Tropics, 1865–1873*, 1967, pp. 121–51; Kimble, pp. 259–60.

only to stifle opposition'. Kimberley also ruled that the Fante Confederation could not be recognised in its present form as an intermediary between the chiefs and the government, but that the chiefs could form their own Confederation for the defence of their frontiers. However, just before this despatch was to be posted, news of the Asante invasion reached London. Kimberley therefore decided to postpone deciding on the future government of the coast until 'a more favourable juncture of affairs presents itself'.

It was the long period of waiting, from April 1872 to March 1873, and not the threat of an Asante invasion, that finished off the Confederation. At the start of the Asante invasion the King of Mankessim did write to the Administrator asking for support and this was, of course, promptly turned down. When another request came in May 1873 it was from 'the Fantee Kings and Chiefs' and not from the Fante Confederation. By then the Confederation was indeed non-existent. So disunited had the Fante become by the beginning of 1873, thanks to the intrigues and machinations of the local British officials, that not even the full-scale Asante invasion could pull them together. But it should be obvious from the above account that the Confederation did not die a natural death. It was, by and large, deliberately killed by the British.

Brief though its history was, the Fante Confederation left an indelible imprint on the history of Ghana. It was a nationalist movement embracing non-Fante as well as Fante and its influence extended as far east as Accra. Though it was killed, the ideals it stood for, the attainment of self-government and 'the achievement of the social improvement of our subjects and people, of education and industrial pursuits', survived it and were to inspire future generations. It was the first movement in which the traditional rulers and the educated élite joined together and maintained their co-operation despite all the British intrigues. It was the precursor not only of the Aborigines' Rights Protection Society but also of the National Congress and the United Gold Coast Convention. Mankessim now lost its significance as the religious and political centre of the Fante nation and never recovered it, while the educated class acquired an odious reputation in the eyes of the British which lasted for many years to come. Finally, there is no doubt that the early, spectacular success of the Confederation hastened the withdrawal of the Dutch while its collapse led to the entrenchment of the British position. It is one of the ironies of our history that the Confederation ended by compelling the British to abandon the idea of self-government until 'a more favourable juncture of affairs presents itself', which was not to be until 1957, eighty-four years later.

Establishment of a Crown Colony

The withdrawal of the Dutch and the collapse of the Fante Confederation were not the only considerations that led to the establishment of a Crown

Colony by the British in 1874. The third, and probably the most decisive reason, was the crushing defeat of the Asante in 1874.[420] This event left the British once more with the decision whether to withdraw or entrench themselves, or, as one Under-Secretary put it, 'complete annexation or total abandonment are I fear the only second alternatives'. This, as usual, led to a flurry of minutes in the Colonial Office and a heated debate in Parliament.[421] The majority opposed abandonment because they were convinced that the sole guarantee against further Asante attacks was a strong government on the coast and a strong defensive grouping under the British. As one of them stated, 'unless we directly govern up to the Prah, we can have no guarantee against wars and disturbances'. Secondly, it was felt that the British had a moral obligation to protect the southern peoples, an obligation in no way minimised by the recent defeat of Asante. E. Fairfield said:

> The duty arises from the fact that our presence renders the protected tribes less able to defend themselves, whilst our peculiar policy has exposed them to the undying hatred of their most powerful enemies, although, of course, the cost of providing such defence is primarily chargeable to local resources, and can, it may be hoped, be usually defrayed, in full from that source.

There is no doubt that the new Secretary of State, Lord Carnarvon, was influenced by these considerations in what he described as 'a very evil choice' between annexation and abandonment. His decision in favour of remaining, as he told the House of Lords in May 1874

> is simply a sense of obligations to be redeemed and duties to be performed. But . . . as long as we do stay there, whether it be long, or whether it be short, we should exercise an effective control—a control beneficial to the natives themselves, and worthy of the history and position of this country.

Two months later, on 24 July 1874, Letters Patent and Order were issued in London by which the British settlements were transformed into the Gold Coast Colony and together with Lagos were separated from Sierra Leone and constituted a separate administration with its own governor and its own Legislative Council. A month later, the Legislative Council was empowered to legislate for the adjacent Protected Territories. Thus, having gained their independence from Asante in 1826, the southern states lost it to the British in 1874.

Alarmed not only by the revival of the Asante Empire but also by the activities of the French to the west and the Germans to the east of the Colony in the 1890s, and spurred on by the new spirit of imperialism in

420 See above, p. 202–3.
421 Metcalfe, *Great Britain and Ghana*, docs. 300, 301, 302.

Europe, the British arrested the Asantehene in 1896 and exiled him first to Sierra Leone and then to the Seychelles Islands. They then went on to annex Asante and to declare a Protectorate over Northern Ghana between 1900 and 1901. The British Colony of the Gold Coast with its three separate but related territories of the Colony at the coast, Asante and the Northern Territories came into existence on 1 January 1902. It was this colony which was to attain independence fifty-five years later under the new name of Ghana.

CHAPTER 7

The Atlantic coast and the southern savannahs, 1800–1880[1]

YVES PERSON

THE SITUATION AT THE BEGINNING OF THE NINETEENTH CENTURY

During the eighteenth century there were a number of upheavals in the area comprising the southern savannahs and the Atlantic coast, one of the most spectacular being the rise of Fouta-Djallon, discussed in the preceding volume. Such changes were a natural consequence of expansion in long distance trade and developments in new political structures. These two factors are closely linked and their geographical relationship is such that the entire area may be divided into three distinct zones running from west to east. These are the trade routes joining the Upper Niger to the estuaries of French Guinea (Rivières du Sud), the hinterland of the great Kru forest and the routes from the Middle Niger to the Gulf of Guinea between the Bandama and the Volta.

In this vast region where the only appreciable relief is the Fouta-Djallon plateau and the Upper Guinea Highlands, the vegetation zones appear in fairly regular longitudinal bands and this feature is echoed in the parallel position of the Gulf of Guinea. For this reason, trade has traditionally followed a north–south direction, except in the west where the mountains and the coastline takes it towards the Rivières du Sud and Senegambia.

The general absence of ocean traffic until the sixteenth century meant that the trading network gradually spread all over the region from the Sudan to the fringes of the forest and in some cases even as far as the coast. Salt and horses from the north were bartered for kola nuts and slaves from the south and commerce in general was much stimulated by the output of gold from the Volta basin and the fringes of Fouta-Djallon. Trade was usually in the hands of the Dyula, and had been since the middle ages. These Mandingo-speaking converts to Islam had formed a useful balance to the pagan peoples in the society of the western Sudan. The area from

1 The Atlantic Coast, in this region, is also known as Upper Guinea, see Rodney, W., *History of Upper Guinea*, Oxford, 1970. This chapter was written in 1966 and the study of the sources is mainly limited to this date.

the Upper Niger to the Bandama was subject to a massive infiltration of the Malinké, a warrior people still holding the old animist beliefs, who spread over the entire region to the edges of the forest. The Dyula were basically a specialised Malinké group who fulfilled a religious and, more particularly, a commercial function which made them quite indispensable to their fellow countrymen. A different situation prevailed in the East, where the basic population was racially different, speaking Voltaic, Akan or southern Mande languages. Here the name Dyula was synonymous with a closely-knit ethnic minority group which often played a leading role in society, usually on account of the socio-professional nature of its talents.

In the sixteenth century the old order changed, with the establishment of European merchants along the coast who soon became deeply involved in the slave trade. The old south-bound trading routes were duplicated by others, in the same direction but from the Sudan to the sea. Thus even more slaves, gold and ivory found their way to the Portuguese, Dutch, English and French trading posts. The Europeans provided salt, spirits and various manufactured objects. Textiles and more particularly firearms were carried to the interior of the continent and the growth of these new demands increased the influence of the Dyula who took control of this type of trade.

Although slave trading was made illegal at the beginning of the nineteenth century the export of slaves went on until the third quarter of that century in isolated sectors of the coast. Legitimate commerce also continued to spread European influence in every part of the area.

Inevitably, this process played an essential role in modifying the structure of those societies involved in the new patterns of trade. The influence of the trade routes varied in intensity with local geographical conditions, and the resulting conflicts explain the main events under consideration.

The west

In the west, the historical axis of Sudanese civilisation stretched to Kankan, with relatively easy access to the sea between the Upper Niger and the Rivières du Sud. Although the forest belt here is quite narrow and varies in density, the mountainous nature of the region has affected communications. The Somba or Quoja invasion took some Manding to the coast in the sixteenth century, but Sudanese commerce did not succeed in securing a regular outlet to the sea. Until the eighteenth century, the people of the Upper Niger often had to travel to the Gambia in order to reach the sea, an enormous detour made to avoid the insecurity prevailing in the Fouta-Djallon highlands.

However, the establishment of the Fulani theocracy in this region changed the situation from 1727 onwards and Dyula-controlled trade soon stretched as far as the estuaries of what would later become French Guinea,

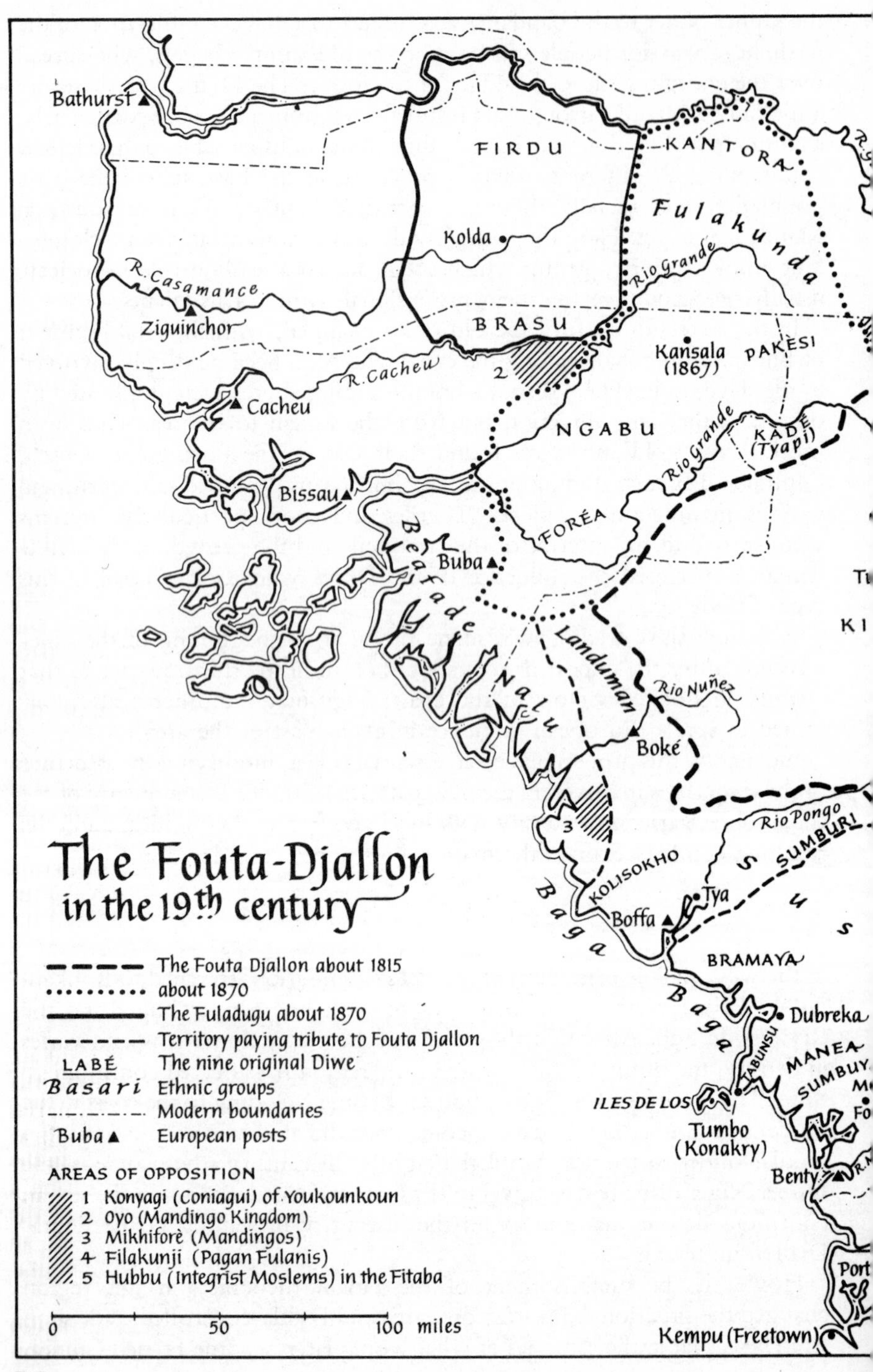
The Fouta-Djallon in the 19th century
The Fouta Djallon about 1815
about 1870
The Fuladugu about 1870
Territory paying tribute to Fouta Djallon
LABÉ The nine original Diwe
Basari Ethnic groups
Modern boundaries
Buba European posts
AREAS OF OPPOSITION
1 Konyagi (Coniagui) of Youkounkoun
2 Oyo (Mandingo Kingdom)
3 Mikhiforè (Mandingos)
4 Filakunji (Pagan Fulanis)
5 Hubbu (Integrist Moslems) in the Fitaba
0 50 100 miles
Bathurst
FIRDU
KANTORA
Fulakunda
Kolda
R. Casamance
Rio Grande
Ziguinchor
BRASU
Kansala (1867)
PAKESI
R. Cacheu
Cacheu
NGABU
KADE (Tyapi)
Bissau
FORÉA
Beafade
Buba
Landuman
Nalu
Rio Nuñez
Boké
Rio Pongo
SUMBURI
KOLISOKHO
Baga
Tya
Boffa
BRAMAYA
Dubreka
TABUNSU
MANEA
SUMBUYA
ILES DE LOS
Tumbo (Konakry)
Benty
Kempu (Freetown)

7:1

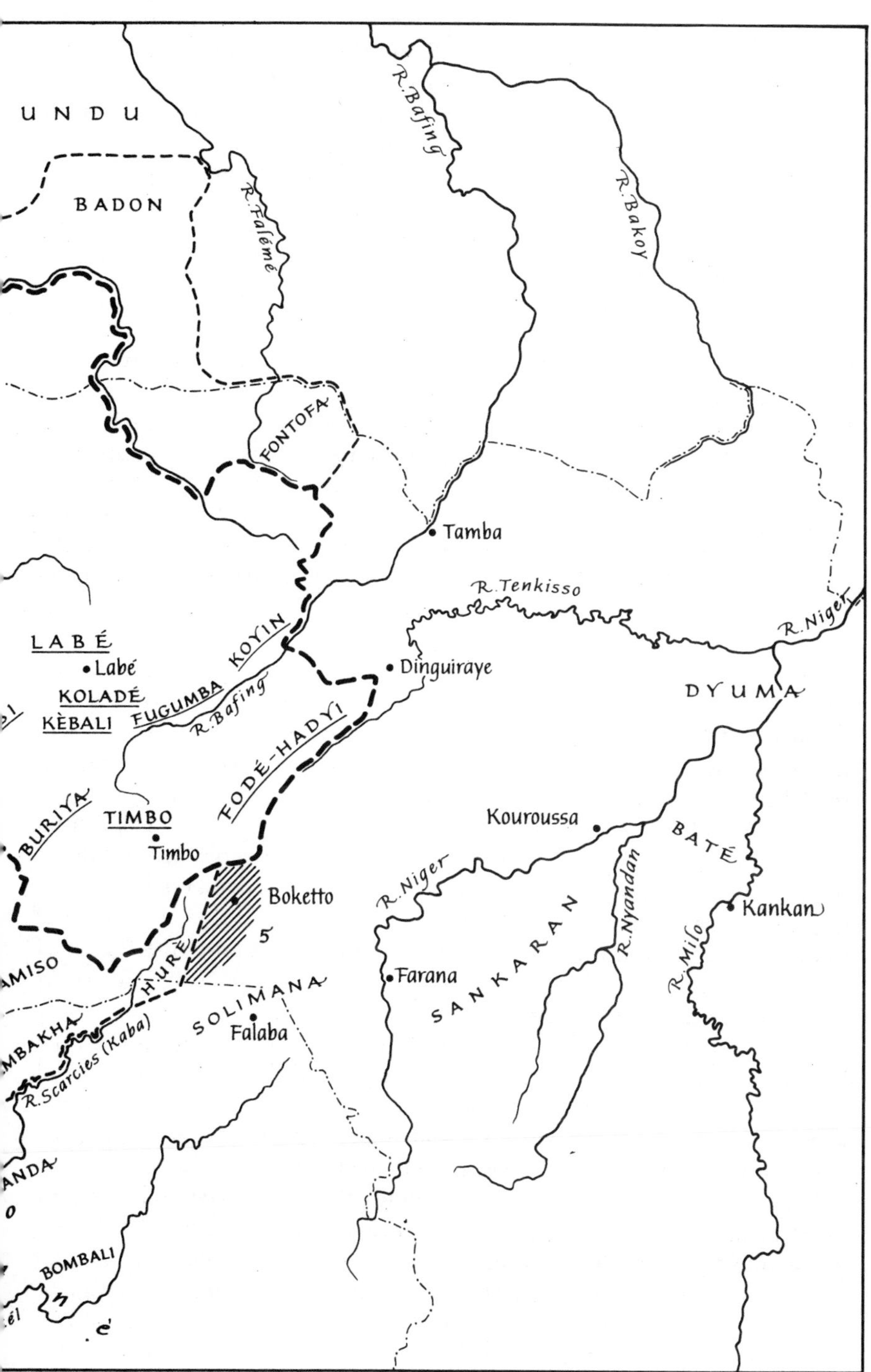
UNDU
BADON
R. Falémé
R. Bafing
R. Bakoy
FONTOFA
Tamba
R. Tenkisso
R. Niger
LABÉ
Labé
KOLADÉ
KÈBALI
FUGUMBA
KOYIN
R. Bafing
Dinguiraye
DYUMA
FODÉ-HADYI
BURIYA
TIMBO
Timbo
Kouroussa
BATÉ
Boketto
5
R. Niger
R. Nyandan
Kankan
SANKARAN
R. Milo
Farana
HURÉ
SOLIMANA
Falaba
R. Scarcies (Kaba)
BOMBALI

and to Freetown in Sierra Leone. European products soon became widespread among the Malinké of the south and the role of commerce became more and more important as its influence increased. This was the great era of slave trading and there was a constant and tragic procession of human beings from the Upper Niger to the Americas.

This region was of considerable importance for long distance traffic as it was the meeting point for three trade routes. The traditional route for kola nuts from the Kisi, the Toma and the Guerze came along the Milo, the Nyandan and the Niger towards Bamako, below the zone where these rivers meet. Products from regions on the edge of the forest were collected here to be transported to the European trading posts at the river estuaries by way of Fouta-Djallon or a more southerly route through the land of the Dyalonké people, and Limba country. The Valley of the Niger's Milo headstream was the logical site for an important commercial centre and it was in fact for these reasons that the Dyula city of Kankan was established there.

The stabilisation of the Fulani aristocracy in the Fouta-Djallon area after 1788 had important ethnic consequences. After the defeat of the Upper Niger tribes led by the Wasulunke Kondé-Brèma, the autochthonous Dyalonké, an archaic Manding people with little political cohesion, were brought completely under Fulani influence. One branch of these people, the Susu, were pushed towards the coast, where they assimilated the original inhabitants of the area who spoke variants of the Mel language (Baga or Mani). The Susu borrowed much from the Atlantic forest peoples before coming eventually under the Islamic influence of the Fouta.

The Fouta-Djallon exercised considerable control over the weak coastal populations, using them as a means to spread the influence of the Sudan. It was at this time that the flow of trade from the hinterland established the importance of Sierra Leone as a commercial centre, and thus the sphere of influence was moved away from the Susu-dominated estuarine region of French Guinea. Thus, during the eighteenth century, the way was paved for the creation of the British colony.

Many traders who were originally Malinké, Fulani or Dyakhanké (an Islamic trading minority from Upper Senegal and Fouta-Djallon) moved into the area and married into the local population. They soon became assimilated but remained Muslims, and were instrumental in shaping the new political structures. These traders had already undergone some cultural changes due to their contact with the slave trade and this new racial synthesis developed some very unusual characteristics. The Temne gained most from the intercultural exchange; they took control of the coast between the Scarcies and Sierra Leone. Under the influence of Portuguese settlers in the seventeenth century, they had become somewhat Christianised, but they turned towards Islam as they gradually assimilated the

Fulani, Susu and Malinké settlers. Port Loko became the base for this cultural expansion.

The principal losers in the cultural exchanges were the Loko, forerunners of the Mende, who had come from the eastern forest, probably as an outcome of the sixteenth-century Somba invasion. Stubbornly opposed to Islam despite their association with peoples of Sudanese origin, they were gradually infiltrated by the Temne and the Limba. Assimilation was inevitable and did not cease until the third quarter of the nineteenth century.

The Bunduka from Upper Senegal were among the most active traders during the period of these changes. They were of Dyakhanké and Toucouleur origin, and were hawkers who travelled constantly about the country. They later turned to slave-trading when it became illegal in the nineteenth century and practised it on an inter-African basis up to the start of the colonial period.

South of Sierra Leone and along the Rokel river the situation was completely different. The Sudanese type of long-distance trade could not operate in the enclosed world of the forest where tribes still lived and worshipped in the purest animist tradition. Instead of a network of long-distance professional traders, goods were carried from place to place through a system of relays dominated by local big men. The increasing activities of Europeans along the coast in the eighteenth century slowly had an effect and it was during this period that the Mende completed their move westwards towards the sea, overwhelming the Bulom and some of the Temne. The movement started much earlier with the Kpa Mende and they were soon in active communication with the trading posts of Sherbro Island.

In this way the great Mende people isolated the two Manding groups who were settled before the sixteenth-century Somba. These were the Kono, who lived high up and alone, panning for gold in the mountains and the Vai, who held the coast almost as far as present-day Liberia and who were actively involved in the slave trade. The Vai had trading routes through the Toma forest to the lands of the Malinké from Konyan country, but traffic along this route remained negligible until Monrovia was founded.

Events took a different turn in the southern savannahs which lie above and beyond the coastal regions. Here the Dyalonké people were not assimilated by the Fulani, but were pushed towards the east and the south where they met the Malinké. Their civilisation therefore remained purely Sudanese and they gradually integrated with their neighbours. They nevertheless retained their language until the beginning of the twentieth century except in the gold-bearing Sudanese region of Buré. The threat of the Fouta-Djallon became particularly menacing in this region at the end of the eighteenth century when Fulani armies reached the upper Niger, crossing it repeatedly in order to reach the Kisi region. Some of the Dyalonké of the east began to form military states as a first step towards resistance. The greatest opposition came from the Sulimana, led by the Samura clan, who fortified Falaba and, despite several defeats, victoriously resisted the assaults of the Fulani until the middle of the nineteenth century. They even on occasion entered into open alliances with them. Further east, the Kamara people of the Firiya formed an alliance with a Malinké clan, the Ulare from western Sankaran, and this chiefdom fortified the ford of Farana, closing off the route to invaders.

Much farther north, between Tenkiso and Bafing (in Upper Senegal), the Dyalonké of Tamba submitted to the Sako, descendants of assimilated Dyula, and organised a state which was destined to become powerful. The Fouta-Djallon therefore saw its eastern frontiers blocked at a time when internal dissensions stemmed the tide of its expansion. Its military and political influence in the Upper Niger steadily declined, but its cultural and religious prestige remained considerable throughout the nineteenth century.

In the Upper Niger, the most important factor in the eighteenth century was the growth of Kankan. This enormous village had been founded by the Kaba at the beginning of the seventeenth century. It stood in the territory of the Konde, a people of animist beliefs and traditional rulers of eastern Sankaran. The Kaba were a Sarakhole lineage from Dyafunu, west of Nioro in the Sahel area. Dyula people came to settle with them when trade routes to the southern estuaries had been established. Kankan had all the importance of a city from the middle of the eighteenth century onwards and the Muslim settlements gathered round it, eventually forming a *kafu* that took the name of Baté ('between rivers'). The city supported the holy war of the Fouta-Djallon which controlled its outlets to the sea. The city was destroyed by Kondé–Brèma's animists, but when the Fulani were victorious around 1788, Kankan was rebuilt and threw off the Konde yoke to form a sort of merchant republic. The city was ruled for over a quarter of a century by a holy man, Alfa Kaabinè Kaba, whose reputation spread to the edges of the forest, but who limited his military policy to defending the city's freedom. He maintained the supremacy of his spiritual and commercial influence until his death in 1810 and his successors respected this political legacy until the Tijanniya of Al Hadj-Umar set the whole region in a ferment.

Kankan was the principal staging post between the trading centres on the coast and the kola-producing regions of the forest on the one hand, and Buré gold-washing areas and the middle Niger markets on the other. Developments in the Sudan were therefore of great importance to the town's existence.

In the second half of the eighteenth century the stretch of the Niger River from Kouroussa to Kangaba was ruled by the Keita of Dyuma, whose most famous leader was Kasa-Musa (*c.* 1770–1800). He was an ally of the Segou emperors and helped them fight the Fula of Wasulu, a Fulani people linguistically assimilated by the Malinké. The Fula were, however, allied to the people of Kita who, led by Basi Dyakité, managed at one point to command the river area from Bamako to Kangaba. Their capital, Samayana, was eventually destroyed in 1808 by Daa Dyara, the sovereign of Segou. Daa Dyara gained control over the Buré, reached Kouroussa and crushed the eastern Sankaranké, some of whom accepted his leadership. His aim was to control the trade routes to the southern river estuaries as far as the confines of the Fouta-Djallon. He succeeded in doing this but only for a short period, from about 1808 to 1818. The Bambara mandate in the Upper Niger collapsed shortly after the Masina rebellion and Segou control never again went beyond Kangaba.

The control of the area up-river was taken by the Dyalonké, then at the height of their power under the leadership of Tamba-Bwari (Bukari-Sako) (*c.* 1810–40). He conquered the Buré, crushed the Keita of the Dyuma and pushed east of the river as far as Wasulu, with the help of his Kundyan allies. The city of Kankan refused to join him and in order to crush it he tried to redirect the caravan route from Segou to the Fouta-Djallon via the Buré river and the Tamba. The Dyula people neither forgave nor forgot this manoeuvre. It was Tamba-Bwari's wars which prevented Caillié from travelling down the Niger River and directed him instead along the kola routes.

Tamba supremacy crumbled soon after the death of Bwari. Al Hadj-Umar wiped out the Dyalonké as an ethnic entity soon after settling the Toucouleur at Dinguiraye, in the heart of their territory, during the period 1849–50.

The forest fringes

As the kola trade routes leave Kankan and wind their way towards the great forest to follow its outer fringes in the direction of the Bandama, the countryside changes radically. The southern savannahs stretch for over three hundred miles and meet the impenetrable wall of the great forest which separates them from the sea. This dense mass of huge trees is a vast complex of primeval forest and great rivers which traverse endless series of rapids before they finally reach the Atlantic. Waterside settlements are usually found only on navigable sections near the river mouths and local people are dependent on dugout canoes. The area is sparsely populated, although a few favourable districts have been heavily settled.

This country was the territory of the Kru people (Basa, Wènyon, i.e., Gérè and Wobè, Bété and Dida) who lived in the area from Monrovia to the Lahou lagoon, and further north, near the southern savannahs. Various Mande peoples also lived in the region, and they spoke either western-type languages (Mano, Dan, Wen or Tura and Guro). The first group of peoples lived a totally forest-type life, their society being framed by the hunting associations. Manioc and rice were cultivated fairly intensively in some areas by certain peoples (the Bété of Gagnoa, the Dan of Man and the Guro of Zuenula). The coastal peoples engaged in a good deal of fishing and trading, but the slave trade itself often bypassed these desolate areas, which rarely saw slaves from the interior.

By the middle of the nineteenth century, when steamships made regular trips backwards and forwards to Africa, the 'Krumen' had remodelled their social structure to such an extent that they could provide crews to work on board the tramp steamers for the coast-hugging journeys from Liverpool to the Congo.

Successive waves of southern Mande peoples forced the Kru towards the south and they lived in overlapping societies from the sixteenth to the eighteenth century. Their culture showed markedly Sudanese characteristics as they were great farmers who kept in close contact with the Malinké dealers operating in the markets of the forest fringes.

It is extremely difficult to write about the history of these peoples for they lived separate, self-absorbed existences. There was some political cohesion, for example such districts as the *zu* from the Toma, and the *se* from the Northern Dan and Guro, but it was rare to find important leadership among them. Even when, as in the case of the Guro, careful research has produced detailed information, and it is possible to reconstruct the migrations of the last centuries, it is still difficult to find a middle way between vague historical generalisations and a depressing welter of confused detail. Pending further research, we will not spend much time on these forest peoples.

So far as this study is concerned their main importance lies in their position in the development of trade in old Africa, and this is significant only along the northern fringe, where they produced kola nuts. Due to the presence of the forest wall, broken only by the trading routes to Monrovia (themselves of relatively little importance), this region remained a blind alley until the French conquest. The only long-distance trade which concerned it was that of kola nuts, salt, and slaves and this had not changed since the late Middle Ages. The influence of European settlements on the Rivières du Sud and the Gold Coast was barely felt, although the opening of the Fouta-Djallon and the rise of Kankan increased that influence at the end of the eighteenth century.

Inside this area, between the Niger River and the edge of the forest, was a solidly-rooted population of Malinké animists. Its oldest element was formed by the Kamara people of Konyan, known as the Dyomandé further east in the Mau of Touba. This victorious clan had taken part in the Somba drive to the sea in the sixteenth century, coming from the east where they had settled on the cool, high, open plateaux which soar up above the heavy dampness of the forest (Beyla–Kérwané region). The Kamara drove the

Toma and the Guèrzé people southwards, though some of the Dan, who were also involved in this forced migration, were left behind in isolated groups. These small groups accepted their conquerors in the Touba region.

The main descendants of the Kamara from the Konyan have a semi-mythical ancestor, Fèren-Kaman (sixteenth century?), who led the migration and established the customary law. His old residence of Musadugu (Musardu to the Liberians) was entrusted, together with neighbouring villages, to the Dorè, Bèrèté and Kuruma Muslims, so that no one family would make use of it for personal advancement. During the seventeenth and eighteenth centuries, descendants of this victorious line abandoned the high plateaux and went deep into the Toma forest where they set up a number of states and intermarried with the people of the region. From here, intermittent communication was established with the coast in the Vai region to the west of Monrovia. The Toma people were thus absorbed as vassals into the society of the newcomers and were not exterminated.

In the eighteenth century the Kwen ruled the Upper Konyan. The capital was Sondugu, at the foot of the enormous Tibe peak. Some dissidents broke away and sought their fortune in the high Dyani valley, which was completely cleared of the local Toma, to become the new kafu of Buzye. This kafu became very powerful and destroyed the Kwen people in about 1790, with the aid of related tribes from the Simandugu in the high Dyon valley on the other side of the Gbe range. The Simandugu lineages, who were to maintain their supremacy until the appearance of Samori, nevertheless found themselves faced with a new threat on the western border of the Konyan during the eighteenth century. On this western side, the Sankaran (Ularè, Kondé) and the Toron peoples (Konaté, Kuruma) had been comfortably settled for centuries, and had rarely troubled the Kisi or Toma tribes from the forest. They were a wealthy people but much given to long domestic wars.

Right at the end of the sixteenth-century migrations, a Malinké group broke away and settled down as the Kuranko, far to the west, between the Dyalonké of the Solimana and the Kono people. This warlike and unstable nation, led by the Mara clan, included small Kuruma and Mansarè groups and maintained a very pure Sudanese civilisation, despite a strong substratum of forest peoples of different origin. Towards the end of the seventeenth century, the Kuranko thrust eastwards in an extraordinary burst of energy, driving along the forest fringes as far as the Nyandan, and overwhelming the Kisi, many of whom were captured. At the beginning of the eighteenth century, a force led by two brothers, Musa-Fin and Musa-Gbè Mara, crossed the river, drove back the western Toma people, and reached the Baulé, a tributary of the Milo river which marks the border of the Konyan. It was here that they met the Kamara who prevented their further expansion although the two sides fought until the middle of the nineteenth century. Their capital, Soghomaya, became the centre of a ruling system which was to be relatively stable.

In the east, on the other flank of the Konyan, various lineages spread along the edges of the forest as far as the Bandama river, repulsing the Guro in the south during the seventeenth and eighteenth centuries. They overwhelmed and occupied the Worodugu of Seguela (Dyomandé, Sumauru, and Donzo clans), and the Koyara of Mankono (Donzo, Fofana, Karamogho and Dyabaghaté clans). In the Koro, they mingled with various Dyula lineages who had come by way of the city of Kong, and they took part in the early development of a maritime outlet on the Bandama river. It was in this region that they founded the ancient city of Boron which was to change its site many times after its destruction in the eighteenth century.

The extraordinary expansion of the southern Malinké people had taken them close to the Sénufo who still held the Odienne region. As trade became more organised, various Dyula settlements appeared along the trading route which passed through this area from Seguela to Bamako. There settled a lineage of Dyarasuba, leaders of a Bambara army which came from Segou, towards the middle of the eighteenth century, drove away the original inhabitants, and established the Nafana kingdom.

Further north, in the Wasulu and among the southern Bambara (around Bougouni and Kolondyeba), the people lived in a complex of small self-contained kafu and this remained the general pattern, in spite of the fact that the kola traffic along the trade roads from the south led to the birth of important trading stations (Maafélé, Gbaralo, Ntèntu). The organisation of Tyendugu was something of a curiosity; this was a warrior settlement founded to the south-east of Bougouni in the mid-eighteenth century by a Kondé lineage of Sénufo origin, the Gonkoro-Bi. Their influence along the right bank of the Baulé was considerable. It was within this strongly animist Manding world that the Muslim Dyula came to organise the routes for the kola trade. Any venture in kola was bound to be speculative because the nuts are perishable and the main areas of consumption were scattered along the Niger waterway far away from the actual areas of production. Prices varied considerably from one end of the trading route to the other, depending upon the time of the year and the volume of traffic.

The organisation of kola supplies was the prerogative of Dyula agents permanently installed in the large southern markets. The main ones were Fèrèdu in the Kuranko country, Beyla and Musaduga in the Konyan, Touba, founded in the nineteenth century, and Séguéla and Mankono. The Dyula had the monopoly of dealing with the forest peoples, and acted as *dyatigi* or 'hosts' and compulsory middlemen for the traders who came from the north with salt, cattle or horses. The traders would travel as fast as possible along the southern trails, calling on the dyatigi at each market-town they visited. Caillié described in vivid detail how the caravans hurried from place to place, constantly inquiring about prices in an attempt to dispose of their merchandise with the least loss of profit.

This trading system was based on quick returns and a perishable commodity; it demanded great speed of operation and its proper organisation gave rise to an interesting society. The main dyatigi combined their brokerage operations with caravan-running on their own, and they were often equally involved in slave-trading, agriculture and local handicrafts. By the end of the nineteenth century, some of them had become very well-known, Nana-Fali Kamara, for instance, who was uncrowned king of the Buré. He controlled all the local gold-panning and trade relations with Freetown. Another successful man was Bubu Silla, who operated in Kankan and was in charge of most of the dealings between the forest edge and the coast. All this suggests the existence of a social class, and we may reasonably speak of a capitalistic group within a traditionalist society.

As this Dyula society stressed the importance of the individual and viewed trade in the light of a modern economy, they formed a perfect complement to the Malinké world in which they lived. They expanded the subsistence economy of the old kafu and even though they operated in small quantities of produce they nevertheless had a very powerful cultural influence. This was all the stronger as they were a small group of Muslims living in a vast animist society. The location of the brokerage centres was such that they lived mainly to the south, a long way from the Muslim Sudan. Inevitably, they were not therefore very enlightened Muslims and their culture was poorly developed, but they lived in a close community and kept together. Their relations with the infidels and especially with the local chiefs who gave them protection were usually excellent and the idea of a holy war was quite foreign to them. They were in no way interested in converting their neighbours, in fact they much preferred to maintain a state of affairs which worked entirely in their favour. It is noticeable, however, that the relationships between Muslims and animists were not so relaxed in the intermediate regions, where the former limited their activities to the trading-stations along the south-bound tracks and where they were numerically much less strong.

The region between the Bandama and the Volta

Between the Bandama and the Volta lies an extraordinarily complex region, fundamentally different from the immense stretch of dense forest. There is still a belt of forest here, but it is broken up by savannahs of the Baulé gap between the Bandama and the Nzi, so that in places it is barely sixty miles wide, a feature which strongly tempted the people of the north to travel down towards the sea. It is curious that this tendency, which seems to have started in the sixteenth century and gone on well into the seventeenth, was to be abandoned altogether during the period now under consideration.

Further east, on the Comoé River and south of the Volta, the forest is nearly two hundred miles wide but human initiative operated in such a way that the barrier it offered was by no means insurmountable. This was the home of the Akan peoples who seem to have originated along the northern edge of the forest. At an early date the search for gold brought the Dyula from Djenne to this region where they founded the city of Begho and by the fifteenth century were established on the Gold Coast. Their influence, combined with that of the great Benin civilisations, accounts for the rise of the Akan kingdoms in the sixteenth and seventeenth centuries, a development culminating in the foundation of the Asante confederacy in the 1690s. The forest in this region was no longer an absolute obstacle to human relations and it is not by sheer chance that the Portuguese established themselves on the Gold Coast at the end of the fifteenth century and that European forts were built all along the coast from the seventeenth century onwards. In this way a permanent link was established between the Middle Niger and the Gulf of Guinea, a link which lasted until the colonial conquest.

The area now to be considered lies slightly to the west of this Niger–Gulf of Guinea axis. On this side, between the Bandama and the Pra Rivers, the forest had long been the home of an ethnic family, of which the Akan were an early offshoot, far removed from the influence of Sudanese culture. They were forest farmers and lagoon fishermen living in a society based on age-sets and matrilineal hierarchies. The principal group, the Akyé, are probably long established in the Camoé region, although their oral traditions have recently been re-examined. The present distribution of the others is relatively recent and due to readjustments following pressure from the Anyi in the east (the Aburè from Bassam) or the Dida in the west (the Adyukru from Dabou). The peoples settled between the lagoons and the sea; the Alladian and Avikam (Mbrignan from Lahou), came from the east in the seventeenth century, almost certainly drawn by a desire to play some active part in the trade movements between the interior and the European centres on the coast.

While the shores of the Kru region (the Grain Coast, the Côte des Mal Gens, the Côte des Dents) did not attract European ships, this western neighbour of the Gold Coast developed as a trading area quite early in the seventeenth century. Dapper and Barbot noted that trade in this area was linked to goods brought down the Bandama River from the hinterland. Textiles which were carried there in great quantities by canoes from Lahou could only have come from northern craftsmen. In fact, a careful study of the present Baulé region makes it clear that at this time the area was settled by various peoples, but was dominated by those of Sudanese origins. There were various southern Mande people, in particular the Guro and some small allied groups, large numbers of Sénufo and many small Dyula settlements scattered all over the area. The Dyula groups were made up of southern Malinké, both heathen warriors and Muslim traders, who had come from the north-west along the forest edge, and real Dyula who had come from the north-east along the Djenne–Gold Coast route and through the home territory of the Sénufo. It is probable that they came to pan for gold (Kokumbo), but their importance lay in their movement towards the coast. Their descendants are still to be found spread as far as Tyassale on the lower Bandama. Other groups were probably autochthonous and Bauler speaking.

This complex society was completely transformed at the end of the seventeenth and the beginning of the eighteenth centuries by a massive westward movement of Akan lineages. This was the decisive factor in settling the present location of the Anyi and Baulé peoples. These successive migration waves started from an epicentre in the heart of modern Ghana, a region which was in a state of ferment during the entire seventeenth century. The increase in the volume of trade from Djenne through the Kulango kingdom of Bouna was obviously the main reason for this tremendous change.

The first migration wave began during the second half of the century when the new Dankyira empire extended its power to the present Asante region. The local inhabitants fled north and expelled a chiefdom related to the Akwamu empire which dominated the hinterland of Accra at that time. This chiefdom retreated towards the north-west, conquering the original Kulango and Malinké from the Barabo, as far as the banks of the Comoé River. They finally set up the Abron kingdom in the third quarter of the seventeenth century. Dyula refugees from the recently destroyed city of Begho settled in this new state and founded the merchant city of Bondoukou. The Abron kingdom had its roots in both forest and savannah; it existed under the rule of a minority group observing Akan traditions and following a system of government similar to that of the Asante. Interest was, however, directed northwards in the direction of the trade caravans, for it was from here that most of its subjects had originally sprung. The kingdom's history was troubled by an interminable duel with the Kulango from Bouna and by the Asante invasion of 1739. The Abron lost their independence and had to pay tribute to Kumasi until the fall of that city in 1873. They revolted twice in 1801 and 1818, but on the latter occasion King Adingra was killed and most of the nobility were massacred.

At the end of the nineteenth century, people breaking away from Dankyira found protection with Ano Aseman, king of the Agri kingdom of Adwin. When the Asante crushed the Dankyira, in 1701, new refugees fled west, some pushing beyond the Nzi River, where they began the conquest of the Baule country; they are the Alangyira. When Asante subjected Aowin, after 1710, others left Ano Aseman to found the Ndenye kingdom around Alengourou (Ivory Coast). Finally Amalaman-Ano settled between the Tano and Comoe River, around the Aby Lagoon, where he founded the Sanwi kingdom (1720?).

Another wave of migration was caused by the civil wars which came with the disputed succession to the Asante throne in the period 1717–20. Those who were defeated, led by the legendary Queen Poku (Aura Poku), left Kumasi and joined the Alangyira between the Nzi and Bandama Rivers. These were the Asabu people. These two Baulé groups united to set up an unusual and extremely vigorous society. The original inhabitants were assimilated in spite of forming the majority, but their language and traditions (wood carving, a new type of weaving and yam cultivation) were of vital importance to the new culture. A new civilisation was created, uniting in admirable synthesis the contribution of the Asante and the surviving local traditions.

Baulé society never overcame a strong tendency to break up into small groups and the early political unity was no more than a memory by the end of the eighteenth century. In spite of this, the natural vigour of the society was undiminished and the people were greatly feared by all their neighbours, especially the Guro whom they had driven back beyond the Bandama. Right up to the European conquest the Baulé people (Boule people) displayed a strong propensity for expansion and for excluding foreigners. They also excluded the traditional network of Sudanese long distance professional traders, being satisfied with a system of relays carrying goods from place to place under the control of local chiefs. Thus from the first quarter of the eighteenth century the Bandama route was closed to Sudanese trade and remained so until the colonial conquest. On the coast, Lahou fell into a decline and ships stopped calling there. It was only in the nineteenth century that traffic in oils from the Adyukru made the position of the Alladyan (Jacqueville) a valuable one and at last revived European activity on this coast.

Many of the Dyula previously settled in the Baulé went back to the far side of the Upper Nzi River, on the borders of the Dyimini Sénufo, and there, under the rule of a Watara lineage, they established the Dyammala. In this way they drew nearer to the Djenne–Gold Coast trade route, which was to take all future trading traffic.

The Empire of Kong

In the seventeenth century, the southern Malinké, migrating from the west, met some Dyula groups on the Bandama. These groups had come from the Middle Niger centuries before, originally in search of gold, but later in pursuit of the Gold Coast European trade. The area they had settled in, a region stretching from Segou or Djenne as far as the Black Volta, was inhabited by an extraordinarily complicated mixture of ethnic groups, unusual even for this continent. The population was basically of Voltaic stock, but varied greatly from area to area with the exception of the two main groups, the Mosi in the east and the Sénufo in the west. Apart from the Mosi, whose kingdoms had taken on a definite form during the fourteenth and fifteenth centuries, these Voltaic peoples had no tradition of political centralisation and their basic culture, which formed a constant substratum throughout the area, was of an archaic type which Baumann called 'palaeonigritic'. Sudanese influences masked this to some extent among the Sénufo, a numerically impressive people, whose tribal system seems to have taken its first shape at a very early date along the Korhogo–Bobo–Diolasso axis. On the other hand, the ancient culture flourished strongly among such widely differing groups as the Bwamu or Bobo Ulé, the Dagari and the Lobi, the Gwen of Banfora and the Kulango from Bouna and Bondoukou. There were also ethnic groups with similar cultures but of very different origins. There were for instance such old Mande tribes as the Samo of Tougan and the Samogho, some of whom are autochtonous in the Sikasso region, or the Sambla and Bobo from around Bobo-Diolasso. There were even isolated Kru, such as the Seme of Orodara, and also a number of ill-defined ethnic minorities.

From the middle ages onwards, Dyula migrations complicated this picture still further. Their earliest migratory waves left behind small units of people who integrated to a greater or lesser extent with the local population. There were for instance the Bobo-Dyula, the Blé and the Nigbi of old Begho. Their own culture seems to have adapted itself to that of the Voltaic peoples with whom they settled and they were able to fit comfortably into the centralised states which were established all along the trade route. This was the case in Bouna, well before the advent of the Abron kingdom. Bouna was the capital of a kingdom founded at the start of the seventeenth century by a Dagomba dynasty in the heart of the Kulango people who were still at the palaeonigritic stage. At about the same time, some Dyula, or more accurately Malinké, founded the Gonja kingdom, after they had integrated with the local tribes of the Begho region. In these states traders of Manding origin soon encountered their Hausa counterparts and also met the Zerma later in the nineteenth century.

Beyond this great north-west to south-east route it is possible to discern an old track going towards the Bandama. From this sprang some old Dyula villages in Sénufo territory near Korhogo and in the present Kong region. It is not without relevance that there are traditionally traders of Mosi origin and tribes from the Dafin, that is to say from Manding colonies, in the Tougan region of the Upper Volta. Their close relations with the ancient city of Boron are symbolic of the encounter between the two Manding waves of migration.

This region was the very heart of the Dyula empire of Kong, born in a sudden and violent upheaval at the beginning of the eighteenth century. Local tradition is the only source of information concerning these happenings. They seem to have resulted from the growth of Dyula influence, the development of trade along the Gold Coast route in the seventeenth

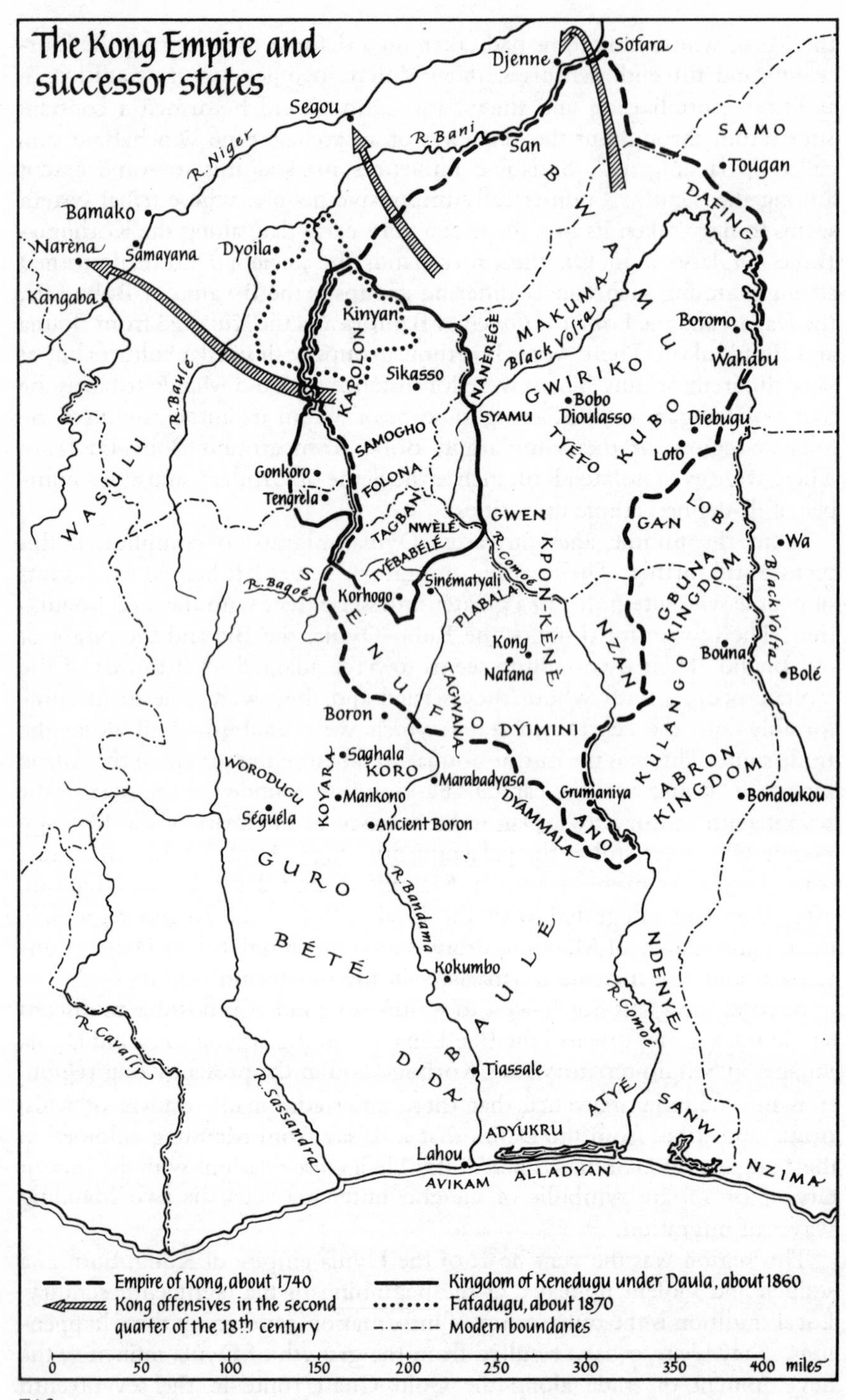
The Kong Empire and successor states
Djenne
Sofara
Segou
R. Niger
R. Bani
San
SAMO
Tougan
Bamako
Narèna
Samayana
Dyoila
Kangaba
Kinyan
KAPOLON
NANERÉGÉ
MAKUMA
Black Volta
Boromo
Wahabu
Sikasso
R. Baulé
GWIRIKO
Bobo Dioulasso
SYAMU
SAMOGHO
Diebugu
Loto
TYÉFO
KUBO
DAFING
BWAMU
WASULU
Gonkoro
Tengrèla
FOLONA
TAGBAN
NWÈLÈ
GWEN
LOBI
GAN
Wa
TYÉBABÉLÉ
R. Bagoé
Sinématyali
R. Comoé
KPON-KÈNÈ
GBUNA KINGDOM
Black Volta
Korhogo
PLABALA
Kong
Nafana
NZAN
Bouna
Bolé
SENUFO
TAGWANA
Boron
DYIMINI
KULANGO
Saghala
KORO
WORODUGU
KOYARA
Marabadyasa
Grumaniya
ABRON KINGDOM
Bondoukou
Mankono
Séguéla
Ancient Boron
DYAMMALA
ANO
GURO
R. Bandama
BÉTÉ
BAULÉ
NDENYE
Kokumbo
R. Comoé
R. Cavally
R. Sassandra
DIDA
Tiassale
ATYÉ
SANWI
ADYUKRU
Lahou
AVIKAM
ALLADYAN
NZIMA
Empire of Kong, about 1740
Kong offensives in the second quarter of the 18th century
Kingdom of Kenedugu under Daula, about 1860
Fafadugu, about 1870
Modern boundaries
0
50
100
150
200
250
300
350
400 miles

7:2

century, and the spread of firearms, which now appeared in large numbers for the first time. The Dyula were in the best position to obtain them, and they planned to force the local inhabitants to become involved in free trade as soon as they had the means to do so. This manoeuvre came shortly after the formation of the Asante confederation, which established control over the southern sector of the route and was more or less contemporary with the rise of the Baulé kingdom which closed the Bandama River to commerce. The establishment of Kong was not unconnected with these crises. The trade route from Kong usually crossed the Asante to reach the coastal settlements, but the Dyula also planned to develop the Comoé route. This would secure a free outlet to Bassam and, at the cost of a detour, to the forts of the Nzima and Fanti coasts. The basic purpose of the Dyula was in fact to bring a new system to the southern caravan routes up to the Niger. With the aid of their rifles and swift ubiquitous cavalry it was easy for the Dyula to triumph over the unorganised bowmen of the palaeonigritic peoples. In only a few years they managed to establish their supremacy over the immense territory they coveted; as traders they already controlled the trade routes throughout the area.

Séku, who was the founder of this empire, was born at the end of the seventeenth century. His ancestors were traders who claimed descent from the Kèita but had been settled in the area for several generations and had adopted the name of Watara. During the early years of the eighteenth century the young chief had no difficulty in either eliminating or winning over the Dyula groups who had conquered Nafana (Kumbala) and Kong. In Kong, which he made into his capital, Séku organised a powerful army by recruiting large numbers of Sénufo, usually commanded by Dyula officers. His authority spread rapidly southwards to the Dyimini and to the Ano, a country in a forest zone below the Comoé River. To the north, the Bobo-Dyula gave him command of the Gwiriko and the loop of the Black Volta. From here the Kong armies pushed on to Sofara on the shores of the Niger, opposite Djenne, which was attacked in 1739. To the north-west the whole of the Sénufo and Bambara territory, as far as the right bank of the Bani, was ruled by Kong.

The vast empire had barely been established when its limitations began to appear. The Watara's assault on Segou was repulsed by Biton Kulibali, who asked the Malinké and the Fula for aid. The defeat of the Abron by the Asante brought the armies of Opoku-Ware to the walls of Kong in 1739. Séku seems to have extricated himself from danger only by handing over the refugees. He died soon afterwards, probably in 1740, bequeathing to his heirs the largest empire that had ever existed south of the Niger River. After Séku's death the weaknesses of the Dyula empire became only too apparent. Established by a minority, stretching endlessly along the trading routes, it lacked the necessary elements of cohesion. Not only was its ethnic diversity impossible to alter, but the political and commercial aims it pursued excluded all possibility of religious unity. The payment of tribute and the opening of trade-routes were imposed, but never conversion to Islam. The most notable feature of the empire was the establishment of a warrior caste around Séku's family. Its origins were Dyula or Sénufo, but the members were to all intents and purposes animists, even if they were of Muslim descent. They were known as the Sohondyi or 'sacrificers' to distinguish them from the Silama, who were authentic Dyula who stayed faithful to commerce and Islam. It is significant that the Sohondyi abandoned the city of Kong to the Silama after Séku's death; each prince went back into the bush with his warriors and set up his own domain.

In such circumstances as these and in the absence of a coherent system of territorial government, the dismemberment of this immense domain was inevitable and indeed took place very quickly. As Séku had founded the empire, succession had to follow Malinké custom and pass to his children. Séku's brother Famagan refused to obey his nephews, seized power and established himself in Sya (Bobo-Dioulasso), taking the Gwiriko, which at that time included the entire northern part of the empire. The southern part, which took the name of Kpon-Kènè, remained under the rule of Samandugu, Séku's eldest son, and later passed to his brothers and their sons. This cleavage was nevertheless not a total break. Towards the middle of the century, the entire southern part of the empire revolted under the leadership of Demba-Koroko Watara, chief of the Dyula of Dyammala, and the city of Kong was threatened. The Faama Kombi asked his uncle Famagan for help and it was with his aid that the Dyammala were crushed. In the nineteenth century, the people of Kong reciprocated and went to the aid of Gwiriko which was threatened by Sikasso. This basic separation into two main blocs was not the only division in the empire. Branches of Watara's family settled more or less everywhere, establishing small kingdoms which were to prove extremely viable, for instance the Nzan east of the Comoé on the border with Bouna and the Kubo around Djébougou. They attempted to conquer the breakaway Lobi and Dagari and to hold the Volta fords that led to the Mosi. The Makuma also became organised north of the Gwiriko, on the Djenne trade route.

There were rather more disturbing disruptions during the third quarter of the century when Sohondyi from other clans began to carve out their own personal kingdoms. A Sénufo called Nangen who was one of Séku's old lieutenants organised the Tyèbabélé (Kyèmbagha) around Korhogo, and the old Malinké state of Boron also tried to impose its authority below the Bandama River. Further north, Kasina-Traore, a 'Sénufised' Dyula whose ancestors had lived near Banfora for many years, conquered the Folona. His successors were to rule the Sikasso region where the Sénufo had fought to regain their independence. In this way, the kingdom of Kénédugu took shape during the eighteenth century. The final break-up of the Watara empire came when the paleonigritic tribes succeeded in freeing themselves. During the latter years of the century the Plabala (Pallaka), local Sénufo from the Kong region, revolted and dug themselves in less than sixty miles north-west of the city. For nearly a century they fought off all attacks and did not yield until the coming of Samori.

At the start of the nineteenth century the immense Watara empire no longer existed as a cohesive political entity. Kong retained the prestige of a commercial capital and intellectual centre, but was no longer the seat of a great power. The alliance of Kpon-Kènè and Gwiriko still kept some semblance of unity and managed to keep open the southern trading routes, but elsewhere Dyula and Sénufo hierarchies organised small self-contained states and gradually the local populations moved towards independence.

THE WESTERN ROUTES

Throughout the eighteenth century the growing influence of the European tradings posts was felt along the trade routes and eventually disturbed the interior. Revolutions of considerable size and importance were stimulated at the two extremities of the region under consideration; in the Fouta-Djallon and the Kong empire. The central area, on the other hand, remained intact and untroubled behind its forest walls. It was precisely this area which, in the nineteenth century, found itself torn by the internal dissensions known as the Dyula revolution, while the east and the west became steadily more disturbed and less able to discover a new stability.

The Côte des Rivières

It would take many pages to unravel the tangled skein of the nineteenth-century history of the Rivières du Sud. On the coast the different states, usually small self-contained units, were subjected to European influences of every kind, firstly from the forts of Portuguese Guinea in the north and also from Sierra Leone in the south. In between lay the Rivières du Sud, the complex of waterways which eventually formed French Guinea. This was a region where clandestine slave trading was carried on in friendly competition with the Senegalese settlements until the authorities in St Louis decided to intervene (Boka, 1866; Benty, 1867). In the north, the coastal peoples were subject to pressure from the Fouta-Djallon and some of them acknowledged their suzerainty. In the south, the entire area was ruled by Dyula families who had long since settled with the local inhabitants. They were, however, continually engaged in civil war, for there was a constant struggle between more or less Islamised chiefdoms for control of the trading routes, and monopoly of local middlemen. The wars were bitter and hard and grew worse as European rivalries increased on the eve of colonial conquest.

The old Malinké kingdom of the Gabu, in the Upper region of the Rio Grande (Portuguese Guinea), as well as those of Casamance and the Gambia, were the home of many pagan Fulani (Fulakunda and Pulli) who had come to the area during the rule of Koli Tengela. These people were deeply attached to their blood brothers elsewhere and many were converted to Islam. The conditions were ripe for a holy war.

The conquest began in the first quarter of the nineteenth century. Grey has noted an Almami campaign here in 1816 but two years later Mollien commented that the Peuls' authority did not go beyond Kadé. There were, however, many battles, especially from 1830 onwards, for which two successive Alfa-Mo-Labé, Ibrahima and Yaya Maudo, were responsible. The king, Yangi Sayon, was killed by Almami Umaru in 1849. The siege of Cansala, in 1866–7, finally brought the collapse of the Malinké of Gabu, many of whom emigrated to the Gambia. The war continued for a further three years on the border of the Braso, which remained unconquered. The success of the war was due to the revolt of the indigenous Fulani. They went on to establish several kingdoms, such as that of the Forea who settled south of the Corubal around 1850. Beyond the Malinké zone, the Byafada and the Nalu peoples were driven back to the river lowlands where they settled down with the help of the Portuguese at the forts of Geba and Buba.

The movement gained strength in the north after the fall of Cansala, and the Fulakunda from the Braso and the Firdu under Alfa Molo's command eliminated the Malinké from the downstream areas of Casamance and the Gambia. Musa, son of Alfa Molo, broke away from the Fouta-Djallon some twenty years later and began to court the French.

Further south, the Landuman of Boke were subject to the Labé's domination and were not able to break away in spite of massacring the Fulani in 1840. Their kings, Sara and Tongo, fought on opposite sides during the period 1844–9 and their weaknesses were cruelly exposed. Their Nalu enemies controlled the mouths of the rivers and they too were divided, between the Dapellom and Towelia tribes. The latter claimed Malinké origin and enjoyed a brief moment of fame at the beginning of the French colonial period with the advent of the celebrated Dina Salifu.

In the Rio Pongo region (Boffa), the Susu, led by the Damba (Kati) clan, had driven the Baga into a few coastal enclaves and had established the Tya kingdom. Clandestine slave trading resulted in racially mixed dynasties and the suzerainty of the Fouta-Djallon stirred up domestic quarrels. In 1865 these disturbances reached a head with the 'Kolisokho' or 'Mulattos' War. The Lightburns and Fabers supported the young Manga Kulum against Bala Bangu, who was himself supported by the pro-slavery party of Curtiss and the Fulani. Kulum finally triumphed and was recognised by the Almami, but the country remained in a state of general weakness and anarchy despite the French occupation of 1868.

Further to the south, away from the influence of the Fouta-Djallon, the Tabunsu kingdom (Kaporo, Conakry), inhabited by the Baga, battled constantly against the Susu from the Manéa (Koya) and the Sumbuya. The latter people were involved in a series of dreadful wars which ravaged the Morea (Melakori/Mallacoree) region. A number of the Turè, who had originated from the Baté but had become assimilated by the Susu, seized power in the middle of the eighteenth century under the command of Fode-Bakari, the founder of Forèkaria. The Turè gave their chief the title of Almami and strengthened their position with the help of the Solimana.

After civil war had been simmering, a major conflict erupted in 1865. On one side was the proselytising Muslim Almami Bokari, who had been educated in Timbo, and on the other side was Maliki-Gbeli, whose supporters took the name of 'Maliguists'. This conflict, further complicated by French, British and even Samorian intervention, lasted until the colonial era. By then, all the neighbouring countries had been drawn into the struggle. Notable participants were the Bènna, from above the Scarcies, the Tambakha, whose chief, Karimu, became famous for his resistance to the English, and the Temne of Kambia. The latter were usually headed by the Satan-Lahay family, who were also descended from the Turè of Baté. This group came to play a major part in the border rivalries of the colonial period. The dissensions were to last well beyond the 1889 agreements.

The Loko people under the leadership of chiefs of Malinké origin fought for survival to the south of this area. They were attacked on all sides by the Temne of the Sanda and the Bombali. The greatest of these Loko chiefs was Pa-Koba who gained the upper hand by his support of Samori in 1884. He kept his people secure until the establishment of the protectorate in 1889.

In Port Loko, now the starting point for all the caravans travelling from the hinterland to Freetown, the Susu seized power at the end of the eighteenth century, only to lose it again in 1815 to the local Temne, who were already ardent converts to Islam. The Alkarli, who had made an alliance with Britain in 1825, managed to achieve some sort of stability, but each new accession to the throne brought further trouble from power-seeking immigrants and took the country eventually to the brink of civil war.

On the Rokel another Temne group, the Yoni, led by a family of Fulani origin, found themselves cut off from the trade routes and trapped by the Kpa-Mende of Madam Yoko, and their brothers from the Masimera tribe. The Yoni struggled to break free, and a British military expedition had to settle the problem in 1886.

These skirmishes show a confusion more apparent than real. They also show the archaic peoples facing the gradual rise of perils they had never known before and which they were quite unable to overcome. Bravery and initiative were never lacking, but they were not enough. The weakness of the neighbouring Fouta-Djallon was equally great and could not be concealed by the brilliance of its Muslim culture or the success of its feudal system.

The drama of the Fouta-Djallon

After the murder of Almami Saadu, the powerful Alfa-Mo-Labé, Modi Sulèmani, interfered in the affairs of the crown in 1799 and installed as Almami Abdulay Bademba, the third son of Karamogho-Alfa. Thus commenced the rule of the biennial alternation between two Almamis, members of both dynasties, the Alfaya and the descendants of Ibrahima Sori who took the name of Soriya. This rule was to apply not only to Timbo but to all the other *diwal* (provinces) except for Fugumba, which was reserved for clerics who were in charge of the investiture. The solution was an essential step in the stabilisation of Fulani society. The curious system it set up was to function, for better or for worse, until the French conquest. It suited the aristocrats well enough, for they were anxious to avoid any catastrophic clash of arms and at the same time it provided a way of limiting the powers of the Almami. Thus, there was peace for the next ten years. This tenuous unity allowed the country to fight off new attacks from the Sankaran and to lead a campaign against the Dyalonké from the Solimana region. However, Modi-Sulèmani died in 1810 and the Soriya Almami, Abd-Ul-Gadiri, was then sent into exile in the Pakèsi area (Portuguese Guinea). He did not return until four years later when he formed an alliance with the Manga of the Solimana, defeated Abdulay in Kétigiya and killed him with his own hand. Alleging the violation of the biennial rule, he was able to impose his claim on the Fugumba priests and was duly proclaimed Almami in 1814. He was to stay alone in power for eight years.

Whilst the ruling parties quarrelled among themselves the society which had sprung from the eighteenth-century holy war began to settle down. Fighting did continue in the north-west towards the Rio Grande and Gambia; it was chiefly the work of the Labé, helped from time to time by the Timbo people. Here, the last pocket of resistance held by the Malinké from Gabu, Cansala, fell in 1867. Elsewhere the frontiers were quiet and stable from the beginning of the century onwards, for any attempts at expansion were impeded by the powerful Dyalonké kingdoms of Tamba and the Solimana. The Almami still raided in this area, for example in the Sankaran in 1818, but without any positive plans for conquest. Towards the south-west, their control spread along the Freetown route as far as the Susu of the Tambakha. Towards the west, the diwal of Timbi drove back the Susu from the Rio Pongo in 1830 and established its control over the Tya (Boffa) kingdom. This area was, however, soon torn by civil war between the supporters of the Labé and those of Timbo. It never recovered from this conflict. The Susu from the south (Dubreka-Forèkaria) were never to pay tribute to the Fulani.

The Fouta-Djallon had exhausted its conquering strength but its cultural and religious prestige remained enormous. This was largely due to the Dyakhanké minority, whose main centre was Tuba, west of the Labé. Its influence in the Upper Niger, and particularly on the city of Kankan, was shaken only by the rise of the Toucouleur. In each *miside* (parish or sub-district) the aristocracy crushed the Dyalonké masses, who soon lost their ethnic individuality. There was no longer any fear of attack from outside, but in spite of this relative peace, the Fouta-Djallon proved incapable of establishing a stable political system. Although the succession was often peaceful in each party, the biennial rule, however, was often broken, and this led to civil war. The struggles had repercussions in each diwal, for each ruling family was split between the two parties. Much of the trouble was the responsibility of the Labé, who occupied a good half of the country and were constantly seeking to extend its frontiers. When Modi Sulèmani became Alfaya, he ousted his brother Mohamadu Dyaw who was supported by the Soriya party.

Abd-Ul-Gadiri died in 1822 and the Alfaya Bubakar Zikru, an old man, followed him eight months later. The Soriya attempted to keep themselves in power through their new Almami Yaya, a brother of the dead man. However, they soon quarrelled among themselves and after civil war Yaya was replaced in 1827 by his nephew Amadu. The fighting allowed the Alfaya to regain power in the person of Almami Bubakar, Maudo ('the great') son of Abdulay Bademba, and for about twelve years (*c.* 1827–39) he ruled Timbo with a firm hand. He spent much of his time supporting Ibrahima Maudo and Yara Maudo, the two Alfa-Mo-Labé, in their campaigns in the Kantora of Upper Gambia. He was badly defeated by the Malinké from the Braso and his quick decline seemed to be confirmed by an earthquake and an eclipse of the sun in the year 1253 after the Hegira (1836–7).

The aged Almami Yaya had retired with his nephew Umaru to live in exile in Bundu, after the triumph of the Alfaya, but he returned later and began to intrigue against his colleagues. The civil war which followed ended in a compromise; the ruling Almami gave place to his old rival, promising to respect henceforth the rule of alternation (1839). Yaya died one year later and was replaced by his nephew Umaru, and he gave way two years later to Bakari, who nevertheless tried to retain power when he had served his term. Thus a new civil war followed, which was very bloody, and was still going on in 1844, when Al Hadj-Umar, who had arrived in the Fouta, back from his pilgrimage, offered his mediation and restored the rule of alternation.

Thus Bakari took over, but he died in 1845, giving place to Umaru. In 1846 the latter was replaced, without incident, by his colleague Ibrahima

Sori Dara, and the agreement was now to be generally respected. Umaru was nevertheless threatened by his colleague in 1851, when the French traveller Hecquard arrived and reconciled the disputants. Making the system work was not easy but alternate succession finally began to work properly in 1853 when the threat of a far-reaching internal revolt brought the two reigning families together. After this the system was respected, apart from minor violations, until the French conquest.

The unusual danger which brought Alfaya and Soriya together was the famous Hubbu revolution. This movement has not yet been studied in depth but it seems to have had all the characteristics of a social revolution of a type which was quite exceptional in pre-colonial Africa. The founder of the sect, Modi Mamadu Dyué, was a resolute Kadiri who had studied in Mauritania with Sheikh Sidia, and his reputation as a scholar was such that the Almami Umaru allowed him to teach his children for a time. He was a So (Pèrèdyo, Sidibé) who belonged to one of the families who had come to the Fouta many years before and who were closely linked to the Dyalonké, with whom they shared territorial rights. They had, however, allied themselves with other Fulani and had eagerly taken part in the holy war. As the new society became established, many Fulani, and particularly those descended from tribes long resident in the region, found themselves constantly repressed by the established aristocracy. Their territorial rights, like those of their old Dyalonké allies, were in jeopardy, and in the face of this injustice it was natural that the most literate among them should seek the egalitarian judgements of Muslim law. Circumstances soon offered them a religious justification for their complaints. Al Hadj-Umar's sojourn in the Fouta (1844–8) had some surprising results. With the exception of a few Dyakhanké intellectuals and other marginal elements, the entire Fulani aristocracy was ready to adopt the new Tijanniya teachings. They were drawn more by its promise of glory than by its declarations of equality. The Tijanni leader founded Dinguiraye, in Dyalonké territory, in 1849 and two years later contingents from the Fouta and Kankan followed him when he made war on his host, King Imba Sako, and destroyed the old hegemony of Tamba (1851?).

Mamadu Dyué refused to accept the Tijanniya heresy and decided to form a separate group. Those who joined him were on the whole drawn from the ranks of the disaffected, people rejected by the Fulani hierarchy and members of the Dyalonké and Malinké minorities. These people flocked to join Mamadu Dyué in the diwal of Fodé-Hadyi which marks the eastern frontier of the Fouta on the Upper Tenkiso and they then took the name of Hubbu; *hubbu rassul allay*, 'those who love God'. For a time Umaru's attitude remained somewhat equivocal; he may well have been distracted by the growing influence of the Toucouleur, and the difficulties caused by the turbulence of the nobility. When the Almami came to understand the full implications of the Hubbu movement, he instantly decided to destroy it.

The struggle which began in 1849 was waged sporadically for about a quarter of a century. The astonishing fact was the strange powerlessness of the Fulani, in spite of their overwhelming material superiority. The Hubbu captured Timbo at least twice. In time, however, they were forced to retreat to the lonely Fitaba region south of the Upper Tenkiso. Here they were on the frontiers of the Dyalonké of the Firiya, some of whom decided to become their allies. Mamadu Dyué died shortly after this but his son Abal was to prove himself an admirable war leader. In his small community there reigned an atmosphere of mystic exaltation which was all the more formidable for its stark simplicity.

Despite a victorious campaign in about 1865 Almami Umaru was abandoned by his troops, who seem to have been won over by enemy propaganda. Umaru abandoned the campaign and went to help the Alfa-Mo-Labé, Yaya Maudo, in his fight against the Gabu. He had already killed their king, Yangi Sayon, in 1849. The fall of Cansala marked the end of this war, in 1866–7, but the Brasu were still fighting in the west. The Almami died of an illness on the way back from this country in 1869–70. His brother, Ibrahima Sori Dongol Fella (1870–90), replaced him as the Soriya Almami, but his colleague Ibrahima-Sori-Dara took control. The latter was anxious to take up the fight against the Hubbu again but he was killed with many members of his family while assaulting Boketto in 1871.

This disaster was never avenged. Amadu Dara, the new Alfaya, was denied power by Ibrahima Sori, his colleague, and took over only in 1875, through a coup d'état which restored the alternation. He went to war with the Susu and in the Gambia only. He, like his Soriya colleague, was a former pupil of Mamadu Dyué and seems to have had a soft spot for the intransigent Hubbu. The latter was ruler, however, when Samori reached the Fouta frontier in 1879 and it was at his request that the conqueror had the Fitaba heretics wiped out in 1884.

The successes of the Hubbu can be explained only by the appeal they held for large sections of Fulani society. If the aristocracy seemed reluctant to wage war with their blood-brothers, who were a living reproach, it was doubtless because they could not be sure of their troops. When Umaru asked the Fugumba council for an army, he received a blank refusal, and ten years later, at the height of victory, his men abandoned him at Boketto.

There were Hubbu groups more or less everywhere in the Fouta, but they were all wiped out, except for those of the Timbi who fled to Susu country. Fulani society was apparently breaking up, and was able to exist only by exercising great care. The excellent relations it maintained with Samori from 1879 to 1893 helped to maintain it, and the Hubbu revival of 1889 was short-lived. Despite their pride and hostility to white men, the Fulani seemed incapable of offering any serious military resistance. Divided as they were, all they could do was gain time by skilful delaying tactics. The surviving Hubbu joined the Shadeli from the Wali of Gumba and under

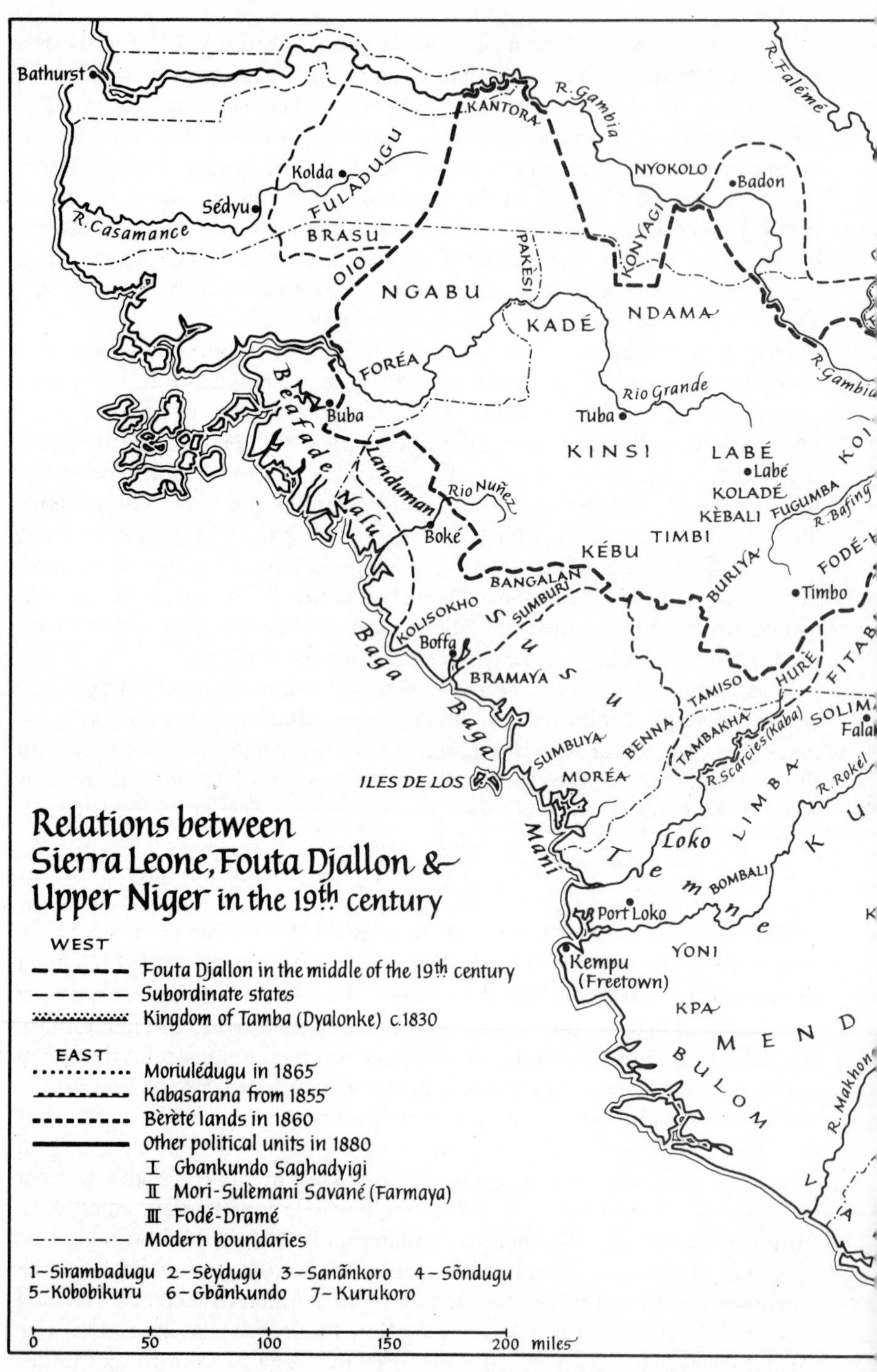
Relations between Sierra Leone, Fouta Djallon & Upper Niger in the 19th century
WEST
Fouta Djallon in the middle of the 19th century
Subordinate states
Kingdom of Tamba (Dyalonke) c.1830
EAST
Moriulédugu in 1865
Kabasarana from 1855
Bèrèté lands in 1860
Other political units in 1880
I Gbankundo Saghadyigi
II Mori-Sulèmani Savané (Farmaya)
III Fodé-Dramé
Modern boundaries
1–Sirambadugu 2–Sèydugu 3–Sanãnkoro 4–Sõndugu
5–Kobobikuru 6–Gbãnkundo 7–Kurukoro
0 50 100 150 200 miles
Bathurst
R. Gambia
R. Falémé
KANTORA
FULADUGU
Kolda
Sédyu
R. Casamance
BRASU
OIO
NGABU
PAKESI
KONYAGI
NYOKOLO
Badon
KADÉ
NDAMA
FORÉA
Rio Grande
Tuba
Buba
Beafade
Landuman
Nalu
Rio Nuñez
Boké
KINSI
LABÉ
Labé
KOLADÉ
KÈBALI
FUGUMBA
R. Bafing
TIMBI
KÉBU
BURIYA
Timbo
BANGALAN
SUMBURJ
KOLISOKHO
Boffa
Baga
BRAMAYA
Susu
TAMISO
HURÉ
TAMBAKHA
R. Scarcies (Kaba)
SOLIM
BENNA
SUMBUYA
MORÉA
ILES DE LOS
Mani
Loko
LIMBA
R. Rokél
Temne
BOMBALI
Port Loko
Kempu (Freetown)
YONI
KPA
MEND
BULOM
R. Makhon

7:3

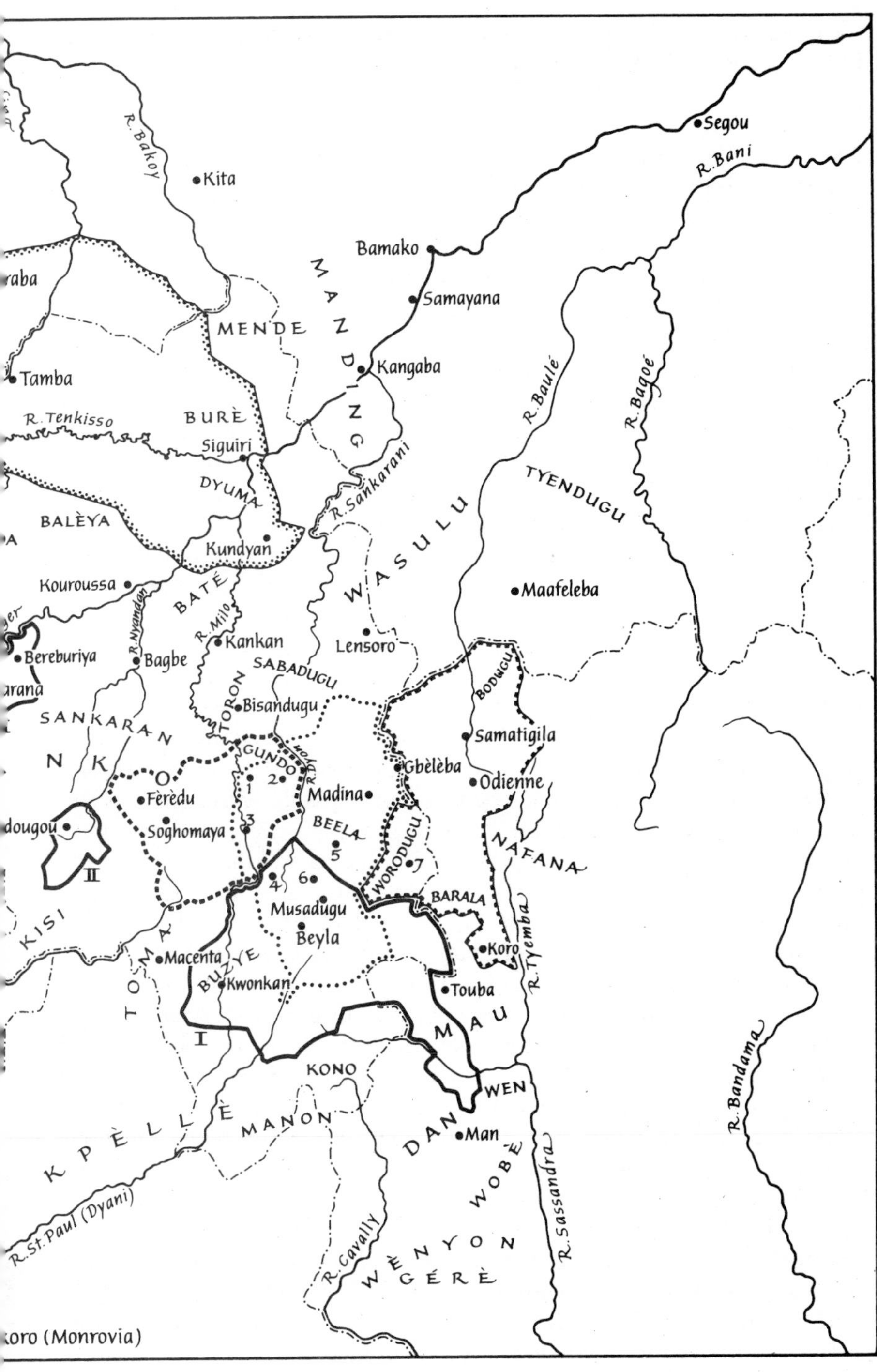
Segou
R. Bani
Kita
R. Bakoy
Bamako
Samayana
MANDING
MENDE
Kangaba
Tamba
R. Tenkisso
BURÈ
Siguiri
DYUMA
R. Sankarani
BALÈYA
Kundyan
Kouroussa
BATÈ
R. Milo
R. Nyandan
Kankan
Bereburiya
Bagbe
Lensoro
WASULU
TYENDUGU
R. Baulé
R. Bagoé
Maafeleba
SABADUGU
TORON
Bisandugu
SANKARAN
GUNDO
Fèrèdu
Soghomaya
Madina
BEELA
BODUGU
Samatigila
Gbèlèba
Odienne
WORODUGU
NAFANA
BARALA
Musadugu
Beyla
KISI
TOMA
Macenta
BUZYE
Kwonkan
Koro
R. Tyemba
Touba
MAU
KONO
WEN
MANON
DAN
Man
KPÈLLÈ
WOBÈ
R. St. Paul (Dyani)
R. Cavally
WÈNYON
GÉRÈ
R. Sassandra
R. Bandama
Koro (Monrovia)
I
II

their command took part in one of the last anti-French revolts in 1910. Others resettled the Fitaba during the colonial period.

THE DYULA REVOLUTION

The Upper Niger and the imperialism of Kankan

During the crisis which disrupted the Upper Niger Malinké in the nineteenth century, the influence of Al Hadj-Umar played a more important role than it did in the Fouta-Djallon. The Fulani had withdrawn inside their own frontiers and left the animists to their own devices. The influence of the Fouta was, however, maintained through the city of Kankan, which kept its intellectual and commercial links with Timbo, though its direct relations with the coast were equally vigorous along the Solimana route.

Content to be fully independent, the Dyula capital followed a policy of great prudence under Mamadu-Sanusi (Koro Mamadu Kaba), a nephew of Alfa Kaabine (*c.* 1810–50). The expanding Dyula population and the growth of trade enhanced the prestige of the city well beyond the forest fringe. Its only serious handicap was the temporary closure of the downstream area of the Niger during the hostilities against Tamba-Bwari. There was not, however, much internal harmony within the city. The prestige of the Kaba was eclipsed by that of the Sherifu, who had come from the Middle Niger in the eighteenth century and had maintained a close link with the Moorish countries from where they had received the *wird* of the Kadiriya. This difficult situation, together with the established tradition of the Dyula, made Kankan unwilling to take up arms against the animist chiefs, who were its neighbours, even though life became increasingly unpleasant for the wealthy trading class who had to put up with local laws and traditions. This somewhat precarious balance was upset by Al Hadj-Umar. He visited Kankan in 1842 and persuaded the Kaba and the families of the political class to adopt the new religion, whilst the Sherifu and a majority of marabout families remained faithful to the Kadiriya. Alfa Mamadu, a grandson of Sanusi, followed Al Hadj-Umar as a *taalibu* and in 1850 or 1851 took a contingent from Kankan to fight in the siege of Tamba.

This young man became the Mansa of the Baté in the following year and thanks to his powers of persuasion so inflamed the Dyula that they decided to try to impose their law by force of arms. They started a series of small wars against the Kondé of Sankaran and the Konate of Toron, their aim being to extend their frontiers, shatter their neighbours' pride and force them to respect the trading routes. So the traders became warriors and since the crisis came at the same time as Al Hadj-Umar's exploits in the Western Sudan, the threat they offered was considerable. Their pagan neighbours did not accept the situation meekly; the Fula of Wasulu, for example, reacted with swift and astonishing violence. A young vassal chief of the Baté, Dyèri Sidibé of Ulundu, saw in a dream a spirit who revealed to him a dance of possession which would allow him to pick out those warriors who would be invincible. This 'war of the sons of dreams' *(sugo den-kele)* became an absolute tidal wave which in a few months swept over the whole of the Wasulu and reached the very gates of Odienne and the land of Bougouni. A few years later (*c.* 1855) Dyèri attacked Kankan. The city's downfall would have strengthened his leadership, but he died during the battle and his army broke up immediately afterwards. The Wasulu fell into a state of anarchy which persisted until two great chiefs, Kodyè-Sori of Lensoro and Adyigbè of the Dyètulu restored order.

Alfa Mamadu took advantage of the situation to throw himself boldly into a holy war. It is clear that he wished to imitate Al Hadj-Umar but the latter's victories had carried the conqueror northwards and the Dyula from the Milo had to rely on themselves. Disappointment soon followed. They first tried to crush the growing power of Adyigbe, but the coalition of the Wasulu and the Lower Toron (then united behind Nantenen-Famudu Kuruma) proved too strong for them. The eastern route was then closed and it was in vain that the Kaba turned against the Toron in the hope of joining the Sisé and forcing a way through to the forest. Alfa Mamadu then turned northwards, towards the Toucouleur outposts. The latter held the Bure, but Alfa Mamadu was unable to overcome the resistance of the Kèita from Dyuma, as they were quickly reinforced by their kinsmen from Kangaba. These events occurred between 1864 and 1865. Soon afterwards Al Hadj-Umar died and the Mansa was not long in following him. The discouraged Kaba decided to give up their desire for conquest.

Alfa-Mamadu's son Karamogho Mori (Dyinabu-Fatima-Mori) was a pious man, but his brothers were ardent supporters of the new imperialism and they were soon involved in more fighting. Umaru-Ba took up the struggle again towards 1870, in order to open up the Kisi kola route. This entailed a preliminary defeat of the Konde from the Sankaran, but after some successes, he died at Bagbe in 1873 with most of his army.

Up until now the Kaba had been overwhelmingly successful but they suddenly found themselves in desperate straits, for every kaju along the banks of the Niger from the junction of the Milo to Kouroussa united against them. The Dyula capital was blockaded and the city was soon in the most alarming situation it had known since the time of Tamba-Bwari. It was saved only by the intervention of Samori, whose influence had by now spread all over the southern lands where the main events were taking place.

The Moriulédugu and the imperialism of the Sisé

Although the Kaba war was indicative of general developments in human attitudes along the Upper Niger, it was not a focal point of the great upheaval usually called the 'Dyula Revolution'. This had begun much earlier, near the borders of the forest. The first centre of this movement developed in about 1835 along the borders of the Toron and the Konyan, that is, south of the intermediate zone and on this side of the pre-forest strip, well away from the commercial routes. It was founded by a holy man, Mori-Ule-Sisé, who came originally from the Baté but studied for a long time among the Dyakhanké of the Fouta-Djallon. He left them in the 1825–30 period and it is not without significance that he did not return to his country, Kankan being at that time a peaceful city, but turned instead towards the forest in the company of a fellow student who was returning home. Shortly afterwards he founded Madina, well away from the local animist villages and in a bleak wilderness belonging to the Kuruma from the Toron. Here he gathered around him a motley following of devout believers, professional adventurers, and bankrupt Dyula as well as a few hunters (elephants were plentiful in this area).

By the time his pagan neighbours realised their danger it was too late, and the Muslims forced themselves and their religion on the neighbouring kafu. Mori-Ule-Sisé, who took the title of Faama, at first practised a policy of forced conversion but soon abandoned it and found allies among his neighbours. He organised a strong army based on the extensive use of the rifle and horse, and his military superiority was unquestionable. The new state was, however, organised in a very sketchy way and for this reason had to limit its territorial conquests.

As soon as he decided to abandon his dream of a holy war in favour of a more concrete reality, Mori-Ule directed his conquests towards the north–south trading routes. It then seemed as if he wished to organise a new kola route in competition with those to the west and the east, which were out of his reach. He first extended his domain towards the south at the expense of the Kone from the eastern Konyan, but was unable to capture the mountains of the Upper Konyan which separated him from the forest. He then turned towards the north, where the Kuruma of the Sabadugu stood between him and Kankan. He was unable to subdue them and this act of aggression encouraged the Toron people to develop a powerful state under Nantenen-Famudu's leadership. When Kankan finally declared itself in favour of a holy war it was too late, and the two Muslim centres were not able to unite until after the victory of Samori.

This unstable empire nearly collapsed in 1845 when Mori-Ule, again trying to push towards the south, attacked the Worodugu. He was killed assaulting Kuro Koro, by a co-religionist, the young Vakaba Turè, who was fighting in his first important battle. Nevertheless, a compromise on the part of the two Dyula powers saved the territory of Moriulédugu from total extinction. It was, however, much reduced in size and was obliged to live in peace and circumspection, whilst the pagans returned to their old ways and its warriors fought under Vakaba's leadership (*c.* 1845–9).

When Sérè-Burlay Sisé, the new Faama, returned to Madina to take command, the movement quickly revived. The local tribes were easily subdued and the conqueror, barred from the east by the Turè and from the north by Nantenen-Famadu, turned towards the west, crossing the Dyon to reach the great kola route from the Toma which followed the Milo to Kankan.

Here the Sisé came up against a Dyula group which was firmly established in the Gundo (Upper Toron) and which defended the Konaté animists. The defeated Kalogbe-Bèrèté was obliged to surrender in about 1849, and the rebellion of Saran-Swaré-Mori, his nephew, was crushed after the long siege of Sedugu, in 1852 or 1853, during which the Sisé captured Samori's mother. As the Bèrèté sought exile in Kankan, the conquerors remained masters of the Upper Toron and turned to attack Nantenen-Famudu, but again with no success. The new burst of religious intolerance from the Sisé soon provoked a rebellion among the animists. Sérè Burlay died at Kobobikoro in the Beela (Lower Konyan) whilst trying to crush the revolt in about 1859. His brother Sérè-Brèma soon took control and managed to maintain his supremacy by a campaign of repression and pacification. It took him only a short while to extend his authority over the whole of the Upper Konyan to the Guèrzé frontier. Meanwhile, Saran-Swarè-Mori Bèrèté had returned to the Gundo; he built himself a well-fortified headquarters in Sirambadugu and spread his rule over the two banks of the Milo. With the help of Nantenen-Famudu and the young Samori, Sérè-Brèma managed to defeat him and utterly destroyed the power of the Bèrèté (1865). He then turned against Samori, who fled to the forest, so the Sisé people were left in control of the Konyan in 1866–7.

The structural weaknesses of the Sisé state were such that this high peak of achievement could not last. His eagerness to stop the Kabasarana from seizing the Wasulu led Sérè-Brèma into long and sterile wars (1867) which took his forces towards the north. Samori took advantage of this situation and recaptured the Milo valley, while Gbankundo-Saghadyigi organised animists to drive Sérè-Brèma out of the Upper Konyan. The Wasulu was free from Odienne, and even one group remained in the Sisé orbit but the Sisé were forced to abandon the territorial expansion they had aimed at. In order to escape from this isolation they were obliged to come to an understanding with Samori, who gave them part of the Sabadugu after defeating Nantenen-Famudu in 1873–4. This was but a slender gain and Sérè-Brèma was now obliged to observe the irresistible rise of his former soldier.

It would of course have been wise to give up all ambitions but the old man could not suppress the warlike passions of his nephew, Morlay, who was very popular with the military caste. Once Samori was well out of the way, the Sisé crossed his territory, defeated the Sankaran and took the area up to the frontier with Sierra Leone in 1878–9. This move provided Samori with an excellent *casus belli* and scattered the Sisé's forces so widely that Samori found it extremely easy to wipe them all out in the space of a few months. At the beginning of 1881, after taking Kankan and capturing Sérè-Brèma, Samori amalgamated their former 'sofas' with his army and in this way laid the foundations for a military superiority no one in the southern regions could possibly challenge.

The Turè of the Kabasarana or Kabadugu

The founding of Kabasarana, or Kabadugu of Odienne, was in some ways similar to that of Moriulédugu, and both date from approximately the same period. Both were joint undertakings of Dyula and Muslims, but with the difference that these were families who had their roots in the area and indeed had been settled for many years on the southbound route from Bamako to Touba and Seguela, a route they were most anxious to control for themselves. The founder, Vakaba Turè, belonged to a family of weavers who originated in Sidikila on the lower Sankarani, and who, according to tradition, were distantly related to Samori's lineage. After many years of trading and some training in the art of fighting from Mori-Ule, Vakaba started his new career (*c.* 1842–3) by subduing the Fani pagans from the eastern Toron on behalf of his native village, Samatigila. He then went to the Worodugu to help his mother's family fight the troops of Mori-Ule, but he was magnanimous in victory and had the good sense to spare the Sisé. It was then that he enlisted the services of their army, commanded by Sérè-Burlay, in about 1845. With the aid of these reinforcements, he found it easy to bring down the Dyarasuba kingdom of the Nafana, which had controlled the Odienne region ever since the mid-eighteenth century. The defeated people withdrew to an uninhabited area in the south-east on the borders with the Seguela country and here they lived until the French conquest of 1896–7.

Vakaba's task was not, therefore, the unification of a region which had been split into small units since time immemorial, and unlike the Sisé, he was neither a foreigner nor a ruthless master. He was above all a native of the region and he left the existing centralised system of authority exactly as it was, giving the state the name of Kabasarana, after his mother, Kaba-Saran. He did, however, change the nature of the group in whose hands the ultimate authority lay, replacing the pagan Bambara with Muslim Dyula. The framework of the state stayed as it was with its nucleus of slave villages (Sofadugu) all at the disposal of the Faama and its ring of vassal kafu. The introduction of guns and horses gave the army a new superiority and allowed the new kingdom to spread beyond its borders. A belt of fortified villages was built to protect the frontiers.

As one might have expected, this Dyula rule spread along a kola route, following the north–south axis. After protecting its eastern frontier by making peace with the Nafana and the Sénufo from the Noolu, Kabasarana easily took control of the one hundred and twenty-five-mile stretch from the great Muslim centre of Koro to the Bodugu. After the fall of Dyèri Sidibé, who had seemed dangerous at one point, Vakaba seized the Folo with the large market of Maninyan in about 1855. His next plan involved expansion as far as the forest in the Mau, but the defeat he suffered at Borotu put an end to the undertaking and he died soon afterwards in the year 1857. The Kabadugu was never to spread beyond the frontiers its founder had given to it. Vakaba's eldest son, Vabrema (Ibrahima), tried to expand eastwards but he died before the walls of Korumba, a town he was unable to capture despite help from the Sisé (Fuladugu).

Vamuktar (1858–75), who replaced him, was a popular military leader. He liked fighting and made the military caste the strongest force within the state. His efforts were not destined to be successful, however. Like his brother, he tried to march towards the east, no doubt to capture the Mankono–Séguéla route, but strong resistance from the Bambara and the Sénufo (*c.* 1860–5) prevented him. He turned instead towards the forest and also failed to take Borotu, though he was successful in subduing Mau territory (Touba) by playing an active part in the civil wars of the Dyomande Sakuraka (*c.* 1866–7).

The Turè then turned towards the north and seem to have planned to found a new Dyula kingdom in the Wasulu in order to clear the Bamako route for their own use. The campaign was undertaken by Bunu-Mamèri, a cousin of the Faama, who was well known for his fighting courage. This first Wasulu war failed, however, in the face of the resistance offered by Adyigbe Dyakité, aided by the Sisé, between 1868 and 1870. A second war in 1873 was even worse. It is true that Adyigbe was killed, but Bunu-Mamèri, cut off from Odienne and driven right back to the Toucouleur frontier, was forced to flee to Segou where he lived for many years. Vamuktar, who had hastened to his aid, died of wounds received at Mafeleba, at the beginning of 1875.

A third brother, Mangbè-Amadu, became the next Faama but was unpopular with his brother's warriors. He was almost immediately confronted with a gigantic rebellion of all the animists and this all but engulfed the kingdom. After the secession of an important Kèlètigi, Vakuru Bamba, the situation grew desperate. Vakuru returned to his native region, the Barala, where he recruited a private army which he used to start a new revolt. It took Amadu five years of fighting to recover his kingdom and he was never able to crush Vakuru, who by this time had formed an alliance with Borotu and had taken over the eastern Mau. It took even longer to rebuild an army half-ruined by general desertion. It is not surprising that this weak ruler welcomed the arrival of Samori on his frontier at Gbeleba, after the capture of Sérè-Brèma. Samori was already married to his sister and Amadu married the conqueror's daughter. The Kabasarana was thus absorbed into the Dyula empire in May 1881. The two countries were not dissimilar in structure but Kabasarana preserved a measure of autonomy. Its leader fought under Samori's orders until his capture in 1898.

The forest fringe

The intermediate zone was turned upside down in the early nineteenth century by elements seeking to establish a new political order, but the land around the edge of the forest stayed undisturbed until much later. This is not surprising because the excellent relations between the local chiefs and the Dyula middlemen made any possibility of revolution most unlikely unless it were prompted by outside influences.

Those disturbances which did occur were therefore of little importance. There was a rising of the Dyakhanké Fodé Dramé in the western Sankaran a long way to the north, which had certain similarities to the rise of the Sisé. Its leader belonged to the small priest caste of the Fouta-Djallon and had fought with the Hubbu before founding Bereburiya among the Ulare people, north-east of Farana. He gathered a body of disciples from among the bankrupt traders and wandering adventurers who roamed the area. It was only in about 1875 that he began to fight the pagans, but the land he won was small in area. In spite of support from the Hubbu his forces were so weak that he was glad to ally himself with the Sisé when their army marched into the area in 1878. It meant the loss of all autonomy and it was whilst fighting for Samori that he died in 1884.

A unique and surprisingly successful enterprise was launched in Kisi country by the Bunduka Mori-Sulèmani Savané. He was a prosperous merchant who had been living in the area for over twenty years and who enjoyed close relations with the local inhabitants. In spite of his great age, he decided to use force of arms to impose his rule on a highly anarchical region where there were no markets and where risks to traders were therefore very considerable. Unlike Fodé Dramé and the Sisé in their early days, he had no intention of eliminating the existing society; he simply wanted to give it a proper form and structure. Indeed, he used local warriors (Kisi or Kuranko) for all his fighting. His use of existing dissensions set him firmly in power and he then decided to extend his activities. A brisk excursion against some opponents in the south was followed by a campaign against the Farmaya who were given a new chief, Kaba Lèno, otherwise called Kisi-Kaba, who was a local man. The enterprise was not easy and the civil wars which followed lasted until 1881 and were terminated only by the intervention of Samori. Mori Sulèmani allied himself to the new empire and served it well and faithfully. His work of acculturation of the Kisi was not wasted, for after his death they found themselves able to carry on with the constructional processes he had begun.

There was no other attempt at imposing Dyula leadership from the Kisi region to Sassandra before Samori's conquest. Everywhere, the old kafu either formed alliances or started unconvincing struggles for local supremacy. The Mandinka went on attacking the Toma around Macenta and the Mauka whittled away Dan country between Touba and Man. The threat of Sisé imperialism had nevertheless broken the rules of the game quite early on, at least in the Upper Konyan. When Samouri was helping Sérè-Brèma he expelled Saghadyigi-Kamara from the Simandugu. The latter was a young chief whose family was related to that of Samori's mother as well as to the people who had taken the Buzye from the Toma. This daring man joined a group of rebels who had entrenched themselves in the inaccessible Gbankundo mountains and organised a powerful army (in 1865) by encouraging the local tribes to revolt. In 1869 most of the Upper Konyan including some Muslim villages decided to support him.

From that time onwards Saghadyigi became the defender of the old traditions against Muslim imperialism, but he had the sense neither to pursue the defeated Sisé nor to oppose the growing power of Samori. He directed the forces at his disposal firmly towards the south and for the next ten years took more and more Toma territory with the help of the Buzye. During the same period he moved eastwards against the Guerze. The Karagwa and the western Mau recognised his authority and he found himself master of a real empire. He gave it no permanent form but his army kept firm control of it. The existence of this great state in the heart of the Konyan presented Samori with a challenge, for it offered his compatriots an alternative way of safety. Even though the conqueror directed his energies northwards, it was obvious that in time he would have to confront it. Saghadyigi realised this, for in 1881 he offered the Sisé military aid. His move was, however, in vain, and by 1883 Samori was free to march against Gbankundo. His campaign succeeded after an eventful siege (February–November 1883). By incorporating many of the defeated into his own army, he established himself as sole defender and unifier of the Konyan.

Samori's early career (1861–84)

This rapid review of the Dyula movements gives a brief indication of their nature and significance and throws light on Samori's tremendous undertaking. He was to take complete control of all of them and give the revolution unpredictable breadth.

In contrast to the great upheavals in the north, there was no question here of the mass conversion of an entire ethnic group. It does not seem that the Fulani exercised any direct influence, although the Dyula world was in close contact with them, both in the Masina for the kola trade, and in the Fouta-Djallon on the trade routes to the Rivières du Sud. The vision of a triumphant Muslim society was no doubt encouraging, but the example shown by the Fouta-Djallon, where revolutionary fervour had grown rather thin, provided no incentive at all. It is significant that Kankan, which should have been affected first, did not react until much later, and then only on the initiative of Al Hadj-Umar. In these circumstances, it would seem that the idea of a holy war appealed only because the Dyula happened at that moment to be in a reasonably receptive frame of mind. The real driving force was surely the economic situation and the tremendous increase in trade since the eighteenth century. The Dyula were by this time both wealthy and numerous and, being conscious of their indispensability, became steadily more unwilling to put up with their minority status. Doubtless they grew tired of supplying the animists with firearms and began to think of using them for their own ends.

Whatever the case, it is clear that there were two kinds of revolt. The first was led by foreigners such as Mori-Ulé or Fodé-Dramé and most of their troops were outcasts and soldiers of fortune. In this case, the pretext of religion and holy war was constantly put forward and the object of the enterprise was to destroy the indigenous society as impure. In the second kind of revolt the leaders usually had roots in the area, as did the Turè, the Bèrèté or Mori-Ulé's children, and troops were recruited locally. The religious pretext disappeared or was played down because the objective was the reform and not the destruction of the existing society. It seems in fact that the first kind of revolt, the product of an aggressive minority, was never able to be properly resolved. It either degenerated completely as in Fodé-Dramé's uprising, or else it turned quite swiftly into the second type of revolt; and this was more or less what happened with the Sisé. Aggression was transformed into revolution.

This seems to have been the basic problem. The old traditional animist society of the Malinké was apparently too set in its ways to evolve or to face the Dyula threat. When this began to take on the aspect of a holy war, the mass of the unconverted could not accept the fate which threatened them and must have felt in some obscure way that they did in fact have the answer if only they could find a way of adapting their way of living. If the old order could not survive, then it was far better that any transformation it might undergo should be the work of a local leader rather than some foreign invader. It was logical that such a leader should be sought among the local Dyula, a people well enough aware of the outside world to know how to resist it, but yet sufficiently close to the local people to remember that they were kith and kin, not enemies to be destroyed. The Turè from Odienne and the Bèrèté all faced the problem in this way. It was Samori who showed just how successful the method could be and it was he who embodied the whole of the Dyula revolt in his great achievement.

Samori Turè came from the Lower Konyan. He was born in about 1830 in Manyambaladugu of a family who had been settled there for a long time and was so closely allied to the Kamara animists that it had practically abandoned Islam. The young man's strong personality soon drove him to break his family ties and he worked as an itinerant trader for several years. This broadened his outlook and inevitably led him back to the religion of his forefathers (*c.* 1848–53). This was a time when his native region was threatened by the Sisé and, after the capture of his mother, he joined their army where he learned the craft of war and began to distinguish himself during the period 1853–9. He left them shortly after the accession of Sere-

Brema, with whom he was on bad terms, but as he was still uncommitted to any policy, he lent his services to the Bèrètè, who were then at the height of their power in about 1859–61.

Samori soon broke away from the Bèrètè and took advantage of their first defeats to impose his leadership on his mother's kinsmen, the Kamara of the Lower Konyan. His personal career thus began in 1861 in a paradoxically anti-Muslim guise. He did not disown his readopted religion, but he placed his skills at the disposal of his animist relations to help avert the spread of Sisé imperialism, now a serious threat to his country. His army was still weak but he had extraordinary tactical ability, knowing instinctively how to divide his enemies, attacking only when he was sure of himself and retreating whenever it was necessary to do so. From the *tata* of Sanankoro, where he established his headquarters in 1862, he extended his authority over the Dyon and Milo valleys and was reconciled with the Sisé. He then helped them crush the Bèrètè in 1864–5. After this Sérè-Brèma turned to attack him and Samori avoided an unequal fight by fleeing in 1866 to the Toma in the Kononkoro region on the edge of the forest. Taking advantage of the Wasulu war, he returned to Sanankoro in 1867 and discreetly began to build a strong army on Dyula lines. Every warrior in this force felt a personal loyalty to him, and this made him finally independent of his 'uncles', the Kamara.

In 1871 Samori decided to move away from the Upper Konyan, then held by Saghadyigi, and he made an agreement with the weakened Sisé. He thrust northwards, inaugurating the great era of expansion. It took him only three years to destroy Nantenen Famudu's kingdom completely; he then controlled the lower Milo valley and the whole of the Toron as far as the outskirts of Kankan. In 1873, he settled in Bisandugu, in a region he had only just conquered, which showed that he intended to work for himself from then on and not for his 'uncles', the Kamara. He had already exchanged the title of Murutigi for that of Faama in 1868, and this made it clear that he was set on sovereignty of a military nature.

At this point Samori reversed his policy, and entered into a Muslim alliance by going to the aid of Kankan, in order to fight the local tribes whose blockage was ruining the city. From 1875 to 1879 he crushed the Sankaran for the benefit of the Kaba, conquered the entire Niger valley from Siguiri to Kouroussa, compelled the Burè to pay tribute-money and, in the west, conquered the Balèya and the Ulada. These achievements gave him a common border with the Fouta-Djallon and the Toucouleur of Dinguiraye. He concluded a treaty with the Toucouleur and, thanks to the Fulani, gained his first access to the Rivières du Sud and its great warehouses of modern weapons.

Kankan was now completely ringed by Samori's conquests and the Muslims in turn grew uneasy. The invasion of the Sankaran by the Sisé, at the end of 1878, made Samori decide to find new allies. As the Kaba refused to help him he turned to the animist kafu again, to ask them to fight their oppressors, and in one single campaign he destroyed the Sisé empire and seized Kankan (December 1879 to April 1881). Madina was then destroyed and its population transferred to Bisandugu and to the Kabasarana, so that it was integrated into the new empire. The conqueror's army, swelled by hundreds from the ranks of the defeated, acquired an overwhelming superiority. Samori now surrounded himself with Muslim advisers chosen from those who had just submitted. Another move towards Islam had started.

Dyula conquests always moved along meridians, following the axis of the kola routes, and this was also the direction taken by Samori. Now, however, two new features came into play; the necessity to pursue the fugitives from Kankan and Samori's desire to seize the Niger fords near Bamako and thus control the Sahel markets. At the end of 1881 he marched northwards and captured Kènyéran in February 1882, in spite of unexpected intervention from the French. He then became master of the Manding and Bambara regions, from the right bank of the Niger to the frontier with the Toucouleur of Segou. The year 1883 was devoted to conquering Saghadyigi and this enabled Samori to place his southern frontier along the fringe of the forest, a region he had no intention of penetrating. As the French invasion of Bamako had just closed his northern frontier, Samori could advance only in an easterly or westerly direction.

This change of direction took place in 1884, and it was then that the empire reached its peak of territorial expansion, spreading across countries which were broken into small units and could offer no effective resistance. In the west, Langaman-Fali destroyed Falaba and took only a few months to reach the British outposts of Sierra Leone. In the east Tari-Mori advanced as far as the Bagoé, to the outposts of Kénédugu. The proximity of this powerful state was a source of anxiety for Samori at a time when the French threat in the Niger was growing, and he was afraid of being trapped in a pincer movement. As soon as his relations with the Europeans appeared to be in order, he thrust eastwards in 1887, determined to break through Sikasso's defences. In spite of his skill and obstinacy, the war ended badly for him, and French intrigues instigated the Great Revolt in 1888, which brought the empire to its knees.

Samori thought he had found a successful formula for a new society by taking the title of Almami in 1884, and by setting out to convert all his subjects to Islam. This new policy was instrumental in causing the crisis of 1888 and Samori soon realised this. Faced with the threat of French infiltration, he devoted himself exclusively to the construction of a military state capable of doing battle with the French. He held his ground until 1898 but in his final years, when his sole aim was to resist as long as possible, he did nothing to advance the tide of African achievement.

In spite of the theocratic period (1886–8) and although Samori, himself a sincere Muslim, worked hard to acquire the Koranic learning he had been denied as a child, his empire always retained a strikingly lay character, unlike the Fulani whose holy wars were steeped in religious implications. From 1879 onwards, he worked to establish a system of military rule and centralised administration quite unlike any to be found elsewhere in the region. This made it possible for him to exercise his sway over vast areas of territory, which his predecessors had never been able to do, and to impose a state of law and order which was of great benefit to trade and other aspects of life. If an ill-advised search for perpetuity had not led him down the blind alley of theocracy, he might well have solved the crisis of the Malinké world. He was the incarnation of the Dyula revolution and his impact on West Africa was so strong that the new colonialism was never able to blot it out entirely.

THE DECLINE OF KONG AND THE EASTERN TRADE ROUTES

The eastern regions had their 'Dyula revolution' a century before those of the Upper Niger and, in the nineteenth century, they experienced the constant decline of the old Kong empire, and the development of new political structures better suited to the area.

Kpon Kènè

Kpon-Kènè was by now only a small state, which had lost control over the Dyimini and the Ano (Grumanya) in the south, and had to tolerate constant provocation from the Pallaka (Plabala) at the very gates of its capital. In the north, this state still controlled the Gwen and Tyéfo areas, that is, the Tyèrla of Banfora, a flat region below the famous escarpment. The Watara kings usually acceded to the throne when they were very old men and as a result succeeded each other rapidly, but in spite of this, the commercial and religious elements of the city kept their prosperity. The caravans regularly came from Djenne through Bobo or from Segou through Sikasso. The trade route still went down to the Gold Coast through Bondoukou, then a satellite of the Asante, but a great deal of trade was sent for preference along the Comoé, through the middle of the forest, as far as the French outposts of Bassam and Assinie, where the 'Bambaras' had already begun to trade in 1842.

Bobo-Dioulasso

The Gwiriko still formed a powerful and cohesive unit around Bobo-Dioulasso at the beginning of the century. Caravans from Djenne passed through unhindered, although a large part of the vast Bwa (Bobo-Ulé) country was no longer under Dyula control. There was a different situation in the south-east along the route to Bouna, as the untamed Lobi were gradually taking over the principality of Kubo (Diébougou) and the northern part of the old Kulango kingdom. It was, however, from the west that the main threat came. Here the Kénédugu began to expand along the route to Segou, and the Makuma chief, Suramana Watara, came into conflict with them in about 1850, over the possession of the populous border lands of the Samogho and Nanèrègè. He was helped by Dyori, the Faama of Gwiriko, and later received aid from Kong, so that the whole of the old Dyula empire found itself involved in a struggle with the renegade Traorè. The conflict lasted for many years and was often very dangerous; the young Tyèba, the future Faama of Sikasso, was even held captive for a short while.

In the end, neither party gained an inch but the vassals of the Watara had played such a vital role in the struggle that they now demanded a better status. In the last quarter of the century, all the local peoples, once subdued by Séku, regained their freedom, even the Bobo-Dyula of the Sano clan, who seized the main trading centre from the control of the last Faama. They maintained, nevertheless, a certain solidarity and in 1892 they united under the leadership of Amoro, the Tyefo chief, to stop Tyèba's last offensive. It was with the Bobo-Dyula and not the Faama that Samori had to deal after the collapse of Kong in 1897.

Growth of centralised states

Everywhere else, throughout the enormous area abandoned by the Watara, the nineteenth century saw the growth of centralised states established by local peoples or by foreign conquerors who took advantage of the interregnum to impose their rule. The most characteristic feature was the growth of warrior states among the Sénufo. This type of development was contrary to their usual traditions, and up until the French conquest many tribes still remained organised in large villages, in the vague framework of customary unions or *tar*. The new development showed the influence of Kong culture, which had affected Sénufo society in various other ways as well. Almost all these warrior states were grouped along the loop of the Upper Bandama which seems to have been the western province of the Watara empire.

The same sort of thing happened to the Tyèbalélé, who had been established around Korhogo by Nangen's descendants, and to the chiefs of Sinématyali, who settled at a later date in Nafagha country. Further north, the Tyèbabélé from Nwèlé (Nielle) and the Tagban from Bengé followed the same pattern. Established in the eighteenth century, all of them went through a period of rapid consolidation at the beginning of the nineteenth century and expanded at the expense of their less organised neighbours before finally coming into conflict with each other.

Kénédugu kingdom

Although it started off in the same way, the Kénédugu kingdom merits further attention because its achievements were so much more considerable.

At the beginning of the nineteenth century, the Traorè who lived on this distant frontier, along the route from Segou to Kong, had practically become Sénufo Muslims. Islam was in fact limited to some 'aristocratic' families who claimed the name of Dyula and it posed no political problems. The army, whose organisation had been inherited from Kong, was recruited exclusively from the local population, and it eventually took on a quasi-national character. It was the eighteenth Faama, the second Daula, who fought the Gwiriko in the middle of the century and he managed to establish a real empire. His authority extended from Tengréla and Nwèlé near Korhogo, to the borders of Segou and in 1862 he allied himself with Al Hadj Umar to subdue the Bambara who were fighting against the Toucouleur.

After his death in about 1865, a series of short reigns weakened the kingdom considerably, whilst a formidable enemy emerged on its northern border. Supporters of the Toucouleur broke with Sultan Amadu and joined with Bambara from the Baninko to establish a powerful military state around Kinyan. This kingdom, called Fafadugu after its founder, Fafa Togola, seized the northern and western borders of the Kénédugu and also certain Sénufo areas such as the Kampo (Kapolondugu). Between 1865 and 1870 Fafa played an important part in the civil wars which rent the Traorè, and when the young Tyèba came to power shortly after 1870 all he had left was a very small territory at the junction of the Lotyo and Farako Rivers. He fortified the village of Sugokaa, the home of his maternal ancestors, and gave it the Dyula name of Sikasso. A brief respite then followed and in about 1875 the Faama began a systematic reconquest of his kingdom. He began with fierce battles against Fafadugu, which the Toucouleur army from Segou opportunely attacked from the rear. Tyèba's power was at its height when his vanguard reached the Bagoé and came up against Samori's troops in 1884.

The merciless struggle which followed and culminated in the famous siege of Sikasso, which lasted fifteen months in 1887–8, had all the characteristics of a national war. Tyèba, however, refused to become a mere pawn of the French and would not sacrifice the ambitions of the Kénédugu to them. Fighting between the Samorians and Sénufo went on until the beginning of 1895. Tyèba's brother and successor, Babèmba, eventually established better relations with the Almami but died at the hands of a French army in 1898. The original framework of the Kénédugu had its roots set in Sénufo society and this meant that it could not join the Dyula revolution. It is probable that Samori would never have attacked if the French had not closed their northern trade route to him. This absurd and fruitless struggle was one of the repercussions of colonial aggression.

Military states

Various racial minority groups also set up military states on the Black

Volta in the northern part of the old Kong Empire. The Fulani from Masina endeavoured to subdue a number of the Bwa by pushing southwards along the route from Djenne to Bobo. This was the origin of the Barani kingdom, whose most important ruler, Widi, was in power at the end of the century. After the Toucouleur victory, other Fulani fled from Masina and went further east and founded the Dokuy kingdom in about 1865. Dafin Muslims from the Tougan area began the conquest of the Boromo-Wahabu region in about 1850 and before the end of the century had become masters of the Volta fords and controlled the trading route from Bobo to Ouagadugu. From 1887, Al-Kari framed the kingdom of Bousse against the Samo.

On several occasions, the Dafin came up against the Zabèrma (Zèrma) who had come from the east as mercenaries for the Mamprusi and who in about 1865 had begun the conquest of the Gurunsi. These bands of men, whose most famous chiefs were Gadyari and Babatu, do not come within the scope of this study. It should, however, be noted that Zèrma or Hausa traders had settled further west, in Bouna, in the Dyimini and even as far as Saghala on the Marawé (White Bandama). It was one of these, Mori Turè, who started to ravage the Tagwana and the Dyimini in about 1885. He was finally defeated and withdrew to found the village of Marabadyasa in 1891 on lands ceded to him by the Baulé. Mori Turè was to ally himself with Samori in 1893 but the net result of his activities, all so fearfully destructive, seems in the end to be completely negative.

One can in fact say the same thing about most of these ventures because they were basically conducted for gain and aimed particularly at destroying the local populations. They were totally different from the Dyula revolution and the developments in Sénufo territory after the decline of Kong.

For all these widely differing countries, the nineteenth century was a period of history so filled with change and upheaval that any kind of unity is barely discernible. But one can perceive some sort of general guideline in the operation, and this is what distinguishes them both from the enclosed lands of the south (Kru, Baulé), where life went on in its usual slow way until the European intervention and from the open savannahs of the north, where the Fulani preached their holy war.

The southern savannahs were divided between the age-old influence of the Sudan and that of the European trading posts, which was not felt until the eighteenth century. The development of their history is closely related to the ever-increasing part played by this second sphere of influence. European products spread gradually all over the land, starting at the Rivières du Sud and the Gold Coast, and ending finally at the closed world of the Great Kru forest.

The traditional societies of these countries, weakened in many ways by the new influences, were often unable to find an effective means of revival within themselves. They, therefore, fell by the wayside. As early as the eighteenth century the two extremities of the area under consideration collapsed. In the west, the Fouta-Djallon changed its ethnic and religious individuality and the coastal peoples were converted to Islam. In the east, the great Kong empire grew in strength and size, but it never sought to destroy the personality of the peoples it conquered. When it eventually fell into decline, its subject peoples took advantage of its weakness to assert themselves by setting up self-sufficient states which reached their fullest development in the nineteenth century.

There was a similar process in the central zone, although change and upheaval came relatively late in the nineteenth century. The local populations took the Dyula revolt as part of their own, and they kept as much as possible of their traditional structures. Their ultimate achievement was incarnated in the spectacular venture of Samori, son of the Konyan.

SOURCES OF INFORMATION

The reconstruction of the history of this immense region depends as much on ethnographic and linguistic researches as on the archaeological investigations which have only just begun on the northernmost fringes. For the nineteenth century, it is dependent on three sources:

(a) The oral traditions, which are in urgent need of collection, because the drastic changes currently taking place in traditional society will soon obliterate them. This work has been carried out seriously only in a few areas and it is a mistake to think that such areas as the Fouta-Djallon have been studied most thoroughly. We have carried out this work ourselves in the Milo valley, in the Niger valley, above Siguiri, and also in the Konyan (Beyla) and the Kabasarana (Odienne). The Guro are exceptionally well documented, thanks to the works of Meillassoux and Ariane Deluz.

(b) Documents written by Africans, especially in the Fouta-Djallon. This field is now being studied by Ibrahima Sow.

(c) Archives of European origin; *P.R.O.* in London; *Section Outre-Mer des Archives Nationales* in Paris; *Anciennes Archives du Gouvernement Général* in Dakar; *National Archives of Guinea,* in Conakry, *of Mali,* in Koulouba, *of the Ivory Coast* in Abidjan, *of Sierra Leone* in Fourah Bay College.

In addition to documents which enable readers to reconstruct European activities in the region and which, for Sierra Leone, cover the entire nineteenth century, there are also many monographs and unpublished studies. These often contain important collections of oral traditions, but as the latter were not gathered under proper conditions they must be used with care. (For example: CO 879/25 at the *P.R.O.* for the history of Moréa and the coastal peoples north of Freetown.) The following selected bibliographic guide attempts to give the best available coverage of the subject.

RIVIÈRES DU NORD (Guinea–Bissau). See the bibliography in Texeira da Mota, *Guiné Portuguesa*, Lisbon 1954. There are three basic works for the history of the Fulani conquest, and these are also valuable for the Fouta-Djallon. A. Carreira, *Mandingas da Guiné Portuguesa*, Bissau 1947; J. Vellez Carroco, *Monjur*, Bissau 1948, J. Mendes-Moreira, *Fulas do Cabu*, Bissau 1948.

RIVIÈRES DU SUD (Republic of Guinea). There are some bibliographic indications in Houis, *La Guinée Française*, Paris 1952. The only work devoted entirely to the subject is the old somewhat confused book by A. Arcin, *Histoire de la Guinée Française,* Paris 1911. There are a number of interesting articles in the following reviews: *Etudes Guinéennes* (I.F.A.N., Conakry, from 1947 to 1958) and *Recherches Africaines* (I.N.R.D. Conakry, from 1958 onwards).

SIERRA LEONE See the bibliographies by C. Fyfe, *A History of Sierra Leone,* London 1962, and *Sierra Leone Inheritance*, London 1965. These refer mainly to colonial history. An outline of the ethnography of the area can be found in M. McCulloch, *Peoples of Sierra Leone Protectorate*, London 1950. Many essential articles are to be found in the long series of *Sierra Leone Studies.* A particularly important one is that by J. Spencer Trimingham and C. Fyfe: 'The early expansion of Islam in Sierra Leone' in *S.L. Bulletin of Religions,* ii, 2, 1960.

LIBERIA This country remains almost unexplored from a historical point of view. For the hinterland peoples see the bibliography of Schwab, *Tribes of the Liberian Hinterland,* Cambridge, Mass. 1947.

IVORY COAST See the bibliographies of Atget, *La France en Côte d'Ivoire,* Dakar 1962; and Holas, *Arts de la Côte d'Ivoire,* Paris 1966. The first is historical and the second ethnographical.

FOUTA-DJALLON The history of the Fouta-Djallon appears now in a new light, thanks to the translation of Fulani literary texts by M. Alfa Ibrahima Sow, *La Femme, la vache et la foi*, Paris 1966 and *Chroniques et récits du Fouta Djallon*, Paris 1968. These are first-rate documents with an excellent linguistic presentation, although the historical apparatus is somewhat shaky. The historian can find a good analysis in the work of Thierno Diallo, *Les institutions politiques du Fouta Djallon au XIXe siècle,* (Fii Laamu Alsilamaaku Fuuta Jaloo)—a research thesis doctorate, submitted in 1968. The future researches of this young historian should prove of great interest. There is still much of value in Tauxier, *Moeurs et histoire des Peuls,* Paris 1938. This author collected most of the eye-witness accounts published in the nineteenth century, but he was not able to make much use of them. He constantly refers to the accounts of bygone travellers (Gray and Dorchard, Mollieu, Hecquard, Noirot, Bayol, Sanderval) and to the works of Guebhard and Arcin. The work of G. Vieillard on the Fulani of the Fouta-Djallon is essential to a study of the people, but he died before he could start a history of the area (*BIFAN*, ii, 1 and 2, 1940, 85–210). Paul Marty, *L'Islam en Guinée*, Paris 1921, made use of many unpublished monographs.

KANKAN AND THE UPPER NIGER Apart from A. Arcin and many unpublished monographs (Conakry and Dakar archives) there are also the following: Humblot, 'Du nom propre et des appellations chez les Malinké de la vallée du Milo', *B.C.H.S.A.O.F.*, Dakar 1918 and 1919. Humblot, 'Kankan, métropole de la Haute Guinée', *Renseignements Coloniaux*, 1921.

UPPER IVORY COAST Some old monographs were published in the volume *La Côte d'Ivoire*, on the occasion of the colonial exhibition in Marseilles (1906). Others were used by Paul Marty, *L'Islam en Côte d'Ivoire*, Paris 1922. This book is not easy to follow and should be used with care.

SAMORI Most of the works on Samori deal exclusively with his struggles against the French and are not serious studies. Two of the three volumes of the author's dissertation on this subject are now available: *Samori, une révolution dyula,* i and ii, Dakar 1970–71. There are old standard works which are worth re-reading: Peroz, *Au Soudan français,* Paris 1889 (imaginative and full of understanding) and an admirable account of African exploration by Binger, *Du Niger au golfe de Guinée,* Paris 1892 (important both for Kong and Kénédugu). Among modern authors, the best account is that of Delafosse in Hanotaux, *Histoire des Colonies françaises,* Paris 1931. General Duboc, *Samory le Sanglant,* Paris 1947. This author has made use of some original documents.

THE KONG AND KÉNÉDUGU EMPIRES For all the Sénufo the bibliography by B. Holas, *Les Senoufo*, Paris 1956, is a good ethnographic outline, although it neglects the historical problems. For the Kong empire: E. Bernus, 'Kong et sa région', *Etudes Eburnéennes,* viii, Abidjan 1960. This excellent study has a most exhaustive bibliography, giving an important place to historical problems. However, it is basically the work of a geographer and leaves many problems unanswered. The same author published 'Notes sur l'histoire de Korhogo', *B.I.F.A.N.*, 1–2, 1960. The history of the Kénédugu was to be the subject of a major work by Father de Bengy; unfortunately he was killed in a car accident in 1970. The present writer will use the missionary's papers to augment his own research, and will write a history of Sikasso and the Kénédugu in the near future. At present the best study is the relevant section of the Quiquandon report, 'Dans la boucle du Niger', *Bulletin de société de géographie commerciale de Bordeaux,* 1892. Perron, 'Précis chronologique de l'histoire de Sikasso', *B.C.H.S.A.O.F.,* 4, 1923; and Collieaux, 'Histoire de l'ancien royaume du Kénédougou', *B.C.H.S.A.O.F.,* 1, 1924, have made but little use of the traditions they collected. Méniaud, *Sikasso,* Paris 1935, is of value for the history of the conquest.

CHAPTER 8

The Republic of Liberia

ABEODU BOWEN JONES

ORIGINS

Liberia was originally the forest and coastal home of the Mande, Kru and other Melle-speaking peoples, among whom the process of state formation had begun before the arrival of the Portuguese in the fifteenth century.[1] Hemmed in by mountains and forests, the area did not come under the influence of the savannah empires of Ghana, Mali and Songhai; yet its protective geographic features allowed the creation of states, many of them small and widely scattered but with sufficient contact to form confederations occasionally, for the purposes of defence and trade.[2] By the nineteenth century several of these confederations were in existence, among them the Grebo United Kingdom and the Dey representing the Kru confederacy; the Gola of the Melle-speaking and the Kondo of the Mande-speaking people.

Oral tradition supported by Portuguese accounts suggests certain migratory trends. It is generally believed that the disintegration of the Mali Empire in the fifteenth century led itinerant Malinke traders and warriors to move from the savannah into the kola plantations of the forest, bringing with them their merchandise as well as Islam.[3] Waves of Mande people

1 C. Fyfe, *A Thousand Years of West African History*, eds. J. F. A. Ajayi and Ian Espie, London, 1964, pp. 149–59; Charles Huberich, *A Political and Legislative History of Liberia*, New York, 1947, i, pp. 1–5; Department of Information and Cultural Affairs, *Oral Traditions of Liberia*, Monrovia, Liberia, 1967.

2 *Oral Traditions of Liberia*; Jane Martin, *The Dual Legacy: Government Authority and Mission Influence among the Grebo of Eastern Liberia*, doctoral dissertation, Boston University, 1968, pp. 1–73; C. Abayomi Cassell, *Liberia: History of The First African Republic*, New York, 1970, p. 87.

3 Ajayi and Espie, p. 149; Dr Nehemia Levtzion, Lectures on 'Islam in West Africa', University of Liberia, 8 and 9 February 1971; James Hopewell, *The Spread of Islam in the Guinea States of Sierra Leone, Liberia and the Ivory Coast*, doctoral dissertation, Columbia University, New York, 1963; Walter Cason, *The Growth of Christianity in the Liberian Environment*, Ph.D. thesis, Columbia University, New York, 1963.

are known to have descended in the middle of the sixteenth century from Upper Guinea into the southern regions of Liberia and along the coast to Sierra Leone.[4] One of the most important of these migratory routes seems to have been the one emanating from Guinea and entering the north-western regions of present-day Liberia through Voinjama and Lofa County, Zorzor, Zolowo, Bella Yella and finally Bopolu, a popular trade and market centre which lay at the crossroads of trade routes between the savannah and the coast.[5] Bopolu provided easy access to the coast through Mesurado and Cape Mount. It was also a major centre in the Kondo Federation. The impact of Islam was sufficiently great to have led to the invention of an indigenous alphabet among the Vai, a subgroup of the Mande, in the early decades of the nineteenth century. This alphabet, employed as a medium for general communication, became famous and was used during the Second World War as a German secret code.[6] The Mande preserved law and order through powerful secret societies, the *Poro* for men and *Sande* for women, which were also vehicles for the transmission of culture and traditions from one generation to the next.

By 1461, trade on this windward coast had attracted sufficient attention for it to be styled the Grain Coast by the Portuguese and by the sixteenth century it had become widely known to European traders as an important centre of international trade.[7] European trade with those confederations having outlets to the coast was well established, and the Kru were notorious for providing assistance to slave traders while they themselves avoided capture. The trade in tropical goods was now replaced by a new trade.

The modern state of Liberia was an offshoot of slavery and the anti-slave-trade movement of the nineteenth century and was made possible by a combination of fortuitous circumstances. The makers of the American federal constitution in 1787 had called for the gradual abolition of the slave trade by 1808, and in 1807 the Congress of the United States passed an act which prohibited the further importation of slaves and provided for the return to Africa of slaves illegally imported.[8] The question then arose as to the means of their return to Africa. This was at first considered to be within the jurisdiction and responsibility of the state in which the slaves were found, but since few states had the necessary resources, the federal government had to intervene. Thus began a series of legal battles which plagued the whole scheme attending the founding of Liberia.

4 Ajayi and Espie, p. 153.
5 Cassell, p. 87.
6 Department of Information and Cultural Affairs, *Handbook on Liberia*, Monrovia, 1968, p. 5.
7 Ajayi and Espie, pp. 151–6; Huberich, i, pp. 1–5.
8 Alfred H. Kelly and Winifred A. Hardison, *The American Constitution: Its Origin and Development*, New York, 1955, p. 932.

In January 1819 Congressman Charles Mercer of Virginia, aware of the colonising sympathies of two fellow Virginians and Episcopalians, Thomas Jefferson and President James Monroe, presented facts to Congress on the continuation of the slave trade by United States citizens in violation of the anti-slave-trade act of 1807. Two months later a new bill was passed by Congress and signed by Monroe which stated that all slaves illegally imported or taken at sea should be held in custody of the government of the United States until they could be returned to their homes in Africa. The bill authorised the President to appoint agents to reside on the west coast of Africa to receive them and appropriated the sum of one hundred thousand dollars to cover the expenses. The bill, however, did not authorise the purchase of land in Africa by the United States Government.

As the debate over this matter continued, the number of rescued slaves in government custody increased steadily. Monroe was earnestly entreated to interpret the bill not merely as contemplating the dumping of rescued slaves on the shores of Africa but as envisaging the creation of a special colony for their rehabilitation. Ultimately, this view prevailed and it was the necessity of placing the rescued slaves, now a federal responsibility, which drew the attention of Monroe to the work of the American Colonization Society. The Society had been established in 1816 with a view to ending slavery in the United States by emulating the British experience in founding Sierra Leone nearly four decades before. The federal government was concerned in ending the slave trade rather than slavery. In cooperation with the Society, Monroe agreed that if the Society would accept the responsibility for buying African land, he would provide ships and money to transport both rescued slaves and those freed from slavery in the United States (the latter already numbered nearly sixty thousand). Motivated by the expediency of the whole transaction, he thus allowed a broad interpretation of the bill and gave an impetus and importance to the founding of Liberia which otherwise would have been long in coming.[9] He disowned, however, all idea of an empire. He stressed that the primary objective of the agents was to make the necessary and amicable arrangements with the government of the area they selected in Africa, paying special heed not to exercise any power or authority founded on the principles of colonisation.[10] This later fostered a two-nation *ad-hoc* policy.

Thus, it could be said that Liberia was born out of a great American controversy focusing on the best solution to the Negro problem. Abolition and colonisation were two hotly debated topics, and in the early years immediately preceding the founding of Liberia, few American politicians

9 J. P. Staudenraus, *The African Colonization Movement*, New York, 1961, pp. 48–58; B. Sunderland, 'On Liberian Colonization', *Liberia Bulletin (L.B.)*, no. 16 (*Liberia Bulletin* succeeding the *African Repository* in 1892), February 1900, pp. 24–5.

10 Huberich, i, pp. 73–5. This clause which seems to discountenance any possibility of the African populations being made subservient to immigrant rule was to prove very troublesome later.

dared risk their political fortunes by taking a strong stand on either of them. People were so sensitive to the numerous legal niceties attending colonisation in Africa that the government of the United States received no expression of gratitude in any of the proceedings of the Society and the impression was allowed to persist that the founding of Liberia was the work of the American Colonization Society alone.[11]

In his annual message to Congress in December 1819, Monroe announced that he was sending two agents to Africa with teams of 'laborers and mechanics' provided by the Society to prepare the government's African station so as to receive rescued Africans under the anti-slave-trade act of 1819.[12] Accordingly, the *Elisabeth*, a United States naval vessel, sailed out of New York harbour on 31 January 1820, with eighty-eight immigrants chosen by the Society. They took with them a wagon, several wheelbarrows, ploughs, a sawmill, a seine net, a variety of farming utensils, arms and ammunition. In all the United States Government spent thirty-three thousand dollars on this first expedition. Later, the cost was to be borne jointly by the Society and the government. The immigrants could remain in Africa, but there was an implied agreement by the government to provide for their transport back to America if they so desired and a few immigrants did in fact return. Once the immigrants had arrived in Africa the government was prepared to go no further than declare a moral protectorate over them. The Society failed to obtain any legal status for Liberia whether as an American colony, or as a state under the protection of the United States, or as a nation whose neutrality would be guaranteed by the United States, Great Britain, France or any other great power.[13]

While anticipating congressional aid, the Society in 1818 had dispatched the Reverend Samuel Mills and the Reverend Ebenezer Burgess to seek land suitable for the settlement of immigrants as close to Sierra Leone as possible. The land deal with local African chiefs was a fiasco. John Kizzel, who was recommended by Governor Charles MacCarthy of Sierra Leone to assist Mills and Burgess, saw an opportunity to make a fortune and erected a number of huts on his land at Sherbro. The site was inhospitable, malaria-ridden, and even lacking in fresh water, so nearly all the immigrants from the *Elisabeth* died. Sherbro was abandoned and temporary shelter was sought at Fourah Bay, while 1820 and 1821 were spent in a desperate search for a new site. Finally in December 1821, when patience was running out, Captain Robert F. Stockton of the United States Navy, commanding U.S.S. *Alligator* in the anti-slave-trade campaign, threatened bombardment and forced the rulers of Mesurado to cede some of their land to the

11 *Ibid.*, p. 77. The U.S. Government did not seek recognition for its rule in the Liberian venture.

12 Staudenraus, p. 56. Once in Africa, however, the tables were turned and the servicemen, so-called, provided leadership in all things.

13 Huberich, i, p. 258. This subsequently became one of the most burning issues of the period.

immigrants.[14] On 7 January 1822, the immigrants were removed from Fourah Bay and brought to Cape Mesurado. In 1824, the land they occupied was named Liberia and the capital Monrovia. It was the white agent of the United States Government who raised the American flag over the settlement[15] and it should also be noted that it was white Americans and not the Liberian immigrants who took the lead in the negotiations for land.

THE COLONIAL EXPERIENCE

Under the constitution of 1820, full authority for the government of Liberia was delegated to the American Colonization Society. The immigrants were, however, led to believe that in Africa they were entitled to all the rights and privileges enjoyed by citizens of the United States.[16] Since these rights and privileges were not qualified or limited, the immigrants began to demand them in their entirety and were reluctant to believe that the promises could have been mere inducements for them to leave America. The most important of these rights and privileges were the franchise and eligibility to hold any office, including that of head of government, though this office continued to be occupied by white men, a fact which they said reminded them of slavery. The immigrants disliked the idea of the white agent's representing both the Colonization Society and the government of the United States, and they resented his veto powers. There was also dissatisfaction over the distribution of town lots and the rationing of food. There were frequent riots and on several occasions the white agent had to flee the colony in fear for his life.

In 1824 the Liberian immigrants petitioned the Society for a general redress of their grievances and were provided that same year with a new constitution which marked a step towards the transfer of power to them. The Society delegated more power to the white agent resident in Liberia. Subject to his veto powers and the ultimate approval of the Society, laws were to be made by a Colonial Council, all of whose members were to be nominated by the settlers. The settlers also had the power to elect the agent's deputy and certain colonial offices such as that of sheriff and justices of the peace were open to them through appointment by the agent.

Other moves in the agitation for self-government were the Shepherd Petition of 1830, which asked that the offices of the agent of the Society and of the United States Government be separated, and the sending of a delegation to Washington in 1833 requesting greater constitutional powers for the Colonial Council. The Society refused the Shepherd Petition, expressing regret that any individuals in the colony should be so dissatisfied with the existing form of government and should harbour desires for political offices and distinction. It was, however, prepared to

14 Staudenraus, pp. 59–68.
15 Huberich, i, pp. 253, 258, 263. The act was unauthorised but not condemned.
16 Huberich, i, p. 146.

concede that the time had arrived for the people to have greater powers and more privileges than they had hitherto enjoyed.

The Society was riddled with problems of its own both at home and abroad. By 1830, their Congressional political opponents were beginning to look over the Anti-Slave-Trade Act of 1819 and Amos Kendell, auditor of the United States Treasury, when taking inventory after the inauguration of Andrew Jackson, discovered that the Act had indeed been stretched too far and that in only eleven years, Congress had spent $264,710 on two hundred and sixty rescued slaves who had been sent back to Africa under the management of white Americans. All officials, both those of the Society and those appointed by the United States Government to care for the repatriated Africans, were on the payroll of the United States Government. Kendell urged a drastic revision of policy but Jackson postponed action and only became hostile towards colonisation when Henry Clay, an influential and powerful congressman and a supporter of colonisation, showed strong inclinations towards occupying the White House. In 1834 Andrew Jackson's Secretary of the Navy, Levi Woodbury, ordered a much reduced budget for the American agent in Liberia. Jackson's administration did not disband the colony for the sole reason that it was still useful to the Navy as a depot for the rescued slaves.[17]

Meanwhile dissensions erupted within the parent body of the Society over financial matters and the method of selecting immigrants for Africa. The financial situation became acute as several members of the Society felt no obligation to send money to a central agency for disbursement. New American colonies began to mushroom along the Liberian coast between Mesurado and the Cavalla River in Cape Palmas. Colonisation societies or companies in New York and Pennsylvania joined to found a colony in Bassa between 1833 and 1835; another society from Maryland planted its own in Cape Palmas in 1834; yet another from Mississippi placed one in Sinoe in 1838 and an Indiana society sent immigrants to Cape Mount in 1855. Monrovia, the oldest seat of established government, argued that it should police the new settlements, but all the other colonies denied its jurisdiction.

Wars between native Africans and the immigrants due to a disturbance of the old pattern of confederations were rampant, though sporadic. Those colonies which were imbued with the spirit of the temperance movement then prevailing in America, and consequently most anxious to impose a moral code on their neighbours, suffered most in terms of conflict with Africans. Heavy casualties were sustained on both sides, with almost total depopulation in certain areas, Bassa in particular. The protection and defence of scattered colonies became a real problem, but notwithstanding their separatism, Monrovia always went to their rescue in the name of the anti-slave-trade movement.

17 Staudenraus, p. 178.

Meanwhile the Colonial Council took steps to regulate coastwise trading and in so doing came into direct conflict with both African and European traders. The most serious opposition came from the British who had hitherto regarded the Liberian coast as a no-man's-land. They reacted to the Liberian laws by questioning the international status of the colony and its right to impose customs regulations on trade which had been conducted for centuries without undue interference from the Africans.

In 1836, Monrovia, with the cooperation of the Colonization Society, began to press for a federation of the colonies, and in 1838 a commonwealth constitution was drafted—the Monrovia Draft—which brought into existence the Commonwealth of Liberia. The colony of Maryland, however, on advice from the Maryland Society, refused to join the new political arrangement. Supported by an annual grant of ten thousand dollars, it was the most prosperous of the colonies.

Under the Commonwealth, a few changes were made in the laws then existing. As could be expected the Liberians took the occasion to curtail the powers of the white agent, whom they now called a Governor and who, though still appointed and paid by the Society, was no longer the agent of the Society but rather chief executive of the colony. Legislative powers were now vested in a local legislature. The veto powers of the Governor were abrogated but all laws were still subject to the ultimate approval of the Society. The Monrovia Draft reflected the high level of political thinking which had been reached in the colony in barely seventeen years of existence. Africans were officially recognised as wards of the government.

In 1841 Thomas Buchanan, the first and only white Governor of the Commonwealth, died and was succeeded by J. J. Roberts, who had been lieutenant-governor. Roberts soon pressed the idea of self-government on the Commonwealth, using as a major argument the failure of European traders to observe Liberian customs regulations. Britain and France were aggressive and the United States Government wished to avoid a confrontation, especially with Britain, with whom it had already begun to exchange diplomatic notes on a definition of the United States relationship with Liberia.

The United States Department of State and the American Colonization Society agreed to advise the settlers to seek independence as a way out of the dilemma. Negotiations over the transfer to them of lands bought by the Society with American money caused some delay; the Society at first wished to keep the lands but later agreed that portions be set aside as future homes for rescued Africans, inasmuch as the anti-slave-trade movement was then still active.

The vote for independence taken in 1847 was carried by a simple majority of the Liberians, most of whom resided in Monrovia. The people of Bassa and Sinoe, who were known to have serious doubts over the wisdom of severing connections with the Society, took no active part in the voting. Their leaders, however, signed the Declaration of Independence in their capacity as elected officers already serving in the Colonial Council. Following this example, Maryland too declared its independence in 1854 but lost it two years later in 1856 in a war with the Greboes. In the peace settlement which followed with Monrovia as mediator, it decided to join Liberia as a county.

Meanwhile, under President James Polk, the government of the United States took scant notice of Liberia and withheld recognition of its independence. However, by 1856, every major state in Europe had accorded it recognition. Successive governments in the United States, however, continued to delay recognition for fifteen years. The people of Liberia never forgot the insult.

PARTY POLITICS: 1834–78

The first step towards party politics in Liberia was taken in 1834 when the Colonial Council assumed the power to divide the colony of Liberia into electoral districts according to population. Dissensions soon arose not only over the outcome of the divisions, but also on the question of whether elections to the Council were to be won by a plurality or by a majority of votes. So as to restore public confidence in its deliberations, the Council in 1835 repealed the Act of 1834 and resolved that each settlement should have equal privileges of suffrage and that the outcome of elections be determined by a majority of votes cast. This action seems to have met with public approbation and set the pattern for future elections.

The first general elections were accordingly held in 1835 for the nomination of councillors, but no sooner were they completed than complaints of racial discrimination were raised. Settlers of the area called New Georgia, who were Negroes from Brazil, charged that they were discriminated against in the elections and had not been accorded full citizenship. In fact, Negroes not from the United States frequently voiced this complaint.[18] The Society in 1836 affirmed the right of the New Georgians to vote if they had fulfilled the requirements for citizenship, but it failed to grant similar assurances to native Africans.

By the time of the next general elections in 1840, party politics had taken shape along clearly recognised lines of class and interest. There was the grouping of Negroes from the United States, who were predominantly mulattoes and wealthy, opposed to Negroes from other lands who were black and poor. There was an alignment of Monrovia against the St Paul River districts, Bassa and Sinoe; and also a pro-government party led by Joseph J. Roberts, a mulatto who had emigrated from Virginia in the United States, with an opposition led by the Reverend John Seys, a black

18 Christopher T. Minikon, 'History of Liberian Political Parties', *Liberian Historical Review (L.B.H.)*, i, 1, 1966.

West Indian. The opposition, it appears, had not been won over to complete acceptance of the commonwealth and it agitated for a loose type of federation in which each interest group, the Methodist Episcopal Church of which Seys was pastor included, would be granted special privileges and make its own laws. The idea of tolerating an opposition was never popular, so although the pro-government party won the elections of 1840, it took steps to cripple the opposition by passing a law stating that no native Africans, unless they had been brought to Liberia under the authority of the American Colonization Society, or had been resident in the colony for three years in a civilised Western fashion, should be entitled to the franchise.

By the time of the first presidential election in 1847, party labels had been chosen. The pro-government party was known as the Liberian Party, later to become the Republican Party, while the opposition chose to be called the True Whig Party.[19] The opposition lost in 1847 and continued to lose every election until 1869 when it emerged victorious under Edward J. Roye, who was both black and wealthy. In 1871, however, after less than two years in office, Roye was deposed and later assassinated following clouded political charges. These included the embezzlement of loan funds, although he was known at the time to be the wealthiest Liberian of the day; and of aiming to become a dictator, since he won the by-elections in favour of an extension of the term of presidential office from two to four years.[20] J. J. Roberts, leader of the Republican Party, was soon President again for two additional terms from 1872 to 1876. The Whigs, however, made a successful return in the elections of 1878 and have since remained in power. On 8 January 1964, they laid the cornerstone for their headquarters and named it the E. J. Roye Memorial Building.

TERRITORIAL EXPANSION

From the beginning, the American Colonization Society in its numerous petitions to the United States Congress had brought forward arguments to show the benefits to American trade from an African settlement, especially trade arrangements with the local populations which could secure for American merchants African trade then monopolised by European merchants. It was through this insistence on the commercial potential of Liberia that the first successful white agent, Jehudi Ashmun, was able to give the settlement a degree of stability. Earlier humanitarian arguments that Liberia would be the harbinger of Christianity and civilisation in West Africa failed to enthuse members of the Society and the work of evangelisation was left entirely to separate missionary societies.

19 The names were an attempt to identify with similar political labels then in vogue in the United States so as to attract to their ranks new immigrants arriving in Liberia. The True Whig Party posed as a people's party to unite blacks, mulattoes, recaptives and Africans.

20 Cassell, pp. 267–93. Repayment of the English loan of five hundred pounds was to plague every Liberian President from 1871 until redeemed by Arthur Barclay (1904–8). Liberia benefited little, if at all, from this loan.

There was little land for expansion around Cape Mesurado. Barely a mile wide and three miles long, the land was rocky and steep, making agricultural production and the division of town lots difficult. Its commercial potential, however, was high and the immigrants found a flourishing trade in the area. The area of the colony was gradually increased through treaties with the chiefs and outright grants by weaker tribes who sought from the settlement protection against their more powerful neighbours who tried to sell them as slaves to European and American slave-traders.[21]

On the whole the period 1822–68 was a time of great territorial expansion. After nearly losing hope when the Sherbro venture proved a fiasco and Governor MacCarthy of Sierra Leone gave them limited time to remain in Fourah Bay, the immigrant Liberians finally got a foothold in Africa. Elijah Johnson's celebrated remark, 'For two long years have I sought a home, here I have found one and here I will remain', was long recalled on the observance of Pioneers' Day on 7 January and Independence Day on 26 July. It was a clarion call to courage and determination in the struggle for survival. In 1823 when the British intervened to provide aid for the weakened colony ravaged through wars with the neighbouring Africans, Elijah Johnson refused foreign assistance, declaring: 'We want no flagstaff put up here that will cost us more to take down than it will to beat the natives'.[22]

Land for settlement, trade and survival was what they wanted, and the Liberian declaration of independence later proclaimed that it was not greed for territorial expansion that had brought them to the shores of Africa, but the search for an asylum from the most grinding oppression. However, when the Africans attempted to regain Mesurado it took a major battle fought on 1 December 1822 to establish a permanent foothold. The kings of Mesurado were defeated and the immigrants hailed the victory as a triumph for the regeneration of civilisation in Africa. In the peace settlement, Chief Sao Boso of the powerful Kondo confederation pronounced the famous decision that the Africans, having sold their land and accepted payment, should accept the consequences.[23] The chiefs of the Dey and the Gola confederacies were also present.

Becoming the first African friend of the immigrants, Sao Boso also showed them the path into the interior which lay through Bopolu, the seat of his chiefdom, then still a popular trade route for Mandingo middlemen. The interior and the various items of tropical merchandise brought down to the coast by the traders fascinated the settlers. The anti-slave-trade campaign emphasised the need for legitimate trade in tropical produce to replace the trade in human beings. Jehudi Ashmun, using American

21 Cassell, pp. 91–293.
22 *Liberia Bulletin*, no. 7, November 1895, Washington, D.C., p. 1.
23 Jehudi Ashmun, *History of the American Colony*, Washington City, 1826, p. 12.

money, acquired much land under the pretext of discouraging slave-trading in the vicinity of Monrovia. He was also agent for the Baltimore Trading Company of Maryland, United States, and was thus anxious to make the trade routes into the interior safe for commerce. He complained that the American flag flown over the colony also gave protection to the neighbouring territories occupied by slavers. He was therefore anxious to consolidate the entire area under his rule. By a gradual extension of its coastline the Liberian settlement rendered it more and more difficult for slave traders to find outlets for their cargoes.

Friends of the American Colonization Society in America also felt that Liberia was ideally situated geographically to garner the unfathomed interior trade and New York merchants were urged to export the Bible along with trade articles. Certain that southern immigrants would carry southern tastes to Africa, Virginia anticipated a brisk demand for its tobacco and flour. Notwithstanding the limitations set by the Monroe Doctrine in 1823, Liberia was hailed by the Society as the first step in building an American empire which would flourish and expand until it overshadowed the continent of Africa and would lead to a steady stream of ships and colonists linking the African and American continents.[24]

The Society was apparently to reverse the tide of empire from west to east, and Americans were called upon to reclaim western Africa and open the resources of an immense continent to the enterprise of the civilised world. Supporters urged the Society to work for the day when it would control a thousand-mile area extending from the Senegal to Cape Palmas. In their vision of empire, they saw the United States supplanting in time all its European rivals in Africa because it could pour forth thousands upon thousands of Negro pilgrims for whom the African climate was as congenial as was that of New England to their own Anglo-Saxon forefathers. Even Edward Blyden believed this vision of an American empire to be the whole purpose of creating a Negro diaspora.

No more than fifteen thousand immigrants arrived in Liberia during the period 1820 to 1865 and this imposed a limitation on how much land the Liberians could occupy by their own efforts. But following the early example set by Ashmun, each new agent thought it necessary to justify his position by stretching the boundaries of Liberia, much of this being done at random with the deeds forwarded to the Society. Ashmun especially induced Africans to sell their lands and become 'Americans'.[25] At the time of independence the territory owned by Liberia was some thirteen thousand square miles with nearly three hundred miles of coastline from Cape Mount to Grand Sesters and a population of not more than a hundred thousand including eighty thousand native Africans, seven thousand

24 Staudenraus, pp. 157–8.
25 *Ibid*, pp. 157–66.

immigrants, and fifteen thousand rescued slaves and wards. The annual value of exports was over five hundred thousand dollars.[26]

Expansion continued after independence, but the problem of safeguarding the territorial boundaries involved Liberia in further dependence upon the United States. Benjamin K. Anderson, explorer and government agent, was given much credit for opening up Liberia's interior to the immigrants. In 1868 he travelled as far as the Guinea highlands and down into the San Pedro valley claiming lands and signing treaties with African chiefs on behalf of the Liberian Government.[27] The following year a Department of the Interior was created by the government of President James Spriggs Payne. Soon E. J. Roye, succeeding Payne in 1870 as President, was saying to the Liberian people:

> In a country like ours, destitute of large navigable rivers or canals penetrating the interior where indigenous and spontaneous wealth covers the ground, the necessity of railroads must be at once evident. I believe that the erection of a railroad will have a wonderful influence in the civilisation and elevation of the native tribes. The barriers of heathenism and superstition will disappear before the railroad as frost and snow dissolve before a summer's rain. This is one of the most efficient means which God's promise made concerning Africa, is in my opinion to be fulfilled. The natives will readily consent to do all the manual labor in the construction of railroads for comparatively small pay and kind usage. After the completion of the roads, they will become the best of customers to bring the camwood, palm oil, ivory, Mandingo, gold, cotton country cloths, peanuts, iron ore, hides, bullocks, sheep, goats, rice and other things too numerous to mention to the Liberian markets on the seaboard, and thus multiply indefinitely the exportable products of the country.[28]

The northwest territory was lost, through dispute with a British trader, John Myer Harris. Moreover, the interior lands claimed by Anderson and recognised by the Berlin Conference of 1884–5 were soon disputed by the British and French. The African chiefs were often in the habit of selling their land to more than one party or putting it under the protection of two overlords, thus creating disputes.[29] In 1892 long-standing disputes with the British were settled, the territories claimed by Liberia being greatly reduced and the United States Department of State advising Liberia to accept whatever the colonial powers thought it should have. The interior lands adjacent

26 Cassell, p. 162.
27 *Ibid.*, pp. 226, 260–1; Jakema V. Fahnbulleh, former district commissioner of Western Province in Liberia, 'Liberia's Birth, Growth and Progress', Monrovia, Liberia (unpublished manuscript).
28 Cassell, p. 270.
29 R. L. Buell, *The Native Problem in Africa*, New York, 1928, p. 10.

to Sierra Leone were acquired much later, in the first decade of the twentieth century, by agents of the Liberian interior service.[30]

The pattern was the same with the French, the dispute with them lasting until the independence of Guinea in 1958. Then, the Liberian Government simply stated that it felt satisfied that the lands contended for were now in the hands of an African government, and that it would no longer press for claims over them.[31] The greater part of the mountains of Nimba, laden with some of the world's richest supply of iron ore, remained part of Guinea as a result of this agreement. Much land too was lost to the Ivory Coast when the Cavalla River was accepted as the boundary with Liberia. In all the Liberian Government lost nearly half of the territories it originally claimed.

Policing even the reduced area of forty-three thousand square miles and keeping peace among a population which, by the turn of the century, was a little more than a million inhabitants were heavy responsibilities involving frequent wars, some of which threatened the physical survival of the Republic. French conquest had removed Samory Toure who threatened the northern boundaries. Along the coast, the United States Navy continued to offer protection until 1916. No major wars were fought in the interior although there were Gola and Kpelle uprisings until after the First World War. In the early decades of the twentieth century the United States provided white commissioners to assist in the administration of the interior lands and to keep a watchful eye over the treatment of the African population. American assistance was, however, always subject to the stipulation included in the treaty of belated recognition of Liberian independence signed in 1862 by Abraham Lincoln:

> The United States government engages never to interfere, unless solicited by the government of Liberia, in the affairs between the aboriginal inhabitants and the government of the Republic of Liberia in the jurisdiction and territories of the Republic.[32]

Liberia survived through nominal though positive protection of the United States.

ECONOMIC PROBLEMS

When the early pioneers came to West Africa from the United States, an industrial outlook on national development was conspicuously absent. This

30 Cassell, pp. 355–82; *Liberia Bulletin*, 1910; Fahnbulleh, p. 283. The Liberian army at the time contained mostly troops from the border areas, Lofa and Grand Gedeh especially.

31 Republic of Liberia, *Statement on the Recognition of the Independence of the Republic of Guinea*, Monrovia, 1958, p. 2.

32 National Archives, United States Department of State, 'Act of Recognition of the Independence of the Republic of Liberia, 1862'.

was a reflection on the stage of industrial growth of the United States at that time. Before industrialisation had made its impact, Liberia was created and declared a republic governed by black men.

From the beginning, even before the birth of Liberia as an independent nation in 1847, the cause of the Colonization Society was never widely popular in the United States. Its plan for colonisation always suffered from shortage of funds since it was dependent on meagre financial contributions raised at rallies in fraternities, churches and during July Fourth independence celebrations. The pattern was thus set for inadequate financial backing and this plagued Liberia throughout the nineteenth and early decades of the twentieth centuries. No effective currency was ever established until the introduction of the U.S. dollar during the Second World War. The Society had experimented with minting its own coins but it failed when an indifferent United States Congress curtailed its financial support.

The pioneers had said that they came to Africa to prove that black men were capable of self-government, and politics dominated their life. For example, when Liberia College was founded in 1862 there emerged a heavy and early emphasis on politics and the training of statesmen, but little enthusiasm was shown for economic development.

The lack of concern with industrial enterprise in those early days reflected the ideas of Jeffersonian democracy emanating from an agrarian southern United States. These were interpreted by the white agents to mean making Liberia a nation of small farmers. The immigrants in their first years in Africa could not adjust to the climate, soils and crops and continued to live on an American diet. Even their style of dress and building were attempts to transplant a little bit of the Old South to Africa. They had some success in planting coffee but generally they rebelled against any attempts to make them agriculturalists. To the great dismay of their white governors, they soon began to turn to commercial pursuits where gains could be made quickly. Engaging at first in barter trade with neighbouring Africans in the vicinity of Monrovia, they gradually extended their contacts into the interior and exported, often in their own ships, such tropical products as palm kernels, palm oil, piassava, sugar cane, coffee, cocoa, dyes, kola nuts, ivory and gold. It was a flourishing trade which lasted unfortunately only until the middle of the 1870s. For a while, coffee was all-important, but then the Brazilians took away the Liberian monopoly of the world's finest coffee when Liberian scions were introduced into Brazil from the United States. These scions had been taken to the American Fair of 1876 in which Liberia participated in honour of the first centenary of the independence of the United States. By then sailing ships had become old-fashioned and European firms with steamships drove Liberian merchantmen out of the coastal trade. German synthetic dyes and other European inventions undermined the Liberian export market. By the 1890s, compared with other West African areas growing cotton, groundnuts and cocoa, Liberia exported insufficient produce to attract much European shipping to its shores.

Obtaining money to run the affairs of the nation then became a problem. The government at first considered arranging money loans from abroad. E. J. Roye, the first True Whig President, initiated this policy to finance the building of a railway into the interior in 1871. After his assassination, the loan was diverted to other projects and no railway was built until after the Second World War when an American company constructed the line to the Bomi Hills iron mines. Foreign loans proved unproductive and dangerous. One of the best known examples was the Harry Johnston loan of one hundred thousand dollars under President Arthur S. Barclay in 1907, in return for rubber concessions and British control of the Liberian Frontier Force, customs duties and other resources. It provoked French territorial aggression and led to a grave political crisis including a mutiny of the Frontier Force which threatened the very existence of the Republic. Loans were always given reluctantly, at very high rates of interest and for other considerations which put the economic and political independence of Liberia in jeopardy.[33]

The government then tried other ways of raising money. They contracted with foreign governments to tax migrant labour recruited to work in European colonies in other parts of Africa. This was at first a popular business venture in which many Europeans participated, but it proved to be an ill-regulated enterprise which, in the end, brought much suffering and embarrassment to the Liberian Government and led to the intervention of the League of Nations in investigating charges of forced labour and slavery. The Spanish Government, the most important agent with whom Liberia had entered into labour contracts, was never implicated nor was its role brought to light. In Fernando Po especially, where most Liberian workers were sent without the protection of a consul, the Spanish overlords failed to honour their own commitment to care effectively for the workers and on returning home these never failed to narrate tales of horror on the cocoa island.

Finally, taxes were imposed, the chief one being a hut tax. However, far from solving the government's financial problems, this led only to widespread riots and wars against the authority of the government. Not until the Tubman administration, which started in 1944 with the proclamation of an economic Open Door Policy for investment by private concessionaires from abroad in the natural resources of Liberia and in local businesses, was the situation altered radically for the better. Total revenues rose from five hundred thousand dollars in 1847 to over seventy-one million dollars in 1971.[34]

33 Harry H. Johnston, *Liberia*, New York, 1906; Raymond W. Bixler, *The Foreign Policy of the United States in Liberia*, Buell, ii, pp. 786–802.

34 Cassell, p. 162; Republic of Liberia, *The National Budget, 1971*, p. 1.

FOREIGN MISSIONARIES AND THE PROBLEM OF NATIONAL INTEGRATION

Shortly after their arrival in Africa, the immigrants began to discuss the position which native Africans should occupy in the state. They considered various alternatives and approaches, always bearing in mind the injunction not to exercise any authority founded on the principles of colonisation. The white agents were present to see that the injunction was obeyed.[35] The work of foreign missionaries and the idea of a home mission were encouraged and widely championed at the time. Basle missionaries were particularly welcomed but inevitably American missionaries became predominant. By 1837 there was a great deal of missionary activity, centred on the coast.[36]

Among the immigrant Liberians the view was prevalent by 1850 that a generation or more should pass before there could be a perfect union with the native Africans and that this would result naturally from the apprenticeship system then in vogue.[37] They extolled the role which apprenticeship could play in assimilating the African into Liberian civilisation. African children were brought into Liberian households, and were introduced to Christianity, Western family life, dress habits, material needs, the rudiments of education, and general acquaintance with the values of immigrant Liberians. An increase in the wants and desires of the Africans, it was argued, would lead to habits of industry, work, commercial transactions and agricultural operations. The implications were that the civilised man from abroad, white or coloured, was a white man to the African and that every American Negro was a missionary without necessarily being aware of it.

Opportunities for citizenship, however, had been made available to the Africans as early as 1841 during the Commonwealth. Citizenship could be acquired through ownership of a town lot with a house built on it, and at least two acres of farm land under cultivation. The African must in addition have subscribed to 'civilised' habits for at least three years by living decently clad, and there should be witnesses to testify that he had fulfilled these requirements.[38] These rules, by and large, remained in force until 1944 when, activated by certain liberal recommendations of the League of Nations and by the principles of the Atlantic Charter, mass suffrage was granted.

The Independence Constitution of 1847 evaded the issue of national

35 Edward W. Blyden, 'The Future of Liberia', *Liberia Bulletin*, no. 26, February 1905, p. 45.
36 Huberich, i, p. 519.
37 *Ibid.*, p. 48. The white agents had opposed the institution of apprenticeship in compliance with the instruction not to exercise powers based on the principle of colonisation. The Commonwealth Constitution of 1838 allowed the settlers a freer hand, but apprenticeship was encouraged only at the end of the period of white rule.
38 Huberich, iii, pp. 851–1012.

integration entirely. In view of the delicacy and complexity of the problem, and also the dangers of overextending to others *en masse* the newly won rights and privileges of the immigrant Liberians, the constitution gamely stated under its miscellaneous provisions that:

> the improvement of the native tribes and their advancement in the arts of agriculture and husbandry being a cherished object of this government, it shall be the duty of the President to appoint in each county some discreet person whose duty it shall be to conduct regular and periodical tours through the country for the purpose of calling the attention of the natives to these wholesome branches of industry and of instructing them in the same, and the Legislature shall as soon as it can conveniently be done, make provisions for these purposes by the appropriation of money.[39]

Shortage of funds, therefore, seems to have dictated a course of delayed action. When after 1847 native Africans complained of neglect, the usual justification was that resources for the survival of the immigrant population were inadequate and that there was a real risk in spreading them too thinly. After 1900 representation in the National Legislature was extended to those ethnic groups which could qualify through payment of the registration fee of one hundred dollars annually. Only a few of them, the Krus and Greboes in particular, who were living on the coast and had been long accustomed to the advantages of foreign trade, were ever represented in this way. Even so, some politicians were soon expressing fears of being overshadowed and outvoted by legions of African tribes in the National Legislature.[40]

A policy of gradualism in recognising the rights and privileges of the Africans in the state was therefore considered wise. The African peoples remained largely wards of the government. They were rebellious when various taxes were imposed upon them or when the Liberian Frontier Force, taking justice into its hands, as it frequently did under the early regulations which gave it a share in the government of the interior, disturbed their centuries-old carefree existence and pressed them into service as migrant labourers.[41]

Rather than hasten integration with Africans, the government placed a great deal of reliance upon immigration from abroad, even after 1847. The government was encouraged in this by the fact that at this time the Colonization Society was emphasising emigration as the most popular prescription for ending Negro slavery in America in place of colonisation which

39 *Ibid.*, p. 864.
40 *Ibid.*; Liberia Department of State, *Inaugural Message of President Hilary Wright Johnson*, Monrovia, 1884.
41 Nathaniel R. Richardson, *Liberia Past and Present*, p. 32.

was being severely attacked by abolitionists.[42] The big chance for mass emigration should have come during the American Civil War, but the leaders of the Society and the Liberian Government watched in utter amazement as Abraham Lincoln's government spurned Liberia, and toyed with one emigration plan after another as a solution to the presence of masses of emancipated slaves roaming the United States. Lincoln agreed privately with President Daniel B. Warner of Liberia that Liberia was the logical place for American Negroes to be sent, but continued to consider in a more serious vein Central American alternatives.[43] In desperation, Edward Blyden persuaded Warner to appoint him commissioner for immigration. Blyden, himself a West Indian, then quickly seized the opportunity in 1865 to bring three hundred and forty-six black immigrants from Barbados to Liberia, with money from the Society. It was an unprecedented act but so committed were Liberia and the Society to fostering immigration that the end and not the means was what counted. This immigration from the West Indies, however, introduced an important new element in Liberian politics which was long dominant. As they arrived virtually at the end of emigration by United States Negroes, the West Indian immigrants overshadowed in the passage of time the older streams of settlers from the old country and set up their own hierarchy in government. By 1904 they had successfully assumed the highest political leadership with the election of Arthur Barclay. This leadership was to last until 1943, and even today their influence on Liberian politics is unquestioned.

But there was danger in continuing to depend on foreign manpower, and Marcus Aurelius Garvey seems to have driven this lesson home. Garvey, who was of West Indian origin, organised a back-to-Africa movement after the First World War and contemplated Liberia as the base of his operations in Africa. Diplomatic pressure on the United States from the colonial powers then surrounding Liberia succeeded in bringing about a total collapse of the Garvey movement. There was fear that his real intentions were to supplant the Liberian Government by force and to assume control.[44] Up to that time, importing trained manpower from abroad seems to have been preferred to seeking collaboration with neighbouring Africans in the task of building the nation.

Schemes for the integration of native Africans with immigrant Liberians continued to be discussed intermittently. The journeys of B. K. Anderson into the interior in 1868 and 1874 in particular stimulated serious discussions. First, integration through trade and commercial expansion was suggested. Liberian agents were appointed to advise the chiefs on how to

42 Staudenraus, pp. 224–51.

43 *Ibid.*, p. 248.

44 *Inaugural Message of President C. D. B. King*, 1924, Monrovia, p. 2. This was a real setback to the Garvey movement, attributed by many to the influence of the colonial powers, especially since soon afterwards Liberia opened her doors in 1926 to grant a concession to the white American Firestone Rubber Company.

engage their people in agricultural production and cooperate in trade and commerce with the immigrants. With the creation of the Department of the Interior in 1869, government, law and taxes were extended into the hinterland. Alexander Crumwell described the new trend as one which recognised the full and equal ability of the African 'to become all that we are, and capable of doing all that we can do'.[45] Still, little was achieved.

It was then proposed to develop settlements of Liberian immigrants in various parts of the interior. Again there was fear of a general relapse into barbarism if the human and material resources of the Liberian settlers were spread too thinly. This, coupled with an unwillingness to go too far from Monrovia, the centre of civilisation and the source of livelihood and protection, frustrated the scheme.[46] The idea, however, remained popular to the extent that at one time force was considered necessary to spread democracy by erecting new towns in the interior, and compelling Africans to live there. This method had been tried with limited success by the Protestant Episcopal Missions, who had planted several Christian villages in Maryland which were intended to be centres for disseminating civilisation and Christianity in eastern Liberia. Government resources, however, were limited and the new towns which developed were additional homes for fresh immigrants.

Intermarriage was also suggested by Edward Blyden in the last quarter of the nineteenth century when immigration was dying out, as an ideal method to achieve integration between the immigrants and native Liberians. But as many Liberians then saw the basic antagonism to be between Christianity and heathenism, the differences in points of view on every aspect of life and existence—social, civil, political and religious—seemed irreconcilable and there was no noticeable acceptance of intermarriage.

The American Colonization Society, which derived revenue chiefly from numerous churches and denominations while refusing to participate actively in the evangelisation of Africans, extended equal facilities to the various denominations to establish schools and missions in Liberia.[47] Probably the classic experience of the effect of Christian missions in integration was the work of the Protestant Episcopal Church which began in Maryland in 1836. The policies of Johann Gottlief Auer, Bishop of Cape Palmas and parts adjacent in the 1860s and 1870s, stressed Grebo initiative and education in the spreading of Western culture among their people. The immigrants watched this development with misgivings.

For the evangelised Greboes, the problem was one of assimilation into one of two irreconcilable communities—their traditional societies or the Westernised Liberian communities—and they were often forced to choose.

45 Jane Martin, *How to Build a Nation*, Boston University and the University of Liberia, Monrovia, 1968, p. 10.
46 Martin, *Dual Legacy*, p. 265.
47 Huberich, ii, p. 893.

They found the experience diverse and conflicting and said they were like the bats, neither beast nor bird. The use of new knowledge gained at mission stations was consequently another problem. Some graduates became teachers, catechists, stevedores, or worked with individual missionaries to build new stations, or simply continued their education as independent scholars studying grammar, philosophy, theology and history. Others just relapsed into the old tribal ways. The mission worried over this unanticipated aspect of their labours and the answer seemed to be provided by the erection of Grebo Christian villages at Cavalla, Cape Palmas, Hoffman Station, Fishtown and Rocktown. But often there was antagonism between the new Christian communities and the traditional towns. Those in the Christian villages were isolated from their own people and their influence on the whole was negligible.

By 1871 Christianised Greboes began to play a more forceful role within the indigenous communities. In December 1873, they called a conference and organised the Grebo Reunited Kingdom. It was the renewal of a confederation which had formerly existed before the arrival of immigrant Liberians in the area in 1834. The conference pledged that the gospel of Christ would be freely taught in the Reunited Kingdom. At the end of the meeting, prayers were offered on behalf of the President of Liberia and their own Grebo kings, and the *Gloria in Excelsis* was sung. The Grebo knew of the existence of the Fante Confederation in the Gold Coast and they read the *African Times* which carried reprints of its constitution and frequent and optimistic reports of its operations. They knew too of the existence of the German Confederation and they tried to impress upon the government the fact that the Grebo kingdom had existed before the Republic of Liberia. Any reference to their being under the supremacy of the Liberian Government was bitterly resented by the Cavalla people.[48]

In company with the Krus, who were also long accustomed to diplomatic and commercial contacts with foreigners, the Grebo Reunited Kingdom began to show its opposition to the Liberian Government, particularly in the taxes levied on their goods and services as stevedores and on the closure by the government of certain ports of call along the coast, which adversely affected Grebo economic interests.

The Foreign Committee of the Episcopal Church disapproved of the idea of the confederation and informed the Grebo organisers that it recognised no government within the limits of Liberia other than the Liberian Government. In June 1875 President Roberts sent James Spriggs Payne to Maryland to discover the causes of the disloyal feelings held by the Africans that led them to form themselves into a kingdom within, and yet free and distinct from the government of, Liberia, notwithstanding their Christian outlook. Investigations later disclosed that the coastal Greboes believed

48 Martin, *Dual Legacy*, pp. 250–65.

that too much neighbourliness was not to their benefit, that the government had taken away their trading rights as middlemen between the Europeans and the interior peoples and had sided with the latter when wars broke out. They also thought that too much of their land was being set apart for the use of new immigrants from America. They resented being called derogatory names and being excluded from activities and offices in the Liberian settlements, especially when they had accepted civilised standards and their competence was widely acknowledged.

Meanwhile, Payne's efforts to call a meeting of the Greboes proved futile. All invitations and other documents on the investigation were returned. Then the confederation heard rumours that Superintendent James Tuning, the local governor, had called the interior peoples to a meeting in an attempt to force the Greboes in the Christian missions to give up certain aliens working among them without the permission of the government. This was a right which the Greboes had long denied the government, they had even gone so far as to write to the British Governor at Cape Coast and Sierra Leone as well as to the Earl of Caernarvon in Great Britain denying the jurisdiction of Liberia over them and asking to be made citizens of Her Majesty's Government. Although the British Government refused to grant their request, it was believed that if they were successful in a war against the immigrants, the British would make a treaty with the Greboes.

The Greboes then declared war on the grounds that they did not want to be caught unawares as had been the case on previous occasions. They put nearly seven thousand men in the field against the one thousand men of the armed forces of Liberia. Using Schneider repeater rifles and well provided with powder and guns from their trade with Europeans, they soon defeated the Liberian forces who, as they retreated, were taunted by the Greboes again singing the *Gloria in Excelsis*.

News of the defeat of government troops spread from Cape Palmas to Cape Mount, and fear of a general African uprising was rampant. Plans were made to put Monrovia in a state of defence. In February 1876, the U.S.S. *Alaska* commanded by A. A. Semmes arrived to aid the Liberian Government at its invitation, in keeping with the Lincoln treaty of recognition. Although the Liberians insisted on a renewal of the war, Semmes thought otherwise and persuaded them to desist. At Cape Palmas, he convinced the Greboes that the deeds to territories on which the Liberian Government based its claims to land would be accepted as authentic in any court in Europe and that they had been amply rewarded, more so than the American Indians had been for their territories. Besides, forty years' residence had given the Liberians a right to possession of the land, and he asserted that if he had been given his way he would have commenced some distance up the coast to shell the Greboes out of their towns. Semmes further impressed on the Greboes that he had been sent out by the

President of the United States to aid the Liberians in suppressing the insurrection and, pointing to the *Alaska*, he said that it was only one of a hundred ships that the United States could send. P. K. Valentine, an educated Grebo who was present at the peace talks, recalled that Semmes asked them which they preferred—peace or war. Again pointing to the big guns on board the *Alaska* he asked them if they had anything like them, stating that he had no desire to fight poor helpless people who could not respond to the roar of thunder from his ships. Despite these threats, both sides later expressed satisfaction with the terms of the peace settlement. Semmes in the end proved a calm, just and impartial umpire who suppressed unreasonable desires by both sides and restored peace. The Grebo Reunited Kingdom from then on ceased to be an active opposition, though the feeling of Grebo nationalism remained and was later to be a major factor in the movement led by Prophet William Wade Harris in Liberia and the Ivory Coast.[49]

The war of 1875 brought to a head the antagonism which had long existed between church and state in Liberia. The government was always suspicious of foreign, particularly white American, missionaries and barely tolerated them. The war which was led by mission-educated Greboes aggravated tension between the two groups. The Protestant Episcopal, the Presbyterian and Methodist Episcopal Churches aroused the greatest criticisms. There was no fear as long as missionary activities were confined to the coast but when they moved inland, the government criticised them for developing stations in the remote interior beyond its power to reach, protect and control. Even along the coast, mission-educated Africans often proved disloyal to the government in times of crisis. They often wrote to the United States Department of State, holding the government that had assisted the emigration of the settlers responsible for their alleged misdeeds in Liberia.

President Payne, who had tried to avert the war, remarked sadly that years of untiring Christian missionary efforts among the Greboes as preferred objects of the Presbyterian and Protestant Episcopal Churches had made them hate rather than admire Christian civilisation. He believed that the ability of the Grebo to make mischief had been heightened by the spread of mission education. As retaliation, the duty-free privileges of the missions were taken away and restored only at will by the government.

There was also a matter of conscience involved. The Liberians were prone to view every national catastrophe as a chastisement from God. The Grebo war of 1856 when Maryland lost its independence, the assassination of President Edward J. Roye in 1871 in his bid to extend the presidential term from two to four years, and the Grebo war of 1875, when the combined armed forces of Liberia were defeated by the Greboes, were a few

49 *Ibid.*, pp. 265–86.

of the soul-searching occasions which pointed towards the need to speed national integration. After the war of 1875 the 'native' question became especially prominent in every state paper, legislative act, public address and sermon. In the treaty of 1876 after the war, the Liberian Government promised to give the Greboes equal rights as citizens of one and the same country.

Those who had previously regarded the relationship of immigrant Liberians with Africans as the exclusive sphere of foreign mission boards now began to say that it was a responsibility the nation must assume. There was increased concern for the building of a nationality which embraced the native Africans. In 1879 the government created its first honour—the Liberian Humane Order of African Redemption—which had as its object the civilisation of the African peoples throughout the territory of Liberia. There followed discussions on the creation of an independent national church and a home missionary society for the building of a Christian state out of American and African material. But this remained only a hope for the future.

The European scramble for Africa forced the Republic to move into the interior and show evidence of effective occupation of the areas claimed and effective control of the African populations there. Liberian troops were posted in every major town in the interior—Jene Wonde, Bopolu, Gbarnga, Sanniquellie, Tappita, Zwedru, Kolahun and Salala. Gradually, some administration based on the powers of the traditional chiefs emerged. President Arthur Barclay attempted in 1908 to put this indirect rule administration on a regular footing with a hierarchy of provincial and district commissioners accompanied by the Liberian Frontier Force to uphold and strengthen governmental authority. Paramount, clan and village chiefs were appointed or approved by the government to be responsible for the collection of the hut tax, the recruitment of labour for road construction, the enforcement of the labour law, and the maintenance of law and order.[50]

The activities of the Frontier Force often provoked resentment and frequent disorders. Efforts of the government from time to time to control the activities of the Force in remote areas achieved no permanent success. There were not sufficient funds to maintain a comprehensive and efficient administration. Liberia's colonial neighbours, taking advantage of her inability to demarcate and police effectively her international boundaries, offered to assist her. Troops from Sierra Leone were stationed in Monrovia under the command of a British soldier, Major Cadell, who took his orders from the Governor of Sierra Leone. He encouraged insurrection among the men because of alleged arrears in salaries and nearly achieved a *coup d'état*. A timely intervention by the government of Theodore Roosevelt at the

50 Liberia Department of State, *Inaugural Message of President Arthur Barclay*, Monrovia, 1908, p. 10.

urging of Booker T. Washington that the United States should take some interest in the fortunes of her foster-child forestalled Cadell's intentions and those of the French and Germans. The First World War also helped to avert the designs of the colonial powers. It was necessary in the coastal areas as late as 1915 for the United States Navy to intervene against the Kru. It was on this occasion that the Kru summed up the African complaint against the Liberian Government in this extraordinary message to solicit British intervention:

> We are that whole race in general which knows . . . what our fellow brethren suffered in time of slavery in West Africa. . . . We were exempted from them and are proud of it; and it is that freedom we want and mean to keep up from our ancestors. . . . We want the Liberians to respect our race. We admit and obey everything they say but they are not satisfied. They want every particle that belongs to us nearly every three months. While we work hard on the foreign navy and merchant ships . . . they do nothing. But we must work and they must have the benefits of our blood. They place men or military escort among us if we do not give anything.
>
> The Frontier Force will chain our chiefs and elders and even go so far to shoot women and children if we have nothing to give. . . . We contend for our rights and have persuaded them that they were placed among us of one race. The very people that brought them to us do aid them in landing . . . on our shores with ammunition and then aid them with their cutter and use machine guns on us who have nothing in the way of arms. We pray you in God's name that your government should intervene if not the Kru race will be out of existence. We want you to do all you can and ask the United States of America Government about this act they have done towards us. We can stand them if we have ammunition.[51]

European imperialism in Africa had, it seemed, led to certain adverse and protectionist tendencies.

THE LEAGUE AFFAIR: 1930–5

Certain atrocities committed by the Liberian Frontier Force were reported abroad by the opposition in the 1928 presidential campaign. These led to a bad international press, a sore election issue, and finally the intervention of the League of Nations in 1930 at the invitation of the Liberian Government. These happenings did more to accelerate the pace of national integration than any other factor.[52]

51 'The Kru Remonstrance', National Archives, Department of State, United States, 1915, quoted in Abeodu B. Jones, *The Struggle for Political and Cultural Unification in Liberia, 1847–1930*, Ph.D. thesis, Northwestern University, Evanston, Illinois, 1963.
52 Huberich, ii, p. 899.

The intervention of the League of Nations to investigate charges of domestic slavery in Liberia was not an isolated occurrence. Liberia was a member of both the League of Nations and the Anti-Slavery Convention of 1926 and as such its doors were open for inspection in compliance with the principles of membership in international associations. When in 1928 the People's Party, in opposition to the elections won by the dominant True Whig Party, charged that unfair political activity and repressive measures had characterised the Liberian domestic scene and that they would like to see these corrected, the Charles D. King government, in good faith and on the recommendation of the United States, officially requested the intervention of the League. Matters seem to have worsened when disgruntled citizens, taking advantage of the presence of the League, testified to the charges of slavery as being true. They also spread rumours that Liberia was to become a mandate of either the United States or Britain. A fresh uprising broke out in Sasstown in the 1930s, serious enough to cause all foreign diplomatic missions to withdraw from Monrovia, and this appeared to support the rumour that Liberia was about to lose its sovereignty. Indeed, Liberia's colonial neighbours were already vying with each other at the League's headquarters as to who should possess her as a mandated territory. The Firestone Rubber Company also entered the fray, arguing that it was the most logical contender for the prize since it had lent five million dollars to the Liberian Government in 1926. The British particularly resented the American presence represented by Firestone and this was a factor in the instigation of the League to intervene in Liberia's internal affairs.[53] The United States had undermined British rubber monopoly.

The findings of the Commission sent by the League disclosed that slavery as defined by the anti-slavery convention of 1926 did not exist in Liberia but that the means employed in recruiting labour for public and private purposes and for shipment to Fernando Po and Gabon were associated with slavery because the method of recruitment entailed compulsion. It was noted, however, that the government of Liberia had never sanctioned the recruitment of labour with the aid of the Frontier Force but that persons holding official positions had illegally misused their offices as agents for labour contracts with foreign governments (Spain, Portugal and France) by recruiting with the aid of the Force.[54]

The Commission recommended a radical reconstruction of native policy and a more humane treatment of the Africans. It further suggested that barriers between the 'civilised' and 'uncivilised' be broken down through the extension of education to all, regardless of cultural origin. More especially it asked that pawning, the shipment of labourers, road programmes and the activities of the Frontier Force be curtailed, and it called

53 *Ibid.*

54 Nnamdi Azikiwe, *Liberia in World Politics*, London, 1934, pp. 193–4; Huberich, ii, pp. 910–14.

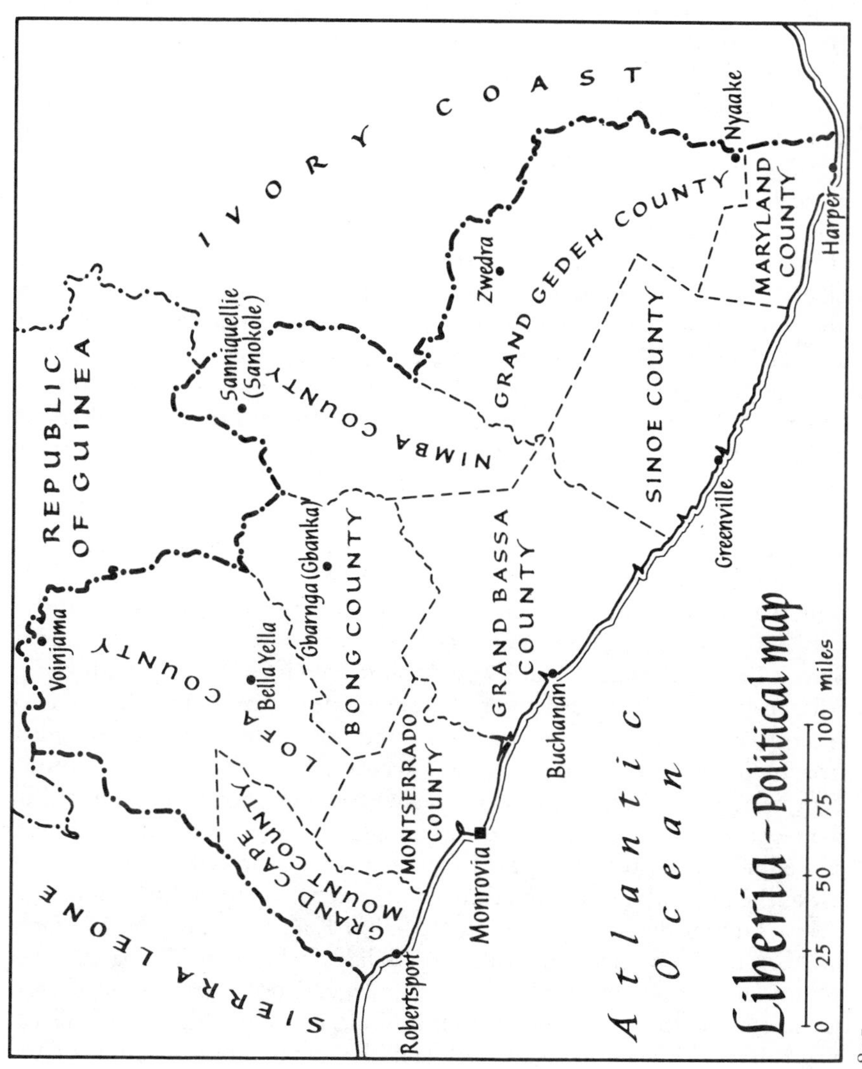

8:1 Liberia – Political map

for a complete reorganisation of the administration of the hinterland into counties with the creation of a civil service and that Liberian district commissioners be replaced for a while by Americans or Europeans.

The recommendations of the League received, on the whole, the approval of the Liberian Government, which only raised objection on principle to internal supervision of its affairs by the big powers. Moreover, the activities of the League in Liberia were soon overshadowed by the emergence of Adolf Hitler and the gathering of a much bigger storm on its doorsteps in Europe. But the League did bring Liberia into the limelight of world events and thinking, thus marking a turning point in Liberian history.

For decades following independence it was argued that the constitution of 1847 was ordained and established by the people of the commonwealth of Liberia—the word *people* being used in a technical and juridical sense and not in the popular meaning. It was not synonymous with inhabitants and did not include white residents, native Africans, or rescued slaves, unless they had acquired citizenship, any more than the Indians were included under the designation of people in the American constitution.[55] Fears of imperialist powers, boundary pressures and the risk of losing large numbers of the African population in Liberia in time brought about major constitutional changes. The government of Arthur Barclay in the first decade of the twentieth century recognised the Africans as citizens although they were not granted the franchise until after the Second World War and this right was not exercised until the presidential elections of 1952. Previously, only Liberian immigrants residing in the coastal counties of Mesurado, Sinoe, Bassa, Maryland and Cape Mount could vote along with those Africans who had fulfilled the requirements for citizenship.[56] There also emerged the annual convocations of the interior people and chiefs presided over by the President and known as Executive Councils. There, the President in personal diplomacy negotiates with the interior peoples, listens to their complaints and grievances, usually against District Commissioners, and adjudicates. President David Coleman set the precedent in 1900, but it was then considered a premature move, and he was forced to resign.

In a famous decision in the case of *Ballah Karmon versus John L. Morris* in 1919, the Supreme Court ruled to abolish dual rule in Liberia between county and province, thus giving the independence constitution of 1847 jurisdiction in the hinterland and not limiting it solely to the coast.[57] It was this ruling of the Supreme Court and the later county division of the

55 Huberich, ii, p. 865.

56 Richard A. Henries, *Statement on the Occasion of the Silver Jubilee of the Presidential Administration of William V. S. Tubman*, Monrovia, 1969, p. 6.

57 Huberich, ii, pp. 1207–20. Tribal citizens entered the coastal or county areas near Monrovia and took stolen goods into the interior. A legal battle ensued as to whether the criminal laws of the county areas applied equally in the interior.

hinterland as recommended by the Commission of the League of Nations which led to the creation in 1964 of a uniform political and administrative structure of county government in Liberia. A deep respect for law and order meant stability, survival and progress.

EDUCATION

In treaties ceding their lands the Africans had always asked that the government establish schools among them but it fell to the lot of missionaries to provide them at first.[58] Higher education was of special interest to the Protestant Episcopal Church. Liberia College, now the University of Liberia, was first planned by Bishop John Payne in 1848 to be located in Cape Palmas among the Greboes. The government of J. Roberts thought otherwise and took an active hand in placing the college in Monrovia, where it began operation in 1862. Until 1950, the chief beneficiaries were the Liberians of settler origin. Not satisfied with this state of affairs, the Protestant Episcopal Church in 1889 established Cuttington College and Divinity School in Cape Palmas where it remained until 1929 when its doors were closed by the world economic depression. In 1948 the college was re-established in the interior of Liberia in Bong County where it has had far-reaching effects in the training of hundreds of Liberians from all cultural backgrounds.[59] The Holy Cross Fathers were the first Christian missionaries to penetrate the remote interior in their attempt in 1922 to counter the further spread of Islam from Guinea into Lofa County. They were followed in the late 1920s by Methodist missionaries who opened their school in Ganta, Nimba County. As late as 1944, of the two hundred and fifty-one schools with a student enrolment of nearly twenty thousand, two-thirds were mission institutions. By 1970, however, of the one thousand and eighty-four schools with an enrolment of one hundred and thirty-four thousand, only one-third were mission operated.[60]

MODERN TRENDS AND ISSUES

The coming of the Firestone Rubber Plantations Company to Liberia in 1926 made a great difference to Liberia's financial means for national development. Besides procuring through Firestone its first major loan of five million dollars, there was the provision of infrastructural and industrial services, and thousands of native Africans and immigrant Liberians were for the first time introduced to a wage economy. The stationing of American troops during the Second World War necessitated the undertaking of public works to accommodate their operations. Other factors also helped

58 Martin, *Dual Legacy*, p. 48.

59 Abeodu B. Jones, 'Of Things Episcopalian', *Liberian Age*, xxv, 8, 5 February 1961, p. 1; Melvin J. Mason, *The Role of Cuttington College and Divinity School in the Development of Liberia*, Ph.D. thesis, Michigan State University, 1965, p. 15.

60 'Annual Report of the Department of Education', Monrovia, 1970, p. 288.

to speed up changes in Liberia's outlook for economic and industrial development. The observance of the first centenary of independence in 1947 highlighted the need for suitable modern facilities to accommodate foreign guests. Similar facilities were erected for construction workers in the Free Port of Monrovia and the numerous concessionaires from abroad, who had arrived since the war, tried to enforce international standards of living in Liberia by undertaking the building of infrastructural facilities. Finally, many Liberians returned from universities abroad having acquired foreign habits of living. Moreover, under the Open Door and National Unification policies vigorously pursued by the Tubman government after 1944, thousands of Liberians migrated to Monrovia from rural areas and were engaged in gainful industrial employment. Industrial and technical education were given major emphasis in educational development plans, and all foreign concessions were required to provide technical education for their Liberian workers. Liberia no longer placed a premium on the cry that the aim of its existence was to prove that black men were capable of self-government. Since the Second World War, the emphasis has been on industrial development and enlightened and active participation in world affairs.

Yet, important political issues remain unresolved. For a long time there was controversy about the term of office of the presidency. From 1847 to 1940 the period was two years but this was widely considered too short a duration leading to lack of continuity, frequent elections and inaugurations, too little time spent on national development and too much spent on politics. Arthur Barclay, of West Indian origin, succeeded where Roye had failed in 1871 and won an extension of the term of President to four years. In 1932, Edwin Barclay, cousin of Arthur Barclay, had the term extended to eight years without benefit of re-election. Finally, William Tubman, on succeeding Barclay in 1944, subscribed to the eight-year term but with the benefit of re-elections to serve additional terms of four years at a time.

Meanwhile the native African populations seemed to be patiently biding their time. While representation in the National Legislature had been allowed them by the end of the nineteenth century, cabinet positions were not open until 1924 during the presidency of Charles Dunbar King. In 1950, Didwo Tweh, a Kru, organised the Reformation Party ostensibly to protest against the alienation of Kru land during the construction of the Free Port of Monrovia. His action resurrected old conflicts between tribe and government, created the first effective opposition to Tubman's government, and significantly was also the first time a native African had made a bid to become President of Liberia.

Tubman, in reply, denounced the distinction Tweh tried to make between Americo-Liberians and aborigines: 'H. R. W. Johnson, Daniel Howard, Charles Dunbar Burgess King and William V. S. Tubman are

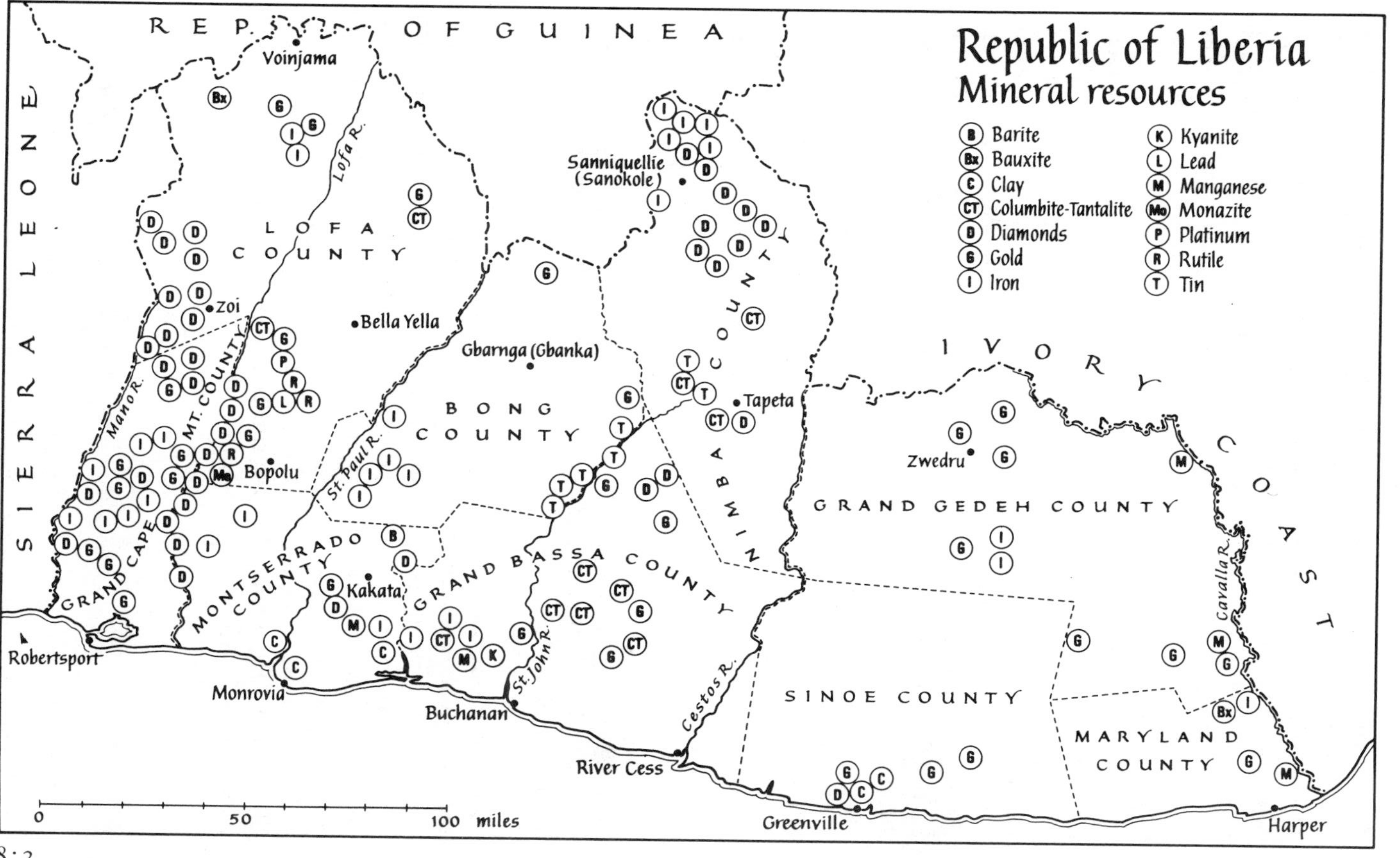
Republic of Liberia
Mineral resources
B Barite
Bx Bauxite
C Clay
CT Columbite-Tantalite
D Diamonds
G Gold
I Iron
K Kyanite
L Lead
M Manganese
Mo Monazite
P Platinum
R Rutile
T Tin
REP. OF GUINEA
IVORY COAST
SIERRA LEONE
LOFA COUNTY
BONG COUNTY
NIMBA COUNTY
GRAND GEDEH COUNTY
SINOE COUNTY
MARYLAND COUNTY
GRAND BASSA COUNTY
MONTSERRADO COUNTY
GRAND CAPE MT. COUNTY
Voinjama
Zoi
Bopolu
Bella Yella
Gbarnga (Gbanka)
Sanniquellie (Sanokole)
Tapeta
Zwedru
Kakata
Robertsport
Monrovia
Buchanan
River Cess
Greenville
Harper
Lofa R.
Mano R.
St. Paul R.
St. John R.
Cestos R.
Cavalla R.
0
50
100 miles

8:2

all aborigines and indigenous people of this country, for we were all born, bred, and raised here,' he said. He further pointed out that since independence in 1847, H. W. Johnson (inaugurated in 1892) was the first Liberian-born President and D. E. Howard, inaugurated twenty years later in 1912, was the second and that '. . . as true as night follows day, the tribes will produce a president who would be elected by the people of Liberia and not by a single tribe or number of tribes'.[61] Thus, integration would ultimately be achieved by an evolutionary process.

It is not without significance that the decades which spanned the establishment and departure of colonial rule in Africa saw nearly parallel developments in Liberia. In 1908 indirect rule penetrated the interior areas which became emancipated in 1964 by an act of legislature creating counties out of the provinces. Moreover, during the decade of the 1960s there was greater emphasis on identification with things essentially African. Culturally and socially, Americo-Liberians became more Africanised through the wearing of African dress and the institutionalisation of African music, dance, arts, crafts and village society.

Of equal importance too has been Liberia's contribution to the creation of the Organisation of African Unity by assembling in Monrovia in 1961 those African states which took a less militant view of solving African problems and by assuming in 1963 a leading role in the preparation of the charter of the Organisation. After decades of war, the philosophy of goodwill for national unification promulgated by William Tubman in 1944 ensured political stability by extending equal rights to all citizens. And at his passing in 1971, William V. S. Tubman, Liberia's greatest President, seemed to have achieved a degree of unity and stability unsurpassed in the annals of his country.

Tubman's successor, William R. Tolbert, promises to raise Liberian living standards 'from mats to mattresses and to higher heights' by calling for the total involvement of all citizens to work for a wholesome functioning society. In activating the National Planning Council, an integrated approach to implement planned development is employed, which in essence aims at a fair distribution of the national wealth to all parts of the country.

Thus, western civilisation as known in Liberia was transplanted there by black men, a fact which the government probably considered its greatest contribution to Africa and the world.

61 E. Reginald Townsend, *Tubman Speaks*, London, 1959, p. 95.

TABLE 8:1

LINGUISTIC CLASSIFICATION OF THE TRIBES OF LIBERIA*

Mande	West Atlantic	Kwa (West Akans)
Northwest	North	South (coastal)
Mandingo	Kissi	Grebo
Gbandi	South (coastal)	Kru
Kpelle	Gola	Bassa
Loma		Dey
Mende		
Vai		
East central		East central
Mano		Belle
Gio		Krahn

*This classification excludes the English-speaking non-tribal descendants of the Negro settlers from America who inhabit the coastal and southern territories of Liberia.

TABLE 8:2

RELIGIOUS AFFILIATIONS IN LIBERIA*

Religion	Estimated number of adherents	Comments
Christian	88,500	Roman Catholics are the single largest group of approximately 20,000. Next are the Episcopalians, Methodists, Baptists, Lutherans, Pentecostals, African Methodists, Seventh Day Adventists, Jehovah's Witnesses, Church of the Lord, Presbyterians.
Moslem	100,000	The majority live in the north, northwest and in the capital city of Monrovia.
Tribal religions	Over 800,000	These are to be found largely in rural areas and remote interior lands.

*The figures were collected from oral field investigations.

TABLE 8:3

POPULATION OF LIBERIA*

Area	Population
Grand Cape Mount County	32,472
Montserrado County	168,139
Marshall Territory	14,523
Grand Bassa County	100,990
River Cess Territory	5,332
Sinoe County	44,845
Sasstown Territory	9,565
Maryland County	39,912
Kru Coast Territory	21,153
Grand Gedeh County	
Lofa County, Bong County, Nimba County	562,432
Total	999,363

*National Planning Agency: 1962 Census.

TABLE 8:4

PRESIDENTS OF THE REPUBLIC OF LIBERIA

President	Years	Party Affiliation	Comments
Joseph Jenkins Roberts	1848–1856	Republican	Guided the colony to independence as the Republic of Liberia. Later opposed because of mulatto complexion.
Stephen Allen Benson	1856–1864	Republican	Liberia College was opened.
Daniel Bashiel Warner	1864–1868	Republican	First migration to Liberia from the West Indies to increase number of U.S. immigrants which the Civil War had caused to decline.
James Spriggs Payne	1868–1870	Republican	Beginning of interior explorations.
Edward James Roye	1870–1871	True Whig	Deposed. First True Whig President.
Joseph Jenkins Roberts	1872–1876	Republican	Strove for restoration of Republican domination of Liberia or Mulatto rule.
Anthony William Gardiner	1878–1883	True Whig	Real beginning of True Whig rule in Liberia. Resigned over border protests with the British in Sierra Leone.

TABLE 8:4—*continued*

President	Years	Party Affiliation	Comments
Alfred F. Russell	1883–1884	True Whig	Completed Gardiner's term of office, border problems continued with British and French.
Hilary Richard Johnson	1884–1892	True Whig	First president born on Liberian soil.
Joseph James Cheeseman	1892–1896	True Whig	Introduced first Liberian currency. Died in office.
William David Coleman	1896–1900	True Whig	Completed Cheeseman's term of office, subsequently elected in his own right but forced to resign over his liberal policies towards native African population.
Garretson Wilmot Gibson	1900–1904	True Whig	Completed Coleman's term of office, elected in his own right. End of two-year presidential term.
Arthur Barclay	1904–1912	True Whig	Start of four-year presidential term of office. Start of Indirect Rule in the interior of Liberia among the African populations. End of boundary disputes which had lingered after official treaties in the 1890s.
Daniel E. Howard	1912–1920	True Whig	Enforced Indirect Rule. Led opposition The People's Party upon retirement.
Charles D. B. King	1920–1930	True Whig	First to win three terms under new four-year term for the presidency. Forced to resign on charges of forced labour by the opposition The People's Party. Led opposition United Whig Party upon retirement.
Edwin James Barclay	1930–1944	True Whig	Defended Liberia on forced labour charges, completed King's term of office, advocated new presidential term of eight years only. Led opposition on retirement—The Independent True Whig party.

TABLE 8:4—*continued*

Year	Event		
William V. S. Tubman	1944–1971	True Whig	Announced and promulgated major policies of national unification and the open door for foreign investment. Abolished territorial divisions and made them full counties. Extended suffrage to women tribal citizens. Died in office.
William R. Tolbert	1971–	True Whig	Pledged to carry on policies of William V. S. Tubman.

TABLE 8:5

MAJOR EVENTS IN THE HISTORY OF LIBERIA IN CHRONOLOGICAL ORDER

Year	Event
1816	The American Colonization Society was organised to send free American blacks to Africa.
1819	Annual Message of President James Monroe in which the U.S. Congress was called upon to give much needed financial assistance without which Liberia would not have been founded.
1820	First Negro colonists sail for Africa; first constitution was provided by the American Society.
1821	Land for settlement bought by the United States Navy during Anti-Slave-Trade Movement.
1822	First town planned on Cape Mesurado; first major tribal war.
1822–1828	Jehudi Ashmun first effective governor of the colony, a white American.
1824	The Colony was named Liberia and its capital Monrovia in honour of President James Monroe.
1825	Constitutional changes provided for limited self-government.
1830–1838	Four additional settlements were established independent of Liberia by New York and Pennsylvania, Mississippi, Maryland and Indiana.
1838	Three of the independent settlements joined to found the Commonwealth of Liberia-Montserrado, Bassa, Sinoe.
1847	Declaration of Independence, a new constitution, the Commonwealth now named the Republic of Liberia.
1847–1878	Period of Republican rule in Liberia characterised by mulatto domination of black citizens.

TABLE 8:5—*continued*

Year	Event
1856	Second major tribal war, Maryland lost its independence and joined Liberia as fourth county.
1862	Lincoln Government recognised the independence of Liberia and treaty of friendship signed.
1871	Beginning of True Whig rule on defeating the Republicans. First loan obtained from foreign source.
1875	Third major tribal war and beginning of liberalisation of Liberian laws to include African tribes upon the defeat of the Armed Forces of Liberia by Grebo soldiers.
1885	Berlin Conference which ignored Liberia but required it to obey the rules governing the appropriation of African lands to which she was not a signatory power.
1895	Liberia-Ivory Coast boundary demarcated through treaty with France.
1904	Beginning of Indirect Rule among African tribes in Liberia.
1911	Final boundary lines settled between Liberia and the British and French colonial territories.
1912	Liberia in bad economic straits, subjected to an international receivership of customs receipts for the servicing of foreign debts.
1914	Liberia entered First World War on the side of the Allies and nearly lost her independence.
1919	Liberia signs League of Nations Charter.
1926	First major loan received from the United States. First U.S. investment established in Liberia—Firestone Rubber Company.
1930	League of Nations on reports by Liberian opposition party intervenes in Liberia's domestic matters to investigate charges of forced labour.
1942	Franklin D. Roosevelt visits Liberia, Defense Areas Agreement signed with the United States.
1944	William V. S. Tubman inaugurated president, proclaims new deal for Liberia in promulgation of national unification and Open Door policies.
1971	Death of William V. S. Tubman, end of the Tubman era.

CHAPTER 9

The Western Sudan and the coming of the French, 1800–1893

B. ỌLATUNJI ỌLỌRUNTIMẸHIN

Before the middle of the nineteenth century the Western Sudan consisted of a number of states and principalities. These included large empires, such as Ka'arta under the Massassi Bambara, Segu Bambara and the Fulani state of Masina, and, in the Senegambia, such small states as Jolof, Futa Toro, Futa Jallon, Bambuk, Bondu and Khasso. These states had different political systems, in some cases accentuated by ethnic and religious differences, and had established relations between themselves and with the outside world, including some European countries. In the Western Sudan itself relations between states were facilitated by the political and cultural influence of the Mande group of people and the use of their group of languages over virtually the whole area.[1]

Socio-political instability, the existence of a lively tradition of proselytising and religious revivalism and the presence of European, especially French, settlements and trading posts in the Senegambia, are among the dominant factors which must be recognised before studying the emergence of the Tukulor empire and the relations of the French with it and with other states in the Western Sudan.

Jolof and Futa Toro were situated in the most fertile parts of the Senegal Valley. Jolof's territory was very suitable for the cultivation of gum, an important article of trade in the area. The significance of this lies in the fact that the fertility of the area easily attracted migrant populations in search of a settlement where they could grow their crops and tend their cattle. Conflicts often arose as population groups settled in areas coveted by other land-hungry peoples. The influx of peoples in search of fertile land led to the population of Jolof becoming heterogeneous. Notable with his disciples to Futa Jallon where the Almami, Bubakar, allowed him

1 H. Labouret, 'Les Manding et leur langue', *Bulletin du Comité d'Etudes Historiques et Scientifiques de l'Afrique Occidentale Française (B.C.E.H.S.A.O.F.)*, 1934; Jack Goody, 'The Mande and the Akan Hinterland', in *The Historian in Tropical Africa*, eds. J. Vansina, R. Mauny and others, Oxford, 1964, p. 193; also *The Historian*, 'The Influence of the Mande in West Africa', pp. 91–3.

among the settlers were the Moors who, by the beginning of the nineteenth century, had become the most important carriers in the gum trade, especially with the Europeans. The struggle for land similarly affected Futa Toro which up to the nineteenth century was a federation of several principalities each of which was dominated by particular clans competing with one another. The prominent lineage groups in Futa Toro were Dimar, Toro, Lao, Irlabé-Ebiabe, Bosséa, N'Guénar and Damga.[2]

The struggle for the possession of land was often connected with the struggle for the control of political power. From the Middle Ages to the nineteenth century, the political history of Futa Toro was largely an account of the establishment of successive empires. In the search for fertile land various peoples, notably the Fulani, Tukulor, Wolof, Moors, Bambara and Sarakole, had come to live in Futa Toro and each group had, at one time or the other, taken part in the struggle for political power until, from the eighteenth century, the Tukulor became dominant in terms of power and numbers.[3] The need to possess political power as a prelude to exercising control over the economy of the area must have been obvious to the contestants. The dissolution of the Jolof empire resulted from developments similar to those in Futa Toro.[4]

Instability sometimes resulted from naked struggle for political power between rival groups in the ruling dynasties. This was the case in Futa Jallon and Khasso.[5] In both states the people were divided into feuding groups up until the time that al-hajj 'Umar emerged to establish the Tukulor empire. Ka'arta and Segu Bambara were relatively stable in spite of the insurrections from subject peoples, like the Diawara and Kagoro, which plagued the former, and the wars of the latter with Masina.[6]

Another important factor in the history of the Western Sudan was the tradition of proselytising and religious revivalism. In Futa Toro, the tradition of Islamic proselytisation dated back to the establishment of Tekrur in the eleventh century.[7] The imamates in both Futa Toro and Futa Jallon resulted from the eighteenth century Islamic revolutions

2 Felix Brigaud, *Histoire Traditionnelle du Sénégal*, Saint Louis, 1962, pp. 9, 55–60; V. Monteil, 'Le Dyolof et Al-Bouri Ndiaye', *Bulletin de l'Institut Fondamental de l'Afrique Noire*, xxvii, Sér. B., 3–4, 1966, p. 596.

3 Abdoulaye Bara Diop, *Société Toucouleur et Migration*, Dakar, 1965, pp. 12, 15.

4 Brigaud, *Histoire Traditionnelle*, pp. 53–4; Monteil, 'Le Dyolof . . .', p. 602.

5 André Arcin, *Histoire de la Guinée Française*, Paris, 1911, pp. 101–2; Archives Nationales, Section d'Outre-Mer, Paris (A.N.S.O.M.), Sénégal III, 8, 'Rapport de M. Hecquard sur son voyage dans l'intérieur', 1852; A.N.S.O.M. Sénégal I, 41b/402, 'L. Faidherbe à Son Excellence, Monsieur le Ministre de la Marine et des Colonies', Camp de Médine, 1 October 1855.

6 Archives du Sénégal, Fonds de l'A.O.F. (A.S.A.O.F.), IG195/9, Lt. Sagols, 'Notice générale sur le Soudan—Administration', May 1897; in IG124, 'Pays du Ka'arta: Historique'.

7 Brigaud, *Histoire Traditionnelle*, p. 11.

led by Suleyman Bal and Ibrahim Nuhu respectively.[8] Tukulor rule in Futa Jallon resulted from the eighteenth-century 'Islamic' revolutions very much alive at the beginning of the nineteenth century. The religious leaders of Futa Toro and the main agents for the propagation of Islam in the Western Sudan were the *toroBé*. They were renowned for their role in toppling the non-Muslim Denyanké rule in Futa Toro and for aiding the revolutions in Futa Jallon and other adjacent territories.[9] 'Umar's family belonged to the *toroBé* group and was part of the religious leadership. In Masina, the Cissé successfully led a revolution against the non-Muslim dynasty in 1818 and thereafter established an Islamic theocracy.[10] The wars between Masina and Segu resulted in part from the desire of the Cissé to spread Islam and their own rule to neighbouring states.

Thus, it is clear that political instability was widespread in the Western Sudan by mid-nineteenth century. In the Senegambian states of Futa Toro, Futa Jallon, Jolof and Khasso, there was political instability owing largely to struggles between competing economic and political interest groups. The Bambara empires of Ka'arta and Segu were fairly strong militarily; but they were weak in the face of external invasion because of their internal problems. Ka'arta had for long been grappling with insurrections from subject peoples, and in Segu, the Bambara had had to struggle for survival against their aggressive Muslim neighbour, Masina. The latter was weakened by external military involvements. Given the mass movements of people which the events in the various states involved, the situation appeared fluid and susceptible to exploitation by external forces. Throughout this period, however, normal relations continued both between states and with the outside world.

INTER-STATE RELATIONS

The strong cultural ties existing between the states, the majority of which belonged to the Mande group, were supplemented by an extensive trade. Mungo Park, who visited parts of the Western Sudan between 1796 and 1797, reported on the relations between the Senegambian states. In an account of his journey to these states he recorded the existence of trading activities that were regulated by the exaction of customs dues. In relation to Bondu state in particular, he noted that '. . . from its central situation between the Gambia and the Senegal rivers . . .' it became a place '. . . of

8 Abdoulaye Diop, *Société Toucouleur*, p. 14; M. J. De Crozals, 'Trois états foulbés du Soudan occidental et Central, le Fouta, le Macina, l'Adamaoua', *Annals de l'Université de Grenoble*, viii, 1, 1896, p. 296; H. F. C. Smith, 'The Islamic Revolutions of the 19th Century', *Journal of the Historical Society of Nigeria (J.H.S.N.)*, ii, 1, 1961; Alphonse Gouilly, *Islam dans l'Afrique Occidentale Française*, Paris, 1952, p. 67.

9 Abdoulaye Diop, *Société Toucouleur*, pp. 49–50.

10 A.S.A.O.F., IG158/2, 'Notes sur l'histoire et la situation actuelle du Macina', Le Résident, Ségou, 1 March 1892, pp. 1–2. See A. H. Ba and J. Daget, *L'Empire Peul du Macina. I. 1818–1853* (I.F.A.N., Centre du Soudan, 1955), for the history of the Cissé dynasty.

great resort; both for the slatees *(sic)*, who generally pass through it, in going from the coast to the interior countries; and for occasional traders who frequently come hither from inland countries to purchase salt,' and that '. . . heavy customs were levied on travellers' in almost every town.[11] European influence was also already in evidence in the economic and political life of these states. Some of the articles of trade like Indian baft, muskets and gunpowder were imported by European trading agents. Some of these articles were used in paying tributes, and these European articles, particularly the arms and ammunition, were becoming sources of strength for the ruling elites in these states who often used their new weapons for purposes of state-building and for fighting external wars.[12] Many of the rulers behaved like that of Bondu who '. . . is well supplied with arms and ammunition; a circumstance which makes him formidable. . . . The state was strongly and aggressively organised. . . .'[13]

Some of these wars were caused by conditions already explained above. But others were directed at religious conversion,[14] an exercise which often involved territorial expansion. In general, however, the confusion resulting from wars within and between the states often provided the opportunity for external forces to operate to their own advantage. For instance, the internal political struggle in Walo and other Jolof states has been noted as a cause for incursions by the Moors, subsequent border fights and political intervention by the French. Allying with one party or the other within the Jolof states, the Moors successfully exploited the situation and established their settlements on the Senegal. In the same way, the Tukulor from Futa Toro were able to establish themselves in Jolof.[15]

THE EMERGENCE OF THE TUKULOR EMPIRE

The movement which led to the creation of the Tukulor empire was headed by ʿUmar b. Sa'id Tall, commonly known as al-hajj ʿUmar. He was born in the village of Aloar (Halwar) in Futa Toro around 1794.[16] His father was a very learned marabout who had brought up his children in strict Islamic traditions. As has been mentioned, his family belonged to the *toroBé* group, religious leaders from among whom were chosen the Almamies of Futa Toro. Like the other Muslims of the period, 'Umar's family belonged to the Qadiriyya brotherhood and 'Umar himself grew up as a Qadiri.

11 Stephen Gwynn, *Mungo Park and the Quest of the Niger*, London, 1934, pp. 57–8.
12 Gwynn, *Mungo Park*, pp. 58, 68, 76–9.
13 Gwynn, *Mungo Park*, p. 58.
14 Gwynn, *Mungo Park*, p. 72.
15 V. Monteil, 'Chronique du Walo Sénégalais ((1186?–1855) par Amadou Wade (1886–1961)', *Bulletin de L'Institut Français d'Afrique Noire (B.I.F.A.N.)*, Sér. B., 3–4, 1964, pp. 440–1, 445–7
16 Mohammadou Aliou Tyam, *La Vie d'El-Hadj Omar, Qacida en Poula*, trans. H. Gaden, Paris, 1935, p. 6; A. Le Chatelier, *L'Islam en Afrique Occidentale*, Paris, 1899, p. 167.

He must have received the usual education in Islamic theology and jurisprudence. It appears that he attained a remarkable degree of excellence as a religious scholar though not very much is known about his youth.

While preparing to perform the pilgrimage, 'Umar visited Saint Louis, Senegal, in 1825, to look for means of carrying out his plan. It was usual for intending pilgrims to ask for alms to supplement their own resources for undertaking the pilgrimage. Saint Louis was at this time the main commercial centre in Senegal and was a converging point for Muslim traders from adjacent territories. It was natural, therefore, for a *torodo* marabout, such as 'Umar had obviously become at this time, to go there to preach and appeal for alms. We have no information on the extent of his following at this time, either in Futa Toro, or in Saint Louis. He, however, received generous donations from the Muslim population of Saint Louis and with these and other resources at his disposal, he was able to leave for the pilgrimage in 1826.[17] He was apparently the man whom Clapperton met at Sokoto and prematurely described as al-hajj 'Umar on 7 November 1826.[18] On his way to Mecca, he was said to have stayed for about seven months in Sokoto and two months in Gwandu, both in the empire created by 'Uthman b. Fudi.

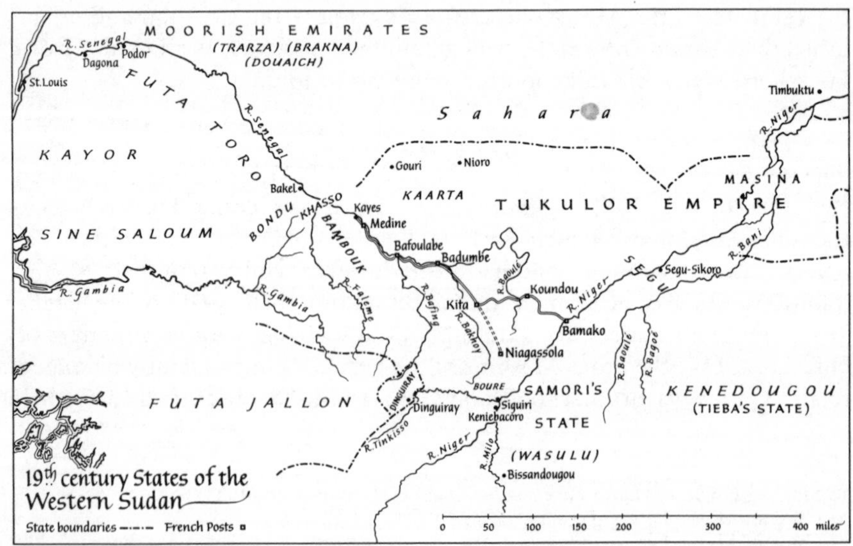

9:1

17 Frédéric Carrère et Paul Holle, *De la Sénégambie Française*, Paris, 1855, p. 192.
18 Hugh Clapperton, *Journal of a Second Expedition into the Interior of Africa*, London, 1829, p. 202.

'Umar, who had been born into the Qadiriyya brotherhood, then the predominant Islamic sect in the Western Sudan, was probably initiated into the Tijaniyya Movement during his pilgrimage. One tradition, however, claims that 'Umar's initiation into the Tijaniyya had come earlier through Sidi 'Abd al-Karim al-Naqil of Timbo.[19] In any case, 'Umar deepened his knowledge of Islam and gained a wide experience of other lands through his travels. He is said to have received instruction from Muhammad al-Ghāli Abu Tālib in Medina. It is also claimed that the founder of the Tijani brotherhood, Shaikh Sidi Ahmad al-Tijāni, appeared to Muhammad al-Ghāli and ordered him to initiate 'Umar and to invest him with powers of a *muqqadam* and appoint him the Khalifa of al-Tijani in the Western Sudan. He claimed subsequently to have been given the injunction to wage a jihād in the Western Sudan.

By 1832, he had reached Sokoto on his return from the pilgrimage. He was very generously treated by Muhammad Bello who provided him with the means of settling down for the seven years he stayed in Sokoto and also gave him one of his daughters, Mariam, in marriage. 'Umar also married a Hausa girl, who later became the mother of his eldest son, Ahmadu.[20] In December 1838 'Umar left Sokoto on his journey towards his home in Futa Toro. By this time he had become wealthy in terms of slaves and property. During this homeward journey he spent about nine months in Masina as a guest of the ruler, Shaikh Amadu. He also reached Segu, the capital of the Segu Bambara empire, some time in 1839. The *fama* of Segu at first detained him; but released him soon afterwards. He arrived in Futa Toro in 1840 but did not settle there. Instead, he moved with his disciples to Futa Jallon where the Almami, Bubakar, allowed him to establish his *Zawiya*[21] at Dyegounko, near Timbo. Here, while recruiting and teaching his disciples Tijaniyya ideas, he took to large-scale trading, using his disciples as agents.

Through his disciples, he ran a large trade in guns and gunpowder with the British in Sierra Leone and the British and French agents in the trading posts in Rio Nunez and Rio Pongas (in the area known by the French as the 'Rivières du Sud') north of Guinea. He also had trading connections with the French settlements in Senegal. With time, he became rich and popular; and with the arms and ammunition at his disposal, he also became powerful and influential. Eventually, he became involved in the politics of Futa Jallon, where he is known to have participated in finding a solution to the Alfaya-Soriya dynastic tussle for power.

19 Jules Salenc, 'La Vie d'Al-Hadj Omar, Traduction d'un manuscrit arabe de la Zaouia tidjaniya de Fez', in *B.C.E.H.S.A.O.F.*, 1918, p. 410.
20 Al-Hajj Sa'id, in *Tedzkiret en Nisiam fi Akhabar Molouck Es-Soudan*, trans O. Houdas, Paris, 1901, p. 308.
21 A Zawiya was a religious and militarily-oriented community, a nursery of men for the expansion of the faith, if necessary by force of arms.

'Umar was a learned scholar who was respected as the leader of the Tijani community in the Western Sudan.[22] 'Umar's teaching clearly emphasised his discontent with the existing religious systems in the Western Sudan. He was dissatisfied with the practice of Islam where this was the religion of the people, as in Futa Jallon and Futa Toro where a long tradition of Islam had existed. His dissatisfaction must have been due in part to lapses observable among Muslims in these areas. But this must be seen also in relation to the fact that the Tijani believed themselves morally and spiritually superior to all other groups of Muslims. 'Umar's desire was to change the Muslims in these areas, mostly Qadiri, to Tijani. As for the areas not under Muslim rule, like Ka'arta, 'Umar's detestation was even greater, and like most Muslim leaders, he wished to see these states accept Islam, but specifically under the Tijaniyya brotherhood.

Needless to say, 'Umar's discontent with the religious systems in the states of the Western Sudan necessarily implied a condemnation of the existing social and political order, as well as of the ruling elites. In his own state of Futa Toro, for example, the religious leaders were the same as the political leaders of the communities. These were mostly Qadiri. Their acceptance of the Tijaniyya brotherhood would involve the acceptance of 'Umar's claim to be the Khalifa of Tijani in the Western Sudan. This, of course, would mean a renunciation on their part of their positions as politico-religious leaders of their communities in order to give place to 'Umar. The position was the same for the Almamies of the Senegambia as well as for the Cissé rulers of the theocratic state of Masina. As it turned out, the existing ruling elites refused to make such a heavy political sacrifice only to change their identity as Muslims from being Qadiri to become Tijani. The prospects must have appeared more frightening to the ruling groups in the non-Muslim states. 'Umar's condemnation of all showed him to be a revolutionary and he was recognised as such by those with interests in the preservation of the existing social and political system.

The situation seemed ripe for revolution. As already indicated, many of the states were at this time racked by tension and strife arising from social, political and religious conflicts. 'Umar profited from the existence of socio-political disorder, the causes of which could be attributed, rightly or wrongly, to the existing rulers. It is arguable that without these social and political conditions 'Umar's propagation of Tijaniyya brotherhood could not have resulted in a revolution. One might recall that the fact that he was a Tijani did not lead to friction between him and his hosts in Sokoto, all Qadiri, during his long stay there from 1832 to 1838. He certainly tried to convert people to the Tijaniyya sect as the apocryphal claim that he converted Muhammad Bello and other leaders of the Sokoto empire

22 See Jamil Abun-Nasr, *The Tijaniyya: A Sufi Order in the Modern World*, Oxford, 1965, for a discussion of Tijani ideas and the role of Tijaniyya communities in West and North Africa.

indicates. He also had the ambition to rule in Sokoto after the death of Muhammad Bello. This ambition must have been responsible for the strained relationship between him and Atiq who succeeded Bello. The evidence from 'Umar's descendants is that his rift with Atiq was not settled until after he had left Sokoto in December 1838.[23]

Some writers, notably Jean Suret-Canale, have tried to explain 'Umar's prestige and success in terms of the supposed differences between Qadiriyya and Tijaniyya brotherhoods. The Qadiriyya leaders who ruled before the emergence of 'Umar are said to have perpetuated in Islam the privileges of the tribal aristocracy and thereby created and maintained barriers between themselves and the ordinary believer. It is then said that 'Umar's strength came in part from the fact that the Tijaniyya which he preached broke these barriers and created a new value system based on individual talents. This theory is logically concluded by the claim that the Tijaniyya had a revolutionary character and was relatively more democratic than the Qadiriyya.[24]

This interpretation is based on certain false assumptions. It assumes that the Tijaniyya brotherhood was a new phenomenon introduced by 'Umar but, as we have pointed out, evidence exists to the effect that the brotherhood had spread to the Senegambia even before 'Umar went on his pilgrimage and that he himself was probably initiated into it earlier in Futa Toro. There is also the assumption that all or most of those who took part in the revolution understood the doctrinal points of differences between the Qadiriyya and Tijaniyya, as well as the supposed socio-political ideas that separated them. In a sense the fact that 'Umar's opposition came from the ruling elites seems to support the socio-political interpretation. The claim that the Tijaniyya was more democratic presupposes that 'Umar's movement indeed led to the establishment of a revolutionary socio-political organisation from the point of view of the ordinary people.

'Umar was a *torodo* and a member of that class of religious aristocrats who were supposed to have created barriers between themselves and the ordinary believers. It is also necessary to appreciate that the *toroBé* were accepted as leaders by their society. There is hardly any evidence that the common people understood the differences between the Qadiriyya and Tijaniyya so that it is difficult to maintain that their mobilisation to action derived from such an understanding. On the other hand, one would be on firmer ground in suggesting that 'Umar's movement was significant from the social and political point of view mainly because it offered a new context within which men of the same politically important group could carry out their usual competition for power. This explains why, although he

23 See oral evidence collected from Shaikh Madani Muntaga Tall, the religious head and patriarch of 'Umar's descendants at Segu on 12 May 1965. This is recorded in B. Ọlatunji Ọlọruntimẹhin, *The Segu Tukulor Empire*, London, 1972, ch. 2.

24 J. Suret-Canale, *Afrique Noire—Occidentale et Centrale*, Paris, 1961, pp. 191–2.

was repulsed by those actually exercising power, his immediate assistants came from members of the same social class who were still aspiring to positions of power and to whom his movement offered opportunity for new status. Those who joined him in Dinguiray were already religious leaders in their own right. Some, like the Almami of Bosséa, joined him at the head of their own militia. The composition of his army clearly reflected the differences between the aristocratic Tukulor and the rest as well as the different lineage groups to which the Tukulor themselves belonged. In the organisation of the empire, the political leadership, though not exclusively Tukulor, emphasised the dominance of the old aristocracy from Futa Toro. Indeed, there is a sense in which one can regard the administration and politics of the Tukulor empire as an extension of the Futa Toro system.[25]

It is difficult to assess how far the mobilisation of the people to action was due to religious factors. But if one were looking for the sources of strength and success for 'Umar's revolution one could regard the religious factor as the trigger or the occasion for action. What is very clear is the contribution made to his initial success by the antecendents of his revolution—the confused social and political situation in the various states, which one can regard, using Eckstein's language, as '. . . unallocated political resources . . .'[26] which he capitalised upon and used for his own purpose. His strength lay in the fact that his movement represented a common umbrella for people with various ambitions to fulfil, and varied degrees of attachment to his religious concepts.

'Umar's ambition to carve out a new empire by uniting the existing states under his own rule necessarily brought him into contact with the Europeans, especially the French, who were already in the Senegambia. He toured the Senegambia to enlist support for his plans and during one such tour in 1847 he revealed his ambition to the French. Probably as a way of winning their sympathy, he affirmed his friendship to them and promised to stop the wars which were raging in the Senegal valley and which were ruinous to commercial interests.[27] The basis of 'Umar's relations with the French at this time was their common concern for trade.

THE FRENCH BASE IN SENEGAL

As for the Europeans, their interests were at first mainly commercial. At first the Europeans, especially the British and the French, confined their activities to their 'factories' on the coast and dealt with the interior through agents. The trade in the gum of the Senegambia with the coast

25 See B. Ọlatunji Ọlọruntimẹhin, *The Segu Tukulor Empire*, for details on the army of 'Umar and the administration and politics of the empire.
26 Harry Eckstein, 'On the Etiology of Internal Wars', *History and Theory*, iv, 2, 1965, for an analysis of the various elements in revolutions.
27 Carrère et Holle, *De la Sénégambie*, p. 195.

was carried mainly by the Moors and Arabs from North Africa. The Mandingo traders also dominated the carrying trade between the Western Sudanese states and the British settlements of Sierra Leone and Gambia, as well as the French establishments in Senegal. The Diulas were also prominent in the kola nut and salt trade between the coastal states and the interior.[28] Up to the early nineteenth century gold dust was still traded between the Western Sudanese states and the French on the Senegal Coast.[29] With lucrative trade also came the first European settlements on the coast, as well as rivalry between the different European nationalities. Notably, the French and the English competed with each other from their different settlements for the trade of the interior.

The French first established themselves at Saint Louis (Senegal) in 1637. Later they founded other settlements at Gorée Island, Rufisque and Joal. Britain obtained these colonies from the French by the Treaty of Paris in 1763. However, they were restored to France by the Treaty of Versailles of 1783. But the struggle for these settlements continued until during the Napoleonic wars of 1809 when they were recaptured by Britain who turned them into dependencies of her colony of Sierra Leone. By the Treaty of Paris (30 May 1814) Britain agreed to return them to France. The final papers for the handing over of these colonies were, however, not signed until 24 January 1817 in respect of Saint Louis and 15 February 1817 in respect of Gorée Island.[30] The struggle for these colonies was protracted because the two rivals saw them as gateways, the control of which was essential to controlling the interior market. Having regained control of their colonies, the French, though continuing their traditional pattern of trade through indigenous agents, took additional measures to maximise the profits accruing to them from their trade connections with the interior. These measures necessarily involved the expansion of French activities from the coast to the interior.

This expansion did not result from any studied plan of the French Government. Rather, it was conceived, and sometimes executed by local French officials and merchants in the Senegal settlements who later persuaded their government to acquiesce in the situation created. It is this characteristic of French expansion that has led a writer on French colonial policy to claim that '. . . dans l'histoire des origines de notre empire colonial le calcul des hommes d'Etat entrèrent pour beaucoup moins que le caprice d'un colon audacieux, ou l'initiative intelligente de tel officier'.[31]

28 G. Mollien, *Voyage dans l'Intérieur de l'Afrique aux sources du Sénégal et de la Gambia fait en 1818,* Paris, 1822, vol. ii, pp. 193, 199; E. Mage, *Voyage dans le Soudan Occidental 1863–1866,* Paris, 1868, pp. 120–7, 187–8.

29 L. Faidherbe, *L'Avenir du Sahara et du Soudan (annoté par L. Faidherbe),* Paris, 1907, pp. 12–13.

30 Jean Darcy, *France et Angleterre: Cent Années de rivalité Coloniale: Afrique,* Paris, 1904, pp. 214–19.

31 Darcy, *France et Angleterre,* p. 222.

Among the steps taken by the French in Senegal to expand their economic activities in the Western Sudan was the negotiation of treaty agreements with several of the states. Such treaties usually put them in more direct contact with the sources of raw materials, and consequently reduced their dependence on agents. In pursuance of these objectives, the Schmaltz Treaty (1819) was concluded with Walo. The treaty guaranteed to the French a certain proportion of agricultural products of Walo against the payment of fixed 'tributes' to the Walo chiefs.[32] This treaty, in effect, sought to exclude the Moors, the traditional carriers of the gum trade, at the same time as ensuring the French adequate supply. Efforts were made to establish similar relations with the other states in the Senegambia.

Generally, however, both the indigenous populations and the trading agents whose interests were threatened took steps to protect their interests against the French. The alliance entered into by Walo and the Moors in 1819 could be seen in this way. Between 1819 and 1835, the alliance endured, fostered by the dynastic marriage in 1833 of the daughter of the Brak of Walo to the emir of the Moors, Muhammad El-Habib, and the common hostility of the two peoples to the French expansion. Wherever the French felt able, a combination such as that in Walo was usually countered with force. In this way by 1835 they were able to force the Moors to renounce their alliance with Walo.[33] Such successes, however, were usually ephemeral and could not guarantee protection for French interests. They therefore had to resort to such steps as sending exploratory and peace missions to the states concerned.

Largely at the instigation of the French settlers in Senegal, the French Government sent missions to various parts of the Western Sudan beginning with the Senegambian states. It also gave approval to the formation of the *Compagnie de Galam* to trade mainly with the Senegambia. In 1818, under the leadership of G. Mollien, an official mission was sent out, charged with finding out routes linking the Senegal with the Gambia and with ascertaining the quantity and quality of the gold deposits of Bambuk.[34] Also, M. Duranton was sent on a commercial mission between 1828 and 1838 to several states in the Senegambia. This mission resulted from the report of the Governor of the Senegal settlements, Roger, on the richness of the trading post of Bakel. The Ministry responsible was pressed to intensify its efforts for exploration 'with a view to commercial expansion'. Another consideration that weighed heavily with the ministry was the need to forestall the British in the penetration of the Western Sudan. Thus the Duranton mission set out in June 1828 not only to explore the commercial possibilities of the Upper Senegal region, but also with the objective

32 Monteil, 'Chronique du Walo', p. 448.
33 *Ibid.*
34 Mollien, *Voyage dans l'Intérieur*, p. 97.

'. . . à devancer les Anglais dans la pénétration du Soudan'[35] so that the commercial potentialities of the Sudan as a whole could be monopolised by the French. In his eagerness to establish the French presence in the states visited, Duranton got involved in their politics. The hostility which Duranton aroused against the French by his involvement, notably in Khasso where he participated in the wars of Awa Demba (of Medine) against Logo and Bambuk in 1829, marred any chances of success that he might have had. The Compagnie de Galam lost some of its trade as a result of hostility incurred through Duranton's activities. The same company later used its influence with the colonial government in Senegal to get Duranton discredited, after accusing him of never stopping '. . . de fomenter la division entre les divers peuples du Haut-Sénégal'.[36]

The Duranton affair showed clearly the close connection between the French concern for commercial interests and for political affairs. It also demonstrated how political ineptitude on the part of a commercial agent could damage both the political influence of the government and the commercial interests of the trading companies. In this case, Duranton was offered as a sacrifice to save French political and economic interests. Subsequently, the political aspect of French relations with these states was properly regulated with the establishment of a political affairs division with its own officer in the Governor's administration in Saint Louis. The missions sent out later confined themselves to commercial activities and left the regulation of political relations to the Saint Louis authorities. Exploratory missions to various parts of West Sudan were undertaken for French commercial bodies by Anne Raffenel (1848–9); Commandant Bouet Willaumez (1849) and M. Hecquard (1849–50).[37] Hecquard, in particular, was charged with the task of verifying the information given by previous agents. He was also to find out the caravan routes, especially those used by Bambara traders, and other direct communication links between Senegal and the Sudan. This clearly shows that the French were already thinking of expansion to the Western Sudan. On the political plane, the officer responsible for external affairs was sent to Futa Toro in 1851, with instructions from the Governor to foster the good understanding between Saint Louis and the states in Upper Senegal. The Governor also instructed him to take measures to protect the persons of French merchants in the area.[38] This mission appears to have succeeded as the Almami of Futa, Muhammad, his Chiefs and the people of Lao were reported to have protected French boats trading in their areas of authority. In appreciation of this act of friendship, the director of

35 Georges Hardy, 'Un épisode de l'exploration du Soudan—L'Affaire Duranton, 1828–1838', *Annuaires et Mémoires du C.E.H.S.A.O.F.*, 1917, pp. 413–20.

36 Hardy, 'Un épisode . . .', pp. 420–36.

37 A.S.A.O.F., IG 22, 'Voyage à Timbo de M. Hecquard, 1849–51'.

38 A.S.A.O.F., 13G33/1 (Q-a-34), 'Le Gouverneur à M. le Directeur des Affaires Extérieures', Saint Louis, 1 December 1851.

External Affairs later recommended the Governor to present the almami and his people with '. . . 85 pièces et 3 fusils doubles'.[39]

In the period under discussion a feature of French relations with the Western Sudan was the development of fortified trading posts, notably along the River Senegal. At first these posts were regarded as trading centres or 'factories' where protection and security were guaranteed for the persons and property of the merchants. Posts that were built to satisfy these objectives included Bakel, which was built in 1820, Dagana (1821), Merinaghen (1822), Lampbar (1843) and Senoudebou (1845). The plots of land occupied by these posts were usually rented from the rulers concerned. Rents were paid as regulated by treaty stipulations, which also included the understanding that these posts were mainly to further commercial interests by being used as 'factories' and ports of call and refuelling for French ships.

The growth of posts along the Senegal was in line with the general pattern of French colonial activities in this period when the government discontinued the old policy of centralised colonial administration and gave a free hand to enterprises to develop as they liked. The commercial houses developed what was generally known as the 'système des points d'appui' which took for granted that the establishment of posts along the waterways would guarantee to their owners the control of the trade of the area concerned. There was hardly any effort by the government to control or coordinate their activities and involvement in political commitments was generally ruled out as an objective. The posts generally depended on the goodwill of their neighbours rather than on their military forces to guarantee their commercial interests.[40] This might account for the trouble taken to cultivate and even compensate, as noted above, the friendship of some of the local rulers.

What has been said so far sets out the general pattern; but it by no means tells the whole story. As has been said earlier, the determination of the French to expand their commercial activities, especially by reducing the participation of traditional agents, earned them trouble in some quarters. Sometimes, as in Walo and with the Moors, it was difficult to reconcile the conflicting interests of the indigenous populations and the French. In some of these cases, the French had recourse to force to establish their own predominance. Thus, for instance, in 1832 the French in Senegal fought the Trarza Moors to establish their control over the gum trade. The same situation applied in the relation between the French, the Moors and the Jolof state of Walo in 1835.[41] Military involvement of this nature was

39 A.S.A.O.F., 13G33/1 (Q-e-42), 'Monsieur le Gouverneur', Saint Louis, 25 February 1853.
40 A.S.A.O.F., 13G23/13, 'Notes sur les postes militaires du Sénégal et de sa dépendance', 19 July 1867; Georges Hardy, *Histoire de la Colonisation Française*, Paris, 1928, pp. 169–71.
41 Hardy, *Histoire de la Colonisation*, p. 170; Monteil, *'Chronique du Walo'*, p. 448.

often protracted and indecisive in its results, since it also implied reprisals usually taken against the French by the offended parties and was necessarily disruptive to good commercial relations. The volume of trade was usually reduced and commercial activities became unremunerative. This situation continued until about 1850 when commerce was already in a miserable state with complaints that '. . . Les ennemis de notre expansion, les peuplades indigènes . . . étaient aussi prompts à gêner notre action'.[42] Between 1845 and 1850, commercial activities were said to have been reduced to about half of the former volume.[43]

The dwindling stature of Senegal as a commercial centre was due in part to the absence of any controlling political authority to regulate commercial and political activities. The July Monarchy had left the colonies almost uncontrolled and during this period Senegal was said to have remained an '. . . archipelago of isolated posts where life cannot but be miserable'.[44] With the coming of the Second Republic, however, came the desire to revive the colonies and make them part of a greater France. The revival of Senegal and the reorganisation of its relations with West Sudan came at the same time. In 1850, an Inter-Ministerial Council was set up to examine the problems of the French colonies in Senegal and propose solutions. The commission concluded that the colony was capable of a rich and fertile development provided that the free movement of the gum trade could be re-established, the privileged companies suppressed, commercial expansion linked with the establishment of spheres of influence and lastly, the re-establishment of security be undertaken and an energetic policy adopted towards the neighbouring indigenous populations. In effect, the commission wanted a new policy of commercial and political expansion to be vigorously undertaken by the government itself. The realisation of this programme was the task to which the Government of the Second Republic set itself.

The reorganisation of Senegal began with the appointment of Protet as Governor in 1850. With his appointment, one of the main causes of past stagnation and failure—lack of a directing mind and the consequent instability of the local power—was removed. Protet set about curing what he considered to be the ills of the past. In 1852, he put the colony under the same law, suppressed most of the existing regulations and organised the gum trade on a new basis. He wanted to continue treaty relations with the neighbouring states; but he emphasised that to be effective he needed force to back up his policy. He stated that '. . . une simple politique de traités, sans manifestations de force, n'aboutira comme par le passé, qu'à

42 'The enemies of our expansion, the indigenous people, . . . were also quick in counteracting our activities'.
43 Hardy, *Histoire de la Colonisation*, p. 201.
44 Hardy, *Histoire de la Colonisation*, p. 201.

une pacification toute apparente . . .'[45] and demanded to be furnished with a column of artillery and two war vessels. He also asked to be permitted to march on the Trarza Moors immediately.[46] In this way, Protet launched a policy of aggressive expansion which eventually led to war with the neighbouring states. The construction of new posts serving as strong military stations began as part of this policy. In 1854, Protet sent an expedition to construct a new post at Podor. The construction, under Faidherbe, took a month and the post was used later as a base for waging wars on the neighbouring peoples for the purpose of winning control of the gum trade for the French. Posts of the new type were later constructed at Medine (1855), Matam (1857), Saldé (1859), Aéré (1866), Ndiagne (1866), Klur-Mandoumbe-Khary (1867), Khaoulou (1867) and Talem (1867).[47] Henceforth these posts ceased to be mere trading stations; they were conceived as serving to maintain French authority in territories along the River Senegal.[48]

This aggressive imperial policy necessarily provoked greater hostility towards the French from the indigenous populations than hitherto. In many areas French posts were subjected to attacks. For instance, the French in addition to fighting the Moors, were said to have faced the Tukulor in war. At Dialmath alone, they had been confronted with about two thousand Tukulor forces in 1854.[49] The hostile reaction of the indigenous populations was understandable. From being the protectors of the French, they were now being told to accept French authority. With the spread of hostility against them, the French soon found that they were ill-equipped to execute the policy launched by Protet. The policy failed almost inevitably and the need for a change seems to have been accepted by the French Government who replaced Protet with Louis Faidherbe later in 1854. By the time Faidherbe assumed office, as Governor, French commercial and political relations with the states of the Western Sudan had fallen into chaos. The states had virtually paralysed the commercial activities of the French.

L. FAIDHERBE

The state of war between the French and the neighbouring states did not change with Faidherbe's appointment. It was clear from his instructions that the French Government had not renounced its imperial objective of establishing spheres of influence over the areas where it traded. Faidherbe

45 '. . . a policy based simply on treaties without any show of force will result, as in the past, only in a superficial pacification . . .'. Hardy, *Histoire de la Colonisation,* p. 201.

46 Hardy, *Histoire de la Colonisation*, p. 202.

47 A.S.A.O.F., 13G23/13, 'Notes sur les postes militaires du Sénégal et de sa dépendance', 19 July 1867.

48 A.S.A.O.F., 13G33/4, 'Considérations générales sur l'état des postes du fleuve au mois d'août 1875'.

49 Georges Hardy, *Histoire de la Colonisation*, pp. 202–3.

was told that his appointment was not meant to mark the beginning or continuation of a bellicose era; but rather that his main mission was to further the peaceful expansion of commerce through the exploration of new avenues. This was a realisable task which was, however, made impossible by the political task which Faidherbe was asked to perform. He was to impress upon the people among whom the French traded that the French were always ready to use force to back up their claim to political authority in their countries, especially in the areas bordering on the Senegal from Saint Louis to the Cataracts of Felou.[50] In effect, the French were claiming political authority over the territories adjoining the River Senegal.

Faidherbe's attempt to realise these objectives necessarily involved a continuation of the state of war left over by Protet's administration. Faidherbe, who had commanded the battalion of troops stationed at Saint Louis under Protet, continued the war against the Moors, and some of the other riverain peoples. For their own part, the indigenous populations continued to base their relations with the French on the earlier treaty agreements and they fought to preserve their independence of the foreigners. They continued to force the French to pay tributes over the land used for building the posts.

Apparently, Faidherbe then sought to deal separately with the states in the hope that he would thereby be able to defeat the Moors. With this aim he entered into negotiations with the Almami of Futa Toro over the payment of tributes, while continuing the war against the other states. He was shocked to discover that there was cooperation between the states in their efforts to preserve their independence and rights against the French. To the request that Futa Toro should deal with the French separately, Faidherbe was told that Futa chiefs were having consultations with the chiefs of the Trarza and Brackna Moors, the Almami of Futa Jallon and others. Confronted with a united front in this way, Faidherbe concluded despondently that '. . . cela semblerait annoncer un commencement de concert, de ligue contre nous'.[51] The power of the French was based on naval boats operating during the high waters of the Senegal. They could, therefore, not afford campaigns against hostile neighbours who were united during the dry season. This situation had hindered the progress of the French and by the end of 1854, they were in a stalemate in their relations with the Senegambian states. The situation was also being complicated by the activities of the Tukulor under 'Umar's leadership, who were creating a new empire for themselves in the area.

The beginnings of 'Umar's empire building campaigns came at a time

50 A.N.S.O.M., in Sénégal I, 41C, 'Le Ministre a M. le Chef de Bataillon du Génie, Faidherbe à Sénégal', Paris, 9 November 1854.

51 A.N.S.O.M., in Sénégal I, 41b, 'L. Faidherbe à M. le Ministre de la Marine et des Colonies', N. 551, Saint Louis, 26 February 1854.

when the commercial activities of the French colony of Senegal were almost at a standstill. The colony was beset with problems. It was regarded as a *simple comptoir* and was seen, a little despondently, as being '... toujours comme en 1817, ce maigre archipel de postes isolés, où la vie ne pouvait être qu'inquiète et misérable'.[52] Part of the problem with the colony had been that it had to deal with several rulers, most of whom were hostile. This was the period before the reorganisation carried out under Protet as explained above. The French could see at least two good reasons for welcoming the emergence of 'Umar and his followers as a strong political force, capable of imposing a single authority over the existing states. 'Umar's commercial interests were complementary to theirs, since a good deal of his transactions were with the French in whose goods he traded. They could have seen him as a likely agent for the expansion of French commerce in the area under his control, especially if the existing friendly trends continued. Secondly, on the political plane, it would be easier to deal with a single political power than with many. Apparently, 'Umar was astute enough to appreciate these points. He promised to liberalise trading activities as soon as he established his rule and called for French help in building a fort to be used against the existing states. The French were naturally favourably disposed to 'Umar's plans. By this tactful handling of the situation, he was also free to deal directly with the states against which his campaigns were soon to be launched.

It was after this initial understanding between the French and 'Umar that the former launched their programme under Governor Protet for the realisation of their ambition to expand their commerce and spread their political influence in the Western Sudan. But, as has been said earlier, the rulers of the Senegambian states easily discovered the imperialist content of the French activities under the governorship of Protet and Faidherbe. Their reaction against this resulted in hostilities and war against the French. It was an attempt to safeguard their states against French imperial ambitions that resulted in the alliance reported by Faidherbe. Their awareness of the danger from external forces also meant that the ruling élites in these states had become sensitive, and in some cases hostile, to 'Umar and his following and this had obliged the latter to withdraw to Dinguiray where the campaigns against the Senegambian states were begun at the end of 1852.

THE CLASH OF RIVAL IMPERIALISMS

Broadly, by the early 1850s the French and the Tukulor led by 'Umar had emerged, independently, as forces with identical imperial ambitions. Some years earlier, their ambitions had seemed complementary, at least to the French. However, their competitive nature was soon to emerge. Strug-

52 '... always, as in 1817, a barren archipelago of isolated posts where life cannot but be disturbing and miserable.' Georges Hardy, *Histoire de la Colonisation*, p. 201.

gling against both forces were the indigenous states who were anxious to protect their own independence. Of the two enemies, most of the ruling élites found the French less dangerous, partly because it had become clear that they were not really equipped to fight a general war, especially when confronted by an alliance of the states. ʽUmar and his Tukulor army appeared to present the more immediate threat. This was due to a number of reasons. ʽUmar and his following belonged by birth to various Senegambian states. He himself, as a *torodo*, belonged to the ruling class in Futa Toro. He could be seen, therefore, as an indigenous revolutionary seeking to overthrow the existing socio-political system. Naturally, his movement had adherents from among various interest groups who for different reasons were against the existing power structures. These groups could easily be accommodated in ʽUmar's all-embracing programme, which involved changing the existing religious, as well as the socio-political, system. Whereas the ruling groups could rally the support of all sections of their communities against the French as foreigners, they could not do the same with ʽUmar's movement. It was for many an internal revolution creating a fruitful milieu for competition between many social and political interest groups. It therefore divided the states, and left the ruling groups and their supporters fighting to preserve the integrity of the states as they were, but only as one of several competing parties. In addition to the above, the early successes achieved by ʽUmar in the military campaigns against some of the states convinced the other rulers that he could easily destroy them unless they had better and greater forces at their disposal.

ʽUmar also seems to have been encouraged by his early successes. By 1854 he had demanded payment of 'tributes' from the French as a condition for their continuing to trade in his emerging empire. We may recall that it was the desire of the French not only to discontinue paying these tributes, but instead to establish their political influence over areas where they traded, that had resulted in wars between the French and the Senegambian states. This move by ʽUmar, which amounted to an assertion of political authority over them, annoyed the French and made imminent a clash between the two. Thus, from the beginnings of the Tukulor empire to its fall, we can describe Franco-Tukulor relations as a clash of imperialisms.

The subject peoples played a vital role in determining the course of the history of the Western Sudan. They trusted neither the French nor the Tukulor, against both of whom they continued to struggle to retain their independence. Since they were not militarily as strong as either of the two competing imperialist forces, their tactics consisted mainly in exploiting Franco-Tukulor conflicts to satisfy their own interests. Their ruling élites felt the greatest detestation for the Tukulor under ʽUmar and believed that theirs was the party against whom concerted effort was needed. The French forces were fully committed in the struggle against ʽUmar and they were for this reason seen as the lesser evil by the ruling groups. For their success against their rival, the French even needed the sympathy of the local people. They, therefore, posed as champions of the people's independence against the invading Tukulor and this tactic worked, not because they were trusted but because the people too needed allies against 'Umar. This situation explains some of the alliances contracted between the French and the states in the Senegambia between 1854 and 1860. In certain cases, however, some of the ruling élites joined forces with 'Umar against the French. These were usually in areas where the ruling groups had had direct experience of the French and had concluded that it was better to be part of a new political system, represented by 'Umar, than to admit French domination. The outcome of the network of alliances was the polarisation of the conflicting forces into two main blocks—the French and the Tukulor—each of which represented a welter of interests.

In spite of the heavy odds created by the French and the ruling élites who were opposed to him, 'Umar continued the military campaigns for the establishment of his empire. Between 1852 and 1854, he had successfully established control over parts of the Senegambia, notably Diallonkadougou, Bambuk, Bondu and Khasso. But both the rulers who were displaced by these exercises and Faidherbe, who in December 1854 assumed office as governor in Senegal, worked hard to reverse the situation. Faidherbe was convinced that 'Umar was more dangerous to French interests than any of the enemies they had been fighting hitherto. He therefore tried to turn these erstwhile enemies into allies against 'Umar. Those who were being overthrown by 'Umar found this a convenient arrangement and entered into negotiations with Faidherbe and his agents. Thus, by the summer of 1855, Futa Toro chiefs, hitherto hostile to the French, concluded a commercial agreement, with a political understanding, with the French to ward off 'Umar and his men.[53] Similar agreements were reached with Sambala, King of Khasso and his chiefs who, in return for an assurance of French protection against 'Umar and his men, allowed the French to construct a fort at Medine. Faidherbe entered into identical arrangements in several other states. He then took immediate military steps to contain 'Umar's activities. Among these was the construction of a garrisoned fort at Medine in September 1855. Other posts, built from 1855 onwards, had the same objectives as the Medine forts. Faidherbe also created a battalion of African sharpshooters, the *Tirailleurs Sénégalais* in 1857.[54]

While the French and their allies were busy organising themselves against the Tukulor, 'Umar was leading further campaigns in the neighbouring state of Ka'arta. But all along, his allies, among whom were chiefs in the various states, tried to consolidate their positions in the places already con-

53 L. Faidherbe, *Le Sénégal*, Paris, 1889, pp. 165–7.

54 Commandant Chailley, *Les Grandes missions françaises en Afrique occidentale.*, I.F.A.N., Dakar, 1953, p. 45.

quered. Thus it could be said that both sides spent the period 1855–6 organising their forces in the Senegambia in anticipation of further clashes. By early 1857 'Umar had completed the conquest of Ka'arta and, after putting his lieutenants in charge of the administration, returned to the Senegambia with the objective of securing his hold over the area by wiping out his rivals—the French and their collaborators among the indigenous populations. The struggle over the Senegambia began with the siege of Medine (April to July 1857) and lasted until 1860 when the destruction of Guémou took place.[55] After Guémou, negotiations between 'Umar and the French took place with a view to concluding a definitive peace treaty. Although no treaty emerged, an understanding resulting from the desire of the two sides for peace was reached. Faidherbe even promised to send envoys to 'Umar to negotiate further. These were not, however, sent until 1863.

Briefly, the events of 1857 to 1860 marked the failure of 'Umar to oust the French from the Senegambia. This fact was significant as it involved the frustration of 'Umar's ambition to incorporate the area in his empire. After 1860 although he succeeded in maintaining Tukulor control over a few districts like Logo and Natiaga (in Khasso) and Dinguiray, a *de facto* situation did arise whereby the idea of the river Senegal being the boundary line between the French in the Senegambia and the Tukulor was more or less accepted. For all practical purposes, the French had established their supremacy over the Senegambia and later on, after consolidating their position, they were to use the area as a springboard for the attack on, and conquest of, the Tukulor Empire, and the establishment of the colony that was known as the French Sudan. The 1860 *entente* was dictated mainly by the need to guarantee non-intervention between the two imperialists. It gave both sides an opportunity to consolidate and possibly expand their various domains.

THE ESTABLISHMENT OF SPHERES OF INFLUENCE

Before 'Umar died in February 1864 in a campaign in Masina, he and his men had succeeded in conquering the Bambara Segu kingdom and the Fulani empire of Masina.[56] 'Umar's death led to a struggle for the leadership of the empire. The decentralised form of administration that emerged was partly due to the effects of this struggle. In spite of internal problems, however, the Tukulor managed to hold the territory comprising the former states of Ka'arta, Segu, Masina and the greater part of Beledugu under their rule until 1893. But because the administration of the empire was not centralised, the relations of the empire with the outside world were

55 Ọlọruntimẹhin, *The Segu Tukulor Empire*, ch. 3, for a discussion of the events in this area.
56 *Ibid.*, ch. 5, for a discussion of the situation in Segu and Masina, and Umar's campaigns. campaigns.

determined largely by the responses of the individual Tukulor rulers affected.[57]

While the Tukulor were expanding eastwards, the French were busy making good their hold over the Senegambia. The first problem was what to do with their allies who had helped to rout 'Umar mainly because they thought they would thereby regain the independence of their states. Many of these were opposed to any domination, either Tukulor or French. The other problem concerned what to do with former allies of 'Umar. Many of these had migrated with 'Umar and were participating actively in the campaigns for the extension of the empire east of the Senegal. It was, therefore, easier for the French to deal with those left behind. They represented the minorities in their various states and, by the defeat of 'Umar, had been subjugated. Among those conquered during the campaigns in the Senegambia was Walo, which was occupied and annexed by Faidherbe in 1856.[58] The pockets of 'Umar's supporters in the other states could easily be left to the care of the dominant pro-French rulers. As for these rulers themselves, they easily accepted French protection and were left in their chiefdoms and principalities. They acted, as in the case of Sambala (Khasso), under the direction of the French administration because this was the main guarantee of their existence against 'Umar. Thus, in trying to gain redress for his grievances against 'Umar's Tukulor empire, Sambala constantly referred the measures he proposed to take to the French for approval. In this way it was possible for them to prevent Sambala engaging in war with the Tukulor in 1860–1.[59]

By approaching the problem in this way the French were able to maintain their control over the Senegambian states at the same time as they maintained a semblance of peace with the emerging Tukulor Empire. The next line of action was to leave 'Umar alone to continue his wars of conquest and devote their energy to consolidating their positions in the Senegambia. This policy was announced by Governor Jauréguiberry who in 1861 analysed the task of his administration as being to conserve '. . . avec vigeur, l'autorité et la préponderance si glorieusement acquise au drapeau de la France sous le Gouvernement de M. le Colonel Faidherbe',[60] and having done this, to stimulate agricultural production in the areas under their control as a means of boosting commerce.[61] Politically, the new policy was conservative as far as relations with the emerging Tukulor empire were concerned. It was simply to maintain the status quo as defined by the

57 *Ibid.*, ch. 6, for an analysis of the administration and politics of the empire.
58 Monteil, 'Chronique du Walo', p. 448.
59 A.S.A.O.F., in 15G64, 'De Diougou-Sambala à Monsieur le Gouverneur Faidherbe', 6 February 1861.
60 '. . . with vigour, the authority and the preponderance so gloriously acquired for the French flag under the governorship of Colonel Faidherbe.'
61 A.S.A.O.F., 13G23/8, 'Avis Officiel', Le Gouverneur Jauréguiberry, Saint Louis, 14 December 1861.

1860 *entente*. But Faidherbe had begun a number of campaigns in the Senegambia and these needed to be continued. These campaigns were aimed at establishing French political authority, primarily in areas not yet involved in the Tukulor movement. With this object, he had subjugated N'Diambour (1856–8) and Casamance (1860). This programme was continued by Jauréguiberry who extended the campaigns to Futa Toro in 1863. Faidherbe himself returned to Senegal from his Algerian mission in 1863 and continued the expansion programme in this area. The French later on succeeded in establishing their political control over the Sérères in Saloum (1865) and Kayor (1871).[62] Having established their political authority in the areas adjacent to Saint Louis, it was easy for the French to control the exploitation of their commercial activities. The next problem which Faidherbe tackled on his return from Algeria was how to extend French commercial and political influence in the emergent Tukulor empire.

By the *entente* of 1860 the Tukulor and the French had managed to establish a *modus vivendi* which accommodated their competing interests. The 'live and let live' policy which was thus adopted continued to be the basis of their relations for the next fifteen years. The two sides accepted the situation for different reasons. The French needed all their available resources for their programme in the Senegambia, an area which appeared to be a natural adjunct to their establishments on the Senegal. They rarely had the means of engaging in greater expansionist activities and therefore found it necessary and useful to maintain an arrangement that delimited, at least for some time, the respective spheres of influence of their rivals and themselves. As for the Tukulor, after the setback in the Senegambia (1857–60), they found it necessary to concentrate on expanding their territories east of the Senegal. Hence, there was hardly any time for consolidation, much less for engaging in further battles with the French, before their leader, 'Umar, died. The Tukulor leaders were later engulfed in their own internal problems, including a succession struggle that remained largely unresolved in spite of the many civil wars that they fought. In their continuous struggles, the Tukulor leaders all needed the arms and ammunition which the French could easily supply. Apart from this, they were eager to avoid external complications. In the prevailing circumstances, both the French and the Tukulor found it expedient to maintain relations to avoid hostility and safeguard their interests.

After the *entente* of 1860 further channels for negotiations were opened by the offer of peace made by Tukulor emirs in Ka'arta and the acceptance of the same, with a promise to send a mission to Segu, by the French officials.[63] Although no emissaries were sent at this time, the friendly trend continued and was fostered by the way French affairs were conducted under

62 Hardy, *Histoire de la Colonisation*, p. 207.
63 A.S.A.O.F., in 15G64, 'Mīn amir Musa ilā Kumadau Madina', marked 'Parvenu à Médine le 5 7bre 1860'.

Faidherbe and Jauréguiberry, both of whom limited their activities to the Senegambia. On his return from Algeria to Senegal in 1863, however, Faidherbe initiated a new move to exploit the existing friendly relations with the Tukulor.

Obviously, with his Algerian experience behind him, Faidherbe was more convinced than before that the Tukulor, representing the forces of Islam, were the greatest obstacle to French interests both in the Western Sudan and in the Sahara. He considered several possible ways of dealing with the Tukulor, including the establishment of a French military presence in the Sudan and the Sahara, a measure which he concluded was inexpedient as long as ʿUmar lived. He believed that the latter, much travelled and sophisticated, was realistic enough not to seek to engage in further military acts against the French. He believed that unless a less experienced and intelligent successor came after ʿUmar, it would not be necessary for the French to employ force to prosecute their interests in the Western Sudan, which he regarded as a natural, vast hinterland which was to be managed for the expansion of French commercial and political influence from the Senegal.[64] To lay the foundation for the eventual exploitation of the resources of the Western Sudan by the French, Faidherbe sought to cultivate the friendship of the Tukulor rulers by sending a mission to ʿUmar in 1863.

MISSION OF MAGE AND QUINTIN

In a letter dated 7 August 1863 Faidherbe appointed Captain Mage to go on a mission to Segu. During this mission, Mage was to undertake a study of the nature of the available raw materials and the density of population in the various parts of the Tukulor empire. The mission was to explore the possibility of establishing a line of communication to join the French establishments in the Senegambia to the rest of the Western Sudan, including the Tukulor empire, especially Bamako, which he thought marked the nearest navigable point on the Niger before the Bussa falls. This exercise was to be a groundwork for the eventual creation of a line of garrisoned posts between Medine and Bamako, as well as at any other convenient points in the territories adjacent to the Niger. Faidherbe at this time foresaw that these posts would be used subsequently to suppress Moroccan trade relations with the Western Sudan. The Moroccan trade, Faidherbe claimed, was beneficial only to the British and eliminating it would bring a double advantage to the French—the elimination of British competition in this way would assure the dominance of French political and commercial influence from two sectors, Algeria and Senegal. Faidherbe appreciated that the achievement of this grandiose objective depended on the cooperation of ʿUmar. He told Mage that: 'The support of an all-powerful chief of a large empire in the Central Sudan, such as al-hajj

64 Faidherbe, *L'Avenir du Sahara*, pp. 17–21.

'Umar is today, is needed for the realisation of this project . . . it is therefore in the capacity of an ambassador to al-hajj 'Umar that I am sending you.'[65] Mage and Quintin set out for Segu in October 1863, and arrived there on 28 February 1864.

Mage reached Segu at a time when 'Umar and his men were still engaged in the Masina campaigns. It was, therefore, 'Umar's eldest son, Ahmadu, who was holding the fort at Segu, who received the mission. News of 'Umar's death soon spread, however, and the succession struggle ensued. In the confusion Faidherbe sought to increase French advantages by cleverly claiming that certain agreements guaranteeing territorial possession to the French along the Niger existed with the Tukulor empire under 'Umar, and asked Ahmadu to honour these. In effect, Faidherbe wanted to exploit the confusion to achieve the objectives for which Mage's mission was designed. However, Ahmadu not only rejected such moves[66] but kept Mage and Quintin in Segu until early 1866. In the meantime, while awaiting the return of Mage's mission, Faidherbe continued to extend French influence in the Senegambia and even opened direct relations with some Tukulor emirs in the Ka'arta sector.

By May 1864 the commandant of the French post at Bakel was already on a mission to conclude commercial agreements with local rulers, including the chief of Koundian, the Tukulor enclave in Bambuk. Efforts were also made to sound the Tukulor emirs in Ka'arta on the possibility of entering into relations with the French, independently of Segu. The commandant at Matam engaged in similar activities under the control of Bakel.[67] Earlier in 1864 a convention had been signed with the ruler of Natiaga giving the French permission to establish new fortified posts at Bafoulabé and Fakhotala.[68] Altiné-Sega, the ruler, had been defeated by 'Umar in 1857 but was able to regain his position through alliance with the French. The convention, guaranteeing French military presence in his territory, must have been seen by him as a way of preserving his country against the Tukulor menace.

More significant was the mission which Faidherbe sent to Ka'arta to negotiate with the Tukulor emirs in December 1864. The mission comprised M. Perrand, commandant at Medine, and M. Béliard. The two men were to go to Koniakari, headquarters of the emirate of Diomboko, and there to persuade Tukulor rulers to enter into commercial relations with the French. The mission was received by Tierno Musa, the emir, and his

65 A.S.A.O.F., 3B75/2, 'Année 1863—Mr Mage, Lieutenant de Vaisseau, Mission pour le haut Sénégal auprès d'El-hadj Omar', L. Faidherbe, Saint Louis, 7 August 1863; see also Mage, *Voyage*, pp. 12–17, 167.

66 Mage, *Voyage*, pp. 325–6.

67 A.S.A.O.F., IG32/3(4), 'Le Commandant à M. le Gouverneur', Bakel, 19 May 1864.

68 A.S.A.O.F., 15G108/12, 'Convention avec Altiné-Sega, roi de Natiaga', Médine 15 January 1864.

military chief, San Madi. The Tukulor leaders agreed to have commercial relations with the French but they raised a political aspect to the question. They regarded French activities in Khasso as unfriendly. Specifically, they accused the French of using their cannon to protect Sambala at Medine, thus enabling the latter to continue undermining the Tukulor regime by encouraging rebellion in, and receiving refugees from, Diomboko. Khasso, particularly Medine, occupied a strategic position in the Franco-Tukulor struggle for the control of the Senegal Valley. This was illustrated by the bloody siege of 1857, during which the Tukulor were routed and forced to confine their hold to the Logo district. Since then, Medine had remained a sore point with the Tukulor who were convinced that without French support Sambala could easily be toppled. But the French could not withdraw support for Sambala because he was really the agent through whom French imperialism had been established and assured in that part of the Senegambia. The French envoys knew the Tukulor claim would not be entertained but they still wanted to achieve their goal without arousing antagonism over Khasso. They therefore tried to separate the political question by promising to raise it with the governor who alone was capable of giving an answer.[69] This episode illustrates that the competition between the two imperialisms continued, even though the two were now adopting subtle means because of the necessity to maintain the delicate equilibrium that guaranteed the existence of both.

In undertaking the activities described above, Faidherbe must have doubted the possibility of realising, through Ahmadu, the objectives he had hoped to achieve when the Tukulor empire was still under 'Umar's control. The alternative activities were aimed at achieving the same goals, if only on a limited scale. After much delay Mage's mission left Segu early in 1866. Before leaving, a treaty agreement, of doubtful validity, had been concluded with Ahmadu on 26 February 1866.[70] The Tukulor made this treaty the basis of their relations for the next decade; but from Mage's report and later events, it is clear that it was a disappointment to the French.

Conscious of the need to prevent the empire from being consumed by French imperialism, Ahmadu had rejected the vital points in the proposals sent by Faidherbe. He had rejected the French request to establish posts along the Niger and had refused permission to the French to live in any part of the territories ruled by him. After an assessment of the situation in the empire, and of the attitude of Ahmadu to the French, Mage argued against the establishment of a line of posts as envisaged by Faidherbe. He maintained that the wars which had been raging since the establishment

69 A.S.A.O.F., In 15G108/13, 'A Monsieur le Commandant Supérieur de l'arrondissement de Bakel', Médine, 22 January 1865.
70 Mage, *Voyage*, p. 589.

of the empire had ruined the area economically. In the circumstances, it would be unprofitable and costly to establish posts. Moreover, the construction of the posts would involve constant wars with the Tukulor since Ahmadu did not consent to it. He concluded by saying that the posts were not a *sine qua non* for the expansion of French commerce. The peoples of the Western Sudan, he said, were in need of French commercial goods and they would on their own come to deal with the French in the established posts.[71] In other words, Mage recommended that French relations with the Tukulor empire should be carried out within the existing framework.

Faidherbe left Senegal in 1866 and the new Governor, Pinet-Laprade, could not accept the treaty brought by Mage for ratification by the home government. In spite of the failure of Mage's mission, however, Franco-Tukulor relations were not unduly strained. The two sides still found it necessary to maintain the equilibrium upon which their relations hung. Ahmadu was still faced with a crisis within his empire and he still needed French arms to use against his rivals. This explains why he had sent a letter through Mage to the Governor asking to be supplied with two cannon, which apparently Mage had promised him to facilitate his negotiations at Segu. He did not get what he wanted from the French partly because the latter were not willing to strengthen him, a rival, by supplying him with heavy weapons. Ahmadu needed these cannon mainly to give him an advantage in the succession struggle. It was, however, made possible for Ahmadu, as well as his rivals, to buy other arms from the French. It was in his interest therefore not to be openly hostile to the French. He continually assured the French of his friendship; but also continued to press for the cannon he needed.[72]

After Mage's mission the French continued to maintain the status quo because the alternative would have been a war for which they were ill-equipped. The Minister for Colonies made it clear to the Governor in Senegal that the government found it absolutely impossible to increase the colonial force, military or naval, in Senegal. He was consequently requested not to engage in any operations for which the existing forces were inadequate. In the light of the ministerial instruction the Governor made a policy statement to his staff emphasising that the French would not engage in any wars, unless absolutely necessary, and that all existing treaties between the French and the indigenous peoples would be respected. He reserved the authority to himself to decide what, if any, expeditions would be embarked upon in the Senegal and in the rest of the Western

71 A.S.A.O.F., In IG32, vol. 1866, 'Relation du voyage par Mage—Notes diverses', 21 June 1866; A.N.S.O.M., In Sénégal III, 9, Mage I, 'Note sur le voyage d'explorations de MM. Mage et Quintin dans l'Afrique occidentale de 1863 à 1866', Paris, 21 July 1866.
72 A.S.A.O.F., IG32/5, 'Rapports avec Ahmadou et son vassal Mohammed Bassirou de la région de Nioro, 1866–1869'.

Sudan.[73] Since the French would not fight, Ahmadu's attitude had to be tolerated.

The failure of Mage's mission, rather than leading to bitterness against the Tukulor, appears to have encouraged another trend in the development of French imperial ambitions. This was the Faidherbe policy of consolidating the French hold on the neighbouring states of the Senegambia—an area to be effectively blocked to the Tukulor, politically their most dangerous neighbours. This policy also aimed at establishing a French monopoly against their British rivals. From 1866 to 1876 the French engaged in protracted negotiations with the British over Gambia. They wanted Britain to cede Gambia in exchange for the French settlements of Grand Bassam, Gabon and Assinie. The negotiations were complicated by the rivalry between the two powers in the riverain areas of Rio Nunez, Rio Pongo and Mellacorée. The British, who had controlled the French settlements in Senegal with their Gambian settlement, knew the economic and political opportunities promised by an unfettered control of the Senegambia. They could not afford to give their French rivals the advantages which the surrender of Gambia would concede and consequently the exchange plan was rejected in 1876.[74]

Towards the end of the negotiations with the British, the French made an ill-fated attempt to achieve similar objectives through the Tukulor. After the false rumours that spread to Saint Louis early in 1874 to the effect that Ahmadu, the Tukulor ruler, was dead Governor Valière tried to reopen direct negotiations with the head of the Tukulor empire. This was the first attempt since the return of Mage's mission. The opportunity came with the arrival at Saint Louis of an emissary, Tambo, from the Tukulor ruler. Tambo told Valière that he had been sent to reopen friendly relations with the French. To this end, it was claimed, Tambo was empowered to negotiate a treaty of alliance and commerce similar to that already concluded through Mage but which, as has been noted, was not followed up by Pinet-Laprade, then Governor of Senegal. In the renewed negotiations Ahmadu again asked for the supply by the French of twelve cannon. It is useful to note that Ahmadu was at this time involved in a civil war in Ka'arta where he was trying to establish his supremacy over his half brothers—Abibu and Moktar. In fighting this war, both Ahmadu and his rivals had access to French arms and ammunition. Ahmadu's desire to buy cannon emphasised his eagerness to possess weapons which were superior to those of his rivals.

Valière reciprocated by sending a draft treaty in which he tried to

73 A.S.A.O.F., 13G33/3 (No. 77 Circulaire-T-b-23), Gouvernement du Sénégal, Saint Louis, 10 June 1867.

74 Archives du Ministère des Affaires Etrangères, Paris (M.A.E.), in Mémoires et Documents: Afrique 48, 'S. Excellence Le Marquis d'Harcourt à Londres', Versailles, 27 March 1876; also a cutting from *The Times* marked '*Times* de mars 76'.

resuscitate French ambitions as embodied in Faidherbe's policy for Mage's unsuccessful mission. But nothing came of this new attempt at treaty making because Tambo died before he reached Segu,[75] and with him perished all the papers embodying Valière's proposals.

BRIÈRE DE L'ISLE AND FRENCH EXPANSION

Hitherto, French activities in Senegal had conformed to the general French colonial policy as dictated by the aftermath of their defeat in the Franco-Prussian war of 1870–1. In the period following the war France was politically unstable and the government was mainly concerned with repairing the country's damaged image and re-establishing its position in Europe. On the colonial front, therefore, expansion projects were generally abandoned, and efforts were concentrated on preserving what was already held. The *prudence passive* which characterised this policy was to be cast away and replaced by a new dynamism in the Western Sudan in 1876, when Colonel Brière de l'Isle replaced Valière as the Governor of Senegal. Brière de l'Isle, actively devoted to the cause of French imperialism, introduced greater force into the task of consolidating French position in the Senegambia. But more important was the fact that he revived Faidherbe's political and commercial programme in relation to the Tukulor empire and the rest of the Western Sudan. He tried to overcome the anti-military attitude of the local commercial classes by arguing that it was in fact in the interest of rapid commercial expansion to employ force against anti-French groups in the Western Sudan. The wars for the expansion of French spheres of influence and, later, conquest of the Western Sudan could be said to have been made inevitable by Brière de l'Isle's conduct of French policy.

In his effort to assure the complete dominance of the French in the Senegambia, Brière de l'Isle launched military campaigns against the few remaining Tukulor footholds in the area. In this process, he intervened militarily on the side of Sambala, enemy of the Tukulor and protégé of the French, to crush Logo during the crisis between the two sectors of Khasso between 1876 and 1878. In this way he secured French control over the whole of Khasso. By the destruction of Sabousire (headquarters of Logo) in 1878 he deliberately provoked a new era of armed conflict between the French and the Tukulor.

Brière de l'Isle also tried to establish the French position more firmly in Jolof and Kayor. Here, a resurgence of religious proselytism and military expansion had led to the establishment of Tukulor political rule between 1870 and 1875. During this period, Ahmadu Seku (Shaikh Ahmad?), after conquering the Buurba Jolof, proclaimed himself Almami and forced the *tiedo* (the warrior class) to accept Islam at the same time as he put his own

75 Y. Saint-Martin, 'Les relations diplomatiques entre la France et l'Empire toucouleur de 1860 à 1887', *B.I.F.A.N.*, xxvii, Sér. B., 1–2, 1965, pp. 192–6.

brother, Bara, at the head of his *talibés*. The French reaction was to help Ali Bouri N'Diaye to overcome Ahmadu Seku and become the Buurba Jolof. At the same time the French sustained the Damel of Kayor, Lat Dior N'Goné Latir Diop, by killing Ahmadu Seku and routing his forces during his expansionist bid in Kayor. Thereafter both Ali Bouri, the Buurba Jolof and Lat Dior of Kayor were controlled, indirectly, by the French. Brière de l'Isle, however, tried to turn both Jolof and Kayor into direct French protectorates[76]—a move that was resented by the ruling elites in the two states and eventually led to war with the French.

Brière de l'Isle's effort to establish direct French control over Kayor and the neighbouring states came with a railway project which he conceived in 1877. By 16 September 1877, he told the Damel of Kayor, Lat Dior N'Goné Latir Diop, that he intended to construct a railway line from Dakar to Saint Louis through Kayor territories. On 23 January 1878 he put his plan forward to the Minister for Colonies, Admiral Pothuau. He tried to convince the latter that the projected railway line was essential to tap the increasing agricultural products of the Senegambia. He cited the enormous harvest of groundnuts which was traded in the previous season as evidence of the increasing productivity of the area. The railway line, he argued, would improve security for commercial activities in the area and would serve 'the cause of the metropolitan dominance and of the development of its maritime commerce'.[77] The Ministry was convinced by Brière de l'Isle's case and under the new Minister responsible for Colonies, Admiral Jauréguiberry, himself a former Governor of Senegal, he was asked, in July 1879, to put forward a military and logistic plan aimed at linking together the navigable waters of the Senegal and the Niger.

This involved more than Brière de l'Isle's first proposal but it was no doubt inspired by it. Jauréguiberry believed that if linked together by a railway line, the two waterways could provide the necessary communication links with Timbuktu. From economic and political viewpoints, Jauréguiberry's new idea would provide double the advantages argued for Brière de l'Isle's original plan.[78] A good imperialist, Brière de l'Isle loved his assignment and even proposed a new project in addition, namely, the construction of a telegraph line across Futa Toro (Senegalese Futa) from Saldé via Matam to Bakel on the Senegal. The French parliament later voted money for the building of the railway to link the Senegal with the Niger, and thus began a protracted, arduous and costly task which culminated in wars of conquest in the Western Sudan.

While preparations for the building of the railway line were afoot the indigenous rulers fully recognised the imperial ideas which gave birth to

76 Monteil, 'Le Dyolof et Al-Bouri Ndiaye', pp. 615, 626.
77 G. Ganier, 'Lat Dyor et le Chemin de fer de l'arachide, 1876–1886', *B.I.F.A.N.*, xxvii, Sér. B., 1–2, 1965, p. 226.
78 *Ibid.*, pp. 223–6.

the project. Consequently, they struggled to protect their states against French imperialism. Many of them had fought against Tukulor domination and they had no intention of surrendering to the French. As far as the Damel of Kayor was concerned, the proposal to build a railway across his country was correctly seen as an act that would lead, inevitably, to French intervention in the affairs of the country and an undermining of his own authority and the independence of the state. But the Damel, Lat Dior N'Goné Latir Dyop, could not resist the French for long. He owed his position to them since it was French forces which in 1874–5 helped him to repulse the Tukulor invader, Ahmadu Seku. He succumbed to French propaganda and, on 30 September 1879, signed the convention permitting the French to build the railway line across Kayor. But although he succumbed in this way his people did not. They believed that the Damel had put his personal interests above the common good in signing the convention. The people's discontent resulted in a general rising of the principal chiefs, led by the Damel's nephew, Samba Laobé, in October 1879. The Damel was threatened with deposition in favour of his seventeen-year-old nephew. With considerable diplomacy, however, Commandant Boilève, sent by Brière de l'Isle, was able to mediate successfully to stave off trouble in January 1880.[79] But the evil day was merely postponed.

Brière de l'Isle's 1879 proposal to construct a telegraph line across Futa Toro provoked hostility similar to that in Kayor. The Almami, Abdul Bubakar, made it clear that he '. . . saw in this project only French intervention in Futa affairs'.[80] The Almami immediately tried to rally round the various sections of Futa for a common action against the French. But the French, employing 'divide and rule' tactics, were able to thwart his efforts for a while. They had had dubious treaty relations with some of them and they were able to claim that Lao, for instance, was a state which had been placed under French protection by the protectorate treaty of 24 October 1877. In this way they were able to weaken the resistance to their penetration in the state.[81]

The campaigns for the building of the Senegal–Niger line began with the sending of Captain J. S. Gallieni on a mission to Segu early in 1880. When this was held up at Nango, after being attacked by the Bambara at Dio, the French appointed another officer, Borgnis Desbordes, to take charge of the project. The assignment was seen as a way of preparing for the eventual conquest of the Senegambia and the Tukulor empire. It was also, as always, part of French ambition to keep the British out of the area. To this end, Gallieni was to negotiate a treaty with Ahmadu, such as would give the French control over navigation on the Niger waterway and,

79 *Ibid.*, pp. 228–9.
80 *Ibid.*, p. 227.
81 A.N.S.O.M., in Sénégal I, 63a, 'Brière de l'Isle, Gouverneur du Sénégal, à Monsieur le Ministre de la Marine et des Colonies', Saint Louis, 23 April 1879.

in addition, give the French exclusive commercial privileges within the Tukulor empire. Although the treaty which Gallieni concluded with Ahmadu in 1881 failed to give the French what they wanted, they were generally successful with their railway project. By 1883, Desbordes' mission reached Bamako, where a garrisoned fort was later built.

French success could be explained with reference to the political situation in the Tukulor empire in this period. Like the Senegambian states, the Tukulor and their Bambara opponents detested French imperialism. To the Tukulor, the French were competitors. The Bambara at first suspected them and were even hostile. But they were soon won over by French diplomacy. They had, since the founding of the Tukulor empire, been engaged in wars of resistance through which they hoped to regain their independence. The French knew this and exploited their hostility to the Tukulor. They took care to persuade the Bambara and other anti-Tukulor groups that the French had the same goal as they—the destruction of the Tukulor and the reinstatement of the dispossessed ruling groups. All that interested the French, it was often stressed, was the liberalisation of commercial activities in the area that would revert to the authority of the Bambara after the defeat of the Tukulor. This approach to the problem was determined by Gallieni who deliberately chose the line Bafoulabe–Kita–Bamako for building French military posts because the French would find in the area '. . . a series of states hostile to the Tukulor . . . whose rulers would counter-balance the power of the Segu rulers . . .'.[82] From this, it is clear that while anti-Tukulor groups thought the French were sincere allies who could be used against the Tukulor to achieve their independence, the French were in fact using them against the Tukulor to achieve their own imperial ambition. The desire to crush the Tukulor was a common objective; but the competing nature of their interests would not be revealed until after the fall of the Tukulor empire.

In the meantime, various anti-Tukulor groups cooperated with the French, and even signed 'treaties' with them.[83] This assured the French of a friendly atmosphere for their operations against the Tukulor, whose territories were penetrated until 1889 when open war seemed inevitable. Ahmadu clearly recognised the subversive nature of French activities but he was unable to do anything against them for a long time because of the internal political situation of his empire. First, he had to contend with resistance wars mounted by his non-Tukulor enemies in active collaboration with the French. But his second problem handicapped him almost more than the first. The Tukulor ruling groups were divided and some were

82 A.S.A.O.F., IG50/20, 'Gallieni, Le Chef de la mission du haut-Niger à M. le gouverneur', Nango, 7 July 1880. See John D. Hargreaves, *The Tokolor*.
83 For a list of treaties concluded by the various missions between 1879 and 1889, see A.S.A.O.F., in 15G1, 'L. Archinard, Traités rélatifs à la Sénégambie—', Kayes, 4 February 1889.

in arms against him. The most serious of the civil wars which marked the intra-Tukulor political struggle was that between Ahmadu and his half brother, Muhammad Muntaga, in Ka'arta in 1884 and 1885. Hence, whereas Ahmadu could fulminate against the French for their hostility towards his empire, he really could not go beyond mere complaining. Not only was it impossible for him to take arms against the French while he was fighting his civil wars but he in fact still needed French arms to use against his rivals. Consequently, he had to negotiate with his imperialist competitors, the French, in order to get arms to settle his internal problems, hoping that afterwards he would be able to tackle the French. This explains why after all the complaints he listed in 1880, he still negotiated with Gallieni, demanding large quantities of arms and ammunition as part of the treaty agreement between him and the French. Of course, the French were unwilling to strengthen their rival, and so they did not follow up the treaty.[84]

Before the French were able to prosecute their plans in the Tukulor empire by exploiting the political situation, they had had to contend with obstacles in the Senegambia. All the states learnt their lesson from the experiences of Kayor and Futa. They realised they were weak if they faced the French individually and they entered into defensive alliances against French imperialism. Throughout 1880 and 1881 there were agitations against the French in Kayor, Futa Toro and Jolof. In this area, the situation was already getting out of control for the Governor, Rear-Admiral de Lanneau, before he died during the epidemic of yellow fever which ravaged Saint Louis in July 1881.[85] The new Governor, Colonel Canard, assumed office in October 1881, and found to his dismay that Kayor was already negotiating alliances with the neighbouring states of Walo and Mauritania in addition to strengthening the bonds between it and Futa Toro.[86] The Damel of Kayor then showed open hostility by banning the transportation of agricultural products from Kayor to the French in Saint Louis. Similar anti-French activities, including pillaging, continued throughout 1881 and by early 1882, they were already paralysing French activities in the Senegambia. The chief participants in the anti-French alliance—Ali Bouri, the Buurba Jolof, Abdul Bubakar of Futa Toro and Lat Dior himself—were already mobilising their troops for action and were holding consultations with the ruler of the Mauritanian tribe of Trarza. Governor Canard was frightened at the possibility of a war in the Senegambia at this time. The French would not be able to cope since all its forces in Senegal were already engaged in the Niger mission. Canard

84 A.N.S.O.M., In Sénégal IV, 75(A), 'Ministre de la Marine et des Colonies—"Note"', Paris, 11 March 1882.

85 Ganier, p. 229.

86 A.N.S.O.M., in Sénégal I, 66b, 'Le Gouverneur Canard au Ministre . . .', 8 August 1881.

emphasised how shaky the French military strength was, when he remarked to the Minister that 'All is engulfed in the Upper Senegal'.[87]

At this time French operations in the Senegambia were deadlocked. As Jauréguiberry's letter pointed out in February 1882, the ministry gave priority to the Niger project in the Tukulor territories. The Governor was advised to temporise in relation to Kayor and avoid any new pretext for hostilities. He was to avoid unnecessary intervention or meddling in the affairs of Kayor, military display, and any attempt to send military columns to Kayor borders. For the Minister, it was emphasised, '. . . The important thing is to prepare the way for the campaign which should be decisive in the Upper Senegal . . .'.[88]

But Governor Canard thought differently from the Minister. He was convinced that the most urgent task was the rail project from Dakar to Saint Louis across Kayor and that the penetration into Tukulor territory was not only premature but hazardous. He was worried by the movements of troops from Kayor and Jolof to a camp at Keur Amadou Yala, from where they could take off against the French. He reported this to the Minister in April 1882. But the reaction of Kayor and its allies came swiftly in May. Lat Dior refused any permission for the construction of the railway line and announced the determination of his allies and himself to fight the French to prevent the construction. Governor Canard, however, held on to his belief that the best thing was to speed up arrangements for the line. Canard later left Senegal, but his successor, Vallon, shared his ideas about the strategy for prosecuting French policy.[89]

Jauréguiberry, the Minister, also opposed the alliance in the Senegambia; but he preferred to deal with it not by military action, but by subtle diplomatic moves. He maintained that the Niger line was the priority project for which the parliament voted money. The Minister saw the Niger project not just as a way of conquering the Tukulor but also as a way of barring the British from the Western Sudan. To achieve this objective, he insisted, 'We must reach the Niger and establish ourselves there in the coming season: this is the view of Parliament . . .'.[90] The Niger mission reached and established a post at Bamako early in 1883; but hostilities had also broken out in the Senegambia in December 1882.

THE WARS OF CONQUEST

The French attacked Kayor from several points, contrary to the expectations of the Damel and his allies. They broke the Damel's resistance before his allies could assist him and by 16 January 1883 the Damel had fled and

87 Ganier, p. 230.

88 A.N.S.O.M., in Sénégal I, 66c, 'Le Ministre à M. le gouverneur du Sénégal', Paris, 9 February, 1882.

89 Ganier, p. 231.

90 A.N.S.O.M., in Sénégal I, 67c, 'Le Ministre à M. le gouverneur du Sénégal', Paris, 27 September 1882.

A statue commemorating collaboration between the French army and African 'mercenaries' in French imperial expansion

Samba Yaye, a relation of his who had collaborated with the French, was installed Lat Dior by the commander of the French operations, Colonel Wendling. This was under the governorship of Servatius.[91] After this initial success, however, the French faced more determined military opposition from an alliance led by Ali Bouri of Jolof and Abdul Bubakar of Futa Toro. These campaigns continued with varied fortunes from 1883 to 1885 when the French achieved their first significant success against the alliance. This came in April 1885 when Ali Bouri of Jolof was forced to sign a peace treaty with the French. It was reported that the Buurba Jolof agreed to the construction of the railway line and promised to send his eldest son to France for education.[92] But the campaigns had been ruinous to

91 Ganier, pp. 232–5.
92 Monteil, 'Le Dyolof et Al-Bouri Ndmye', p. 628.

French commerce and the commercial houses in Saint Louis had been critical of the government policy. By 1885, commercial concerns must have been hoping for better times. But the treaty proved to be only a truce.

With the successes of the Niger expeditions, the French appeared to have been encouraged to embark upon wars of conquest in the Senegambia. New hostilities broke out in 1889 when the French converted into a Protectorate treaty the peace treaty earlier concluded with Jolof. This was on 1 July 1889. The French explained their action by claiming that Ali Bouri had, since January 1889, been in contact with Ahmadu in addition to forming fresh anti-French alliances with a view to repulsing the French from the Senegambia. The French attack on Jolof was later extended to other Senegambian states. It soon became clear that the French campaigns were a part of a common plan for the conquest of the Western Sudan. The campaigns against the Tukulor began towards the end of 1889.[93]

It then became clear to all that the French were their common enemy, and consequently all the ruling groups who were eager to fight French imperialism entered into an alliance headed by Ahmadu of the Tukulor empire. This alliance included among others Ali Bouri of Jolof, Abdul Bubakar of Futa and Samory Ture, founder of the Wasulu empire, who had been at war against the French since the latter's establishment at Bamako in 1883. The alliance was dictated purely by self-interest.

By August 1890, Ali Bouri had lost many battles and seems to have abandoned the hope of defeating the French in Jolof. Confronted with superior French arms, he had no future and could not even be sure of his personal safety if he remained in Jolof. He then withdrew and went with his forces to join Ahmadu in Ka'arta to fight against the French. Henceforth, Ali Bouri, with several other contingents from the Senegambia, especially Futa Toro forces led by Abdul Bubakar, took an active part in all the campaigns against the French in the Tukulor empire.[94]

In the French campaigns against the latter, Archinard achieved successes against the Segu sector in 1890 and by January 1891 Ahmadu and his allies had been forced to evacuate Nioro, headquarters of the Ka'arta Sector. Ahmadu and his lieutenants thereafter escaped to Masina in 1891.

Hitherto, the French had had the support of the Bambara and other anti-Tukulor forces as faithful allies in the fight against the Tukulor empire and this accounted for part of their success. But with the fall of the Tukulor in Segu and Ka'arta, the Bambara interest in the struggle ended. They wanted to see the fruits of their efforts by being reinstated over their former territories now captured from the Tukulor. Their hopes were, however, disappointed when, by the new administrative arrangements, the French made it clear that they were out to achieve their own imperialist ambition

93 *Ibid.*, p. 628.
94 *Ibid.*, pp. 630–2.

and that the Bambara and others had merely been used as instruments in the process. The Bambara had no intention of replacing the Tukulor with another imperial master. They realised they had fallen victims, as the Tukulor had, to French imperialism. Henceforth, until 1894, there was a new turn in the war. The Bambara leaders decided to join forces with the Tukulor in a fresh attempt to drive out the French.

From 1891 to 1894, Ahmadu, with great diplomacy and determination, became the nerve-centre of a network of alliances comprising various opposed forces. In the new anti-French coalition, there were, in addition to those already mentioned, Tieba, ruler of Kenedougou and formerly a tool in the hands of the French against Samory, Samory himself and the Bambara. Its members saw the coalition as the only way left to achieve their varied and conflicting interests. In spite of all their efforts, however, they lost Masina to the French in 1893 and by early 1894 Timbuktu was lost. Thus, they were all consumed by French imperialism, which was established in the whole of the Western Sudan thereafter.[95]

We might conclude by noting that at the beginning of the nineteenth century the Western Sudan was made up of several states, some of which were in commercial relations with the French. But from trade the French developed political ambitions in the Western Sudan, though these were at first confined to the Senegambia which they were eager to hold against their British rivals. They were just taking the first steps to achieve their ambition when another imperialist force appeared on the scene. This was the Tukulor revolution, led by al-hajj ‘Umar. Their conflicting interests proved irreconcilable and wars ensued. While the struggle lasted, the existing ruling groups in the various states, who would in the end lose, no matter whether it was the French or the Tukulor who won, tried to save themselves by allying with one or the other of the imperialist rivals. The first stage of the struggle ended in 1860 with a delicate equilibrium established between the French and the Tukulor. This gave the two their respective spheres of influence—the French in the Senegambia and the Tukulor, east of the Senegal. But the real struggle was only postponed by this arrangement. The French used the interval to persuade the anti-Tukulor forces that they had a common interest in destroying the Tukulor. In this they were largely successful. The Tukulor, on the other hand, were bedevilled by internal wars. These factors were exploited to their ruin later on. But the anti-Tukulor forces discovered too late that by allying with the French, they had merely been pursuing a mirage. An alliance, embracing assorted anti-French groups in the Western Sudan, later emerged. It came too late to be effective and all concerned fell victims to French colonisation.

95 B.Ọlatunji Ọlọruntimẹhin, ‘Anti-French Coalition of African States and Groups in the Western Sudan 1889–1893’, *Odu*, iii, April 1970, pp. 3–21.

CHAPTER 10

Economic change in West Africa in the nineteenth century

J. E. FLINT

The revolution in the writing of African history which has taken place since 1945 has hitherto been concentrated on political and religious history and has been much less pronounced in the sphere of economic history. For this there are many reasons, not the least being the difficulty of the subject. Economic historians[1] are trained in systems of analysis based on the experience of European and American capitalism, cash-based economies which can be measured statistically using quantity measurements of weight and volume, and cash measurements of value. These techniques are of little use in African history, except for that of quite recent times, and can be misleading if improperly used, for all the evidence we possess that is amenable to orthodox forms of economic analysis relates only to the trade with Europeans, the only portion of African economic life which was in any way systematically measured at the time in records that have been preserved for us. The trade with Europeans, however, though of great historical importance, was in reality but a very minor part of the normal economic activities of the mass of West Africa's population during the nineteenth century. Thus, if the volumes and values of African exports and imports to Europe were rising, this does not necessarily mean that Africans were producing more, or that the economy was 'healthier'. Conversely, if European trade with West Africa fell, this did not necessarily indicate an 'unhealthy' or depressed economic situation. European traders, missionaries and officials commenting on the rises and falls of trade at the time, however, usually thought in these terms, and it is they who have passed on the bulk of the evidence we possess.

1 Of whom this writer would not claim to be one, hence this chapter cannot be regarded as a full outline of the economic history of West Africa in the nineteenth century; a theme which still awaits its proper historian. In this chapter the author tries merely to indicate the problems awaiting research and analysis, and to make suggestions for interpreting the more obvious changes.

FALLACY OF THE SUBSISTENCE ECONOMY

Once the historian accepts that the sector of the West African economy in the nineteenth century which can be measured was only a minor part of the overall economic picture, he must then avoid the temptation to withdraw from all further attempts to understand the workings of West African economic life by characterising the rest of the economy as 'mere subsistence'. The concept of a 'subsistence economy' is virtually meaningless, except perhaps in the cases of a few unfortunate human groups battling with desert or arctic conditions who in certain periods of history have won barely sufficient food to maintain, without expanding, their populations. The history of the Negro peoples of West Africa from the remote period of the development of agriculture is one of a colonising race, steadily expanding its numbers and its territorial extent whether in savannah or tropical forest. This could not have occurred if the economic life of such peoples had been based merely on a sufficient production for 'subsistence'; to colonise, multiply and expand a population necessarily demands the production of surpluses greater than the immediate needs of the community. What we know of the history of African social organisation and religious and political life also reduces any concept of a 'subsistence economy' to absurdity. The existence of towns, merchant communities, rulers and their officials, artisans and craftsmen, priests and healers, the mounting of military expeditions, the collection of tribute and taxes, even the mere exchange of goods—all presuppose the existence of organised economic surpluses which could be used for such purposes.

The creation of such surpluses depended basically upon the society's system of agriculture, for the first task of every economy is to feed its population. When an economic unit—family, village, tribe or region—is able to produce surpluses these can then be exchanged in the form of trade with other units to create a more varied diet, or to import manufactured goods or raw materials not available locally. Alternatively, labour can be withdrawn from agriculture and used to make local manufactures, or even withdrawn from material production altogether and used for purposes the society considers worthwhile or necessary, for religious observances, artistic production, defence and warfare, or the maintenance of a ruling bureaucracy.

FOOD PRODUCING SYSTEMS

The first essential, therefore, in attempting to build up an understanding of the development of the West African economy in the nineteenth century is to try to examine the basic food-producing systems used in the area. The efficiency of these agricultural systems depended on the interaction of several factors, some of which were not under human control. Productivity depended on the natural fertility of the soil, rainfall adequacy, the technical

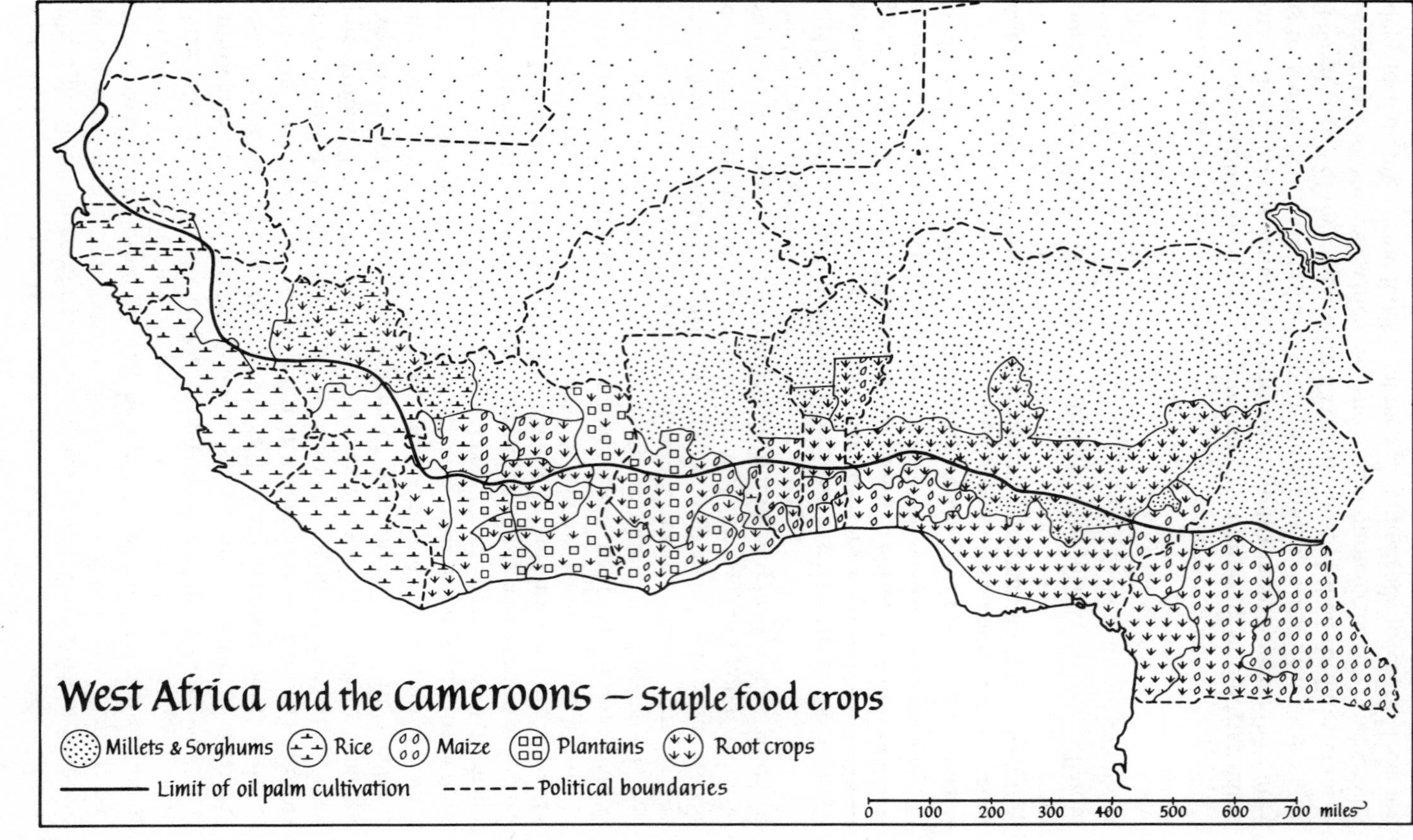
West Africa and the Cameroons – Staple food crops
Millets & Sorghums
Rice
Maize
Plantains
Root crops
Limit of oil palm cultivation
Political boundaries
0
100
200
300
400
500
600
700 miles

10:1

skill of the farmers, and their ability (or luck) in finding the most successful crops to grow in given conditions of soil and rainfall. Naturally, therefore, the success of West African farmers varied greatly from area to area.

To support life human beings must take in four essential substances: energy-giving starches, proteins, vitamins and salts. Virtually all[2] West Africans prefer to take meals in the form of a bulky paste or mash of energy-producing starches, to which is added a tasty sauce to give protein, vitamins and salt. The protein elements in this sauce are often contributed by plants such as palm oil, shea butter and groundnuts stewed for their oil. Animal meat and fish are much richer sources of protein and when these are available most West Africans will eat and enjoy them. But many areas, especially those remote from the sea or large rivers, suffered acute shortages of meat and fish. Similarly there were areas of salt shortages. Communities which were naturally rich in salt or fish could thus, even in very remote times, develop trades in these commodities over considerable distances.

If the 'mash and sauce' meal was general throughout West Africa, the actual foods used varied a great deal, especially in the staple food content of the 'mash'. The starchy element could be provided either by cereal crops, such as rice, sorghum or millet, or maize, or it could be supplied by the tubers of root crops, such as yams, cassava or forms of the potato. An additional staple which was neither a root crop nor a cereal was plantain, which became a dominant staple in East Africa but which in West Africa attained only local importance in the eastern area of the Ivory Coast.

In the remote period during which the Negro peoples of West Africa created their agricultural revolution, they had evolved a specific agricultural complex of their own, involving the domestication and improvement of local wild plants. The first area to be effectively colonised was the northern savannah, where wild grasses were ennobled perhaps six thousand years ago into the millets and sorghums. Such cereals would not grow in the tropical rain forest area, so that here the indigenous guinea yam was domesticated as the staple food, perhaps four thousand years ago. Within this southern area of high rainfall, however, the yam staple extended westwards only to the Bandama river, in modern Ivory Coast. West of the Bandama, and roughly up to the Gambia, an African rice was domesticated as a basic cereal crop, some three thousand five hundred years ago.

These ancient regional divisions in West Africa—the savannah sorghums and millets, the southern and south-eastern yam complex, and the south-western rice area—remain to this day as distinct agricultural regions. However, West Africans were adept in seizing opportunities of intensifying food yields by adopting new crops from outside. These new crops

2 The exceptions may be the 'purely' pastoral peoples who in remote times may well have used a radically different dietary combination.

came in two distinct waves: the first from about the time of Christ was the complex of new plants from Asia brought by Indonesian colonists to East Africa and transmitted gradually across the continent by Africans. The most important of these was the Asian yam, and of lesser importance in West Africa, plantains and bananas, which enriched the rain forest area. A second great wave of agricultural crop borrowing came after A.D. 1500 when the Portuguese began in an unsystematic way to bring across to their island and coastal African settlements the products of the unique and separate agricultural crops of the American Indians, developed in isolation from the Old World. West Africans in turn diffused these new food crops from the Portuguese. Several of these foods were items now of major importance in West African diet; in particular maize and cassava.[3] Maize at first spread much more rapidly than cassava, for it provided a cereal crop which could be grown in the forest area, and also in the intermediate area between savannah and forest, to supplement the normal root staple of yams. By 1800 it was well established as a secondary crop in these areas and had become an almost equal staple with yam in the eastern Ivory Coast.

The spread of cassava was much later and slower. Cassava, to be used as a food, could not be simply 'borrowed', for cassava tubers often contain quite lethal doses of prussic acid. The earliest African experiments in eating the crop must have had tragic results, and earned a sinister name for the food. The American Indians of the Amazon basin had developed the techniques for removing the poison and taught them to the Portuguese. The crop could thus spread only to the accompaniment of quite elaborate instruction in its preparation as food or *gari* flour, amid considerable local suspicion. The first evidence of its use in Africa comes from the Congo mouth and from Benin, both centres of Portuguese activity, at the end of the sixteenth century. By 1700 it was well known in São Thomé, Fernando Po, and Warri, all of them strongly under Portuguese influence. It began to spread more rapidly after 1800, especially near coastal towns which felt the presence of returned African ex-slaves from Brazil. By this time 'manioc flour' (gari) was an established article of diet among the Negroes of Brazil, and it was these men who spread the knowledge of its preparation from centres such as Lagos, Grand and Little Popo, Whydah, Porto Novo and Cotonou. In Dahomey the influential 'Brazilian' slave trader F. F. da Sousa is known to have taught the art of gari preparation in and around the king's court.[4] Even by 1900, however, there were extensive areas of southern West Africa where cassava was virtually unknown, or regarded

3 The diffusion of cassava, or manioc as it is called in America, has been discussed in detail in W. O. Jones's excellent monograph *Manioc in Africa*, Stanford, 1959. A useful discussion of the diffusion of cassava, maize and other African food crops is contained in B. F. Johnston, *The Staple Food Economies of Western Tropical Africa*, Stanford, 1958.
4 A. Le Herissé, *L'Ancien Royaume de Dahomey*, Paris, 1911.

with suspicion, or even despised as beneath the dignity of man to cultivate.[5]

Even today there is much prejudice against cassava as a food; it is often regarded as inferior food for use by poor social groups and not of great nutritional value. However, it has powerful advantages as a crop. Its yield is very high, and it flourishes on poor soils such as those partially exhausted by too much yam cropping. It survives drought and insect pests and is proof against locusts. It was not until 1890 that cassava suffered from any plant diseases.[6] Although it is poor in human nutriments if eaten alone, it contains a high quantity of starch, and therefore gives a quick energy yield, and it provides some vitamin C and calcium. Thus it spread rapidly after 1900, as populations strained the resources of the soil and overcropped yams, and it spread most rapidly in the areas of thickest population, among the Yoruba, Ibo and Ibibio of Nigeria.

West Africans on the whole seem to have shared basic techniques and systems of agriculture, though with regional variations determined by the staple crop, the fertility of the soil, and the availability of land. In general West Africans had not developed much of a corpus of knowledge of fertilisers, mainly because land was plentiful and it was easy to find new soil, and also because in the forest areas cattle and horses, from the effects of whose dung the processes of fertilisation may be observed, were rare because of the prevalence of the tsetse fly. In the savannah areas of the north agricultural groups like the Hausa welcomed cattle-herding peoples on to their land after the harvest to fertilise it and developed a symbiotic relationship with them, a fact which does much to explain the spreading influence of the Fulani. Elsewhere, however, there were only two fertilising techniques available; the fallowing of areas for a period of years once the crop yields began to drop, and the burning of the covering of bush and grass on new soils before cultivation, both to clear the land and to deposit fertilising ash. Fertility itself, and the germination of seeds, was naturally regarded as an awe-inspiring and deeply spiritual process and a great deal of the ritual and ceremonial of religion was designed to ensure fertility and germination upon which the society utterly depended.

The land itself was not usually regarded as property, except in areas close to European traders where rental payments and leases, even by Africans, became increasingly common in the nineteenth century, or in the northern savannah areas affected by the Muslim jihād where concepts akin to 'feudal' overlordship began to develop after 1810. Instead land was regarded as a natural resource, in control of him who put it to use. Its distribution was generally controlled by the village authorities, whether headmen or elders.

5 Jones, pp. 60–79.
6 *Ibid.*, pp. 4–24.

Cultivation was by men, women and children, often with quite specialised division of labour as between tilling, hoeing, weeding and cropping. The basic tools in use were the iron hoe and the iron-bladed matchet, which in themselves created the need in every agricultural district for blacksmiths and forges. Deep digging was not practised, and was indeed hardly desirable, for without fertilisers deep digging or ploughing would have produced extensive soil erosion.

POPULATION

We do not possess population statistics for nineteenth century West Africa but from the general observations of travellers it is possible to pick out certain striking concentrations of thick population which a combination of local agricultural skill and soil fertility had created over a long period of time. The most intensive areas of settlement in all West Africa were in Nigeria, where there were three thick clusters of settlement; the Ibo-Ibibio area in the south-east, the Yoruba group in the south-west and an intensive cluster in the north around Kano and Katsina. Not quite so dense, but still intensively settled, was the coastal belt from Ivory Coast to Dahomey, thickest in Ghana and Togo. More isolated pockets of thick population occurred in the coastal area of Sierra Leone, and at the mouth of the Senegal river.

In general it appears that the areas of forest and coast supported human life and reproduction more abundantly than the savanna interior, with the exception of the Kano–Katsina cluster. It cannot be demonstrated that this had any relationship to the presence of European traders in the south (except in the restricted sense that quite large trading ports gathered considerable numbers of people at places like Porto Novo, Whydah, Lagos and Calabar, which helped to raise average densities nearby). The reality was that European traders, whether looking for slaves or legitimate goods, naturally gravitated to dense areas of population where surpluses for trade, whether in men or things, were naturally greater.

During the nineteenth century it appears that there was a fairly steady drift of the forest populations in a southerly direction towards the coasts. In some areas political pressures helped to intensify this movement; Yorubaland moved its centre of gravity southwards as a result of Fulani pressure in the 1840s, and pressure from Asante and Dahomey caused a tighter packing of settlement in areas to their south. But more basic geographical and economic factors were also at work. With the rapid decline of the Atlantic slave trade between 1840 and 1860 slaves who would have been exported were put to productive work in agriculture or trade in many areas on or near the sea coast. This in turn helped to increase population in these areas. Increased security of life nearer the coast may well have attracted people southwards in this period, but there is also considerable evidence that the forest vegetation itself was retreating southwards as a result of farming. The northern parts of the rain forest received considerably less rainfall, with the result that cultivated areas abandoned for fallowing were not able, like those further south in wetter conditions, to replace the tree cover quickly. With a rising population in the nineteenth century, grass fires at the end of the dry season became more frequent and these destroyed the young saplings which had sprouted. The savannah thus crept southwards year by year, and there is much evidence to show that certain areas we now regard as 'savannah' ought to be described as 'derived savannah', e.g. in the northern Ibo areas of Nigeria.[7]

EFFECT OF EUROPEAN COMPETITION ON NATIVE MANUFACTURES

It has been shown that West African agriculture was able to support growing populations without great difficulty and to overcome problems arising from population growth and soil exhaustion by utilising fresh land and new crops. The food supply was also quite adequate to allow a considerable ancillary activity in manufacturing. Much of this was non-specialised, and remained so in the nineteenth century. Women engaged in the manufacture of their household pots, baskets and clothing as part of their normal household duties. Other crafts had become more specialised, especially the metalworking industries. Blacksmiths were specialists, with 'mysteries' of their craft concealed from the common people. Not every village possessed its blacksmith group and some might, therefore, need to travel some distance to purchase farm implements or weapons. Some towns became centres where blacksmiths had congregated in cooperative guilds numbering several dozen craftsmen and became known as places where iron products of greater variety and superior workmanship could be obtained. In general such centres did not organise the export of their goods but relied on nearby peoples coming there to buy.

In the nineteenth century it would appear that the smithing trades were on the decline, especially in the southern areas close to the coasts. By 1800 the West African iron mining and smelting industries, upon which the blacksmiths had once had to rely completely for their raw material, were almost at an end, ruined by the competition of cheaper and purer iron bars imported from Europe. Throughout the nineteenth century African blacksmiths had to face the competition of increasing quantities of cheap European imports of iron basins, matchets, knives, hoes, wire and other metal goods turned out by the expanding mass-production techniques of the industrial revolution. The English cities of Birmingham and Sheffield were especially active in this respect, and by the mid-nineteenth century had become an important political lobby in matters affecting British policies in West Africa.

7 R. J. Harrison Church, *West Africa: A Study of Environment and Man's Use of it*, 4th ed., London, 1963, pp. 73–84.

The manufacture of cloth was a larger, more elaborate West African industry which exhibited a greater degree of specialisation and division of labour than any other, and seems to have faced the challenge of European competition with far greater resilience. The market for cloth was large. In the savannah area the more open exposure to the sun, the dusty dry season and the influence of Islamic culture created patterns of dress which demanded voluminous clothing. In the forest zones where this was not necessary, male and female vanity nevertheless asserted themselves among those with the means to embellish their bodies, and on the coast new styles of dress emerged from contact with European traders and Africans from the North.

In 1800, despite three centuries of European cloth imports, West Africans still produced the bulk of the cloth that was consumed—in a bewildering variety of fabrics which utilised almost every substance from which cloth could be made, including animal hairs and the bark of trees as well as cotton. Much of it was plain stuff produced by village women in the household, but for the high quality cloth specialised centres had developed exporting their goods over long distances. These centres produced cotton cloth almost exclusively and each would have its distinctive shapes, sizes, colours and pattern designs. In such places a definite artisan class, often female, maintained itself almost completely by the cloth trade, and the professions of weaving, spinning, dyeing and embroidering were often quite distinct. In particular, colours and patterns were highly specialised, and the regions of West Africa each had their own rooted preferences (many of which have been carried through to the present in the characteristic colours and patterns of the 'national' dress among Ghanaians, Ibo, Yoruba, Mende, etc.).

The most highly developed centres of cloth manufacture naturally arose in the towns which lay in the thickest clusters of population. Thus in the savannah, Kano, Timbuktu and Kuka were famed centres of the industry, as was Old Ọyọ before its destruction. The new urban centres arising from the Yoruba wars developed cloth manufacturing on an extensive scale, and Ibadan in particular became the centre for the production of the famous Yoruba blues. Smaller towns would also have their own industry, but often the goods would be simpler and less specialised. The larger centres developed long-distance trades selling goods in quantity to merchants who would sometimes transport the cloth for hundreds of miles. Cloth from Bornu, for instance, could be purchased not only in Hausaland, but in Ilọrin and further south in Yoruba country.

It was the specialisation of pattern, design and colour, and the excessive 'choosiness' of the West African customer, which gave the indigenous cloth industry its ability to resist European competition. In the seventeenth and eighteenth centuries European cloth, being mostly woollen and heavy and itself the product of handicraft industry, was unable to compete in price or quality, and thus the slave traders had turned to India for a supply of exotic and richly patterned lightweight cottons and silks for the African markets. The Indian cloths were naturally expensive and furnished only a limited demand among African rulers and coastal merchants with a taste for the unusual. After the mid-eighteenth century the factories of Lancashire began producing cheap cotton cloth on an ever-increasing scale at lower and lower prices. Before 1880, however, although Lancashire goods increasingly entered the market, they were unable to displace the local cloth. It took years of experience in the trade for the European merchants to obtain a close understanding of local patterns and prejudices, and these sometimes varied so greatly from village to village that the factory system of Lancashire could not adapt its production for such small units. Lancashire cloth was also at first much inferior in quality to the hand-made local product,[8] and where Africans dressed more for embellishment than for warmth quality was more important than quantity. Macgregor Laird, on his Niger expedition of 1830–2, found that he was unable to sell Lancashire cottons in Nupe, despite the fact that his cloth was cheaper than Nupe cloth.[9] In the far north, where white plain cloth was more frequently worn, the high cost of transportation from the coast delayed the triumph of Lancashire cloth until the coming of the railway after 1911.

If cotton manufacturing was able to resist European competition, other substantial trades and products in West Africa were totally unaffected by European competition in the nineteenth century. One of the most important long-distance trades was that in kola, the only stimulant permitted to Muslims. A large trade in this commodity continued to flourish between the forest zones producing it, especially those of the Asante region, and the Hausa states, with large caravans coming from the north to collect kola, as well as gold. The gold trade and gold production also seem to have been very little affected by European trade until the beginnings of mining in the southern Gold Coast after 1880. Gold-producing areas continued to export the precious metal both into the savannah and down to the coast throughout the century. The important complex of trades and crafts arising from leather-working similarly faced little or no European competition; the savannah zone not only produced all its own requirements of tanned skins, footwear and saddlery, but exported cured skins to North Africa and the Atlantic coast, and sent high quality leather goods across the Sahara for re-export to Europe where the so-called 'Moroccan' leather commanded high prices.

8 There was much discussion of the problems of exporting cotton cloth to West Africa in 'Report of the Select Committee on the State of the British West African Settlements', *Parliamentary Papers*, v, 1, 1865.

9 M. Laird and R. A. K. Oldfield, *Narrative of an Expedition into the Interior of Africa by the River Niger in the Steam Vessels Quorra and Alburkah in 1832, 1833, and 1834*, London, 1837, i, p. 103.

EFFECTS OF EUROPEAN TRADE

It has sometimes been argued that the development of European trade on the West African coast, both during the slave trade period and after, had the effect of diverting trade from the trans-Saharan routes to the Atlantic coast, and thereby weakened the economic position of the savannah states. It has also been argued that this economic decline lay at the root of political upheavals, such as the Fulani jihād, so characteristic of the nineteenth century in the north. The evidence at our disposal, however, will not support these conclusions. In some areas of the savannah there was undoubtedly economic dislocation in the nineteenth century but this was not general to the whole region and represents pockets of depression caused by local political upheavals. In general the creation of the Sokoto empire, and the maintenance of Bornu as a large unified state provided increased security for traders and allowed considerable commercial expansion. C. W. Newbury, using the evidence of French explorers and diplomatic missions and the records of the Moroccan and Tripoli archives, has suggested that in the nineteenth century the trans-Saharan trade actually increased to reach a peak of about £1,500,000 in 1875 and that much of the ivory, ostrich feathers and gold dust thus exported from West Africa was ultimately destined for European markets. The real decline in the trans-Saharan trade after 1880 was caused initially not by European economic competition but by the political effects of the 'Scramble', the setting up of colonial frontiers and tariffs, and accentuated later on by the creation of the colonial railway systems feeding the coastal ports.[10]

Much of the rest of this chapter will be taken up with a discussion of the European commercial presence and the changes, revolutionary in their implications, which came over the very nature of European trade in the nineteenth century. At this point, however, it might be well to stress the fact, which the above analysis of West African agriculture, industry and trade should have made abundantly clear, that European economic contact with West Africa, at least before 1880, was essentially peripheral to the West African economy. It was the local economy which dominated the lives of the people, created the surpluses needed to provide for traders and craftsmen, the maintenance of kingship, ruling oligarchies and the state, and the support of priests and religious oracles. This economy was one of considerable complexity, one which had long before passed the stage of simple barter, in which buying and selling, especially in the long-distance trades, was conducted with the aid of complicated systems which utilised easily stored commodities such as metal bars, foreign coinage, gold

10 C. W. Newbury, 'North Africa and the Western Sudan trade in the nineteenth century: a re-evaluation', *Journal of African History (J.A.H.)*, vii, 2, 1966, pp. 233–46.

dust, or even alcoholic liquor to serve as 'currencies'.[11] Sometimes prices were calculated abstractly in terms of 'bars' or some other unit which was not actually exchanged at all.

END OF THE SLAVE TRADE

Though the economy of West Africa may not have been basically changed by economic and commercial contact with Europe in the nineteenth century, it is nevertheless a vitally important feature of West African history that gradually after 1807 and more or less completely by 1860 'legitimate commerce' replaced the trans-Atlantic trade in slaves.

The anti-slave-trade movement in Europe was itself, directly and indirectly, a product of growing European industrial capitalism. The evangelical religious revival, which produced under the leadership of William Wilberforce the extraordinarily powerful political movement that forced the British government to outlaw the slave trade by British subjects in 1807, was itself in origin a movement of religious revival aimed at the near-pagan cities created in England by industrialisation. That the anti-slavery movement was able to crush the vested opposition of powerful slave-trading interests in Liverpool, Bristol, London and the British West Indies was partly the result of new economic attitudes towards West Africa which were growing in Britain. British industry increasingly needed tropical products, dyes, gums and vegetable oils, and it was thus stupid, as well as immoral, for British slave traders to be permitted to denude Africa of the labour with which these commodities could be produced. Moreover, British industry was an exporting industry, and men were turning to the belief that the Negro, left a free man in Africa and given purchasing power through 'legitimate' commerce, was of more value than the Negro, transported as a slave to the West Indies, sullenly working in response to the stimulus of the whip, with no cash income to spend on British goods.[12]

Nevertheless the abolition of the British slave trade in 1807 was initially a blow to the British commerce already established and there was little experience of 'legitimate' trade which could be used to build up substitutes for slaving. The initial attack on the trade was mainly political; penalties for the crime of slave-trading were intensified until in 1822 the offence carried the death sentence. The West African Squadron of the Royal Navy was created to patrol the coast from its new base in Freetown, taken over from the bankrupt Sierra Leone Company, and the

11 On currency see A. H. M. Kirk-Greene, 'The Major Currencies in Nigerian History', *Journal of the Historical Society of Nigeria (J.H.S.N.)*, ii, 1, December 1960; G. I. Jones, 'Native and trade currencies in Southern Nigeria during the eighteenth and nineteenth centuries', *Africa*, xxviii, 1, January 1958; A. Hopkins, 'Currency in Southwestern Nigeria in the late nineteenth century', *J.H.S.N.*, iii, 3, December 1966.

12 On the relationship between economic theory and contemporary attitudes to the Negro see P. Curtin, *Image of Africa: British Ideas and Action 1780–1850*, Madison, 1964.

British Government conducted a diplomatic offensive to persuade other European states to outlaw the trade. The French slave trade was made illegal in the 1815 peace settlement and thereafter the Spanish and Portuguese, in return for British loans and financial subsidies, increasingly restricted the scope of their slaving interests. The result was a rapid change in the balance of the trade: after 1810 the British slave trade ceased to exist, by 1820 the French had ceased to participate, but the demand for slaves continued from Brazil, Cuba and the southern states of the U.S.A. which required labour for expanding coffee, sugar and cotton plantations. Brazilians increasingly took over the trade, building up influential positions with African rulers whose states lay at some distance from areas of British influence and off the regular patrolling routes of the West African Naval Squadron. As a result, the slave trade actually increased from 1807 until about 1830, despite British and French abolition.

GROWTH OF 'LEGITIMATE' TRADE WITH EUROPE

In these years most of the French and British slave-trading firms were ruined, despite the attempts of some of them to build up a legitimate trade. A few new, but small, trading concerns began to appear, mainly from Liverpool and Marseilles, looking to develop new produce trades for the chemical and soap industries of France and England. They established viable, if small, trades, exporting groundnuts from the Senegal and Gambia, picking up a little gum, indigo and peppers from the same region, helping to provision the British station at Freetown and the moribund forts on the Gold Coast from which a trickle of gold was still obtained. By 1830 it had become clear that palm oil was the most lucrative commodity to be obtained from the West Coast and that the Ijo and Efik areas of the Niger Delta, once the centres of slave-trading activity, were the most productive areas from which oil could be obtained. Dike has discussed in detail the way in which the coastal Africans of this area adapted their social and political organisations to the European demand, and preserved their powerful middleman position.[13]

The French in these same years were faced with even more serious problems than the British, for they lacked the highly developed industrial base from which to build up a legitimate commerce. At first the French attempted to tackle the problems raised by the cessation of the slave trade by setting African labour to work in Africa itself and in 1818 the French government attempted to establish white settlers and a plantation economy in Senegal. It was hoped that the scientific farming methods of the French planters would be observed and imitated by the nearby African chiefs, who would themselves start plantations, grow cash crops, and become a

13 K. O. Dike, *Trade and Politics in the Niger Delta, 1830–1885*, Oxford, 1956; G. I. Jones, *The Trading States of the Oil Rivers*, London, 1963.

consuming class for French manufactures. Even when white settlement proved a disaster the French continued to hope that plantations would be established either by Africans or by men of mixed French–African descent. Much capital and effort was expended on these ventures but by 1830 the only result was an annual export of some twenty tons of cotton. By 1840 the plantations schemes were abandoned, and the French began to concentrate, like the British, on the development of trade with the interior, attempting to use the Senegal river as a highway for the groundnut trade.[14]

The precise effects of the impact of legitimate commerce are very difficult to delineate, but from what we know it is clear that the humanitarian hope that the development of trade in things would *ipso facto* create radical new social attitudes of an egalitarian kind was naive. Paradoxically areas such as the Oil Rivers, where the quantitative expansion of legitimate trade was largest, experienced social change which, although significant, was essentially gradual. Though the overseas slave trade was virtually at an end by mid-century, domestic slavery expanded to become an important element in the economy, and African traders, heads of houses and noblemen measured their prestige not only in profits of trade but in numbers of 'boys' in their control. Similar developments occurred in the areas further west which experienced commercial penetration from Europe; great chiefs in Ibadan or Dahomey became 'capitalists' expressing their prestige in the size of the farms they held and the number of slaves working for them. In these developments there was a kind of continuity from the slave trading period, whilst at the same time African appetites for European goods were being stimulated and expanded, thus laying the foundations for a later cash-crop export-oriented economy.

Elsewhere in the coastal towns more subject to direct European cultural and political impact, though trade often languished after the suppression of slaving, new groups of Africans with new economic attitudes had begun to emerge, more as the result of political and administrative changes than of any economic forces. Freetown, though uneconomic and costly to the British, had been taken over by the Crown as a naval station for the Navy's anti-slave-trade patrols, and it was here that the Navy discharged its cargoes of liberated Africans from the captured slave ships. These Africans were of many tribes and language groups; flung together in a British colony they of necessity began to adopt English as a *lingua franca* and to adopt British ways, and the local missionaries made converts among them. From the local Creole settlers[15] they soon acquired the understanding

14 For a discussion of economic development in Senegal see P. Cultru, *Les Origines de l'Afrique Occidentale: Histoire du Sénégal du XVe Siècle à 1870*, Paris, 1910; G. Hardy, *La Mise en Valeur du Sénégal de 1817 à 1854*, Paris, 1921.

15 For a discussion of the economic and social significance of the Creole population see A. T. Porter, 'Family histories and West African social development: the role of the Creole in nineteenth century Sierra Leone', *Historians in Tropical Africa*, Salisbury, 1962, pp. 302–16.

that the way to wealth, security and influence lay through literacy and education. Economically this liberated African group increasingly had to operate within a framework of the cash economy, working for wages or even aspiring to salaried posts, or accumulating cash savings through the profits of trade. This type of African society was in fact beginning to arrange its social hierarchy according to European, or indeed even English, concepts of 'class' based upon wealth.

The same phenomenon, on a less extensive scale, was also present in other foreign-controlled urban centres on the West African coast. French attempts to develop plantations in Senegal, though a failure, had resulted in an inflow of government capital to build up St Louis and later Dakar, whose African populations expanded in response to employment and trading opportunities and were increasingly involved in a cash economy. The colonisation of Liberia after 1816 by free American Negroes produced similar effects in Monrovia.

EXPLORATION AND TRADE

More immediately spectacular results began to affect the patterns of trade after 1830 as a result of geographical exploration by Europeans. Though the European explorers were usually men of genuine scientific bent, in a broader sense they nonetheless represented an agency whereby Europe collected economic and commercial intelligence of the African interior. By the 1820s most European explorers were quite systematic in the way in which they noted down and assessed the possible export crops of a region, or the potential of an African people as consumers of Lancashire cotton cloth. In West Africa, from the beginnings of exploration of the interior by Europeans in the eighteenth century, the problem of the course and termination of the Niger obsessed travellers. The obsession was not simply natural curiosity, but lay in the commercial possibilities which the discovery of a highway into the Muslim savannah would open up. By 1830 experiments in the use of steam propulsion for ships in England had created a vessel independent of winds and currents which, unlike the sailing ship, could ascend any river deep enough to float it.

In 1830 Richard Lander dramatically solved the problem of the location of the Niger mouth by the simple method of letting the current of the river carry him down to the sea.[16] It was thus demonstrated that the Niger emerged at exactly the point where European commerce had already discovered the most plentiful resources of palm oil, in the Niger Delta, whose mouths they had hitherto regarded as independent small rivers. The oil-producing regions behind the coastal middlemen thus lay open to penetration and, with the middlemen cut out, the profits of the trade upon

16 R. Lander and J. Lander, *The Niger Journal of Richard and John Lander*, ed. R. Hallett, London, 1965.

the lower Niger promised to be even higher than those at the coast. Moreover, there might be lucrative export and import opportunities by direct contact with the Muslim emirates of the north. Macgregor Laird, a younger member of a Liverpool shipbuilding firm which was pioneering the use of iron and steam propulsion in shipbuilding, seized on Lander's discovery to test out the world's first ocean-going steamship constructed entirely of iron, and in this and other ships Laird with Lander ascended the Niger in 1832–4.[17] But the result was a disaster in which a third of the Europeans involved died from fever. In addition the expedition had incurred violent hostility from the African middlemen traders of Brass, who naturally resented this intrusion into their palm oil markets and were supported by the Liverpool traders whom they supplied.

Laird, and those traders who had interested themselves in the prospects of interior penetration via the Niger, now abandoned the venture as too dangerous to health. Others were less cautious. The humanitarians in England, fresh from their political triumphs in abolishing slavery (1834) and its successor status of 'apprenticeship' (1838) in the British colonies, now turned to consider the outrageous fact that the slave trade with West Africa had actually increased, despite British abolition, the British treaties with European nations, and the activities of the West African Naval Squadron. They concluded that the policies of political and naval suppression had failed, and that new policies would have to be found. In 1840 their leader, Thomas Fowell Buxton, published *The African Slave Trade and its Remedy* in which he resurrected the original policy which had lain behind the foundation of Sierra Leone by arguing that the slave trade could be eradicated by the forces of free competition in legitimate commerce. If only Africans in the interior of the continent could be shown that by keeping their labour at home they could produce cash crops which would sell at a higher profit to the community than the export of slave labour, then the slave trade would dry up at its source for want of supply. The means to achieve this lay to hand in the steamship and the highway of the Niger, and Buxton proposed a great British expedition to Lokoja where the Niger and Benue flowed into each other. Here his plan was to establish a model farm to teach agricultural science. Treaties would be made in which local African rulers would agree to abolish slave-trading and be compensated with promises of equivalent 'legitimate commerce'. Missionaries would teach the Gospel and the arts of 'civilisation', and trade would flow down to the sea.

In 1841 the British Government decided to test Buxton's ideas and the largest expedition of Europeans yet seen in West Africa ascended the Niger. The result was a disastrous failure in which forty-eight Europeans lost their

17 Laird and Oldfield, *Narrative*.

lives and the humanitarian influence on British policies suffered a blow from which it never completely recovered.[18]

The disaster of 1841 emphasised that West Africa was the 'white man's grave' and resulted in a definite withdrawal of white personnel from the interior. This withdrawal coincided with an upsurge of ambition among the 'liberated African' class from Sierra Leone, a movement whose most spectacular effects were to be seen in the mission field with the organisation of the C.M.S. Niger mission under Crowther,[19] but which had equally important economic aspects. Liberated Africans in this period were in fact developing not only as wage-earning artisans and salaried professionals; a nascent group of small capitalist traders was also emerging, quite unlike the traditional middlemen of the Oil Rivers. During the late 1830s these men began trading contacts with Lagos and Badagry from which developed the 'return of the exiles' to Abeokuta.[20] From these coastal beginnings they penetrated inland along the trade routes of Yorubaland and in the 1850s they had begun to settle as traders in Porto Novo, Whydah and some of the ports of the Oil Rivers, making no attempt to restrict their activities to areas of their own language or tribal group.

During the 1850s and 1860s there was renewed British governmental support for British traders, British shipping and for Niger penetration; a support motivated by several considerations. One was the political support given by the French government to its traders, and the establishment of French posts at Assinie, Grand Bassam, the Popos, Porto Novo and in the new French colony of liberated Africans at Libreville. Another was increasing anxiety among British cotton manufacturers about the availability of raw cotton supplies in the future and the dependence of industry upon supplies from the U.S.A. The British Government made efforts to encourage cotton cultivation in its West African settlements; a Cotton Supply Association was formed which distributed seeds and ginning machinery in Sierra Leone, the Gold Coast, Lagos and Abeokuta.[21] Macgregor Laird showed renewed interest in West Africa and received considerable government support for his schemes. In 1849 he had formed the African Steamship Company, in itself a significant addition to the economic equipment of West Africa. Hitherto British traders had employed their own shipping in the trade, with no fixed schedules, and naturally such ships gave no facilities of passage or freighting to their rivals. African traders naturally found great difficulty in shipping both themselves and their goods around the coast. Laird's ships were now thrown open to

18 For the 1841 expedition see C. C. Ifemesia, 'The Civilising Mission of 1841', *J.H.S.N.*, 1962, pp. 291–310; J. Gallagher, 'Fowell Buxton and the New African Policy, 1838–42', *Cambridge Historical Journal*, 1950, pp. 36–58.

19 J. F. A. Ajayi, *Christian Missions and Nigeria: the Making of a New Elite*, London, 1965.

20 S. O. Biobaku, *The Egba and their Neighbours, 1842–1872*, Oxford, 1957.

21 These developments are well documented in C. W. Newbury, *British Policy towards West Africa, Select Documents, 1786–1874*, Oxford, 1965, especially Chapter II, Section B.

all comers, providing regular connections between Freetown, Accra, Lagos and Calabar for small African traders. In 1854 the African Steamship Company was given the official British mail-carrying contract, and regular postal services may be said to have originated from that date.[22]

Laird was also active on the Niger, promoting Baikie's famous expedition in 1854 which by using quinine ensured the survival of all its members. After 1857 the Foreign Office began subsidising, on a decreasing scale, annual expeditions up the Niger and regular trading stations were established at Abo, Onitsha and Lokoja. Violent opposition from the delta states of Brass and New Calabar, encouraged by the Liverpool traders, was overcome by naval protection and the bombardment of offending villages in the 1860s.[23]

EFFECTS OF EUROPEAN TRADE

By 1870 it may be said that British, French, German and liberated African traders had succeeded in establishing 'legitimate commerce' in West Africa. The slave trade overseas was no longer significant. Along the coast, from St Louis to Libreville new urban centres, or expanded traditional ports, not only acted as the entrepôts for the exchange of European for African goods, but had begun to create new classes of African wage-earners, professional men and small capitalists whose impact on traditional society was to prove revolutionary in the long run. But the successful implementation of legitimate trade had not been the result of economic and commercial forces alone. Naval blockade, government subsidies, interference by European consuls and naval officers, and the foundation of colonies like that of Lagos in 1861[24] had all played a vital part. Moreover, the effects of the changeover to legitimate commerce were limited in their extent and disappointing to the exaggerated hopes of doctrinaire humanitarians and free traders who had imagined that a free trade in things replacing the trade in slaves would revolutionise and revitalise African society. Except in the coastal 'colonial' towns such as St Louis, Freetown, Lagos or Monrovia, legitimate commerce had produced little fundamental economic change. More than 90 per cent of West Africa's produce remained totally outside the cash-based economies of the coastal towns. Seen from the viewpoint of African traditional society at the time the triumph of legitimate commerce was less than revolutionary. In general, African societies provided the European traders with new commodities without undertaking any major changes in their own internal economic structures. The palm oil trade is the major illustration of this phenomenon; palm oil was already a major item

22 'Papers Relating to the Grant of a Subsidy to the Company of African Merchants', *Parliamentary Papers*, xli, 1864; Newbury, *Documents*, pp. 114–15.

23 Newbury, *Documents*, pp. 121–2, Minute by Wylde on the Niger Trade, 8 January 1862; Dike, pp. 171 ff.

24 Economic factors behind the occupation of Lagos are discussed in J. F. A. Ajayi, 'The British occupation of Lagos, 1851–1861', *Nigeria Magazine*, 69, 1961, pp, 96–105.

of African diet before its export to Europe began, and the large increase in its production and processing which occurred in the nineteenth century was accomplished by traditional methods and traditional means of transport, extended in scale and size. The cultivation of groundnuts also expanded, and farmers now enjoyed European imports in exchange which were an addition to income, but the groundnut trade seems to have made little impact on the actual system of production, tenure of land, or utilisation of labour. Only in the attempts made to develop cotton cultivation, where, especially in the Yoruba interior, small-scale African capitalists tried during the 1860s to develop plantations employing wage labour and producing entirely for the export market, did fundamental agricultural change occur, and in the aftermath of the American Civil War these projects either collapsed, or remained minor in their impact on the economy.

If the character of the export trade did little to stimulate the traditional economies of West Africa, the nature of the imports which paid for African produce did less. A few printing presses and cotton ginning machines, replacement parts for locally used steamships, the books imported by the missionaries and the tools imported by mission-trained African carpenters and coopers is a list which almost exhausts those imports which may be said to have had any effect in increasing productivity. The vast bulk of imports before 1870 consisted of a small number of items which were by their nature quickly consumed and contributed almost nothing to the quality of the economy or to its development; alcoholic liquor, guns and ammunition, tobacco, beads and not a very large proportion of cotton cloth together formed the major import items.

The ending of the trans-Atlantic slave trade must undoubtedly have produced important economic effects in the entire West African area. The mere retention of labour which would otherwise have been exported must have increased the total scale of both production and consumption, and there were no doubt many other subtle effects of the change which further research and investigation will one day illuminate. Nevertheless the impact was not dramatic nor productive of fundamental change. The institution of slavery remained little affected by the abolition of the oceanic trade, except in certain coastal areas. The slave trade continued to exist along internal routes where Africans imported or exported labour as of old. The spread of legitimate commerce even, in some areas, worsened the lot of the African slave, traditionally a domestic slave integrated to the larger family structure, by creating a demand for cash crops like cotton which could be produced by plantation slaves working under the whip.

During the 1870s European traders began to experience difficulties of their own. For some time the actual volumes of produce exported from West Africa had remained static, whilst competition intensified between a growing number of firms, British, French and German. The competition of small African capitalists, now assisted by the facilities of the African Steamship Company, was also difficult to face, as African traders operated on smaller profit margins than the Europeans with their more elaborate organisations and expensive European personnel. After 1873 these difficulties were compounded by falls and fluctuations in produce prices caused by slumps in the European demand for produce, especially in the palm oil trade. African middlemen in the Oil Rivers naturally objected to any attempt to reduce prices, and powerful kings like Jaja of Opobo could play off one firm against another to try to maintain prices, or even take steps, as Jaja did, to ship oil directly to Liverpool and thus cut out the European trader's profit altogether. European firms also sought to reduce costs by dispensing with the African middleman and obtaining direct access to palm oil producing regions. The Liverpool firm of John Holt was to provoke a bitter quarrel with Jaja by its attempts to penetrate his markets in the Qua Ibo river. Firms from Glasgow, Manchester and London, unable to penetrate the closely developed middleman system of the Oil Rivers, began extensive trading on the Niger in the 1870s, but here again over-competition was soon evident, with each firm setting up competing stations at the same places on the river. There was also violent opposition from Brass and New Calabar to Niger penetration, as well as active competition from African traders from Lagos.[25]

PARTITION

The *laissez faire* attitude to trade seemed in fact to have worked itself out, the trade seemed stagnant, competition too fierce, and African middlemen too strong.[26] Increasingly some of the British traders began to turn to ideas of monopoly as a way out of their difficulties. From this time forward there were constant negotiations among the more powerful Liverpool traders, designed to secure price-fixing arrangements whereby African middlemen could be presented with a united front. The larger traders also began to make arrangements with the shipping lines for cheaper rates on bulk freight and dearer rates on small packets so as to exclude African competition from Lagos and Freetown. On the whole, such arrangements were none too successful; there were too many small traders left out of the rings, who would take advantage of disputes with African middlemen to jump in with offers of higher prices of their own. In these new conditions trade was beginning to take on a quasi-political aspect; groups of traders might see advantages in negotiating special privileges for themselves with African rulers, or even be tempted to use the power of British consuls and gunboats to force the coastal middlemen to

25 J. E. Flint, *Sir George Goldie and the Making of Nigeria*, London, 1960, Chapter II.
26 For a detailed description of a local situation of this kind see H. J. Bevan, 'The Gold Coast economy about 1880', *Transactions of the Gold Coast and Togoland Historical Society*, ii, 1956, pp. 73–86.

allow them direct access to the markets in search of cheaper produce. Some began to feel that colonial rule could serve to 'open up' producing markets. French traders, faced by similar difficulties of competition from British, African and German competitors, began to consider the advantages which a French administration and a protective tariff might bring.[27] Such moods and ambitions were to provide an important commercial background to the beginnings of the partition which was to follow in the 1880s. The transition from commercial to political ambitions is neatly illustrated by events on the Niger, where George Goldie Taubman (later Sir George Goldie) succeeded in 1879 in persuading the four main trading firms to amalgamate under his leadership in order to force down produce prices and share profits. In 1880 the United African Company, as the amalgam was now called, made an exclusive agreement with the Emir of Nupe whereby, in return for supplying the Emir (and him alone) with arms and ammunition, he agreed to grant the company a local monopoly. But he broke the agreement in 1882 by admitting French traders who were offering higher prices than the U.A.C. Goldie and the U.A.C. now realised that they could not go on indefinitely admitting newcomers into their amalgamated group and conceived the long-term aim of reviving the colonial device of the chartered company and themselves securing the power to administer the Niger, and thus exclude their commercial rivals, whether white or black. As this coincided with the beginnings of the 'Scramble', Goldie and his associates were in fact able to achieve this ambition with the chartering of the Royal Niger Company in 1886 after the Berlin Conference.[28]

The political partition of West African territory among the European powers may also be seen as a movement closely related to commercial and economic ends.[29] The initial phase, from 1882 to 1890, was, with the exception of the French occupation of the Western Sudan, one in which France and Britain succeeded in establishing political control over the coastal areas which during the nineteenth century had been the main centres of their economic and commercial activity, except in the Cameroons and Togo which fell to Germany in Bismarck's diplomatically motivated intervention of 1884–5. In the 1890s it was France which forced the pace in the African interior, attempting by military means to secure the maximum hinterlands for the coastal colonies in the Ivory Coast and Dahomey around which tariff frontiers could be thrown to create new markets for French industry. The British had no need for

27 P. Atger, *La France en Côte d'Ivoire de 1843 à 1893*, Dakar, 1962; B. Scnapper, *La Politique et le Commerce Française dans le Golfe de Guinée de 1838 à 1871*, Paris, 1961, pp. 259–62.
28 Flint, *Goldie*, Chapters III and IV.
29 This is not the view of R. Robinson and J. Gallagher in *Africa and the Victorians*, London, 1961, where strategic motives involving Egypt and the Suez canal are thought to have outweighed any local West African considerations for the British.

protective tariffs, but responded by seizing as much of the hinterlands of the Gold Coast and Nigeria as could be preserved for free trade. In this contest, although the French succeeded in acquiring a vast territorial empire, the British, who secured all the major clusters of thickest African population, were clearly the economic victors. They had first preserved and then expanded their territories according to a rough and ready strategy which attempted a real assessment of the economic potential of the areas taken over.

The partition of West Africa may thus be related, indirectly at least, to the failure of 'legitimate commerce'—the West African version of the pervasive mid-nineteenth century British doctrine of free trade—to set in train an economic revolution in West Africa and to integrate the West African economy into the expanding cash-based economies of Europe. Conversely, colonial rule after the 1880s may be regarded, from the economic point of view, as a system designed to overcome the earlier limitations and failures of *laissez faire*, and to orientate the mass of Africans into the production of commodities needed in world markets and into consumption of the products of Europe's industrial system. Even before the partition the development of the small coastal colonies such as Freetown or Lagos had demonstrated that it was European administration which created the maximum of economic change and went farthest to reshape African society along 'class' lines by its close involvement in a cash-measured economy. The establishment of colonial administration over larger territorial units after the partition itself proved to be the catalyst of fundamental economic change. Colonial administrations needed officials, great and small, black and white, to administer the domains; these had to be paid in cash, housed and transported about the country. Cash revenues were, therefore, needed and a military force had to be there to insist in the last resort on the taxes being paid (and incidentally to consume more revenue). A large and ever-expanding working force, salaried or wage-earning, therefore arose from the very fact of colonial administration. By 1895, revenue collection and the accumulation of revenue surpluses became a major source of the capital, with which the colonial governments began laying down railways, harbours, bridges and feeder roads to swell the export of cash crops and the yield of the revenues. And now the mass of West Africans began to respond not only with huge increases in the production of such traditional crops as palm oil or groundnuts, but with the cultivation of cash crops like cotton and cocoa which entailed new methods of agriculture, changed systems of landholding, novel attitudes to property and an increasing dependence upon a cash income which was determined almost entirely by impersonal forces outside their control. The African revolution had begun.

CHAPTER 11

The European partition of West Africa

J. D. HARGREAVES

In 1880 the European presence in West Africa was largely confined to coastal districts and the valleys of the navigable rivers—Niger, Senegal, Gambia. Apart from the southern Gold Coast, which Britain had annexed as a Crown Colony in 1874, the French colony of Senegal (which still did not effectively control the whole of its future territories) and the still more restricted Colony lands of Sierra Leone, foreign flags still flew only over isolated forts and settlements and European power hardly extended beyond the range of naval artillery. Twenty years later the whole of West Africa (saving only the Republic of Liberia, whose independence seemed precarious) had been divided out by diplomatic agreement among the European claimants, whose armed representatives were everywhere advancing to make their rule effective. One of the great revolutions of African history had taken place with surprising speed. Considering the area involved and the proud independence of the peoples there had also, on the whole, been surprisingly little violence.

Sudden and dramatic though these changes were, they did not occur without preparation. The 'new imperialism' of late nineteenth-century Europe was linked to its predecessors by the varied expedients of 'informal empire'. Earlier chapters of this work have provided examples of the methods by which French and British Governments tried to secure for their subjects some of the fruits of empire, while limiting their own military and financial responsibilities. The logic of these earlier Afro-European relations already indicated increasing African dependence, in some form, on the insurgent, technological, individualist societies of industrial Europe; the far-sighted could already anticipate an increasing use of European power. But few expected this power to extend so suddenly and brutally.

African rulers who received foreign merchants or missionaries everywhere expected to retain their sovereignty; Al-Hajj 'Umar's letter to the French Governor in 1854 expressed assumptions common to them all:

The whites are only traders: let them bring merchandise in their ships,

> let them pay me a good tribute when I'm master of the Negroes, and I will live in peace with them. But I don't wish them to erect permanent establishments or send warships into the river.[1]

Such assumptions were not necessarily as naïve as they later came to seem. The maintenance of African sovereignty was perfectly acceptable to many mid-Victorian statesmen—on condition that African sovereigns collaborated in programmes presented to them for the advancement of civilisation and commerce, or the 'peaceful radiance' of French institutions. Localised annexations such as those of Dakar (1857) and Lagos (1861), naval bombardments of states which obstructed the expansion of foreign commerce, did not imply abandoning hope that the natural result of cultural and commercial contacts would be to expedite the evolution of African societies along paths already trodden by favoured nations of Europe. Nor did experience in their few established colonies encourage European governments to lay ambitious plans for formal empires.

African attitudes towards these processes of informal empire varied, as they did among Europeans. Elites in the coastal colonies might identify themselves quite positively with the aims of their rulers. In Freetown Creole leaders joined with Europeans in demanding stronger government action on the colonial frontiers, although in 1885 one angry young man publicly dissociated himself from the measures they proposed to expand a demoralising 'rum and gin civilisation'.[2] Attitudes of independent rulers differed similarly. George Pepple of Bonny adopted a wide range of Anglo-Saxon attitudes and sought European support in spreading Christianity, civilisation and commerce among his subjects; Jaja of Opobo, although anxious to develop export trade under his own control, was still more strongly 'determined on preserving indigenous religion and institutions at all costs'. Jaja saw no contradiction here; the record of consular influence in the Niger delta, and indeed the attitude of Lord Salisbury himself, encouraged confidence that in the last resort African independence would be maintained. From a Nigerian viewpoint it was thus the arbitrary deportation of Jaja by Vice-Consul Johnston in 1887, and Salisbury's failure to overrule him, which signalled the change in the spirit and purpose of British policy.[3]

For Lat-Dior, Damel of Cayor, the change arrived a little earlier. Having resisted Faidherbe's prematurely expansionist policies in the 1860s, he found a basis for somewhat uneasy collaboration with the less active governor Valière. Lat-Dior's policy was to strengthen his authority within his hitherto disorderly state by simultaneously encouraging Islamisation and the cultivation of groundnuts; his success in the latter

1 P. Cultru, *Histoire du Sénégal*, Paris, 1910, p. 337.
2 Sierra Leone Association. Paper by Samuel Lewis, Freetown, 1885. The phrase quoted is from the contribution to the interesting discussion by A. J. Sawyerr, p. 30.
3 E. A. Ayandele, *The Missionary Impact on Modern Nigeria*, London, 1966, pp. 77 ff.

policy made the former more acceptable to the French. Yet his very success in commerce encouraged the French to drive the Dakar–Saint Louis railway through Cayor, hoping to increase production still more; in 1882 Lat-Dior again resorted to arms to resist this obvious harbinger of French control.[4] But elsewhere it was often some years later before African rulers understood how grave a threat to their independence was developing in Europe.

It was not only Africans who were troubled by the implications of the change in Afro-European relations during the last two decades of the century. But early European protests came mostly from missionaries or traders whose own interests were in some way involved; West African affairs were still very much the concern of specialised minorities. Contemporary historians and commentators could overlook the breaks in continuity and see the conquest as involving simply 'the visible entry of Africa into the empire of civilisation'—'the one barbarous continent parcelled out among the most civilised powers of Europe'.[5] It was only later, as imperialist rivalries seemed to have grown important enough to cause revolutions in European diplomatic alignments and Britain's South African entanglements began to exercise profound and disturbing influences on her own internal politics, that more critical analyses began to be made. Because of certain rather obvious features of the South African situation, and also of deeper-lying tendencies which could be discerned within European society, these usually laid great emphasis on the influence of capitalists and their desire to maintain the rate of profit by finding lucrative and secure outlets for investment overseas. The interpretative essays of Hobson and Lenin, two men passionately concerned about the effects of imperialism on developments in Europe, were to have long-lasting effects on writing about the partition of West Africa. The very existence of their hypotheses encouraged many enquirers to begin with a search for financial capitalists and monopolistic trusts (or with a denial of their existence) rather than an examination of the changing activities and demands of Europeans on their frontiers of influence in Africa.

NEW FRONTIERSMEN AND NEW PROGRAMMES

It is certainly possible to detect the arrival in West Africa during the late 1870s of new sorts of 'frontiersmen', commanding more capital than their predecessors and with more far-reaching requirements for diplomatic or armed support in determining the conditions of their commerce.

4 Vincent Monteil, 'Lat-Dior, Damel du Kayor (1842–1886) et l'Islamisation des Wolofs', *Archives de Sociologie des Religions*, 16, 1963, pp. 77–104; Germaine Ganier, 'Lat-Dyor et le chemin de fer d'arachide, 1876–1886', *Bulletin de l'I.F.A.N.*, Ser B, xxvii, 1965, pp. 223–81.

5 Emile Banning, *Le Partage Politique de l'Afrique*, Bruxelles, 1888, quoted in R. F. Betts, *The Scramble for Africa*, Boston, 1966, p. 1; J. Scott Keltie, *The Partition of Africa*, 2nd ed., London, 1895, p. 1.

Goldie, reorganising the Niger traders, enforcing his monopoly with help from departments of the British Government, and ultimately obtaining a Charter authorising him to govern in the name of the Crown, is an obvious example. He is hardly typical; but neither is he unique, even at this period. C. A. Verminck of Marseilles, moving into shipping and the processing of vegetable oils after nearly forty years in West African trade, was laying the foundations of a great commercial company of widespread interests, the future C.F.A.O.; the Rochellais trader Arthur Verdier, charged with maintaining the French flag on the Ivory Coast since 1870, began in the 1880s to develop exploration and inland commercial penetration, to prospect for gold, and to experiment with coffee plantations. Problems arising from such extensions of 'legitimate commerce' of course affected African participants as well as Europeans, and Jaja's response to market conditions was no less vigorous than Goldie's. But the Europeans' greater power to control the course of events by involving their governments could make the change in their attitudes of more decisive importance.

Yet even now it was not usually full territorial sovereignty that traders demanded; the *pax* they wanted to impose on their African customers might go beyond the conditions characteristic of the 'imperialism of free trade' without involving the full apparatus of colonial rule, customs services and all. Chartered company rule, for example, need imply little more than the right to exclude or control the activities of foreign competitors and to punish Africans like the Brassmen, who rejected the dependent roles prescribed for them; to adopt the terminology which Goldie introduced only when chartered government was nearly at an end, commercial interests in nineteenth-century West Africa rarely required *landeshoheit*, and even when *oberhoheit* was claimed it did not necessarily press very hardly upon Africans except in questions affecting external market conditions.[6] Traders still demanded full imperial control only when the expedients of informal empire seemed manifestly exhausted. The Europeans who made the most immediately far-reaching and novel demands upon Africans were not capitalists, financial or commercial, but officers of the Marine Infantry, who from 1879 spearheaded the French advance from Senegal towards the upper and middle basins of the Niger.

Although some sort of scramble for political control of African territory was no doubt the logically inherent consequence of the rivalries among European traders on the coasts, this French initiative did much to determine the tempo and the climate of the European conquest. It seems to have owed little to pressure from established French commercial interests, who were in general content to accept the limitations imposed by existing patterns of trade in the lower Senegal basin. The ambitious

6 G. Goldie, Introduction to C. S. Vandeleur, *Campaigning on the Upper Nile and Niger*, London, 1898, reprinted D. Wellesley, *Sir George Goldie*, London, 1934, pp. 175–6.

schemes of railway construction launched in 1878–9 as counterpoise or complement to the still more grandiose trans-Saharan project certainly implied extended military control, but these schemes, even if they invoked visions of ultimate economic development, were essentially political in character. These were government railways into which no prudent investor would have put his capital and from which only contractors could possible hope to draw financial profit. The real incentive was to reach the countries of the Niger valley, subject of imaginative projects and source of imaginary riches; and to do so before the British, whose somewhat desultory interest in the 1870s was greatly magnified in the minds of patriotic Frenchmen.

These projects were set in motion by Colonel Brière de L'Isle of the Marine Infantry, Governor of Senegal from 1876 to 1881. Between 1854 and 1865 Faidherbe had pursued a more or less deliberate policy of consolidating French power in the lower Senegal valley, by armed force when necessary, while expanding French influence towards the upper Niger peacefully, by diplomacy and commerce. But in 1869 Governor Valière was appointed to inaugurate a period of peace, retrenchment and cautious restriction of direct responsibility; these tendencies became still more marked after France's defeat by Prussia in 1870. Brière, reacting against such pusillanimous measures, harnessed to his new forward policy the resurgent patriotism of young French officers humiliated by Sedan, and sometimes further disturbed by the implications for their country and their class of the establishment of the Third Republic. In the lower Senegal he sought to re-establish Faidherbe's empire of direct control, founded on military power and held together by the Dakar–Saint Louis railway (the one element in the whole programme with immediate commercial significance). In the Sudan, Brière's governorship saw the establishment of the autonomous military command of Haut-Sénégal-Niger, based on Kayes; young colonels and majors, impelled by vague concepts of a natural civilising mission or simply by personal and professional ambition, were to use the largely African forces of this command as the engine of imperial advance, culminating in the 1890s in outright war against the Muslim states of the Tukolors and of Samori.[7]

This private Sudanese empire of the French Marine Infantry, it may be argued, was an exceptional case, providing no exact parallels with the conquest of Nigeria, or even of Dahomey. It is nevertheless an important case, both because of its influence on policies and attitudes elsewhere in West Africa, and because it directly embraced a great part of the total area of the region. The methods by which these inland areas were brought under military control differed from those used in the expansion of the old coastal colonies of Britain and France, and had far-

7 A. S. Kanya-Forstner, *The Conquest of the Western Sudan, 1879–1899,* Cambridge, 1969.

reaching effects on the future of the inhabitants. It may or may not be important for the general debate about the significance of the new imperialism in world history to note that this most vigorously aggressive group of imperialists had little connection with any sections of French society which could be held to represent a mature and aggressive capitalism. What mattered to the peoples they brought under French rule was their supreme confidence in their right to take control in the name of their country, and their readiness to deal ruthlessly, sometimes brutally, with those who impeded them.

Other agents of European expansion might act less violently than the French military; but increasingly they shared the arrogance of their approach. Changing European attitudes, some of which have been described in other chapters of this volume, at once underlay the changing ambitions of African 'frontiersmen' and were profoundly influenced by their activities. Increasingly common among them was the confident belief that the superior culture of Europeans (sometimes, though not invariably, identified with racial superiority) gave them a right—even a duty—to determine what should happen in Africa. This outlook, which was to obstruct the achievement of human contact between European and African in the twentieth century, may be detected not only among commercial and military empire-builders but, for example, among the young University-educated missionaries who on their arrival in the Niger mission in 1887 assigned themselves the duty of rectifying the shortcomings of Bishop Crowther and his clergy.[8] The intellectual background of the new European imperialism is complex and full of contradiction; among its rising themes are the muddled racial theories favoured by early anthropologists, the secularised service ethic of the English public schools, the ethnocentric paternalism of many missionary societies and the shrill chauvinism with which many political leaders tried to entrance enlarged electorates. As impressions and prejudices derived from African experience were fed back into Europe a new 'image of Africa' began to emerge, as an incorrigibly barbarous continent, more suited for civilisation by conquest than by the old agencies of Christianity and commerce. As one Liberal Minister who regretted the trend noted, an increasingly common prescription for African disorder was the introduction of white men to take care of the black men.[9]

To justify more active government policies it was frequently alleged that African authority was breaking down, so that European intervention would be necessary to save the continent from 'tribal war'. It was with such arguments that Freetown's leading European and Creole citizens

8 J. F. Ade Ajayi, *Christian Missions in Nigeria, 1841–1891*, London, 1965, ch. 8; Ayandele, *Missionary Impact*, pp. 207–17.

9 Minute by Kimberley, 4 December 1882, quoted by D. Kimble, *A Political History of Ghana 1880–1928*, Oxford, 1963, p. 277.

urged stronger government action in the 1880s to enforce order within Sierra Leone's established sphere of commercial interest. These assumptions were questionable. Hostilities were certainly taking place in several quarters of the future Sierra Leone protectorate; indeed the expansion of the commercial economy, and the social changes thereby produced, often exacerbated old rivalries, and produced new causes of conflict. But the prime cause of the Colony's economic problems (perhaps also of some of the 'tribal wars' themselves) was not the collapse of African authority but the instability of agricultural prices in international markets.[10] Nevertheless, where Africans were fighting, Europeans now seemed increasingly moved to join in. During the period 1885–96 the British government (aware also of the growing dangers of French intervention) accepted the need for more direct control and authorised the creation of new instruments of frontier policy such as the Sierra Leone Frontier Police, formed in 1890.[11] These in time became expansionist forces in their own right. Active militia officers seconded from colonial service sought opportunities for gallantry in the service of Queen, country, civilisation and promotion; Creole traders, following their flags, opened new markets and new rivalries; African resistance to frontier justice or interference with slavery helped to justify the establishment of the formal Protectorate of 1896. But this time an operation originally undertaken reluctantly to prevent deterioration could be presented to the British public as a national mission of highly constructive purpose: a stage (to quote the title of a book by a Frontier officer with Lugardian ambitions) in *The Rise of Our West African Empire*.

The belief that Africa, unless preserved by European intervention, would relapse into anarchy and tribal war was at least as much the product of this new imperialistic outlook as its cause. Where something of the sort might seem to be happening (as in the Sierra Leone and Guinea hinterlands and to some extent among the Yoruba) it was largely due to the effects on existing feuds and contradictions within the African political structure of new tensions arising directly from the very successes of European civilisation and commerce. But in some states at least these tensions were being resolved in ways which produced structures and rulers much better adapted to existence in the commercial environment. It was the growing strength and confidence of Jaja and Lat-Dior, not their weakness, which provoked European intervention against them. Dahomeans and Tukolors, in different ways, were trying to use resources and ideas derived from intercourse with the West for the strengthening and renewal of African institutions. In the 1880s Samori, with the help of home-produced and European firearms, was building a strong empire of a new type. The

10 J. D. Hargreaves, *A Life of Sir Samuel Lewis*, London, 1958, pp. 40–58.
11 N. H. R. Etheridge, *The Sierra Leone Frontier Police*, M.Litt. thesis, University of Aberdeen, 1967.

reception of reforming ideas and techniques from the Atlantic seaboard, as well as from the heartlands of Islam, was bringing about drastic re-distributions of power within Africa. A sort of internal partition was already in progress, involving the emergence of rulers with modernising attitudes, and definite ideas of how to serve the interests of their states in dealings with foreigners. In mid-century some Europeans had been prepared to welcome and even encourage the emergence of such 'strong native powers' and to seek means of collaborating with them for mutual advantage. If such attitudes became much rarer and less effective in the 1880s and 1890s this was partly due to apprehension that some strong native powers might grow too strong, and partly to the development of tribal feuds and international tensions within the continent of Europe.

EUROPEAN GOVERNMENTS BEGIN TO SCRAMBLE

The French campaigns in the Upper Senegal which began in 1879, although prompted in part by fears of some great British enveloping strategy, at first provoked little reaction in London. The area of Futa Jalon and the upper Niger basin had never lived up to the optimistic commercial appreciations periodically formulated in Freetown and Bathurst, and a reconnaissance by V. S. Gouldsbury in 1881 produced a particularly depressing report; the highest hopes of spreading civilisation and commerce remained attached to the Sokoto Caliphate. But in 1880 the Secretary of State could still assume that 'We shall get to the Hausa country via Lagos long before the French'; Britain's limited interests in West Africa could hardly be much affected if the French chose to squander resources on military campaigns and wild railway projects in the Senegal valley.[12]

French activity in the lower Niger was another matter. By 1882 Goldie saw the enterprise of Count Semellé's *Compagnie Française de l'Afrique Equatoriale*, and of Verminck's Senegal Company, as threats to the monopoly on which he based his grand design for the Niger (whatever this may have been), and was actively preparing to assert political control in the name of the British government. But unexpectedly he encountered resistance not merely from these companies but from the French Government, which had hitherto shown only desultory interest in the affairs of the lower Niger. Forces in Paris were now working for a more assertive defence of French interests and claims in West Africa generally (whereas the pressures behind the advance into the Sudan were in large part locally generated). These forces, largely political in orientation, help to explain the

12 C.O. 87/116, Minute by Kimberley to F.O. to C.O. 4 August 1880; also Minute by Meade on Rowe to C.O. 20 August 1880. For additional reference and amplification of the argument in this section, see John D. Hargreaves, *Prelude to the Partition of West Africa*, London, 1963, ch. 5.

immediate origins (as against the underlying significance) of the sudden scramble to appropriate African territory which began in 1883–4.

It was not mere coincidence that this was precisely the period when Anglo-French relations began to be dominated by the dispute arising out of the British occupation of Egypt during the summer of 1882. Distinguished scholars have suggested that the latter event led very directly to new initiatives south of the Sahara. Seeking for 'adequate compensation', or even 'quickened by the hope of prising the British out of Cairo, the French drove deep into west Africa'; the British response was in turn subordinated to 'the greater game for the vital interests in the Mediterranean and the valley of the Nile'.[13]

As an explanation of the partition of West Africa this view is oversimplified (as its authors seem to recognise when they seriously get to grips with the evidence relating to the area). For one thing, signs were apparent before 1882 that the diplomatic restraints upon Anglo-French colonial rivalry might be weakening. Many officials had long been discontented with the enforced passivity in foreign affairs which followed France's isolation and defeat in 1870. The Tunisian protectorate of 1881 was largely prompted by a feeling that inertia in the face of Italian encroachments represented both a danger to French prestige among North African Muslims and 'a menace to our whole foreign policy'.[14] Henceforth, however, Gambetta, potentially the strongest of the Republican leaders, identified himself with this new doctrine that France's position in Europe could be strengthened by a vigorous defence of her overseas interests. During his short ministry of November 1881 to January 1882 colonial policy was in the hands of two provincial businessmen with overseas interests, Maurice Rouvier and Félix Faure, who gave clear indications of their desire to 'accentuate' the defence of France's local West African interests against Britain.[15] When the collapse of the joint Anglo-French Egyptian policy which Gambetta had encouraged resulted in a British occupation which was clearly likely to be long-lived, strong resentment of British perfidy was expressed by the French public (or at least by newspapers and parliamentarians anxious to justify their patriotism to the public). These repercussions of the Egyptian question finally released the braking mechanisms in Paris which had held back patriotic naval officers and consular agents on the frontiers of French influence in Africa. The first reaction came on the Congo, where the implications of King

13 R. Robinson and J. Gallagher, *Africa and the Victorians*, London, 1961, pp. 162, 189.

14 *Documents Diplomatiques Français,* 1st series, iii, no. 405; Noailles to St Hilaire, Pte. 19 March 1881; cf. no. 406 St Vallier to Noailles, Pte. 21 March 1881. For the Tunisian affair, Jean Ganiage, *Les Origines du Protectorat Français en Tunisie*, Paris, 1959.

15 Ministère des Affaires Etrangères, Mémoires et Documents, Afrique 57, Rouvier to Gambetta, 3 January 1882; Hargreaves, *Prelude*, pp. 278 ff.; J. P. T. Bury, 'Gambetta and overseas problems', *English Historical Review*, lxxxii, 1967, pp. 277–95 deals with the background, though not with the effects in West Africa.

Leopold's activities were beginning to appear, and a suitable representative of the nation was already present; the unusual decision of the Duclerc government to seek formal parliamentary ratification of Brazza's treaty with the Bateke showed appreciation of the political advantages which might reflect upon them from the exploits of an attractive popular hero. Shortly afterwards the French Government moved in the Niger region also; several related decisions by the Ministry of Marine in January 1883—to establish protectorates at Porto Novo and in the Popos, to 'develop our influence alongside that of the British' in the Niger and especially the Benue, to seek treaties in the Cameroons—showed a readiness to cross lines previously respected.[16] There was not necessarily a considered intention of establishing a permanent French presence in all these areas but any compensation which might eventually be obtained by abandoning these initiatives was likely to be found elsewhere in West Africa, not in Egypt.

Some British policymakers responded quickly to these signs of change in Paris; but not all. One Colonial Office official apparently connected them at once with the advance already begun from Senegal. 'The French put us to shame by their activity in these regions,' [specifically, Senegambia] wrote Hemming in February 1883, 'and if we do not take care they will obtain their reward by diverting to themselves and their colonies the greater portion of the West African trade.' To this Derby, the inert Secretary of State, returned what he thought a conclusive answer: 'I think the French will soon get tired of throwing away their money.'[17] French activity in the Niger and Congo areas, however, forced even Derby and his colleagues into responses of some sort. There were growing fears of serious fiscal discrimination in French colonial territories and both Goldie and those interested in the growing Congo trade were actively demanding more security for their operations. In the Congo, Britain reacted by an elaborate attempt to renovate the somewhat shaky edifice of Portuguese sovereignty under guarantees for free commercial access. For the Niger delta and the Cameroons, after lengthy ministerial discussion and haggling over small expenditures of public money, a system of protectorates was authorised, to be based upon treaties with African rulers which would not only exclude the French but would permit the development of plans formed by Consul E. H. Hewett for 'altering the whole system of British policy as regards the native chiefs of the West Coast of Africa and the abolition of the monopolies which are destructive to trade and productive of endless

16 Jean Stengers, 'L'impérialisme colonial de la fin du XIXe siècle; mythe ou réalité', *Journal of African History (J.A.H.)*, iii, 1962, pp. 469–91; Henri Brunschwig, 'Les origines du partage de l'Afrique occidentale', *J.A.H.*, v, 1964, pp. 121–5.

17 C.O. 87/121, Minutes by Hemming, Derby, on F.O. to C.O. 24 February 1883. Derby made an identical comment on F.O. to C.O. 7 July 1883.

squabbles and wars'.[18] From the point of view of London (though not necessarily of the British on the coast) even this was a reluctant extension, accepted partly in the hope that further responsibilities might be limited by some fiscal or territorial agreement with France. But this possibility was now reduced by the unexpected claims of a third major European government.

Bismarck's motives for bringing the German Government into active competition for African territory during the spring of 1884 have been much discussed. Although German missionaries and traders from Hamburg and Bremen had been active on the West African coast since the 1830s, it seems that the decisive clue to his actions is not to be found by detailed scrutiny of their interests but somewhere in the complexities of international or domestic German politics. The immediate result, however, is clear. Not only did Germany acquire protectorates in Togo and Cameroons (which soon became a more direct responsibility of the Imperial government than Bismarck had intended); but—and this Bismarck certainly had intended—the various local Anglo-French disputes became involved in a much more complex pattern of inter-European relations. The international conference which met in Berlin from November 1884 until February 1885 was called to regulate these relations, not to 'partition Africa'. Its actual results were much more technical and limited than legend suggests—agreements to secure customs-free access for European trade to the whole Congo basin and (supposedly) free navigation of the Congo and Niger rivers, enunciation of some rather sanctimonious general principles and the definition of a formula to minimise inter-European friction when future colonies or protectorates should be established on African coasts. (By the time the treaty was signed, few African coasts remained unappropriated; all the future colonies of the West African seaboard already existed in embryo, although boundaries might remain indefinite, and occupation was usually still very far from effective.) Yet in a more general sense the legend is not so very misleading. Africa had become clearly a matter of greater concern to European chancelleries than ever before; the repercussions of inter-European rivalries now reinforced the expansionist tendencies of those forces of civilisation and commerce which were already operating on the frontiers. The result was to accelerate the conquest and partition of the continent as a whole, including inland districts which had hitherto seemed safely remote.

INTERLUDE AND CONQUEST

Immediately after the Berlin Conference there was something of a pause in European activity. It is tempting to suggest economic explanations for this; but it seems premature to press any single explanatory model too

18 P.R.O. 30/29/135 (Granville Papers), Memo by Lister 4 January 1882.

far, at least 'until a history of West African trade and commerce has been written'.[19] During these years West African trade passed through a crisis which affected both African and European participants. From 1885 the international economic depression reduced the prices obtainable for most African products,[20] sharply cut the value of seaborne commerce, and so restricted the budgets of colonial governors. The political consequences varied. On the one hand falling rates of profit might encourage Europeans to press for strong action against African merchants (as Liverpool members of the African Association did in the case of Jaja). Yet Johnston's handling of Jaja was hardly typical of government policies in 1887; elsewhere even Johnston favoured a more negative approach. 'So long as we keep other nations out we need not be in a hurry to go in.'[21] Even where merchants were demanding military action as the remedy for their ills, funds could not easily be found and the old tendency to seek interim solutions by negotiation with African rulers persisted. Continuity with old European policies of limited commitment was not broken. If the tide of political imperialism had really been flowing strongly in European capitals governments would surely had been more ready to help their African colonies hit their way out of trouble by spirited policies of territorial advance.

In fact the political enthusiasm which had mounted during the hectic scramble of 1884 cooled during the next few years. Bismarck, having achieved his immediate purposes, seems to have been somewhat disturbed to discover the extent of the responsibilities he had acquired in Africa; his efforts were directed less towards the inland expansion of influence and supervision over African middlemen—which German traders in the Cameroons wanted—than to inducing those traders to bear the costs of administration.[22] Indeed Bismarck soon became ready to consider bargaining new colonies away.[23] British politicians turned to more absorbing

19 C. W. Newbury, 'Victorians, Republicans, and the partition of West Africa', *J.A.H.*, iii, 1962, pp. 498–501. Cf. A. G. Hopkins, 'Economic Imperialism in West Africa: Lagos 1880–92', *Economic History Review*, xxi, 1968, pp. 580–606.

20 *Parliamentary Papers*, 1890–1, lxxxix, 'Statistical Abstract of the U.K.', calculates average prices of palm-oil imported into U.K. as follows (shillings per cwt):

1877	35.62	1884	33.50
1878	34.80	1885	26.90
1879	30.52	1886	20.92
1880	29.43	1887	19.48
1881	29.09	1888	19.83
1882	30.49	1889	21.17
1883	35.11	1890	22.90

21 Johnston to Anderson, Pte. 13 November 1886, quoted in J. C. Anene, *Southern Nigeria in Transition*, Cambridge, 1966, p. 72.

22 *Kamerun under Deutscher Kolonialherrschaft*, ed. H. Stoecker, E. Berlin, 1960, pp. 81–6; H. R. Rudin, *Germans in the Cameroons*, New Haven, 1938, ch. iv.

23 For early suggestions of readiness to withdraw from Togoland see C.O. 879/28, C.P. African 356, no. 6, Griffith to Holland, Conf. 21 March 1888.

problems of domestic and Irish policy; the return of a Conservative government brought no sudden shift to a new imperialism, although Salisbury did begin to acquire a coolly practical understanding of British interests in Africa. In France the general mood of disenchantment with colonial adventures which set in when Ferry was defeated in the Chamber in March 1885 over a military set-back in Indo-China was reinforced by awareness of specific difficulties and dangers in the Senegal–Niger area.

In February 1882 French forces on the upper Niger clashed with those of Samori, a ruler of whom they had previously known little or nothing. Borgnis-Desbordes and his officers clearly regarded him from the first as a rival claimant for military power in the Sudan, and saw in his defeat a possible source of further military glory; but their hold on the Niger was still precarious and for a time the French had to temporise with him, as with Amadu, while building up their forces. Three years later Colonel Combes felt strong enough to invade Bouré, a gold-bearing region of great economic importance to Samori; but he underestimated his opponent and had to withdraw in June 1885. His small force, at the end of long and difficult lines of communication, suddenly seemed in jeopardy from all sides. Lat-Dior continued to resist in Cayor until October 1886; before he had been defeated, Mamadu Lamine (a Muslim Sarakule who had performed the *haj*) raised his people in rebellion in the upper Senegal valley; meanwhile the railway project went forward slowly, and at many times the expected cost. From Paris especially, the danger of a new Lang-Son disaster in Africa seemed alarmingly great.

Had such a disaster occurred, it would certainly have restrained French expansion, or even halted it; and, since France was now the pacemaker, would have greatly affected the speed and structure of the whole European partition. But France's enemies failed to combine. The man who might most clearly have seen the need for African unity was Amadu, their oldest antagonist, who was indeed reorganising his states in preparation for such a conflict; but he seemed equally determined to resist the challenge to his authority represented by rival Muslim empires. He had formerly held Mamadu Lamine in prison and had no wish to encourage Sarakule nationalism within his own sphere of power and influence. Nor would he make any move to cooperate with the upstart empire of Samori. Strangely enough it was Samori who responded more positively to the French menace, by attempts to win British friendship and to secure his supply of firearms through Sierra Leone; but even Samori gave priority to his feud with Tiéba of Sikasso over the needs of his relations with France and did not seek collaboration with Amadu until too late.[24]

Nevertheless, when Gallieni returned as commander of the Upper

24 Yves Person, 'Samori et la Sierra Léone', *Cahiers d'Etudes Africaines*, vii, 1967, pp. 5–26; John D. Hargreaves, 'West African States and the European Conquest', *Colonialism in Africa*, eds. L. H. Gann and P. Duignan, Cambridge, 1969, i, pp. 199–219.

Senegal–Niger district between 1886 and 1888 there was some temporary reassessment of French objectives. While concentrating on the military defeat of Mamadu Lamine, Gallieni signed treaties which established French protection, but not French control, over the states of both Amadu and Samori; and although some of his officers certainly regarded these merely as cynical expedients to buy time, there is evidence in Gallieni's earlier and later career to suggest that he may not have excluded the possibility of advancing French power in collaboration with the Tukolors rather than by military operations against them. Other treaties with Tiéba, Aguibou, and the Almamies of Futa Jalon completed a system designed to exclude British influence from the Sudan rather than to consolidate French power. Gallieni, convinced that 'the richest parts of French Sudan were to be found to the southward, in the upper valleys of the Niger and its tributaries and in the lands separating them from the Southern rivers',[25] began the process of swinging the French centre of gravity away from the Senegal basin. He placed the chief emphasis on the future colony of Guinea; but after his departure Captain Binger, whose journey he had authorised, revealed the possibilities of the Ivory Coast by arriving at Grand Bassam after traversing Mossi during his long march along the *dyula* routes from Bamako.

As Europeans in the coastal colonies became more aware of possibilities for development, so their governments felt the need to extend towards the interior those lines of partition which had so far usually been drawn only on the coastline. Entrepreneurs like Verdier (whose agents represented French authority on the Ivory Coast until 1889) wanted to develop agricultural or mining operations, or long-range trade, and sometimes their ambitions overflowed into areas of interest to their European neighbours. Freetown merchants not only sent precision firearms and other goods to Samori but sought to substitute a British protectorate for the one France claimed; Edouard Viard, a former agent of C.F.A.E., acted with Catholic missionaries in exploiting Egba dissatisfaction with the British, thus hoping to expand the French protectorate from Porto Novo to the lower Niger.[26] There was renewed talk of a comprehensive Anglo-French agreement which might involve the exchange of the Gambia; instead there was a more limited agreement on 10 August 1889, regulating existing colonial boundaries in the Gambia, Sierra Leone, Assinie and Porto Novo (and thus making them all the harder to modify).

But now a new and stronger wave of French expansionism was developing. The first initiative came from Archinard, the ruthlessly ambitious commandant in Upper Senegal–Niger. His attack on Koundian in February 1889 marked the beginning of a drive to destroy the Tukolor

25 J. S. Gallieni, *Deux Campagnes au Soudan Français*, Paris, 1891, p. 620.
26 Ayandele, *Missionary Impact*, pp. 49–51; E. Viard, *Au Bas-Niger*, Paris, 1885.

empire (and to secure promotion for himself); just over two years later Archinard's attack on Kankan marked the end of coexistence with Samori and the beginnings of a seven-year campaign against that still more formidable opponent. Although the French suffered only one outright defeat, at Timbuktu in January 1894, the Sudan was not conquered easily. The logistic difficulties of maintaining and controlling forces operating far from their bases, the tenacity of African resistance, and the difficulties of establishing alternative African authorities to work under French orders in administering the conquered territories, all slowed down the pace of advance.

The military commanders also encountered problems in their relations with the politicians in Paris. The foundation in 1890 of the *Comité d'Afrique Française* as an expansionist pressure group and the tenure of the Under-Secretaryship for the Colonies from 1889–92 by one of its patrons, the Algerian-born Eugène Etienne, may be taken to mark the emergence of African imperialism as a constant and organised force in French politics. Two years later Etienne became chairman of a 'colonial group', which claimed the support of ninety-one Deputies drawn from all parts of the Chamber; in 1893 bankers and merchants with interests in Africa and Indo-China came together in the *Union Coloniale Française*.[27] Although supporters of these bodies shared the desire of the military to establish French control over large areas of the African land mass, they dissented sharply from them over priorities and methods. While Archinard saw the Sudan essentially as a battlefield on which to win laurels for France and himself, the colonial party remained convinced that it contained rich and populous territories which could become the heartland of a continental empire. They favoured speedy reconnaissance and treaty-making expeditions towards Lake Chad, the pivot which could unite Algiers, Saint Louis and Brazzaville into a single bloc. In 1890, even before the *Comité d'Afrique* was constituted, its members had provided support and financial backing for three probes towards Lake Chad—Crampel's expedition by the Ubangui, Monteil's from Saint Louis, and Mizon's voyage up the Benue to Adamawa, testing the Royal Niger Company's professed acceptance of the principle of free navigation. Archinard on the other hand did not take treaty-making seriously; impatient under the restraints imposed from Paris and Saint Louis, he was convinced that it was necessary 'to get finally rid of these two customers, Samori and Amadu'.[28] Again, while the colonial party hoped to see civilian administration introduced as quickly as possible so as to prepare the way for regular

27 H. Brunschwig, *Mythes et Réalités de l'Impérialisme colonial français,* Paris, 1960, ch. 8. Kanya-Forstner points out however that the 'colonial party', far from reacting against the Anglo-French Treaty of 4 August 1890, welcomed it as leaving open access to Lake Chad.
28 Archinard to Desbordes, 24 October 1889 in J. Méniaud, *Les Pionniers du Soudan*, Paris, 1931, i, pp. 434–8. See also Kanya-Forstner, *Conquest.*

business activities by French firms, the soldiers abandoned control of their conquests only slowly and reluctantly. There was nevertheless sufficient concordance between the objectives of politicians and military to give the impression that French power was advancing according to a set plan.

Although grand designs for Africa as a whole tended to turn on Lake Chad, the upper Niger valley would be the point of union for French activities in West Africa. For Guinea the necessary link was secured by the progressive tightening of French control over the Futa Jalon area. On the Ivory Coast the French began to expand northwards from Assinie, somewhat hesitantly at first, after Binger had proved that no 'mountains of Kong' stood in the way; but before the French could consolidate their power in the savanna to the north of the Ivory Coast they had to deal with Samori's brilliant rearguard action in this area. The expansion of the French protectorates of Porto Novo and Great Popo raised still greater problems. Gelele had defended the independence of the Fon state of Dahomey with tenacity and skill, but latterly his independence had been increasingly menaced. French soldiers occupied Cotonou, his busiest port, and claimed a protectorate over Porto Novo, which he considered as a revolted fief; French missionaries were also increasingly undermining the traditional religion of his subjects. But the French Government hesitated to launch what might be a difficult military campaign until Behanzin, who became king in 1889, asserted his independence with a militancy no longer tolerable. In 1892 French troops (under the command not of Archinard but of the Senegalese General Dodds) occupied Abomey; two years later Behanzin was finally deposed and his kingdom broken up.

This operation opened up the possibility of a northward advance to reach the lower Niger below the Bussa rapids. Since Mizon's second filibustering voyage of 1892–3 the French had begun to realise that the monopoly claimed by the Royal Niger Company rested largely on a confidence trick. In southern Nigeria the British were only now moving to make their power effective; Jaja's deposition had merely been a first step in the transformation of Consular authority into more effective controls. After 1891 changes in the organisation and title of the Oil Rivers Protectorate were matched by piecemeal, but increasingly forceful, measures to subdue the rulers of the southern Nigerian states.[29] In Yoruba, Governor Carter's attack on Ijebu in 1892, and the treaty-making journey which followed it the next year, signalled a new readiness to 'resort to arms . . . in the interests of civilisation'[30]—or in other words to open Western Nigeria to fuller penetration by British trade and mission-work,

29 On these see Anene, *Southern Nigeria.*

30 Speech to Legislative Council, 27 November 1891, quoted in Ayandele, pp. 60–1. See also A. B. Aderibigbe, 'The Ijebu Expedition, 1892', in *Historians in Tropical Africa*, Leverhulme Conference Report, Salisbury, 1962, and his doctoral thesis, *The expansion of the Lagos Protectorate, 1863–1900*, London, 1959.

and so exclude the last danger of French infiltration through the Egba. But to the north of these infant protectorates the French would have only the paper tiger of the Niger Company to deal with if they decided to try and reopen the question of navigation on the lower Niger, and to test the claim to Hausaland which Goldie had not yet tried to convert into effective control. The famous 'race to Borgu' between Captain Decoeur and Captain Lugard in 1894 represented a French challenge to the heart of Britain's most substantial concentration of West African interests. Here—and only here—European governments envisaged a distinct, though distant, possibility that the scramble for West Africa might need to be settled by armed conflict among themselves.

At the diplomatic level there were signs that the struggle for the Niger might become a triangular affair. In both Togoland and the Cameroons German frontiersmen had begun to expand their influence inland and Bismarck's successors, less strongly resolved than he to maintain the priority of European over colonial interests, found themselves drawn into Anglo-French rivalries. In 1893 the British, worried by Mizon's activities on the Benue, tried to interpose a German buffer between the French in the Congo basin and Bornu (and also, less hopefully, between the French and the Upper Nile); this plan was upset by the Franco-German agreement of February 1894 which gave France a foothold on a navigable tributary of the Benue within the German sphere.[31] France and Germany seemed to be discovering common interests in opposing the Niger Company's navigation regulations, on which their monopoly rested; in March 1894 Dr Kayser, spokesman of colonial interests in the German Foreign Office, warned the British Ambassador that this might become 'the next great international question'.[32] There was even talk of France and Germany agreeing to give a north-easterly turn to the expansion of Togoland and Dahomey, so that both colonies might touch the Niger below Bussa.[33] German West African policy continued to be governed by wider considerations, so nothing came of these schemes; but, together with French attempts in the middle 1890s to reopen the Egyptian question by establishing an expedition in the upper Nile valley, they reflected the growing concern of European governments with African problems.

In Britain this was symbolised by the appointment to the Colonial Office in 1895 of the Unitarian businessman Joseph Chamberlain, symbol of the most forcefully expansionist elements in British public life.

31 Flint, *Goldie*, pp. 179–86; G. N. Sanderson, *England, Europe and the Upper Nile*, Edinburgh, 1965, pp. 107–10.

32 Malet to Kimberley, 7 March 1894, in T. A. Bayer, *England und der Neue Kurs*, Tübingen, 1955, pp. 117–20.

33 F.O. 64/1358, Malet to Kimberley, 117 African, 30 October 1895; F.O. 64/1373, D.M.I. to Sanderson, 4 October 1895.

Demands for a more vigorous defence of claims pegged out by frontiersmen were increasingly being heard—from Chambers of Commerce in provincial cities, zealous supporters of missionary endeavour, imperialist lobbies like the Royal Colonial Institute, idealists appealing now to the charter of enlightened expansionism which the powers had agreed on in the Brussels Conference of 1890. For some time the political effectiveness of these pressure groups had been contained by the old-fashioned anti-imperialism of Gladstone, the sceptical Toryism of Salisbury, or the tendencies to inertia of the 'official mind'; but during the 1890s radical opposition (often more concerned with economy than morality) tended to be shouted down while the 'disjointed and incoherent utterances' of African protest[34] could rarely secure a hearing at all. During Gladstone's last government the contest over the Uganda protectorate showed which way the cards were running. Chamberlain at the Colonial Office was the joker in the Imperialist hand.

His advent had important consequences for economic and administrative policies; but except on the Niger he arrived too late to affect the outcome of the partition of West Africa. The frontiers of Sierra Leone and the Gambia had already been substantially settled, and it remained only to organise the protectorates. To the north of the Gold Coast belated efforts to secure treaties with states in the savannah belt had been made under his Liberal predecessor; but although most of Gonja, Mamprusi and Dagomba were thus saved for eventual British control the French military were already poised for control of Mossi and Chamberlain could only struggle to get relatively favourable boundaries. In Asante he authorised the expedition of 1896 which provided the occasion for Prempeh's deposition and exile; yet even this was simply a logical outcome of the ambiguities in earlier Anglo-Asante relations. But on the Niger, once his attempts to arrange a comprehensive deal with the French fell through, Chamberlain insisted on a very tough line, and a risky policy of replying to French encroachments in kind. Although increasingly at odds with Goldie over the future of the Niger Company, Chamberlain supported his insistence on keeping the French away from the navigable Niger below Bussa; 'Nigeria with its 30 millions of Hausa and its civilisation and wants' was to him an undeveloped estate to be secured at any risk. This 'Birmingham Screw Policy' (as Lugard, its reluctant agent, called it)[35] was risky to the point of irresponsibility; but the French Government knew their position to be vulnerable internationally and at home, and in the agreement of 14 June 1898 Chamberlain's view, in its essentials, prevailed.[36]

34 A term used to describe Jaja's deputation in F.O. to Johnston, 30 September 1887 in *P.P.*, 1888, lxxiv, no. 53.

35 *The Diaries of Lord Lugard,* iv, eds. M. Perham and M. Bull, London, 1963, pp. 342–3, 334.

36 M. Perham, *Lugard: the Years of Adventure*, London, 1956, part vi; Flint, *Goldie*, ch. 12; Sanderson, *Upper Nile*, pp. 317–23.

Although a few minor frontier adjustments remained to be made and several areas (including most of the Sokoto Caliphate, and virtually the whole of Mauritania) were still not effectively occupied the European partition of West Africa was for most purposes complete. The attention of French and British Governments shifted even more decisively to the north—first to Marchand's small detachment at Fashoda and the underlying conflict over Egypt which had taken him there, then to the possibilities of diplomatic reconciliation through a businesslike partition of interest in northern Africa. Delcassé, transferred from colonial to foreign affairs, increasingly appreciated what such an agreement could mean for French policy in the Mediterranean and in Europe; the conversion of Etienne and his colleagues in the *parti colonial* to this scheme of priorities made it politically possible.[37] The Anglo-French agreements of 8 April 1904 represented that principle of 'comprehensive dealing' which British and French statesmen had intermittently hoped to apply to West Africa, ever since the exchange of the Gambia was first mooted in 1865. But the change in international alignments which permitted the application of this principle to Egypt and Morocco came too late to permit the creation in West Africa of large compact colonies with rational frontiers. The partition had been achieved with too much difficulty to be easily remade; Britain's cession of the Isles de Los to France, and some minor frontier rectifications in Northern Nigeria and the Gambia, merely removed some of the more troublesome defects of those often arbitrary boundaries by which remote negotiators had tried to compound the conflicts of French and British frontiersmen.

THE INITIAL IMPACT

As far as the European participants were concerned, the partition was completed without any major fighting among themselves: an interesting fact whose implications lie somewhat beyond the scope of the present volume. From an African point of view the absence of violence is less striking. True, much of the expansion of the coastal colonies was carried out by peaceful, if not always very frank and fair, negotiations and even when military campaigns were needed most colonial governments tried to carry them out with a minimum of force. It was after all present customers and future subjects who were being coerced and in any case the forces available were rarely adequate for carrying out extensive punitive campaigns. The Europeans' decisive weapons—such as the celebrated Maxim gun—might cause heavy casualties in battle but they could also serve to bring a campaign to a speedy conclusion. In the Sudan on the other hand the *Infanterie de la Marine* tended to use more consistently violent methods; even if psychopaths like Voulet and Chanoine (who in 1899

37 Sanderson, *Upper Nile*, pp. 374–80.

sought to turn themselves into African chiefs and carve out an empire in Hausa country by methods of sheer terror)[38] were in no way typical, this empire was taken by force and initially often needed to be held by fear. There was talk of liberating slaves from their oppressors; but for the individual slave the consequences of liberation might be to find himself detailed to serve as a carrier or even assigned as war-booty to a Senegalese soldier or an African auxiliary. Indeed, many features of the conquest most repugnant to European consciences were results of the conquerors' dependence on dubious and ill-controlled African collaborators.[39]

If the European invasion did not immediately strike with equal violence everywhere, neither were its effects on African institutions and structures everywhere the same. This was not simply because the invaders were moved by different ideas of empire—the French concerned to level and eventually assimilate, the British to preserve African aristocracies under indirect rule. The contrast between the fate of Samori's empire and Sokoto is certainly striking; but if Asante is compared with Abomey, or Mossi with Dagomba, no such clear antithesis stands out. Unlike the military empires of the Sudan, the French coastal colonies expanded under civilian direction and with regard for established commercial interests: hence, initially at least, without needlessly violating the independence of rulers who were prepared to cooperate with them. It was indeed precisely the complaint of G. L. Angoulvant, who became Governor of the Ivory Coast in 1908, that for the previous fifteen years the colony had attempted to penetrate inland by peaceful and commercial methods, which he held to be incapable of establishing French authority over the diverse states of the forest belt in their 'independence and anarchy'. What was needed, claimed Angoulvant, was the strong hand of a colonel or general; this he proceeded to apply, with painful effects for the people of the Ivory Coast.[40]

Since African resistance could not indefinitely withstand the Maxim gun, it might appear that the storms of the European partition would be best weathered by those African rulers who were most flexible in responding to the advent of European power, most ready to collaborate with this inexorable tide of history. Such leaders, it has been suggested, if far-sighted, and well-advised by Europeans,

> might well understand that nothing was to be gained by resistance, and much by negotiation. If they were less far-sighted, less fortunate, or less well-advised, they would see their traditional enemies siding with the invader and would themselves assume an attitude of resistance, which could all too easily end in military defeat, the deposition of chiefs, the

38 J. Suret-Canale, *Afrique Noire*, i, Paris, 1958, pp. 236–44.
39 Paul Vigné [d'Octon], *Au Pays des Fétiches*, Paris, 1890, pp. 70–6.
40 G. Angoulvant, *La Pacification de la Côte d'Ivoire, 1908–1915*, Paris, 1916, part I.

loss of land to the native allies of the occupying power, possibly even to the political fragmentation of the society or state.[41]

It is doubtful whether such a distinction is truly relevant; the choices which African rulers had to make were more complex than that. It has been suggested above that, in coastal areas where Afro-European contacts were long-established, collaborative bases of relationship were usually abandoned only slowly—sometimes too slowly—as it became clear that European demands were developing in such a way as to put African independence at imminent risk. But neither was collaboration rejected out of hand by rulers of inland states for whom the appearance of the Europeans was a relatively new development. For Amadu and Attahiru, Samori and Wobogo, military resistance represented no blind rejection of a European programme for development, more a policy consciously accepted as the last desperate hope of preserving a way of life. In practice 'resisters' and 'collaborators' were usually the same men. Virtually all West African states made some attempt to find a basis on which to coexist with Europeans; virtually all had interests which in the last resort they would defend by resistance or revolt—some conception of a rudimentary 'national cause', anterior to, and distinct from, the national loyalties demanded by modern independent states. Whether it was judged necessary to defend this national cause by arms, and at what stage, depended on variables on both sides of the Afro-European relationship—on African statecraft as well as European intentions.[42]

As far as immediate survival was concerned, armed resistance was usually a forlorn hope. Except among the martial peoples of the desert marches in Mauritania, Mali and Niger (where the French had to maintain military government into the 1920s) it could prolong the period of political independence for only a few years, at best, and its collapse might be followed by severely repressive measures. A later chapter in this volume will consider the reasons for the relative ease with which the Europeans usually took over. In the short run, it is true that negotiation and collaboration could bring concrete advantages, not only for the individual ruler but for the 'national cause' of his state. 'Loyal chiefs' like Momo Ja and Momo Kai Kai of Pujehun, in return for their services in the subjection of Mende country and their support of the Sierra Leone Government in the 1898 insurrection, might see their small states develop as favoured centres of trade and administrative activity.[43] Tofa of Porto Novo, who on his accession in 1874 chose to work with the French in order to further commercial development and to secure their

41 R. Oliver and J. D. Fage, *A Short History of Africa*, Harmondsworth, 1962, p. 203.

42 For development of this argument, see my paper 'West African States and the European Conquest', (fn. 24 above).

43 C. H. Fyfe, *A History of Sierra Leone*, London, 1962, pp. 451, 480–4, 584–5, 604.

support both for the dynastic claims of his lineage and the territorial claims of his state, fared better than his more powerful enemy Behanzin; he remained ruler of Porto Novo under the French protectorate until his death in 1908, though with diminishing privileges. The French administration even gratified one of his ambitions by nominating his son to succeed to his functions.

But historians can hardly content themselves with counting such immediate gains in privileges, perquisites and status. The second generation of collaborators would not necessarily enjoy the same credit, either with the colonial authorities or with their own people. Forms and titles might be preserved by leave of the conquerors; but, as the historians of Porto Novo put it, the old monarchy came to an end with Tofa's funeral.[44] To preserve a moral basis of African identity through the period of colonial rule the intransigence of the resisters, as well as the flexibility of the collaborators, would continue to be necessary. And whether the cause of African independence could in future be represented by traditional rulers and structures might depend largely on how such rulers were judged to have defended their 'national cause' during these twenty years of European conquest.

44 A. Akindélé and C. Aguessy, *Contribution à l'étude de l'histoire de l'ancien royaume de Porto Novo*, Dakar, 1953, p. 89.

CHAPTER 12

The establishment of colonial rule, 1900–1918

A. E. AFIGBO

By 1900 the partition of West Africa amongst the colonising nations of Europe—France, Britain, Germany and Portugal—was practically over. It may be true that the detailed demarcation of the colonial boundaries was still to be achieved but the general outlines were already clear and agreed upon, thus removing the danger, which had once appeared very real and loomed very large, of armed conflict amongst the colonial powers. But considering the total process by which European rule was imposed on West Africa the partition, largely an affair of the European powers and therefore basically an aspect of European diplomacy, was only a first and introductory step. The second, more important and more difficult, stage lay in actually establishing colonial rule, that is in bringing colonial rule home and making it acceptable to the different West African peoples and their rulers, most of whom were not consulted either before or during the partition. The accomplishment of this task was tedious and protracted, in some places lasting over two decades. It required the application of both brutal and subtle methods of coercion and persuasion.

If in the partition of West Africa the colonising powers fell just short of armed conflict amongst themselves, in the actual establishment of colonial rule force of arms, or the threat to use it, occupied a prominent place. Military conquest not only impressed colonial peoples with the might of their new masters, but usually cleared the way for the application of more lasting methods of consolidation. But these reasons hardly explain why the partition was peaceful and the imposition of the colonial regime largely violent and bloody. In spite of the feverish competition which marked the scramble, the interests of the powers involved were not necessarily in conflict to an unnegotiable degree. The two powers which came nearest to war were France and Britain, but they stopped short of it for their interests and claims were not utterly irreconcilable. France, led by her soldiers, concentrated on the Sudan to whose ancient glory and romance they had fallen victim. The commercially minded British, for the most part, wanted the forest regions which were rich in cash crops. The only real crisis between the two powers was precipitated by what might be termed British intrusion into the central Sudan. Even in such cases, the absence among the contestants of any sentimental attachment to the disputed areas left the way open for negotiation and territorial adjustments.

Such easy accommodation was not possible when it came to the imposition of colonial rule on West Africans who were in sentiment and all other ways attached to every inch of area they claimed as their own. Their individual and collective well-being and self-respect were closely tied to the inviolability of their sovereignty over their time-honoured patrimony. Furthermore they had always regarded the Europeans as their guests and felt responsible for their safety and comfort. The post-partition reversal of the traditional roles by which the former 'protectors' became the 'protected', with all its implications for the sovereignty and independence of the people, was totally unacceptable and was contested at every turn with all the means at their disposal. Consequently, in spite of initial claims that treaties of protection with West African chiefs constituted the basis of the colonial regime, European rule derived ultimately from military conquest and to the end depended on the universal awareness amongst West Africans that military might could always be called up in its defence in times of emergency.

Military conquest as a means of imposing and consolidating European presence and rule in West Africa was not a post-partition discovery. It went back to a much earlier time. For the French it was Lieutenant-Colonel (later General) Louis Faidherbe who inaugurated it with what S. H. Roberts has described as his 'delightfully simple policy of "Peace or Powder"';[1] while with the British its origin can be traced back to the 'gunboat politics' of Consul Beecroft, the first notable result of the policy being the bombardment of Lagos in 1851 and its annexation in 1861.[2] Through the use of this swift, if bloody, method, the establishment of colonial rule to some extent went hand in hand with the partition. The overthrow of King Jaja of Opobo in 1887, of the Ijebu in 1892, of the Tukulor empire in 1893, of Behanzin of Dahomey and Nana of Itsekiri in 1894, of Benin, Bida and Ilorin in 1897, of Bai Bureh in Sierra Leone and of Samory Toure in 1898 were only some of the more important milestones marking the progress of the establishment of alien rule in West Africa. In a sense the wars which sought to maintain West African independence were fought and lost in the nineteenth century.[3]

1 S. H. Roberts, *History of French Colonial Policy*, Oxford, 1929, i, p. 302.
2 K. O. Dike, *Trade and Politics in the Niger Delta*, Oxford, 1956, pp. 128 ff.; J. F. A. Ajayi, 'The British Occupation of Lagos 1851–1861', *Nigeria Magazine*, 69, August 1961.
3 *West African Resistance*, ed. M. Crowder, London, 1971; A. S. Kanya-Forstner, *The Conquest of the Western Sudan*, Cambridge, 1969. These two books contain the most detailed and latest accounts of the military side of the scramble in West Africa.

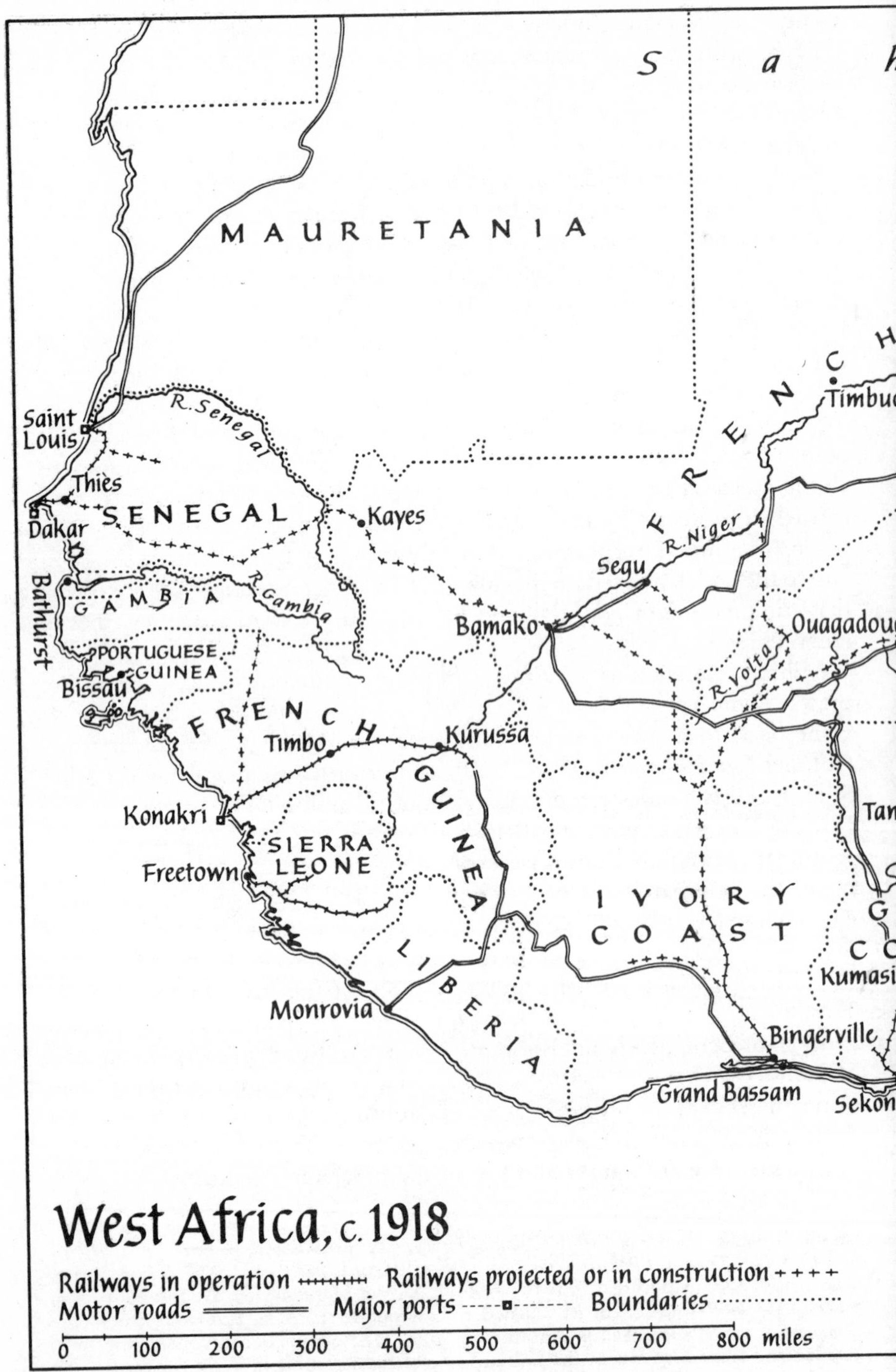
S a h
MAURETANIA
F R E N C H
Timbuc
Saint Louis
R. Senegal
Thiès
Dakar
SENEGAL
Kayes
Bathurst
GAMBIA
R. Gambia
Segu
R. Niger
Bamako
Ouagadou
R. Volta
PORTUGUESE GUINEA
Bissau
FRENCH
Timbo
Kurussa
GUINEA
Konakri
SIERRA LEONE
Freetown
Tam
IVORY COAST
LIBERIA
Monrovia
Kumasi
Bingerville
Grand Bassam
Sekon
West Africa, c. 1918
Railways in operation
Railways projected or in construction
Motor roads
Major ports
Boundaries
0 100 200 300 400 500 600 700 800 miles

12:1

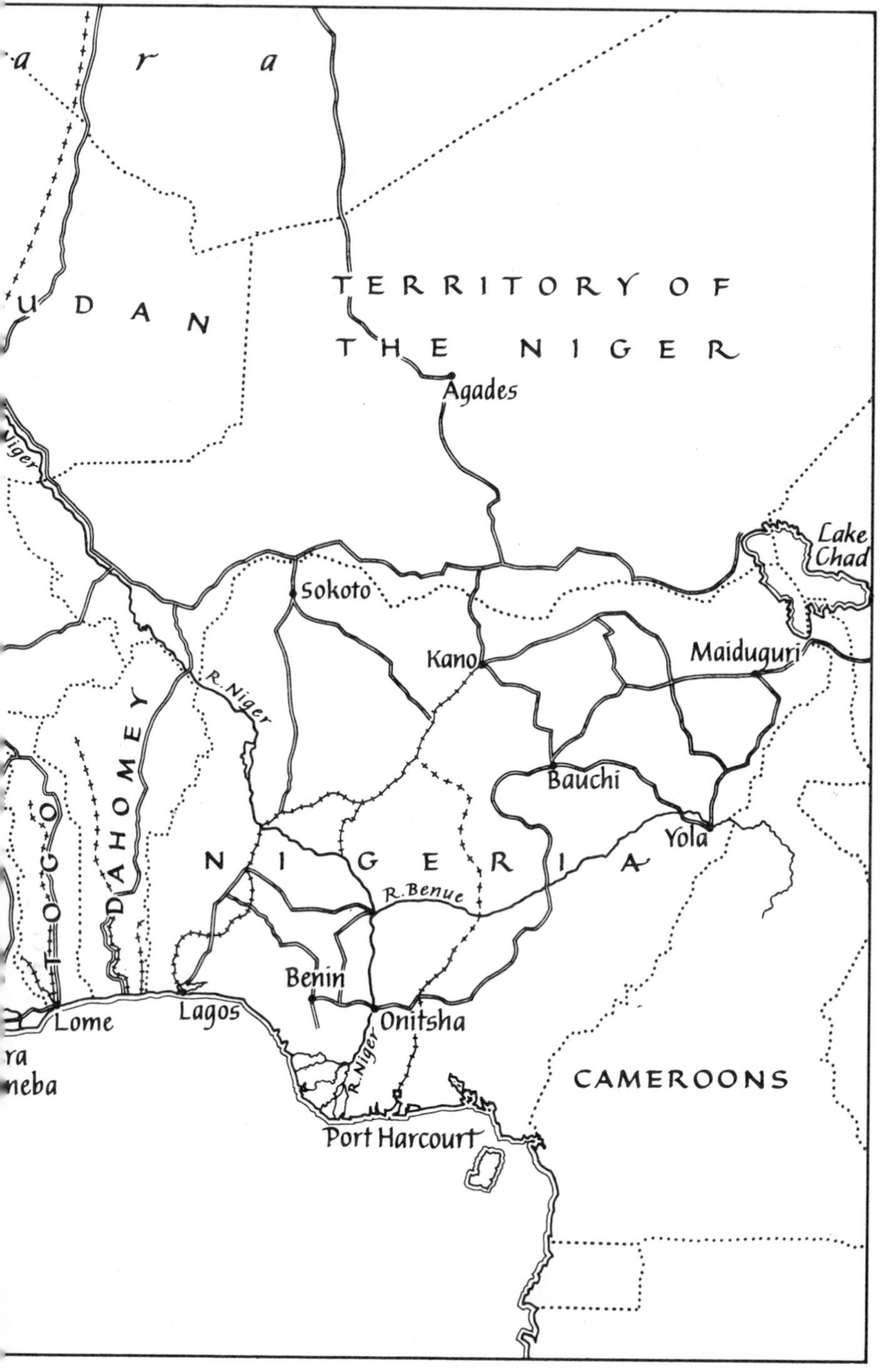
TERRITORY OF THE NIGER
Agades
Lake Chad
Sokoto
Kano
Maiduguri
R. Niger
DAHOMEY
TOGO
Bauchi
Yola
NIGERIA
R. Benue
Benin
Lagos
Lome
Onitsha
R. Niger
CAMEROONS
Port Harcourt

But one cannot say that by 1900 the military side of European occupation was complete. In both British and French territories large areas remained either to be conquered for the first time or to be reconquered. 'The stereotyped image of a conquest,' writes Professor Jan Vansina, 'is Caesar's *veni, vidi, vici*, but Caesar's books show well enough that this is only part of the story, in fact only the first chapter. The following chapters dealt with revolts.'[4]

In the Gold Coast (modern Ghana) the once formidable Asante Confederacy remained resentful of the 1896 incident which had led to the exile of Prempeh, the Asantehene. In 1900 the Governor, Sir Frederick Hodgson, by his arrogant indiscretion in demanding that the Asante hand over the Golden Stool to him, provided them with a *casus belli* which precipitated the futile nine months' war of Asante independence. The following year the confederacy was annexed to the British crown.[5] The greater part of Nigeria was still largely independent in 1900. British power in Hausaland remained precarious despite fourteen years of so-called rule by the Royal Niger Company and the capture of Ilorin and Bida in 1897—victories which Goldie in his usual style had refused to press home to the point of permanently altering the political order. Sir Frederick Lugard had to begin the military occupation of Northern Nigeria from scratch. Starting with Bida and Kontagora in 1901, he took Bauchi the following year, then Kano and Sokoto in 1903.[6] This was by no means the end of armed resistance to British rule in the Sokoto Caliphate, for three years later the British had to face and defeat a Mahdist uprising at Satiru.[7] In Eastern Nigeria there was no single state or power whose defeat would put the whole region into British hands. At first Aro Chukwu was thought to be the key but the Aro expedition of 1901–2 disproved this and there followed military operations to subdue or overawe the village groups in what later became Bende, Aba, Umuahia and Ikot Ekpene divisions.[8] The rest of the region was left unconquered until subdued by a series of small military expeditions which scoured the former Eastern Region of Nigeria between 1905 and 1919. Among the more important of these

4 J. M. Vansina, *Kingdoms of the Savannah*, Wisconsin, 1966, p. 18. Most colonial military commanders were familiar with Caesar's boastful and arresting situation report, but not with the story which, according to Professor Vansina, the subsequent chapters of Caesar's books tell. See for instance A. E. Afigbo, 'The Eclipse of the Aro Slaving Oligarchy of Southeastern Nigeria 1901–1927', *Journal of the Historical Society of Nigeria (J.H.S.N.)*, v, 4, December 1971.

5 See chapter 7.

6 See chapter 2.

7 M. Perham, *Native Administration in Nigeria*, Oxford, 1937, pp. 46–7.

8 A. E. Afigbo, 'The Aro Expedition of 1901–1902: An Episode in the British Occupation of Iboland', *Odu, A Journal of West African Studies*, new series, 7, April 1972; S. N. Nwabara, *Ibo Land: A Study in British Penetration and the Problems of Administration*, Ph.D. thesis, Northwestern University, 1965; J. C. Anene, *Southern Nigeria in Transition, 1885–1906*, Cambridge, 1966; J. C. Anene, 'The Southern Nigeria Protectorate and the Aros 1900–1902', *J.H.S.N.*, i, December 1956.

were those against Ezza of Abakiliki and the Ahiara in 1905, the people of the Onitsha hinterland in 1904 and Awgu in 1919.[9]

The Federation of French West Africa also witnessed after 1900 a series of military actions designed to break the back of indigenous resistance to French rule. In Guinea, in spite of the 'settlement' of February 1897 by which the *imamate* of Futa Jallon became a French Protectorate and the *imams* pensioners of France, resentment against French presence remained popular and persistent. In 1900 this led to an open revolt which was put down with a heavy hand, the leaders being executed. Yet the spirit of the people remained unbroken and militant until 1906 when the last manifestation of armed resistance was destroyed. In the Ivory Coast until 1908 only a narrow strip of the coastal territory was under effective control. The rest of the country defied every effort at negotiated 'settlement'. Then came Governor Angoulvant who believed in a policy of blood and iron or what was described as a 'méthode rigoureuse'. After seven weary years French arms triumphed. In Niger no such complete victory was achieved by the French until 1921, three years after the period covered by this chapter.[10]

Why, one might ask, did West African resistance to European conquest prove generally so ineffective? This question raises the whole problem of the nature and organisation of indigenous resistance to alien rule in the late nineteenth and early twentieth centuries. The colonial powers, especially the British and the French, sometimes attributed their easy success to what they described as the 'isolation' of West African rulers of the time. Convinced that such opponents as Oba Overami of Benin, Samory Toure, Ahmadu Seku, the Fulani oligarchy in Northern Nigeria and so on were soulless Leviathans who mercilessly ground under their heels the overwhelming majority of their peoples, the colonial powers believed that the traditional rulers did not always enjoy the support of their own subjects in their wars of resistance to alien rule. In the light of this misconception the powers saw themselves as the liberators of the oppressed peoples of West Africa. 'The majority of the inhabitants of this large town [Ijebu],' wrote Sir Gilbert Carter before the Ijebu expedition of 1892, 'would welcome the advent of the British flag.'[11] In the same vein Lugard claimed in 1902 that 'the misrule of the Fulani had rendered them so hateful to the bulk of the population, who would welcome their overthrow'. On the overthrow of Kano he also claimed that the British troops entered the town unopposed by the common people who looked on as though the matter did not immediately concern them.[12] No doubt paraphrasing the opinions of the French colonels and generals who conquered French West Africa, Raymond L. Buell wrote in 1927:

9 Nwabara, *Ibo Land*; Anene, *Southern Nigeria*.

10 J. D. Fage, *A History of West Africa*, Cambridge, 1969, pp. 175–6; *West African Resistance*, p. 1.

11 C.S.O., 1/1/13, Carter to C.O., 20 April 1892.

12 *Colonial Reports Annual, Northern Nigeria*, 409, 1902, pp. 83–7.

> When the French undertook the occupation of West Africa they were confronted with a number of native tyrants who cruelly exploited their subjects. Life and property were insecure; slavery and human sacrifice prevailed everywhere. In a few cases local Almamys had imposed a form of discipline maintained by terrorism upon thousands of unwilling subjects.[13]

Thus, it came to be fashionable to conclude that the feeble resistance of West African states to European rule was mainly due to the indifference of the masses to the fate of their rulers or even to their willingness to accept new masters.[14]

This explanation of the weakness of the resistance movement in West Africa is suspect for a number of reasons. Firstly it was to the interest of local commanders to represent the small but costly wars in West Africa as wars of liberation, crusades for humanitarianism and civilisation in order to justify military measures before home governments and to parry criticisms by humanitarian societies in Europe. Secondly, there was contemporary European ignorance of the political organisation and tradition of most West African peoples. West African rulers were portrayed as despots. But research has shown that their governments were more democratic than their hostile detractors knew or would allow. In fact some of the so-called autocratic chiefs were mere puppets of their councils which in turn depended on popular support for their powers and authority. Furthermore the traditional military system of West African rulers, with few exceptions such as that of Samory Toure, depended wholly or mainly on the principle of the general liability of all adult males to carry arms. Under this type of political and military system it was difficult for most West African rulers to impose unpopular war policies on their subjects. The trouble with the armies of resistance was not numbers but training, experience and arms.

Thirdly, West African society was not as rigidly stratified as the European explanation implies. There was a distinction between free men and those of servile origins, but among free men there were no class distinctions so rigid as to hamper the pursuit of common political goals and unified military actions. With a few exceptions here and there, for instance on the West Atlantic coast, there was no class stigma to hamper social mobility indefinitely. This was true even of Northern Nigeria which would seem to provide a classic example of such rigid social

13 R. L. Buell, *The Native Problem in Africa*, London, 1965, p. 987.

14 Incidentally the view that the imposition of European rule meant the dawn of a new era of liberty, equality and fraternity for the so-called oppressed classes of Africa has entered into popular tradition, at least in Northern Nigeria. 'We had,' said Madam Baba of Karo, 'wanted them [the Europeans] to come, it was the Fulani who did not like it!' See F. M. Smith, *Baba of Karo*, London, 1954, p. 67.

stratification between Fulani and the Habe. The *talakawa* were not necessarily all Habe; some Fulani were commoners unless they were rich or held hereditary or appointive offices. Furthermore, miscegenation was already blurring the ethnic line between the conquering Fulani and the conquered Habe by the time the colonial powers were threatening the independence of the Sokoto Caliphate. Above all, the classical concept of a Muslim community was that all members identified themselves with any struggle against a non-Muslim people. Division in the body politic could therefore hardly explain the weakness of the resistance movements at least in the case of Muslim communities.

Moreover, this usual explanation for easy European success misrepresents the attitudes of the different groups in West Africa to the pre-colonial order. To the Europeans the political, economic and cultural situation in West Africa at the time of the scramble and after depicted nothing but squalor, chaos and barbarism which African masses were anxious to give up. But this was the European's view of African society, not the African's. Though some of the masses might have had grievances against the established order, they could hardly have questioned the fundamental basis of society. If they expected change it was within the context of what they knew. They were not in a position to say whether the colonial regime which they had not yet experienced, would make their situation better or worse. Baba of Karo is supposed to have given the Habe reason for 'welcoming' the Europeans as the fact that 'if you worked for them they paid you for it, they did not say like the Fulani: "Commoner give me this! Commoner bring me that!"'[15] This story was certainly a latter-day myth. In 1900 the Habe had no way of knowing that Europeans would be better taskmasters than the Fulani.

The ultimate explanation for the ineffectiveness of West African resistance to colonial conquest lay in the superiority of the colonial armies in arms, experience and strategy as well as in the fact that at the time of the wars of independence West Africa was in political and diplomatic disarray. The armies of resistance, even those commanded by Samory Toure, were, compared with the armies of the colonial powers, poorly trained, poorly armed and poorly officered. Hilaire Belloc's cynical observation:

Whatever happens we have got
The Maxim gun and they have not

was unfortunately true for West Africa. In these wars a society which for the most part had a primitive and naïve conception of warfare was confronted with another that had institutionalised war in its modern form and was taught the bitter lesson that in modern warfare enthusiasm is no substitute for professionalism. The story of the British conquest of the Fulani empire of the Central Sudan, which reads like an account of a triumphal procession,

15 *Ibid.*, p. 67.

illustrates the pathetic futility of West Africa's traditional war machinery in the modern world. The account of Samory Toure's defiance of French might for over a decade gives a tantalising impression of how entirely different the story would have been had West Africa adopted professionalised militarism on a wider scale.

To military weakness West Africa's resistance leaders added a culpable political myopia which made it impossible for them to take a united stand against the colonial invaders. One might have expected at least the new Islamic states of the Sudan to unite against their common enemy. But Samory Toure did not cooperate with Ahmadu Seku against the French, and Ahmadu also failed to cooperate with the Caliph at Sokoto to whose ancestors his empire owed so much and in whose court he spent his weary last days. If states with so much in common would not help each other one could not expect such traditional rivals as Asante, Dahomey and what remained of Oyo to make common cause against the French and the British. The reasons for this failure require further investigation but on the basis of already available evidence, the scholar cannot but be dismayed by the inability of West Africa's different leaders to appreciate the immensity of the new challenge, the fact that they were all in the same plight and the need for a radically new approach to the problem of preserving their sovereignties.

Apart from a general lack of diplomatic insight and ability each of the then existing states had its own peculiar internal problems. Throughout the Sudan the appeal of Islam as a rallying ideology had been paradoxically gravely weakened by the jihads which had sought to strengthen it. In Ahmadu Seku's empire the tradition of Tijani exclusiveness bequeathed by Al-Hajj 'Umar had left a legacy of deep-rooted divisions not only between the Tijani hosts and the pagans whom they treated abominably, but also between them and those Muslims who belonged to such older established brotherhoods as the Qadiriyya.[16] Consequently these other religious interests saw the French threat to the Tukulor state as an opportunity for settling scores with the Tijani brotherhood. In the Central Sudan the Sokoto Caliphate was, from its inception, in conflict with the Kanuri empire of Bornu, thus ruling out any possibility of cooperation between the two. The Sokoto Caliphate which was large enough by itself to put up a stubborn resistance eventually turned out to be a colossus with feet of clay. Under its old military system the Caliphate had no central army, and defence tended to be left to the frontier provinces at a time when the new challenge required an all-out assault on the British by the whole might of the Caliphate. This military conservatism was coupled with an olympian fatalism which nothing could shake. Even after Bida, Kontagora, Bauchi and Bornu had been taken, the Caliph continued to write to

16 J. M. Abun-Nasr, *The Tijaniyya*, Oxford, 1965, pp. 138–9.

Lugard that 'There is no power or strength save in God on high' as if nothing had happened. Only when the Sokoto armies were finally dispersed did the Caliph realise that there were 'power' and 'strength' in seven-pounders and maxims also.[17]

Among the Guinea states the problem was political disarray. The Benin empire had long become a mere shadow of its former self. The Oyo Yoruba empire had fragmented early in the nineteenth century while the successor states had entered a long-drawn-out internecine struggle for ascendancy and leadership in Yorubaland. This struggle had led to a military stalemate from which only British intervention could free the participants.[18] The highly militarised Dahomey kingdom had blunted its weapons in indecisive conflicts with rival powers in Yorubaland. In the Asante state divisions within the Confederacy had opened the door to interference from without. By the time of the scramble that once formidable kingdom was fighting a losing battle against disintegration.

Whatever the reason for Europe's easy conquest of West Africa the fact remains that the lesson went home especially among the traditional élite who were most stunned by the failure of the military and magical systems on which they had until then placed so much reliance. There is abundant evidence that the traditional magico-religious powers were also called up against the Europeans and found wanting. Dr E. P. Skinner reports that 'when evil befell Mossi country in spite of the annual sacrifices to the ancestors and to the earth deity, the Mogho Naba had to offer additional sacrifices'. This was true of most West African communities. The Muslims would offer prayer to Allah. But with disconcerting consistency this supernatural source of strength failed to stem the tide of European conquest. According to Dr Skinner:

> It is generally believed by the Mossi, that when the French attacked Ouagadougou, the deposed Mogho Naba Wobogo made sacrifices to earth shrines. Tradition has it that he sacrificed a black cock, a black ram, a black donkey and a black slave on a large hill near the White Volta River, beseeching the earth deity to drive the French away and to destroy the traitor Mazi whom they had placed upon the throne.[19]

The failure of this and innumerable similar appeals[20] went further to

17 D. J. M. Muffett, *Concerning Brave Captains*, London, 1964; D. J. M. Muffett, 'Nigeria—Sokoto Caliphate', *West African Resistance*; Lady Lugard, *A Tropical Dependency*, Nisbet, 1905.

18 See chapter 5.

19 E. P. Skinner, *The Mossi of Upper Volta*, California, 1964, p. 133.

20 There were many other such cases of the leaders of the resistance to colonial rule seeking to invoke supernatural forces against the colonial powers. Mr Robert Cudjoe, a Ghanaian who came to Nigeria as a carpenter but later became an interpreter and who accompanied an expedition against the Ezza of Eastern Nigeria in 1918, recorded the following in his reminiscences: 'It was during this time (1918) that an accidental fire, caused by Lightning,

(continued overleaf)

convince the traditional élite of the power of the Europeans. According to Dr Busia:

> The most notable and impressive lesson (learned by Africans), was that of the power of Europe, a power based on superior technology. It is an unforgettable lesson. Older Africans, those who fought in the wars against European countries, when asked about their most vivid impressions of the wars, invariably say, 'The White man is powerful'.[21]

Military conquest was the preliminary to the actual organisation and consolidation of colonial rule in West Africa. 'An empire had been won,' writes S. H. Roberts about French West Africa, 'rich but only partly organised, and with its resources practically undeveloped. . . . Having achieved the difficult task of conquest, the French were thus confronted by the far more difficult task . . . of consolidation.' It would be a mistake, he continues, to think that any final result had been achieved by conquest. On the contrary 'the problem was just being posed and only the preliminary obstacles, those due to the opposition of the native kingdoms, had been removed. As France soon perceived, the destruction was far the easier task. It was at the stage of rebuilding, when political and economic organisations were just commencing, that the real difficulties emerged.'[22]

The colonial powers recognised that the grim lessons of military conquest were likely to be quickly lost on their newly won subjects if steps were not immediately taken to convince the people that European rule had come to stay. The need to realise the economic ambitions and social reforms which inspired the powers reinforced the above argument. Consequently the French, the British and the Germans had to take steps to strengthen their hands against subsequent threats to their position and to promote the work of economic exploitation and social change. These measures eventually brought about far-reaching transformations in the

(20 *continued*)
set fire to the soldiers' Barracks at Abakaliki, and destroyed some of the houses. A wife of a soldier and her daughter of 10–12 years of age were burnt to death. . . . The following morning chief Okanu Opehu and 5 others came to the station. The chief shook hands with me and said "Urobot" (meaning Robert) "things are moving well". I became surprised and asked "What is the matter?" He replied "Ezeke Una, the founder of the Ezzas is not asleep." I burst into a fit of laughter but the chief hushed me up saying, "There is no joke . . . for the white man is about to go. . . . Don't you know he has begun to burn down the soldiers' barracks." I explained the whole story of the accident to him, but he would not be reconciled to this as it was contrary to what he had been assured by the Native Doctors and Diviners.'

R. Cudjoe, 'Some Reminiscences of A Senior Interpreter', *The Nigeria Field*, xviii, 4, 1953, p. 159.

21 K. A. Busia, *The Challenge of Africa*, New York, 1962, p. 56.

22 Roberts, pp. 306–7.

West African scene and led to the emergence there of modern nation states.

COLONIAL ADMINISTRATION

After military conquest, the establishment of colonial rule involved two closely related issues—the relationship between the administrations of all the colonies under each colonial power and the relationship between the new colonial administration and the traditional machinery of government of the West African peoples concerned. In dealing with these questions the several colonial powers often adopted methods which on detailed study prove to be, in spite of the theoretical conflicts between different national publicists, broadly similar to one another.

When the noise and whirling dust of the scramble died and settled down by the 1900s the new political map of West Africa showed that France had seized by far the largest area, holding as she did 1,800,000 square miles of territory. Britain came second with about 480,000 square miles, Germany third with 33,000 squares miles and Portugal last with 14,000 square miles. For Germany and Portugal, whose colonies were widely isolated, there was no difficulty over deciding to link together or separately rule their holdings in West Africa. But the French and the British had this problem and in seeking a solution they oscillated between federating their colonies and allowing each to develop on its own. The manner in which the two powers changed positions in this matter makes an interesting study.

Up to 1886 it was the British rather than the French who showed a tendency to federate their holdings in West Africa. As long as British territories in West Africa were mere coastal footholds the sea was a more effective link than it could be after the 1890s when the British had pushed their holdings into the interior and the French held extensive areas between one British territory and another. In the earlier period, therefore, the British attempted first to rule three of their West African dependencies (Gambia, Sierra Leone and Gold Coast) from Freetown, and then to rule Gold Coast and Lagos from Accra. The main argument for this was economy in administration. The French on their side did not consider federating their West Africa at the time. In the first place, while French rule was being many reasons for the French tendency towards administrative separatism in West Africa at the time. In the first place, while French rule was being imposed along the rivers and trade routes, any attempt at centralised control would have hampered the initiative of local military commanders. In the second place the colonial department of France was up to 1894 merely part of the Ministry for the Navy. The department, which was over-burdened with problems from other parts of the colonial empire, was not able to undertake a proper organisation of France's West African empire. Thirdly, argued C. N. Newbury, 'this administrative separatism was assisted by the relatively independent economic life of the coastal territories when the bulk of imports into Senegal, Guinea, the Ivory Coast and Dahomey came from Britain and Germany rather than from France'.[23]

But as the scramble progressed the two powers rethought and changed their positions. The first to move were the British. The 'Federation' of Lagos and the Gold Coast colony was very unpopular with the élite of the former territory who resented subordination to Accra. Secondly, as the scramble gathered momentum the British realised that their interest in the interior of Lagos colony would be more effectively protected by a Lagos administration that was independent of Accra. The British, therefore, abandoned 'federation' or amalgamation for their mid-nineteenth century West African holdings but revived it later for their various territories in the area subsequently named Nigeria.

The French on their part had strong reasons for choosing to federate their own territories. For one thing, unlike the British territories, the French colonies formed a continuous block, a first administrative prerequisite for effective amalgamation. 'The one obvious fact at this stage,' writes S. H. Roberts, 'was that [French] West Africa was a single geographical and economic whole, and as such had to have its problems attacked by one directing agency. In other words the necessary prelude to any organisation or development was the establishment of a single government for all [French] West Africa.'[24] There were other reasons. By 1895 the scramble was nearing its end, the former French coastal enclaves had all been connected and French territorial ambitions in West Africa seemed more or less secured. The chief argument for a many-pronged attack had therefore lost validity. In this new phase military policy in French West Africa required centralisation not only in order to gain closer control over expenditure but also in order to mobilise France's West African forces for a final onslaught against the elusive and redoubtable Samory Toure. This centralisation of military policy was particularly necessary as civilians were replacing military governors and, above all, it was considered necessary to avoid futile boundary disputes amongst the various colonies. In 1894, that is a year before the first step towards federation, there were two boundary disputes involving the Sudan and Senegal. In the first, Governor Lamothe of Senegal took punitive action against certain villages around Futa Jallon, only to be greeted by a protest from Governor Grodet of the Sudan who felt piqued that Senegal should interfere with people under him without consulting him first. In a second incident Grodet refused to send military reinforcements to Lieutenant-Colonel Monteil after the latter had been repulsed by Samory Toure because he felt that the Senegalese authorities had once again

23 C. W. Newbury, 'The Formation of the Government General of French West Africa', *Journal of African History (J.A.H.)*, i, 1, 1960, p. 112. I owe to Newbury's article the main facts about the origin of the Federation of French West Africa. See also M. Crowder, *West Africa Under Colonial Rule*, London, 1968, pp. 174–82.
24 Roberts, i, p. 307.

exceeded their authority.[25] These two episodes underlined for the French in West Africa that the only way out of unproductive rivalry lay through administrative amalgamation.

The formation of the Federation of French West Africa was achieved in four stages, the first being the decree of 16 June 1895 which brought into existence a central authority for French West Africa. At this stage the centralisation was half-hearted and, to a great extent, ineffective and unsatisfactory. The decree created a Government-General for French West Africa headed by a Governor-General. To this officer, or rather to a Commander-in-Chief under him, the Lieutenant Governors of the coastal colonies surrendered their control of military questions. The Governor-General was also the only officer in French West Africa who could communicate directly with the French Minister of Colonies. The interior colony of Sudan, however, was subordinated to the Government-General to a greater extent than any other colony. Its budget, for instance, was to be approved by the Governor-General. These were, more or less, the only elements of centralised control the decree of 1895 was prepared to concede. For the rest the new arrangements were full of contradictions and difficulties.

Not all French West Africa was brought into this tentative effort at federation. Dahomey was left out and the coastal colonies remained largely autonomous since they retained their own separate budgets, and approval for these budgets lay not with the Governor-General but with the Minister of Colonies. Secondly, the Governor-General was also Lieutenant Governor of Senegal. In addition the Government-General had neither its own separate budget, nor an independent source of income, nor a separate civil service to carry out its policies and schemes. For revenue it depended on grudging contributions from the coastal colonies included in the Federation. Buell justifiably described the Government-General set up in 1895 as 'merely an addendum to the Government of Senegal',[26] and Roberts said the Governor-General was a 'shadowy officer'.[27] As if these difficulties were not enough, the Governor-General's confidential instructions assigned to him more far-reaching powers and duties than the decree conceded. He was expected to arbitrate between all French colonies in West Africa, to 'coordinate their native policies and to share out the expenses of the Government-General between them'. Then a decree of September 1895 set up a Government-General Council to advise the Governor-General in his arbitration between the colonies but, absurdly enough, only Senegalese officers were included in the Council. The other territories were not represented.[28]

25 Newbury, 'The Government General', p. 114.
26 Buell, i, p. 926.
27 Roberts, i, p. 308.
28 Newbury, 'The Government General', pp. 115–18.

Not surprisingly this rickety arrangement failed to work. Sudan was not amenable to control, especially in its military policies, while the coastal colonies resented contribution to the Federation's budget. Consequently the old problems remained unsolved. For instance, the Sudan continued to suffer from chronic budgetary instability because of reckless spending on military expeditions. The terms of federation were reviewed in 1899 and a new decree issued to remove some of the patent irregularities of the first decree. Among the innovations were the inclusion of Dahomey in the Federation, the partition of the Sudan between Senegal, Guinea, Ivory Coast and Dahomey and the introduction of centralised control of France's West African forces which were commanded by an officer subordinate to the Governor-General.[29] The other failings of the Federation remained, however, especially the financial impotence of the Government-General. The federal authority had not the financial means to promote capital investment or to coordinate a badly needed programme of public works.

This structure was slightly modified by the decree of 1902 which sought to free the federal authority from the stranglehold of Senegal by shifting the headquarters from Saint Louis to Dakar. The Government-General Council was reconstituted to include the Lieutenant-Governors of the five colonies, and given the function of meeting once a year to ratify the budget of each territory and to 'discuss general administration'. Yet the financial problem remained since the federal centre still had no independent source of revenue while the Ministry of Colonies and the Governors-General continued to assign powers to the centre beyond what the decrees would allow or what the inadequate central administrative service could handle.[30]

The decisive step in the organisation of the Federation was taken in 1904 with the decree giving the new colony of Upper Senegal–Niger its own Lieutenant-Governor, thus freeing the Government-General from responsibility for the administration of any particular colony. The federation got its own independent source of revenue from customs duties and its own personnel to carry out its policies. The Federation of French West Africa thus emerged in the form known throughout the greater part of the period it was under French rule.[31]

In Nigeria the British were faced with similar problems and adopted an analogous solution. The idea of federating Lagos and the Gold Coast colony had scarcely been discarded when amalgamation was taken up as the best solution to the administrative problems posed by Nigeria. At that time what later became Nigeria was divided into three separate administrations, the Niger Coast Protectorate, the Colony and Protectorate of Lagos

29 *Ibid.*, p. 121.
30 *Ibid.*, pp. 124–5.
31 *Ibid.*, pp. 125–6. See also Buell, i, pp. 925–8; Roberts, i, pp. 307–9; Crowder, *West Africa Under Colonial Rule*, pp. 180–2.

and the territories of the Royal Niger Company. Each of the three administrations had its own fiscal policy and was guided by different considerations in its administrative policies. Whereas, for instance, Lagos and the Niger Coast Protectorate were regular governments, the Royal Niger Company was concerned to make profits for its shareholders. Furthermore the existing arrangement imposed hardships on the British and the African alike. Traders from the Niger Coast Protectorate, for instance, were denied free entry into the Niger Company's territories.

Amalgamation as a solution to these problems was mooted by Sir Ralph Moor of the Niger Coast Protectorate in 1896[32] and in 1897 the Niger Committee was appointed by the British Government to look into the question. Reporting the following year, the Committee accepted eventual amalgamation in principle but ruled against immediate amalgamation on grounds of poverty of communications and lack of a man of the requisite experience and with the staying power to administer so large a territory in so difficult a climate. For the time being, the Committee advised a piecemeal approach, dividing the country into two provinces, coastal and inland, which became in fact in 1906 the Southern and Northern Provinces of Nigeria; the first drawing together the Colony and Protectorate of Southern Nigeria, the second consisting of the former territories of the Royal Niger Company.

With the partial amalgamation of 1906 the problems created by Northern Nigeria's land-locked position and general poverty became more glaring than before. Southern Nigeria was rich enough to embark on the extension of the Lagos railway, to balance its budget, to pay and house its officers well and to subsidise Northern Nigeria. The latter administration was so poor that the revenue it raised locally and what it got from Southern Nigeria were not adequate to balance its budget, finance its railway and maintain its officers. It was dependent therefore on subsidies grudgingly made by the imperial exchequer. What was more the two administrations wasted their energies in unprofitable rivalry over railway policy and boundary disputes. The Southern Administration expected Northern Nigeria to accept the humble position of a 'mere hinterland' of Southern Nigeria and allow its products to be carried on the Lagos railway at rates fixed by the Lagos authorities. But the Northern Administration wanted to assert the separate existence of Northern Nigeria. Thus with money borrowed from Southern Nigeria it began to build a railway that would help the export of Northern produce via the Niger. On the boundary issue both administrations claimed Ilorin and Kabba Provinces, as well as Igala and Idoma territories. At one point in the boundary dispute the Northern Authorities threatened to use force

32 T. N. (formerly S. M.) Tamuno, *The Development of British Administrative Control of Southern Nigeria,* Ph.D. thesis, London, 1963, pp. 69–70.

to eject the Southern Authorities from what they considered Northern Nigeria's territories.

By 1911 it had been decided to solve some of these problems and to evade the others by amalgamation. Lugard was sent back to Nigeria in 1912 to carry out the scheme. In October that year he amalgamated the Lagos and Northern railways and renamed the line the 'Nigerian Railway'. On 1 January 1914, the formal amalgamation of the two Nigerias was proclaimed. In actual fact this meant little more than the amalgamation of certain government departments, but all the same it marked another stage in the establishment of British rule. By 1914 the British were sufficiently sure of their position in Nigeria to bring under one man territories which earlier they thought would be more effectively controlled if partitioned among three administrations. Having decided on this administrative federation the British stuck to it in spite of increasing difficulties. But the French reverted to 'administrative separatism' as the independence of their West African colonies drew near.[33]

As has been mentioned the other aspect of administrative consolidation related to local government. Here again, in spite of conflicting ideologies and attitudes there were broad similarities between the solutions found by the different powers. The system adopted in the coastal colonies which were already in existence by 1875 will first be considered. These colonies included the communes of Senegal and the four British coastal colonies of Gambia, Sierra Leone, Gold Coast and Lagos. The communes of Senegal stood in the same relation to the rest of French West Africa as each of the four crown colonies stood to the Protectorates immediately inland of them. Using the terminology loosely, the communes constituted the 'crown colony' and the rest of French West Africa the Protectorate.[34]

The communes and the four crown colonies were acquired by France and Britain in the age of optimism, romanticism and liberalism, when it was believed that Africa and Asia could easily be transformed by a touch or two of European commerce, European technology and western Christianity. At this time both the French and the British were in their own different ways cultural imperialists who sought to impose a substantial part of their cultures on those under their rule. The French were at the time convinced of the universal applicability of French civilisation to all peoples, no matter what their provenance. The English, while not subscribing to any such idea as regards their own culture, still considered it necessary and possible to endow their colonial peoples with aspects of it. To both parties the emergent westernised élite, aided by and cooperating with the colonial aristocracy, were to be the agents of this civilising

33 See ch. 19.

34 For a time the 'distinction between annexed and protected territory' was made in Guinea, the Ivory Coast and Dahomey, but this was soon abandoned.

mission. This was the time when members of this class were given respectable and responsible positions in church and state in Sierra Leone, the Gold Coast and Nigeria. Even Britain believed that for such people only institutions embodying the fundamental principles of western civilisation could be suitable. Thus, Sir Herbert Richmond Palmer, a colonial civil servant and an 'indirect ruler' of note, writes:

> During most of the nineteenth century the ideal of colonial administration was to create colourable counterfeits of the Motherland, its institutions, its idiosynacrasies and its pecularities. There was a governor to represent the king, a judge to represent the Rule of Law and a chaplain to represent the Established church. Land tenure—regardless of what actually it had been—was made to simulate the time-honoured characteristics of English Freehold. Dim rows of street lamps attested the high degree of civilisation reached by the river ports, and the Beadle said 'Oyez', 'Oyez' when His Honour the Judge bowed to his be-wigged bar and his be-wigged bar bowed to him.[35]

And in any case the French and the British had to take into account the fact that by the mid-nineteenth century each of these colonies had a long history of direct and intimate contact with Europe and had developed a fairly westernised urban community. Lord Hailey has described them as 'comprising the coastal region(s) in which European influences have been established for a considerable time'.[36] Because of the all-pervading foreign influence the indigenous institutions of government had been seriously weakened. Not surprisingly therefore the French and the British proceeded to establish alien institutions for administering these coastal colonies.

In the Senegal the process began in 1848 with the extension of French citizenship to the inhabitants of the old towns and the grant to them of adult male franchise. At the same time Senegal was given the right to elect one of the twelve colonial deputies allowed in the French Chamber of Deputies. French local government institutions were introduced in 1872, when Saint Louis and Gorée were created communes. In 1880 and 1887 two others, Rufisque and Dakar were added. Each of these had an elective council, the *commune de plein exercice,* presided over by a mayor exercising powers similar to those of the communes in France. In 1879 the creation of the *Conseil-Général du Sénégal* gave Senegal the equivalent of the council of a French *département*. The members were also elected.

The British on their side did not export to their crown colonies the English system of local government, but instituted in each a legislative council made up of official and unofficial members to advise the

35 H. R. Palmer, 'Some Observations on Captain R. S. Rattray's Paper: "Present Tendencies of African Colonial Government"', *Journal of Royal African Society*, xxxiii, 130, 1934, p. 37.

36 Lord Hailey, *An African Survey*, Oxford, 1937, p. 171.

Governor on the administration of the colony. This political concession took place in Gambia and the Gold Coast in 1888, in Sierra Leone in 1863, and in Lagos in 1862. In addition to the legislative councils the British also introduced municipal government into Sierra Leone.

Both the French and the British recognised the practical impossibility of entirely ignoring the traditions and the culture of these colonies. The inhabitants of the Senegalese communes for example could be *citoyens français,* while in civil matters retaining their Muslim status and coming under Muslim tribunals which dealt with their civil cases. In the Gold Coast, where the indigenous institutions still retained a substantial measure of vitality, the authority of the chiefs and their councils was recognised and encouraged by the British in the Native Jurisdiction Ordinances of 1876 and 1883, which gave the chiefs and their councils defined judicial, executive and legislative functions.

For a time these measures won for the colonial powers the general support and cooperation of the Western-educated élite in Senegal and the British crown colonies. In Bathurst, Freetown, Accra and Lagos people prided themselves on being British subjects. At the beginning of the scramble the argument in certain circles amongst this élite was whether it was better to be under French or British rule. By the end of the partition, however, this attitude had changed.[37]

In the protectorates the problem of administrative consolidation was slightly different but equally straightforward. Here the question was how to forge a link between the new colonial government and teeming un-westernised populations with a view to minimising friction and obstruction and maximising harmonious and fruitful cooperation. Every imperial power in human history has had to tackle such a problem and the solution has generally been to use in *some form* at *some stage* the traditional rulers and institutions of the conquered territory as a bridge for reaching the people. The arguments which led to the adoption of this method by the French, the British and the Germans were both philosophical and practical.

On the philosophical plane the earlier assumption by the French and the British that all men were equal and that the African was a benighted brother who could be led to the level of European achievement through the light of European civilisation was increasingly brought under question and in fact rejected by a responsible section of European opinion. With quasi-scientific arguments from evolutionist biology and pre-scientific anthropology, such people argued that the African was of a different biological genus from the European or, as they put it, was no more an undeveloped European than a mouse was an undeveloped rabbit. In any case, these men maintained, even if the African and the European were the

37 See chs. 13 and 14.

same species, they were of different kinds and at different levels of cultural development, and that this difference must be respected. Therefore whatever the apparent advantages of endowing Africans with European culture and institutions, such a course must be avoided for it would be throwing pearls before swine. It would also, it was argued, do untold harm to the Africans.

On the other hand there were many practical problems which made the extension of the mid-nineteenth-century policy of liberalism, assimilation or partial assimilation into the interior impossible and unadvisable. In the first place applying that policy to a few thousand already westernised subjects along the coast was one thing, applying it to millions of unsophisticated villagers was quite another. In the interior the colonial powers came across such nomadic groups as the Tuareg and cattle Fulani. How, people asked, could one make European institutions function among such groups? The westernised élite along the coast were too few to be used in working such alien institutions among the peoples of the protectorates. And even if the élite were available in sufficient numbers, there was the fact that the earlier understanding between them and the colonial powers had broken down by the end of the partition. In the British colonies those of them who held positions of responsibility were being systematically humiliated and removed.[38] The French tightened the conditions for becoming a French citizen.[39]

However, while the powers could not and would not use the new West African élite, their own citizens could not do the work either. Until railways and roads were built, slow and inadequate communications limited the effectiveness of field officers. Then, there remained the fact that Europeans still had to reckon with serious health problems despite the increased use of anti-malarial measures. Debilitated and leave-hungry Europeans resulted in troublesome administrative discontinuities. And in any case Europeans were costly, and colonial governments with budgets as small as their economic ambitions were large could not afford many of them.

All these arguments dictated the use of an African agency at some point in the administrative chain. The traditional rulers were, at least at first, an obvious choice. For one thing they were already familiar with their people and had the necessary experience of government. For another, using them helped to simplify the work of consolidation. Since they invariably led the resistance movements to the imposition of European rule, conciliating them and getting them involved in the affairs of the new regime would leave the resistance movement leaderless and demoralised. The rulers themselves, it was hoped, would soon realise that their interest lay in supporting the colonial administration which had graciously decided to continue to keep them in being. 'To bring pressure on the chiefs,' said

38 See ch. 14.
39 See ch. 13.

Captain Rottier, 'the best means was to show them by tangible proofs that their interests were rather with us than against us.'[40] If this were achieved the traditional rulers would serve the colonial powers as very sensitive instruments for reading the political pulse of the people.

In fact the choice which the colonial powers first had to make on conquering the interior of West Africa was not whether or not to rule through the indigenous system but where to effect the link with that system, whether at the level of the paramount chief, where such existed, or somewhere lower down. The general pattern eventually adopted by the powers would seem to have been to break the paramount chief into chiefdoms where they encountered severe opposition to their rule and to effect the junction at the district or village level. The British did this in Benin and Asante, the French in the Western Sudan where they encountered strong men like Umar, Ahmadu Seku and Samory Toure as well as in Dahomey. But where the resistance was weak or relatively so as in Mossi, Hausaland and Yorubaland the French and the British retained the paramount chieftaincy. The French were more inclined than the British to adopt the former policy which they described as 'the progressive suppression of the great chiefs and the parcelling out of their authority',[41] but in general both powers applied each policy wherever it suited best.

In effect either method of handling paramount chiefs could be described as a variation on the ancient principle of *divide et impera.* Whether the paramount authority was altogether set aside and a direct relationship established between the colonial administration and his former vassals, or whether an all-powerful Resident or Commandant stood across the channels of communication between a formally recognised suzerain and his subordinate chiefs the result was the same; the chiefdom had been divided to be ruled.

In areas where there were no chiefdoms but instead autonomous villages or village-groups, each with populations of not more than two or three thousand souls, the problem presented itself upside down and had to be tackled differently. To use only the village head and his council as the point of highest contact between the administration and the people would require a large army of European officers if effective supervision of government business was to be achieved. To avoid this unpalatable, because costly, solution the colonial powers, especially Britain in the former Eastern Nigeria and Germany in Togo, sought to build up 'super-chiefs' who would be able to exercise political control over territorial units larger than the traditional village or village-group.

40 Quoted in Buell, i, p. 920, fn. 1.
41 J. Suret-Canale, 'Guinea Under the Colonial System', *Présence Africaine*, English edition, i, 29, p. 29.

The Abomey Protectorate[42] perhaps provides one of the most striking examples of the application by the French of the policy which favoured destroying the paramount authority and effecting the linkage between the new colonial administration and the indigenous political organisation at the level of the district (canton) and village. The kingdom of Abomey, which embraced most of the southern half of what is now known as Dahomey, was throughout the nineteenth century probably the most effectively centralised and militarised state in West Africa. Its resistance to French aggression had, not surprisingly, been stiff. For this reason the French soon decided to destroy the centralised control of the paramount authority as a further stage towards the effective consolidation of their rule. The first step in this direction was to depose and exile Behanzin, the King of Abomey who had led the resistance against them. In his place they enthroned a placeman, Gouthchili, who took the throne name of Agoli-Agbo. Because this man was a weak character and owed his accession to the French, the latter hoped to be able to control him without much difficulty. As a second step the French decided to weaken the position of the Fon *vis-à-vis* the other groups in the kingdom whom they had conquered and ruled. Among these were the Yoruba who lived between the rivers Zou and Weme and the peoples to the south of the Fon formerly under the control of the kingdom of Alladah and Whydah. The French decided to give these peoples their independence. The former kingdom of Alladah was constituted into a separate protectorate with Gli-Gla as paramount chief. The Yoruba on their side were asked to move to the region east of Weme where they would fall outside the Abomey Protectorate. Hoping that these two measures would remove the most dangerous threat to their presence and rule in Dahomey, the French were prepared to allow the paramount chieftaincy in Abomey to lead an emasculated existence.

But the place of the Abomey authorities in the new scheme of things was too undefined and too full of contradictions to produce harmonious cooperation between the French and Agoli-Agbo. The king found himself in an impossible economic position. When the French settled themselves at Porto Novo and Cotonou in 1885 the royal court of Abomey lost the revenue from import and export duties which had formed its most important source of income. With military conquest in 1894 the king lost even the right to farm taxes from his dependent provinces to the east of Zou river and south of the Fon. Yet despite this reduction in his revenues the king was expected to maintain a large and expensive court and

42 For the French Administration in Dahomey see C. W. Newbury, *The Western Slave Coast and Its Rulers*, Oxford, 1961, pp. 180–7; Newbury. 'A Note on the Abomey Protectorate', *Africa*, xxix, 2, pp. 148–54; Buell, i, pp. 918–19; Crowder, *West Africa Under Colonial Rule*, pp. 177–89.

entourage as well as to fulfil the traditional duties of his office such as the performance of expensive funeral rites for his deceased ancestors.

Attempts to find new sources of revenue brought the king into conflict with his people and his new masters. In place of the revenue from the rich palm belt to the south from which the French had excluded him, the king tried to tax those Yorubas who had not moved to Ketu or beyond the Weme. The Yoruba resisted; the French refused to help the king and in 1895 actually intervened to stop the activities of royal agents in the Yoruba districts. In return Agoli-Agbo chose the occasion of the French imposition of direct taxation on Southern Dahomey to obstruct them in the Abomey Protectorate. Such political naïvety disqualified him from his office. He was removed and exiled to Gabon in 1900.

The deposition of Agoli-Agbo was not just a personal tragedy, it also meant the end of the Abomey kingship, as the French refused to appoint a successor. But the suppression of the paramount authority at Abomey should not be taken to mean the total destruction of all there was of the indigenous political system of the Fon. At the district (canton) and village level the French showed no intention of breaking with tradition. They divided the kingdom into cantons but in doing so respected lines of traditional village groupings. Furthermore the cantons and the villages under them were left in the hands of traditional rulers who had also been Agoli-Agbo's functionaries. And these officials were aided by native courts or tribunals, whose membership included village and canton chiefs, which tried cases of disobedience to chiefly authority especially in the matter of taxation. An investigation in 1904 revealed that of the surviving male members of the Dahomean royal family only three of them who had particularly offended the French were not holding office in the local administration. Four were chiefs of cantons, while six were ward chiefs. These office holders successfully opposed the revival of the kingship which they feared would bring unwelcome interference from Abomey in their local affairs and bring on them once again the burden of maintaining the royal court. In 1910 when Agoli-Agbo was allowed to come back to Dahomey he was permitted to perform traditional religious functions only and had to content himself with an annual pension of 3,600 francs and exile to Savé.

Likewise in the rest of Southern Dahomey, the paramount authority in the Allada protectorate was quickly rendered powerless by recalcitrant elements within the protectorate as well as by the French. The paramount chief, Gli-Gla, in trying to play his new role as agent of the French Administration, fell foul of his people and lost his authority over them. In 1895 he precipitated a rebellion against his authority by assisting the French in the recruitment of labour for their colony of Madagascar. Gli-Gla forgave the rebels but they never forgave him. When they came back to Dahomey after three years they proceeded to undermine his authority and even instigated another revolt against him. In all these troubles with his subjects the French refused to support Gli-Gla decisively. In fact they further weakened his position by partitioning his territory. Under Jihento, his successor, the Tori area of the protectorate, which objected to control from Alladah, was included in the Whydah District. In Porto Novo the authority of the paramount chief Tofa was whittled away by the French. Here as elsewhere the French regarded the ability of the chiefs to collect direct taxes without incidents as the touchstone of their usefulness. Tofa and his ministers did not pass this test and so their powers were reduced. In 1913 the paramount chieftaincy was suppressed altogether. In Dahomey, therefore, the real contact between the French administration and the people was forged at the district or canton level where the traditional rules of succession to office were respected as long as they did not conflict with French interests.

In the Sudan, where the French had encountered what they considered the worst forms of Muslim fanaticism, they were even more determined to break up the Islamic states which had come into existence by the nineteenth-century jihads. Umar's empire, the least organised and stable of the great caliphates of the nineteenth century, was easy to disintegrate. On the conquest of Segu in 1890 the French brushed aside Umar's royal line and appointed as king a member of the old Bambara royal family of Amkienne. On the conquest of Massina, Aguibu, a brother of Ahmadu Seku, was appointed king under French protection. He held the position until 1902 when he was pensioned off, the kingship suppressed, and French local government thenceforth based on cooperation between the central government and the local district and village chiefs. The provincial and district functionaries created by Umar were removed and their positions given to members of the old royal families they had forcibly superseded. 'The Toucouleur chiefs installed as provincial chiefs by El Hadj and his successors,' it has been written, 'were suppressed by extinction: at the death of each of them, the cantons and villages which they commanded were returned to their original chiefs.' By this policy the French hoped to strengthen their hands by alliance with the communities whom they believed the Tukulor imperialists had oppressed and exploited.

The imamate of Futa Jallon provided another classic case of the application of this policy by the French. As in Umar's empire, the French were assisted by the fact that the political system of the imamate was highly decentralised. Suret-Canale has described it as 'feudal in character', 'an obstacle to real central authority', a 'confederation' in which 'order resulted less from the exercise of real authority than from the (unstable) equilibrium between the forces of rival families'.[43] In their 1897 treaty of protection with the Futa Jallon authorities the French had promised,

43 Suret-Canale, p. 26.

among other things, 'to respect the present constitution of the Futa Jallon' and the arrangement then obtaining under which the rival families of Alfaya and Soriya provided the imam alternately. But no sooner was this pact made than the French proceeded to ignore it by applying the policy now being discussed. Firstly they deposed Imam Oumarou Bademba and put in his place their nominee, Baba Ahmou. Secondly they reduced the imamate to only three provinces, proclaiming the others independent under chiefs acceptable to them. Thirdly when Baba Ahmou died in 1906 they abolished the principle and practice of alternate succession and partitioned the imamate between candidates chosen from the two rival families. Along with this change went the reduction of the imams from paramount to provincial chiefs. For these imams the end of the road was to be reached in 1912 when they were further reduced to the rank of canton chiefs.

Where the French decided to effect the linkage between their new administration and the indigenous system of government at the level of the paramount authority or king, they usually ensured, as did other colonial powers, that the authority was so 'purified' of 'barbarous practices' and so reorganised that it became emasculated, totally dependent on them and therefore incapable of mounting any serious challenge to their regime. This end was achieved in various ways. Structural changes were sometimes effected in the system, territorial rearrangements were carried out within the area of rule of the paramount authority resulting in the alteration of the traditional extent of the local functionaries' responsibility and invariably, as with the British, a new concept, philosophy and purpose of government were imposed.

In the ancient Mossi state of the Volta Valley, for instance, the French applied the policy of preserving the paramount authority and governing the Mossi people through that authority and its subordinate agents. But to ensure that the Mossi authorities had no doubt as to where sovereignty now lay and did not obstruct their rule, the French set out deliberately to weaken the traditional political system before hitching it to their own administrative wagon. First the central authority, then the provincial authorities were undermined. Said Commandant Destenave:

> We have no interest in strengthening the power that is regarded as central, nor in increasing the powers of the various Nabas; on the contrary we must look for points of stress which will permit us to divide the country and thus preclude any coalitions against us. . . . In doing so the authority of the Mogho Naba itself will be weakened because we can easily acknowledge the independence of these great [princes] and free him from their influence. In future we shall seize every opportunity to weaken the authority of these vassals by declaring the independence of the villages under their command.[44]

44 Quited in Skinner, p. 155.

Thus on capturing the kingdom of Ouagadougou the French treated the paramount chief Mogho Naba Boukary Koutou, who had led the resistance against them and had fled his capital when his forces were dispersed, as deposed. They appointed in his place Mogho Naba Sighiri whom they were confident of controlling. But this man turned out to be a passive resister and would tell the French nothing about his kingdom, his people or his predecessor. 'The king of the Mossi,' complained the French, 'is lavish with vows of devotion and fidelity, but these vows are never translated into action. It is impossible to obtain any pertinent information about the country—he tries to give the impression that he knows nothing about it. . . .'[45]

When, therefore, Sighiri died in 1905 the French appointed as Mogho Naba his sixteen-year-old son Saidou Congo, since they reasoned that a child would be more amenable to control than an obscurantist adult. The Mossi kingmakers thought the boy too young to 'keep the country' but to the French this was the strongest argument in his favour. Two years after the elevation of Saidou Congo, the French went a step further to weaken his position in order to make him and his chiefs dependent on them. They reorganised the provincial administration and in the process suppressed certain principalities and districts while enlarging others. At the end of the exercise some chiefs and functionaries saw themselves subordinated to their former equals and members of the royal family were forced to become district chiefs. For instance the Mogho Naba's principal ministers responsible for provincial administration were deposed and their places given to men approved of by the French. These new appointees were given wider powers than their predecessors on the grounds that the young Mogho Naba could not exercise some of his traditional powers. Furthermore their traditional order of precedence was tampered with and the territorial responsibilities of the chiefs determined without reference to traditional practice.

While the executive power of the chiefs was being made dependent on the will of the French Administration, their judicial power was first strictly limited and then completely abolished as in the rest of French West Africa. Under a decree of September 1887 the arbitrary code of law known as the *indigénat* was made applicable to French West Africa. This code provided the French administrator with a rough and ready answer to any subversive behaviour on the part of the *sujets* by empowering officers to discipline criminals without necessarily holding a trial. Against punishments imposed under this code there was no really effective appeal.[46]

45 *Ibid.*, quoted on p. 155.

46 The administrator was required by law to report any such case and the penalty imposed to the Lieutenant-Governor who had the right to alter or annul the punishment. But since the Lieutenant-Governor was usually not on the spot, he could hardly effect any changes before a victim had spent the period of imprisonment awarded. See Buell, i, pp. 1015–17.

Exercising their powers under this code the administrators in Mossi, as elsewhere in French West Africa, undermined the judicial authority of the chiefs over their subjects. Yet cases remained which the administrators could not deal with under the *indigénat*. Therefore in 1905 a hierarchy of village, district and *cercle* courts, run by chiefs at the lower levels and by a joint system under European presidency at the *cercle* level, was created in Mossi. These local courts continued until 1912 when they were swept away by a decree. It was argued then that the chiefs were incompetent and that the functions of the courts conflicted with the judicial powers of the administrative officers under the 1887 decree.

Thus reduced and taught their dependence on the administration, the Mossi king and his chiefs became the obedient servants of the French. By the end of the period covered in this chapter the Mogho Naba had lost most of his powers and all initiative, but was permitted to retain his ceremonial position as a figurehead. He could still appoint district chiefs but subject to French confirmation. Some of the chiefs who had lost by the whole process of reorganisation had tried to create trouble for the French in 1910 but had been forced to reconcile themselves to the *fait accompli* 'when the police burnt their villages, seized their goods and animals' to pay their taxes. 'These harsh measures so shocked the Mossi,' comments Dr Skinner, 'that they remained docile from that time onward, and thereafter the French were able to rule the country with one European administrator for every 60,000 Mossi.'[47]

In Benin and Asante the British found two fairly well integrated political systems which were able to offer stiff military opposition to their rule and, like the French in the cases dealt with above, they decided to do away with the central authorities in these states and rule through what had hitherto been the provincial and district authorities. The general pattern of the new organisation instituted by the British in the Benin kingdom between the exile of the Oba in 1897 and the revival of the Obaship in 1915 is fairly clear. The provinces of the empire lying beyond the capital were treated as if they had never been parts of the Benin empire. Bini officials resident in them were chased out and the local chiefs organised into Native Courts and Councils to carry on the administration of their internal affairs. In Benin itself a council of chiefs was constituted to govern the capital city in place of the Oba under the supervision of the British Resident.[48]

As for Asante the British had been bent on the disintegration of the confederacy since 1874, and the conquest of Kumasi in 1896 merely offered

47 Skinner, p. 156.
48 P. A. Igbafe, 'The District Head System in Benin 1914–35', *Odu*, iii, 2, 1966; Igbafe 'British Rule in Benin 1892–1920: Direct or Indirect', *J.H.S.N.*, iii, 4, June 1967.

them the opportunity to implement this policy more effectively from the centre of the confederacy. Firstly, the Asantehene and his principal chiefs and advisers were seized and exiled and the appointment of successors was forbidden. Thus the lynch pin of the Asante confederal, political and administrative system was removed. In the words of Dr W. Tordoff the events of the fateful day of 20 January 1896 broke 'the link, forged over hundreds of years, which had held together the various states of the confederacy'.[49]

Secondly, the British took steps to ensure that Kumasi, the capital territory of the confederacy, lost her pre-eminent place amongst the member states of the empire. The member states were encouraged to organise and run their internal affairs independent of Kumasi and under the supervision of British Residents. Territories in the conquered parts of the empire whose government was supervised by Kumasi chiefs were removed from the control of these chiefs. This policy played into the hands of those provinces on the northern and southern frontiers of the kingdom which had never taken kindly to Asante rule and in which rebellion had been endemic. Thirdly, within Kumasi itself the British took steps to prevent the emergence of a leading spirit. Thus a *Native Committee of Administration* was created to conduct government and administration within the capital city which was now reduced to a mere province. Even this body was soon allowed to fall into desuetude. In 1905 another body, the *Council of Kumasi Chiefs*, was constituted to take its place. Its chairman was the Kontihene, who in the preconquest days generally presided over the meeting of Kumasi state council in the absence of the Asantehene. But to prevent this official from considering himself the political leader and head in the changed circumstances, it was stipulated that the council should meet in the houses of its twelve or fifteen members in turn. The British used this body as a mere sounding board of popular feeling rather than as the governing authority in Kumasi and after 1918, when they felt that their rule in Kumasi was secure, they allowed it again to fall gradually into decay. The disruptive effect on Kumasi of this policy was such that local government in the division remained in a chaotic state until after 1926 when Prempeh was reinstalled Kumasihene and allowed to undertake the internal government of the former imperial capital.

After forcing 'independence of Kumasi' on the other member states the British did little else to interfere in their internal administration. They were allowed to govern themselves according to their traditional usage as long as they did nothing the British considered barbarous or unjust. Between 1911 and 1913 the colonial administration also tried the idea of

49 W. Tordoff, *Ashanti Under the Prempehs*, London, 1965, p. 81.

chiefs' councils among the states of the Central and Northern Provinces. Through these councils the British hoped to learn the effect of their policies on the people and to inform themselves about the law, traditions and usages of their subjects. As in the case of Kumasi these councils were soon neglected. Ultimately, before the re-creation of the Asante confederacy in the 1920s, the effective link between the colonial administration and the people was made at the level of the traditional authorities of the districts. Elsewhere in their West African territories the British sought to consolidate their rule by using the paramount chiefs and their subordinate functionaries as the agents of their rule or where these paramount chiefs were not in existence the British sought to build them up artificially.

The earliest attempt by the British at a systematic use of indigenous chiefs as agents for extending and establishing their rule was made in the territory known as the Niger Coast Protectorate between 1893 and 1899 and as the Southern Nigeria Protectorate between 1900 and 1905. This colony comprised the region now occupied by the Midwest, East Central, Rivers and South-Eastern States of Nigeria as well as part of the Owo Division in what is now the Western State. Except perhaps in the Benin territory and part of Owo Division the so-called chiefs of this area were often no more than village ritual headmen who had little political status and authority. The British were therefore compelled in these areas to set up councils of chiefs to which they assigned specified executive and judicial powers. Those of these councils which met where they could be supervised and presided over by European political officers were known as Native Councils and given wider powers than were the Minor Courts which sat in district and provincial suburbs where they had to be presided over by one of the native chiefs.

This administrative system came into existence in 1891. By 1899 these councils, each with its rudimentary treasury fed by court fees and fines, directing and financing local projects, represented the main instrument of British administration throughout much of the Protectorate. But there was much in this system that was artificial and unacceptable to the people. First, the members were often chosen in a haphazard fashion without proper inquiry into indigenous political practice and systems of authority. Consequently most members of the councils were not even village heads. Their authority derived entirely from the *Warrant* or 'certificate of recognition' given to them by the government. Secondly, even where they were traditional village heads they were given an artificial position through being made to represent, and supervise the affairs of, villages and village-groups over which they traditionally had no authority whatsoever. For these reasons they came to be popularly known as *Warrant Chiefs*, that is chiefs whose authority derived from their warrants only. Without knowing it the British were foisting a new constitution on the people. This affront to the traditional constitution within the village and to the system of political relationship amongst villages was to become worse after 1912 under Sir Frederick (later Lord) Lugard as Governor-General.[50]

In the neighbouring territory of Lagos Colony and Protectorate the first serious attempt to recruit the traditional élite of Yorubaland into the service of the colonial government was made by Governor MacCallum in 1897 when he undertook the reorganisation of the conciliar government of the traditional Yoruba aristocracy. He launched the programme at Ibadan with the reorganisation of the Ibadan Council of Chiefs. This body was enlarged to include in its membership a European Resident. From being the sovereign authority in the Ibadan city and empire the Council was reduced to the status of a subordinate local authority of the colonial administration with competence limited only to Ibadan city. When Sir William MacGregor succeeded MacCallum as Governor in 1899 this experiment was extended to the rest of Yorubaland. MacGregor tightened government control over the already existing councils, posted Residents to such areas of the Yoruba interior as had none on his arrival and appointed Travelling Commissioners whose work it was to tour the interior constantly in order to supervise the native authorities where they already existed or to organise them where they were not already properly constituted.[51]

On 4 November 1901 the local government system which he envisaged as a means of spreading and stabilising British rule was placed on a definite legal basis. His Native Councils Ordinance of that year created for the Lagos colony a Central Native Council which was to advise the Governor on matters 'concerning the good government and the well-being of the native population'. The Governor himself presided over the council and appointed the local members. By advising the government on Yoruba traditional law and custom touching on such delicate issues as land ownership, divorce, the use of ceremonial umbrellas and crowns, the council saved the government from such blatant contraventions of traditional practices and procedures as might have generated popular disturbance. Through it the traditional rulers of Lagos, though they had lost real power, retained a tenuous right to influence decisions on matters affecting indigenous Lagosians. Through it too they learnt from the Governor much about government policies and intentions. The mere fact that they were considered worthy of consultation for advice must

50 A. E. Afigbo, *The Warrant Chief System in Eastern Nigeria 1900–1929*, Ph.D. thesis, Ibadan, 1964; Afigbo, 'Herbert Richmond Palmer and Indirect Rule in Eastern Nigeria 1915–1928', *J.H.S.N.*, iii, 2, 1965; Afigbo, 'Chief Igwegbe Odum: The Omenuko of History', *Nigeria Magazine,* 90, 1966; Afigbo, 'The Native Treasury Question Under the Warrant Chief System in Eastern Nigeria', *Odu*, iv, 7, 1967; Anene, ch. VII; O. Ikime, *Niger Delta Rivalry*, London, 1970; I. F. Nicolson, *The Administration of Nigeria 1900–1960*, Oxford, 1969, ch. IV.

51 Nicolson, ch. III; J. A. Atanda, *The New Oyo Empire: A study in British Indirect Rule in Oyo Province 1894–1934*, Ph.D. thesis, Ibadan, 1967.

have gratified the self-esteem of these dignitaries. By creating mutual understanding between the chiefs and the government the council reduced the occasions for head-on collision.

For the rest of Yorubaland, MacGregor's Native Council Ordinance recognised provincial and district Native Councils where they were already in existence and authorised their establishment elsewhere. The most prominent chief in the area of jurisdiction of each Native Council was to be its president, while the British political officer in the province or district acted in an 'advisory' capacity. Below the District Native Council was the Village Native Council which was responsible for settling village disputes and applying government regulations to the village. Under the provisions of the ordinance of 1901 the village council was organised by either the Governor or the Provincial or District Council.

In Northern Nigeria the British also sought to convert the indigenous systems of government of the people into instruments for extending and stabilising their rule. There were 'three political areas' in Northern Nigeria in 1900 which the British had to cope with. First, there was the Bornu empire which was the most ancient state in the central Sudan and had a long tradition of ordered and stable government and administration. This stability had been recently upset by the overthrow of the fifty-three-year-old al-Kanemi dynasty in 1893, but the son and successor of Rabeh, the usurping ruler, died at the hands of the French and the British were able to call Abubakar Garbai back to the throne of his fathers—a traditional ruler but at the same time a British creation.

Secondly, there was the Sokoto Caliphate which constituted one religious community under the Sultan, but for political and administrative purposes was divided into two under Sokoto and Gwandu. Here the British adopted a slightly modified form of the French policy of 'the progressive suppression of the great chiefs and the parcelling out of their authority'. The Emir of Gwandu and the Sultan of Sokoto ceased to be paramount chiefs who had the right to expect and receive tribute from the other emirs under them or to have a say in their election and confirmation. These functions were instead assumed by the High Commissioner who was the supreme authority over the Caliphate, had the right to impose and receive tribute, to confirm all elections to the provincial governorship or emirship and could legislate for the Caliphate by proclamation. To make this reduction of Gwandu and Sokoto to provincial political lords effective a Resident was posted in each province. It was this officer who under the guise of 'advising' the Emirs dictated policy and supervised its implementation.[52] The Emirs dealt with the High Commissioner through the

52 J. N. Paden, 'Aspects of Emirship in Kano', *West African Chiefs*, eds. M. Crowder and O. Ikime, Ife, 1970; R. Cohen, 'The Kingship in Bornu', *West African Chiefs*; R. Heussler, *The British In Northern Nigeria*, Oxford, 1968.

Resident rather than through the Sultan or the Emir of Gwandu as would have been the case before 1900. Thus the British utilised for local government not the political system of the Caliphate but the administrative system of the Emirates. The empire disintegrated but the provinces (emirates) were preserved in considerably modified forms.

The third political area which the British encountered in Northern Nigeria was that of the so-called pagan provinces which the Fulani had failed to conquer though they had begun to infiltrate some of them when the British came. The political system of these pagan areas was as fragmented as was that of the Ibo and Ibibio peoples of Eastern Nigeria. The attempt to convert the administrative institutions of these peoples into instruments of British policy led to results similar to those in Eastern Nigeria. Artificial councils of artificial chiefs were created and vested with political and judicial authority over artificial districts and many appointments of warrant chiefs were made.[53]

Three measures harnessed the emasculated provincial system of administration in the Caliphate, Bornu and the pagan areas to the central administration. There were the Native Authorities, Native Courts and Native Revenue Proclamations (later Ordinances). The Native Authorities Proclamation recognised the existing hierarchy of chiefs. Five grades of chiefs, with different insignia of authority, were accepted. At the top were the more important emirs, like the rulers of Sokoto, Gwandu, Kano and Bornu who had their judicial councils and personal police. At the bottom were the district heads of the politically fragmented non-Muslim communities. The right to appoint these chiefs was vested ultimately in the High Commissioner, who had to confirm every election and could refuse to accept the election of a candidate he did not approve of even if the candidate should be the most qualified under the traditional rules of succession to political office. In confirming the appointment of any chief the High Commissioner emphasised that the appointee owed his position to the British and would be deposed or disciplined in the event of incompetence, corruption or disloyalty. In order to ensure the loyalty of the Muslim rulers the government exploited their religious superstitions.

> I [ran the oath of office which they were required to take] swear, in the name of Allah and Muhammad his prophet, to serve well and truly His Majesty King Edward VII, and his representative, the High Commissioner of Northern Nigeria, to obey the laws of the Protectorate and the lawful commands of the High Commissioner and of the Resident, provided they are not contrary to my religion. And if they are so contrary, I will at once inform the Resident for the information of the High

53 D. C. Dorward, 'The Development of British Colonial Administration Among The Tiv 1900–1949', in *African Affairs*, 1969, pp. 316–333.

> Commissioner. I will cherish in my heart no treachery or disloyalty, and I will rule my people with justice, and without partiality. And as I carry this out so may Allah judge me.

For non-Muslim areas the oath was amended to suit their religious beliefs.

The Native Courts Proclamation (1902) gave a legal basis to the existing system of native courts in the Emirates. It recognised and upheld the judicial council of the leading emirs, and then the *alkali* courts. The Native Revenue Proclamation (1904) sought to systematise and rationalise the system of taxation in the Emirates. After making a detailed study of the existing systems of taxation and finding them complicated and capricious, Lugard introduced a single tax which immensely simplified assessment and collection as well as reduced the burden which the *talakawa* endured under the old cumbersome system. Out of the total proceeds of the taxation in an emirate the government took or, as the theorists of Indirect Rule would say, was given a portion which varied from a quarter to a half depending on the status of the emir or the chief in question. What remained was left to the chief for meeting his own salary, the salaries of his subordinates who hitherto were not on fixed salaries and for satisfying some of the public needs of the area under his authority.[54]

For a long time, this money was not formally accounted for by the emir nor was its expenditure controlled by the Resident or the central administration. This must be borne in mind in view of the popular notion that the Native Treasury was fundamental to Lugard's interpretation and application of the old principle of ruling a conquered people through some form of their indigenous institutions. It was only after Lugard had left Nigeria that gradually the idea was evolved by his successors that it was much better and neater to regularise the emir's use of his own portion of the proceeds of direct taxation. Out of this emerged the idea and institution of the Native Treasury as it was understood in Northern Nigeria. Under this system the money was placed in a safe in a public building and systematically administered. The emir, his chiefs and other functionaries got their salaries out of it, while what remained was devoted to providing public needs and utilities in the emirate or province under the supervision of the Resident. The Native Treasury came into official records in 1911. After that date it won great prominence as an indispensable institution in any local government system that claims to use indigenous institutions.[55]

It is popularly believed that in Northern Nigeria the British preserved intact the administrative system of the emirates after purifying it of its grosser forms of barbarism and corruption. Recent research carried out on

54 M. Perham, *Native Administration in Nigeria*, Oxford, 1937, pp. 43–80; M. Bull, 'Indirect Rule in Northern Nigeria 1906–1911', *Essays in Imperial Government*, eds. K. Robinson and F. Madden, Oxford, 1965.

55 *Ibid.*

early British administration in the Kano[56] Emirate reveals, however, that the emirate administration was so drastically weakened and changed that it would be a travesty to describe it as strictly traditional. The government of the Kano Emirate was, like that of other emirates in the Sokoto Caliphate, an autocracy. In Kano this autocracy was buttressed on an elaborate court organisation made up of the emir's personal slaves, hangers-on, favourites and other dependents. Among these there were the *hakimai,* free-born district chiefs. These were usually absentee officials who resided almost permanently at the emir's court. To do the work of the hakimai, there was a host of subordinate slave officials, *jakadu* (messengers), usually under the direct control of the emir and mostly responsible for the collection of the innumerable taxes on which Fulani rule depended. Some specially favoured slaves were raised from the rank of messengers and given titles and district chiefships.

The Emir of Kano took elaborate measures to ensure that his rule was absolute. First, there was the tradition by which all fief-holders were absentees and resided at court where they were closely watched by the Emir. Second there was the excessive dependence on slave officials and messengers for the actual enforcement of the Emir's will in the province. As slaves they knew they owed everything to the Emir and did what they could to uphold his authority. Third, the Emir ensured that the territories under each hakimai were scattered all over the emirate rather than formed one continuous block. Whatever the original history of this arrangement, it ensured that no hakimai could easily raise enough forces from his districts to challenge the Emir's authority. Any administrative reform which tampered with these fundamental props of the Emir's autocracy would of course undermine the traditional system.

The British administration found the system unacceptable for many reasons. They disliked the absenteeism of the district chiefs as well as the large horde of slave officials which they believed gave rise to waste, inefficiency, corruption and oppression. The dispersal throughout the emirate of the territories making up a hakimai's district they found untidy and unreasonable. British rule in Kano throughout the period covered by this chapter was dominated by attempts to remove these features of the traditional system.

As a first step in this direction the British in 1903 removed Aliyu, the ruling Emir who was intelligent, courageous and independent-minded and therefore found submission to British rule uncongenial. They placed in his stead his brother, Abbas, who, they hoped, would be easier to

56 The following account is based on information taken from an unpublished manuscript entitled 'Some Aspects of Early British Administration of Kano' read by Mr C. N. Ubah at a staff Seminar in the Department of History and Archaeology, University of Nigeria, Nsukka, on 18 February 1967. I am grateful to Mr Ubah for permission to use the manuscript. In addition see J. N. Paden, 'Aspects of Emirship in Kano', *West African Chiefs.*

manipulate. In 1904 they replaced the dispersed hakimai territories with eight large districts each of which was territorially continuous. Eight of the most important hakimai were made heads of these units, while all the others were made sub-district headmen, thus introducing inequality and a hierarchy amongst the hakimai who had hitherto been equals, at least in theory. The administration also ruled that thenceforth the district and sub-district heads were to undertake the collection of taxes themselves and bring the proceeds personally to the Emir. These measures undermined the Emir's power at many points. The new administrative arrangement made it impossible for him to create new district heads. Also the authority and position of the jakadu on which the Emir relied for dealing with the free-born office holders were being gradually undercut. For a time something remained of the jakadu system since the British did not take immediate measures against absenteeism. But this too went in 1906 and the district and sub-district heads were sent to their districts to undertake the tax collection themselves. This reform destroyed that direct control which the Emirs of Kano had always exercised over their important functionaries; it destroyed what remained of the *raison d'être* of the jakadu. The innovation offended the hakimai who did not want to be banished from the Emir's court or to undertake the rigorous task of tax collection. The practical consequences of the innovation for the Emir's authority were soon revealed. Some of the hakimai soon came to realise that it freed them from close control by the Emir. One of the immediate results of this was tension between the Emir and the Resident such that both went out of their way to angle for the support of the hakimai and the talakawa. It was a measure of the extent to which British rule had undermined the Emir's authority that the Emir should now have to campaign for the support of his subordinate functionaries and even of the talakawa. In the course of this quarrel, Abbas demonstrated that he was not going to lose the traditional powers attached to his office without a struggle. At one stage the British contemplated the Emir's deposition and H. R. Palmer, a political officer, even suggested what he called 'a more direct administration' and the destruction of the palace organisation.

The Emir's resistance to this progressive subversion of his powers made little impression on the British who went ahead with other measures whose effect was to reduce him to a cipher. In 1908 the Administration abolished the slave districts and carried out a redistribution of the districts without consulting the Emir. They now dealt directly with the district heads. In the same year the administration of Kano City was reorganised and placed under the *MaajinWatari* who soon became a declared opponent of the Emir and would not cooperate with him. The tax system was streamlined and even those taxes sanctioned by the *Quran* were abolished leaving only the direct tax sanctioned by the Native Revenue Proclamation. The Resident, with the approval of the Governor, abolished the *Waziri's* district, reduced the *waziranchi* or the office of the waziri to a city office and contrary to all tradition appointed a slave to it. In the process of doing this, the title of *Dan Rimi* was taken from the Emir's son and given to this slave, the Emir's son being reduced to the office of *Ciroma* from which he had been elevated in 1904. As soon as he received his appointment, the slave allied himself and his following with the British and became arrogant towards his former master, whose authority he now sought to undermine. In the same year, 1908, the Resident superseded the Emir's court which Lugard had earlier agreed to recognise as serving a useful function. After this only the alkali court was left in Kano.

The Emir's power was now a mere shadow of what it used to be and in August 1908 Abbas pathetically appealed to the Resident saying:

> This is to bring before your notice that at present I have nothing to do, and I would like you to give me some proper work to do—formerly I was nobody, before the white man made me a king—now that I am sitting down doing nothing I always feel sick—and as you are the big white man I trust that you would give me some strong work to do that will please the government.[57]

Help for the Emir was at hand. In 1909, Mr C. L. Temple became the Resident of Kano, after serving in Sokoto. He was a staunch advocate of preserving Fulani power intact. He removed the slave waziri and appointed a free man, though not a prince of blood as had traditionally been the case. Also the Emir's judicial council was restored.

In spite of Temple's retreat from the extreme position of the earlier Residents, the Emirship was now a changed institution. The new district organisation was not abolished. The jakadu system was not given official recognition though it persisted as an informal aid to the Emir in his dealings with the district chiefs. It is said that with the rapprochement inaugurated by Temple, Emir Abbas kept the slave messengers out of the sight of the British. Nonetheless the palace clique or *sarkin gidda* declined in prestige, authority and number. In 1900 they were said to have numbered ten thousand. But with the introduction of the Native Treasury which led to the Emir's being placed on a fixed salary it became difficult to keep a large retinue which had no official existence and therefore no vote for their maintenance. This decreased entourage was of course positive evidence of the changed status of the Emir.

Up to 1912, when Lugard came back to Nigeria as Governor-General charged with the duty of amalgamating the Northern and Southern Protectorates, the British made no attempt to achieve uniformity in the method by which they employed or attempted to employ traditional authorities for purposes of local government. In the south the European

57 Quoted in Ubah.

political officer was a member of the Native Authority Councils and Courts, and east of Yorubaland he actually presided over sittings of a Council when present. In Yorubaland there was not even a rudimentary form of Native Treasury, but east of Yorubaland there were Native Treasuries which derived their revenue mainly from court fees and fines.[58] In the south the Native Authority was usually a Council of Chiefs or supposed chiefs but in the Sokoto Caliphate and in Bornu province it was usually a single chief or emir called Sole Native Authority. There was a well-established tax system and after 1911 a well-established Native Treasury. There was also a distinction between the Native Authority and the Native Court; in the south the two were usually one and the same body. In the pagan districts the system was very much like that in the southern protectorate except for the system of taxation and the Native Treasury.

In the course of the amalgamation Lugard decided to extend to the south the system he and those after him had built up in Northern Nigeria. To this end he introduced into the south in 1914, 1916 and 1917 Native Courts, Native Authorities and Native Revenue Ordinances on the Northern Nigeria model. The attempt to carry out these reforms gave the Obas and Warrant Chiefs more powers than they had exercised under the traditional constitutions and so estranged them from their councils, and the latter from their peoples. The introduction of direct taxation into Yorubaland led to widespread riots in Oyo in 1916 and in Abeokuta in 1918. The political officers east of the Niger resisted the plan to extend the taxation to the areas under their control. Lugard was forced to leave Nigeria in 1919 before he had succeeded in building up a uniform system of local government for Nigeria. Throughout the period, the system in the emirates, in Yorubaland, in Eastern Nigeria and the non-Muslim areas of the north differed one from another.[59]

In Sierra Leone the system of using native chiefs in local government was organised in a more haphazard fashion than in Nigeria. There it was based on three ordinances—the Native Protectorate Ordinance (1901) which divided the Protectorate into Provinces and districts under provincial and district commissioners; the Protectorate Courts Jurisdiction Ordinance (1903) which gave legal recognition to the judicial powers of the native chiefs; and the Protectorate Native Law Ordinance (1905) which, *inter alia*, regulated the traditional claims which the chiefs were still entitled to make on their subjects. For the rest no attempt was made to organise Native Treasuries or to organise the chiefs' courts and place them under regular European supervision or to place the chiefs on fixed salaries. 'At

58 A. E. Afigbo, 'The Native Treasury Question under the Warrant Chief System in Eastern Nigeria', *Odu*, iv, 1.

59 Afigbo, *Warrant Chief System*; Ikime, *Niger Delta Rivalry*; Atanda, *New Oyo Empire*; P. A. Igbafe, 'British Rule in Benin 1817–1920; Direct or Indirect Rule', *J.H.S.N.*, iii, 4, 1967.

present,' wrote Professor Buell in 1927, 'the Government pays the chiefs a five percent commission on house tax . . . plus stipends in pursuance of treaties. . . . In addition it makes certain presents to the chiefs.'[60] The Sierra Leone authorities, however, felt that if the chiefs were to continue in the future to be useful as effective instruments of colonial rule, then future chiefs should be educated in order to increase their efficiency. So they opened a school for sons of chiefs at Bo in 1906. The school was given 'a native atmosphere' and the scholars were sent frequently to the village on holidays to avoid being detribalised by their education. In 1927 five graduates of the Bo schools were paramount chiefs and ninety-two were in government service.[61] The exact position which the Sierra Leone chiefs occupied under the new regime is still to be unveiled by detailed and systematic research. But there is no reason to believe that their traditional statuses and powers were preserved absolutely unimpaired.

In the Gambia, where the system was better organised than in Sierra Leone, the first step was taken in 1893 under R. B. Llewellyn, Administrator of the Colony from 1891.[62] In that year the Gambian Government appointed two Travelling Commissioners for the two banks of the river to tour the villages, acquaint the native chiefs with the changes implicit in the declaration of a protectorate and to advise them on the running of local affairs. On giving up the idea of setting up a 'super-chief' to control local government in Gambia through the various districts and village chiefs, Llewellyn enacted the Protectorate Ordinance in 1894 which remained the main legal basis for the local administration until 1913. The Ordinance upheld all traditional laws and customs which were not repugnant to British conscience or to any law of the Gambian Government that applied to the Protectorate. It divided the Protectorate into districts, each of which was placed under a chief for purposes of local administration. Other provisions conferred minor executive powers on the district chiefs and established Native Courts, presided over by district chiefs but supervised by British officers, to settle specified classes of cases. Some of the district chiefs had traditional rights to these powers, but the majority were men artificially lifted out of obscurity to occupy these positions created by the 1894 Ordinance.

The outlines of the local government system were more or less complete when in 1895 the Yard Tax Ordinance introduced direct taxation into the Gambian Protectorate. In 1902, 1906 and 1909 ordinances and regulations were issued to modify the above structure here and there by further defining the executive and judicial functions of the chiefs and the Commissioners as well as the limits of the newly created districts, but the principle remained the same as had inspired the 1894 Ordinance. In 1913

60 Buell, i, pp. 864–5.
61 *Ibid.*, p. 867.
62 H. A. Gailey, *A History of Gambia*, London, 1964, pp. 111–45.

all the provisions made in the various ordinances were gathered together under a definitive enactment. If the exact impact of these innovations on Gambian social and political systems remains to be examined in detail, what evidence there is shows that the effects were no less revolutionary and unsettling than elsewhere in West Africa.

Like the other colonial powers, the Germans saw the necessity for using indigenous chiefs and headmen as local agents of their rule in Togoland. But in this they were as handicapped as the British were in the former Eastern Nigeria by the conspicuous absence of paramount or even district chiefs.[63] Political authority here was highly fragmented, rarely exceeding the limits of villages or village-groups of one or two thousand people. Since at this time no colonial power had the time or the inclination to investigate how such communities governed themselves in the absence of chiefs, the Germans, like the British in Eastern Nigeria, resorted in places to the creation of artificial chiefs. In the District of Misahohe there were only four chiefs of any standing—the *Akpanya* of Buen, the *Dagadu* of Kpandu, the *Plako* of Nyamvo and the *Hosu* of Ho. The Germans recognised these, used them for purposes of taxation and upheld the traditional rule of succession to their titles and positions. In the district of Atakpame the Germans did the same with two of the four ward heads under whom the people were grouped. Here too the line of succession was rigorously upheld.

In the Lomé district the villages were even smaller, sometimes consisting of no more than a few dozen people. Authority was therefore even more fragmented and widely dispersed. Here the Germans for the most part imposed on the people artificial chiefs endowed with wider powers than any single individual had exercised in pre-German times. Some of these functionaries were village heads; others were former clerks or tax-collectors who were appointed and upheld for their intelligence and loyalty to the administration. These men were not only required to collect taxes, they also recruited and supervised labour for the roads as well as exercising minor judicial powers for the settlement of petty civil cases.

Apart from recognising head chiefs where they existed or creating them where necessary, the Germans upheld the traditional system of justice as far as this was possible under a European regime. The head-chiefs were empowered to apprehend and punish certain grades of offenders using their own interpreters and policemen who were accorded official recognition and maintained out of the funds of the native courts. At the village level the Germans also retained and encouraged the village courts for the purpose of composing petty disputes. German officers attended meetings of the village courts where and when they could and on such occasions tried to infuse into the courts European methods of presenting and handling evidence. Certain grades of cases were reserved specifically

63 Newbury, *The Western Slave Coast*, pp. 177–80.

for the village courts. These included all civil disputes between Ewe and other natives resident in the area of jurisdiction of the courts. But all criminal cases as well as cases in which educated natives or Europeans were involved went to the German administrative officer.

Some of the cases dealt with in the village courts eventually came to the German officer on appeal. In order to handle such cases without blatantly transgressing traditional law and custom the German administration at the beginning of this century commissioned Drs Henrici and Asmis to produce systematic and scholarly studies of the Ewe law and custom for codification. This, it was hoped, would help German political officers to acquaint themselves with the intricacies of Ewe law and thus equip them to handle the many civil cases which came before them. This was one of the earliest attempts in West Africa at such a study of the legal system of colonial peoples. The expected code, however, was never issued. The project was overtaken by the First World War and the ejection of Germany from Togoland in 1914.

Owing to an over-preoccupation with the theory of colonial government, it has become the stock-in-trade of many works on the colonial period of West African history to produce sharply contrasting pictures of British colonial administration on the one side, and French and German administrations on the other, the scale being often weighted unduly favourably on the British side. It is usually asserted that the British employed Indirect Rule, while the others employed Direct Rule with the undertone that the former was wiser and showed more consideration for the interests of the African subjects.

Further research still remains to be done on this aspect of colonial administration. Many of the earlier works written on the topic suffer from over-dependence on official, policy-making documents containing the pious wishes and ambitions of the governments. Experience and recent research have shown that there is often a wide margin between theoretical policy and actual achievement. The present state of our knowledge seems to reveal that neither the French nor the British nor the Germans applied any particular variation of the policy with rigorous logic and consistency. The above analysis would seem to show that strictly speaking there was no French or British style of Indirect Rule[64] if we consider not the theory but

64 For the debate on Indirect Rule see M. Crowder, 'Indirect Rule: French and British Style', *Africa*, xxxiv, 3, 1964; Crowder, *West Africa Under Colonial Rule*; A. E. Afigbo, 'The Warrant Chief System: Direct or Indirect Rule', *J.H.S.N.*, iii, 4, 1967; Afigbo, 'The Native Treasury System'; Afigbo, 'West African Chiefs During Colonial Rule and After', *Odu*, new series, 5, 1971; O. Ikime, 'Reconsidering Indirect Rule', *J.H.S.N.*, iv, 4, 1968; *West African Chiefs* (see especially the Introduction); P. A. Igbafe, 'British Rule in Benin 1897–1920'; I. F. Nicolson, *Administration of Nigeria*, ch. II–VII; C. Coquéry-Vidrovitch, 'French Colonisation in Africa to 1920: Administration and Economic Development', *Colonialism in Africa*, eds. L. H. Grann and P. Duignan,, i, 1969; J. E. Flint, 'Nigeria: The Colonial Experience 1880 to 1914', *Colonialism in Africa*.

the practice of administration. The administrative imponderables like differences in local situation, the temperament of the European political officer on the spot, and the effectiveness of control from the centre, were so varied that uniformity within the French or British areas over any reasonable length of time was out of the question. The departures from the so-called British and French styles were too many to be regarded as mere local or temporary aberrations. The situation in Nigeria was a case in point. Not until 1914 was an attempt at a Nigerian style of Indirect Rule made and then with ill-success. Not until after Lugard had published his *Dual Mandate* was any attempt made to extend to Ghana, Sierra Leone and Gambia the system propounded there which was based on the experience in the Muslim areas of Northern Nigeria.

Furthermore the two terms Direct and Indirect Rule are vague and uninformative.

> I have [wrote Lord Hailey, a great authority on this matter, in 1942] purposely avoided discussing methods of native administration in terms of 'direct' and 'indirect' rule. There is now little advantage in the use of those terms. They have no claim to precision. . . . The use of these terms conveys the erroneous impression that there are two opposing systems of rule. This is not of course the case. All African administrations are dependent to a greater or lesser extent on the use of native authorities as agencies of local rule.[65]

On this matter H. R. Palmer warned in 1934 that 'there cannot be and [there] is not one universal sealed pattern of Indirect Rule. . . . [All that Indirect Rule] was intended to mean was some variety of local administration of African antecedents which was not direct Crown Colony of the old type.'[66] One cannot clearly distinguish between what are called 'direct' and 'indirect' rule. But there were certainly differences in the degree to which traditional local authorities were used by the different colonial powers in different places, at different times.

The consequences of colonial rule for the status, prestige and power of traditional local institutions were usually similar under all the different colonial powers. It is platitudinous but perhaps necessary to point out that the preconquest states and kingdoms of West Africa had their own ideals of government which were inspired primarily by the local religion as well as by sheer tradition. These were necessarily different from those evolved in the industrialised and mechanised societies of the colonial powers. One of the most widely advertised aims of the colonial powers was the introduction into Africa of 'civilised', that is, European ideas of good government. Inevitably this meant supplanting the traditional ideas and

65 Lord Hailey, *Native Administration and Political Development in British Tropical Africa 1940–42*, H.M.S.O., 1951.
66 H. R. Palmer, 'Some Observations on Captain Rattray's Paper', *loc. cit.*, pp. 38–43.

ideals. The traditionality of indigenous systems of government derived not only from their structural arrangement but also, even more fundamentally, from the philosophy of government and society that animated them. When that idea of government is lost the system ceases in a very real sense to be traditional even if it does not suffer structural disfiguration. The total cultural situation in which that institution developed also helped to determine its nativeness. If that cultural situation is undermined either by the preaching of Christianity or the abolition of so-called barbarous practices, the traditional political system is also undercut.[67] On this matter Miss Mary Kingsley said towards the end of the nineteenth century that there was a difference between governing Africans through African institutions but according to European principles and governing Africans according to African principles. Not even in the Muslim emirates was the latter the case.[68]

Finally, under the French, the British and the Germans the local authority, whether or not he had any shadow of traditionality, enjoyed a new status by the very existence of the colonial regime. *Vis-à-vis* the colonial administration or its representative he was entirely powerless, depending as he did for his continued existence on the policy or whims and caprices of that administration or its representative. In this situation there was not, for the local authority, any real difference between 'advice' or 'command' coming from the political officer. But *vis-à-vis* the local population the local authority was more than ever before an autocrat. He was not bound to consult his people in anything he did. What the people thought about him mattered little as long as he retained the confidence of his makers. In this sense the most insignificant village head under the French or the Germans enjoyed a power relationship with the colonial administration analogous to that of, say, the Sultan of Sokoto. But with big chiefs like the Sultan the traditional pomp and ceremony associated with their offices were retained to give them and their people the false impression that nothing had really changed and to delight the nostalgic administrators who thought that through these relics of the traditional culture they were having a glimpse of life in feudal Europe. In places the chiefs so treated were successfully hoodwinked and not only became staunch supporters of the colonial regime but, worse still, later constituted exasperating obstacles on the road to modernisation.

ECONOMIC AND SOCIAL DEVELOPMENTS

A very important aspect of the problem of consolidating colonial rule had

67 Afigbo, 'West African Chiefs During Colonial Rule and After'. Here it was argued, *inter alia*, that Indirect Rule was *only 'an attempt'* to rule through indigenous institutions because the colonial powers were accursed with something akin to the Midas touch; but instead of turning every indigenous institution they touched into gold, they debased it into an alien machinery of control, pp. 105–6.

68 M. H. Kingsley, *West African Studies*, third edition, London, 1964, pp. 401–2.

to do with the imposition of a new economic and social regime on West Africa. The effective exploitation of the region's human and material resources to the advantage of Europe was the most important single factor accounting for French, British, German and Portuguese presence there. To a very great extent, therefore, the methods of consolidation already discussed were designed to lead to the successful achievement of this goal through ensuring the maintenance of law and order on a scale acceptable to the powers and conducive to their interests.

In another sense, however, the imposition of a new economic regime was by itself a method of extending and consolidating European rule. For instance, fundamental to this new economic order was improved communication by water, road, railway, telegraphy and telephone. But these means of transport and communication were also important politically and militarily.

> Politically [argued Dr Allan McPhee] speedy and efficient transport means that troops and stores can be moved more readily, that therefore troops can be reduced in numbers with a simultaneous diminution in costs, that outbreaks and rebellions can be more easily and expeditiously quelled, that officials can supervise a greater area, that therefore their numbers and expenses can be decreased.[69]

It was as much in consideration of the administrative implications of efficient communication as of its direct economic uses that many French and British officials in West Africa agreed with Sir Walter Egerton's famous statement that 'if you ask what my policy is, I should say "Open means of communication" and if you would wish for additional information, I would reply "Open more of them!"'[70] Furthermore it was hoped that through these new lines of communication European trade, ideas and way of life would reach even the remotest village and that those who benefited from trade and contact with Europe would have a stake in the security and continuation of the new regime.

There were two sides to the creation of the new economic regime. The first dealt with the setting up of the infrastructure, that is modern means of communication and transport, the second with the direct promotion of the production of the desired economic or cash crops.

Before the imposition of European rule, means of communication and transport in West Africa was very poor and primitive indeed, relying as it did mainly on head porterage and canoe transport in the forest zone and on pack animals in the savannah belt. These traditional means might be useful and effective for short distances and limited quantities of

69 A. McPhee, *The Economic Revolution in British West Africa*, London, 1926, p. 126, see also p. 111 and fn. 4 on p. 128.
70 On Egerton's policy regarding communication see I. F. Nicolson, *Administration of Nigeria*, and Tamuno, *The Development of British Administrative Control*.

merchandise, but for trade and communication on the scale sought for by Europe it was unpromising. With the intensification of European commercial enterprise in the second half of the nineteenth century it was sadly discovered that West Africa's main rivers, the Gambia, Senegal, Volta and the Niger, did not provide efficient routes for all-the-year-round heavy traffic, a situation which derived partly from their small volume, partly from their tendency to develop rapids and partly from most of them entering the Atlantic in small streams and rivulets. The Gambia was navigable by ocean-going vessels for a distance of about one hundred and fifty miles, and then by light launches for a further one hundred and forty miles. For two months in the year, August and September, the Senegal was navigable up to Kayes by medium-sized steamers. The Volta, perhaps the least navigable of the rivers, could be plied for limited distances only by light launches and canoes. The Niger was navigable by shallow-draught vessels between Burutu and Jebba, a distance of five hundred and forty miles within Nigeria. In French territory it was navigable between Asongo to the east of Timbuktu and Koulikoro which was the nearest point to Kayes on the Senegal. Not surprisingly therefore, from the 1880s mercantile interests in France and Britain clamoured for energetic and imaginative railway policies in West Africa as a solution to the problem of easy and dependable access to the West African interior.

The French led the way and their approach to the problem was dictated by their overall grand strategy for West Africa in which the Senegal was seen as the main route of advance into the 'rich' and romantic interior. But the usefulness of the river as the highway for evacuating the resources of the Sudan was seriously limited by the inadequacy of Saint Louis, the main town and harbour near the mouth of the Senegal, as a port. Dakar to the south had a good natural harbour and was strategically placed on the ocean routes. Early French railway policy in West Africa therefore sought to bring together the advantages of the Senegal river beyond Saint Louis and the ocean port of Dakar.[71] Thus the first French West African railway ran from Saint Louis to Dakar and was completed by 1885. With the end of the scramble and the French acquisition of much of the Sudan it became necessary to build a line from Kayes to Bamako on the Niger so that the Senegal and the Saint Louis–Dakar line would serve as the main exit for the products of the whole Sudan. This line was completed in 1904. Between 1903 and 1923 the two lines were linked and through traffic from Dakar to Bamako became possible, thus removing dependence on the occasionally wayward Senegal river. Also as part of the drive to avoid excessive dependence on the Senegal the French built other railway lines from their colonies on the coast further south into the interior. These lines, which ran from Conakry in Guinea, Abidjan in

71 Roberts, *History of French Colonial Policy*, p. 328; C. Coquéry-Vidrovitch, 'French Colonisation in Africa to 1920: Administration and Economic Development'.

Ivory Coast and Cotonu in Dahomey, sought to reach the Niger. But only the Central Dahomean line later succeeded in reaching one of the tributaries of that river.

The British launched their own railway policy in the 1890s. In Sierra Leone, beginning in March 1896, they built a Y-shaped railway system which linked the palm-producing regions of the protectorate to the coast. In the same year the Lagos Administration began what later became part of the western line of the Nigerian Railway. The eastern line from Port Harcourt was not begun until 1913. In the Gold Coast (modern Ghana) railway construction began in 1898 with two lines, one from Takoradi to Kumasi serving the gold mines, the other from Accra to Kumasi, the cocoa-producing regions of the colony. By 1932 Sierra Leone had 311, Ghana 510 and Nigeria 1,903 miles of railway.

The Germans were equally alive to the economic and political importance of improved means of communication in general and of railways in particular. They built three lines. The first, which ran from Lomé to Anecho, parallel to the coast, was opened in 1905. The second line ran from Lomé to Palimé and was completed in 1907. The third and longest from Lomé to Atakpamé was started in 1908 and opened to traffic in 1913.

All the powers also realised that those railways would be uneconomic without good roads to act as feeders. Consequently road building was undertaken by the French, British and German administrations as an important aspect of the bid to provide the proper infrastructure for the new economic regime and to establish their rule. The history of these roads remains to be studied. But in French, British and German territories an important part of the duties of the chiefs was the recruitment of forced labour for the roads as well as the supervision of roadmaking and maintenance. Hand in hand with roadmaking and maintenance went the clearing of creeks and streams, where they existed, to make them suitable for launch and/or canoe traffic. The 'Southern Nigeria' Protectorate in 1903 issued a Roads and Rivers Proclamation (later Ordinance) under which a warrant chief, on instruction from the District or Provincial Commissioner, was required to recruit labour for meeting these needs. Each adult was expected to work for six days in a quarter.[72] In Togo and Dahomey every male was required to spend twelve days a year on this or like work or make a stipulated money payment in lieu.

The exact nature and extent of the impact of these new modes of transport on the economy of West Africa are still to be revealed by detailed research. It is, however, indisputable that these railways and roads opened up new markets in the interior to European business men and that they supplanted the creeks, rivers and footpaths as 'the main avenues of transportation'. Their construction stimulated economic activity. The

72 A. E. Afigbo, 'Sir Ralph Moor and the Economic Development of Southern Nigeria 1896–1903', *J.H.S.N.*, v, 3, 1970.

wage payment of workers on the rail lines boosted the circulation of cash currency. New towns sprang up on the railway lines, while old centres of commerce and civilisation which were unlucky to be sidetracked by the lines declined, unless linked to the rail lines by feeder roads.

The decline of the old centres of trade was not always sudden and dramatic. The case of Lagos and Ikorodu in Western Nigeria is an example. Colin Newbury has shown that on the completion of the line from Lagos to Ibadan many economic analysts expected the speedy decline of Ikorodu and the equally dramatic rise of Lagos as a collecting centre for indigenous produce. But for several years this did not happen. The railway line failed to stifle the route from the interior to Lagos via Ikorodu and Ejinrin. More produce continued to travel by the older route than by rail. Much of what the railway carried consisted of stores for government departments and materials for the further extension of the line.[73] In Dahomey the local businessmen were slow to exploit to the full the opportunity which the railway offered for business in the interior. They tended to confine their activities to the south especially as palm produce remained Dahomey's main export commodity. What was more, in transporting this produce to the coastal ports the local traders long continued to prefer canoe transport on the River Weme to the railway. This conservatism was rooted in economic calculation. The canoe transport, though slower and more tedious than the railway, was cheaper.[74]

There was, however, one area in the economic life and history of West Africa in which the effect of improved communications was dramatic and final. The railways and roads were largely responsible for the final resolution of the four-centuries-old struggle to decide whether the Sahara or the Atlantic should be the highway for the commerce of West Africa. Ever since the Portuguese started business along the West Coast this question had stood out prominently. Slowly but steadily the Atlantic had started gaining the upper hand in the rivalry. Yet in 1900 there was still some trade across the Sahara between North Africa and the Sudan. The last faint echoes of this conflict were heard in what later became Nigeria when the Southern Administration accused Lugard of artificially sustaining the trans-Saharan traffic by his imposition of tolls on trade between Northern and Southern Nigeria. This toll, the Southern Administration alleged, forced Northern traders to sell their products to Arabs who carried them on camel-back across the Sahara. With the linking of the two territories by rail and road and the abolition of the tolls this ancient trade route was rendered irretrievably uncompetitive.[75]

73 Newbury, *The Western Slave Coast*, pp. 144–6.
74 *Ibid.*, p. 146.
75 Webster *et al.*, *West Africa Since 1800*, London, 1967, ch. 5; J. C. Anene, 'Liaison and Competition between Sea and Land Routes in International Trade from the 15th Century—the Central Sudan and North Africa', in *Les Grandes Voies Maritimes Dans le Monde*, Paris, 1966.

The political effects of the roads and railways were also far-reaching. 'Politically,' writes Dr Newbury, 'the railways were enormously influential in binding together into an administrative unity the disparate areas of jurisdiction.'[76] It was the improved means of communication that made possible the effective administration of the large political units which emerged out of the scramble. 'With the means of communication we now have,' reported Mr F. S. James of the Southern Nigerian Political Service in 1912, 'a provincial commissioner is enabled to administer far more thoroughly than of yore the many districts and people in his charge. . . .'[77]

Except in the affair of transport and communication the economic policy of the colonial powers in West Africa throughout the period covered by this chapter was unspectacular and uninspiring. It had three distinguishing features. Firstly it was marked by a determination to introduce European methods of trade in place of indigenous methods which were considered primitive and obstructive of trade expansion. Secondly it sought in a mercantilist sort of way to achieve the economic development of West Africa by encouraging the people's agricultural efforts as a complement to Europe's industrial effort. Finally the policy was marked by timid experimentation in the introduction of new methods of cultivation and new breeds of certain cash crops.

In spite of centuries of contact between West Africa and Europe, in 1900 one main method of trade transaction in our area of study was still barter supplemented by cumbrous currencies like manillas, iron rods, copper wire, cowrie shells, slaves and so on. The colonial powers considered this system inefficient, pernicious and vexatious and saw the answer in the introduction into their colonies of their own currencies which were generally lighter to carry. By 1900 coastal anchorages of European powers like Lagos, Accra and the communes of Senegal already had these European currencies. The problem was to extend their circulation to the interior. Through the payment of carriers and labourers on the railway lines, through the institution of direct taxation and court fees and fines which could be paid only in the new currencies, and then by proclamations, ordinances and decrees declaring indigenous currencies no longer legal tender, this ambition was achieved.[78] By 1916 sufficient progress had been made in British West Africa for paper currency to be introduced into Nigeria in July and later the same year into Sierra Leone and the Gold Coast (modern Ghana).

For the rest, the economic policy of the powers aimed at encouraging the production of those vegetable and animal raw materials required to keep the industries of Europe working. For the most part attention was focused on the production of those cash crops indigenous to West Africa

76 Newbury, *The Western Slave Coast*, p. 146.
77 Quoted in Tamuno, *The Development of British Administrative Control*, p. 365.
78 Afigbo, 'Sir Ralph Moor'.

and favoured by the geography of the region. In consequence of this many of the colonies became committed to dangerous overdependence on monoculture. Senegal depended on groundnuts, Ivory Coast on timber and Dahomey on palm oil and kernels. Among the British territories, Gambia depended almost entirely on groundnuts, the Gold Coast on cocoa, Nigeria on palm produce, cocoa and rubber. Through the creation of botanical gardens, for instance in Lagos and at Calabar, and experimental plantations along the Niger both in British and French territory some effort was made to improve cultivation. From 1903 the French tried to shift the emphasis in the Sudan from the cultivation of groundnuts to the cultivation of cotton in a bid to make French textile industries independent of cotton supplies from potential rivals or enemies. To this end improved breeds of cotton, the so-called 'improved Allen' were introduced from Nigeria. The *Association Cotonnière Coloniale* supplied the seeds and at times encouraged farmers by buying their products. This effect was, however, not enough to persuade the local farmers to change over to cotton for groundnuts continued to fetch better prices. The most dramatic agricultural development in West Africa, the Gold Coast cocoa revolution, was primarily the work of the peasant farmer and owed little to government initiative or encouragement.

In Nigeria experimental cotton and coffee plantations were established by the Protectorate government at Ishan in the present Mid-West State, and at Onitsha, Asaba and Nkissi on the Niger. The government supplied improved breeds and the equipments and at times bought the yields at artificial prices to encourage the farmers. But the experiment failed. Ultimately the cotton exported by Nigeria came from Northern Nigeria, especially the Sokoto and Zaria provinces. There the British succeeded in supplanting the indigenous breeds with *Allens Long Staple Cotton* which is an American variety. Apart from these timid experiments which were supported neither by sufficient zeal and determination nor by enough money, the colonial powers issued instructions designed to preserve the natural produce of West Africa. Pamphlets were issued and laws made to regulate the tapping of rubber, the felling of timber, the collection and drying of cocoa seeds and the preparation of palm oil. A moderate policy of forest reserves was also pursued. In places, however, these came late. In Guinea and the Gold Coast, for instance, reckless 'devastating and exhausting exploitation' quickly led to the almost total extermination of the wild vines on which these two colonies depended for their export of rubber. For Guinea this was a serious matter since between 1890 and 1914 rubber was the main export crop.[79]

A noteworthy feature of the economic policy of the colonial powers in West Africa was that, unlike the policy pursued in the Congo,

79 Suret-Canale, 'Guinea Under the Colonial System', pp. 46–9.

Cameroons and East Africa, it left the exploitation of the resources of West Africa, mines excluded, in the hands of the West Africans themselves, not in those of concessionary companies. It was not that concession hunters were lacking but that they were discouraged. In French West Africa, for instance, up to 1927 the administration made only about forty agricultural concessions of more than two hundred hectares each. Out of these about twenty-eight had by then either lapsed or been abandoned as unprofitable. German Togo came nearest to falling into the grips of huge concession hunters through the activities of the Austrian geologist, Friedrich Hupfeld, and his supporters. But then the schemings of this man were frustrated by the opposition of the small German firms and other interests on the coast who feared losing their business to a monopolist company. From about 1917 big British business also clamoured for concessions in West Africa, among the most notorious of these claims being Lord Leverhulme's in the years after the First World War. As elsewhere in West Africa these demands met with the staunch opposition of the local administration.

That West Africa escaped being overrun by concession-mongers is to be attributed to geography and demography. On the one hand the climate of West Africa ruled out European settlement. This still left the possibility of huge plantations run by European management but worked by disinherited West Africans. But the forest regions where this policy could be profitably pursued were thickly populated and already heavily farmed. Furthermore, the failure of the mid-nineteenth century British experimental farm at Lokoja and of the French experimental farms on the Senegal during the reign of the restored Bourbons were chastening experiences.

> In West Africa [warned Miss Mary Kingsley late in the nineteenth century] the most valuable asset you have is the native: and the more prosperous those natives are, the better for you; for it means more trade. All the gold, ivory, oil, rubber, and timber in West Africa are useless to you without the African to work them; you can get no other race that can replace him and work them; the thing has now been tried, and it has failed.[80]

The French, the British and the Germans heeded this warning.

It was not only by brute force, unequal 'alliance' with traditional rulers and subtle appeal to the generality of the people through the imposition of a new economic regime that the colonial powers sought to extend and establish their rule in West Africa. They also depended to some extent on the propaganda and the western education and civilisation peddled by Christian missions. These bodies were themselves manifestly aware of their dependence on the colonial administration for the

80 Kingsley, *West African Studies*, p. 337.

maintenance of the law and order in which alone their work could flourish. They were also dependent on the commercial interests who were among the staunchest supporters of the colonial regime for the stores they needed either for their work or to keep body and soul together. The missionary bodies therefore saw their interest and the will of God in West Africa as bound up with the continuation of colonial rule. They not only openly preached obedience to metropolitan governments but through the education they spread they won adherents to the western way of life. Educated elements who had acquired European tastes were among the most avid consumers of European produce. They were also employed in the administration as junior clerks and inspectors, and thus helped to uphold the colonial bureaucracy.

The missions served the imperial interests of the powers in other ways. They were usually nearer the people and often came closer to winning their confidence than the political service ever did. They therefore served as an effective means of gauging the reaction of the common people to the policies of the powers. The missionaries also had considerable knowledge of the customs, traditions and languages of the people. This knowledge they placed at the disposal of the administration in the form of books, pamphlets and oral advice. The monographs on indigenous languages and culture produced by missionaries like Crowther, Schön, Miller and Basden in Nigeria were particularly valuable to the administration as sources of information about Nigerians. Examinations in Hausa and Ibo, for instance, were conducted for the administration in Nigeria by the Rev. Mr Miller and Archdeacon Basden respectively.

What was more, the missionary usually began by selective conversion of a section of any community in which he settled. In many places this section was made up of the outcasts and the underprivileged such as slaves. To these they preached the equality of all human beings before God as well as social justice and the like. The upshot usually was that the community was irreconcilably split into two camps—the 'progressives' or Christians and the conservatives, with the former often regarding themselves as instruments of the new regime and therefore looking to the colonial administration for protection against their numerically superior opponents. This split weakened the indigenous resistance to colonial rule and the new way of life. It meant that wherever there was a mission station indigenous resistance had to reckon with a potential fifth column.[81] There was much truth in Sir Harry Johnston's opinion on the role of missionaries in the whole process of colonisation:

> The missionary is really gaining your experience for you [the colonial power] without any cost to yourself.... They strengthen our hold over

81 A. E. Afigbo, 'The Eclipse of the Aro Slaving Oligarchy of Southeastern Nigeria 1901–1927', *J.H.S.N.*, v, 4, December 1971.

> the country, they spread the use of the English language, they induct the natives into the best kind of civilisation, and, in fact, each mission station is an essay in colonisation.[82]

Not surprisingly the colonial powers, Britain in particular, encouraged the missionaries in their colonies, especially where there was no fear of conflict with Muslims. The French were less forthcoming in lending positive help to the missions, especially after 1903, thanks to the streak of anti-clericalism which had been present in French policy ever since the Revolution.

The tendency of the colonial powers to adopt, after military conquest, a policy of caution and compromise was reflected not only in their attitude to traditional rulers but also in the policy they adopted towards the vexed question of domestic slavery. Radical humanitarian opinion would have favoured the *immediate* abolition of slavery but practical considerations forced the colonial powers to adopt a lenient line towards that institution which for some time was a necessary evil. Closer acquaintance with West Africa taught the powers that until the universal acceptance of European currency in West Africa, the construction of roads and railways and the introduction of wheeled traffic, had been brought about, the immediate abolition of slavery would seriously harm business. Slaves were not only currency, they were also largely responsible for the collection of West Africa's produce and the transportation of it to the markets and trading posts where European businessmen and agents could get at it. There were also other aspects of the problem. The slaves were not usually badly treated. Thus most of those amongst them who had lost their kith and kin were none too eager to be torn away from the families to which they had become attached not only economically, but at times also sentimentally and by marriage. Then there were various forms of the institution including some which escaped the notice of the colonial authorities. Overwhelmed by the difficulties of a radical and sudden reform these powers resorted to gradualism.

In French West Africa the problem of rooting out slavery was further complicated by two other factors peculiar to the French colonial regime there. Firstly, in the period of conquest the French authorities had sought to please their African soldiers and other collaborators by allowing them to take *captifs* from subdued villages and communities. These *captifs* were converted into slaves by their captors, and became personal property. To kill the institution of slavery in all its manifestations would involve the alienation of these supporters and beneficiaries of the French regime.

82 Quoted in R. Oliver, *The Missionary Factor in East Africa*, London, 1965, p. 128. For the role of the Christian missions in the British occupation of Nigeria see E. A. Ayandele, *Missionary Impact on Modern Nigeria*, London, 1966. See also A. E. Afigbo, 'The Background to the Southern Nigeria Education Code of 1903', *J.H.S.N.*, iv, 2, 1968.

Secondly the administration was guilty of retaining slaves in order to meet its manpower needs for labour. In 1887 Gallieni had brought into existence the misnamed *village de liberté* or Liberty village. In theory these were homes to which slaves rescued from maltreatment by their African masters were sent for rehabilitation by a kind and humanitarian government. But in practice they were the government's reservoirs of forced labour.

> It is quite certain that the motive which generalised the 'liberty villages' was that they were an excellent solution to the problem of porters and manpower, far superior in any event to the system which consisted of going to look for people, weapon in hand, whenever they were needed, to see them flee at the first relaxation of attention, a slow, complicated and hazardous system.[83]

The condition of the inmates of these villages was generally such that many of them escaped, but this was usually followed by recapture and subjection to further hardships and punishments. Not surprisingly in the (French) Sudan the liberty villages were popularly known as 'the commandant's captive villages'.[84]

With such a record, the French avoided coming to grips with the problem. In 1901 Governor Ponty issued a circular which forbade masters to pursue and recapture escaped slaves but the purpose of this circular was often set at nought by slavemasters and at times by the government agents. According to Suret-Canale, unless a fugitive slave had first bought his freedom through the payment of 150 francs, he was handed back to his master 'with the connivance of the authorities', if recaptured. In 1905, a new government decree promised two to five years' imprisonment to parties to any arrangement which had as its purpose or end result the alienation of the liberty of a third person. In 1908 the African 'friends' of the French regime were asked to liberate the *captifs* they had been allowed to take during the years of military conquest. Then between 1905 and 1910 the administration embarked on the liquidation of the liberty villages. For the rest the French authorities closed their eyes to what remained of the institution. 'In fact up to the eve of the Second World War,' writes Suret-Canale, 'hut [domestic] slavery remained, at any rate in Futa, a reality under the discrete protection of the authorities.'[85]

In this matter, if British performance was less cynical, it was not radically different. In Northern Nigeria, for instance, the government issued in 1901 a Proclamation which abolished the legal status of slavery. This meant, among other things, that all children born after April 1901

83 Suret-Canale, 'Guinea Under the Colonial System', pp. 50–1.
84 *Ibid.*, p. 51. See also Crowder, *West Africa Under Colonial Rule*, pp. 182–7.
85 Suret-Canale, 'Guinea Under the Colonial System', p. 52.

were free, that it was illegal to enslave or claim property rights in anybody, that a slave could desert his master without the risk of recapture and that a slave became legally accountable for his actions and therefore could give evidence in court.

At first sight this law gave much; on closer examination the concessions were very limited. Slave owners were not forced to liberate their slaves. On the contrary, government measures limiting vagrancy undermined the right of the slave to leave his master at will. For instance it was the policy of the government not to grant land for settlement to runaway slaves unless they had purchased their freedom from their masters. If a runaway slave took employment under the government and this was found out, it was the duty of the Resident to arrange for the ex-slave to pay to his former master part of his salary until he had completed the sum accepted in local law for manumission. Then the Native Court would issue him with a certificate of freedom. Thirdly the government left the trial of slavery cases to the Native Courts and where a slavery case got to the Resident he settled the matter between the slave and his master informally.

The reason for all these limitations on the operation of the law was that the British took cognisance of the fact that the Fulani rulers regarded slaves as property and that a sweeping and unmodified law of emancipation would amount to expropriation. They also feared alienating the Fulani aristocracy. In spite of these administrative obstacles many slaves succeeded in leaving their masters. Some of these went back to their kin especially where they were recent captives and had not forgotten their homes. Some took up work under the government. But for freed slave children the government established Freed Slave Homes where a genuine effort was made to rehabilitate and prepare them for independent and useful life in the community.[86]

In broad outlines at least the methods of administrative consolidation adopted by the French, the British and the Germans were similar, but the reactions that these evoked from the different groups in West Africa varied widely at different times owing largely to varying group interests. A thorough study and analysis of these reactions still awaits its historian but certain features are already fairly clear. For the purposes of this chapter a distinction will be made firstly between the reactions of the old political and economic élite and the masses, secondly between those of the Muslims and the adherents of West Africa's traditional religion popularly described as paganism.

The traditional élite were, no doubt, among the most shocked, the most bewildered and the most dismayed by the failure of their armed resistance against the European powers. They were also among the first to be called

86 G. O. Olusanya, 'The Freed Slave Homes: An Unknown Aspect of Northern Nigerian Social History', *J.H.S.N.*, iii, 3, 1966. The first of these Homes was opened at Zungeru on 1 January 1904.

upon to cooperate with their conquerors, in establishing and protecting the new regime. How exactly they felt is not yet, and may never be, known in detail since most of them left no records and are now beyond the reach of the collector of oral tradition. It can, however, be assumed that their cooperation was at first grudgingly given, some of them no doubt hoping that the gods of their fathers would soon come to their rescue. But as time went on and the gods failed to intervene and they came to learn their utter dependence on the colonial administration, silent resentment probably gave way to active and willing support. By 1906 in Northern Nigeria, for instance, only two emirs were inclined to support the abortive Mahdist uprising at Satiru. It is true that most of the emirs were at the time British appointees, but they were nonetheless Muslims and heirs to the empire of Uthman dan Fodio.

In those societies where traditional authority or political leadership was not vested in an individual or a narrow oligarchy and therefore where it was easy to sidetrack the traditional authorities in favour of individuals who gave some promise of dynamic leadership, the situation was a little different. In the former Eastern Nigeria most of the titular village heads seemed to have escaped appointment as warrant chiefs either because they were old and uninspiring or because not every village head could be made a chief under the new regime in any case. Also other institutions such as title lodges and secret societies which shared political authority with the councils of elders were sidetracked either because their true functions were not appreciated or because in any case they were associated with those 'barbarous' practices which the colonial powers felt they had a duty to abolish. In such cases the tendency was for these traditional repositories of authority and power to operate underground in opposition to the new agencies set up by the government. Throughout the period dealt with in this chapter, and in fact beyond, political officers constantly uncovered the secret activities of these bodies. Mr Horace Bedwell, a political officer in the former Eastern Nigeria, in 1916 described the continuance of these practices as 'really an attempt to usurp government functions and rule the country as secret societies of all sorts did previous to the advent of [the colonial] Government'.[87]

One group which could, in a vague way, be classified along with the traditional economic and political élite were the coastal middlemen. It is now the standard view that the establishment of colonial rule saw the final destruction of their entrenched position and that for this reason they were uncompromisingly opposed to the new regime. It does appear, however, that this theory is too simplistic to be correct and that the situation as well as their reaction to it was much more complex.

The fact is that both the missionaries and the colonial administrations

87 Afigbo, *The Warrant Chief System in Eastern Nigeria*, pp. 404–7.

were first established amongst these coastal middlemen and did not penetrate the interior to any great extent, except probably in Yorubaland and a few other places, for many decades after. These two arms of European rule—the secular and the spiritual—were keenly interested in raising a class of subordinate native agents to help them in their work. In pursuance of this policy they established schools for the education of local youths. The coastal peoples benefited from this policy with the result that within a few decades they had produced a small but significant class of ambitious young men who were literate enough to serve as clerks, interpreters, church agents and teachers and so on. By the time the interior was effectively assaulted by the forces of European colonialism, the coastal peoples were in a position to staff all sorts of junior positions in the government and missions. In the event the expansion and consolidation of European rule meant the expansion and consolidation of a new kind of coastal dominance over the interior. The result was that the tribes of the interior found themselves in a quandary. They were opposed to all kinds of external domination and so had to oppose the missions, the colonial administration and its native subalterns. Yet they were aware that the establishment of the missions and the government amongst them would bring along with it education and general sophistication and so equip them to deal with the coastal peoples on equal footing. Yet, for a time at least, the coming of the missions and the colonial government would subordinate them to the many coast-native agents of these bodies. One immediate result, therefore, of the establishment of colonial rule was not the liquidation of the influence of the coastal tribes, but its transformation into a new kind of influence unwittingly bolstered by their former European rivals.

Yet this did not mean that the coastal tribes supported colonial rule unequivocally. They also found themselves in a quandary. If missionary and government penetration of and establishment in the interior offered them new avenues for power over their hinterland neighbours, they knew that it also had its dangers. For one thing missionary education and government influence would in time make the interior peoples whom they despised and cheated as good as themselves. It would also undermine what remained of the economic advantages they enjoyed in the interior for government penetration meant penetration by the European firms who were richer and better organised than they were. These calculations led the coastal tribes to adopt the paradoxical posture of at once opposing and encouraging European penetration and the establishment of colonial rule.[88]

The reaction of the common people is even more difficult to determine.

88 For a detailed investigation of this matter for the Cross Rivers area of Nigeria see A. E. Afigbo, 'Trade and Politics on the Cross River 1895–1905', *Transactions of the Historical Society of Ghana,* viii, 1, 1972.

Some of them did what they could to benefit from the new regime by enlisting as messengers, policemen, soldiers and the like. Some tried to exploit the dislocation and confusion which preceded the effective establishment of colonial government by impersonating government agents and levying blackmail wherever they could. Throughout the Southern Provinces of Nigeria, for instance, British officers complained of the ravages of 'consulmen' who, arming themselves with discarded government uniforms, pages torn from any printed book as evidence of their authority, water-bottles and revolvers, terrorised and exploited unsophisticated villagers. The same practice obtained in Northern Nigeria and no doubt elsewhere in West Africa. The annual reports of Northern Nigeria from 1900 to 1911 carried complaints against this practice.

Some others, however, gave their services to the government when conscripted, attended the new Native Courts when dragged there by their more litigious neighbours, but for the most part went about their business as much as possible in the same way as their fathers had done. It was probably this group that developed the tendency to invent myths which by and large reconciled them with a situation they felt themselves incapable of changing, especially by asserting that their learned men had predicted the dawn of the era of colonial rule.

> Ever since we were quite small [claimed Baba of Karo] the *malams* had been saying that the Europeans would come with a thing called train, they would come with a thing called motor-car, in them you would go and come in a trice. They would stop wars, they would repair the world, they would stop oppression and lawlessness, we would live at peace with them. We used to go and sit quietly and listen to the prophecies.[89]

The Mossi of Upper Volta have a similar myth explaining their conquest by the French. According to them a holy man warned Mogho Naba Koutou, the ruler of the Mossi state of Ouagadougou, that an evil child would bring about the ruin of his kingdom, but the Mogho ignored the warning. 'He made no sacrifices and a child who had "two teeth like the tusks of an elephant" was born into his house. This child Boukary Koutou was disrespectful of his father and made war against his brother and was eventually defeated by the French.'[90] In short the French conquered the ancient Mossi state not because they were stronger but because a Mossi ruler had ignored the warnings of the gods. This was some consolation.

Yet by and large, the majority of the people remained unreconciled to colonial rule, as shown by the numerous riots which took place in this period and after, especially in the matter of taxation. Direct taxation was indeed the most important single measure for determining the extent of the

89 Smith, *Baba of Karo*, p. 66.
90 Skinner, *Mossi of Upper Volta*, p. 153.

people's acceptance of European rule. This was largely because it involved the whole question of overlordship and tributary status. Also in places it contravened certain traditional practices and taboos especially where associated with a census of taxable adults. The last and most famous riot against such imposition was the Women's Riot of 1929–30 in Eastern Nigeria during which the women, among other things, asked the British to leave the country so that the people would run their affairs as they had done in the days of yore. To this group the past remained the rose 'whose faded petals are for ever sweet'.[91]

The Muslims were more inclined than were the pagans or former pagans to support the colonial powers or to adopt an attitude of indifference towards the colonial regime.[92] The reasons for this were many and varied. Though with the establishment of colonial rule most of the Muslim states, especially in French West Africa, disintegrated and the Muslims lost the opportunity of seizing political power by *coups d'état* under the guise of religion, they soon discovered that the new regime had its compensating advantages. The French and the British clearly realised the force of Islam as a religious, social and political ideology and thus did not make the mistake of trying to root it out. In fact they believed that, more than paganism, Islam was an advanced stage along the road to civilisation. For these reasons the colonial powers treated Muslims with more consideration than they thought necessary for pagans. Muslim or Quranic schools were established and given official encouragement. The French in particular assiduously wooed the Muslim elements in their territories. In the schools they built for sons of Mohammadan chiefs they taught not only French but also Arabic. To convince the Muslims that these schools did not undertake anti-Islamic propaganda they allowed Muslim school masters to be present during classes. They also undertook the publication of a bulletin in Arabic, which along with the Arabic newspaper *Al Mobacher* they distributed to leading Muslims in the Sudan, especially at such strategic centres as Timbuktu, Jenne and Nioro. These publications of course contained pro-French propaganda.

In the second place an effect of colonial rule was to favour Islamic proselytisation. The law and order which French and British rule ensured guaranteed the safety of Muslim preachers throughout West Africa. The imposition of colonial rule therefore opened a new era for Islamic propaganda. Pagans were confronted with the choice of becoming either Muslims or Christians as paganism seemed to have no future. In places the

91 Afigbo, 'Revolution and Reaction in Eastern Nigeria'.

92 For the reaction of Muslims to Colonial Rule see the following: E. D. Morel, *Affairs of West Africa*, London, 1902, pp. 208–24; J. S. Trimingham, *Islam in West Africa*, Oxford, 1959, ch. 9; Trimingham, *The Influence of Islam Upon Africa*, London, 1968, ch. 5; J. H. Fisher, *Ahmadiyyah*, Oxford, 1963, pp. 25–6; J. Abun-Nasr, *Tijaniyya*, ch. VI; S. H. Roberts, *History of French Colonial Policy*, i, pp. 313–16; T. Hodgkin, 'Islam and National Movements in West Africa', *Journal of African History*, iii, 2, 1962.

pagans preferred conversion to Islam. For one thing under Islam they would be able to retain much of their traditional culture—for instance, polygamy and a certain measure of their ancestral religion. Furthermore the Islamic preacher appeared a humbler man than the white missionary. He was usually about the same complexion, had lived long in the area and shared the everyday life of the people.

In return for these benefits derived under colonial rule the Muslims for the most part preached submission to French and British authority. Alfa Hashim, a nephew of Umar who died in either 1930 or 1931, for instance, preached submission to French rule. A former Grand Mufti of Ahmadu Tall's empire, he had fled to Medina on the advent of the French. There he acted as a liaison between Muslims from West Africa and the other parts of *Dar-al-Islam*. During the First World War he is said to have called on Muslims in West Africa to support the British. Another great Tijani leader, al-Hajj Malik, whose centre was Tivouane (Senegal) and who exercised influence over a motley community of followers drawn from Tukulor, Wolof, Lebou and Sarakolle, was also pro-French. He is said to have accepted French rule and to have exhorted his followers to do the same. During the First World War he held regular prayers with his disciples for the success of French arms. The French reciprocated this loyalty by publishing in 1914–15 at their own expense about twenty-two of his linguistic and theological writings.

The third reason why the Muslims remained largely unopposed to colonial rule would seem to be the fact that the imposition of European rule saw the exodus from West Africa of most of the fanatically committed Muslims. This 'mass' exile remains to be studied in detail with a view to determining the number involved and its weakening effect on Islam in West Africa. According to popular tradition, however, the phenomenon would seem to have taken place throughout the Sudan. It is the view of Dr Jamil Abun-Nasr that after the conquest of Ségou and the other Tijani states in West Africa, many of Umar's descendants emigrated to Mecca to escape French reprisals. There they were maintained with grants from the Ottoman government. Some others, however, fled only as far as Sokoto and other parts of Northern Nigeria where they were free to preach the Tijani way of life. But nonetheless they were displaced persons. They could not oppose the French administration effectively from Northern Nigeria, while on the other hand they had to be grateful to the British administration which allowed them to settle.

In the fourth place there were certain tendencies in Islam itself which tended to militate against radical nationalism, and political activism. Among these was the inclination to fatalism. Once European conquest became a *fait accompli* certain Muslim elements tended to think that nothing could be done or that only a Mahdi or some supernatural intervention could restore to them their lost independence. Some extracts from poems written by Muslim Fulanis in Futa Jallon between 1900 and 1910 illustrate this point clearly. One extract reads 'The Mahdi, whose hour we await, will come and bring succour to religion; he will take arms against the devourers of taxes [the French] and drive them into the next world.' Another from a poem entitled 'Let us Accept the French' reads 'This world is a camp; a camp is not a home, so many others before the French pitched their tents, and went on their way.'[93] This attitude of mind could not but engender passivity and indifference.

Thomas Hodgkin has also shown that *'ilm* and the *tariqas* tended to hinder radical reformism or nationalistic activity. 'Ilm or 'education in the traditional Islamic sciences' with its 'emphasis on Maliki fiqh', he says, 'tended to produce an attitude of mind that is resistant to the ideas both of Muslim reformism and of secular nationalism'. The tariqas on their side came under attack by the reformists because they were inclined to 'maraboutism', tended to be anti-rationalist, sectarian and prepared to preserve their class privileges through cooperation with the colonial powers. Finally in places like Northern Nigeria where the political position of the Muslim leaders was preserved in some form this was an extra reason for cooperating with the colonial masters.

Yet not all Muslims were in favour of cooperation with the colonial rulers and of those who were hostile some were prepared to go beyond passive resistance or mere withdrawal. Among such groups were the *Muridiyya* in Senegal led by Ahmadou Bamba (*c.* 1850–1927) and the *Hamalliyya* in what later became the Mali Republic led by Shaikh Hamallah (*c.* 1883–1943). Hamallah founded the movement named after him on the teachings of his great master Sidi Muhammed b. Ahmad b. 'Abdullah. The Hamalliyya was a splinter group of the *Tijaniyya* tariqa and was considered a heretical sect by the parent body. The two groups regarded each other with great animosity and it is not unlikely that much of the anti-French activity indulged in by the Hamalliyya owed something to its jealousy of the cooperation which existed between the French and the Tijaniyya. The conflict between the two groups did not, however, take the form of open breach of the peace until 1924, that is beyond the period covered by this chapter.

Generally speaking, opposition to colonial rule would seem to have been fiercer and more dogged in the pagan areas, especially amongst pagans who later turned to Christianity. The reasons for this were many. The colonial powers regarded so-called paganism as one of the most degraded and primitive practices of mankind. They did not even consider it as a religion but as a superstition. They therefore saw it as their appointed task to drag the West Africans immersed in it on to a higher plane of civilisation through focusing on their society the full blast of secular and

93 R. Delavignette, *Freedom and Authority*, p. 102.

missionary propaganda. Lugard for instance thought that the Christian missions were more necessary in the pagan areas than in the Muslim emirates and so advised them to concentrate there. Furthermore the pagan areas, especially in the forest regions of West Africa, came under the full impact of European economic activity. The forest areas were richer in sylvan produce than the Sudan. They were also more thickly populated and thus had the manpower for the exploitation of their resources.

For these reasons the social systems of the pagan peoples were more easily fractured than were those of the Muslims. Also, partly because they had the money and partly because they were less conservative, many more of them sent their children to the mission and government schools where they came into contact with an exciting body of new knowledge which equipped them to do battle with the colonial governments. In this sense at least some of the methods of consolidation—western education, missionary activity and the creation of a new economic regime—were self-defeating.[94]

94 For a detailed treatment of this development the reader is referred to chapters 13 and 14.

CHAPTER 13

The 1914–1918 European war and West Africa

MICHAEL CROWDER

Few aspects of West Africa's colonial history have received such scant attention in relation to its importance as the impact of the First World War on French West Africa. The large-scale revolts against recruitment for the European front, which nearly lost France control of her West African empire, have gone largely unnoticed.[1] While the war had much more impact on French West Africa than on the four British colonies, it nevertheless marked for them too an important stage in their colonial history; their direct involvement in the European politics of their colonial masters. This war too was a turning point in colonial history; the eventual suppression of the wide-scale revolts against the French during the war marked the end of the era of pacification and the beginning of the era of intensive administration. The same is true for British West Africa, which also had its revolts during the First World War, though it is difficult to attribute their cause directly to the war; after 1918 the British colonies were little troubled by armed opposition from their colonial subjects. From then, until the outbreak of the Second World War, the colonial rulers were able to consolidate their administration, free from armed opposition by the traditional political leaders and untroubled by opposition from the modern political leaders, whose voices, when heard, rarely criticised the colonial governments.

This study of the impact of the First World War on West Africa will consider first the military role played on African soil by the West African colonies in what has been described as the Great European Civil War.

THE EUROPEAN WAR IN AFRICA

Once war had been declared in Europe, the governments of colonies

1 A recent treatment of the revolts against the French in Dahomey appeared after this chapter was written: see Luc Garcia, 'Les Mouvements de résistance au Dahomey (1914–1917)', *Cahier d'Etudes Africaines*, x, 37, 1970, pp. 144–78. See also M. Crowder, *West Africa under Colonial Rule*, London, 1968, ch. 2, pt. IV.

belonging to the Allied powers—France and Britain—had to decide what to do about the Germans in Togo and Kamerun. Strategically Kamerun possessed one of the finest deep water ports in West Africa, Duala, whilst in Togo the Kamina wireless transmitter was vital for communications with the South Atlantic. From the German point of view, Sierra Leone, with its deep water port and coaling station, used as a staging base for troopships and foodships from Britain to the Antipodes and Far East, was the most important strategically.[2]

The Germans in Togo were in no position to defend themselves. They had pathetically small forces at their disposal in relation to their immediate neighbours, the Gold Coast and French West Africa. What was more, Nigeria's capital, Lagos, was within easy sailing distance of Cotonou or Accra, and just over a week's march away. In the Kamerun, even though the troops at the disposal of the Germans were greatly outnumbered by those of the Allied territories surrounding them—Nigeria, French Equatorial Africa, and effectively French West Africa, with Belgian Congo nearby—the lack of roads and mountainous nature of the terrain favoured them.

It has become common currency that the Germans militarised their African colonies. This is certainly not true of either Togo or Kamerun. In fact, in West Africa, only against France can this accusation be justly levelled. France alone before the outbreak of the war had used West African troops outside their homelands, notably in Guiana, Mexico, the Franco-Prussian War, Algeria, Morocco, the Congo and Madagascar. During the war France used African troops on the European front, while Britain used West African troops and carriers in their East African and Mesopotamian campaigns. No Togolese or Kamerun troops were used outside their colonies.

A comparative table of the forces at the disposal of the contestants in West Africa at the beginning of the war and at the end, shows that Germany had 'militarised' neither Togo nor Kamerun (see Table 13.1). Before the outbreak of the war as far as the British and Germans were concerned their African troops were maintained primarily to keep law and order in the newly acquired colonies; once their 'pacification' of a colony had been completed, the colonial armies increasingly adopted the role of auxiliary police forces which, in times of rebellion, could return to their basic military function. They also had a ceremonial role as representing the pomp, circumstance and might of the colonial powers. The different approach to the role of their colonial armies by the British and Germans on the one hand and the French on the other is clearly indicated by the disparity in their size (see Table 13.1).

Militarily Togo presented no problem to the Allies. This was recognised by the German Governor of Togo, Von Döring, who only had Polizei-

2 *The Red Book of West Africa*, ed. Allister Macmillan, 2nd ed., London, 1968, p. 234.

TABLE 13:1

TROOPS STATIONED IN WEST AFRICA AT THE OUTBREAK OF WAR

Territory	European Officers	European N.C.O.'s	European Troops	African Troops	Total
French West Africa	389	← 2182 →		14,785	17,356†
British West Africa	244	115	—	7,201	7,560★
Togo	← 300 →			1,200	1,500⁑
Kamerun					
(a) According to Heywood:	← 200 →			3,200	3,400★
(b) According to Rudin:	30 (Polizeitrüppe)			1,200 (Polizeitrüppe)	2,915‡
	185 (Soldiers)			1,500 (Soldiers)	

FRENCH WEST AFRICAN TROOPS ABROAD: ON OUTBREAK OF WAR†

Morocco: 12 Battalions   Algeria: 1 Battalion   Madagascar: 1 Battalion

Sources

★ A. Haywood and F. A. S. Clarke, *The History of the Royal West African Frontier Force*, Aldershot 1964, pp. 89, 37, 105.

⁑ Robert Cornevin, *Histoire du Togo*, Paris 1962, p. 207.

† General Maurice Abadie, *La défense des colonies*, Paris 1937, p. 217.

‡ Harry F. Rudin, *The Germans in the Cameroons 1884–1914: A Case Study in Modern Imperialism*, Yale 1938.

trüppe, as distinct from regular soldiers, at his disposal. His men were equipped with rifles dating back to the Franco-Prussian War of 1870, and he had no artillery. Though the French and British together mustered only 1,738 troops against his 1,500, the fact that theirs were regular soldiers and had artillery at their disposal meant that the Germans had little potential for long-term resistance. Consequently, Von Döring suggested to his French and British counterparts that Togo be neutralised so that Africans would not witness the spectacle of Europeans fighting each other. Actually, though the West African campaigns were, as in the case of the original colonial occupation, largely a matter of African troops fighting each other, the German Governor was right in diagnosing that the spectacle of European powers fighting each other could be damaging, and in the event dangerous, to the prestige of the colonial powers.

Neither the Acting Governor of the Gold Coast—both the Governor and the Officer Commanding the Gold Coast Regiment were on leave at the time—nor the Lieutenant-Governor of Dahomey heeded Von Döring's appeal and British and French forces invaded Togo from all sides (see map). They met no opposition on the Coast and indeed Lome, the German capital, was occupied without a shot. Realising that the small forces at his disposal could not defend the whole of Togo, Von Döring had decided to concentrate on the defence of the vital Kamina wireless station; indeed he abandoned all territory a hundred and twenty miles

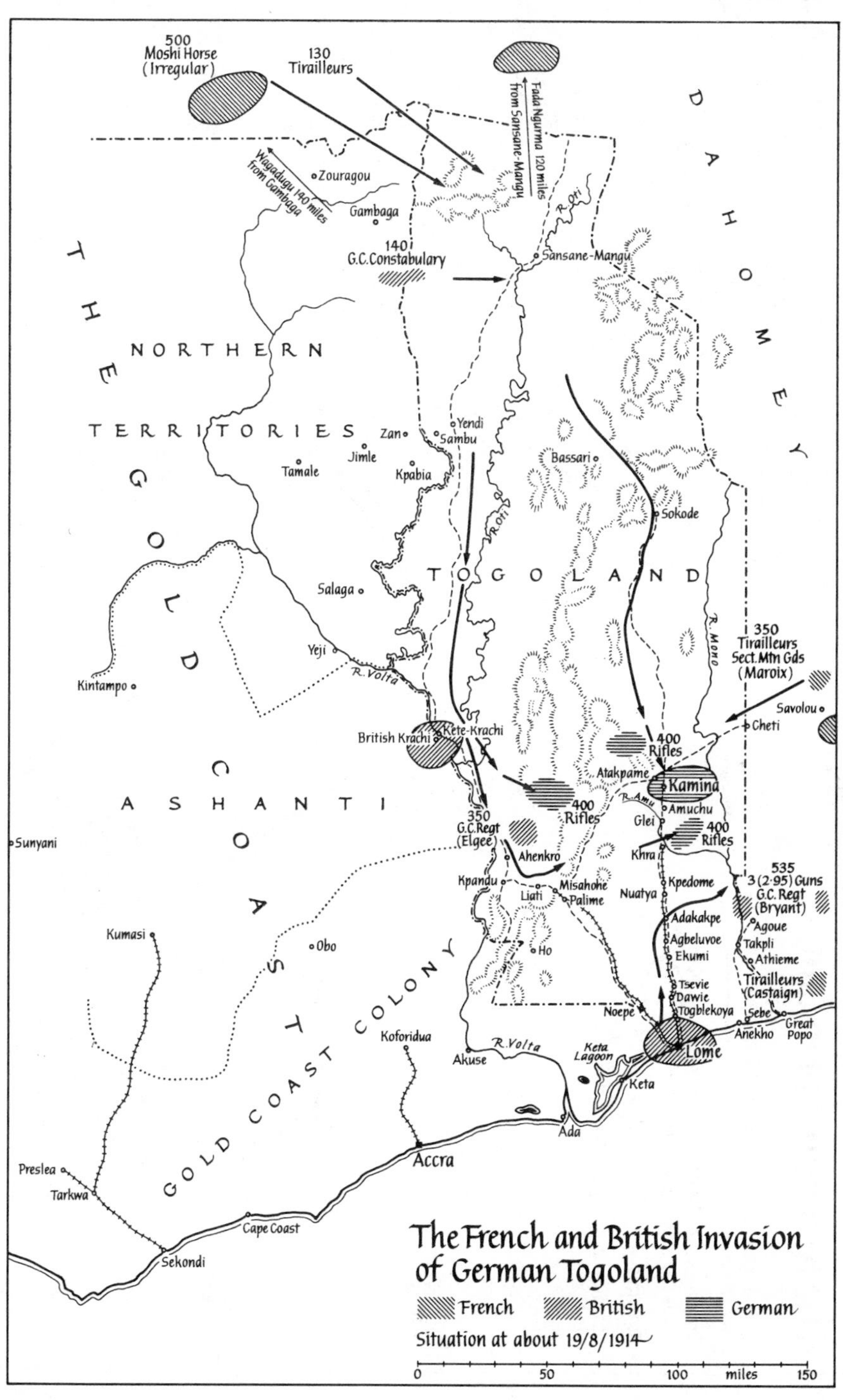
500
Moshi Horse
(Irregular)
130
Tirailleurs
Fada Ngurma 120 miles
from Sansane-Mangu
Wagadugu 140 miles
from Gambaga
Zouragou
Gambaga
140
G.C. Constabulary
Sansane-Mangu
R. Oti
DAHOMEY
THE GOLD COAST
NORTHERN
TERRITORIES
Zan
Yendi
Sambu
Jimle
Tamale
Kpabia
Bassari
Sokode
TOGOLAND
Salaga
Yeji
R. Volta
Kintampo
R. Mono
350
Tirailleurs
Sect. Mtn Gds
(Maroix)
Savolou
Cheti
British Krachi
Kete-Krachi
400
Rifles
Atakpame
Kamina
R. Amu
Amuchu
Glei
400
Rifles
400
Rifles
ASHANTI
350
G.C. Regt
(Elgee)
Sunyani
Ahenkro
Khra
Kpandu
Liati
Misahohe
Palime
Nuatya
Kpedome
535
3 (2·95) Guns
G.C. Regt
(Bryant)
Agoue
Adakakpe
Agbeluvoe
Ekumi
Takpli
Athieme
Kumasi
Obo
Ho
Tsevie
Dawie
Togblekoya
Tirailleurs
(Castaign)
Noepe
Sebe
Anekho
Great
Popo
GOLD COAST COLONY
Koforidua
Akuse
R. Volta
Keta
Lagoon
Lome
Keta
Ada
Accra
Preslea
Tarkwa
Cape Coast
Sekondi
The French and British Invasion
of German Togoland
French
British
German
Situation at about 19/8/1914
0
50
100
miles
150

13 : 1

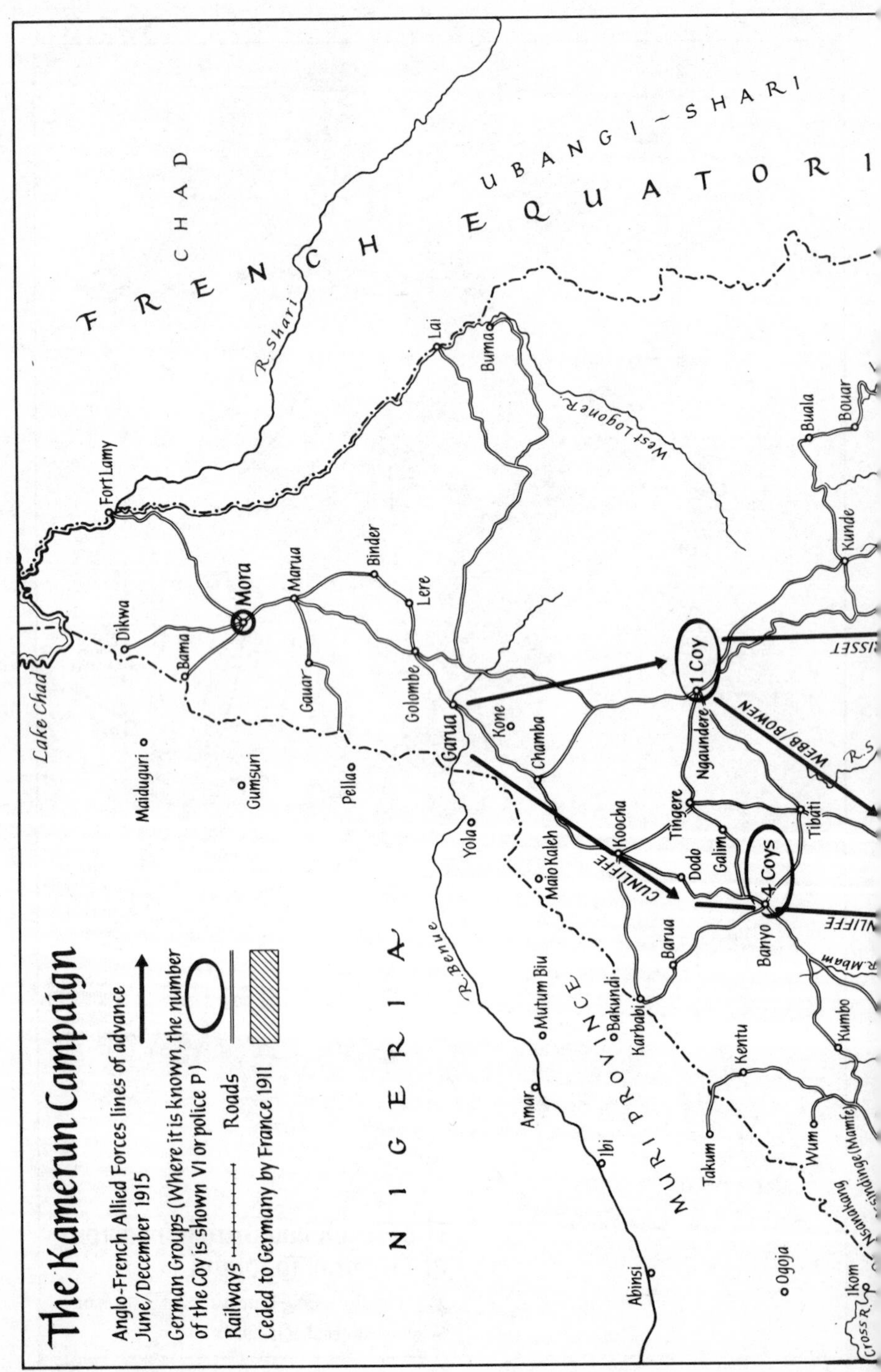
The Kamerun Campaign
Anglo-French Allied Forces lines of advance June/December 1915
German Groups (Where it is known, the number of the Coy is shown VI or police P)
Railways
Roads
Ceded to Germany by France 1911
FRENCH EQUATORIAL
UBANGI-SHARI
CHAD
NIGERIA
MURI PROVINCE
Lake Chad
R. Shari
West Logone R.
R. Benue
Fort Lamy
Dikwa
Bama
Mora
Marua
Binder
Lere
Golombe
Gauar
Maiduguri
Gumsuri
Pella
Garua
Kone
Chamba
Lai
Buma
Buala
Bouar
Kunde
1 Coy
Ngaundere
WEBB/BOWEN
Tingere
Galim
Tibati
4 Coys
Koocha
CUNLIFFE
Dodo
Yola
Maio Kaleh
Barua
Banyo
Bakundi
Karbabi
Mutum Biu
Amar
Ibi
Kentu
Takum
Wum
Kumbo
Abinsi
Ogoja
Ikom
Cross R.

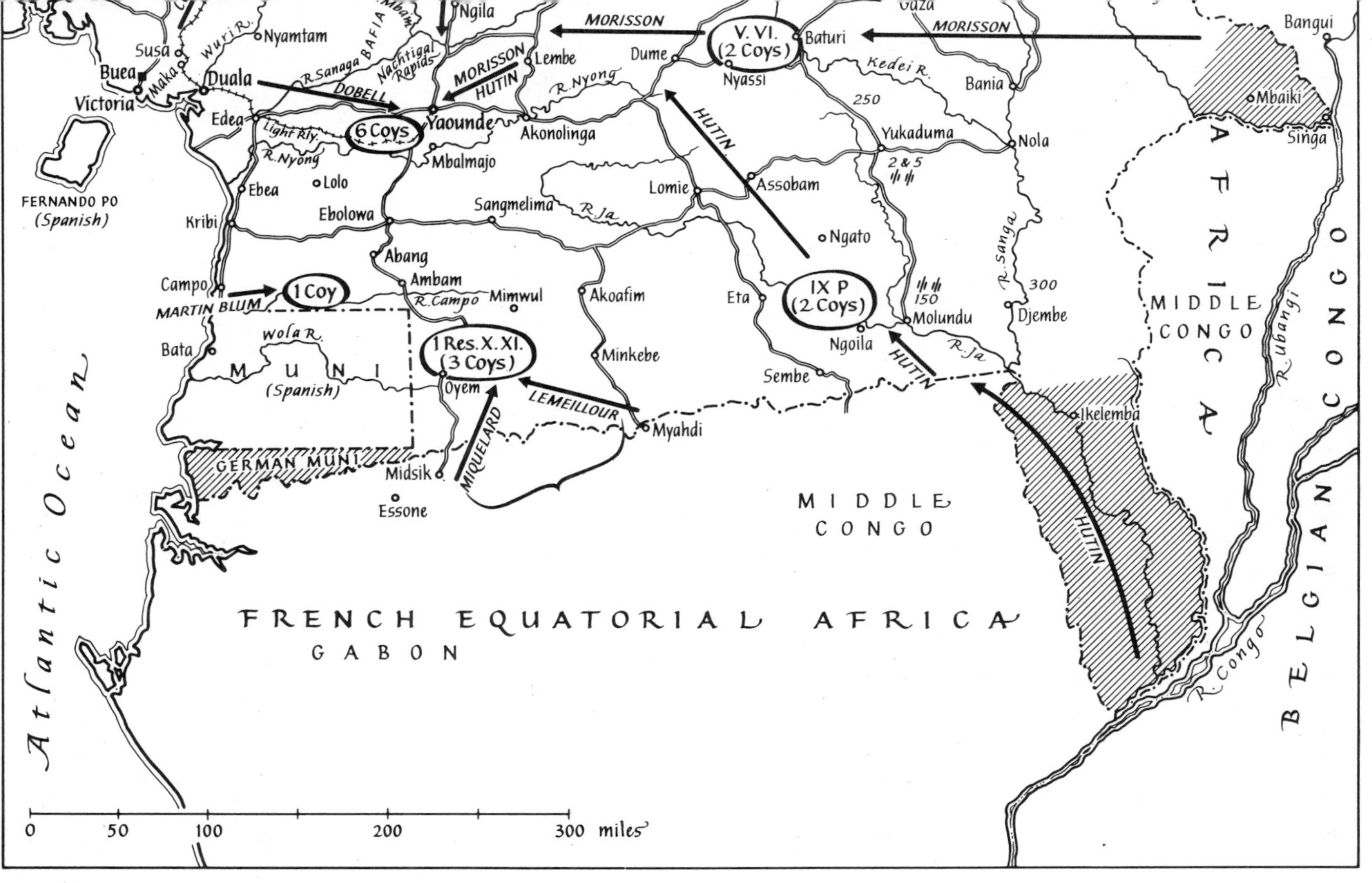
BELGIAN CONGO
FRENCH EQUATORIAL AFRICA
MIDDLE CONGO
GABON
GERMAN MUNI
MUNI (Spanish)
FERNANDO PO (Spanish)
Atlantic Ocean
R. Ubangi
R. Congo
R. Sanga
R. Ja
R. Nyong
R. Sanaga
R. Campo
Wola R.
Wuri R.
Kedei R.
Bangui
Mbaiki
Singa
Nola
Djembe
Ikelemba
Bania
Yukaduma
Molundu
Ngoila
Ngato
Baturi
Nyassi
Assobam
Eta
Sembe
Dume
Lomie
Myahdi
Akoafim
Minkebe
Lembe
Akonolinga
Yaounde
Mbalmajo
Sangmelima
Mimwul
Oyem
Midsik
Essone
Ebolowa
Lolo
Abang
Ambam
Ebea
Edea
Kribi
Campo
Bata
Dula
Maka
Susa
Buea
Victoria
Nyamtam
Ngila
BAFIA
Nachtigal Rapids
Light Rly.
MORISSON
HUTIN
DOBELL
LEMEILLOUR
MIQUELARD
MARTIN BLUM
V. VI. (2 Coys)
IX P (2 Coys)
1 Res. X. XI. (3 Coys)
6 Coys
1 Coy
250
2 & 5
150
300
0
50
100
200
300 miles

13:2

inland to the invading Allies. Even his hope of defending the wireless base was unrealistic. On 12 August the Allied and German forces engaged each other along the railway line.[3] On 14 August the advance on the Kamina Wireless base began. Ten days later, as the Allied forces converged on the base, the Germans blew it up and on the 25th the Germans surrendered unconditionally. Their surrender was accepted on the morning of 26 August. The destruction of Kamina was a great loss to the Allies, but 'to the Germans its importance was a great disaster; without its direct means of communication with German ships in the Southern Atlantic they would be completely out of touch with current events and handicapped accordingly'.[4]

The war in Togo was a simple one compared with that in Kamerun. Apart from the military importance of the destruction of Kamina and the disadvantage to the Allies of not capturing it intact, the campaign showed that in times of stress France and Britain, who normally conducted their colonial administration as though their colonies were islands, not adjacent territories, could cooperate. From the point of view of the African colonial soldiers this was the first time they had come under the machine-gun fire which had proved so demoralising and effective when used by them against African states during the colonial conquest. According to Haywood the effect on them was very demoralising, but at least it prepared them for the Kamerun campaign.[5]

Kamerun, larger than Germany itself, with thick forest and many rivers in the south and plateaus difficult of access in the north, presented an immense strategic problem to the Allies. Their campaign against the Germans lasted from August 1914 until 18 February 1916 and their difficult progress is spelt out in detail by Colonel Haywood in *The History of the West African Frontier Force.* Though written by a British officer, it is a fair account, giving the Germans full credit for their skilful defence with a small army expanded from some three to six thousand during the eighteen-month-long campaign, facing combined British, Belgian and French forces vastly superior in number, as well as an Allied naval blockade along the Kamerun coast.

The principal objective of the Allies was Duala, which apart from its potential military value was also Kamerun's main supply line from the outside world. Second to this was, of course, the port of Victoria and the capital, Buea. In the north, the main objective was Garua, a river port on the Benue. Here the British forces which crossed into Kamerun from Northern Nigeria suffered an early reverse which according to Haywood

3 It is believed that the first shot fired in the Togo campaign, by R.S.M. Alhaji Grunshi, D.C.M., M.M., was also the first shot to be fired by a 'British' soldier in the First World War.

4 A. Haywood and F. A. S. Clarke, *The History of the Royal West African Frontier Force*, Aldershot, 1964, p. 103.

5 *Ibid.*, p. 103.

caused considerable unrest among certain tribes in the Benue valley, though the Emir of Yola remained loyal.[6]

Duala itself was captured on 27 September 1914 and Buea fell on 15 November 1914. Garua was not taken till 10 June 1915 and Yaounde held out until January 1916, after being surrounded by a net of Allied forces. The bulk of the German forces then retreated to their base at Ntem, near Spanish Muni, and crossed the Spanish–German border on 15 February 1916 to seek asylum. However, in the North the German garrison at Mora, which had moved to a hill above the town, held out until 18 February 1916 and surrendered only on hearing of the retreat of their main force to Spanish Muni. Mora itself had proved impregnable to the Allied forces despite repeated assaults on it.

This bald outline of the principal Allied successes can do little to indicate the incredible hardships and suffering of the troops on both sides. At Nsanakang Lt Milne-Holme wrote in his diary of 6 September 1914: 'I have never known such hell and pray God I will never see or feel like it again. The men were magnificent.'[7] Three days later he reported that '60 out of 121 of my company are killed or missing, and about 10 wounded. About 45 of "G" [coy] have survived out of 100.'[8]

Throughout the campaign the main problems for both sides were sickness and the difficult terrain linked by the most rudimentary communications. In July 1915, for instance, of the 3,470 troops under General Dobell's command, 723 were sick.[9] To keep the Allied forces supplied and on the move, Haywood reports that the vast number of twenty thousand carriers were imported into Duala.[10]

European officers who have recorded their memoirs of the Kamerun and East African campaigns seem to be unanimous in their praise for the West African troops under their command. Colonel Haywood, joint author of the history of the West African Frontier Force, even admired the fact that the German African troops in the Kamerun 'stuck to their officers in a most surprising way until the position became hopeless'.[11] Brigadier-General Crozier, who fought in Lugard's conquest of the Sokoto Caliphate, records meeting an officer from West Africa in a train during the First World War who told him 'of the magnificent way in which the WAFFs behaved in the Cameroons'.[12]

6 *Ibid.*, p. 112.
7 *Ibid.*, p. 116.
8 *Ibid.*, p. 118.
9 *Ibid.*, p. 150.
10 *Ibid.*, p. 173.
11 *Ibid.*, p. 175.
12 Brigadier-General F. P. Crozier, *Five Years' Hard*, London, 1932; F. H. Mellor *Sword and Spear*, 1933, pp. 280–1, wrote, 'There can be no doubt, however, concerning the fidelity of the smart and extremely efficient soldiers of the Nigerian Regiment, Royal West African Frontier Force, whose qualities in war and smartness in ceremonial compare favourably with any troops in the world.'

WEST AFRICAN TROOPS ABROAD

The East African campaign

With the conquest of Kamerun completed by the beginning of 1916, British West African troops became available for the campaign in German East Africa (Tanganyika). French African troops, however, were primarily reserved for the European front and for quelling the extensive rebellions against French authority that had broken out in 1915 and continued through 1917 (see below, pp. 496–500). Though the Germans only had a small army in East Africa the British forces there in 1916 could not deal with them, so the W.A.F.F. were diverted to East Africa. Partly this was because of lack of British or other European Allied troops, partly because West African troops had shown themselves highly capable in bush warfare and finally because of the need for carriers, a task European soldiers were incapable of. As Haywood points out: 'Experience had shown that the carrier played a most important, indeed one might almost say vital, part in our African campaign where wheeled transport could not be used owing to the absence of roads.'[13] Furthermore the East African porters available to the British had proved largely useless so that it became worthwhile to go to the expense of shipping porters, mainly from Sierra Leone, all the way to East Africa and later to Mesopotamia.

The details of the East African campaign belong to the history of that part of the continent. What is important for the history of West Africa is the number of troops who participated in a campaign so far from their home. The Gold Coast Regiment sent 1,423 effectives (fifty-one of them Europeans), arriving in East Africa in July 1916. The Nigerians were remobilised with greater difficulty because many had already been discharged before the order for them to go to East Africa came; in December 1916 four infantry battalions together with the Gambia Company reached East Africa. Sierra Leone provided as many as seven thousand carriers, to which task, the Governor of Sierra Leone wrote to the Governor-General of French West Africa, they were better adapted 'than for providing troops of the line'.[14]

The British campaign in East Africa owed much of its success to both carriers and troops from the W.A.F.F. Since military campaigns are often seen in terms of the bravery of the few, recognised by the award of medals, it is significant that in the East African campaign fifty-two Nigerians won

13 Haywood and Clarke, p. 188.

14 National Archives of Senegal (NAS), Série D 4D75, Governor of Sierra Leone to Governor-General, French West Africa, 30 March 1918. The figure for the carriers from Sierra Leone given as seven thousand by N. A. Cox-George (*Finance and Development in West Africa: The Sierra Leone Experience*, London, 1961, p. 182) is higher than that given by Sir Charles Lucas, *The Gold Coast and the War*, London, 1920, p. 18, who puts it at 3,500 for East Africa in addition to those sent to Mesopotamia for the Inland Water Transport Service.

the Distinguished Conduct Medal and thirty the Military Medal, while twenty Ghanaians won Distinguished Conduct Medals, two with bars, and twenty-three won Military Medals. Of the Gold Coast Regiment, which was in almost continuous active service throughout the First World War, and was about to be sent to Palestine at the end of hostilities, 215 were killed in East Africa, 270 died of disease, 725 were wounded, thirteen missing and 567 were invalided. Three hundred and forty Nigerians were killed, 263 died of disease and 852 were wounded.

It is not often realised what a major role West Africans, so recently subjected to colonial rule, played in the winning of the war for their European masters, nor with what fortitude those under German colonial rule fought for their own colonial rulers. As Major-General Sir Charles Dobell wrote of the West African soldier in his despatch to the War Office at the end of the Kamerun campaign: 'To them no day seems to be too long, no task too difficult. With a natural aptitude for soldiering, they are endowed with a constitution which inures them to hardship.'[15]

French West Africans on the European front

If the British West Africans played a major part in the war—some thirty thousand participated from British West Africa—the French West Africans, from a more sparsely populated country, provided 137,000 soldiers over and above those already in service at the beginning of the war, not only for the Togo and Kamerun campaigns but also for the European front.

In principle, compulsory recruitment was forbidden in British West Africa, though in Northern Nigeria some chiefs, in spite of warnings, sent in batches of men who were 'unwilling to enlist';[16] in French West Africa compulsory recruitment was effectively the order of the day and, as we shall see, produced immense social upheavals.

French West Africans, unlike their British counterparts, had been used to military service in very different lands from their homelands: Nigerian troops had of course served in the Ashanti Campaign, but none, like the Tirailleurs Sénégalais, had served in such different and distant climes as Morocco, Guiana, Madagascar and Europe itself where their opponents came of different racial stock. General Archinard estimated that as many as 181,000[17] Black Africans (including those from French Equatorial Africa) were used on the European front against the Germans. Estimates of the number killed vary from thirty-five to sixty-five thousand.[18]

15 Haywood and Clarke, p. 257.
16 N[igerian] A[rchives] K[aduna]/SOKPROF/2/10/555/1918, Sokoto Province Annual Report No. 14 for 1917.
17 General Louis Archinard, 'Les troupes indigènes', *L'Empire colonial Français*, Paris, ? date, pp. 219–38.
18 For instance Edouard de Martonne, who once commanded a Senegalese battalion, puts the figure as high as sixty-five thousand in an article in the IFAN-Dakar collection of articles on the Army in French West Africa (no date, no source).

As early as 1910 General Mangin in his *La Force Noire* had called for the greater use of Africans as soldiers in the French army.[19] His idea was that Black African troops should be used to 'police' Algeria, thus releasing thirty-two thousand French and Algerian soldiers to serve on the Franco-German border.[20] However, up to 1912 Black African troops were recruited theoretically on a voluntary basis. In 1912 recruitment, still in principle voluntary, became based on *voies d'appel* for four years for men between twenty and twenty-eight years old.[21] During the First World War these *voies d'appel* became so regular and called for such a large number of men that they constituted effective obligatory military service. Indeed the French talked in terms of recruiting one million men.

Many of the troops, like so many of their European counterparts, were used as 'cannon-fodder' in the grim trench warfare of the First World War. The impact of the long days in trenches in bitterly cold weather on West Africans is not difficult to imagine. But the Senegalese, as the West African troops were collectively but incorrectly known, did not, of course, spend all their time in the trenches. They mixed with the local inhabitants in the towns and villages near their camps. The social problems of these soldiers formed the subject of a book by Lucy Ameillon who was concerned with

19 General Mangin, *La Force noire*, Paris, 1910.
20 *Ibid.*
21 General Maurice Abadie, *La Défense des colonies*, Paris, 1937, p. 215. A. I. Asiwaju in his thesis on *Western Yorubaland under French and British Rule 1880–1945*, Ibadan, 1971, has shown that in practice 'voluntary recruitment' before 1912 was more like a press-gang.

French African troops doing 'occupational' duties in the Ruhr

their welfare.[22] Paul Catrice, writing in a Catholic review some dozen years after the war when African troops were still being recruited into the metropolitan French army, expressed fears about the impact of their metropolitan experience: 'These native soldiers are in direct contact with French people. They enter into their habits, their smaller sides, and especially their vices. They very quickly come to have doubts about the superiority of the coloniser.'[23] Such fears were expressed by administrators in West Africa during the war about the return of discharged or wounded soldiers from the front who could apparently, and did, give their people a bad impression of France.[24]

Some 'Senegalese' troops were commissioned as officers. Blaise Diagne, Senegalese Deputy in the Paris Assembly and High Commissioner for Recruitment of Troops in Black Africa, on his recruiting campaign in Haut-Senegal-Niger took with him Africans who had become officers. That colour prejudice existed in mixed battalions is clear. Capitaine Obissier, writing in the *Revue des Troupes Coloniales*, remarked that European sergeants considered 'native' sergeants as their 'subordonnées et leurs suppléants'.[25]

After the war African troops continued to be recruited into the French army and some of those who had fought at the front were used to occupy the Ruhr which in the context of racial prejudice of the times was considered in some quarters to be an excessive attempt to humiliate the vanquished. 'It not only furnishes Germany with material for an effective appeal to outside sympathies, but it places men and women of European race under surveillance of a particularly galling character. . . . And though Germany, in the light of her barbarities during the war, deserves small consideration . . . nothing can, in our opinion, justify the militant enforcement of coloured soldiers on an unarmed and conquered White Population.'[26]

Given that some one hundred and eighty thousand Black African troops had served in France, and as many as sixty thousand had died, this meant one hundred and twenty thousand able-bodied men, out of a population of not more than twelve million, or one per cent of the population, and a most impressionable and potentially influential section of it, had served in the Metropolis. The impact of this extraordinary experience on French West Africa in the post-war years remains to be studied.

22 Lucy Ameillon, *Des Inconnus dans la Maison*, Paris, 1919(?).
23 Paul Catrice, 'L'emploi des troupes indigènes et leur séjour en France', *Etudes: Revue Catholique d'Intérêt Général*, 20 November 1931, p. 401.
24 See the files on the revolts against the French in Dahomey in the National Archives of Senegal where reference is made to the bad influence on the local population of soldiers returning from the European front.
25 Le capitaine Obissier, 'Notice sur les Tirailleurs Sénégalais', *Revue des Troupes Coloniales*, Paris, n.d. (IFAN-Dakar collection on the Army in French West Africa).
26 Editorial in *The African World and Cape-Cairo Express*, London, 17 March 1923, p. 251.

THE IMPACT OF RECRUITMENT ON FRENCH WEST AFRICA

In the famous 1914 election for the deputyship of the Quatre Communes of Senegal (see ch. 15 below), Blaise Diagne, the first African to be elected, took as one of his platforms that the *originaires*[27] of the Quatre Communes were just as fully citizens as their metropolitan counterparts and opposed recent attempts to deprive them of these rights.[28] In 1911 Senegalese serving in metropolitan units of the army in West Africa under the same conditions as Frenchmen were discharged. In the election Diagne advocated the securing of a law for obligatory military service by the *originaires* without distinction as to race or colour. Thereby he would ensure that their rights as French citizens would be entrenched.

Diagne kept his promise to his electors and by the law of 29 September 1916 gained from the French Government the confirmation that 'the natives of the *communes de plein exercice* of Senegal are and remain French citizens subject to military service as provided for by the law of 15 October 1915'.[29] Using France's need for troops in the war, Diagne had been able to extract this as the price of obligatory military service for his voters.

He told the Minister of War in a debate in the French Chamber of Deputies: 'That group which sent me to this Chamber has the right to consider that the mandate which has been given me will not be complete except on condition that you place this group in the same situation as that group of the population which sent you to Parliament. . . . The question is this: if we can be here to legislate, it is because we are French citizens; and, if we are such, we demand the right to serve in the same quality as all French citizens.'[30]

The law was popular with the *originaires* for it secured them the rights of citizens, that is the opening up for them of a full-scale assimilation policy which for the most part they desired. Governor-General Clozel expressed the hope that the *originaires* called up to fight for France would understand that 'the privileges and rights granted them entail certain duties, and that the very first of their duties is that of shedding their blood for their fatherland'.[31]

The *originaires* were willing volunteers for the war. But the *sujets*, from whom the mass of the two hundred thousand soldiers who served during

27 The *originaires* were also known as *habitants* or *citoyens*, i.e. those with the right to vote in the elections for the Municipal Councils, the Conseil-Général and the Senegalese Deputy.
28 For a full account of this election, see G. Wesley Johnson, 'The Ascendancy of Blaise Diagne and the Beginning of African Politics in Senegal', *Africa*, xxxvi, 3, 1966.
29 See Crowder, *West Africa Under Colonial Rule*, pp. 414–19 for the general background. Also Crowder, *Senegal: A Study in French Assimilation Policy*, revised edition, London, 1966.
30 Cited in Charles Cros, *La parole est à M. Blaise Diagne, premier homme d'Etat Africain*, Aubenas, 1961, p. 29.
31 Gouvernement-Général de l'A.O.F. Discours au Conseil du Gouvernement prononcé par M. Clozel, Gouverneur-Général de l'A.O.F. Session ordinaire de nov. 1915. Gorée, Imprimerie du Gouvernement-Général, 1915.

the First World War were recruited, were reluctant to the point of revolt. One of the principal causes of the open rebellions against the French that followed recruitment clearly was the belief of the French military authorities that Black Africa was an inexhaustible reservoir of men: as we have seen the fantastic figure of a million men had been projected. There was not that number of able-bodied men available in a population of not more than twelve million, especially given that France also demanded increased production of foodstuffs, for the cultivation of which these young men were in large measure responsible.

Before the major recruitment drive of 1915–16 for fifty thousand men, which provoked open rebellion against the French, there had already been three recruitment drives. It is clear from the correspondence between the lieutenant-governors of the colonies and the Governor-General over the 1915–16 drive that the three previous recruitments had largely exhausted the supply of real volunteers and that even with forced recruitment the supply of men in physical condition satisfactory for military service was limited.[32] For the 1915–16 recruitment the constituent colonies of the Federation were required to recruit the fifty thousand on the following basis:

	Required	Actually Recruited
Senegal	7,500	6,520
Guinea	7,000	5,905
Ivory Coast	6,500	6,833
Dahomey	3,800	3,584
Haut-Senegal-Niger	23,000	13,971
Mauritania	1,000	1,069
Niger Military Territory	1,200	1,916
Total	50,000	39,798

Of the 39,798 actually recruited, it was officially estimated that only 7 to 8 per cent were truly volunteers.[33] In Madoua, for instance, on the night of 14–15 November 1915, sixty out of two hundred and forty recent recruits broke out of their camp and only four were recaptured.[34] Chiefs

32 NAS Série D 4D45, (Recrutement indigène: Execution: comptes rendus des colonies: 1915–16: Recrutement des 50,000 tirailleurs Rapport No. 117 de janvier, 1916) hereinafter referred to as NAS/4D45(a) *and* NAS Série D 4D45 (Recrutement indigène: Execution: comptes rendus des colonies: 1915–16—1 Dossier, Affaires Civiles) hereinafter referred to as NAS/4D45(b).

33 NAS/4D45(a), Governor-General Dakar to Minister of Colonies, 28 January 1916.

34 *Ibid.,* Rapport sur les opérations du recrutement au Territoire Militaire du Niger pendant le mois de novembre 1915.

were given financial benefits in accordance with the number of recruits they presented to the recruiting teams. The chiefs naturally forced men from servile classes, rather than those from the free or noble families, to engage themselves. Strangers, visiting a village, were always in danger of seizure and presentation to the recruiters.[35] A famous novel, *La Randonnée de Samba Diouf*, by Jerome and Jean Thiraud, winners of the Grand Prix de Littérature of the Académie Française in 1919, concerned the capture of a young Senegalese by the chief and villagers of a village which had to provide five recruits.[36]

Of those actually presented for service, by force or of their own free will, only 25 per cent were found to be fit. The Lieutenant-Governor of Dahomey complained that if they did not want to remove the entire physical élite from the country, they would have to lower standards for acceptance of recruits.[37]

The safest recourse for many of those who did not wish to be conscripted was flight across the border into a neighbouring non-French territory. From Senegal, cut in half by the Gambia, with consequent borders of three hundred miles on the northern and southern side, and a similar length border with Portuguese Guinea, flight was relatively easy. In February 1916 the Governor of Gambia reported to the Secretary of State that a large number of Senegalese, in particular Diola (Jola in Gambia) from Casamance, had crossed into his territory.[38] From Guinea, with long borders with Sierra Leone and Liberia, many crossed the borders, in particular the Guerze and Kissi.[39] The Gold Coast was a refuge for escapees from what is now Upper Volta and for escapees from Ivory Coast which also had a common border with Liberia. Governor Angoulvant of the Ivory Coast reported that many men had escaped across these frontiers.[40] Nigeria served as a refuge for men from Dahomey and Niger. In particular, nomads in Niger were difficult to track down for the recruiting teams since they slipped easily across the frontier.[41] In Nigerian Borgu the district officers recorded influxes of inhabitants from Dahomeyan Borgu, not always without satisfaction

35 See NAS/4D45(b), where the Lieutenant-Governor of Ivory Coast reports that in the District of Tiassale, Cercle de Baoule Sud, of the 25 presented only 10 were from the District itself. The other 15 were all strangers who had been seized in a 'rafle'. Angoulvant to Governor-General, 18 December 1915.

36 Jérome and Jean Thiraud, *Le Randonnée de Samba Diouf*, Paris, 1919.

37 NAS/4D45(b), Lieutenant-Governor of Dahomey to Governor-General, 2 September 1915.

38 Gambia Archives, 3/26 Colonial Secretary's Office. Conf. No. 175, 'Immigration of Natives from French Territory to Escape Military Service'.

39 NAS/4D45(b), 'Extrait du Rapport Politique du 2ème Trimestre, 1915'.

40 *Ibid.*, Angoulvant to Governor-General, 18 December 1915.

41 NAS/4D45(a), 'Rapport sur les opérations au . . . Niger . . .', November 1915.

since this increased their sparse population for tax purposes.[42] To curb this exodus of population, disastrous not only to the recruitment campaign but also to the economy in loss of farmers and taxable adults, the French appealed to the British, Portuguese and Liberian authorities to send back refugees. The British authorities cooperated in a half-hearted manner. In the Gambia the Gambia Company marched through the Protectorate and arrested deserters for return to the French.[43] In Ashanti an ordinance was passed permitting the Commissioner for Ashanti to deport refugees.[44] But during the 1915–16 and subsequent recruitments the French received the impression that the British were lukewarm in their pursuit of deserters. Angoulvant, as Acting Governor-General, wrote to the Minister of Colonies in March 1918: 'Up to this year we have obtained only courteous promises without any real or effective result, giving the impression that the Franco-British alliance is unknown in British West Africa'.[45] In March 1918 Lugard did however promise to recruit soldiers for his own army along the Nigerian–Dahomeyan border.[46] Lugard also emphasised to the Governor-General in Dakar that 'so long ago as the autumn of 1915 I gave emphatic orders to all officers charged with the administration of districts bordering on the French frontier to send back all fugitives and energetic action was taken with this object by the officers concerned'.[47] But judging by the records and annual reports for Borgu, the French were right in their assessment of British action.[48]

Apart from flight across the border, villagers could hide in the bush. Recruiters sometimes found whole villages deserted. From Indienne in Ivory Coast villagers with all their goods took to the forest.[49] The same was reported from Senegal, Guinea, Dahomey, Haut-Senegal-Niger and Niger Military Territory. Self-mutilation was another means of avoiding service, reported in all the colonies. At Tiebissou in Ivory Coast young

42 NAS/4D79, Lieutenant-Governor p.i. Dahomey to Governor-General of Nigeria, 2 March 1918 and NAK SNP/10/89 p/1919. The Kontagora Annual Report for 1918, however, recorded: 'French recruiting operations in May caused a large number of French subjects to cross the border and endeavour to settle in Borgu. Immediate action was taken and all permission to settle refused, the Native Administration of Kaiama handing back the men to their tribal chief across the border.'

43 Gambia Archives 3/26, 'Immigration of Natives from French Territory, Report of Travelling Commissioner Hopkinson to Colonial Secretary', 30 December 1916.

44 *Ordinance for Ashanti*, 'An ordinance to provide for the Deportation of Certain Persons liable to French Military Service', 29 December 1915.

45 NAS/4D75, 'Correspondance avec les colonies voisines'.

46 *Ibid.*, Lugard to Lieutenant-Governor of Dahomey, 11 March 1918.

47 *Ibid.*, Lugard to Governor-General of Dakar, p. 1, (Angoulvant), 21 May 1918.

48 NAK, Borgu Reports, where there is rarely any mention of action taken to ensure people did not cross the border. The exception is cited in fn. 42. This is also borne out by Asiwaju, *Western Yorubaland under French and British Rule*, where he found little if any evidence of instructions being put into practice.

49 NAS/4D45(b), Lieutenant-Governor of Ivory Coast to Governor-General Dakar, 18 December 1918.

men were taking special medicines and little food so that they would not be considered 'fit for service'.[50]

In some areas such was the resentment that the people took up arms against the French, instigated by rumours that all the French had been killed by the Germans or by the belief that the French were now in a weak position because of their exodus to France for military service. Large-scale revolts broke out in Haut-Senegal-Niger, particularly in Koudougou, Dedougou, Bobo and Ouagadougou; in Dahomey among the Somba, the Yoruba and the Bariba.[51] In Niger Military Territory, the Tuareg seized Agades. While in Ivory Coast, only recently 'pacified' by Governor Angoulvant, and in Senegal and Guinea there were no large-scale revolts, it is clear from all reports that the administration had lost its grip in many areas and feared further recruitment would provoke such revolts, so much so that Van Vollenhoven wrote to the Minister of Colonies begging him to halt his demands for further recruits from French West Africa: 'on le mettra à feu et à sang et on le ruinera'.[52]

By 1918 recruitment, either by provoking revolts or by so weakening French administrative control that revolt by dissidents was always a possibility, had tied down in West Africa large numbers of troops who would otherwise have been sent to the European front. Thus the French had to abandon the suppression of the Somba revolt in Dahomey and use troops destined for it against the more serious Senussist-inspired revolt of the Oullimenden Tuareg led by Kaossen.[53] Indeed, so serious was this latter revolt, and so great was the shortage of troops, that the French in Niger had to call on the British for assistance, which was given.

On top of this, the French government continued to press its demand for more African troops despite Van Vollenhoven's insistence that further recruitment would lead not only to even wider scale revolts but to a decrease in production of food so vital to the feeding of France. In face of Van Vollenhoven's impassioned warnings of disaster, the French Minister of Colonies made what at the time seemed an astonishing decision: to enlist the services of the Senegalese Deputy Blaise Diagne to secure the required troops. Diagne, who was appointed High Commissioner for the Recruitment of Troops in Black Africa with status equal to that of the Governor-General himself, was a willing, and as it proved, an effective tool.

This decision was made in January 1918, by which time Van Vollenhoven had resigned himself to further recruitment with the proviso that

50 *Ibid.*

51 For the latter revolt see 'The 1915–16 Revolt against the French in Dahomeyan Borgu' in Michael Crowder, *Essays on Colonial Rule in West Africa* (forthcoming).

52 NAS/4D75, Recrutement Indigène 1918, Van Vollenhoven to Maginot, July 1917.

53 See my forthcoming paper on the revolt in French Borgu; the article on 'Les sénoussistes pendant la guerre 1914–18' in *Les troupes coloniales pendant la guerre 1914–18*, Paris, 1931, pp. 483–98; and NAK SNP/8/177/16, 'Agades (French Territory) Siege of by Tripolitan Senussi Forces'.

the cooperation of the neighbouring British colonies would have to be secured in carrying out a parallel recruitment along their borders with the French territories.[54] However, without consulting Van Vollenhoven the Minister appointed Diagne as Commissioner of the Republic for the Recruitment of Troops in Black Africa with status superior to that of the Governor-General in matters concerning recruitment.[55] This situation Van Vollenhoven was not prepared to accept and resigned rather than share power.[56] Angoulvant succeeded him as Acting Governor-General and received Diagne cordially in Dakar on 12 February 1918.

Diagne agreed to recruit on behalf of the French in return for reforms at the conclusion of hostilities. These were: relaxation of the *indigénat*, increased facilities for acquisition of French citizenship, improved medical facilities, and reservation of jobs for those who served in the army. Furthermore, after the war, France would be required to initiate development programmes designed to improve the lot of her African subjects generally.[57]

Diagne arrived in Senegal after intensive 'political' preparations had been made for his reception and for the recruitment campaigns in the various French West African colonies. His charge was to recruit a further forty thousand men. To do this he toured extensively in Haut-Senegal-Niger, but did not visit Mauritania, Ivory Coast, Niger Military Territory and only briefly Dahomey. His visits to Guinea and Senegal were directed only to certain areas. Nevertheless the recruitment campaign yielded 63,378 men, 23,378 more than were demanded.[58] His detractors pointed out that colonies he did not visit nevertheless produced men;[59] for instance Dahomey, despite the cursory nature of his visit, produced 3,340 of its quota of 3,500.[60] However, the largest single batch of recruits, 21,000, came from Haut-Senegal-Niger, the seat of the most serious revolts against the French Administration, and as Diagne proudly pointed out, where this had been declared impossible by Van Vollenhoven and Poiret, the recently replaced Lieutenant-Governor of Haut-Senegal-Niger, he had recruited this number in less than two months.[61]

54 NAS Série D 4D73, Ministre des Colonies au Président du Conseil, Paris, 12 January 1918.

55 *Décret portant organisation d'une mission chargée d'intensifier le recrutement*, Paris, 11 January 1918. By Article 3 he was given powers equal to those of the Governor-General and by Article 5 he had the right to demand the communication to himself of any instructions and measures taken by any authority from the Governor-General downwards.

56 *Une âme de chef: Le Gouverneur-Général Joost van Vollenhoven*, Paris, 1920, 'Letter to Minister of Colonies', pp. 265–6.

57 NAS Série D 4D73, Minister of Colonies to Governors-General of A.O.F. and A.E.F. accompanied by series of decrees dated 11 January instituting some of these reforms.

58 NAS Série D 4D77, *Recrutement 1918*.

59 *Les Annales Coloniales* of Saturday July 1918 (two separate issues n.d.) in NAS, Personal File of Blaise Diagne.

60 NAS Série D 4D74.

61 NAS Série D 4D74, Diagne to Colonies, 20 June 1918.

Diagne travelled in the style of a Governor-General, demanding all the courtesies due his rank from both African and European. In his 'retinue' were three African officers as well as bemedalled N.C.O.s, a visible sign to the Africans that their recruits could obtain promotion at the front. In Mossi he used an African lieutenant rather than a European administrator as his recruiting agent. The presence of African officers caused some difficulty since they were not entitled to first class passages on the railway. However, Angoulvant solved this problem by giving them the second class tickets to which they were entitled but allowing them first class compartments.[62]

If Africans were impressed by this demonstration, Europeans were not. In *Les Annales Coloniales* in July 1918, a bitter article censored in part and signed anonymously 'Un Administrateur' denounced Diagne.

> The readiness with which he welcomed, and without control, the petitions of the natives against their European masters, and saw in every administrator a bandit, the thoughtless criticisms to which this state of mind led him in his conversations both in public and private, the off-hand manner with which he treated certain Europeans, the pomp with which he surrounded his mission, all this had no other result than to lessen in the minds of the natives the prestige of the Europeans of France. M. Diagne will have made himself, perhaps unwittingly, the prophet of a new era during which our rule will have to suffer heavy attacks. . . . The natives are raising their heads; they display on all occasions a detestable attitude towards Europeans.[63]

A later issue of *Annales Coloniales* reported the fact that lower-grade European functionaries had been displaced from their houses for Diagne's suite. It quoted a speech by Diagne at Bamako in which he allegedly told African recruits: 'On your return, you will replace the white men in the administrations (sic). You will receive decorations and earn the same salaries as the whites who are here.'[64]

If critics of Diagne believed that the recruitment was largely achieved by the local administration without his help, this was not the view of Governor-General Angoulvant nor that of Brunet, the Lieutenant-Governor of Haut-Senegal-Niger,[65] who had replaced Poiret. Of Diagne's campaign in Senegal, Angoulvant, who confessed to anxiety as to the outcome of the recruitment,[66] wrote: 'A great part of this result

62 NAS Série D 4D74, Angoulvant to Diagne, 19 March 1918 and General Commanding Troops in French West Africa to Angoulvant, 11 March 1918.

63 'Un Administrateur' in *Les Annales Coloniales,* (?) July 1918. [Exact date not clear from clipping.]

64 *Les Annales Coloniales,* (?) July 1918. [Exact date not clear from clipping.]

65 NAS Série D 4D77. See also NAS Série D 4D74, 'Correspondence between Diagne and Angoulvant (March)', for comments on Poiret's critical views.

66 NAS Série D 4D74, Notes by Angoulvant on the Recruitment of 1918.

can be credited to the personal action of M. Diagne, to his moral authority over his fellow countrymen, to the popularity which he has acquired with them, and to the vibrant and patriotic appeals which he addressed to them. . . .'[67] The Administrator of Casamance also wrote warmly of the cooperation and assistance he received from Diagne and one of his officers, Lt Gomis, as a result of which the 1,045 recruits demanded from this hitherto disturbed area were recruited.[68] Similarly Diagne and his lieutenants recruited successfully in the major dissident areas of Haut-Senegal-Niger.

The war came to an end soon after these recruits had been trained and sent to France. Diagne continued in his position of Commissioner until 1923 since recruitment on a more modest scale was continued after the war for France's peace-time army.

The French broke faith with Diagne. The promised reforms were never effected, one reason being the depression that followed so soon after the cessation of hostilities. In the vocabulary of African politics Diagne has been seen as a stooge who sent African cannon-fodder to the European war. A more generous view would be one that gave credit to Diagne for performing a task Europeans were unable to do, for presenting to the African an image of African equality with the European and finally for attempting to obtain social reforms in return for recruiting men who one way or another would have been press-ganged into the army.

THE ECONOMIC CONSEQUENCES OF THE WAR

Germany, before the outbreak of the war, had been a major importer of crops from British West Africa. Similarly she had been a major exporter of processed goods. She purchased over 80 per cent of the Sierra Leone palm kernel crop. In 1913 she had purchased 131,885 tons out of a total export of 174,718 tons of palm kernels from Nigeria.[69] In Gold Coast she was a major importer of cocoa. Though in Gambia, France had been the major importer of groundnuts, the early German occupation of North-east France, where they were processed, left Gambia with the same problem of finding new markets as the other British West African territories.

Since Sierra Leone, Gold Coast and Nigeria were agriculturally self-sufficient, the diminution in imports as a result of the ban on trade coupled with the shortage of shipping space made little difference to their populations except in the rise in prices of those imported goods which did reach them. For the Gambia, whose people used imported rice as their main

67 *Ibid.*, Angoulvant to Service d'Afrique 'Recruitment in Senegal 12 March–6 May', Dakar, 18 July 1918.
68 *Ibid.*, Rapport Général sur le recrutement 1918 en Casamance, Administrateur de Ziguinchor, 25 June 1918.
69 *Red Book of West Africa*, p. 43.

subsistence food, the consequences were serious, since the French shipping line ceased to call at Bathurst and since British traders at the beginning of the war controlled little over 10 per cent of the total trade of the Colony.[70]

The main economic consequence of the war in British West Africa was to entrench British firms as the major importers and exporters in the British colonies. British trade in Gambian groundnuts increased from 4 per cent of the crop in 1912 to 48 per cent in 1916.[71] The French, who even in their own territories, because of mobilisation of expatriate company officials, actually lost ground to foreign traders, provided no competition for the British in the latter's own colonies. So with the Germans eliminated as rivals, the French ineffective and with a monopoly of shipping through the notorious West African Shipping Conference, the British traders were able to secure for the first time the major share of the trade of their colonies. In certain cases the wartime system of licensing further encouraged this British monopoly. For instance in the Gold Coast, licences were not given to farmers to ship their goods direct,[72] until the farmers' leaders threatened to retaliate against this system.[73]

While Nigeria, Sierra Leone and Gambia had little difficulty in selling their palm products and groundnuts once economic reorientation had taken place, the Gold Coast had problems in selling her cocoa crop. The export figures for Gold Coast cocoa are deceptive in this respect:

1913 = £2,489,218
1914 = £2,193,749
1915 = £3,651,341
1916 = £3,847,720
1917 = £3,146,851
1918 = £1,796,985[74]

In Gold Coast, cocoa farming was a young and rapidly expanding enterprise undertaken largely on the initiative of the farmers themselves. Whilst the export value of the crop did rise initially by over £1 million this did not represent the purchase of existing supplies. The Gold Coast Director of Agriculture wrote in 1918 of the problem of booming production and no markets: 'I have estimated that as regards the 1917–1918 crop probably not more than three-fourths of it was gathered, and only half sold; as regards the 1918 crop I imagine I am safe in saying less than three-fourths of it will ever reach the market.'[75] By contrast, Nigeria and Sierra Leone had no problem in selling their palm products

70 P. H. S. Hatton, 'The Gambia, the Colonial Office and the Opening Months of the First World War', *Journal of African History*, VII, 1, 1966, pp. 123–131.
71 *Ibid.*
72 David F. Kimble, *A Political History of Ghana 1850–1928*, Oxford, 1963, p. 49.
73 *Ibid.*, p. 49.
74 *Red Book of West Africa*, p. 161.
75 *Ibid.*, p. 159.

and where in Gold Coast sales of cocoa dropped nearly 45 per cent from 1917 to 1918 and total exports dropped in those years from £6,364,925 to £2,641,927[76] in Nigeria palm oil rose from £1,663,648 (1917) to £2,352,642 (1918) and palm products from £1,882,297 (1917) to £2,704,446 (1918).[77] As Sir Charles Lucas wrote, 'the Gold Coast had cause to regret that palm-oil and kernels had been so completely supplanted by cocoa'.[78]

Economically then the results for British West Africa were initial dislocation of the import–export trade, followed by reorientation in favour of the British commercial houses which began to establish the monopolistic situation which became the major feature of the economy in the inter-war years. Edible oils found a ready market with the Allies, but their export was hampered by German submarine activity. Import costs rose while prices for export goods did not rise accordingly, indeed in the case of cocoa they fell, as the Gold Coast Chiefs complained to the Governor.[79]

Over and above its impact on the import–export trade, the war became a burden to the budgets of the four West African colonies. They had larger military forces to maintain. Allister Macmillan reckoned that the cost of the war to Nigeria was £1,495,000.[80] Even tiny Gambia, apart from maintaining the Gambia Company, contributed £10,000 to the Prince of Wales' National Defence Fund from its budget of £122,225 for 1914.

French West Africa became a major supplier of foodstuffs for France during the war. Though at first the war caused serious economic dislocation, as it had in British West Africa, because Germany had been a major importer of French West Africa's cash crops, France soon not only bought all that Germany had purchased, but even demanded more. This, as we have seen, posed a problem for the wartime Governors-General of French West Africa. The very farmers who could increase production were being siphoned off for service in France. Furthermore, the French employees in the European companies which handled the export of these crops were also being mobilised for the front. In his speech to the Council of Government in November 1915 Governor-General Clozel drew attention to the grave consequences for the economy of the withdrawal of government and commercial employees, which he describes as 'a little excessive'.[81]

The period of reorientation of the economy to the *ravitaillement* of France also meant an enormous diminution of government revenues,

76 *Ibid.*, p. 162.
77 *Ibid.*, p. 43.
78 Sir Charles Lucas, *The Gold Coast and the War*, p. 31.
79 Kimble, p. 49.
80 *Red Book of West Africa*, p. 249.
81 Gouvernement-Général de l'A.O.F., Session ordinaire du Conseil de Gouvernement de nov. 1915. Gorée, Imprimerie du Gouvernement-Général, 1915, p. 4.

since these were principally dependent on import and export duties.[82] But by far the most important impact of the war was the removal of the most productive element in society to the army. As early as December 1915 the Administrator of the Cercle of Ourodougou was writing:

> A new exodus of men would be disastrous for the economy and for the future of the Cercle. We would have to abandon all development projects, such as roads, experimental crops, contribution to railway work, all those tasks which had already been made difficult by the preceding call-ups.[83]

Over and above the removal of prime manpower, the wide-scale revolts provoked by the recruitment reduced trade in many parts of the Federation.

By 1917 Van Vollenhoven was seeing the economic impact of recruitment and the mobilisation of French public and commercial officials for the front as a disaster for the Federation. He reported that 75 per cent of the personnel of French trading companies had left, and as a result foreign-operated trading posts had actually increased from 746 before the war to 851 by 1 January 1918. Public works were almost at a standtill. He reckoned that the neighbouring British colonies had gained some 61,500 able-bodied men from French West Africa, calculating that

20,000 had left Dahomey
15,000 ,, ,, Ivory Coast
5,000 ,, ,, Guinea
15,000 ,, ,, Senegal
6,000 ,, ,, Haut-Senegal-Niger
500 ,, ,, Military Territory of Niger.

Meanwhile, apart from gaining new tax-payers and producers, the British had not had to close down trading posts or halt public works since their European officials had not been mobilised.[84]

No notice was taken of Van Vollenhoven, for France wanted both men *and* food.

The Minister of Production (Ravitaillement) wished to purchase the whole of the harvest for 1917: sorghum, millet, maize, paddy-rice, groundnuts, palm-oil, beans, yams and manioc.[85] One of the reasons Van Vollenhoven put forward for the low level of production of crops, over and above recruitment, was the low prices paid for them. He

82 *Ibid.*, p. 5.
83 NAS Série D 4D45(b), Angoulvant, Lieutenant-Governor of Ivory Coast to Governor-General Dakar, Bingerville, 18 December 1915, encl. report of Administrator of Cercle of Ourodougou.
84 NAS Série D 4D73, Recrutement indigène (1918), Rapport et correspondance du Ministre des Colonies et du Ministre de Guerre. Reprise de recrutement: mission Diagne. 1917–18 'Projet de Recrutement'.
85 'Circulaire au sujet du ravitaillement', 7 June 1917 in *Une âme de chef*, p. 122.

maintained that higher prices would bring forth greater production. He looked with favour at the creation of a consortium of commercial houses to buy up, transport and export all produce. 'This consortium,' he wrote, 'has so few aspects of a monopoly that an explicit provision of the constitutive act prescribes that the agreement shall be submitted to other business houses of French West Africa, and that they will be permitted to adhere to it.' This consortium was to be administered from France with a representative in Dakar working, with full powers, alongside the Governor-General.[86] Van Vollenhoven saw this consortium as a strong weapon against foreign competition, and a means whereby French commercial houses could identify their own interests with those of their country.[87]

By December 1917, however, Van Vollenhoven told his Council of Government once again that the 1918 recruitment would have to be a limited one if it were not to 'injure the vital resources of this colony'.[88]

Van Vollenhoven's successor, Angoulvant, however, described the economic consequences of recruitment as imaginary[89] though this hardly tallies with the various reports of Lieutenant-Governors and Administrators, some of which he himself had forwarded to Dakar as Lieutenant-Governor of Ivory Coast.

In retrospect Van Vollenhoven rather than Angoulvant was justified in his gloomy assessment of the economic consequences of the war which hit French West Africa much more severely than British West Africa. The comparative statistics for external trade and government revenue bear this out.

SOCIO-POLITICAL CONSEQUENCES OF THE WAR

Only some twenty years after they had conquered West Africa, and in many areas even less than twenty years, the colonial powers found themselves in a potentially weakened position. We have seen the way in which the French Africans had risen up against their masters in reaction to their recruitment campaign. Actually the recruitment was the trigger cause of a rebellion which had as its other causes resentment of forced labour, the *indigénat*—or summary administrative justice—taxation and the break-up and reorganisation of traditional political units.[9] Most significant of all, those in revolt responded to the temporary weakening of

86 *Ibid.*, p. 128.
87 *Ibid.*, p. 128.
88 'Allocution prononcée par le Gouverneur-Général Van Vollenhoven à la clôture de la session du Conseil du Gouvernement de l'Afrique Occidentale Française (24 décembre 1917), *ibid.*, p. 265.
89 NAS Série D 4D74, Report by Governor-General Angoulvant to Inspector of Colonies, 1918.
90 See my forthcoming paper, 'The 1915–16 Revolt against the French in Dahomeyan Borgu'.

French administrative control by attempts to rid themselves of an unpopular regime and sought to regain their lost independence. As Inspector-General Phérivong wrote in his report on the Borgu rebellion: 'the Bariba peoples of Borgu and Kandy knew, and there is no doubt about this, of our difficulties in Europe, and the time seemed favourable to get rid of the whites and recover their independence'.[91]

In British West Africa there were revolts against the British during the war, but only those of the Kwale Ibo in Asaba Division of Nigeria, in the Benue valley, and of the Niger Delta Prophet, Elijah II, seem to have been directly connected with the war. In the case of the latter, the temporary depression in the palm-oil trade that set in at the beginning of the war affected the Delta badly. Elijah, profiting from the resulting discontent, promised his followers the independence which he said the Germans would give them in return for their help during the war. The end of the depression finished his movement.[92]

More serious were the Iseyin-Okeiho and Egba risings in Western Nigeria. While the Iseyin-Okeiho rising has often been put down to resentment at the imposition of taxation, Atanda has pointed out that the motives were much more complex. In the first place taxation was not introduced into the area until 1918. Rather the causes were reaction to the introduction of British administration, and in particular the Northern Nigerian system of native authorities, which 'brought about changes which were in conflict with the existing political and social systems of the Yoruba people in Iseyin-Okeiho area'.[93] In particular they resented their subjection to the central authority of the newly created 'native authority' at Oyo under the Alafin of Oyo. Furthermore, they resented the levying of forced labour and the new judicial structure imposed on them. The British believed that the planners of the rising were members of the educated élite in Lagos who wanted to see an end to British rule in Yorubaland and Nigeria as a whole. It was even said that they urged the rising during the war since the British were then at a disadvantage.[94] While the élite may have had such an influence, and Atanda does not dismiss this possibility, the real cause of the rebellion was general discontent with the new Native Authority system.

The Egba rising of June 1918 was potentially more serious for the British, both by virtue of its scale and because if it had happened a little earlier, the Administration would have been hard pressed to find the troops to

91 ANSOM, Ministère des Colonies, Inspection des Colonies, Colonie du Dahomey, Mission 1919. Rapport fait par M. Ch. Phérivong, Inspector General de Ière classe des colonies, 'Affaires des cercles de Djougou et du Borgou', 30 January 1919.

92 Information supplied personally by Professor Robin Horton.

93 J. A. Atanda, 'The Iseyin–Okeiho Rising of 1916: An Example of Socio-Political conflict in Colonial Nigeria', *Journal of the Historical Society of Nigeria*, iv, 4, June 1969.

94 *Ibid.*, pp. 505–6.

deal with it, for the Nigerian forces would still have been on the high seas returning from the East African campaign. A thousand men had to be sent to crush the rebellion and some five hundred rebels were killed. The cause of the rising was, concluded Eric Moore, the Nigerian barrister who investigated it, resentment against the new Native Authority system. He did not see the war as a cause.[95]

While in Dahomeyan Borgu the 1916–17 rebellion had been inspired not only by resentment against the French administrative system but also by a desire to regain a lost independence, the rebellion against the British appointed Native Administration of Bussa in Nigerian Borgu a year earlier seems not to have been motivated by anything other than the desire to rid Bussa of the slave district head, Turaki, who had replaced the deposed Emir of Bussa in 1915, after which Bussa was subjected to its nineteenth century rival Yauri, which also took over some of Bussa's traditional lands. There were similar small-scale revolts in the South-Eastern Provinces, and in the Kabba, Nassarawa, Kano, Bornu and Bauchi Provinces of Northern Nigeria. These were either inspired, as in the case of the Bussa rebellion, by the people's resentment at the imposition by the British of alien systems of rule on them, or else by the absence of the British military and administrative presence.[96]

While the revolts against British authority in West Africa were on a much smaller scale than in French West Africa, the British did, however, keep a wary eye on Muslim northern Nigeria, fearing that the entry of Turkey into the war might elicit support for the Axis from the Northern Emirs. However, even in 1917 when Kaossen took Agades, the Emirs remained loyal, though Gowers, Resident of Kano, had earlier telegraphed the Lieutenant-Governor of the Northern Provinces of his fears that 'if Agades is taken by enemy, news likely cause alarm Katsina, Kano. Presence troops will assist restore confidence.'[97] Arnett, the Resident of Sokoto, also considered the situation potentially dangerous in his province.[98] The Lieutenant-Governor himself wrote to the Resident Zaria that he should not draw too heavily on the Zaria garrison to assist the French 'until the political situation in the North has improved'.[99]

However, despite the fact that Tuareg rebels fleeing from the French entered Northern Nigeria from Niger and were given temporary shelter by some Nigerian chiefs, the situation remained under control. Nigerian

95 See Michael Crowder, *The Story of Nigeria* (Revised Edition), London, 1966, p. 249

96 See my forthcoming *Revolt in Bussa 1915: A Study in British 'Native Administration'* in Nigerian Borgu, *1902–1935*, London, 1973.

97 NAK/SNP/8/177/16, 'Agades (French Territory) Siege by Tripolitan Senussi force', Gowers, Resident Kano to Lieutenant-Governor Northern Provinces, Zungeru, 23 December 1916.

98 *Ibid.*, Arnett, Resident Sokoto to Lieutenant-Governor Northern Provinces, Zungeru, 4 January 1917.

99 *Ibid.*, Goldsmith, Lieutenant-Governor Northern Provinces, Zungeru to Resident Zaria, n.d. (approximately first week in January 1917).

troops were sent to reinforce the French along with contingents from Dakar and Sierra Leone which travelled to the front via Lagos and Kano. The Emir of Daura, for instance, actually captured four Tuareg rebels for the British.[100]

In Sierra Leone and the Gambia the interior remained quiet, except for the problems, already referred to, in the Gambian Protectorate when Senegalese sought refuge there from the French recruiting teams. In the Gold Coast too, all remained calm except in the farthest districts of the Northern Territories. In these only, according to Sir Charles Lucas, 'were there signs of unrest, as one British official after another were called away from their outposts, and the impression gained ground that the whiteman was leaving'.[101]

At the level of the educated élite in British West Africa, despite British suspicions about their part in the Iseyin-Okeiho rising, there was surprising loyalty. This was affirmed in the several legislative councils by the African unofficial members. The overwhelming majority of the educated élite believed in the justice of the British cause and considered themselves good patriots. They certainly did not see the temporary disadvantage of the British as an opportunity for gaining independence. At that time their ambitions were limited to advancement within the constitutional framework laid down by their colonial masters. Even where, like Dr J. K. Randle and many other Lagosians at the time, they were discontented with the British Administration, they would echo the sentiments of his speech at Tinubu square on 4 August 1916 when he moved the resolution of the Central Committee of the National Patriotic Organisation in London in support of Great Britain and the Allies on the second anniversary of the war:

> The fact must however not be disguised, even here, that in recent years the administration of the Government of this colony has not given the people entire satisfaction. The people see that the Government is not carried on in their interest. But, however painfully true this is, let us not forget the wider principle that we are citizens of the British Empire.[102]

Similarly the African Press in the Gold Coast supported the British cause, in particular rejoicing in the liberation of their brothers in Togo from what they, like the British, considered the German yoke.[103]

However, in expressing their loyalty to the British during the war, and indeed in giving them support through contributions to the various war funds, the élite believed that their reward would be a more significant

100 NAK/SNP/8/96/1917, 'Agades: Reinforcements received by Kaossen', report of Capt. J. Dare, General Staff Officer, to Sec. Northern Provinces, 24 October 1917.
101 Lucas, *The Gold Coast and the War,* p. 14.
102 Pamphlet published by C.M.S. Bookshop, Lagos, 1916, p. 1.
103 La Ray Denzer, 'The National Congress of British West Africa', University of Ghana, M.A. Thesis 1964, pp. 27–30.

place in the colonial decision-making process and a role in the Imperial reconstruction discussions.[104] At the Versailles Peace Conference the Accra Committee of the National Congress of British West Africa hoped to prevent the return of the former German colonies to Germany. Above all they hoped that the time had 'passed when the African peoples should be coerced against their will to do things that are not in accordance with their best interests or for their benefits'.[105]

Like Blaise Diagne in Senegal, the British West African élite were to be frustrated in their hope that reforms would follow the war. The decrees issued by the French government at the time Diagne was commissioned to recruit troops were never implemented and it took African assistance in another world war to gain from France the promises made in 1918.

In French West Africa the educated élite had constitutional outlets for political expression only in the Quatre Communes of Senegal. Elsewhere the élite, unless they held the status of citizen—and there were only some two hundred non-Senegalese citizens—were subject to the *indigénat*. Only in Dahomey, indeed, was there an educated élite to compare with that of the Quatre Communes or the capitals of the British West African colonies. This resulted from the educational activities of Catholic missionaries during the nineteenth century in the coastal towns of Dahomey. There, to add to the problems of the French in suppressing the rebellions in the interior, the administration was attacked by a Dahomean businessman, Tovalou Quenum, aided by a *métis* Senegalese lawyer, Germain Crispin. Their attacks on the administration of Lieutenant-Governor Charles Noufflard, coupled with his inefficient handling of the rebellions, led to his removal by the French. Similarly the administration was harassed by Louis Hunkanrin, who founded a branch of the *Ligue des droits de l'homme* in 1914 and was so politically active that he had to flee to Nigeria. In 1918 Diagne, whom he had helped in the 1914 electoral campaign for the Senegalese deputyship, offered him an amnesty by recruiting him into the army and finding him a desk job in Paris. There he continued his anti-colonial activities.[106]

One consequence of the war for the educated élite in French West Africa was the opening up in government and the commercial companies of posts that had hitherto been reserved for the French. The exodus of French officials and commercial employees to the war-front forced the government and trading companies to employ Africans. Furthermore the government made plans as early as 1916 to train African candidates for the posts of agricultural supervisors, postal officials, customs officers, mechanics and surveyors of public works. In particular a medical school to train

104 *Ibid*., p. 34.

105 *Memorandum of the National Congress of British West Africa, 1920*, F. W. Dove, delegate of Sierra Leone.

106 See J. Ballard, 'The Porto Novo Incidents of 1923', *Odu*, ii, 1, July 1965, pp. 52–75.

assistant doctors was established together with an agricultural school. As Angoulvant wrote, the wartime shortage of European staff made it necessary to associate 'more closely than ever the natives with our work of colonisation'.[107]

The main social problem at the end of the war was the reabsorption of soldiers into civilian life. In French West Africa, despite some initial unrest, this was largely solved by the maintenance of a large standing army. Indeed military service continued after the war, initially as we have seen under the direction of Blaise Diagne, and continued as a feature of French colonial policy until independence. All young men between nineteen and twenty-eight were liable to three years' military service, with certain exceptions such as that which stipulated that only one son of a family need serve.[108] In British West Africa, where the army was reduced roughly to its pre-war level, the problems of resettlement and rehabilitation were not very great. Those who had acquired technical skills or literacy were absorbed by commercial companies or government. The main problem was for illiterates. A land settlement scheme was tried in Nigeria but failed, since the land had little appeal for those who had seen the world in the army.[109] In Sierra Leone the 1921 Census report noted that ex-soldiers 'having through force of circumstances seen something of the doubtful attractions of civilisation . . . are at present unwilling to return to their uneventful and peaceful lives . . . in the Protectorate, but prefer to eke out a precarious existence in the crowded capital of the Colony'.[110] Lugard even expressed fears that the ex-soldiers, having learnt how to handle weapons, might prove a threat to the colonial regime.[111] However, the problem of rehabilitation seems to have solved itself without any major social upheaval, with ex-soldiers either finding work in the towns or trickling back to their villages.

In French West Africa, where educational opportunities were much sparser than in British West Africa, ex-soldiers with some education and skill were at a premium. Also the rigid discipline of the *indigénat* meant that there was little opportunity for ex-soldiers to make trouble. Furthermore the universal obligation to pay taxation meant that as a last resort they had to work on the land to earn enough money to pay their taxes.

107 Gouvernement Général de l'A.O.F., Textes rélatifs à la formation et la réorganisation des cadres indigènes en A.O.F., 1916: 'Circulaire rélative à la formation du personnel des cadres indigènes', Dakar, 1 October 1916, pp. 3–4; also 'Circulaire relative à la réorganisation des cadres des agents indigènes de l'A.O.F.', Dakar, 1 October 1916, pp. 27–8.

108 Décret concernant le recrutement des troupes indigènes en Afrique occidentale et en Afrique équatoriale française, 30 July 1919.

109 Nigerian Legislative Council Debates, 16 March 1944, p. 248, cited in G. O. Olusanya, 'The Resettlement of Nigerian ex-Soldiers after World War II. A Guide for the Present', unpublished paper read at the Xth Annual Congress of the Historical Society of Nigeria, December 1969.

110 *Sierra Leone: Report of the Census for the Year 1921*, London, 1922, p. 5.

111 Margery Perham, *Lugard: the Years of Authority 1899–1945*, London, 1967, p. 549.

CONCLUSION

The effects of the First World War have been underrated in previous studies of the colonial period of West African history. In many ways the war marks the end of the period of the establishment of colonial administration and the beginning of its consolidation into a coherent system of government. France, in particular, proved that in exploiting its human and material resources to the maximum, she could weather even rebellions as extensive as those of Haut-Senegal-Niger and Dahomey. The war marked the end of 'pacification': France was untroubled by major revolts thereafter.

Economically the war marked the consolidation of the hegemony of the large trading houses whose monopolistic practices were to characterise the commercial regime of the inter-war years.

Finally the war brought West Africa into the orbit of world politics, and showed that the colonies, far from being mere pawns in international politics, were vital to the European combatants both strategically and materially for supplies of soldiers and food.

CHAPTER 14

West Africa 1919–1939: the colonial situation

J. F. A. AJAYI AND MICHAEL CROWDER

By the end of the First World War, the British and the French, who had strengthened their dominant position in the area by sharing the former German territories of Togo and the Cameroons, could claim to be in effective control of their several territories in West Africa. The Portuguese, too, were firmly established in Guinea–Bissau and the Cape Verde Islands. Resistance continued in a few places, notably Mauritania and other desert, semi-desert and hilly areas where the writ of the conquerors still did not run especially in the hands of tax-collectors. The widespread rebellions during the war and the frequent anti-tax revolts showed that the colonial regimes could be shaken but armed resistance began to die down as the rebellions were usually suppressed with ferocity and the ultimate military superiority of the colonial rulers was hardly in doubt. The period of conquest was over; the few remaining military governors began to give way to civilian administrators, and the various regimes could settle down to the business of administration and economic exploitation.

By 1939 the colonial regimes were still congratulating themselves on the completeness of the conquest, or 'pacification' as they called it. Imperial law and order seemed to reign supreme. The pockets of rebellion had been conciliated or repressed. Colonial officials looked forward to many generations of quiet administration and evolutionary development, with the colonial decision-making processes unfettered by any African initiatives or by international opinion, both of which were to become so important after the Second World War.

The two decades 1919–39 can, therefore, be looked upon as the years most characteristic of European rule in West Africa. The aim of this chapter is to examine the nature of the colonial relationship in that period in terms of the administrative policies and structures, the economic systems and the general impact of colonialism on West Africans.

The colonial regimes and their administrative policies and structures dominated the situation. The bulk of the existing literature on the period has been devoted to studies of these policies and the mechanics of administration based on the records of the administrators. To the obvious technological and military superiority, colonial propaganda has added legends of the moral and social superiority of European civilisation. In consequence, the colonial administrators and their impact on West African society have assumed a larger-than-life proportion. As they moved from the crisis-ridden period of conquest and initial establishment to a more settled period of administration they tried to rationalise their systems of administration. In doing this they tended to emphasise the differences in the attitudes and ideologies of the various regimes with the fervour of denominational evangelists. It is important to understand these differences as they had a significant effect on the roles, fortunes and opportunities open to both the traditional and the emergent Western-educated élites. The response of these élites to the Colonial Situation after the period of conquest had to be within systems created by the foreign rulers. It has therefore been examined in Chapters 15 and 16 in terms of the dichotomy between the English- and French-speaking parts of West Africa. But this dichotomy must not be exaggerated.

In the first place, it must always be kept in mind that the rationalised systems never existed in reality. The settled conditions in which the foreign ruler could do whatever he wished with his African subjects never existed in practice. The maintenance of law and order remained even in this period the constant preoccupation of the administrator. The resources available to him in men, material and facilities for communication did not enable him to achieve an effective day-to-day dominance throughout vast areas of West Africa. The unexpected and unpredictable were daily features of the foreign administrator's life. Even under the most rationalised system pragmatism and laissez-faire, through policy or inertia, remained the predominant approach.

Secondly, it should be borne in mind that the colonial policies and administrative structures were only means to an end. When colonial propaganda about humanitarian intentions and burdens of empire is discounted, all colonial regimes will be found to have shared a common goal in the exploitation of the economic resources of the areas they dominated, to the best of their abilities and with the least possible cost to the metropolitan countries. This overall ambition took precedence over differences in ideology and dictated a greater response, on the part of the foreign rulers and commercial interests, to the realities of the local situation. As we consider these realities and the impact of the social and economic systems of the colonial regimes, not just on the élites but on the mass of West Africans, it becomes more obvious that the long-term effects of the colonial period have been exaggerated. There was in the colonial period not only a common goal of the foreign rulers but also a common West African predicament, involving important African initiatives and continuities with pre-colonial African history.

THE ADMINISTRATIVE POLICIES OF THE COLONIAL POWERS

Before 1919 the administrative policies of Europeans in West Africa had understandably been largely *ad hoc*. During the nineteenth century, when they controlled only small coastal enclaves, the usual approach in governing their African subjects was assimilationist. That is, the framework of reference was the metropolitan model. Thus the Quatre Communes of Senegal were governed through French-style municipal councils controlling their own budgets; the Communes elected representatives to a *Conseil-Général* similar to those of the *Départements* of France and during the Second and Third Republics in France they were allowed to elect a Deputy to the French National Assembly. In Sierra Leone, Freetown was constituted into a municipality with an elected Mayor and the Colony itself was governed by a Legislative Council and an Executive Council, presided over by a Governor, a device which was used in white settler colonies.

Furthermore, Africans or mulattoes could achieve the highest offices in the colonial administrative service. Indeed, a Creole, Honorio Pereira Barreto, was appointed Superintendent of the Portuguese fortress at Cacheu in 1834, and Governor of Cacheu and Bissau in 1837, and he became one of the real architects of Portuguese Guinea.[1] The Quatre Communes of Senegal elected a mulatto *(métis)*, Durand-Valentin, as Deputy from 1848 to 1850. The French Expeditionary force that conquered Dahomey in 1894 was led by another Senegalese mulatto, Colonel, later General, Dodds. When Britain expanded into the hinterland of Sierra Leone its Secretary of Native Affairs, J. C. E. Parkes, was a Creole, while in Lagos in 1875 the heads of police, of posts and telegraphs, and of customs, and the Registrar of the Supreme Court, as well as the Anglican Bishop on the Niger, were all Africans. However, as the European powers expanded into the interior and extended their administration over millions rather than thousands of Africans they found the difficulties of governing their new subjects through metropolitan models overwhelming. It became clear that new administrative formulae had to be devised to govern peoples as disparate in political and cultural organisation as those to be found in a land mass larger than the United States of America. As they settled down to administer the various territories after the shortages and upheavals of the First World War, they tried to rationalise new systems of administration.

Insofar as British administration during this period had a dominant model it was that of 'Indirect Rule' as developed in Northern Nigeria. Indirect rule in the broad sense merely implies governing subject peoples through their own political institutions. In Northern Nigeria it came to have a more precise meaning and it was this systematised Indirect Rule that the British tried to extend to other parts of British West Africa in the

1 R. J. Hammond, *Portugal and Africa 1815–1910: A Study in Uneconomic Imperialism*, Stanford, 1966, pp. 49–50.

1920s and 1930s. Though the administration tried to control the choice of the most powerful rulers as far as possible, they nevertheless often chose them from the lineages that would have ruled in pre-colonial times. As far as possible also, each ruler retained jurisdiction over the area he would have ruled in pre-colonial times, though, for the sake of greater efficiency, small chieftaincies were often federated into larger administrative units and some large chieftaincies were broken up. Chiefs or chiefs in council, or a group of elders, were gazetted as 'Native Authorities'. Like local government councils in Britain, these Native Authorities had legislative powers to carry on the day-to-day government of their areas as well as to implement the wishes of the Colonial Administration. They had the specific duties of collecting taxes, part of which was given to the Colonial Administration and part of which was retained by them for the administration of their chieftaincies. They were required to budget for the amount so retained. Thus taxation and budgeting of expenditure of the revenue so derived became cardinal features of Indirect Rule. 'Native Courts' were also established in which the chiefs administered law according to the prevailing pre-colonial system shorn of those aspects that were repugnant to the Colonial Administration. The chiefs provided the police and prisons to enforce their administration of justice. They controlled the appointment and dismissal of officials and a few large Native Authorities such as Kano even employed British officials.

However, the Colonial Administration severely limited the autonomy of the Native Authorities. The appointment of chiefs was subject to its approval. It was represented locally by a Political Officer whose duty, in theory, was to 'advise' the Native Authorities. In matters that concerned overall policy for the Colony, as distinct from local policy, the Political Officer gave direct instructions to the Native Authorities. Even in his role as adviser, especially in the smaller and less powerful Native Authorities, the Political Officer took a great deal of initiative in the formulation of policy. He was responsible to his superiors for the general maintenance of law and order. The Native Authority budget and decisions of the Native Courts were specifically subject to his review. He supervised, though he did not carry out, the collection of taxes.

It was this system of administration that was gradually extended from Northern Nigeria to Southern Nigeria with the amalgamation of the two Protectorates under Lugard in 1914 and was later introduced into the Gold Coast, Sierra Leone and Gambia. Thus, on the Gold Coast, where the chiefs had always been used as the agents of local administration, the Colonial Administration could talk in the late 1920s of introducing Indirect Rule. What they meant by this was the Northern Nigerian model of administration, and in this case particularly taxation and control of expenditure by the chiefs through a budget.

If the establishment of Indirect Rule or local self-government at the level of pre-colonial chieftaincies can be characterised as the principal theme in British administrative policy in West Africa, it is much more difficult to discern a coherent policy in France's administration of West Africa. Assimilation was abandoned as impractical though it remained possible, but very difficult for Africans to acquire the status of French citizens. By 1939 only some five hundred had in fact become citizens, excluding the *habitants* of the four Communes. In the 1920s, a policy known as *Association* was advocated as most appropriate for French Africa. Essentially this was a policy in which Africans would be administered, and their society reorganised or reoriented, so as to achieve the maximum benefit for both the French and the African. Such a policy has been characterised as the association of a horse and rider since the French would at all times dictate the direction development should take and determine what would be of mutual benefit to themselves and the Africans. The policy of Association did, at least, involve the principle of development for the benefit of the Africans as well as the French and the great dam on the Niger at Sansanding, by which it was hoped many thousands of acres of land would be opened up to cultivation by African peasants, was one of the results. But the policy was never put into systematic practice because the depressions of the twenties and thirties left little money for benign development. The result was that France, in the 1920s and 1930s, did not have an accepted policy for the administration of her African subjects. Some Governors and Governors-General considered the desirability of introducing indirect rule on the British model. But this was never implemented. Though no formal policy was adopted it is clear from a study of French administration in the inter-war years that the French administrators followed a number of guidelines which collectively can be characterised if not as their policy at least as their approach to the administration of their African subjects.

The chief characteristic of French administration in West Africa was its heavy centralisation and consequent lack of scope for local initiative. Thus while the Native Authorities of British West Africa were legislative organs, albeit limited in scope, even the Governor-General of French West Africa had no legislative powers. The principal source of legislation was the French Chamber of Deputies, or, more usually, the Minister of Colonies in Paris who issued decrees. Concomitant with centralisation went a hierarchical and uniform administrative structure. Pierre Alexandre has emphasised the influence on the French administrative system in Africa of the military, who first established the administration of French West Africa.[2] Thus the French equivalent of the British Resident or Commissioner was the *Commandant de Cercle*, and of the District Officer, the *Chef de Subdivision*. The African chiefs were the Sergeants and Corporals of the administrative system. The French chose chiefs, not without some regard

2 See Pierre Alexandre, 'The Problem of Chieftaincies in French-Speaking Africa', *West African Chiefs*, eds. M. Crowder and O. Ikime, Ife, 1970, pp. 24–78.

to their legitimacy, but principally for their potential use as administrative auxiliaries. Competence in the French language or proven administrative ability was thus considered as important as belonging to a ruling lineage. Pre-colonial political units were deliberately broken up to achieve some uniformity in size and population of administrative units and to weaken the more powerful chieftains, thus reflecting the republican, anti-monarchical attitude of most French administrators.

Perhaps the most important distinguishing feature of the French approach is that chiefs were not local government authorities. They disposed of no budget, the taxes they collected being rendered *in toto* to the French administration. Officially they did not exercise judicial functions, had no police force and did not maintain prisons. There was thus no local self-government in the British sense, the chief being not a leader of his people but a mere functionary, supervised, not advised, by the French political officer. He was on paper a member of the French administrative hierarchy and could, in theory, have been replaced by a French official. His power derived little from traditional legitimacy, and was defined in administrative circulars which aimed at uniformity and took little account of pre-colonial differences in political structure. In practice, of course, the French often recognised the advisability of using chiefs with traditional authority to administer a people whose culture was so different from their own. In default of adequate supervision from the French political officers, these chiefs in practice exercised many of the traditional chiefly powers, thus demonstrating the difference between the model and actual practice in colonial administration.

ADMINISTRATIVE STRUCTURES

French West Africa

The administrative configuration of French West Africa was fixed for the next twelve years when in 1920 Upper Volta was excised from Niger, French Soudan and Ivory Coast and made a constituent colony of French West Africa. The French West African Federation, ruled by the Governor-General from Dakar, then comprised Mauritania, Senegal, French Soudan, French Guinea, Upper Volta, Ivory Coast, Dahomey and Niger. The greater part of German Togo was brought under French administrative control, under a League of Nations Mandate, but until 1934 it was governed as a separate entity with its own High Commissioner. From 1934 until 1936, because of the economic exigencies of the Depression, it was administratively integrated with Dahomey to save money. From 1936 until 1945, when it became a U.N. Trusteeship Territory, it was placed under the Governor-General in Dakar, who became its High Commissioner. Even when Togo was administered as part of French West Africa, however, the Mandate meant that certain restrictions were placed on its administration. Forced labour was restricted to four days a year on public works; this could be redeemed by payment of a tax and could not be used for private purposes. There could be no conscription into the army. Togo was kept open to Free Trade and protective tariffs were not allowed. There was comparatively more local participation in the administration than in the rest of French West Africa. Four unofficial members sat on the High Commissioner's Council of Administration, two of them Togolese, two of them French citizens. Though the High Commissioner was not bound by the advice of the Council he rarely acted against it. Between 1920 and 1931 the Council had 231 sessions.

French West Africa proper was ruled uniformly from Dakar, the only exception being Senegal, whose Quatre Communes continued to elect a Deputy to the French National Assembly and to elect municipal councils for Gorée, Rufisque, Saint Louis and Dakar. While the demand of the Conseil Général that its competence be extended over the whole of Senegal was granted, the French administration tried to curb its independence. Governor-General Merlin, who had been appointed to counterbalance the influence of Diagne, the first black Deputy, and other elected representatives of Senegal, reconstituted the Conseil Général as the Conseil Colonial, in which there was to be parity of representation between twenty elected members from the Communes and twenty chiefs elected by fellow chiefs of the Protectorate. Since the chiefs were appointed by the Administration and subject to the *indigénat*—the summary jurisdiction of the French political officer—there was little fear of their opposing the Administration's policy. However, in order to persuade the elected members to pass the budget for the Colony in 1925, the Administration had to concede an increase in the number of elected representatives from twenty to twenty-four at the expense of the chiefs, who lost four seats.

For the rest of French West Africa, the pivotal administrative figure was the Governor-General. He was, as we have noted, not the source of authority, merely its instrument. Only the Minister of Colonies (or in rare cases, the Chamber of Deputies) could issue laws, in the form of decrees that were legally binding in the Federation. The Governor-General and the Governors often drafted these laws but did not *make* them, and they were limited to policy-making within the framework of the laws. This is the most obvious illustration of the highly centralised nature of the French administrative system.

While during the early years of the Federation there had been a natural struggle for power between the Governor-General in Dakar and the Lieutenant-Governors of the Colonies, by the inter-war years the Federation had become stronger than its constituent governments. It has been suggested that the Federation remained a rather weak administrative contrivance and that the *Loi Cadre* which broke it up in 1956 merely recognised the individual personalities of the constituent colonies. This may be an indication of the limited extent to which administrative policies affected the realities of the local situation, but there can be no doubt that the government of the Federation had overshadowed the government of each individual territory. No Lieutenant-Governor had direct access to the Minister, the sole source of authority, except through the Governor-General in Dakar. Circulars and Arrêtés of the Governor-General applied uniformly to all colonies.

Moreover, it was the Governor-General who controlled the economy of French West Africa. All revenues from customs were collected by the Federation and redistributed in the Federal Budget on the basis partly of derivation and partly of need. Roads, railways, telecommunications, ports, health and higher education were all federal concerns. The colonies did not have internal budgetary autonomy. As far as development was concerned only the Governor-General had the right to raise loans. In such circumstances Dakar dominated the colonial capitals. It did so to such an extent that Governor-General Jules Brevié in 1932 advocated a policy of decentralisation,[3] but as Lord Hailey pointed out some five years later, there was little evidence of such decentralisation having taken place.[4]

In a further attempt to decentralise in 1937, Governor-General de Coppet raised the status of the colonial governors from Lieutenant-Governor to Governor and attempted to give them greater initiative.[5] But by 1939 there had been no significant change, and both the Vichy and Free French regimes which ruled French West Africa during the war accentuated the earlier centralising tendencies of the Federal Government.

At the level of the colonies, the Lieutenant-Governors' principal preoccupation was with political affairs. The administrative system devised for the eight colonies may have been more or less uniform, but the people administered differed radically. With his Commandants de Cercle the Lieutenant-Governor had to marry the diversity of the people with the uniformity of the administrative system.

From the point of view of the African peasant, who formed the overwhelming majority of those subject to French administration, the system was one directed to ensuring his production of cash crops for export to France. This may be to look at the system in crude economic terms, but in so far as the French Commandants de Cercles and their subalterns, the Chefs de Subdivisions, impinged on the lives of the Africans, it was mainly in the economic sphere. All Africans, men and women, who were not citizens were taxed directly, though the rates varied from colony to colony, and within colonies from district to district. The minimum age at which Africans could be taxed was often as low as eight or ten, though it was later

3 J. Brevié, *La Politique et l'administration indigènes*, Dakar, 1932, pp. 13–27.
4 Lord Hailey, *An African Survey*, London, 1938, p. 240.
5 Discours prononcé à l'ouverture de la session du Conseil du Gouvernement de l'A.O.F., Dakar, November 1937, p. 9.

fixed at fourteen years. The object of the tax was not only to raise money to pay for the administration but also to force people into production of crops which could earn cash. Such crops were, for the great part, export crops like groundnuts, cotton, palm-oil and cocoa. Where no such crops could be grown in a district, the young men had to migrate to areas where crops could be grown and work as tenant-farmers or labourers on other people's land in order to earn enough money to pay not only their own tax but also that of the rest of the family. In the inter-war years rates of taxation were raised considerably compared with those that had existed in the early days of colonial rule. The principal result of this was a vast increase in the seasonal migration of labourers, particularly from the poorer colonies like Niger, Upper Volta and Soudan to the more fertile coastal colonies. The head tax was supplemented by the *prestation*, a tax in labour, which could, however, be redeemed by cash. All able-bodied males from eighteen to fifty years were subject to this tax of up to ten days' labour in a year, principally for maintaining roads. While large numbers of Africans did redeem their tax in cash, for the vast majority prestation was nothing but the *corvée* of the *Ancien Régime* and was hated as such.

Some idea of the importance of direct taxation to the French West African budget may be gauged from the fact that in 1933 it formed 28.7 per cent of revenue as compared with 18.5 per cent in Nigeria where all were also subject to direct taxation. On the Gold Coast only 0.7 per cent of the budget was of that date made up of direct taxation; for Gambia it was 6.3 per cent and for Sierra Leone 19.4 per cent. In observing that French West Africa's taxation rate was higher than that of British West Africa, we must also note that French West Africa was substantially poorer, sparser in population and that, in addition to his direct tax, the peasant had to work off his prestation.[6] The economic burdens placed on him by the regime did not end here. The compulsory recruitment of African soldiers into the army, begun just before the First World War, was put on a regular basis whereby all adult males were liable to three years' compulsory military service. Though they were paid for the duration of their service, this obligation removed some of the fittest men from the land and further increased the economic burden on those left to cultivate crops for tax payments. Some of the military conscripts did not serve as soldiers but were engaged on public work projects. The Sansanding Dam Project on the Niger in Soudan was built with the help of this class of labour.[7] As if these obligations to the new colonial masters were not enough, peasants were liable for work on the *champs administratifs*—that is, on crops which were deemed essential to the economy. In 1939 Delavignette estimated that obligatory cultivation of cotton in the Upper Volta was responsible for the

6 Hailey, p. 547.
7 Hailey, p. 626.

migration of one hundred thousand peasants to the Gold Coast in search of work.[8]

While the Commandant de Cercle and Chef de Subdivision were responsible for seeing that these various taxes in labour, kind and specie were paid, the agent for their collection or enforcement was the chief. As in British West Africa the chief was the crux of the administrative system. But by the inter-war years in French West Africa the chief had taken on a very different character from that of the chief in pre-colonial times. By the 1930s the French had succeeded in their aim of breaking down large chieftaincies into smaller units, and had established the principle that appointment to chieftaincy depended not so much on traditional claim to the position but on ability. If no one with traditional claims was judged to have the right kinds of ability, an old soldier, *garde cercle* or clerk was appointed. Furthermore, as noted above, the chiefs had lost all legal powers; they did not preside over courts except occasionally in courts concerned with personal law.

The chief, then, had changed very radically from the pre-colonial ruler. He did not head the traditional government of his people as the British West African chief did. He had no legal sanctions over the people whom he was meant to administer, other than unofficial pressure and the implicit threat that those who disobeyed him would incur the wrath of the French administrator. Since he very often did not have any traditional claim to chieftancy and the area which he administered did not necessarily coincide with a pre-colonial chieftaincy, he might even be a complete outsider, not speaking the language of those under his charge. Finally, he had no adequate income of his own—he was paid a pittance by the French government, and was forced to supplement his income by illegal exactions. The chief, as a result of his unrepresentative role and his unpopular duties, lost respect in the eyes of the people to such a degree that in the independent French West African states chieftaincy has become far less significant an element than in British West Africa and in certain cases has been abolished altogether. Even in the 1930s, in the regulation of traditional matters, from customary law to religion, a large number of peasants often dealt not with the chief appointed by the French but with the man who would, according to tradition, have been chief.

While the chief was the agent of the French administration with whom the peasants most often came into direct contact, it was the French administrator who maintained discipline over them. The administration of the law, other than personal law, was at all stages governed by the French administrator. Supplementing the courts, he had powers of summary justice over all African subjects. This system of law, known as the indigénat, empowered the administrator to jail any African without trial

8 Robert Delavignette, *Les Vrais Chefs de l'Empire Français*, Paris, 1939.

for up to fourteen days. This applied equally to chiefs who failed in their duties and many were frequently jailed.

However, because of the comparative durability of his post, the chief had an advantage over the French administrators who, because of the system of *rouage*, rarely stayed in a particular post for more than a year and were thus unable to familiarise themselves with the workings of the chieftaincies under their command.[9] For this reason, the realities of the local situation with its variety of traditional African cultures often proved more operative than the uniformity which the French administration was trying to impose.

British West Africa

The four colonies of British West Africa were separately administered in the inter-war years, practically the only things they had in common being their currency and the West African Frontier Force. Nevertheless by 1936 the 'advantages of a closer relationship between the four British West African colonies and the appointment of a Governor-General for the whole' was, according to *The Times* of London, being repeatedly advocated.[10] The four colonies were each divided into 'Colony' and 'Protectorate', the Colony representing the original coastal enclave in which the inhabitants were technically British citizens, the Protectorate being the vast hinterland in which the inhabitants had the status of British Protected Persons. The Protectorate of Nigeria was divided into two distinct sets of Provinces representing the former colonies of Northern and Southern Nigeria which Lugard had amalgamated in 1914. Nigeria as a whole was administered by a Governor in the inter-war years, with Lieutenant-Governors, later Chief Commissioners, for the Northern and Southern Provinces. The Lieutenant-Governors had a great deal more independence and initiative than their French West African counterparts, particularly in administrative as distinct from economic policy.[11] Indeed administratively Northern and Southern Nigeria were treated almost like two separate countries, with the North–South boundary being often referred to as 'the frontier' in the correspondence between District Officers and their Residents, and political officers spending their lives in the exclusive service of either the North or the South, almost never serving in both.

The principal source of legislative authority was the Colonial Governor,

9 Pierre Alexandre, 'Chiefs, Commandants and Clerks: Their Relationship from Conquest to Decolonisation in French West Africa', *West African Chiefs*, pp. 2–13.

10 *The Times*, 28 December 1936; *The Colonial Empire in 1937–1938. Statement to Accompany the Colonial and Middle Eastern Services*, Cmd 5760, London, 1936, p. 64.

11 This is borne out by a comparison of the detailed information which the Lieutenant-Governor of Dahomey had to provide for the Governor-General in Dakar about the revolt in French Borgu in 1916 and the brief reports the Lieutenant-Governor of Northern Nigeria sent to the Governor-General in Lagos when one year earlier rebellion broke out in Bussa, in British Borgu.

not the Colonial Office. In the case of Nigeria in 1923 the constitution introduced by Governor Clifford made the Governor-in-Council the principal source of legislation for Southern Nigeria while the Northern Provinces were ruled by him through Proclamation. On the Gold Coast the Governor-in-Council legislated for the Colony while Ashanti and the Northern Territories were ruled by Proclamation. The Gambia and Sierra Leone made no distinction for legislative purposes between colony and protectorate, and their legislative councils had competence over the whole territory. However, while a Governor-in-Council could initiate legislation, the Colonial Office could, even after the Governor had given it his assent, disallow it in the name of the British monarch. Similarly, while what may be termed purely domestic legislation was in practice left to the local legislative councils or the Governors, certain classes of legislation involving constitutional changes or overall imperial policy required Acts of the British Parliament or the assent of the British Crown.

The Legislative Councils themselves were composed mainly of the principal and more senior colonial officials. Along with a few non-official Europeans nominated to represent commercial interests there were some African members, most of them nominated, with a few elected in the Gold Coast and Nigerian Legislative Councils. The Africans were always in such a minority that they were limited to criticising bills which they had not the power to reject. Indeed, composed as they were, the Legislative Councils provided little check on the authority of the Governors. Similarly, the Governors' Executive Councils, which were composed of a majority of officials with one or two nominated non-official members, were in practice largely instruments obeying the will of the Governors. Not until the Second World War were Africans nominated to them. Governor Graeme Thomson had in 1930 considered proposing two Africans for membership in his Executive Council in Nigeria but members of his existing Council persuaded him to drop the idea.[12]

That the Colonial Governors were so powerful at this time was also due to the size of the British Empire, the difficulties of communication, and the ignorance of the colonies that prevailed in the Colonial Office. Few Colonial Office officials had ever been to a colony, still less the particular colony on which they could be called upon at any time to advise the Secretary of State. Consequently, as one Colonial Office official told Ormsby-Gore in 1930, the Office's chief duty was 'to select the best man available for any particular job, and send him out to do it, and back him up'.[13] Thus in practice Governors of individual colonies in British West Africa, with their power to initiate legislation, and subject to few checks from London

12 Tekena N. Tamuno, 'Unofficial Representation on Nigeria's Executive Council', *Odu*, new series, 4 October 1970, pp. 54–8.

13 W. Ormsby-Gore, *Comparative Methods of Colonial Administration*, London, 1930, pp. 10–11.

or within the colony itself, were much more powerful than even the Governor-General of French West Africa.

The British Governors were concerned not only with administration but with the coordination of the various technical services and in particular with the economic development of their colonies. Except in matters of major policy, it was their Lieutenant-Governors, Chief Commissioners, or Commissioners who supervised the administration of the African masses. Again, they too, because they had to coordinate the operation of the technical services at the regional or provincial level, had to rely to a great extent on their Residents and District Officers for the conduct of day-to-day administration. It is clear from a close scrutiny of the administration files for one particular division of Northern Nigeria, in this case Borgu,[14] that in a vast area such as Northern Nigeria real administrative power lay with the provincial Resident and his District Officers. The Lieutenant-Governor interfered only on major issues, principal control being through the annual budget exercise. Thus as far as the African was concerned, the main representation of the British administration was the District Officer or District Commissioner. The role of these officials in local administration has already been discussed above.

Portuguese Africa

Portuguese colonial policy in this period leaned towards the French approach of centralised control from the metropolitan capital. Indeed, the Portuguese system as rationalised in the 1930s provided for a greater degree of centralisation than obtained in French West Africa. However, in spite of the relatively greater importance of the colonies as a major national preoccupation of Portugal and a more plentiful supply of low-paid European personnel to draw upon, the internal political and economic problems of Portugal ensured that more persistent bureaucratic interference in the colonies did not result in more effective administrative control.

Under the Republican Regime (1910–26)[15] the colonies were treated largely with indifference. Portuguese settlers and officials, led by the Governors and the colonial legislative councils, were left relatively free to attempt to consolidate Portuguese rule over the African populations, to raise loans for development and exploit the natural and human resources of the Africans. On the grounds of separating Church and State, the subsidy for

14 See Michael Crowder, *Revolt in Bussa: A Study in British Native Administration in Nigerian Borgu, 1902–1935*, forthcoming.

15 R. J. Hammond, *Portugal and Africa 1815–1910:* also his 'Uneconomic Imperialism: Portugal in Africa before 1910', *Colonialism in Africa, 1870–1960*, i, eds. Gann and Duignan; James Duffy, *Portuguese Africa*, Cambridge, Mass., 1959; 'Portuguese Africa 1930–1960', *Colonialism in Africa, 1870–1960*, ii; Lord Hailey, *An African Survey*, revised London, 1957, p. 228 ff.

Portuguese Christian missions was removed in 1911 but restored in 1919 as part of the effort to improve the cruel condition of the African subjects, whose sufferings under oppressive labour codes that were applied even more harshly reached the proportions of the Congo Scandals but were less publicised. There were also frequent reports of mismanagement of funds by colonial officials and of efforts to divert colonial riches away from Portugal, as well as the fear that the degree of local autonomy granted the settlers and officials was likely to produce a disintegration of the Portuguese Empire.

The *Estado Novo*, the nationalist and authoritarian regime that came to power in 1926, was determined to reverse this trend. However, it was not till 1930 that new guiding principles were stated in the Colonial Act, and not till 1933 that these were elaborated in constitutional laws by the Organic Charter and regulations in the Overseas Administrative Reforms. Through these laws and regulations the new regime restored the State's control over the colonies at the expense of the settlers and officials. The Overseas Ministry with the Minister in charge became the main authority over the colonies. Though certain matters like the appointment of colonial governors, concessions to foreign companies and proposed loans had to be approved by the Council of Ministers, and the National Assembly also legislated on major issues, it was the Overseas Minister who directed the administrative, political and financial life of the colonies. The laws or decrees of the Council of Ministers originated from him. No colonial legislation was valid till his Ministry had released it for publication in the *Official Gazette*. The Governor-General of each province and the Governors who were directly responsible to him visited locations and supervised administration; they prepared budgets and advised on legislation; but, as in the French system, they had administrative but not legislative powers. What is more, financial control was so stringent that the Portuguese Governor-General was liable to prosecution if he exceeded or altered the budget passed by the Minister. It was one of the achievements claimed by the new regime that after 1931 colonial budgets were regularly balanced.

As far as the African population was concerned the effective unit of administration was the 'circumscription' under an *administrador*, though in the remote areas, to facilitate communication with the governor, some circumscriptions were grouped together as *intendencia*. Each circumscription was divided into a number of administrative posts, each under a Portuguese official called a *chefe de posto*. The administrador and his officials were directly charged with the political and judicial control of Africans under the *Regimen do Indigenato*. They had the powers of summary trial and imprisonment; they collected taxes and tried to encourage economic development in whatever way they could; above all, they coordinated labour recruitment and supply and dealt with the insatiable demands of private companies and concessionaires requiring labour. The colonial army was also an important feature of the administration, military service being almost as important an obligation on the Portuguese African as on the French African, but with a higher proportion of Portuguese non-commissioned officers.

Even more than in French West Africa, African political leadership had been undermined. While among the Fulahs, Malinke and Mandyako, Muslim religious leaders who were willing to cooperate with the regime were encouraged, it was the administrador and the chefes de posto, the white chiefs, who dominated the day-to-day life of the Portuguese Africans. The African chiefs, called the *regulos*, were uniformed Portuguese officials. Some of them claimed succession through the traditional system but this became increasingly irrelevant. They were appointed by the administrador and were usually former soldiers, policemen or interpreters rewarded for their loyalty and efficiency. They received token salaries and did the dirty work of the administration, especially in the collection of taxes and recruitment of labour. In this, they were assisted by the armed police, who were recruited largely from ex-servicemen and were infamous for their brutal methods.

Portuguese Guinea was different from areas like Angola, Mozambique, the Cape Verde Islands, or even São Tomé, with sizable Portuguese and *mestiço* populations. Guinea-Bissau, as it is now called, remained predominantly an African country with only 2,250 Europeans and 4,500 *mestiços* in 1950 out of a population of half a million. This meant a greater degree of neglect, fewer public works, fewer schools, particularly secondary schools. But the overall colonial policy remained the same. The Estado Novo emphasised the old doctrine of Luso-African solidarity and boasted of Portuguese miscegenation and absence of racial prejudice. It tried to reduce abuses in labour laws, in particular trying to stop private recruitment by channelling all requests through the administrador. But the obligation to work remained the cardinal Portuguese African policy. Beyond that, the ultimate goal that Africans through the regime of enforced labour would acquire Portuguese culture and one day qualify to be regarded as civilised and fit to enjoy the privileges of Portuguese citizenship, remained a distant hope. Rather, it was increasingly stressed that the African had a culture of his own deserving of respect to the extent that his progress to the promised land of Portuguese civilisation should not be hurried with too much education or other interference, except by way of teaching him habits of industry through enforced labour. Meanwhile a few individual Africans, less than fifteen hundred by 1950, about a quarter of one per cent of the population, who were able to prove that they spoke Portuguese fluently, held lucrative jobs and had adopted Portuguese culture, were rewarded with certificates of assimilation, which enabled them to escape from the regimen do indigenato and join the *não-indígenas* in electing one member of the National Assembly.

ECONOMIC AND SOCIAL POLICY

The cardinal principles of the colonial economic relationship were to stimulate the production and export of cash crops—palm produce, groundnuts, cotton, rubber, cocoa, coffee and timber; to encourage the consumption and expand the importation of European manufactured goods; and, above all, to ensure that as much as possible the trade of the colony, both imports and exports, was conducted with the metropolitan country concerned. To facilitate the achievement of these objectives, new currencies, tied to the currencies of the metropolitan countries, displaced local currencies and barter trade. The state of the colonial economy was measured not by the welfare of African peasant producers, manufacturers, consumers, businessmen or taxpayers, but solely by the increase of exports and imports and the proportion of this trade that was conducted with the ruling country.

In theory, the various governments adopted a laissez-faire policy of the state encouraging but not directly interfering with trade. In practice, colonial currencies, banking facilities, navigation, judicial processes, customs regulations and other measures ensured the domination of the economy by a relatively small number of large expatriate firms. These firms took action to ensure—through collusion concerning the prices both of imports and of exports, through agreements not to compete in one another's major spheres of interest and other monopolistic and discriminatory practices—that African businessmen were effectively eliminated from the import/export trade and that the African producer and consumer did not enjoy the benefit of a competitive market in relation to either the price of his exports or the price of the imported goods he bought. Similarly, when his labour was required, the firms instead of paying a competitive price agreed among themselves to fix a low wage. Sometimes the political officer enforced this through recruitment of contract and obligatory labour. Such price-fixing extended all the way to freight charges whereby the West African Shipping Conference, representing the major expatriate firms and having a virtual monopoly of the import/export shipping, imposed higher rates on non-members. Thus it was the European firms who reaped the benefit of the expansion of the colonial economy in the decade 1919–29. The few African merchants who had survived from the pre-colonial period found themselves in increasing difficulties. They tended to hold too much of their capital in goods, without reserves of cash to tide them over the difficult years, and the European banks would not go to their aid. The slump in the economy in 1920 ruined most of them and the Depression after 1929 virtually wiped them out of the export/import business. Their resentment was a major factor in the rise of nationalist politics in the inter-war years.

Similarly, the collection and transport to the coast of export crops and the retail marketing of imported goods, both of which in the nineteenth century had been almost exclusively in African hands, gradually passed to expatriates, not only to the large firms but to Levantines and a few Greeks and Indians as well. The European firms and banks showed greater confidence in these non-European expatriates by extending to them credit and banking facilities which they denied to African competitors. In the areas which produced cocoa, coffee and to a lesser extent palm produce in the Ivory Coast, Ghana, Dahomey and Southern Nigeria, the world price usually brought returns in relation to the labour invested that were sufficiently high to improve substantially the material living conditions of the African farmers. Indeed, faced by regimes little interested in their welfare, these farmers often had enough revenue to finance their own schools, roads, bridges and dispensaries.[16] It was in such areas that there emerged a number of relatively successful African businessmen. In the areas which produced lower-priced export crops like groundnuts, the role of Lebanese middlemen became more prominent. Attempts at African co-operatives in this period were unsuccessful in gaining higher standards of living for farmers of such export crops. The only significant exception was in the Senegal where the marabouts of the Mouride brotherhood preached the heavenly reward of industry on earth and the regular payment of tithes to maintain the marabouts, who were thus enabled to expand the cultivation of groundnuts, accumulate capital and become powerful economic as well as religious and political leaders.[17]

In the inter-war years, therefore, it was in these cash crop-producing areas that West Africans were directly involved in the colonial economy. Even then, their material life was little affected since, apart from the exceptions noted above, the majority of export crops earned low prices and the return to the farmer was so small that after paying his taxes and the usurious rates of interest on advances made to tide him over the cultivation season and other dues, he usually had little left with which to buy imported goods. Moreover, these imported goods were usually only substitutes for goods which had hitherto been manufactured locally—textiles, household utensils, etc.—rather than additions to the range of goods traditionally available to him. The major items imported during the colonial era do not differ greatly from the list of goods Heinrich Barth recorded as being the staples of the Kano market in the 1850s.[18]

The economic life of those West Africans living outside the cash crop areas would have been scarcely affected by the colonial presence but for the obligation to pay taxes—which was not enforced in some places such as

16 Polly Hill, *The Gold Coast Cocoa Farmer*, London, 1956; David Brokensha, *Social Change at Larteh, Ghana*, Oxford, 1966.

17 Donal Cruise O'Brien, *The Mourides of Senegal: The Political and Economic Organisation of an Islamic Brotherhood*, Oxford, 1970; Lucy Berhmann, *Muslim Brotherhoods and Politics in Senegal*, Cambridge, Mass., 1970.

18 Henry Barth, *Travels and Discoveries in North and Central Africa*, London, 1857, II, pp. 118–33.

the Gold Coast till towards the end of the period—and but for the obligatory labour and compulsory crop cultivation in Portuguese Guinea and French West Africa. Money to cover taxes had to be found either from internal trade in selling food crops in cash crop areas, or from the earnings of the younger members of the family who had to migrate voluntarily or through contract labour organisations to areas where cash could be earned. Large-scale migration for this purpose took place for instance between on the one hand Niger and Upper Volta and on the other Ivory Coast and Ghana where work was sought on the cocoa and coffee plantations or in the gold mines. Similarly, seasonal labour from French Soudan worked on the groundnut fields of Senegal and the Gambia. Again, such migrant labour did little more than provide taxes for the workers' families. In many areas throughout this period, barter remained an important form of exchange. In 1935 one British District Officer wrote despairingly on the slow pace of change: 'To describe present day Borgu one could quite easily copy one of Lord Lugard's early reports and give an accurate picture.'[19]

In principle only those areas that had potential for growing cash crops or had mineral resources were opened up by roads and railways financed by the colonial governments. Thus a map of the railways and major roads in the 1930s represents a grid draining the exportable resources of the interior towards the coastal ports. Roads and railways were not built for the specific purpose of developing internal trade. Thus in the 1930s goods from Ibadan to Enugu had to be railed via Jos.

Most economic studies of this period have been concerned only with the export of cash crops and the import of European manufactured goods. Comparatively little is known about the internal trade in subsistence crops and locally manufactured goods such as household utensils, woven and dyed cloth, metalwork, etc. Anthropologists have paid some attention to the economic importance of local trade and manufactures in particular areas. But relatively little, if anything, exists on the overall volume of internal trade in West Africa in such commodities as yams, cassava, cloth, kola-nuts, cattle, poultry, dried fish, etc., or locally manufactured goods such as cloth, hoes, knives, calabashes, jewellery, etc., or the fortunes of trade in these goods. What is clear is that internal trade meant as much to the well-being of the peasant as did the import/export trade, if not more. The depression which hit those involved in the colonial economy so hard in the 1930s probably brought some prosperity to others. French West Africa increasingly in that period and even more during the blockade of 1940–2, which cut it off effectively from European markets, had to rely on local industries and internal trade for textiles and matchets.

19 National Archives, Kaduna/SNP/17/3/24024 Borgu Division, Development of. Resident Ilorin to Secretary, Northern Provinces, 11.4.1935.

In the name of a laissez-faire policy, the various colonial governments did very little to encourage the production of food crops. Yet they did not hesitate to use the resources of the colonial regimes to encourage the production of cash crops. In Portuguese and French West Africa, labour was frequently recruited for European planters. The British in Nigeria did draw the line at the alienation of land to the Lever Brothers for plantations, but the agricultural research stations that were financed with the African taxpayer's money concerned themselves almost exclusively with the cash crops. If anything, as a result of the incessant demand for labour on public works, European farms, cash crop production, military service, etc., food production seems to have declined. Groundnut-producing areas of the Gambia and Senegal began to import rice and to endure annually what came to be known as 'the hungry season' just before and during the cultivation of groundnuts.[20]

Similarly, while France and Portugal and even Britain to some extent were willing to use tariffs and other measures to reserve colonial markets as much as possible for the trade of metropolitan companies, whether or not there were other more economically beneficial customers, there was no encouragement, to say nothing of protection, for local manufactures. No new industries were created to process the cash crops locally. For a long time not even the shells of the groundnuts were removed locally. Rather, cheaper imported textiles and household utensils were allowed to compete with traditional crafts and industries. The full effects of this on the economy deserve further study. While many household utensils began to be imported, farming methods and implements remained unchanged. Cheaper woven cloths were driven off the market faster than the more expensive ones for which there was always some demand. There is no doubt that the cheaper imports improved the standard of living especially in the cash crop areas where there was some money to buy them. But since most of the profit of the import trade was exported, the overall economic advantage to West Africa was limited.

In short the colonial regimes pursued only a one-sided laissez-faire policy. Where the interest of European firms was concerned, the administrations placed their resources at the firms' disposal at the expense of Africans. Where the interest of Africans was concerned, the administrations were indifferent and left the field free for the privileged Europeans to compete unfairly with the Africans. That was the essence of colonial exploitation. An important case in point is public works—roads, railways, harbours, schools, hospitals, agricultural stations, and later the occasional electric plant, pipe-borne water supply, aerodrome, etc.—which were a great source of pride to colonial regimes. It was a cardinal part of the

20 The hungry season in the Gambia continued to be a major problem right up till the late 1950s. See Michael Crowder, 'Better outlook in the Gambia', *New Commonwealth*, 9.12.1957.

philosophy of laissez-faire that colonial regimes should provide only what could be paid for from taxes, customs and excise duties, and loans serviced exclusively from local revenues. There was little conception that metropolitan funds could be invested in colonial infrastructure to stimulate development. The depression of the 1930s provoked one or two very tentative steps in this direction such as the Great Empire Loan in French West Africa for public works to relieve unemployment and the British Colonial Development Fund which made £420,000 available for the four British West African colonies, from which Sierra Leone alone received £250,000. Thus, while the public works represented long-term fixed capital assets, it was essentially the African taxpayers who paid for them but the European rulers who reaped the most benefits from them in the colonial era. Apart from the John Holt and the United Africa Companies which built private ports at Warri and Burutu, it was the African taxpayers who financed the harbours dominated by the monopolistic European shipping lines. They paid for the railways, which were designed for the export/import trade rather than for the expansion of internal trade and industries. They paid for the lighting and the water supply and paved streets of the exclusive European municipalities in West Africa.[21]

In spite of the laissez-faire philosophy, the colonial regimes could be as *étatiste* as they cared to be. In the absence of private companies willing to invest money in them, the colonial administrations operated railways, water works, electricity supplies, etc., thus fortuitously introducing ideas of public ownership of essential utilities. In addition, they not only directed what export crops should be planted, they often used taxation as a direct policy of encouraging migrant labour to assist cash crop production. In Portuguese Guinea, an administrador had the power under the Labour Code to recruit for obligatory work anyone who could not show evidence that he had a job to enable him to pay his taxes, as well as feed, house and clothe his family. Both the French and the British frequently fixed taxes at a rate designed not merely to pay for needed services, but to force as many Africans as possible to increase their production of cash crops or seek paid employment.

However, in matters that concerned the welfare of the Africans, the colonial administrations stuck rigidly to the philosophy of laissez-faire, balanced budgets, apathy and unplanned development. While the French had some notable success in combating major diseases, the Portuguese and

21 For the economy of this period see Alan McPhee, *The Economic Revolution in British West Africa,* London 1926; *Survey of British Commonwealth Affairs,* Volume II 'Problems of Economic Policy, 1918–1939, Part II' by W. K. Hancock, London, 1942; Margery Perham ed., *Mining and Commerce in Nigeria,* London, 1948; The Leverhulme Trust, *The West African Commission 1938–39; Technical Reports,* London, 1943; *Report of the Commission on the Marketing of West African Cocoa,* London, 1938, Cmd. 5845; Constance Southwark, *The French Colonial Venture,* London 1931; Henri Cosnier, *L'Ouest Africain Français*; Department of Overseas Trade, *Economic Conditions in French West Africa* (1928–1930), Report by H.M. Consul-General in Dakar, London, 1930; Jean Suret-Canale, *L'Afrique Noire: L'ère Coloniale, 1900–1940,* Paris, 1964.

British concentrated a disproportionate amount of available resources on safeguarding the health of European officials and paid only limited attention to the health of the African farmers. Similarly, the colonial regimes left the vital subject of education to a laissez-faire policy. Government involvement was limited to a few schools like the William Ponty of Dakar or the Bo School for sons of chiefs intended to produce clerks, interpreters, chiefs and other agents of colonial administration. The colonial administrations supervised and eventually began to subsidise Christian missions to produce more clerks and interpreters needed by the commercial firms or for the grades of staff in the administrative and technical services that were too expensive to import from Europe. Whenever Africans acquired higher education, it was usually through their own efforts or the encouragement of the Christian missions, and the denial of employment in the colonial administrations was calculated to discourage it. By administrative measure, the colonial regimes were willing to restrict the movement of Christian missions, for example in the Muslim parts of Northern Nigeria, or to limit the participation of missionaries from other than the metropolitan country. But they were unwilling to use administrative measures to prevent the consequent unevenness in the rate of economic, educational and social development.

For all the rationalisations of the colonial officials and all the impressive statements of the *Annual Reports*, *Official Gazettes* and other colonial publications, there is very little evidence of any conscious purpose or plan of the colonial regimes to create a new and improved social and economic order for the Africans. The policy of making Africans bear the cost of colonial administration, public works and services while large European combines enjoyed these services and drained the wealth, meant that colonial development in this period consisted of little more than increasing the exploitation of the resources of the forest areas in palm produce, rubber and timber. The forest areas, therefore, rather fortuitously felt more of the colonial impact. Apart from the expansion of groundnuts in Senegal and groundnuts and cotton in Northern Nigeria, the colonial regimes neglected the Sudan belt which at the end of the nineteenth century had appeared the more prosperous part of West Africa. With the seaward orientation of the colonial economy and the consequent total eclipse of the trans-Sahara trade, the Sudan lost its traditional role as the entrepôt of West Africa to the coastal harbours and new capital cities. In the long run, this uneven development was perhaps the most important legacy of the colonial period.

AFRICAN RESPONSE AND INITIATIVES

The West Africans were themselves the first to accept the reality of the colonial situation. It is clearer now in retrospect that, even in the heyday of colonialism in the inter-war years when the threat of revolts and rebellions had largely receded, the basis of the power of the colonial regimes remained shaky and they continued to fear uprisings. They were usually careful to avoid pressing unpopular measures unless they had first taken the precaution of assembling enough force to deal with protests and violent demonstrations. Yet, at the time, not even the most optimistic nationalists could guess that European political control would be so short-lived. No one questioned the reality of the conquest. The experience of some West Africans during the First World War had shown that the Europeans were human after all and not a race of gods but no one questioned their technological superiority and their ability to repress rebellion sooner or later. And no one, before the Second World War, seriously considered that force might successfully be used to overturn their regimes.

Yet, no African people were willing to entrust their fate permanently to the European ruler. He seemed to represent some powerful, unthinking irrational force that argued little and interfered much in matters it did not understand or try to understand. Faced with subjection to such a force, different West African peoples adopted different strategies to ensure their own survival as peoples and in relation to their neighbours. Many peoples, especially if they were outside the cash crop areas, and not close to mines or railway lines, found that they could get away with the minimum of contact. They paid their taxes, tolerated the officially recognised chiefs and courts, but very much went their own way. Others who could not avoid the colonial impact had to seek more positive approaches. Some tried to come to terms with the colonial power and outdo their neighbours in exploiting the regime's ignorance of the region's history in order to improve their own relative status and position. Others sought progress by seeming to cooperate with the colonial powers, by pursuing Christianity and Western education or organising cash crop production, while at the same time taking care to preserve their own identities. These responses and initiatives of the African peoples as ethnic, linguistic and historical units have so far received little scholarly attention other than from some social anthropologists. Most historians have tended to concentrate only on the policies of the colonial regimes and the direct reactions of the traditional chiefs and the rising educated élite. And yet no clear picture of the colonial situation, or the impact of colonialism on West Africa, can emerge without some understanding of these African attitudes.

Perhaps the most immediate problem posed to an African people by colonial conquest and the loss of sovereignty was the question of the most effective unit of organisation in the colonial situation. Given the variety of levels of political organisation in pre-colonial times—villages, village groups, city states, kingdoms, confederations, empires—and the different stages of consolidation of these groups at the time of conquest, the question of the most effective level of organisation became complicated through the apparent vagaries of colonial policy, breaking up existing larger units and forcing smaller units willy-nilly to consolidate. Indeed, the colonial policies seemed in a sense an attempt to create a chaotic situation in which previous historical development mattered less than the changing whims of the political officer in determining the units of local government and the political relationship between different peoples. That chaos did not result was due largely to the constant pressure of the African peoples to assert their wishes and establish some continuity with historical developments. The story is yet to be fully told, but behind the many little struggles against arbitrary amalgamation or the breaking up of historical states—through petitions, litigation, arguments before Tribunals and Commissions of Inquiries and sometimes uprising—was the effort of African peoples to retain some control over their own destinies even under colonialism. It is remarkable how people who in most cases accepted the permanence of colonial boundaries so frequently forced the boundaries of the units of local government to change.[22] However much pre-colonial history was invoked, these struggles were not intended merely to re-create the pre-colonial situation. It was rather that different peoples were seeking to find new units of organisation that would best serve their own interest in the colonial situation by exploiting tendencies shown by the colonial powers, or sometimes by moving against those tendencies. Indeed, levels of organisation unknown in pre-colonial times sometimes came to be preferred, as when the Ewe, Igbo and several peoples in the Cameroons discovered new national consciousness.

At issue in the contest between the consciousness of African peoples and the will of the different colonial regimes was not merely the question of the best unit of organisation, but also the nature of the organisation itself. This, too, was a matter in which the colonial administration had a large stake. Just as they interfered with the units, so they tried to dictate who were to be the new political leaders and what powers they were to wield. Often, as we have seen, colonial powers found it convenient to recognise traditional leaders of the people as their agents of administration. Where, through policy or plain ignorance, the colonial powers insisted on persons without traditional status, African peoples often continued to recognise their traditional leaders and support them through customary dues paid clandestinely in defiance of colonial rulers.[23] They often went behind colonial courts to take their complaints and civil suits to the traditional courts, whose sense of justice they understood better. Sometimes, because colonial police institutions were remote, they continued to operate secret societies which detected and punished crime as understood in pre-colonial society. It was only gradually as colonial power became more effective, or colonial

22 A good example are the Bussawa of Borgu. See Michael Crowder, *Revolt in Bussa: A Study in British 'Native Administration' in Nigerian Borgu*, London, 1973.
23 Delavignette *Les Vrais Chefs de l'Empire*; A. I. Asiwaju, *The Impact of French and British Administrations on Western Yorubaland, 1889–1945*, Ibadan Ph.D. Thesis, 1971.

institutions became more responsive to people's needs, that these traditional law enforcement agencies fell into disuse.

Next to the question of the unit and the system of organisation was that of land tenure. Nothing threatened the African sense of identity and hope of well-being so much as a threat to his land. Colonial policies seeking to invest all land in the control of the metropolitan government, or to set apart uncultivated land as forest reserves or turn them over as concessions to European planters or concessionaires were resisted as fiercely as possible. With such pressures from the colonial regimes, along with increasing urbanisation, expansion of cash crop production, voluntary and enforced migration, the struggle to hold on to one's land and to seek to lay claims to more became for the majority of African peoples a most important feature of the colonial situation.

While most West Africans sought their survival within their own communities by holding on to their land and seeking a preferred unit and system of organisation, others were willing to try their fortunes further afield. The colonial era speeded up pre-colonial patterns of migration and encouraged new ones. This came about not only through the peace the conquerors established by proscribing war among their subjects or through their taxation policies and labour codes. To an extent not always appreciated, many people escaped unpopular measures by fleeing from the rule of one European power to another. The colonial regimes tried to be neighbourly with each other and discourage this but they could never prevent it. Indeed, it may be said that the Africans tolerated the colonial boundaries so well because they were not operated as lines of human divide. This was particularly so for the masses of unknown farmers who moved to and fro, to maintain traditional ties or search for new opportunities by simply walking across the boundaries carrying their belongings on their heads. The boundaries meant more to the élite, who spoke European languages, travelled in lorries or cars and were more conscious of opportunities in the metropolitan capitals than of the nearest markets or the natural outlets for one's goods across the colonial boundary.[24]

Even among those who migrated the communal bond remained strong, either in the sense that they migrated in groups and invoked the communal ties to ease the problems of adjustment in their new homes, or in the sense that migration was only seasonal and the intention to return to their original homes and cultural background was always taken for granted. Thus the problems raised for the West Africans by the colonial situation were not only political and economic. Ultimately involved was a cultural confrontation.

The European ruler was more nervous than most Africans realised at the time, and he did not rely on the superiority of his guns alone. As Lugard

24 *Ibid.*, Chapter IX: Western Education and the Rise of the Educated Elite.

preached, if a single white man was to assert rule over thousands of Africans he needed confidence and a sense of racial superiority more than technical competence.[25] Faced with this attitude, the West African came to rely more and more on his cultural traditions as the most effective armour against the arrogance and nihilism of the European. Some European rulers may have held the mistaken belief that the African was without history or culture and was waiting to be moulded at will by any European innovator. Yet it was precisely his sense of history and the consciousness of his cultural heritage that the West African had to rely upon in his struggle for survival and identity. The best commentary on West African attitudes to colonialism is to be found in the cultural and particularly the religious field.

The colonial regimes maintained an ambivalent attitude towards Christianity in West Africa. Even if some colonial officials were personally non-believers, the cultures of Britain, France and Portugal were officially Christian and the Christian churches were regarded as propagators of European culture. There was little disagreement on that. The ambivalence was to be detected in the extent to which they were anxious to see their African subjects acquire this Christian element of the European culture. Cultural assimilation was said to be the ultimate goal that the Portuguese held out for the Africans, for without it the African was not worthy of the Lusitanian community of which he was declared an involuntary member. To further the cause of assimilation, state subsidy, as we have noted, was restored to Portuguese Catholic missions. Yet, while many officials declared themselves pleased that a few Africans became converts to Catholicism, there was no great hurry to expand missionary activities. Up to 1950, African Christians in Portuguese Guinea numbered fewer than two per cent of the population.[26] It appears that illiterate Christians were welcome because they were thought to be generally loyal to the regimes. But the loyalty of Christians who sought Western education for themselves or their children as the only access not only to the privileges of citizenship but also to the better jobs *ipso facto* became suspect. Missionaries were therefore urged to show some respect for African culture and not hurry the pace of cultural assimilation unduly. The French adopted a similar attitude in restricting the work of missionaries, except that they did not share the added contradiction of offering state subsidy at the same time. The British were not as restrictive, except in predominantly Muslim areas. But they were equally critical of the supposedly adverse effects of missionary work not only on traditional African culture but also on the loyalty of Africans to the colonial regime. It was a firm British attitude that cultural assimilation was to be derided if not condemned, and certainly not applauded.

25 F. D. Lugard, *The Dual Mandate in British Tropical Africa,* London, 1922 (1965 ed., pp. 58–59).
26 Lord Hailey, *An African Survey,* op. cit., p. 238.

Yet the fact that Christianity was the religion of the rulers, and even more that it provided access to Western education, aided its rapid expansion during the colonial era. Many peoples who in the pre-colonial era had refused to accept Christianity began to show new interest. Indeed, the significant fact about the expansion of Christianity in this period was the extent to which it was due to the initiative of African evangelists and the patronage of communities demanding teachers and schools, rather than to any plans of the missions or colonial governments. This in turn meant that West African Christianity was never what the Christian missions or colonial governments wished. The independence movement among churches demanding that control be vested in African lay or clerical leaders remained strong. Such churches were trying to incorporate aspects of African ideas of worship into their liturgies and to show more tolerance for African social institutions like polygamy and the extended family. Other sects arose under African prophets who were anxious to relate Christianity to current West African beliefs in the existence of witches and other spirits which the Christian missions tended to ignore. They offered Christian prayers for the whole range of problems that plagued people in the villages and for which traditional diviners had offered assistance in the form of sacrifices to various gods. In addition, the new churches also showed concern and offered prayers to enable people to cope with the myriad problems of life in the cities—employment and promotion opportunities, good business for the traders, avoidance of accidents for the travellers, etc. In short, the new sects recognised the continued relevance to their followers of traditional beliefs and social practices and were helping them to make adjustments to the new urban situations. Even among those Christians who remained within the churches controlled by the missions, these traditional beliefs and practices died hard. Ethical and aesthetic values continued to be shaped to a large extent more by the traditional values of the old religions than by Christianity.[27]

For all the rapid spread of Christianity, the nagging fact that embarrassed all Christian missions was that Islam was spreading even more rapidly. Many of them sought to explain this by arguing that Islam was an easy religion, making few demands on the convert and being more willing than Christianity to become syncretic and make compromises with traditional African beliefs. They were loath to accept the unpalatable truth that many Africans in the colonial period embraced Islam almost as a form of protest against colonialism, because it offered the wider world-view that Christianity offered but without the indignity of assimilation to the colonial master's culture. Of course, Islam had other advantages. It had been in Africa since the seventh century. The preachers of Islam were

27 See J. F. A. Ajayi, *Christian Missions in Nigeria, 1841–1891: The Making of an Educated Elite*, London 1965; James Bertin Webster, *The African Churches among the Yoruba 1888–1922*, Oxford 1964.

Africans who understood the spiritual needs and social problems of their converts, and no doubt this made it easier for the converts to make the transition from traditional African religions to Islam. Islam had made significant advances in West Africa in the nineteenth century; it had consolidated its position in the Western Sudan, and had begun to advance into the forest areas of Yorubaland, Ashanti and the Senegambia. The colonial regimes quickly recognised that Islam could not be dislodged from this powerful position. They were therefore willing to encourage the spread of Islam provided they could wean Muslim leaders from their initial hostilities into supporting the colonial regimes. They succeeded in this to a large extent, by cutting West African Muslims away from the Islamic centres of the Middle East, by suppressing such universalistic movements as Mahdism, by breaking up the larger Caliphates into smaller, more localised emirates, and by encouraging the Sufi orders provided they limited their organisations effectively to the local level. Many officials who had no doubts regarding the superiority of Christianity over Islam as a way of life were more willing to see the African embrace Islam under those conditions than to see him claim to become heir to European Christianity and culture.

For many West African peoples, particularly those in the Sudan belt, for whom Islam had become a way of life and their cultural heritage, the pressures of the colonial period provided added incentive to hold fast to their religion, even at the risk of shutting out many useful aspects of the invading cultures from Europe. In spite of colonial restrictions imposed on large universalistic and interterritorial organisations considerable movement of ideas went on through simple traders and other channels the colonial regimes could never effectively control. Nor could contacts with the Middle East be controlled as long as even a few Muslims continued to make the pilgrimage to Mecca. Moreover, these centres of Islam in the Sudan exploited the tendencies of the colonial regimes in order to spread their religion among peoples who had hitherto remained impervious to their preachings. It was in these areas where Islam was in direct competition with Christianity that Islam seems to have been preferred. Both religions offered literacy, though literacy in European languages seemed to offer the more immediate relevance. However, many coastal Muslims, partly through the effort of colonial administrations and partly through their own initiative, were able to combine literacy in European languages with Arabic. Both religions appealed to the urban trading communities, Islam more to those interested in the internal chain of markets, Christianity more to those focusing attention on the import–export trade. However, the subordination of Africans within the European orbit increased the attraction of Islam for people who were concerned about their self-respect and self-esteem. Sometimes there were elements of local politics involved: one community would go Christian and their rivals would go Muslim; the Creoles of

Freetown held on to Christianity as their lifeline, while the Protectorate people who began to invade the town preferred Islam[28]; the foreigners in Lagos were Christian and the indigenous community chose to embrace Islam instead. Yet it is remarkable how the majority of the population of Dakar, Freetown, Lagos—those great centres of Christian and colonial activities—entered the colonial period as Christian and emerged as Muslim.[29]

With the rapid expansion of both Christianity and Islam it appeared that the traditional African religions were in danger of imminent extinction. The very act of conquest and loss of sovereignty seemed to have shaken many people's confidence in the old gods and the efficacy of their role in human affairs. The families charged with the rituals of the gods continued to hold their beliefs and warn of the consequences of neglecting the various festivals. But even among those who embraced no other religions, fewer and fewer people in the larger towns were willing to identify themselves as 'pagans'. But this was only a temporary loss of nerve. The vitality of the old religions was scarcely impaired in the villages. The whole cycle of religious festivals, daily prayers and sacrifices went on. Soon, some of the festivals began to move into the cities and the Christian-educated élite who had been taught to be ashamed of them began to rationalise that they were cultural, not religious, ceremonies which every patriot should preserve. The dances and the ceremonies were being increasingly divorced from the religions that produced them. The art that the religions had once patronised also survived and was being revived. The same was true for the oral literature and now it began to be written down. With varying degrees of religious conviction, the oracles and other methods of divination continued to be patronised. Times of stress, such as prolonged illness or unemployment, were apt to reveal that the hearts of many who had formally adopted Christianity or Islam continued to harbour many of the traditional beliefs. Indeed, on the eve of the Second World War, when so much had changed and was changing in West Africa, the remarkable thing was that so much more had been barely touched by European influence. It is easier now in retrospect to see that the impact of colonialism on West Africa has been exaggerated.[30]

28 Edward Fashole Luke, 'Christianity and Islam' in *Freetown: A Symposium*, eds. Christopher Fyfe and Eldred Jones, Freetown, 1968.
29 *Ibid.*
30 See J. F. A. Ajayi 'Colonialism: An Episode' in *Colonialism in Africa, 1870–1960* Vol. I. eds. L. H. Gann and Peter Duignan, Cambridge, 1969.

CHAPTER 15

African political activity in French West Africa, 1900–1940

G. WESLEY JOHNSON

At first glance, the history of French West Africa during the period of full colonial occupation is that of French administrative preponderance in eight colonies of West Africa.[1] The process of conquest, pacification and colonial governance has been studied by French scholars.[2] But the other side of the question, the reactions of Africans and their response—whether covert resistance, overt resistance or political activity—has yet to be seriously investigated.[3] Contrary to popular belief, there was officially sanctioned political participation for Africans. This was restricted to the Four Communes of Senegal, older cities where Africans had been subjected to French assimilation policy over several centuries. Elsewhere, the denial of participation in government or local politics meant that Africans were subjects of imperial France and without recourse. There were a number of individuals, some from the traditional sector but most from the new urban élite, who contested this assumption and as more research is done on France's eight former colonies the range of this effort to oppose the French will become clearer.[4] Newly opened records and a fresh perspective indicate that all was not tranquil during the first forty years of the century, as many official accounts would suggest.

There are three main themes to be examined in this chapter. First, isolated incidents of resistance to French colonisation which continued through the

1 Senegal was the oldest colony, dating from the seventeenth century; the other colonies were acquired during the scramble for Africa: Mauritania, Soudan (now Mali), Niger, Dahomey, Ivory Coast, Upper Volta and Guinea.

2 One of the best examples of the quasi-official view is Camille Guy, *L'Afrique Occidentale Française*, Paris, 1929; also see the works of Georges Hardy, Robert Delavignette and Maurice Delafosse, especially Delafosse's *Afrique Occidentale Française*, Paris, 1931.

3 Here the work of Jean Suret-Canale, despite its inflexible Marxist interpretation, must be acclaimed, *L'Afrique noire, l'ère coloniale, 1900–1945*, Paris, 1964.

4 The work of Professor Terence Ranger and his colleagues for East Africa has not yet been duplicated in West Africa—that is, studying various manifestations of African resistance to the colonial regime. The study of Yves Person on Samori serves as a model for West African resistance studies. (See I.F.A.N. publications, Dakar.)

period. What the colonialists called 'pacification' was a necessity in French West Africa because traditional peoples from Mauritania to Dahomey refused to acknowledge French suzerainty. Only by the 1920s could the French say with assurance that they had gained political control over the vast land area which comprised French West Africa.

Second, African political participation in the colony of Senegal. In the urban coastal areas Africans had received the right to vote in 1848 but did not elect an African candidate until 1914. By 1920 the Africans gained control of local politics and for the next twenty years carried on political dialogue with the French; but they were unsuccessful in their bid to enfranchise their neighbours in the interior protectorate. Politics in a controlled situation meant frustrations and limitations, but Africans organised political parties, gave speeches, held offices and gained valuable experience in Western-style politics.

Third, the refusal of the French to extend local politics to the other colonies caused a number of Africans to write and act against the colonial regime. Some radicals congregated in Paris to question France's activities in Black Africa and to call for emancipation from colonial exploitation; others took their chances and worked in such colonies as Dahomey and the Ivory Coast. And towards the end of the period the young élite turned from seeking political assimilation to confronting the problem: should they continue to be culturally assimilated by the French?

CONTINUED RESISTANCE TO FRENCH RULE

Although the French established a Government General at Dakar in 1902 to administer their holdings in West Africa, the 'era of pacification of the natives' did not draw to a close until 1920.[5] Africans in territories such as Mauritania and the Ivory Coast kept isolated resistance movements alive until the French were preoccupied with the First World War and then intensified them. Soudan and Dahomey were the scenes of violent uprisings during the war—the full story of the extent of these uprisings must wait until the governmental and military archives have been thoroughly explored.[6] One of the most alarming actions to Africans was the French desire to recruit large numbers of men to serve in the French Army during the First World War. General Mangin had visited French West Africa before 1914 and popularised the idea that it was a giant reservoir of manpower for the French military.[7] Recruiting began in 1912 and was stepped up during the war until 1917, when Joost Van Vollenhoven became Governor-General in Dakar. He resolutely stood against recruiting and wrote to the Minister of Colonies that 'The recruiting

5 The Government-General originally was established in Saint Louis in 1895.
6 Suret-Canale has touched upon some of the popular uprisings but detailed work must await archival research.
7 General Charles Mangin, *La force noire*, Paris, 1910.

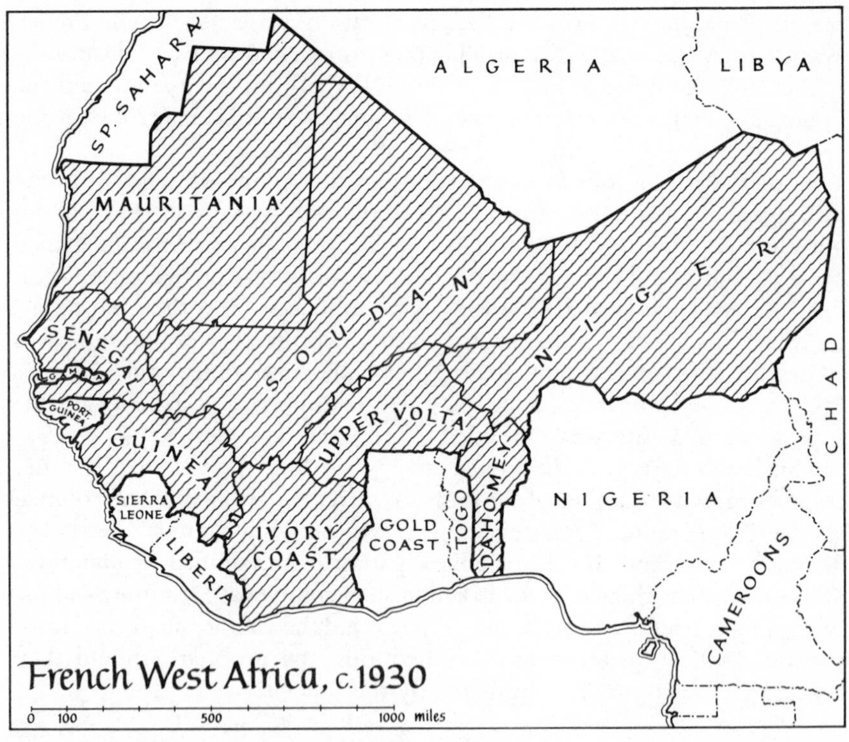

15:1

operations which have taken place from 1914 to 1917 in French West Africa have been excessive both in results and method'. He argued that the Africans were largely indifferent to France's problems in Europe, and that many peoples had revolted rather than serve. Van Vollenhoven reviewed the various areas where major rebellions had broken out: thirteen administrative districts in the Soudan; several in Niger; five administrative districts and seven major cities in Dahomey; an undetermined number of districts in Senegal and Ivory Coast. At least fifty thousand Ivoiriens had fled to Gold Coast and Sierra Leone and many Senegalese had escaped to Gambia and Portuguese Guinea. Despite the Governor-General's warnings, recruiting was taken up again and further instances of resistance were recorded.[8]

After the war, the returning veterans added to the climate of insubordination which caused the French to tighten up colonial rule after 1920. For example, the Governor of Guinea in 1919 complained that returning soldiers were stirring up the local population and inciting them to insub-

8 Archives de la Ministère de la France d'Outre-Mer (FOM); AOF 533-2, Van Vollenhoven to Minister of Colonies, Report for July 1917.

ordination towards their chiefs. Sixty soldiers were sent to put down disturbances and twelve former soldiers were arrested.[9]

The cases of two individuals who did not overtly resist the colonial regime may be examined briefly here to suggest the response of Africans to the dislocation brought about by the imposition of French rule. First was Amadou Bamba, a Senegalese marabout who founded his own order of Islam, the Muridiyya (commonly called the Mourides). Bamba had lived with Lat Dior, one of Senegal's last indigenous chiefs to resist the French. After Dior's death in battle in 1885 Bamba preached a new message of Islam which emphasised work and labour as paramount virtues for the talibé, or follower. Tens of thousands of pilgrims flocked to Bamba's home in central Senegal to see one whom they considered a living saint; the French became alarmed, suspected Bamba of trying to create a theocratic state in their midst and exiled him first to Gabon, then to Mauritania. Finally Bamba was allowed to return to Senegal but to live under house arrest until his death in 1927. Bamba established dozens of new communities in Senegal at a time when the traditional order was in chaos because of the French invasion; many Africans who no longer had traditional chiefs flocked to the Mouride villages where they were exorted to work for their salvation, plant and harvest groundnuts and let Bamba pray for their souls. While Bamba never openly challenged French rule, and during his declining years gave it implicit support, his stance inspired resistance to the French on the part of his followers. He provided a viable alternative to French rule for thousands who were displaced as a result of the French occupation.[10]

Another religious figure, the Liberian William Wade Harris, made an impact upon the people of the Ivory Coast during the First World War. Harris was a Methodist preacher who called upon indigenous populations to burn their fetishes and be converted to an evangelical Christianity. Harris was an older man who dressed in a white turban and a simple garment and carried a large cross in his proselytising effort across Ivory Coast. He was responsible for the baptism of an estimated one hundred thousand Ivoiriens. In Liberia, Harris was not well known; why was he so enthusiastically received in Ivory Coast? One reason is the fact that Ivory Coast's traditional village life, like Senegal's, had been disrupted and reorganised by the French administration. Harris appeared in the midst of chaos and preached a comprehensible message of salvation for troubled and displaced souls. Harris was not trying to start a separatist church, but the French authorities became alarmed and deported Harris to Liberia. Harris died there in obscurity a decade later, but his appearance in the Ivory

9 From AOF 534-10, Governor-General Gabriel Angoulvant to Minister of Colonies, letter of 16 May 1919.

10 See Cheikh Tidiane Sy, *La confrérie sénégalaise des Mourides*, Paris, 1969, for a sociological study which is also rich in historical material. See also Donal Cruise O'Brien, *The Mourides of Senegal*, Oxford, 1971.

Coast led to intensive missionary work and the eventual establishment of the Harris Church.[11]

Harris and Bamba, Christian and Muslim, were both suspected by the French government of separatist tendencies; both denied this. They were threats to French hegemony, however, because they kept the spirit of African initiative alive and gave a will to suffer and resist domination. Elsewhere, especially in Senegal, there were incidents during these years which, if taken together, form a pattern which also suggests a continuing African resistance to colonial rule. There were few mass movements (save for the open rebellions mentioned above when the French, during the First World War, had withdrawn most of their troops to France). Some of these incidents will now be considered.

In 1916 more than thirteen thousand firearms were confiscated from Africans in rural Senegal, convincing the French of the high potential for violence which existed among the Senegalese.[12] In 1920 the employees of the Dakar–Saint Louis railway went on strike and were repressed by French soldiers. The military attempted to run the railway but failed and finally the Governor-General in Dakar met the workers' demands for increased salary and more reasonable hours of work.[13] In Dahomey the famous Porto Novo riots of 1923 caused repression for a decade.[14] Also in 1923 there were serious incidents of unrest on the construction line of the Dakar–Bamako railway. Many Africans from Upper Volta were used in the construction and one of their administrators, Taillebourg, complained constantly to the Governor-General that his subjects were being exploited and mistreated, some being kept from their families for over twenty months. Taillebourg was eventually sacked by the authorities and the affair was hushed up.[15] In Dakar in 1927, telegraph and postal employees struck and would not return to work until their demands were met. Galandou Diouf was called in to negotiate a settlement which the African workers felt did far too little to close the gap between their conditions and those of French workers.[16] These strikes and walkouts, which continued in the 1930s, are significant because in the colonies it was forbidden until 1937 to strike or organise unions—yet the Africans got around this by organising clubs or *amicales*. And they were not afraid to strike.

POLITICAL PARTICIPATION IN SENEGAL

In the colony of Senegal inhabitants of the Four Communes (Dakar, Gorée,

11 I am indebted here to John Zarwan, 'William Wade Harris', an unpublished Stanford senior honours thesis, 1970.
12 Archives de la République du Sénégal (ARS), 2-G-17-5, Report for 1916.
13 ARS, 21-G-134-108, Report on strike.
14 John Ballard, 'The Porto Novo Incidents of 1923', *Odu*, II, 1965, pp. 52–75.
15 ARS 21-G-134-108, Report of Incidents on the Thiès-Kayes Railway, 1923.
16 ARS 21-G-120-108, Report of 7 October 1927. Also see Archives, Dakar, for materials on the 1938 Thiès railway strike.

Rufisque and Saint Louis) had been able to vote and participate in French elections since 1848.[17] Candidates for deputy to the French parliament, town councils, and the *Conseil Général* (the local legislative council) were either French or mulattoes. Few Africans attempted to become candidates until Blaise Diagne, an educated customs official, was successful in becoming Senegal's first black African deputy in 1914. Until then, local political questions were settled by French businessmen and traders (usually agents of Bordeaux firms, which monopolised Senegal's trade) and members of the small mulatto oligarchy of Saint Louis and Gorée. These groups had been successful during the nineteenth century in persuading the French government to set up local government for their benefit so that they would not be subject to total administrative rule. But by 1914 it was apparent that the Africans, who made up about 95 per cent of the electorate, now wanted to seek office in local politics and participate in representative government.

When the First World War was over, Blaise Diagne returned to Senegal from France and led his followers to an impressive victory at the polls in the Four Communes in 1919–20. He was re-elected deputy by an overwhelming majority, forcing his mulatto opponent, François Carpot (the deputy previous to 1914), to retire from politics. Diagne organised one of Africa's first modern political parties, the Republican-Socialists, and brought French electoral campaign tactics to the colony.[18] The Republican-Socialists were opposed by the Bordeaux agents and mulattoes, who fought for political survival. The campaign became bitter; Diagne was accused again (as in 1914) of causing racial strife by his emphasis on 'Senegal for the Senegalese', and 'every African deserves his place under the sun'. Diagne and the Africans, buttressed by returning Senegalese veterans who had fought in France, North Africa and the Balkans, told the electorate that it was now time for Senegalese to run their own local affairs. The African voters responded by giving the Republican-Socialists victory in the town councils of the Four Communes and in the Conseil Général, which ceased to be a preserve for mulatto special interests.[19]

The year 1920 marked the end of the political influence of the mulatto families. Often called Creoles, because they had largely assimilated French culture, religion and style of life, the mulattoes had dominated the Conseil Général, which had certain privileges over the local colonial budget and was often a bulwark against the personal whim of the colonial governor in Saint Louis. The Devès, once the most powerful of the mulatto merchant families, had seen one son, Justin, become mayor of Saint Louis and President of the Conseil Général. Another family head, Louis Descemet,

17 In 1848 only Saint Louis and Gorée participated in elections; the other two communes were given voting rights after the creation of the Third Republic.

18 G. Wesley Johnson, 'Political Parties of Senegal, 1900–1940', unpublished paper given at the African Studies Association, Philadelphia, 1965.

19 ARS 20-G-70-23, Election of 1919 Report.

had also served in both offices; his allies, the Carpots, had served as deputy and council members. But after 1920 only Louis Guillabert, the son-in-law of Descemet, survived the African avalanche—by joining Diagne. The Creoles had made the mistake of allying themselves too closely with the French so that when the African political revolution of 1914–19 came they were compromised.[20]

Local Frenchmen did not have the same continuity in politics, since they served as agents for a few years and returned to France. But during the inter-war period, 1920–40, there were several French liberals who managed to remain in local politics by allying themselves with Diagne or his successor Galandou Diouf.

The first successful African in Senegal's Four Communes politics was Galandou Diouf, who was elected to the Conseil Général in 1909. A Wolof born in Saint Louis, Diouf had practised a variety of trades before finding his penchant for politics. He was the first African to speak up in the council and criticise French policy; he was openly opposed to the attempts of the French administration to disenfranchise the African voters in the Communes between 1900 and 1915.[21] Diouf was a founding member of the Young Senegalese, a political club of young Saint Louis élites, which after 1912 put forward a radical political point of view: they wanted political participation, equal salary for equal work (Europeans were paid higher salaries), better educational opportunities (the scholarships to France were reserved for the Creoles) and equality with Frenchmen. They sought political assimilation with France but did not seek cultural assimilation—an alternative considered only in the 1930s. Diouf was one of their leaders and helped enlist their backing for Blaise Diagne, who returned to Senegal and politics after an absence of more than twenty years in 1914.[22]

During the war both Diouf and Diagne had fought for the Senegalese to have citizenship. To clarify for once and all that Four Communes Africans were legal voters, the French Parliament passed the Diagne laws of 1915 and 1916, which guaranteed Africans French citizenship but allowed them to keep traditional law for marriages, inheritances and family affairs.[23] Diagne and Diouf were hopeful that after the war voting rights could be given to Senegalese of the interior, but this was blocked in the Chamber on several occasions. In fact, the high point of assimilation came with these wartime laws; in return for citizenship the Africans of the Four Communes became liable for military service. More than eight thousand served in the French Army while more than a hundred and fifty

20 For a detailed analysis of the mulatto party in local Senegalese politics, see H. O. Idowu, 'The Conseil Général in Senegal, 1879–1920', unpublished Ph.D. thesis, University of Ibadan, 1966.

21 Michael Crowder, *Senegal: A Study of French Assimilation Policy*, London, 1967 (revised edition), pp. 25–7.

22 ARS 13-G-17/18-17, 'Notice concernant M. Galandou Diouf'.

23 The law of 19 October 1915 and the law of 29 September 1916.

thousand other black Africans served in the French colonial forces during the war. These Africans, from rural Senegal and the other colonies of French West Africa, were envious of the special privileges of the Four Communes; the Four Communes stood for political élitism and assimilation from 1920 until 1945, when France finally gave other colonies political offices.

The French colonial establishment, however, was alarmed by the 1919–20 victories of the Africans in Senegal, and feared the spread of *diagnisme* (which was equated with separatism) to other colonies. Moreover, the Bolshevist regime in Russia now brought forth anti-colonial propaganda which colonial officials thought would corrupt Africans. A new era of tighter governmental controls—creation of secret police, censorship of mail, banning of newspapers from France—affected all of French West Africa.[24] The inter-war years became oppressive for Senegalese and non-citizens alike until 1936 and the Popular Front movement.

Politics in the Communes was dominated by Diagne as deputy until his death in 1934, when Diouf succeeded him in Parliament; he died in 1941. It would be an error to suppose that Senegal was subject to 'boss' rule for these two decades. 'Politics of personal rule' is accurate but still might be misleading. Diagne's political machine was well organised while Diouf's was loosely structured; Diagne was personally domineering while Diouf was personally relaxed. The two dominated the politics of Senegal from 1914 to 1940, but they were surrounded by lieutenants, advisers and opposition politicians which made Four Communes politics more than 'rubber stamp' activities for the French administration. The era of Diagne will be considered first, and then the era of Diouf.

After the consolidation of his party's power in 1919–20, Diagne felt secure and outspoken. In France he talked about Senegalese potential in self-government, opposed the Union Coloniale (a merchants' organisation which favoured tight economic and political control), and gained a reputation as an independent in Parliament. Diagne had recruited African troops for Clemenceau's government during the war and now tried to trade on this political capital for concessions to the colonies.[25] But a new Minister of Colonies, Albert Sarraut, and a new Governor-General of French West Africa, Martial Merlin, were appointed in 1919 and represented a shift to conservative colonial policy which predominated in Black Africa down to 1936. The French became fearful of *diagnisme* and what it might ultimately do to the colonies. Diagne retained his title as High Commissioner of African Troops for several years but his power was slowly eroded after the end of the wartime emergency.

24 This was initiated by Governor-General Merlin and continued by Governors-General Jules Carde and Jules Brévié but finally reversed by Jules Marcel de Coppet after 1936.
25 Diagne refused the Legion of Honour for his successful recruiting, but worked through the Minister of Colonies, Henry Simon, for concessions to French West Africa.

In Senegal, Diagne's followers made important gains in local political affairs, since they now controlled the four town councils and the Conseil Général. Each of the Four Communes had a city hall with some staff positions which were political appointments. These patronage jobs went to Diagnists and helped give stability to the Diagnist political organisation, the Republican-Socialist Party. There was some confusion at first and French officials concluded that many Africans were not adequately trained to take over clerical jobs. Consequently, French staff members remained in the city halls in many important posts although policy was determined by African-led councils. Nevertheless, the Four Communes became the first cities in French colonial Africa to have African mayors, councillors and employees in the urban bureaucracy.[26]

The Conseil Général, which met in Saint Louis in its own palace, was now in African hands although the mulatto President, Louis Guillabert, stayed in office thanks to his political alliance with Diagne. Soon after the election of 1920, Diagne and Governor-General Merlin agreed to reorganise the Council—one of the rare times they were in agreement. Each had his own private reason for wanting the Council changed. Diagne wanted more Senegalese represented and favoured enlarging the Council so that chiefs from the interior could be seated. Merlin feared the growing power of the Africans who were now citizens of the Communes and favoured enlargement as a way of diluting their power—he thought new chiefs could be manipulated. The local populace was divided. To some, it was apparent that non-citizen Senegalese deserved representation. To others, this was simply a plot on Merlin's part to whittle away local powers of control over sections of the budget. Since 1879, the Council had been an effective check on arbitrary powers of the Governor.[27]

The Council was reorganised in 1921 and the sceptics were soon justified: its control of certain budget items was reduced and the administration reserved the right to appoint the chiefs who henceforth composed half the Council.[28] But the chiefs, despite the fact they were in the pay of the French, did not always follow the official line. Some asserted their independence and joined the citizens. Moreover, despite its diminution of power, the Council still remained a powerful voice for local interests in Senegal and continued to serve a valuable function. An American professor, Raymond Buell, visited the Council in the middle 1920s and thought it was the most powerful African assembly in colonial Africa.[29]

Louis Guillabert was retired by Diagne from the Council presidency

26 ARS 20-G-69-23, Report on Municipal Election, Four Communes, 1919.
27 Idowu, chs. VII and VIII.
28 ARS 20-G-74-23, Governor-General Merlin's confidential letter to the Lieutenant-Governor of Senegal, 12 April 1921.
29 Raymond L. Buell, *The Native Problem in Africa*, New York, 1928, i, p. 979.

after several years and replaced by his new first lieutenant, Amadou Duguay-Clédor. For the first ten years in office, Diagne had relied on Galandou Diouf as his most trusted adviser and agent in Senegal, but Diagne's change of policies after 1923 separated him from many of his original supporters.[30] Clédor remained loyal and was named party chief of the Republican-Socialists, editor of the Diagne newspaper, *La France Coloniale*, and president of the Council. Clédor was a journalist by profession and an historian by avocation. He was an ardent defender of the French policy of assimilation. When General Philippe Pétain visited Senegal in 1925, Clédor successfully pressed his demands to be named official African greeter instead of an important African chief. He celebrated the loyalty of the Senegalese to France in his newspaper columns, recalled Senegalese soldiers fighting in the French conquest of West Africa, continually reminded readers of Senegalese contributions to the war against Germany. Clédor believed in the new order of assimilation and presided over the new Colonial Council with the conviction that some day all of Senegal would be assimilated.[31]

Diagne enjoyed the backing of almost all the Africans of the Four Communes in 1919–20 but by 1924 the situation had begun to change. Diagne radically altered his policy of seeking gains for Senegal by opposing the administration if necessary; in 1923 he signed the 'Pact of Bordeaux' and agreed to work with the Bordeaux merchants and the administration for Senegal's future prosperity.[32] Ten years of manipulation and manoeuvre had fatigued Diagne, who now thought cooperation would be more realistic. His younger constituents thought otherwise and interpreted his decision as collaboration. African émigrés and students in Paris and young Senegalese at home in the Four Communes now found a common foe whom they could attack. Direct criticism of French rule was still dangerous, but attacks upon a deputy were fair game.

The most important leader of the growing opposition to Diagne in Senegal was Lamine Guèye, a protégé of Diagne's who had studied law during the war in France and who had won the first law degree to be awarded to a black African. Guèye returned to Senegal in 1922, but soon found Diagne's politics incompatible with his ideas of greater political participation.[33] Guèye met other dissidents in Saint Louis, including

30 After 1923 the Diagne organisation in Senegal was purged of those who would not maintain absolute loyalty to the deputy; he lost many of his original and most dynamic supporters. See ARS 20-G-100-100, Police Reports, legislative elections of 1924; also ARS 17-G-233-108.

31 ARS 3-G-3-108, Report on rift between Clédor and M. M. Gaye; *L'Ouest Africain Français*, 5 March 1927; *L'Ouest Africain Français*, 7 May 1927.

32 ARS 17-G-233-108 describes a number of political meetings Diagne had with party members shortly after negotiating with the Bordeaux merchants.

33 Remarks at a banquet for Lamine Guèye, 16 January 1924, reported in ARS 17-G-233-108; L'A.O.F., 30 April 1925.

Hamet Télémaque Sow, who was to become one of Senegal's leading opposition journalists.[34] Sow and Guèye, however, were young men and lacked the audacity of a younger Diagne; they asked Paul Defferre, a young French lawyer, to be their candidate for deputy. Defferre and his African backers were soundly defeated and Guèye finally understood that, henceforth, Frenchmen were unacceptable as candidates. Guèye entered the municipal elections in Saint Louis in 1925 and won his first electoral victory. He served the next two years as mayor but found it difficult to work for political change when surrounded by Diagnists. Guèye, tired of isolation, tried for a reconciliation with Diagne but failed.[35]

Meanwhile, Diagne faced other defections from his party. Jean d'Oxoby, a liberal French editor, had supported Diagne since 1914 and loyally followed his lead as one of the few French members of the Colonial Council. He broke with Diagne in 1925 and was soon followed by Galandou Diouf, who had been replaced by Clédor as Diagne's chief agent in Senegal. Diouf returned to the world of commerce after suffering defeat in 1925 to the anti-Diagnist forces, losing his job as mayor of Rufisque. Diouf was sought out in 1928, however, by Lamine Guèye, who was now determined to break Diagne's monopoly on power and to open up politics for the younger men. Diouf agreed to stand for deputy although he had fared badly financially; his debts were paid by Cheikh Anta M'Backé, one of the leaders of the Mourides, which had become Senegal's fastest growing Islamic sect.[36]

The election of 1928 put into sharp relief the political situation in Senegal. The 1920s had been a decade of economic growth for Senegal and prosperity had created more citizens anxious to participate in politics and gain access to power, privilege and status. As the French schools grew and expanded, more young men were rising in the system. Not only were members of the political élite anxious for greater participation (those who were citizens of the Four Communes), but the other Senegalese who were in point of law French subjects now took a greater interest in the opportunities local politics offered. Elections from 1928 on in Senegal saw the participation in rallies and campaigns and fund raising of many non-citizens (hence non-voters).[37]

Diouf and Guèye together constituted a formidable opposition to Diagne, who rushed back to Senegal several weeks early to avert political disaster. Diouf conducted a door-to-door campaign emphasising that it was time for the Communes to make a fresh start in politics, since the Diagne regime was atrophied. He criticised Diagne for spending most of

34 ARS 20-G-74-23, Petition of Union Africaine Républicaine, 1 July 1921.
35 *L'A.O.F.*, 30 April 1925; *La France Coloniale*, 23 April 1927.
36 ARS 13-G-17/18-17, Materials on Galandou Diouf for 1928.
37 Télémaque Sow visited many small Senegal River towns to collect campaign funds. See ARS 21-G-133-108.

his time in Paris and rarely returning to Senegal; he objected to new taxes on local goods, which he attributed to Diagne's indifference; he promised that he would help improve Senegal's groundnut industry because he had been a merchant for many years.[38]

Diouf was seconded by Guèye, Télémaque Sow and Kojou Tovalou Quénum. Guèye spoke frequently in Dakar and Saint Louis, reminding voters that Diagne had approved the reorganisation of the Colonial Council, which led to its diminished powers; that Senegal needed cheap public housing, the right to organise labour unions and jury trials for the Assizes Courts.[39] Télémaque Sow argued for the right to strike, the eight hour working day, and the abolition of the chiefs in the Colonial Council 'who are in the pay of the French'. He accused Diagne of not wanting to make contact with the younger Senegalese.[40] Tovalou Quénum was one of the expatriate radicals in Paris who had long opposed Diagne; he was from Dahomey, a member of its élite, and after Lamine Guèye, one of the first Africans trained in the law in Paris. His presence lent a shred of credibility to the main Diagne attack: that Diouf and his entourage were communists.

Diagne himself had often been accused of being a Bolshevist earlier in the 1920s, but by 1928 it was Diagne who raised the spectre of red revolution in Senegal. Quénum, an associate of Lamine Senghor, a Senegalese expatriate in France who had links with the French Communist Party, was denounced by Diagne. The deputy implied that Diouf and his entourage were prepared to sell out to the Bolshevists. Another Diouf supporter, Frenchman Roger Roche, a garage owner in Rufisque, told newsmen, 'We must get rid of the capitalist yoke; forward with the proletariat.'[41] Roche was taken seriously by the police because several weeks before the elections, public assemblies were repressed in Dakar and Rufisque. The African post-office workers, telegraphists and small functionaries in the administration were for Diouf and demonstrated their support in other ways. Moreover, Diouf distributed tracts in Arabic to attract the pious Muslims; this further alarmed the French, who were unusually suspicious of Muslim plots within the colony.

Diagne narrowly won the election on 22 April 1928, but protest immediately arose from the Dioufists. They alleged that in certain suburban voting polls, dead men and absent voters figured on the rolls; in others, troops were kept inside the polling stations, which violated electoral law. Diouf accused the governor of pressuring voters to opt for Diagne and rumours spread through the Communes that the government had

38 ARS 20-G-82-23, Police Reports, Rally for Diouf, Dakar, 9 April 1928.
39 *Ibid.*
40 ARS 20-G-82-23, Police Reports, Rally for Diouf, Saint Louis, 28 March 1928.
41 *Ibid.*

manipulated the elections.[42] Galandou Diouf and Lamine Guèye were convinced electoral fraud had taken place and raised a public subscription, which was easily filled, to sail to France to complain to metropolitan politicians. This was important for later Senegalese politics because, for the first time, Senegalese entered into contact with the S.F.I.O. (Section Française de l'Internationale Ouvrière), or French socialist party, which was sympathetic and interested in the plight of the Senegalese politicians. Lamine Guèye's association with the S.F.I.O. continued down to the 1950s.[43]

Diagne had retained power but his regime was now openly criticised by a number of newsletters and publications manned by young Senegalese determined to keep opposition alive. Télémaque Sow and François M'Baye Salzmann were determined editors who launched several newspapers during the inter-war period, had them closed down, suffered jail terms, but managed to gain freedom and come back to continue the attack. Diouf, discouraged and near bankruptcy after 1928, retreated to a private existence in the face of persecution by the administration, which apparently believed Diagne's charges about Diouf being linked with communists. His unsuccessful trip to Paris, while gaining future friends among Parisian liberals, tarnished his reputation in the eyes of local authorities.[44]

Lamine Guèye, feeling the isolation of remaining in the opposition, made peace with Diagne and was shortly afterwards, in 1931, appointed as magistrate in Réunion and left Senegal for the next few years. Taking advantage of this opportunity to further his legal experience was to cost Guèye support in later years from unforgiving Senegalese who felt he deserted Diouf in the opposition's darkest hour.

Diagne meanwhile remained subject to criticism because of his activities which were increasingly aligned with the government. He supported the renewal of privileges for the Bank of West Africa, for which his opponents alleged that he received a commission. Diagne claimed that more capital was needed in French West African territories and that the Bank was in a position to bring 'equality and economic liberty and well-being for Europeans and natives alike in West Africa'.[45] Diagne, of course, did not reckon with world-wide depression, which he could not have foreseen in 1928, but by the next elections for deputy, in 1932, the situation had changed dramatically.

Senegal was now beginning to feel the effects of depression and economic stagnation. The mood of the times was conveyed in the journals which the anti-Diagne opposition had kept alive and in the activities of the Paris African radicals, described in the next section. In March 1930, Diagne had met the French consortium of merchants in

42 ARS 20-G-82-23, Reports of meetings of 24 and 25 April 1928.
43 ARS 21-G-133-108, Lamine Guèye report on trip to Paris, 21 July 1928.
44 ARS 13-G-17/18-17, Police Report, Dossier Galandou Diouf, 16 September 1929.
45 ARS 20-G-82-23, Cable from Diagne to Clédor, 14 February 1928.

Bordeaux, where he was asked to 'tell the natives we're not responsible for falling prices'.[46] But Senegalese could only perceive that prices paid for groundnuts were steadily going down, while prices asked for manufactured European goods imported into Senegal were steadily going upwards. Housing, food, transportation rose sharply in the colony as the buying power of workers and farmers diminished.

Jean d'Oxoby, editor of the *L'Ouest Africain Français*, an important newspaper in French West Africa, and a former Diagnist, came forward with the idea that overproducing groundnuts would help solve the crisis. Senegal in 1930 produced approximately four hundred thousand tons of groundnuts in the shell; France consumed that year over seven hundred thousand tons. D'Oxoby argued that they needed to produce more to capture part of the balance of three hundred thousand tons now supplied by Nigeria and India. He favoured a stiff tariff on foreign groundnuts and asked for a virtual exclusion of all but Senegalese groundnuts. But d'Oxoby and others, such as Louis Besnard, President of the Rufisque Chamber of Commerce, were not taken seriously by local producers, by Bordeaux, and by the administration.[47] Production went down as the economy ground towards a halt, and the vicious cycle of world market prices wreaked havoc in Senegal. The Governor of Senegal, in his annual report for 1930, tried to sound optimistic: perhaps the crisis would 'help teach the native the virtue of saving'.[48]

In 1931 the Union Coloniale, an association of colonial merchants, came out for free market prices and opposed keeping subsidies, which the Senegalese wanted at all costs to retain. The Union Coloniale was of course speaking to the proposition of keeping prices down in France, but this would leave the Senegalese, with their antiquated and rude facilities, in a less competitive position than Nigeria and India.[49] 1932 became a hellish year for the Senegalese. Groundnut production fell from four hundred thousand to less than two hundred thousand tons as a result of meagre rains, an excess of grasshoppers and reduced demand. Prices tumbled and thousands found themselves out of jobs in the cities and tens of thousands of farmers had no crops to market or no markets for their crops.[50] In the midst of crisis, the voters were called to the polls again to decide whether Diagne and his Republican-Socialists should stay in office or be replaced by a disheartened Galandou Diouf. Lamine Guèye, the idol of the younger Senegalese, was still in Réunion and no younger African had stepped forward to take his place. There were twelve other candidates

46 Report in *L'Ouest Africain Français*, 19 April 1930, of Diagne's meeting with the Syndicat de Défense des Intérêts Sénégalais.
47 *L'Ouest Africain Français*, 5 July 1930.
48 ARS 2-G-30-4, Rapport Politique, 1930.
49 *L'Ouest Africain Français*, 4 July 1931.
50 ARS 21-G-117-100, Cabinet Civil de l'A.O.F., 22 August 1932.

besides Diouf and Diagne, which indicated some lack of confidence in the two old standard bearers. D'Oxoby, Télémaque Sow and Ibrahima Sow (a radical recently returned from Paris) were serious candidates; only Diouf, however, benefited from recognition by a French party, the S.F.I.O., which meant some support from local Frenchmen who were members. As yet, few Africans were affiliated with the S.F.I.O.—one was a member of Diagne's Republican-Socialists, or a follower of Diouf, or an independent.[51]

The thirteen other candidates united in their condemnation of Diagne, some blaming depression, low prices and ruinous markets upon his political policy. Others distributed anti-Diagne material printed in France, which made serious allegations against the deputy, charges which had never been fully aired in the local press because of censorship and lack of contact with France. The radical black community of expatriate West Indians and West Africans denounced Diagne as a 'black Judas' for defending forced labour, still current in most of France's Black African possessions. In June 1930, Diagne had appeared as one of France's representatives to the International Labour Office inquiry into forced labour in Africa. Appearing in Geneva, Diagne supported the use of forced labour for building roads and public projects (but not for private capitalists) and earned the hostility of many European liberals at the conference and radical blacks in Paris. Soon after Diagne became Deputy Minister of Colonies under Pierre Laval, and his alleged defection to the established colonial regime was now proclaimed as absolute. Campaign material circulated in 1932 which warned voters they had many choices in this election, but supporting Diagne would mean 'you will have voted for the ruin of your country and the death of your race'.[52]

Diouf, stretched for funds, finding many of his former supporters now rival candidates and not available to speak for him, issued large posters which told his platform: restoration of an effective Conseil Général, spread of full-government communes to the interior of Senegal, equal access to administrative cadres for Africans, creation of applied agriculture schools, and protection for groundnuts, since bananas, wood and coffee were protected in other colonies.[53] George Barthélemy, a European candidate and former ally of Diagne, promised to help solve the economic crisis afflicting Senegal. He publicised a letter he had received from one of his supporters which suggests the *malaise* in Senegal: 'Our children are educated now, but there are no jobs for them . . . moreover, Africans from neighbouring colonies have increased in the Four Communes and now are

51 ARS 20-G-88-23, Legislative election reports for 1932.
52 ARS 20-G-88-23, Campaign leaflet, 'Appel aux sénégalais: ne votez pas pour Diagne, le Judas noir'. The leaflet was signed by E. Faure, A. Diara and A. Beton, and was printed in Paris.
53 ARS 20-G-88-23, *Profession de foi*, Galandou Diouf.

the cooks and servants which used to be good employment for our wives and children.'[54]

Diagne returned to Senegal with greater fear of losing the election than ever before. He issued bright posters, in garish colours, which contrasted with the sober posters issued regularly since 1914. Diagne told a meeting of four hundred Frenchmen at the Dakar city hall that Senegal's misery was due to the manipulation of international trusts, but that if Senegal could get firm tariff support on groundnuts and preferential treatment, her situation would improve. Diagne claimed that he had not only looked after Senegalese interests, but that he had 'enlarged his mandate' and was concerned about African interests all over French West Africa.[55] He argued that Senegalese interests were increasingly linked with those of other colonies, and that his work in consolidating the position of the Bank of West Africa had helped both Senegal and the other colonies keep a viable currency. Most important of all, Diagne said he was in a position because of his long service in Parliament to help gain the guaranteed subsidy which many now recognised as essential to save Senegal's groundnut crop. This was the deciding factor in the election. The voters, although suspicious of Diagne, could not see how any one of the bewildering array of candidates could command as much political power as the veteran deputy. He was returned to office for the fourth time.

Charges of collaboration with the regime were hurled at Diagne by the angry losers, who said that the French administration would now do anything to keep its puppet in office, especially since he had served in two cabinets under Pierre Laval as the Under Secretary of Colonies. Diagne retreated to Paris to do battle for the groundnut subsidies, in which he was successful. His political organisation was still intact but badly battered; he maintained the support of the veterans of the First World War, most of the Lebou chiefs, and the older African voters. But the younger people, the mulattoes and an increasing number of small French businessmen were looking elsewhere for support. These groups, who now thought of themselves as anti-Diagnist, grew more frustrated after the 1932 victory. New newspapers and newsletters appeared but they were systematically harassed and opposed by the Diagnist press, under Clédor's editorship, and by the colonial regime.[56]

The era of Diagne ended as quickly as it had begun. In the spring of 1934, Diagne visited Africa once again to help dedicate a new monument to the French conquest of the Soudan in Bamako (now Mali). He became ill en

54 ARS 20-G-88-23, Reprint of Agence Havas column by Georges Barthélemy, March 1932.

55 ARS 20-G-88-23, *Profession de foi*, Blaise Diagne.

56 For example: *Le Périscope Africain*, *l'Action Sénégalaise*, *La Sirène Sénégalaise*, *Le Stégomya*, and others.

route, was returned to Dakar and sent to southern France for a rest cure, where he died several weeks later in a small health station in the Pyrenees. The founder of African politics in Senegal and French-speaking Africa was dead; he was mourned not only in Senegal but in other colonies where he was considered as a protector of African interests by much of the population: but since 1923 he had been discredited and opposed by the younger Africans. The question now arose: who would replace Diagne? Who would take over his political organisation?

Insofar as Diagne ever entertained the idea that he might be incapacitated, he undoubtedly thought of his last loyal lieutenant, Duguay-Clédor, as his authorised agent. Clédor headed the party in Senegal, edited the Diagnist paper, and was President of the Colonial Council. But when Diagne died, Clédor's standing fell overnight, and it became evident that his power was based solely on his association with Diagne.[57] When the master was dead, the servant was reduced to nothing. Clédor struggled against this reality but eventually issued a statement inviting his friends and the followers of Diagne to rally round the younger Lamine Guèye, now rumoured to be returning home to contest the succession of Diagne.

The other forces of the opposition now approached Diouf as the candidate who divided them least. He had great appeal for the veterans (since he had fought in the war), the older voters, and the Lebou chiefs, who trusted Diouf. The bulk of Diagnist followers shifted to Diouf, but the Diagnist party apparatus, on the other hand, looked to Lamine Guèye as a younger and more viable candidate and agreed to support him. In fact, Guèye, upon arrival in Senegal, said that he welcomed Diagnists, that he respected the memory of Diagne, but would be his own master with his own party and platform. But there were many voters who had long memories of Guèye and questioned his stability—he had opposed Diagne and become reconciled with him several times, and now returned to oppose his former comrade, Diouf. The younger members of the anti-Diagnist opposition, linked together by a decade of hate for Diagne, were now unclear about what it was that united them. Some remained loyal to Diouf, others favoured Guèye. The bulk of voters preferred Diouf and in 1934 he became Senegal's second African deputy.[58]

Who was going to enact the role of Duguay-Clédor in Senegal for Diouf during his absences in France in Parliament? Diouf had his choice of several candidates for chief lieutenant—his former comrades, Télémaque Sow and François Salzmann, or some of the Diagnists, now released from loyalty to their deceased chief. Diouf's choice created a new situation in Senegalese politics: he chose Alfred Goux, a French small businessman.

57 See announcements about death of Diagne and the question of party continuity in *Franco-Sénégalais*, 17 May 1934.

58 Reports on arrival of Lamine Guèye, 21 and 22 June 1934; ARS 20-G-89-23, Legislative election reports for 1934.

Goux was a strong man in Dakar, where he had wooed the Lebou chiefs and community with promises of seeking retribution for land approriated by the French. Goux had managed a variety of small businesses in Dakar and had lent money to Lebous; he also had influence in the growing community of small French businessmen and Lebanese merchants (who were unable to participate directly in elections because they were not French citizens). Goux rapidly mounted a political machine which had funds and appeared attractive to the French administration. Diouf chose the newcomer Goux as his chief lieutenant rather than a former radical colleague. Goux already had a newspaper to serve as the voice of Dioufism in the colony *(Le Sénégal)*, and he offered bourgeois stability in political management.[59]

But Goux was after all French, and the reappearance on the local scene of a Frenchman in an important political position signalled that Diagne's Africanisation policy had been violated on the morrow of his death. The choice shocked many of Diouf's former radical colleagues, who now opposed Galandou with the same vigour once reserved for Blaise. And Goux, who became Mayor of Dakar, was the forerunner of the events of 1936—the Popular Front—whose net result in Senegal was the bringing of more local Frenchmen into politics and the exclusion of many Africans. The economic depression in France plus a host of assorted political situations produced the Popular Front, France's first alliance of the Communist, Socialist and left-centre parties. The Socialists controlled this coalition under Premier Léon Blum. For the first time, colonial administrators and officials in the African colonies who had liberal sentiments now dared to speak; local Frenchmen from Senegal to Soudan to the Ivory Coast caught the spirit of social reform and openly questioned aspects of colonial rule not previously discussed in public. This new climate in French politics affected all of French West Africa and inspired hope in many Africans that increased political participation as in Senegal might be possible some day.[60]

The situation in Senegal, however, proved to be paradoxical. Whereas in 1934 Goux was the only Frenchman of consequence in local politics, by 1936 the political arena was filled with Frenchmen. None dared challenge Diouf, who became solidly entrenched as a pillar of the colonial regime and accepted the same advantages of parliamentary comfort which had enticed Diagne. But newspapers were now suddenly filled with columns by French writers, who castigated the local Senegalese for not understanding the new issues of the day; several new journals were founded to

59 Goux came to power by intriguing in Dakar's municipal politics, defeating the Diagnist forces which still controlled City Hall, got himself elected Mayor of Dakar, and proved to be an attractive second for Diouf's purposes. See the evolution of municipal politics in 1934–5 in ARS 20-G-91-23.

60 This new spirit was symbolised by the appointment of Jules Marcel de Coppet, an administrator with socialist leanings, as the new Governor-General of French West Africa in 1936.

express Popular Front political rhetoric. The style and content became French and local issues, nurtured since 1914, were pushed into the background by the national issues.[61]

Local African politics did not recede but had their struggle for originality arrested by the Popular Front. The greatest casualty was the Senegalese Socialist Party, founded by Lamine Guèye after his 1934 defeat by Diouf. Guèye spent much of 1935 organising Senegal's first modern political party. The Young Senegalese of 1914 had been a political club; Diagne's Republican-Socialist part of 1919–34 was an unwieldy array of local committees held together by the deputy's prestige and Clédor's watchful eye. But Lamine Guèye hired out large cinemas, printed party cards, designed a party flag, and convinced younger Africans that he was bringing them the latest in political organisation. Lamine Guèye arranged for some of Senegal's first university students studying in France to address the party's annual conference, and one student, Léopold Sédar Senghor, later to become Senegal's first president, thus had his introduction to Senegalese politics.[62]

The first test for Guèye's modern, disciplined party came in the 1936 regular elections for deputy. Guèye faced an entrenched Diouf, supported by Goux, the veterans, loyal Lebou chiefs, and municipal employees of the Four Communes who owed their jobs to patronage. His coalition of young Senegalese, who wanted to move on past the era of personal politics of the old line politicians, realised they would have to go beyond their narrow party basis and try to reach a larger number of voters. Some partisans still called Guèye a 'Bolshevist Socialist' because of the accusations made against him and Diouf in 1928 by Diagne; it was thought by many that Diouf had repented of Socialist leanings while Guèye had remained a staunch leftist. But Guèye's connections to the left were more symbolic than real and his political platform promised little that Diouf was not ready to give. Guèye did favour more scholarships and the right of students to choose the career of their choice (a direct criticism of the French administration which arbitrarily placed the few students it supported). He also favoured creation of a Law School in Dakar to train African barristers, restoration of the General Council and abolition of the Colonial Council, and creation of full municipal privileges for five other Senegalese communities. Guèye did not attempt to push for the vote for all Senegalese of the interior, since Diagne's efforts on this had failed. He sought a compromise for extending the vote to the growing urban areas of the interior.[63]

61 For example *Clarté*, *Le Sénégal*, *Paris–Dakar*, *Le Courrier de l'Ouest Africain* and the venerable *L'A.O.F.*
62 *Clarté*, 25 April 1935; Paul Vidal in *L'A.O.F.*, 8 February 1936; analyses and reports of Senegalese Socialist Party in *L'A.O.F.*, 4 January and 11 January 1936. *L'A.O.F.* was Lamine Guèye's newspaper.
63 See materials on Lamine Guèye in ARS 20-G-92-23.

By the time the voters were ready to go to the polls, a great shift took place among local politicians. Louis Martin, the radical newspaper editor, Amadou N'Doye, the Lebou leader, and other old friends deserted Diouf. On the other hand, Diouf persuaded Duguay-Clédor to back him and to appeal to recalcitrant Diagnists for support. The prestige of office, his appeal as an honest candidate, and his personal magnetism in door-to-door campaigning won Diouf another term as deputy. As in 1928, the vote was close and the defeated candidate, Lamine Guèye, made charges of fraud and left for France to see if the election could be nullified. He thought also that a Parliament with a Popular Front in charge would be sympathetic to his claims, but after two months of protest, Guèye's charges were dropped and Diouf was seated. But Lamine Guèye's efforts were handsomely rewarded later when the Socialist Party called him back to France to serve as a consultant to the Ministry of Colonies on taxation.[64]

In Senegal, however, Guèye's party floundered amidst confusion during the Popular Front and some members deserted to found a Senegalese branch of the French Socialist party, the S.F.I.O. This group in turn was challenged by an exclusively French group, which barred Africans from membership, and which claimed to be the only legitimate S.F.I.O. group in Senegal. Three Socialist parties contesting for influence undermined the unity of the left which Guèye had assiduously built up; these divisions then paved the way for Diouf to remain in absolute control of local politics until the fall of France in 1940.

The last few years of the 1930s witnessed the resurgence of Europeans in politics: Goux continued as Mayor of Dakar and Paul Vidal, a barrister, became Mayor of Saint Louis in 1938. The European-dominated S.F.I.O. group in Dakar was swelled by more Frenchmen coming to the colonies to seek their fortune, but their presence was ambiguous for local politics. On the one hand, most of the arrivals were liberals, who agreed with S.F.I.O.-inspired criticisms of the colonial regime, and helped bring a greater awareness of European class politics, which Senegal had never known. Their influence on younger Senegalese who had never studied in France was therefore great, and one can argue that the political schooling of the politicians of the 1940s and 1950s took place from 1936 to 1940 in the Four Communes. On the other hand, the Europeans pushed many Africans into the background, telling them they knew nothing about how politics should be conducted; they brought in many issues from France which were irrelevant to Senegal and they weakened the Africanisation policy of local politics which Diagne had created and maintained.

Diouf had won stability and prestige by following Diagne's policies of deference to the government and he was courted and flattered by Third Republic politicians. But because he stayed in closer touch with the

64 *Les Annales Coloniales*, 3 January 1938.

Senegalese than had the remote Diagne, Diouf managed to avoid a formalised opposition among the younger Senegalese. (However, if the war had not intervened, the unresponsiveness of Diouf and the local French threat in politics might have led to radicalisation.) The death of Diouf in 1941 during the war cleared the way for 1945, when the deputyship was re-established (Vichy France had abolished local politics) and Lamine Guèye was finally elected.[65]

In retrospect, then, political participation in the Four Communes served mainly as a political initiation for Senegalese; other Africans, for example those studying at the William Ponty school near Dakar, were also affected by local politics—in fact, many of the future leaders of post-1945 West African colonies and states were introduced to politics in this setting. Compared with the other French African colonies, Senegal presented a different picture during the first four decades of the twentieth century. In one sense, politics in the Four Communes was conducted in an artificial vacuum, since the French held ultimate control. But it would be an error to think of the Senegalese as mere puppets; there was enough flexibility in the colonial administration, both in Paris and Dakar, to allow the assimilated Four Communes almost complete political liberty. With the Senegalese preoccupied with their own political problems, then, it is not surprising that we must look elsewhere to find Africans who were denied political participation and who sought other goals: most notably to criticise the colonial regime and to ask the ultimate question, why should France govern African soil?

THE RADICAL CRITIQUE OF THE COLONIAL REGIME

Political participation down to 1945 was officially allowed in French-speaking West Africa only in the Four Communes and unofficially in Senegal's interior. The rural chiefs sat in the Colonial Council after 1920, but their influence was more symbolic than real. The Senegalese hinterland became politicised because of Four Communes citizens who lived and traded there in increasing numbers; they brought politics to the growing urban areas of the interior such as Thiès, Kaolack, Tivouane and Ziguinchor. Non-citizens contributed heavily to the campaign funds of Diagne and Diouf; party newspapers were also bought by people of the interior. Religious groups, such as the Mourides, contributed to sympathetic candidates. Both Diagne and Diouf at various stages of their careers relied heavily upon the Mourides. Candidates from the Communes spent a fair share of their campaign time visiting the towns and cities of the interior. Both Diagne and Diouf had sections of their parties in small towns organised by a local political manager. Thus the interior was not cut off from the Four Communes and exercised a growing influence on political questions

65 ARS 13-G-17/18-17, see 1940 and 1941.

there; but the fact remained that most Senegalese did not have the vote.

Outside Senegal, however, frustration and alienation of Africans who did not have the vote could not be channelled into the lobbying and pressure group activities of rural Senegal. The disparity between the Four Communes and the rest of French-speaking Africa was underscored by the many citizens of the Communes who worked in the French administration in other colonies, or Wolof traders who worked across West Africa. These Senegalese insisted on voting in elections when not in Senegal and their privileges were well known and envied by other Africans, especially the Dahomeyans. Moreover, Blaise Diagne attracted the attention and, generally speaking, the favour of Africans during his recruiting mission and subsequent trips across West Africa. By the mid-1920s Diagne spoke of his mandate as covering all of *Afrique noire*, and Diouf as deputy after 1934 continued this notion that he spoke for all of France's black African subjects.[66] While many Africans were flattered by this idea, that symbolically they did have a champion in the French Parliament, others resented the lack of real political participation. The tighter press and censorship regulations after 1920 made it difficult to print local newspapers or receive papers from France, but there emerged during these two decades many individuals who spoke out against the colonial regime.[67] And there was also a small group of expatriate Africans living in France which attempted to focus criticism at the heart of the colonial system.

Perhaps it is easiest to discuss the expatriates first, since little systematic work has yet been done on African political movements outside Senegal before 1940.[68] This expatriate community was centred largely on Paris, although Marseilles, Bordeaux, Le Havre and Dunkerque also had small groups of Africans. These communities grew up after the First World War as a result of some veterans staying in France and the increased opportunities for Africans to work in French industry, especially in maritime services and shipping. And a few Africans managed to continue their studies in Paris. Some were protégés of Blaise Diagne in the immediate aftermath of the war, but by 1923, when Diagne signed his pact of friendship with the Bordeaux merchants, most of the young Africans viewed Diagne as a traitor. Diagne was originally a socialist who admired Jean Jaurès, but after the war he became more moderate and inclined to the right. The young men in Paris were heavily influenced by the new Communist Party of France and thought that Diagne's political manoeuvring was of little

66 Diouf often bragged he could put over 400,000 black troops at France's disposal. See ARS 13-G-17/18-17, for 1936–40.

67 After the Popular Front, it was easier to publish and distribute newspapers in French West Africa for African editors, but restrictive practices were relaxed only after 1945.

68 The exception to this is the important article by J. Ayo Langley, 'Pan-Africanism in Paris, 1924–36', *Journal of Modern African Studies*, vii, 1, 1969, pp. 69–94. The forthcoming publication of the dissertation manuscript of James Spiegler of Roosevelt University should help clarify this period.

value in getting to the heart of the matter: the injustice perpetrated against the Negro race by the colonial powers. It is difficult to know precisely how these young radicals operated but several of their leaders have been studied recently.

One of the first was the Dahomeyan, Kojo Tovalou Quénum, who studied law at Bordeaux and practised in Paris in the early 1920s. Tovalou started a journal, *Les Continents*, which incurred the wrath of Diagne with its accusation against his activities as High Commissioner for recruiting African troops. Tovalou gained the allegiance of René Maran, the distinguished West Indian writer, whose prize-winning novel of 1921, *Batouala*, had been the first major indictment of the African colonial regime. Tovalou visited New York and Marcus Garvey and promptly became suspected by the French. Tovalou helped found the Universal League for the Defence of the Black Race, but was eventually forced to retire to Senegal, where he aided Galandou Diouf in opposing Blaise Diagne in the 1928 and subsequent electoral campaigns.[69] Tovalou was comparable with Lamine Guèye in education—they were the first Black Africans admitted to the bar—but he lacked the same opportunities since as a Dahomeyan he could not aspire to political office. This inequity caused him to reject a quiet life practising law in Dahomey for political activism at the centre of politics in Paris and Dakar.

Another radical in Paris during the 1920s was the Senegalese Lamine Senghor, who after the war was a student, worked as a mail clerk, and was influenced by the French Communist Party. Senghor was one of the officers of the League for the Defence of the Black Race. He told a meeting of Africans in Marseilles in 1926 that his goal was 'to bring together black people in order to constitute a body capable of aspiring to independence'. Lamine Senghor, although a citizen of the Four Communes, campaigned actively for political rights for all Africans and found himself watched closely by the French secret police until his death in 1928.[70]

Tiémoko Kouyaté from the Soudan was a close associate of Lamine Senghor's and continued his work after his death. Kouyaté served as secretary-general for the League for the Defence of the Black Race, was an editor of *La Race Nègre*, one of several journals published by the association and its successors. Kouyaté sought financial support from the Communists and from W. E. B. du Bois. He told the latter that eventually all the colonial peoples of Black Africa should join to form 'a great black state'.[71] At the same time, Kouyaté wrote to a friend of his, a traditional

69 ARS 20-G-82-23, Police Report, 6 May 1928; *La France Coloniale*, 22 March 1928; Jacques Depelchin, 'Ethnography, Political Radicalism and Conservatism in Dahomey, 1900–1958', unpublished seminar paper, Stanford University, 1970.
70 ARS 13-G-71-28, Dossier on Lamine Senghor.
71 ARS 21-G-33-17, Letter to W. E. B. Du Bois, 19 April 1929.

chief in Senegal, asking him to do his job loyally so that 'we can show them we are capable of governing ourselves'.[72]

Tovalou, Lamine Senghor and Kouyaté are only three of many émigrés who used the relative freedom allowed to Africans in France to mount criticism against the colonial regime. But as the French feared the spread of their ideas, their activities were watched, their journals closed down, or inhibited, and their printed materials prohibited from the colonies, confiscated, or censored. The collective influence of the radicals upon the mass of Africans was not great—only a small number of élite members were in contact with the free-thinking intellectuals of Paris. But this élite took hope and encouragement from the émigrés, and its very existence, despite its impotence, pointed towards political changes in the future.

There were other Africans interested in politics, most of whom were newspaper editors in the various colonies. Some were outspoken, such as Louis Hunkanrin of Dahomey, who served prison terms and suffered exile in Mauritania for his opinions. Hunkanrin was a friend of Mody M'Baye, one of the founders of the Young Senegalese, and both supported Diagne in his earlier radical days. Hunkanrin broke with Diagne after the war, spent time in Paris with the émigré community, and was eventually blamed for the riots in Porto Novo in Dahomey in 1923, even though he was not present. He was singled out by the French as a dangerous agitator and persecuted until the 1950s, when the political climate improved. Hunkanrin was exceptional, however, and other newspaper editors usually avoided his direct rhetoric for veiled barbs and innuendoes.[73]

The Pan-African movement and Garveyism had a limited effect during the 1920s in French-speaking Africa. Blaise Diagne helped W. E. B. du Bois get authorisation to hold the first conference in Paris in 1919. Diagne presided over these sessions and the Brussels and Paris conferences of 1921. But Diagne's conception of Pan-Africanism and du Bois's were quite different; Diagne saw progress best attained by working within the colonial system. Furthermore, as one of the most powerful black men in the world, he assumed that his views carried great authority. Du Bois and the American blacks were more aggressive in their view of the plight of the world's Negroes and after the 1921 conference du Bois and Diagne split. Younger Africans maintained contact with du Bois in the 1920s, but the movement never influenced French-speaking Africans as it did their English-speaking counterparts.

The radicals in Paris maintained contact with both du Bois and Garvey. Tovalou visited Garvey and brought his ideas back to Paris. Of the two, Garvey was more feared by the French because of his allegations that he

72 ARS 21-G-44-17, Copy of intercepted letter to Bakary Bâ, 17 May 1929.
73 See Jean Suret-Canale, 'Un pionnier méconnu du mouvement démocratique et national en Afrique', *Etudes Dahoméennes*, N.S., 4, December 1964, pp. 5–30.

would create new states in Africa. Garvey sympathisers were active in Nigeria and Gold Coast and the French waited for an invasion of Garvey forces. In 1922, the Governor-General announced that Garvey agents were to be sought in all parts of French West Africa. Much to his chagrin the only Garvey stronghold that was found after months of scouring the vast West African domain was a few miles away in Rufisque, where some Gambians and Sierra Leoneans had organised a Garvey cell group.[74] But both Garvey and du Bois remained as courageous symbols to the militant radicals in Paris and the frustrated newspaper editors at home in Africa.

Dahomey began to develop a local African press in the 1920s and by the mid-1930s had several newspapers to match those printed in Senegal. Several African papers appeared in the 1930s in the Ivory Coast and Togo, but Guinea, Soudan and Niger did not yet have a large enough Western-trained élite. The discussion of politics by minor administrative officials, schoolteachers and small businessmen who sympathised with or propagandised for the Popular Front was the introduction to modern political questions for many non-Senegalese. In the Ivory Coast, for instance, before 1936, Africans had little save Catholic-sponsored publications for expressing their views.

The decade of the 1930s also saw the creation of a new university or lycée trained élite group from many colonies which was to have its main impact after the war. These were young men such as Paul Hazoumé of Dahomey, who became interested in ethnography and studied the past of his people; Léopold Senghor, who helped create a new school of African thought, eventually to be called *négritude*; Ousmane Socé Diop, who wrote one of the first authentic African novels in French (*Karim*, the story of a young Senegalese caught in the midst of change from traditional to modern society); and Birago Diop, who collected Senegalese folk tales, proverbs and stories. All except Hazoumé were Senegalese. This new cosmopolitan élite, expanded by the growing number of Africans taking degrees in lycées, was not yet self-conscious. However, a few liberal French administrative officials, such as Robert Delavignette, who became director of the Ecole Coloniale, and Governor Hubert Deschamps, recognised young men of the emerging élites who aspired to positions of political and intellectual leadership.

The concern for cultural matters, which typified the disciples of Senghor, did not manifest itself until after the Second World War. *Négritude* became a form of cultural nationalism for French West Africans, but its antecedents in the years 1900–40 are not apparent because the majority of Western-educated Africans (whether in Senegal, where

74 This bizarre incident thoroughly confused French intelligence. See reports on Garvey sympathisers, ARS 21-G-126-108.

political participation was allowed, or in Paris and the other colonies, where it was not) were mostly interested in political questions.[75] Not until a highly educated university élite, living in Paris and cut off from Africa, raised introspective objections about the dangers of becoming too highly assimilated and isolated from African culture, was the climate of the 1940s prepared for *négritude*, the African personality, and the search for African culture.

The activities of this élite and writers such as Senghor, David Diop, Alioune Diop, Mongo Beti, and others, are usually cited as the first actions of French-speaking Africans to convey their displeasure with Western rule. The point of this chapter has been to suggest that actually there was much political concern before 1940. First, among many nameless Africans who continued resistance to colonial rule. Second, among the Senegalese of the Four Communes, who won citizenship and conducted their own local political affairs and influenced the entire colony. Third, among the excluded Africans, whether in Paris or in the other French colonies. These were courageous individuals who sought political expression both within and without the colonial system and who kept alive the hope of African self-rule.

75 On *négritude* see the discussion by Abiola Irele, 'Négritude or Black Cultural Nationalism', *Journal of Modern African Studies*, iii, 3, 1965, pp. 321–48.

CHAPTER 16

Political activity in British West Africa, 1900–1940

J. B. WEBSTER

OLD AND NEW ÉLITES

Two phenomena of nineteenth-century West Africa continued to dominate West African history throughout the first half of the twentieth century. The first was the consolidation of a number of political entities such as Asante, the Sokoto Caliphate, Bornu, Benin and the Fante and Yoruba states. These polities had developed or were developing an internal cohesion and loyalty among their citizens which later colonial boundaries might ignore but could not destroy.

In Asante, for example, the golden stool which the famous Asante priest, Anokye, described as the soul of the nation, symbolised the loyalty, patriotism and nationalism of the Asante for their kingdom. The development of Asante cohesion encouraged the Asantehene to centralise authority through an appointed bureaucracy—chieftaincies 'in the king's gift'—and a national army. It allowed him to rely upon the direct loyalty of his subjects rather than depend upon hereditary title holders to bring him that loyalty. The maturing of the Asante kingdom can probably be traced to the period 1764–1824 when the royal bureaucracy and the para-military organisation, the *Ankobia*, were developed. It was far easier for the British to conquer Asante and incorporate it into the Gold Coast Colony than to destroy Asante loyalty and make them feel or behave like citizens of the Gold Coast. Furthermore, as will be noted later, while certain economic and social forces in the colonial period favoured the development of a wider loyalty, British political policy did not.

The development of Asante cohesion is stressed only as an example of what was a widespread phenomenon. Bornu and Benin had even longer histories of national unity. The Sokoto Caliphate was a nineteenth-century creation but nevertheless it was very successful in using Islam as a focus of unity in one of the most ethnically diverse areas of West Africa. Beneath the political divisions of the Fante and the Yoruba, there was underlying cultural unity and a conscious effort to build up national cohesion. Because of the dual pressure of the British and Asante on the Fante, the Fante were more successful than the Yoruba in achieving this national consciousness in the nineteenth century.[1]

The second phenomenon of nineteenth-century West Africa was the development of a new élite, adhering largely to Christianity, educated in the Western tradition and speaking the English language.[2] The new élite had loyalties and political goals that went beyond the pre-colonial states and, as we shall see, were in many respects related to those of the African diaspora in the New World. They were largely products of the small British colonial enclaves along the coast and, despite quarrels with the British over methods, they generally held most of the assumptions about the superiority of western civilisation common among the British themselves. A Lagos newspaper summed up the long-term goals of the new élite:

> It is to the steady growth of a College like Liberia College and the hopeful development of Fourah Bay College of Sierra Leone, that we look chiefly for the production of powerful Native forces which, besides operating on the Coast, will urge their way into the interior,commingle with its free and independent tribes, import knowledge to, and receive knowledge from them, and with and by them form intelligent and powerful Christian Negro States.[3]

The British ruled their colonial enclaves directly and their policies were distinctly assimilationist, aiming to turn 'raw' Africans into black English gentlemen. By and large they looked upon the new élite as collaborators in the creation of a new order in West Africa. Despite certain misgivings, the new élite generally supported British imperialism in the late nineteenth century and significantly in terms of 'we'. In 1894 the *Gold Coast Chronicle* urged: 'In plain words we must go straight to Kumasi and occupy or annex it declaring Ashantee a British protectorate.'[4] With the expansion in the British sphere of influence the new élite saw new opportunities for themselves and the fulfilment of their dreams of larger, Christian, and more powerful African states. Moreover the new states had to be English in language and philosophy if the new élite was to benefit. French expansion had therefore to be forestalled. Some of the new élite

1 For a fuller discussion of nationalism in nineteenth-century Africa see J. B. Webster, 'Tribalism, Nationalism and Patriotism in Nineteenth- and Twentieth-Century Africa', in *Empire and Nations*, eds. H. L. Dyck and H. P. Krosby, Toronto, 1969, pp. 195–215.

2 The new élite has been the object of a number of excellent studies. The following is only a selection of some of the best. J. F. A. Ajayi, *Christian Missions in Nigeria*, London, 1965; Ajayi, 'Henry Venn and the Policy of Development'. *Journal of the Historical Society of Nigeria (J.H.S.N.)*, i, 4, 1959; Ajayi, 'Nineteenth Century Origins of Nigerian Nationalism', *J.H.S.N.*, ii, 2, 1961; E. A. Ayandele, *The Missionary Impact on Modern Nigeria*, London, 1966; C. Fyfe, *History of Sierra Leone*, London, 1962; D. Kimble, *A Political History of Ghana*, Oxford, 1963.

3 *Lagos Times*, 11 October 1882.

4 *Gold Coast Chronicle*, 30 November 1894, quoted in Kimble, *Ghana*, p. 285.

became as imperialistic as the British: 'We "black Englishmen" . . . cannot sit still and see her [England] robbed of the well-earned fruits of her sagacity, enterprise and goodwill.'[5]

After the conquest of West Africa the British changed their collaborators from the western-educated élite of the coastal enclaves to the traditional African leaders and rulers. This was partly out of necessity in that the British had neither the forces, the personnel nor the funds to rule without the aid of some of the existing political leaders. It was also because, with the development and spread of racism in the late nineteenth century, British officials came to prefer the former élite who accepted or tolerated them as rulers to the educated élite who claimed equality with them. While they maintained law and order through the traditional élite, they relied on British officials in the modernising sectors to the exclusion of the new African élite. As malaria came to be controlled, it was argued out of sheer chauvinism that Englishmen should enjoy the fruits of office in the British Empire. Lugard later lifted this expediency into a sanctified doctrine called indirect rule. Under it, positions of responsibility formerly open to the educated élite in the missions, the civil service, the army, the medical service and commercial firms were quickly closed. The colonial government became preoccupied with providing amenities for a large influx of European personnel. White reservations, white schools and churches, white clubs and even white cemeteries socially divided the former partners. For the first time western-educated Africans felt the full force of segregation and racial discrimination backed by law. The line was clearly drawn between white conquerors and black conquered, breeding arrogance in the former and resentment in the latter.

Meanwhile, as the British moved inland, they tended to demonstrate their military technology and then offer the traditional élites a salary, enhanced power and greater security if they would collaborate while at the same time threatening them with political and even physical extinction if they resisted. In almost all of the centralised states there were those who counselled collaboration and others who wished to resist. The majority of the Yoruba states and Bornu finally collaborated while Asante, Benin and others resisted. The aristocracy of the Sokoto Caliphate was so divided in attitude that some resisted and some collaborated. Many of the smaller political units and the segmentary societies initially allied with the British through treaties but later when the British sought to tax them, thereby converting an alliance into submission, the so-called tax revolts followed. Among these were the Hut Tax War in Sierra Leone and armed resistance among the Ibo and Urhobo.[6] For the bulk of West Africans, foreign rule

5 *Lagos Times*, 14 February 1891, p. 3.

6 Fyfe, pp. 550–8; A. E. Afigbo, 'Revolution and Reaction in Eastern Nigeria 1900–1929: The Background to the Women's Riot of 1929', *J.H.S.N.*, iii, 3, 1966, pp. 539–57; O. Ikime, 'The Anti-Tax Riots in Warri Province 1927–1928', *J.H.S.N.*, iii, 2, 1965, pp. 559–73.

was an unwanted imposition whether they were resisters or collaborators. The roots of West African nationalism go back to the heroic defence of independence whether the symbol was a Prempeh, a Jaja or a Bai Bureh.

While the western-educated élite admired the heroism of the resisters, they had their eyes focused upon the large-scale African states which the British were creating as the best basis for international recognition of African peoples, for rapid economic development and for social and religious change. While these were laudable goals the new élite failed to realise that the independence which the traditional and Muslim leaders had upheld was an essential attribute of nationhood and prerequisite for sound economic and social change.[7] The new élite at the turn of the twentieth century believed that the Muslims and traditionalists were seeking to preserve political units which were too small and parochial. This only increased the dilemma and division in African society which facilitated British intervention.

ETHNIC NATIONALISM

Where the ruling élite had agreed to submit through treaties or where they could be absolved from responsibility for resistance, the British confirmed the existing royal families on their thrones. This prototype of indirect rule was, however, not always the norm. After the conquest of Asante and Benin, it was British policy as far as possible to break up the states and stamp out traditional nationalism by an attack upon the monarchy. Both kings were deported. The Asante were commanded to surrender the Golden Stool to prevent it becoming 'the rallying point of latent Ashanti nationalism'. The Bini and Asante became preoccupied with the reconstruction of their nations and national institutions, particularly the central institution of the monarchy. This they slowly achieved through manipulation of their British overlords, whose growing commitment to indirect rule, particularly after 1914, made them reluctantly amenable to such manipulation.[8]

Both monarchies were ultimately restored in response to the force of traditional nationalism but the British proceeded cautiously, especially in the case of the Asante. In 1924 the exiled Prempeh I was allowed to return as 'a private citizen'. While the British insisted he be called 'Mister' the Asante called him 'Asantehene'. In 1926, still bowing to the force of traditional nationalism, the British permitted Prempeh's installation as Kumasihene, a title previously unknown to the Asante. Fearful of a revived Asante

7 J. F. A. Ajayi, 'African Resistance to Colonialism and the Development of Nationalism in West Africa', Typescript, Ibadan, 1970.

8 P. Igbafe, 'British Rule in Benin 1897–1920: Direct or Indirect?', *J.H.S.N.*, ii, 4, 1967, pp. 701–18; R. E. Bradbury, 'Continuities and Discontinuities in Pre-Colonial and Colonial Benin Politics (1897–1951)', *History and Social Anthropology*, ed. I. M. Lewis, London, 1968, pp. 193–252.

nation the British attempted to confine his authority to Kumasi division. But to the Asante he was, depite British officials, the Asantehene, King of all Asante. Secret tributes were sent to him from other divisions which when discovered brought down the wrath of the British on those concerned. Frankly admitting the failure of their administration in the other divisions the British finally decided it was easier to support than fight Asante nationalism. In 1935 they officially restored the Asante confederacy and recognised Prempeh II as Asantehene.[9]

The Ibo were one of those peoples among whom there were few signs of ethnic solidarity in the nineteenth century. Traditional political entities were small, often a village group, although there were loyalties to the clan, the totem group or to the secret or title societies which transcended political boundaries. British colonial rule became the spur to the emergence of Ibo ethnic feeling triumphing over, but not entirely replacing, loyalty to the smaller groups. Ibo nationalism first emerged among the 'sons abroad', those Ibos who had become settlers in the cities of the Yoruba, Bini, Hausa, Kanuri and others. In the 1920s the Ibo settlers in these non-Ibo cities began to organise unions of people from the same family or village group. The settler unions organised home branches in their villages in Iboland. The network of unions served two purposes: first to bind the settlers abroad to the homeland and second to convey the progressive ideas of the sons-abroad to the more traditional people at home and thus stimulate the modernisation of Iboland.

Up until 1935, the unions were organised within the social and political framework of the nineteenth century. But between 1935 and 1948 they began to move towards a unification unknown to nineteenth-century Iboland. The result was the birth of a strong and virile feeling of Ibo-ness. In 1935 the Lagos Ibo Union was organised as a federation of the existing Ibo organisations of that city. Since the Ibo settlers of Lagos represented a cross-section of Iboland, the Lagos Union became almost an embryonic federation of Iboland. In 1943 the Lagos Union began the creation of an Ibo federal union seeking to organise the 'Ibo linguistic group into a political unit'. Ultimately an Ibo national anthem was adopted and an Ibo national bank discussed—all trappings of a fully matured Ibo nationalism.[10] The first assembly of the Ibo State Union at Aba brought forth a clarion call to Ibo ethnic solidarity:

> A mighty nation shall resurrect in the west of the Sudan, with a love of freedom in their sinews; and it shall come to pass that the Ibo nation shall emerge to suffer wrong no more and to rewrite the history written by

9 The whole story of the reconstruction of the Asante kingdom is told in W. Tordoff, *Ashanti Under the Prempehs 1888–1935*, Oxford, 1965.

10 J. S. Coleman, *Nigeria: Background to Nationalism*, Berkeley and Los Angeles, 1958, pp. 214–15.

> Ethiopia and Songhay. It is the voice of destiny and we must answer this call for freedom in our lifetime. The God of Africa has willed it. It is the handwriting on the wall. It is our manifest destiny.[11]

The rise of Ibo nationalism was spectacular and later events have drawn attention to it but it was not unique. The Ibo were following the example set by the Ibibio in developing the Ibibio National Union. Other ethnic groups like the Urhobo and the Tiv have also tried to evolve similar 'national unions'.

Yoruba nationalism during the colonial period remained elusive. Memories of nineteenth-century wars and conflicts had tended to entrench the separatism of the states rather than foster the unit of the nation. A major cleavage existed between the Oyo peoples—in theory led by the Alafin but in practice by the Ibadan baloguns—and the other states in the anti-Ibadan alliance. This division was reinforced by the more rapid spread of Islam among the former and Christianity among the latter, giving to each different outlooks, ambitions and ideals. Between 1914 and 1930 the British tried to push the Alafin of Oyo towards a kind of paramountcy among the Yoruba. Rather than fostering unity, this only aroused hostility outside Oyo and the British dropped the scheme after 1930.

Yet some Yoruba continued to work for the unity of the whole ethnic group. Like the Ibos, they began by organising urban unions which reflected the political divisions of the nineteenth century. In the 1930s a federation of these organisations was proposed but it failed and many of the unions themselves died out. The perpetuation of loyalties to nineteenth-century political units was often condemned. Progressive 'Unions' grew up with membership open to all Yoruba, but these in turn often split along the lines of nineteenth-century political divisions. A new effort was made in 1948 with the establishment of the Ẹgbẹ Ọmọ Oduduwa, one of whose aims was to unite around the ancient symbol of Ife 'the various clans and tribes in Yorubaland and generally create and actively foster the idea of a single nationalism throughout Yorubaland'.[12] In reality the Ẹgbẹ represented largely the states of the anti-Ibadan alliance of the nineteenth century and it rapidly became moribund.

The growth of the Ewe feeling of solidarity provides an example quite different from the Asante, the Ibo or the Yoruba. In the nineteenth century the Ewe people (population in 1960 about one and a quarter million) were divided into a hundred and twenty petty states without a paramount chief or other ethnic-wide institution. In 1847 German missionaries, stationed among the Anlo subgroup of the Ewe in Keta, chose the Anlo

11 N. Azikiwe, 'Farewell Message at the First Assembly of the Ibo State Union', *West African Pilot*, 8 July, 1949, p. 2.
12 N. Azikiwe, 'Contemporary Nigerian Politics (3)', *West African Pilot*, 3 December 1947, p. 2.

dialect as the basis for written Ewe and helped to spread it among the Ewe generally. The Anlo thus gained a long headstart in the development of a western educated élite. This élite was reinforced by Afro-Brazilians who became prominent in commerce and identified with the Ewe to an unusual degree. Between 1868 and 1871 the Anlo allied with Asante against the other Ewe, a treachery to the ethnic group which has been neither forgotten nor forgiven. During the partition, the British took over Keta and the Anlo group comprising about 40 per cent of the Ewe, while the rest went under German rule. Since this demarcation followed the lines of hostility among the Ewe it did not then arouse resentment. However, Ewe nationalism was growing with western education, and there was general satisfaction when British forces brought all Ewe under British rule in 1914. After the expulsion of the German missionaries the Ewe operated their missions and in the 1920s formally established the Ewe Presbyterian Church. With its schools and synod the Church became an institution of pan-Ewe unity, a rare case of a European-inspired mission becoming a national church.

Meanwhile, the Ewe were again split in 1919 between French and British mandated territories of the Gold Coast and Togoland, with about 55 per cent of the Ewe under British rule. This was the beginning of Ewe national discontent. In spite of this, the church operated on both sides of the border and the ethnic group operated as an economic unit within the commercial orbit of the Lomé merchants. Protests arose wherever the European powers attempted to make the border effective. In the 1930s when French and British policy moved in this direction there were demands for the return of the Germans, riots in Lomé, and the beginning of a nationalist protest press. The border was sealed in the Second World War causing similar resentment and outcry.

Following the Second World War the Ewe sought to put pressure on the mandate powers through the United Nations. The Ewe Unionist Association in Accra and the *Comité de l'Unité Togolaise* in Lomé were organised for this purpose. Their effort was frustrated by the determined opposition of the colonial powers and lingering suspicion of the Anlo and fear of domination by the large western-educated Anlo élite. Ewe nationalism has so far failed to unite the ethnic group, but there can be no question of its coherent, self-conscious and positive existence.[13]

The growth of ethnic nationalism in the colonial period is a factor

13 The main source for the material on the Ewe is C. E. Welch, Jr, *Dream of a Unity: Pan-Africanism and Political Unification in West Africa*, Cornell, 1966, especially ch. 2, 'The Birth and Growth of Ewe Nationalism', pp. 37–81. Also valuable is H. Debrunner, *A Church Between Colonial Powers: A Study of the Church in Togo*, London, 1965. Debrunner notes also the church competition with traditional religion, which in the early twentieth century created a number of new institutions 'in order to restore the old unity of the tribe on a religious basis', p. 138.

which many historians have tended to ignore, perhaps because it does not fit into any neat pattern. And yet, without it the history of the colonial period in general, or of African politics in particular, cannot be understood. Ethnic nationalism, as we have seen, had its roots in pre-colonial African history. But it also evolved in the context of the colonial situation, the rise of Western education and urbanisation. Even when deeply affected by the colonial situation, and encouraged by many of the new educated élite, its ethos and symbols were usually those of the pre-colonial society. When the British abandoned the educated élite for partnership with the traditional élite, they became the real patrons of ethnic nationalism. The educated élite often called it tribalism and tried to focus attention on the wider political units created by the British.

THE NATIONAL CONGRESS OF BRITISH WEST AFRICA

The first step in the break-up of the partnership between the British and the new élite had occurred in the missions. The reversal of policy from the liberal-humanitarianism of the mid-nineteenth century, with its emphasis upon African personnel and African churches, to the policy of European-controlled and European-orientated missions was dramatically marked in the crisis in the Niger Anglican mission when Europeans tried to discredit Bishop Crowther and succeeded in replacing him with a white bishop. This was symptomatic of the changed thinking in most mission societies. Between 1888 and 1917 a section of the Christian-educated élite, led by Mojola Agbebi, Dandeson Crowther, J. K. Coker and J. G. Campbell, and inspired by the idealism of the mid-nineteenth century, created a number of 'African churches' in Lagos and the Niger Delta in opposition to the 'European missions'.

The African churches of Lagos were designed to be multi-ethnic and West African in membership and scope. They had been inspired by the example of independent churches among the Nova Scotians in Sierra Leone and Afro-American settlers in Liberia. Both Agbebi and Campbell successfully spread out from their Yoruba base to evangelise among the Ibo, Ijọ, Fante, Asante and Freetown Creoles. The educated élite thus resisted the temptation to found ethnic churches after the pattern of the Zulu Churches of South Africa. Because indigenisation meant encouraging the impact of a single ethnic group, they were slow to employ African languages, or local idioms in music and ritual. They wanted an African or multi-ethnic church, much as they wanted an African or multi-ethnic state.[14]

The Christian élite—whether independent churchmen or mission adherents—generally believed that a major bond among the nationalities

14 J. B. Webster, *The African Churches Among the Yoruba 1888–1922*, Oxford, 1964; E. A. Ayandele, *The Missionary Impact on Modern Nigeria*, London, 1966.

of any future West African state would be their common Christianity. Their major political interest, therefore, was in reforms to aid the spread of Christianity. In 1920 J. K. Coker actively encouraged a Christian to succeed to the throne of Abeokuta in the hope that this would induce a mass Christian movement in Egbaland. The period 1890 to 1920 was one of mass conversion to Christianity. Many of the most talented and devoted West Africans were involved in assisting this movement through which they believed they were promoting a basic prerequisite of any modern African state.

Outside the Church, multi-ethnic organisations were not easy to establish. The educated élite were divided in their attitudes to the colonial administration and the missionary societies. The one issue on which they were all united was the need to oppose land alienation. In this they even found common ground with the traditional élite. Land issues thus provided the earliest opportunities for political activities against colonialism on a multi-ethnic basis. When in 1897 in the then Gold Coast the British sought control of land, the old élite led by the chiefs and the new élite led by the lawyers formed a unique partnership in the Aborigines' Rights Protection Society, West Africa's first effective political organisation.

In Lagos, the same issue brought into being the Anti-Slavery and Aborigines' Rights Protection Society. A spectacular victory was achieved in 1920 by the alliance of the House of Docemo—the reigning house in Lagos—and the western-educated leader, Herbert Macaulay, which won from the Privy Council in London the ruling that land was the 'undisputed right of the community'.[15] Macaulay had done what most Nigerians felt was impossible. He humiliated the omnipotent colonial government, forcing it to change course. For people living under a rigid autocracy this brief taste of 'democracy' was an exhilarating experience and Macaulay was fêted in the interior as the 'black whiteman' and has been popularly remembered as the father of Nigerian nationalism. It is not surprising that Britain chose to oppose this incipient multi-ethnic nationalism by patronage of ethnic nationalism.

During the First World War, Britain, France and America had loudly proclaimed self-determination as a solution to the problem of Europe and, after the war, had broken up the multi-national Austro-Hungarian Empire, partly on the principle of ethnicity. In the British Empire colonial nationalism in Canada and South Africa, by insisting on similar rights, rapidly proceeded to 'Dominion' status during the 1920s. Irish nationalism triumphed in independence; Egypt secured limited self-government; and agitation against the British was intensified in India. By contrast, the principle of self-determination was flagrantly violated when the Ewe of Togoland were divided between Britain and France without reference to

15 Kimble, *Ghana*, pp. 330–57; Coleman, *Nigeria*, p. 196.

the people.[16] The Treaty of Versailles, like the earlier conferences at Berlin and Brussels, came dangerously close to dealing with Africans as though they were mere objects. The grievances of the Ewe stimulated the creation of an organisation—the National Congress of British West Africa—of the western-educated élite to demand social, political and economic changes in West Africa. At a conference in 1920 in Accra, representatives from all four British colonies urged the British to lay the basis for future self-government: introduce the franchise, provide for higher education and recruit the senior civil service on the basis of merit.

The Congress was the first political organisation in West Africa to codify in its resolutions the philosophy of the new élite which in fact dated back to the nineteenth century. The Congress brought to the attention of the British what it felt should be the goal of their colonial policy: a multi-national West African state controlled by an élite drawn from all ethnic groups and bound together by a patriotism rooted in a common educational experience, the English language and devotion to the modernisation of African societies. The leaders of the West African educated élite were inspired by the doctrine of self-determination which had broken up the Austro-Hungarian Empire and they named their movement on the analogy of the Indian National Congress, but in fact they spurned nationalism. Neither on the Gold Coast nor in Nigeria did the Congress solicit the support of the traditional élite. It also ignored the governors of the British-created colonial units—much to the annoyance of the governors—and approached the Colonial Office directly on behalf of British West Africa as a whole.

Since ethnic issues were of the greatest importance to the largest number of West Africans, the British, by patronising ethnic nationalism, had little difficulty in limiting the effectiveness of the educated élite in the inter-war period. This was the political goal of indirect rule and its two great architects, Lugard in Nigeria and Guggisberg in Ghana, did not hide their antipathy to the educated élite. It was Guggisberg who at Manchester in 1923 denounced those seeking to revive partnership with the educated élite, and added that if such idealists ever came to power in England they would say to West Africans:

> You are men and brothers with us. Some of you are very finely educated. We will give you self-government and see how you can govern yourselves. If that movement were to come within the next hundred years . . .

16 For Congress concern about the Ewe see J. G. Campbell, 'Facts about the Nigerian Branch of the National Congress of British West Africa', *Lagos Weekly Record*, 10 July 1920, p. 3. For Lagos Togolese reaction see the *African Messenger*, 18 December 1924, where an article signed 'Togolander' included the following, 'Ruthlessly and without the least regard for the people whose tradition it was aimed to break up, the members of the League of Nations signified their approval to the dismemberment of Togoland.'

Manchester merchants would be wise to cut their losses and start trade with some other part of the world.[17]

In place of the multi-national West African state under the leadership of the new élite, British officials were proposing an 'Austro-Hungarian type' of empire where ethnic loyalties would be fostered and the British could pose as the only force capable of holding together the antagonisms inherent in the structure.

The British partnership with the traditional élite was to be cemented by entrenching and enhancing the position of the chiefs, while isolating the educated élite. The old and new élites who had cooperated over land issues in the Aborigines' Rights Protectorate Society were split apart. Ofori Atta, who posed as the spokesman of the Fante traditional élite, accused the Congress and its leader, Casely Hayford, of trying to usurp the chiefs' powers in the leadership of the people. In return *The Gold Coast Independent*, supporting the Congress, castigated the chiefs as 'uneducated and illiterate fetish rulers'.[18]

The shift of British policy from the educated élite to the traditional élite brought new complications in Sierra Leone and the Gambia. There the new élite meant chiefs in the Protectorate. In Nigeria and Ghana, the shift of policy meant a quarrel between chiefs and Western-educated leaders within the same ethnic groups—Fante, Ewe, Yoruba or Ibo. In the Gambia and Sierra Leone, it meant rivalry between the Colony and the Protectorate which took precedence over the solidarity of the educated élite. For example, in the Sierra Leone Protectorate, a small group of Western-educated leaders grew up. They were usually products of the government school at Bo which, in its admission policy, favoured sons and wards of chiefs. Thus, the new Mende and Temne élite was closely related to the chiefly families. J. Karefa Smart and W. C. M. Caulker were both sons of paramount chiefs. Milton Margai was not only the son of a chiefly family, his father's brother was a most influential chief in the Bonthe district. These family ties, when added to the fear of Creole domination of the Protectorate, meant that the new Mende and Temne élite supported their chiefs and the new British policy against the Creoles.

The Sierra Leone branch of the National Congress, a Creole-based organisation, had demanded a larger share for Africans in the government of the colony and an elective franchise. They opposed nominated chiefs sitting in the Legislative Council because 'they have no choice but to back

17 Lugard's antipathy to western-educated Africans is well-known but Guggisberg, because he was connected with the founding of Achimota and because he maintained good social relations with the new élite in the Gold Coast, is often not as closely identified with Lugardian-type prejudice. This quotation is taken from Kimble, *Ghana*, p. 439.
18 Kimble, *Ghana*, p. 393, quoting the *Gold Coast Independent*, 29 January 1921.

the government' to whom they owed their position.[19] This was consistent with Congress policy elsewhere in West Africa and had been the basic cause of friction between old and new élites in Ghana. In Sierra Leone, however, it was the educated élite of the Protectorate who in 1922 founded the Committee of Educated Aborigines to oppose the Congress and protest against the imbalance in the distribution of educational facilities between Colony and Protectorate. Only when the imbalance had been righted, when the 'protectorate can literally march with the colony in every sphere of life' could a common franchise with the Colony be considered. For representation of the Protectorate in the Legislative Assembly, the C.E.A. meanwhile preferred chiefs or Western-educated men who were related by blood to or closely identified with the traditional rulers.[20]

By the 1920s the line of division had been drawn between the Creoles on the one hand and a Mende–Temne alliance on the other. The latter feared political and economic dominance by the Creoles if—as Shorunkeh-Sawyer, a Creole lawyer-politician proposed in 1924—all citizens of Sierra Leone were made equal before the colonial administration. This Mende and Temne alliance was ultimately to become strained because education developed at a faster pace among the Mende than among the Temne. In the meantime, however, British overrule rested firmly on the potential and actual antagonism among Creole, Mende and Temne.

The British used education to support their ethnic policy especially in Northern Nigeria, Northern Ghana and in the protectorate of Sierra Leone and the Gambia. Bo School was specifically designed to strengthen Mende ethnic consciousness. 'Mende pupils will be taught in such a manner as to ensure' that they were 'not educated enough to become senior officials or leaders of any kind of national political movement'.[21] The students of Bo were to be so fashioned that they would never feel sympathy with or be tempted into a partnership with the Creoles of Freetown. Fortunately for the new élite, when the government school at Bo was being set up the missionary-sponsored Albert Academy in Freetown was founded as a place where Mende, Temne and Creole boys could be taught together and prepared for future roles both of 'senior officials' and 'leaders' of national political movements.

The National Congress of British West Africa intended to take over the state structures being created by the British. It therefore could not compete with the British policy of invigorating ethnic cohesion. To win a measure of popular approval it would have had to support distinctly ethnic movements, many of which sought to undermine the solidarity of the

19 M. Kilson, *Political Change in a West African State: A Study of the Modernization Process in Sierra Leone*, Harvard, 1966, pp. 131–2.
20 *Ibid.*, p. 127.
21 'Bo School Prospectus', in C. Fyfe, *Sierra Leone Inheritance*, Oxford, 1964, p. 304.

newly-created states. For example, between 1919 and 1927, the Parakoyi was revived as an instrument to carry forward secret plans for the restoration of the Egba state whose independence had been lost to the British in 1914. In the 1920s, riots and disturbances among the Urhobo were also aiming at a return to pre-colonial political divisions and forms of government. The National Congress did not support or even sympathise with these movements because they ran counter to the wider loyalty which its leaders sought to promote. The wider loyalty ran directly counter also to the basis of British overrule as conceived by Governors like Sir Hugh Clifford who denounced the Congress and made a passionate defence of ethnic nationalism:

> The people of West Africa do not belong to the same stock and are not of common descent; ... [they have] no common language ... no community of religious belief. ... As a matter of fact, the Hausas of Zaria, the Bantu ... of the Benue and, say, the Fantis of Cape Coast are less nearly allied to one another than are, for example, the Scandinavians of the Baltic, the Slavs of Bulgaria and the Semitic peoples of Egypt and Morocco. ... It is the consistent policy of the government of Nigeria to maintain and support the local tribal institutions. ... I am entirely convinced of the right, for example, of the people of Egbaland, ... of any of the great emirates of the north ... to maintain that each one of them is, in a very real sense, a nation. ... It is the task of the government of Nigeria to build up and to fortify these national institutions.[22]

Clifford clearly saw himself as playing the role of the Hapsburg monarch to an Austro-Hungarian empire. He knew that the last thing the Western-educated élite desired was to see the Balkanisation of West Africa. The longer he fostered and fortified ethnic institutions, the more indispensable the British would become to hold together 'this collection of self-sustained and mutually independent Native States'. It is noteworthy too, that while linking 'same stock', 'common descent', and 'common language' as criteria of nationhood, Clifford could in the same breath talk of an Egba rather than a Yoruba nation, a Zaria rather than a Hausa nation. While the forces of nationalism had made it expedient to permit the reconstruction of the Benin and Asante kingdoms, the British often felt safer in focusing people's loyalties on the smallest possible political units. This was the policy in the Sokoto Caliphate where the individual emirates were being encouraged '... to maintain that each of them is, in a very real sense, a nation ...', thus quietly and subtly undermining the unification brought about by dan Fodio in the early nineteenth century. The force of Clifford's arguments lay in the fact that, at the time, the majority of West Africans probably agreed with him and not with the leaders of the Congress.

22 Coleman, *Nigeria*, pp. 193–4.

ELECTIVE REPRESENTATION

Even more fatal to the ambitions of the Congress was the limited introduction of the franchise for elected representatives from Calabar, Lagos, Accra, Cape Coast, Sekondi and Freetown to the various colonial legislative councils. The positive effects of this have been over-emphasised; the introduction of popularly-elected representatives into the sacred precincts of colonial governments, the precipitant in the formation of political parties, and the first step towards self-government. More decisive, however, were the adverse effects it had on the Congress movement. It focused the attention of the intelligentsia 'on the smallest possible political unit'. At the expense of the Congress it fostered municipal parties like Macaulay's Nigerian National Democratic Party in Lagos, formed of small cliques, concerned with even smaller issues and wracked by personal jealousies and quarrels over the spoils of office. While the Congress had had at least four branches in Southern Nigeria, the N.N.D.P. was strictly confined to Lagos.

Then followed the inevitable conservatism which close association with the colonial government produced when Congress men monopolised the elected seats. Influential members who failed to secure election or were from cities without the franchise were also occasionally nominated to sit in the Legislative Councils. By judicious 'buying off', the British assisted the drift to conservatism. The electorate became apathetic and the Western-educated élite, deprived of its leaders, drifted aimlessly. Later generations of independence leaders would have boycotted these inadequate British concessions—as indeed Hayford and the Congress in Ghana attempted to do—because of the predictable consequences. Lastly the Congress by seeking positions in colonial legislatures allowed the British to emphasise the individual colony over a United British West Africa. Henceforth the four British colonies drifted progressively further apart.

Between the early 1920s when the British introduced the elective franchise into certain coastal cities and the mid-1930s, National Congress candidates or sympathisers won most of the seats. In the early days there was considerable popular enthusiasm for election campaigns but it quickly became evident that the elected representatives in the Legislative Councils could accomplish virtually nothing. A Lagos newspaper commented, 'The faint echoes of the protesting voices of elected members have more often than not been drowned in the tumultous uproar of official "Noes"'.[23] The helplessness of the unofficial members against the bloc vote of the official members was so stultifying that even a European unofficial member complained.

> We have heard from the daily press that the eyes of Nigeria are on the unofficial members but we have just had an instance of how perfectly

23 *Lagos Daily News*, 16 October 1928, p. 1.

> hopeless it is, and how perfectly helpless unofficial members are, when it is a question of our coming up against the official vote.[24]

The colonial authorities wanted a show of popular representation but they had little intention of allowing it to influence policy decisions. The Chief Secretary of Nigeria said, 'I have been brought up in the belief that in a typical crown colony or protectorate, the function of the government is to govern'.[25] In plain words the government had little intention of being interfered with by the people's representatives. When members complained that they were given neither the time nor the information necessary to study the financial estimates they were virtually told that they need not bother:

> It is no doubt difficult for unofficial members to study them [the estimates] efficiently for the purpose of criticising them, but you really must trust those of us who have been at this game nearly thirty years not to accept things blindly.[26]

Despite the paternalism and the decorative role the colonial government intended for them, African members were lively, informed critics. In the Nigerian Legislative Council they questioned and criticised all the things which agitated Nigerians then and later. The favourite subjects of questioning were native administration, followed by Africanisation and labour conditions, communications and education. Indeed most of the elected members were a credit to their constituents. Of Adeniyi-Jones it was said, 'the ramifications of his interests were so widespread that no department of government was free from the criticisms involved in his endless flow of questions'.[27]

While the electorate may have appreciated the energy of their members, it became painfully obvious that they could never hope to fulfil even the mildest and most moderate of their election pledges. As might be expected, election manifestoes became more and more subdued in language and popular enthusiasm waned. In the Calabar election of 1938, for example, of the thousand eligible to vote only two hundred exercised their right. Inevitably the colonial government argued that the National Congress had been mistaken in its demands, that the elected members were unrepresentative of the people and that it had been a faulty policy to introduce foreign institutions into Africa (thus indirectly justifying indirect rule). The African mouthpiece of the Nigerian colonial oligarchy stated, '. . . we in West Africa are certainly going too fast rather than slow, as others would have us believe' and 'It has been our contention and conviction that

24 Irving, *Nigerian Legislative Council Debates*, 2 February 1931, p. 89, quoted in J. Wheare, *The Nigerian Legislative Council*, London, 1950, p. 107.
25 *Nigerian Legislative Council Debates*, 5 February 1927, p. 108.
26 Chief Secretary, *Nigerian Legislative Council Debates*, 19 February 1926, p. 119.
27 Wheare, *Nigerian Legislative Council*, p. 63.

the franchise is really a very, very questionable benefit conferred on this country.'[28] Even a press not so given to a colonialist view could agree: 'The franchise was a new gift to the people . . . it aroused much interest at the time, but it was a gift of foreign manufacture and genuine interest in it soon faded.'[29]

The limited and ineffectual franchise was a major cause of the weakening of the position of the leaders of the National Congress. So much did they ultimately become the object of public apathy and press castigation that despite their elected position they appeared to depend more upon the colonial establishment than upon the people for their positions. Inevitably they were swept out of office. In the late thirties candidates of the Youth Movements, who generated a new wave of popular enthusiasm, came to take their place. The Youth Movements, in turn, fell victim to the same system with an accompanying renewal of public apathy.

ECONOMIC NATIONALISM

Had the leaders of the Congress been commercial men rather than mainly professionals they might have had more success in their political demands. If there was any issue over which the largest number of West Africans could have been rallied, it lay in the deteriorating economic situation in West Africa in the inter-war period. Large numbers of African merchants, cash crop farmers, and wage labourers might have rallied to the support of a political movement which became the mouthpiece of their grievances. After all it was an economic issue—land—which had been responsible for the success of the A.R.P.S. before 1914.

Before 1920 there had been two decades of relative prosperity. The spread of new cash crops and generally rising prices contributed to a buoyant and expanding economy. Prosperity was assisted by the infusion of fairly substantial amounts of capital for harbour development, railway and road construction and investment in the gold mines of Asante, Udi coal mines in Iboland and tin mines on the Jos plateau. The demand for skilled and unskilled labour was constantly rising. By 1920 the export-orientated colonial economy had been fixed. While growth might take place within that pattern—more roads, more cocoa, kola, oil and mineral exports—the

28 Sir Kitoyi Ajasa, *Nigerian Pioneer*, 21 May 1926, p. 8, and 18 June 1929, p. 8. At the Anglican Synod in Lagos in 1922 Bishop Jones introduced a motion 'Condemnatory to the granting of the partial franchise'. It was seconded by Ajasa. The *African Messenger* feared that since the synod represented many areas outside Lagos, 'the concurrence of these delegates in the motion could be conveniently and politically used as a representative verdict of those in the interior against the granting of the franchise'. The *Messenger* reported that the motion 'was received with groans' and fortunately failed (18 May 1922, p. 4.) The same newspaper claimed that the Nigerian colonial bureaucracy regarded Ajasa as 'the one straight Negro in Nigeria—always on the side of the government' (24 November 1921, p. 5). In 1924 Ajasa was awarded the O.B.E. having served for over twenty years as a nominated member in the legislative council.
29 *Nigerian Spectator*, 20 July 1931.

colonial government never introduced, and indeed was precluded by its ideology from considering, policies which would create a balanced and modern industrial economy. From 1920 onwards the colonial regime came more and more to appear as an obstacle to the modernisation of society which the Western-educated élite so passionately desired.

The conquest of the West African states and the building of railways permitted the breakthrough of expatriate firms from their coastal bases into the interior markets. The resulting contest between African and European middlemen eliminated the nineteenth-century African merchant princes of coastal cities such as Lagos, Cape Coast and Freetown and the commercial houses of the Delta city-states from the 'commanding heights' of the economy, the import–export trade. With an expanding economy the effect of the disaster was masked by the large number of new opportunities which were opened up to entrepreneurs in the major distributive trades. A few had been displaced at the coast but many more had found opportunities particularly in the interior cities such as Ibadan, Kumasi and Bo. While the heads of the Delta Houses sank into commercial insignificance, a new class of Ibo entrepreneurs around Onitsha and other towns were taking their place.

In the two decades following 1920, the prices of export crops declined and generally remained low. Prices of imported goods, inflated during the First World War, did not decline to the same degree and the wide differential between prices paid for the export crop and prices asked for consumer goods became a continuing grievance. This inflation, combined with a government and commercial policy of low wages based on the myth of the African 'target worker', meant that labourers, junior civil servants and teachers suffered a marked decline in real wages. During the twenties more expatriate firms entered the colonies, supplanting and pushing out African entrepreneurs from the distributive trades they had originally pioneered. The big expatriate firms spread out a network of buying agencies and retail shops from the cities on the railways to displace Africans in the surrounding towns. Substantial African merchants of 1920 had often become petty traders or salaried agents of foreign firms by 1930. What is more there was an almost constant cycle of expatriate amalgamations creating larger and larger firms, eliminating competition in order to maintain, and profit from, the differential price structure. While the European firms monopolised the top, Syrian and Lebanese entrepreneurs broke into petty trade and finally monopolised what was left by the Europeans. The press indicated the plight of the African mercantile community:

> The Native in this country . . . will soon find himself ground to powder between the stones of European and Asiatic economic aggression. . . . What has the government done to improve conditions of the native

> traders with a thousand and one European branch shops at our door, and with all the primary produce market monopolised by European firms and the Syrian seizing the little that could have remained for us?[30]

Between 1920 and 1940 there was little infusion of foreign capital into the country to balance the repatriation of profits to Europe. Foreign banks were accumulating African savings and lending them to expatriate firms to finance their expansion. An attack upon the colonial economy in these decades might have forged farmers, labourers and the entrepreneurial class into a powerful protest group in which the political divisions of the past and the multitude of conflicting ethnic nationalisms could have been submerged under a common economic interest.

As early as the First World War, John Ayew and J. K. Coker, leading farmers of Mampong (near Accra) and Agege (near Lagos) respectively, attempted to organise farmers' unions which bypassed the foreign firms by making collective shipments of cocoa direct to Europe. Both lost heavily in the slump of 1920 but Ayew continued shipments for his Gold Coast Farmers' Association. At the Accra conference of the National Congress in 1920 one resolution called for a British West African Cooperative Association to found banks, promote shipping facilities and establish cooperative buying and selling centres. This was a demand for a share in the colonial economy, a request that Africans should do what the foreigners had done; but it was recognised that since individual African merchants did not have the capital of their European counterparts, the effort must be a cooperative one. The resolution was never followed up because, as professionals, Congress leaders were more immediately interested in political than in economic problems. They also realised that Africans were not likely to get a significant share of the colonial economic system until they had a share in the colonial political system—this is what a later generation meant by 'seek ye first the political kingdom'—but they saw that if protest could be aroused through economic grievances it could be directed towards political ends.

In the early thirties when the Congress was a spent force, Winifried Tete-Ansa, a produce buyer, who later studied commerce and banking in Britain, attempted to implement Congress economic policies in the cocoa-growing regions of Nigeria and Ghana. Through three related companies Tete-Ansa sought to provide an African-controlled alternative to the existing expatriate-controlled firms. The first was the West African Cooperative Producers, to coordinate the farmers' unions and act as their sole buying and marketing agent. By 1930, forty-five farmers' unions had promised support, including Ayew's Gold Coast Farmers' Association and Coker's Agege Planters' Union. The second was the Industrial and Commercial Bank to mobilise African savings for African entrepreneurs and finance the purchase of the farmers' cocoa. The third was the West African–

30 *African Messenger*, 29 May 1924, p. 4.

American Corporation in the United States with an Afro-American directorate to market cocoa and ship back manufactured goods.[31]

The overall aim was to secure a higher cocoa price to the farmers, provide capital for African merchants from an African bank, bypass the expatriate firms and so secure for African entrepreneurs a position in the 'commanding heights' of the economy. The directors included cocoa planters, merchants and professionals who had supported the Congress. The theory was sound but in practice the companies failed very quickly. Capital subscribed was pitifully small, the absence of sound business practice and outright dishonesty among the directors caused bankruptcy in 1930 and again after a second attempt in 1932.

Following the collapse of Tete-Ansa's first set of companies John Ayew and A. J. Ocansey organised a cocoa holdup in Ghana in 1930 to try and force up prices. Tete-Ansa predicted it would fail without the support of a national bank and when it did the foreign firms would force cocoa prices even lower. The prediction came true and at a conference on the boycott failure in 1931 Tete-Ansa persuaded the farmers to suspend action until he could launch a third scheme. In 1934 he outlined a national plan for economic development much like the ones which had previously failed. Although Ayew, the Farmers' Association and some of the chiefs, including the Asantehene, supported the new companies, they collapsed like their predecessors.

The indomitable John Ayew continued organising the farmers and in the 1937 cocoa holdup—the largest demonstration of rural discontent in West African colonial history—he showed the possibility of using economic grievances to unite the people especially when this was combined with the support of the traditional élite. The colonial administrations attempted to divert the militant farmers' organisations which were growing up in the cocoa-growing areas. In the 1930s they began to set up officially sponsored Farmers' Cooperatives as an alternative to Tete-Ansa's. Tete-Ansa and Hayford who had called cooperation 'the greatest word of the century' had popularised the word: the British capitalised on their work. The government cooperative—a perversion of its European counterparts—was designed to be a handmaiden of the expatriate firms rather than an effort to bypass them. Its major 'success' was to direct anger away from the expatriate firms and focus it on the African produce buyers who were also suppliers of credit to the farmers. Instead of behaving as orthodox cooperatives seeking to bypass all middlemen, it rather concentrated its

31 The material on Tete-Ansa is almost entirely from A. G. Hopkins, 'Economic Aspects of Political Movements in Nigeria and the Gold Coast: 1918–1939', *Journal of African History (J.A.H.)*, vii, 1, 1966, pp. 133–52. A biography of Tete-Ansa is in *Lagos Daily News*, xvi, 3, 1932, and Tete-Ansa wrote a number of articles: 'What does West Africa want?', *Empire Review*, xliv, 308, 1926; 'Imperial Cooperation: British West African Development', *Whitehall Gazette*, September 1926; 'The Ottawa Conference and British West Africa', *Lagos Daily News*, xiii, 9, 1932; and *Africa at Work*, New York, 1930.

propaganda against the African middlemen as if they were to blame for the evils of the whole system. Since they were the only group of middlemen with which the average farmer dealt, it was effective propaganda. In Yorubaland the influence of the official cooperative was an important reason why farmers did not effectively support Ghanaians in the 1937 cocoa holdup.

The response to Tete-Ansa and John Ayew indicated that a protest movement based upon economic grievances could draw a larger following over a wider area than the Congress which was asking for political change. Secondly the colonial economic system extended throughout West Africa. The same banks, firms and shipping companies, dominated all of West Africa. A successful attack upon the system had to come from West Africa as a whole. The Congress and Tete-Ansa were the first and last efforts to organise on a West African level. The growing tendency in the thirties was to centre attention on the individual colonies. Political success at the colony level was calculated to make economic success on the West African level more difficult. Lastly, it may be noted that the British dealt with both movements—the Congress and Tete-Ansa's scheme—in much the same way. The leaders were diverted into safe and harmless institutions—the legislative council and cooperatives—institutions basically subversive of the goals they had once espoused.

THE AFRICAN DIASPORA

In addition to the British West African dimension upon which the National Congress and Tete-Ansa had concentrated, and far from the ethnic units on which the British sought to fix African eyes, was the 'Black World' of the African diaspora in America and the West Indies. In the twenties the diaspora was in revolt against racial injustice. This produced sympathy and support in West Africa for a Pan-Negro, Pan-African emancipation movement. Except for the Ethiopian crisis of 1935 the Pan-African movement before the Second World War was a protest of black people from the West African coast, mainly Nigeria, Ghana and Sierra Leone; America, including the West Indies and the United States; and Europe, especially African students and West Indian settlers in Britain. This was, incidentally, the old triangle of the Atlantic slave trade. It was a protest of black people who shared a common origin in West Africa, a common history of suffering, reinforced by the English language as a common means of communication. Contact across the Atlantic had never been broken. Even while the slave ships were still carrying West Africans to America, other ships were carrying Afro-Americans and West Indians to Sierra Leone and Liberia. The West African press always showed a lively interest in Black Americans and West Indians. Agbebi, James Johnson, J. K. Coker and Casely Hayford corresponded with American leaders like Booker T. Washington, and the African Methodist Episcopal Church, an Afro-

American organisation, was welcomed in Ghana. It was therefore hardly surprising that Tete-Ansa sought to enlist American Negro support to break the British monopoly over the sale of West African cocoa to the United States.

London became an important centre for the flow of ideas between America and West Africa partly because direct American–West African traffic almost ceased after the conquest of Africa and partly because American Blacks were viewed with suspicion by the colonial governments in West Africa. In 1900, at West Indian inspiration, a conference was held in London to protest against excesses committed against Africans during the conquest and against the trend of colonial policy in South Africa and Rhodesia. Prominent West Africans from Lagos, Monrovia and Freetown attended. The conference set up a Pan-African Association—the first introduction of the word 'Pan-African'—and a journal. This association was short-lived but it was the father of a series of Pan-African conferences to follow.[32]

In the ferment following the First World War, the Pan-African idea was revived by William Du Bois, an Afro-American intellectual partly of West Indian descent. Du Bois organised the second Pan-African conference in Paris to put the claims of peoples of African descent before the delegates to the Peace Conference also meeting in that city. It was also intended to forge a link between Francophone Africans and their English-speaking counterparts. Despite two further conferences, in 1923 and 1927, the movement failed to become truly Pan-African until after the Second World War.

In order to create a forum for African opinion in Britain and provide a link between the National Congress of British West Africa and the West Indian and American inspired Pan-African movement, Ladipo Solanke organised the West African Students' Union (W.A.S.U.) in 1925. Based in London, W.A.S.U. gave students and Black leaders an opportunity to express their views in its journal.[33] Men who later led the African independence movement were students during the twenties and thirties. Many at one time or the other held executive positions in the Union. There they were exposed to influences from movements of emancipation from all over the Black world.

In 1919 Marcus Garvey (1887–1940), a Jamaican, burst upon the scene in the United States and through inspired oratory created the largest mass movement in American history. By 1923 his Universal Negro Association

32 Pan-African literature is extensive and only a select list appears below. I. Geiss, 'The Development of Pan-Africanism', *J.H.S.N.*, iii, 4, 1967; C. Legum, 'Pan-Africanism and Nationalism', *Africa in the Nineteenth and Twentieth Centuries*, eds. J. C. Anene and G. N. Brown, Ibadan and London, 1966, pp. 528–38; G. Padmore, *Pan-Africanism or Communism? The Coming Struggle for Africa*, London, 1956.

33 Coleman, *Nigeria*, pp. 204–9; Kimble, *Ghana*, pp. 549–50.

claimed six million members. Garvey preached an emotional and inspiring message to the oppressed. While Du Bois had pleaded 'We ask in all the world that black folk be treated as men' Garvey poured scorn on the idea that white men would ever voluntarily liberate Blacks. He demanded that Blacks liberate themselves and behave as men. He called for 'Ethiopia to awake', 'the renaissance of the black race' and 'Africa for the Africans'. Then turning social Darwinism upside down he used it as a powerful weapon against its originators. He condemned the white gods, upheld a black Christ and black Madonna, glorified black skin and preached racial purity. He constantly took inspiration from African history, glamourising its great empires of the past and stressing their outstanding contribution to universal civilisation in an age when Europe was sunk in barbarism.

Garvey suggested that the ex-German colonies in Africa be turned over to New World Blacks. He called for recruitment of a Black army and sent a mission to Liberia to negotiate for land for a mass return of American exiles to their African homeland. He launched a two million dollar fund and a shipping line—The Black Star Line—to carry out the emigration scheme and break the white monopoly of international commerce. His much-quoted statement has proved startlingly prophetic:

> No one knows when the hour of Africa's redemption cometh. It is in the wind, it is coming. One day like a storm it will be here. When that day comes all Africa will stand together.[34]

No movement of American origin had ever before aroused such enthusiasm in West Africa. U.N.I.A. branches sprang up in Ghana, Nigeria and Sierra Leone. An Ibadan citizen wrote,

> Garveyism, one is glad to note, is also gaining ground everywhere in the world. No one can predict what will be the fate of the black races in the near future. Mr Marcus Garvey is trying to build a Negro Empire. The possibility or impossibility of which I am not concerned to say but the downtrodded and oppressed Black knows that his redeemer liveth.[35]

One district officer noted that 'Garvey's manifesto went all through Africa' and related how in remote Borgu the story flashed through villages four days' journey from a telegraph office and eight from the railway, that 'a black king was coming, with a great iron ship full of black soldiers, to drive all the whites out of Africa'.[36]

The Black Star Line was particularly welcomed since it could be expected to provide the direct link to the cocoa markets of America and break the expatriate firms' hold over West Africa's import–export trade.

34 *Marcus Garvey, Philosophy and Opinions*, ed. A. J. Garvey, New York, 1923. The Garvey Movement is also discussed by Padmore in *Pan-Africanism*, i, p. 10.
35 A. Obiseson, Diary, 19 August 1921, Ibadan University Library.
36 Joyce Cary, *The Case for African Freedom*, London, 1944, pp. 20, 22.

Advertisements appeared in West Coast newspapers asking Africans to take out shares.[37] The *Gold Coast Independent* called the Garvey movement 'this great Negro awakening' and went on to link it with the National Congress. 'Negro peoples throughout the world are demanding, not as a favour but as of right, their place among the nations of the earth.'[38] The rigour of the colonial powers' reaction was proof of the popularity of the movement. They warned Liberia against cooperation with Garvey. Lagos mission churches closed their halls against U.N.I.A. meetings. Garvey's book *Philosophy and Opinions* was banned in Ghana as was his journal *Negro World* in Nigeria.

For various reasons the unified movement collapsed when Garvey was imprisoned in 1925 and deported from the United States in 1927. In both America and Africa there was a strong tendency for professional and élite leaders like Du Bois and Herbert Macaulay to look down upon Garvey's movement as one which appealed too strongly to the populace. Most of the Lagos press, for example, scoffed at Garveyism.[39] Many radical organisations, however, arose out of the ashes of the U.N.I.A. and even more came to look back to Garvey for inspiration. Possibly the most publicised of these in modern times is the Nation of Islam, commonly called the Black Muslims.[40] Garvey might very well be called the father of American Negro 'nationalism'. In West Africa his influence was also profound. Kwame Nkrumah, himself very much a Garvey-like figure, used many of Garvey's techniques and provided much the same kind of inspiration to the Black World in the 1950s as Garvey did in the 1920s. Nkrumah not only acknowledged his debt, '. . . of all the literature that I studied, the book that did more than any other to fire my enthusiasm was *Philosophy and Opinions of Marcus Garvey*',[41] but also immortalised Garvey by making his Black Star of African Freedom the symbol of independent Ghana and christening its national steamship company the Black Star Line.

TOWARDS THE NEW NATIONS

By the early thirties the Black emancipation movement had lost its

37 J. Webster, A. Boahen with H. Idowu, *The Revolutionary Years: West Africa Since 1800*, London, 1967, pp. 298–303.

38 Quoted in Kimble, *Ghana*, p. 543.

39 The *Nigerian Pioneer* as might be expected called Garvey 'amusingly distasteful' (12 February 1926). The more moderate paper, the *Nigerian Advocate*, said that the 400 million Negroes 'who would one day drive the Europeans out of Africa' were 'totally mythical, imaginary beings' (5 February 1923). Even H. Macaulay, normally classified as radical, stated, 'This awakening should be, not in the spirit of the rashness of Garveyism, but in the spirit of loyalty and true patriotism until Africa shall take her stand among the British Commonwealth of Nations' (*Lagos Daily News*, 6 July 1928).

40 E. U. Essien-Udom, *Black Nationalism: A Search for an Identity in America*, Chicago, 1964.

41 *The Autobiography of Kwame Nkrumah*, London, 1959, p. 37.

dynamics. The National Congress and Pan-African conferences were dead, Garvey was in exile and Tete-Ansa was bankrupt. In this interlude international Communism sought to form a partnership with Blacks on the assumption that the revolution Blacks sought could not be successful without the overthrow of the world capitalist system. In 1927 the Communists sponsored the League Against Imperialism and the Profintern, the Red Workers International. In 1930 the Profintern organised an international trade union conference of Black workers and began publication of the *Negro Worker*, a journal published in Hamburg which circulated among Black sailors calling at that port. The response in West Africa was negligible but out of these activities emerged two leaders of prominence, the Trinidadian, George Padmore, and the Sierra Leonian Creole, Isaac Wallace-Johnson.

In 1935 Benito Mussolini, an ally of Adolf Hitler, instigated a border dispute with Ethiopia which led to the Italian invasion of that ancient empire. The issues surrounding the crisis were of major world significance. Mussolini chose to base his arguments on the claim of the racial superiority of Italians over Ethiopians. He argued that Italy could bring the benefits of her civilisation and Christianity to Ethiopia, that the doctrine of survival of the fittest dictated that Italy must solve her excess population problem at the expense of Ethiopia. He exclaimed:

> Has the League of Nations become the tribunal before which all the Negroes and uncivilised peoples, all the world's savages, can bring the great nations which have revolutionised and transformed history?[42]

The universal Black protest which followed resulted in riots and disturbances in almost every city in the world with a Black population. Afro-Americans and South African Blacks demonstrated in the streets, West Indians and West Africans offered to enlist in the Ethiopian army. In Nigeria the reaction was unique because it was the first time that her larger ethnic groups—Hausa, Yoruba and Ibo—had ever united in protest.

Ethiopia had come to symbolise all the things which Blacks were constantly being reminded that they did not possess: an ancient history stretching back into the Biblical era, a venerable monarchy, a Christian Church dating from apostolic times, a decisive military victory over a European army during the partition and, most of all, the dignity of independence and international recognition. In fact, in African thinking, the adjective 'Ethiopian' meant not so much a country as a condition of independence. The *Nigerian Daily Telegraph* called the ancient kingdom of Ethiopia 'the sole remaining pride of Africans in Africa, and Negroes in all parts of the world'.[43]

42 *Nigerian Daily Telegraph*, 24 September 1935, p. 3, quoting Mussolini.
43 *Nigerian Daily Telegraph*, 6 July 1935, p. 1.

Britain and France somewhat shamefacedly 'sold' Ethiopia to Italy. This betrayal appeared to indicate that in any clash between the races, whites would stand together regardless of moral issues or international agreements. The betrayal of Ethiopia convinced some Blacks that colour was the unbridgeable gulf and that the great struggle of the future was between black and white. Azikiwe was still bitter as late as 1940.[44] The Ethiopian war was in reality a prelude to the Second World War from which liberal idealism emerged stronger and the doctrine of white or Aryan superiority disgraced, at least temporarily, with the fall of Hitler.

By the mid-thirties the new élite of West Africa was almost voiceless and leaderless. Its political organisations had either collapsed or become so conservative as to be almost unofficial organs of the colonial governments. A new beginning was necessary and it came in the form of four youth movements: the Nigerian Youth League organised in Calabar in 1934 by Eyo Ita, the Lagos (later Nigerian) Youth Movement of the same year, and in 1938 the Gold Coast Youth Conference organised by J. B. Danquah and the West African Youth League under the direction of Wallace-Johnson in Freetown. In aims they differed little from their predecessors of the twenties. The Nigerian Youth Charter called for '. . . complete autonomy within the British Empire . . . a position of equal partnership with other member states of the British Commonwealth'.[45] The Gold Coast Youth Conference echoed Tete-Ansa and John Ayew in warning:

> . . . that the system of trade and commerce obtaining in the country by which its national wealth is exploited for the benefit of the foreigner at the expense of the native is dangerous to the economic stability and permanence of the people of this country . . .[46]

In Nigeria and Sierra Leone, the West African approach of the early twenties was still popular. Eyo Ita established a *West African* Peoples' Institute, Wallace-Johnson a *West African* Youth League which sought self-government for Sierra Leone within a West African federation and Azikiwe the *West African Pilot* which not only in name but in policy stressed a united West Africa: 'So long as we think in terms of Nigeria, Gold Coast, Sierra Leone, Gambia and not as one United West Africa we must be content with a colonial dictatorship. . . .'[47]

For the Creoles of Freetown and Bathurst it was particularly difficult to think exclusively in terms of the individual colonial units. For the whole of their history Creoles had thought of British West Africa as one unit.

44 See Zik's 'Inside Stuff', *West African Pilot*, 1 April 1948 and the *Pilot*'s leader, 31 August 1945, 'UNRRA allocates 50 Million Dollars to Relieve Aggressor Italy. Raped Ethiopia is Given Nothing'.

45 Nigerian Youth Movement, *Youth Charter and Constitution and Rules*, Lagos, n.d., p. 1, quoted in Coleman, p. 225.

46 Gold Coast Youth Conference, *First Steps Towards a National Front*, Accra, 1938, p. 10.

47 *West African Pilot*, 21 July 1938.

Many Freetown families had relatives in Nigeria, Ghana and Gambia and many from these sister colonies had received their university training in Fourah Bay College. Intermarriage with the Efik, Yoruba, Fante and Wolof further strengthened their ties along the coast, rather than with the interior peoples of their own territory. British policy had encouraged a division between the Creoles of Freetown and Bathurst and the people of the interior. Independence for Sierra Leone and the Gambia would inevitably submerge the Creole minority and give political power to the interior peoples. In addition because of their small size and few resources, financial problems were likely to be more acute than in Ghana and Nigeria.

Once Ghana and later Nigeria turned away from the concept of an English-speaking West African state—and this was not clear until well after the Second World War—the Creoles were compelled to accept the individual colonial units as the potential states of the future. It was ironic that the Creoles—politically the most sophisticated West Africans and with the longest tradition of modern political agitation—were the slowest in British West Africa to organise their independence movement.

Ghana was the first to turn away from a West African and Pan-African, to a strictly Ghanaian, approach to politics. Danquah, in contrast to Azikiwe, said, 'When I say "we" I mean the Gold Coast. I do not mean black men. I do not mean Negroes. This is not a question of race at all.'[48] This was not a new attitude in Ghana. As far back as 1920 the A.R.P.S. had urged a Ghanaian approach and condemned the National Congress because it sought to involve all British West Africa. It was easier for Ghanaians to think in terms of the colonial unit than for any of the other peoples of British West Africa because the Akan made up about half of the population and because of the fairly long tradition of cooperation between the chiefly and Western-educated élite. This was strikingly demonstrated in the cocoa holdup of 1937 when the Asantehene, the chiefs, the Fante, Asante and other ethnic groups supported John Ayew and the farmers. The state-wide character of the Cocoa Holdup movement was unique in the inter-war period in British West Africa. A few months later Danquah organised his Gold Coast Youth Conference. The timing could hardly have been more opportune, and as a result the Youth Conference was more truly representative of Ghana than the Nigerian or Sierra Leonian organisations could ever hope to be of their respective territories. The stage was set in Ghana for an independence movement backed by a coherent and comprehensive patriotism relatively free from the ambiguities and fissiparous tendencies so apparent in other British West African colonies.

All the youth organisations sought to become more broadly based than their predecessors, both in terms of ethnic as well as 'class' composition, although none succeeded to the extent of the Gold Coast Youth Con-

48 J. B. Danquah, *Self-Help and Expansion*, Accra, 1943, p. 18.

ference. Wallace-Johnson's Youth League was the first Creole-based political organisation to include Mende and Temne in its membership and to attract the workers and trade unions. The Yoruba-based Youth Movement was the first to include non-Yoruba, especially Ibo, and to concern itself with the problems of lorry owners.

In Nigeria the problem of potentially conflicting nationalisms and their role in a movement towards independence was particularly acute. No one nation occupied the central position which the Akan did in Ghana. There were numerous national groups of which the big three—Hausa, Yoruba and Ibo—presented only the most immediate problem. Both the Youth Movement in Lagos and Youth League in Calabar were conscious of these problems. The Youth Movement set as one of its goals the search for 'better understanding and cooperation and a creation of a common ideal among the tribes of Nigeria'. Eyo Ita stated more precisely the Nigerian dilemma. 'I believe that nation-states are out-of-date and I believe also that empires are out-of-date. . . .' Ita was repeating the philosophy of the new élite since the conquest. The British Empire must be broken up by an independence movement but it could not be done in the manner of the Austro-Hungarian Empire by division into states based on ethnicity.

To his own Efik people Eyo Ita proposed a hierarchy of loyalties, Calabar nationalism—'the love of Calabar must consume us . . .'—Nigerian patriotism—'the whole of Nigeria is a supreme value before God . . .'—and Pan-Africanism—'. . . our mission of service to Mother Africa must become unquenchable'. This was ideal as long as the good of Calabar worked to the good of the Nigerian state and mother Africa. If the interest of Calabar had to be sacrificed to the Nigerian state, there was a danger that the British Nigerian Empire might fragment into ethnic-based states. For Ita who said, 'We are either Calabarians or nothing' and for many other Nigerians, loyalty to the ethnic group held the highest priority.[49]

Between 1919 and 1939 the reform, emancipation and independence movements scored few noticeable successes. Few of the major demands going back to the nineteenth century had been conceded. There had been no real advance towards self-government, no placing of responsibility in African hands in church, state or commerce, no creation of a university to produce skilled manpower, no attempt to right the inequalities of the colonial economy and no effort to attack segregation and discrimination against the African on his own soil. The longer the colonial period lasted the more entrenched and intractable these problems became. It could be argued that British West Africans were actually slipping backwards in the creation of modern societies.

Britain's supremacy appeared unchallenged in her West African colonies

49 Eyo Ita, *Nigerian Youth League Movement*, Calabar, n.d., pp. 6–7.

in 1939. There were too many contradictions in the ranks of West African leaders for a unified assault upon the imperial position to have been predicted. The Western-educated élite was divided as to what should be the arena for such an assault; the ethnic group, the individual colony, British West Africa or the black world. Furthermore the division had not been bridged between chiefly and Western-educated élites. The generation of chiefs in 1939 appeared content with their position, salaries and status. What was more, they seemed to enjoy a monopoly of the loyalty of the masses. It was difficult to predict that the Western-educated élite could, within the foreseeable future, draw this loyalty towards itself. Even the new élite was divided, a large element being collaboration-oriented, cautious and even sceptical about African leadership. Many desired change but moderate, evolutionary change. If the leadership was moderate, the masses were believed to be thoroughly conservative, looking upon Europeans with awe and ascribing to them almost god-like qualities and powers. Could the masses be aroused without playing upon ethnicity and if this was done would it not fragment the colonial state to which the new élite was also committed?

CHAPTER 17

The 1939–45 war and West Africa

MICHAEL CROWDER

Colonial rule between the two world wars can be characterised as a period of social, political and economic stagnation.[1] The prolonged depression of the thirties and the depression following immediately after the First World War meant that little social and economic development could take place in the two decades between the wars. Nigeria, with the most diversified economy of any in West Africa, had a total revenue in 1938–9, the year before the outbreak of war, lower than that of 1925–6 (£5,811,088 as compared with £8,268,928). Her total external trade for these two years was £33,648,510 in 1925 as against £25,957,766 in 1938, though the figures for 1937 had been exceptionally high at £38,143,340. Generally the years between the wars saw little economic development in Nigeria or in any of the other West African colonies. This was largely because the funds were just not available due to the depressed state of the world economy; partly it was because the colonial philosophy of the day held that colonies should be self-supporting and development should not be a charge on the metropolitan budget. However, both France and Britain had taken some hesitant steps towards a philosophy that economic and social aid from the metropolis to the colony would not only benefit recipient and donor but was morally obligatory on the colonial power, in the form of France's Great Empire Loan and Britain's Colonial Development Fund. But the funds made available were small: the four British West African colonies, for instance, received only £419,466.

Politically little change took place in this period: African participation in the process of government in British West Africa was still limited to a few seats on the Legislative Councils. The real seat of power, the Executive Council, was beyond the reach of Africans and in the Legislative Council they were a small minority. In French West Africa the Senegalese had

1 For a detailed study of this period in French and British West Africa see Michael Crowder, *West Africa under Colonial Rule*, London, 1968.

control over the Colonial Council and continued to elect their deputy to the Chambre des Deputés. But in the other Colonies the only concessions made were a few municipal reforms and the inclusion of a few Africans on the Lieutenant Governor's Councils: but they had little effective influence on the process of government. Outside the Quatre Communes the number of Africans assimilated to the position of *citoyen* was small, only 2,136 in 1936.[2]

The major impact of the two colonial powers on West Africa before the war can be said, from the African point of view, to have been largely that of peace-keepers. Vast areas of West Africa remained untouched by economic or social innovation. The colonial powers themselves were controlled neither by criticisms of their politics from within their territories nor from outside: the typical image of that era was of the calm district officer settling 'palavers' or overseeing the collection of tax.

This calm was shattered by the Second World War. In the first place the West African territories changed from colonial backwaters into strategic areas important to the Allied conduct of the war. The question of the loyalty of the African population, hitherto largely assumed, was brought to the fore both by the split in French Black Africa between Vichy and Gaullist supporters and by the propaganda warfare conducted by the Germans to alienate the Africans from their colonial masters. Africans were needed to fight for the Allies: indeed at one time the French army was primarily an African one. Finally, the Americans and Russians made it clear that they were not fighting to preserve the empires of their French and British allies and voiced their criticisms of empire openly. Even within the British and Free French Governments anti-imperialist voices were to be heard.

As a result, apart from the new strategic and economic role of West Africa and the impact of the enlistment of thousands of her men as soldiers, the war heralded a number of reforms which the colonial powers felt obliged to make in the light of the radically changed socio-political situation. The Free French government, based as it was for a long time on French Equatorial African soil, and depending principally on African troops, promised at Brazzaville in 1944, political, social and economic reforms that were to change the whole course of history in French Tropical Africa. Similarly in British West Africa, under pressure from both the Americans and the Labour members of the war-time Coalition Government, the British introduced a series of economic, political and social changes that were to lead logically to self-government and independence a decade or so later.

MILITARY IMPLICATIONS OF THE WAR

The French and British administrations in West Africa had hardly had time

2 Lord Hailey, *An African Survey*, London, 1938, p. 201.

to adjust to the fact of the Second World War before France capitulated to the Germans and on 21 June 1940 an armistice was signed whereby Southern France was left in French hands under the so-called Vichy Government of Marshal Pétain and his Premier, Pierre Laval.

In French West Africa the main feature of the short months before the capitulation of France was the recruitment and training of African soldiers for the French army. In the First World War, some one hundred and eighty thousand Black African soldiers had seen service on the European front.[3] Before the capitulation eighty thousand African troops had already been sent to France,[4] while in French West Africa itself there were some one hundred and eighteen thousand troops at the disposal of the French, the majority of them reservists or recruits enlisted as a result of the 1939 and 1940 recruitments. Of these only twenty thousand were considered sufficiently trained to go into battle.[5] In British West Africa intensive recruitment was carried out and the Royal West African Frontier Force was raised from its pre-war level of eight thousand to one hundred and forty-six thousand.[6] Recruitment was initially on a voluntary basis, compulsory service as in French West Africa never having been introduced by the British. In Northern Nigeria, which supplied by far the largest number of soldiers during the war 'no pressure was employed',[7] even though chiefs were the agents for recruiting, and the borderline between voluntary and compulsory recruitment was sometimes very fine. For example, the District Officer of Borgu in April 1941 wrote to his Resident that he was convinced that the District Head of Yashikera 'has mixed up recruiting with the compulsory registration of adult males and that he proposes to adopt press-gang methods'.[8]

In British West Africa, however, under the Compulsory National Service Regulations of 1940 certain classes of persons such as blacksmiths, bricklayers, carpenters and those with certain educational qualifications were required to register for possible enlistment.[9] As far as the rank and file were concerned these were to be recruited only from 'approved tribes' and in Nigeria never from a line below that drawn from Okuta through Jebba, Keffi, Jalingo and Yola. However, with the increasing

3 M. Crowder, 'West Africa and the 1914–18 War', *Bulletin de l'IFAN*, T sér B. No. 1, 1968, pp. 227–47.

4 Jacques Richard-Molard, *Afrique Occidentale Française*, Paris, 1952, p. 165.

5 Maurice Martin du Gard, *La Carte Impériale: Histoire de la France Outre-Mer 1940–1945*, Paris, 1949, pp. 137–8. N.B. This is an explicitly pro-Vichy account of the war years in the French Empire but has some very useful material despite its bias.

6 *The Colonial Empire (1939–47)*, Cmd. 7167, London, 1947, p. 116.

7 Sir Ahmadu Bello, *My Life*, London, 1962, p. 53.

8 N[igerian] A[rchives] K[aduna]/BORGDIST/DOB/624, 'Recruiting', D. O. Borgu to Resident Ilorm, 18 April 1941.

9 *Ibid.*, Chief Secretary, Lagos to Secretary, Northern Provinces, 10 March 1942.

demand for soldiers, this barrier was abolished and potential recruits were no longer refused on grounds of ethnic origin.[10]

Under the terms of the armistice France's colonies were to be neutralised which meant the disbandment of the majority of France's Black Army, whose level was reduced to twenty-five thousand men. French West Africa ceased formally, like Vichy France, to have a military role in the war. Her reduced forces were merely for self-defence: economically she was limited to providing food for the truncated metropolis, that is if she could get food supplies through the blockade imposed by the British, for whom the establishment of the Vichy regime in Dakar was a potential threat. Shortly before the establishment of the Vichy regime in Dakar the bulk of the Royal West African Frontier Force had embarked for East Africa for the campaign against the Italians in Abyssinia (Ethiopia) and Somalia. This left the four British territories practically defenceless. They had no aircraft except some antiquated ones in Freetown belonging to the Navy. Freetown, vital as a convoy centre, and Takoradi, where planes were off-loaded in crates and assembled to be flown to the Middle East via Nigeria and Sudan, were both within bombing range of French West African air-fields. The British, fearful that the neutrality of the Vichy regime would not survive German pressures to use Dakar as a base for attacks on Freetown, hurriedly increased their forces in West Africa. Three British battalions were shipped to Freetown and part of the Gold Coast Regiment was sent to the Gambia.

Right up to 21 November 1942, when the French West African administration decided to join the Allies the British in West Africa looked nervously across their borders at their French neighbours, especially at Dakar with its large port, harbouring part of the French navy, and its air-base with a comparatively large number of planes. One of the reasons for the return of the West African troops before the end of the East African campaign was anxiety about the security of Freetown.[11]

The fall of France and the setting up of the Vichy Government produced consternation among the overseas French administrators. On the one hand was their loyalty to a legitimate government of the Third Republic presided over by Marshal Pétain, the great hero of Verdun who had successfully defended France against the Germans in the First World War. On the other hand was the call of General de Gaulle to rally to his Free French Government in exile. If de Gaulle could secure the adherence of the French colonies his government would be more than a paper government. It would have land and, more important, men to recruit as soldiers. It could also supply the Allies with much-needed foodstuffs. The administrators themselves, even before de Gaulle's call was made, had considered

10 NAK/BORGDIST/DOB/307, 'Soldiers and Police: Recruiting Instructions'.

11 The above is based on A. Haywood and F. A. S. Clarke, *The History of the Royal West African Frontier Force*, Aldershot, 1964.

whether a bloc comprising France's African colonies might not be formed to continue the struggle against France. Governor-General Léon Cayla even sent a mission to North Africa to find out what General Noguès, General Officer Commanding in French North Africa, planned in this direction but finally decided he had not the forces to resist a German invasion.[12] What seems clear was that if, say, French West Africa had rallied against the Germans, the latter would almost certainly have invaded French North Africa. This certainly was the possibility that was uppermost in the minds of French administrators and is used as a justification by pro-Vichy writers for the policy pursued in French West Africa: by maintaining its neutrality, the Germans were kept out.[13]

As far as the European combatants were concerned, French Black Africa was of strategic importance for two reasons over and above its resources in men and materials. First, Dakar was the third largest port in the French Empire after Marseilles and Le Havre. Secondly, Niger and Chad had common frontiers with the Italian colony of Tripolitania (Libya) while Chad gave access to the Anglo-Egyptian Sudan and the British East African Colonies, all of which were threatened by the Axis.

An important factor in the decisions made by the various French colonial governors about which side to support was the destruction of the French fleet at Mers El Kébir in Algeria. Under the terms of the armistice the large French fleet was to be neutralised. One thing the Allies did possess in their favour at the time was British superiority over the seas. The British were very frightened that the French fleet, though neutral, would be taken over by the Germans whose own fleet was a good deal smaller. The British therefore issued an ultimatum by which the French fleet was offered the alternatives of joining the British fleet, of having its ships disarmed by the British who would guard them in their own ports until the end of the war, or of scuttling their ships themselves. The French refused all three alternatives so the British destroyed the bulk of the fleet in dock at Mers El Kébir, killing over fourteen hundred French sailors.[14] This action, justified by the British on the grounds that they could not afford the remotest chance of the French ships falling into the hands of the Germans or Italians, indicated Britain's lack of confidence in the Vichy regime and was reflected by its administrators in West Africa. The incident embittered many Frenchmen to such a point that they could not accept working with the British under de Gaulle. Furthermore in Dakar the reality of Mers El Kébir and the loss of French lives was brought home when the British crippled the French battleship *Richelieu*, then in port there.

12 Daniel Chenex, *Qui a sauvé l'Afrique?*, Paris, n.d.
13 *Ibid.*; also du Gard, *La Carte Impériale*.
14 According to du Gard, p. 39.

In French West Africa, while de Gaulle had some supporters, the majority of the administration preferred the easier option of neutrality under the terms of the armistice and acceptance of the authority of Cayla's Vichy-appointed successor, Pierre Boisson.[15] It must also be remembered that de Gaulle was a relatively obscure and not very senior French general.

A former Commissioner for Cameroun, Pierre Boisson, himself a veteran of Verdun under Marshal Pétain, where he had lost a leg, was appointed Vichy High Commissioner for Black Africa, that is both French West and Equatorial Africa. His headquarters was Dakar, and he secured the support of all the constituent territories of French West Africa. Not so Equatorial Africa. There, Félix Eboué, the French Guyanese Governor of Chad and the first Negro to be appointed a governor in French Black Africa, rallied to de Gaulle in defiance of Boisson, his High Commissioner, who sent him a telegram: 'By your decision you have betrayed the duties with which you were charged. By taking the initiative in handing over to England the territory confided to you, you have by a deliberate and plotted act, broken the cohesion of the Empire.'[16] This was a major triumph for the Allies, since it gave them a staging post at Fort Lamy through Nigeria to East and North Africa. Shortly afterwards, Cameroun, whose High Commissioner Brunot was hesitating over supporting de Gaulle, was taken over by Colonel Leclerc in a rapid *coup d'état* staged from Nigeria. This secured the Chadian hinterland, for Chad had no territorial frontier with Nigeria, only the Lake. Soon afterwards, Congo-Brazzaville rallied to de Gaulle, and the pro-Vichy administration of Gabon was ejected by force. This gave de Gaulle the much needed land-base for his Free France. Eboué then became Governor-General of the Equatorial federation based on Brazzaville.

Boisson was thus effectively Governor-General of French West Africa alone. Only one administrator of major importance there tried to go over to the Free French: Ernest Louveau, head of Upper Volta, then dependent on the Ivory Coast. In an abortive mission to Dakar to persuade Boisson to support de Gaulle he merely succeeded in getting himself imprisoned.[17] Boisson's policy was clear: to preserve the neutrality of French West Africa to the extent of refusing any German access to it. This policy he succeeded in pursuing and it is quite clear that, despite propaganda to the contrary, no German submarine ever used the port of Dakar, and only one

15 When he was appointed High Commissioner for Black Africa Pierre Boisson had to work carefully at first, on his own admission, because of elements favourable to the Free French cause. Jean Suret-Canale, *Afrique Noire: L'ère coloniale 1900–1940*, Paris, 1964, p. 571.

16 *Paris–Dakar*, 30 August 1940.

17 See his pro-de Gaulle version of the situation in French West Africa at this time: E. Louveau, *'Au Bagne': Entre les griffes de Vichy et de la milice*, Bamako, 1947.

Manning the defences against the unsuccessful Anglo-Free French attack in Dakar

German official ever visited Dakar and that on sufferance and under the close surveillance of Boisson himself.[18]

If Boisson ever doubted that he had the support of the majority of the French for his policy, such doubts were removed by the futile attempt of the Free French and British to invade Dakar. De Gaulle had persuaded Churchill that he had enough supporters in Dakar to take it without difficulty and an invasion force was assembled at Freetown. On 23 September 1940 de Gaulle sent emissaries to Boisson, backed up by part of the British fleet and a Free French force to persuade him to rally to his cause. One group of emissaries was arrested, the other group managed to escape. In the face of Boisson's refusal to negotiate, the combined Free French and British forces attacked Dakar, trying to land troops. Boisson successfully defended Dakar against the assault. The deaths of both French and Africans at the hands of the Allies hardened attitudes against de Gaulle and gave Boisson clearer support for his policy. This was confirmed by an American visitor to Dakar at the time, Paul M. Atkins, who wrote: 'Practically all the

18 This official was a German Foreign Ministry official, Mülhausen, kept under close surveillance by Boisson. The other German to visit Dakar was Dr Klaube, a former representative of Lufthansa in Gambia, who came to supervise the repatriation of Germans interned in Dakar on the outbreak of the war. Mülhausen passed only nineteen days in Dakar in September 1942. Verification of this lack of German activity was given by the American Consul-General in Dakar, Thomas C. Wasson, in an article, 'The Mystery of Dakar', *American Foreign Service Journal*, xx, 42, April 1943. Wasson was Consul during the Vichy period even after the Americans had joined the Allies.

citizens of the city, regardless of their basic sympathies, were proud that the British were successfully repulsed.'[19]

Thereafter Boisson continued his policy of strict neutrality though he restructured the Black African army in such a way that he did not arouse the suspicions of the Germans to the point of intervention. By the time he rallied to the Allies in 1942, he had, according to du Gard, a well-trained African army of one hundred thousand with a half-brigade of the Foreign Legion and a European battalion together with supporting arms, as well as naval vessels and aeroplanes.[20] According to Boisson's supporters, these forces accepted the decision to rally to the Allied cause in 1942 as 'the normal result of a policy whose inspiration, obligations and objectives they had understood. And it was from a well-controlled country that they left for battle.'[21] When the Allies landed in French North Africa in 1942, Boisson decided to rally to them; he thus put over a hundred thousand soldiers at their disposal as well as providing them with the important port of Dakar and vital staging posts for their aircraft. It was the French Government in Algiers under General Gouraud to which he adhered in November 1942 and his decision to do so was not motivated by any concern with de Gaulle. In Dakar, de Gaulle's name was not even mentioned until April 1943, by which time he had become co-president of the government of which he was soon to become exclusive head.[22]

De Gaulle never forgave Boisson for his refusal to turn French West Africa over to Free France in 1940 and he revenged himself by having Boisson imprisoned and nationally disgraced. Boisson had vehement supporters, even after his death, who felt that his policy had been the right one and that de Gaulle had shown gross ingratitude in not recognising that, by preserving French West Africa intact, Boisson had, when the time came, presented Free France with a most important supply of men and provisions as well as a strategically vital part of the Empire.[23]

For the British administration in West Africa the changeover of regime came as a great relief and meant that troops required for defence against a possible attack from French West Africa could be released for the Burma campaign. Furthermore the British blockade of the French ports was

19 Paul M. Atkins, 'Dakar and the Strategy of West Africa', *Foreign Affairs*, xx, 2, 1942, p. 362. A few pro-de Gaulle French were, however, interned at the time and deported, including the Mayor of Dakar, Alfred Goux, and Silvandre, a West Indian, who later became a Deputy and President of the Chamber of Commerce.

20 du Gard, pp. 137–40.

21 *Ibid.*, p. 140.

22 See, for instance, the principal Dakar newspaper of the day: *Paris–Dakar*.

23 See du Gard, *La Carte Impériale*; Chenex, *Qui a Sauvé l'Afrique?*; and Atkins, 'Dakar and the Strategy of West Africa', who quoted Boisson as saying, 'I defended Dakar against the British; I will defend it a hundred times more so against the Germans' (p. 365). *The Times*, 29 January 1943, wrote that Boisson could not 'be regarded as pro-German in sympathy' and was 'in a narrow and rigid way patriotic'.

lifted and West Africa became a fairly safe staging post for the Allies en route to the Middle and Far East.

For the French African the changeover of regime meant a return to systematic recruitment and service in the Allied armies: over one hundred thousand left for the front between 1943 and 1945.

Much play was made by the Gaullists of the 'loyalty' of the African to a Free France; Jacques Stern, Free French Minister of Colonies, in a eulogistic account of *The French Colonies* in which, he declared, no discrimination existed, speculated that the unknown soldier who rests under the Arc de Triomphe might be a coloured Frenchman.[24] But in reality as far as the majority of Africans were concerned the quarrel between Frenchmen—pro-Vichy and pro-de Gaulle—was of marginal interest. Indeed it is one of the myths of French colonial history that the Africans were ready to regard themselves as Frenchmen and fight for France's liberty. As the Camerounian novelist, Mbella Sonne Dipoko, wrote in a recent review about the Free French regime in Cameroun:

> I remember, as a child, seeing Camerounian men being conscripted by non-Camerounian Bambara soldiers [from French Soudan]. Some hid in the bush. Many others who were taken away clearly went against their will, not because they didn't want to fight against Nazi Germany and on the side of France, but simply because they couldn't be bothered one way or the other. . . .
>
> It wasn't Africans who vacillated between the Free French Movement and the Vichy regime. The *right* even to vacillate was denied them because, although French forces had been defeated in France, French-speaking Africans were still a colonised people, there were French-speaking officials on the spot to carry on with the job of autocratic rule. It was these officials who were, with the hand-picked African *notables* associated with them, for or against de Gaulle. And once they had decided one way or the other, all they did was to issue orders for the rank and file to follow suit.[25]

Apart from the potential danger of the Vichy regime in Dakar going pro-German, there was no military action in West Africa apart from Leclerc's spectacular *coup d'état* in Cameroun launched from Nigeria. Early in the war Nigerian soldiers had been sent to the Southern Cameroons in case of any trouble from the Germans working on the plantations there. Apart from severe rationing and shortages of imported goods and the enormous flow of Allied troops, planes and ships, the West African territories felt the war at long distance in North and East Africa and Burma where so many of their troops were sent.

24 Jacques Stern, *The French Colonies*, New York, 1944 (trans. Norbert Guterman), p. 12.
25 Mbella Sonne Dipoko, review in *Journal of Modern African Studies (J.M.A.S.)*, vii, 4, pp. 752–3.

In East Africa the Nigerian and Gold Coast Regiments played a leading part in the defeat of the Italians. In the invasion of Abyssinia, as General Cunningham, the G. O. C. East Africa, reported, the Nigerian soldier led the pursuit of the Italians and 'unaccustomed to cold and damp, fought his way from the hot dusty bush to the west and cold highlands of Abyssinia, where he maintained his cheerfulness and courage in spite of strange conditions and strenuous climbing operations made necessary by the terrain'.[26] The return of Emperor Haile Selassie to take over his liberated country was marked by a guard of honour supplied by colonial Nigerian soldiers.

Britain, desperately short of metropolitan troops at the time, could never have won the campaign in East Africa without the reserve of troops she was able to call on from her British West African colonies.

No less remarkable was the contribution made by the West African soldier in Burma. There conditions were even stranger than in Ethiopia. While popular fiction has visions of Africans living in jungle, none of the West African soldiers had ever experienced anything like the jungle of Burma. From November 1943 to the end of the Burma campaign in September 1945, West African troops fought for the British against the Japanese. This was the first time British West African troops had been used outside their own continent, if one excepts the carriers used in Mesopotamia in the First World War. In Burma the West Africans were indispensable both as soldiers better suited to the conditions of jungle warfare than their white comrades, and also as carriers who could transport supplies over long distances in terrain impassable to vehicles. The West African formations in Burma were as a result much more mobile than the white formations relying on mules or other forms of transport.

Many of the soldiers came from technologically very backward groups—most were illiterate on recruitment. Yet they fought with modern weapons against a fanatically brave enemy in strange and appalling conditions with such success that both Japanese enemy and British commanders had the highest regard for them. A captured Japanese diary noted:

> The enemy soldiers are not from Britain but from Africa. Because of their belief they are not afraid to die, so even if their comrades have fallen they keep on advancing as if nothing had happened. They have excellent physique and are very brave, so fighting against these soldiers is somewhat troublesome.[27]

The Commander of 82 (West Africa) Division, General Sir Hugh Stockwell, wrote of his soldiers:

> The staunch determination of the African soldier in the face of not only a

26 Cited in Haywood and Clarke, p. 347.
27 *Ibid.*, p. 470.

> fanatical and determined enemy, but also terrain in which no self-respecting soldier would ever seek battle, carried us through. . . . They fought splendidly and stood up to much privation and hardship.[28]

THE ECONOMIC CONSEQUENCES OF THE WAR

French West Africa

Before the fall of France, French West Africa assumed its role of the First World War as a source of supply of foodstuffs for the metropolis. Léon Cayla, appointed in August 1939 as Governor-General, established committees at every administrative level to requisition foodstuffs while the possibility of economic cooperation between the French and British colonial administrations was explored by Malcolm Macdonald and Georges Mandel, respectively British and French Colonial Secretaries. For a brief period the rigid political and economic barriers established during the Colonial era between the two sets of colonial administrations were broken down. At the opening of his Council of Government, Cayla announced that priority would be given to construction of roads and railways and improvement of ports and airfields. Eight thousand five hundred kilometres of tracks were to be marked out and work was to go ahead on the second stage of the Mossi railway linking Ouagadougou to Abidjan.[29] This was interrupted in June 1940 by the Armistice. Though French West Africa was still supposed to supply the metropolis with foodstuffs, the British naval blockade brought imports and exports to a virtual standstill. Groundnut production fell from 419,000 tons in 1940 to 114,000 in 1942, the lowest it had been since 1906,[30] while cocoa production fell from 55,185 tons in 1939 to 542 tons in 1943.[31] Though this meant that the peasant was no longer compelled to produce export crops and thus had time to produce subsistence crops, it deprived him of cash income. In certain areas like Senegal, where there was great dependence on imported rice as a staple, the peasant was very hard hit. On the other hand, lack of imported cloth and other manufactured goods gave a new lease of life to the indigenous weavers and blacksmiths.

In April 1943 Boisson presaged the economic significance of the change-over of regime from Vichy to Free French in his declaration of French West Africa's new role: 'We must produce. The next groundnut crop in Senegal must in particular be a great success. . . . I conclude with an equation: Work = Victory = Liberation of the Motherland.'[32] In Senegal under

28 *Ibid.*, p. 471.
29 *West Africa*, 27 January 1940.
30 Discours prononcé par M. Hubert Deschamps, Gouverneur du Sénégal, à l'ouverture de la session ordinaire du Conseil Colonial, 23 August 1943.
31 F. Amon d'Aby, *La Côte d'Ivoire dans la cité africaine*, Paris, 1951, p. 77.
32 *Paris–Dakar*, 21 April 1943.

the Free French a 'Battle of Groundnuts' was undertaken to increase production to pre-war levels.

The call for greater economic output was hampered by lack of transport and equipment as well as the imports which served as a means of exchange for the peasant farmer's export crops. Cut off from metropolitan sources, the Free French had to depend on the Allies for help.[33] To achieve increased production the administration requisitioned fixed amounts of produce and ensured the despatch of seasonal labour from Guinea and Soudan to work on the groundnut fields of Senegal. Forty-five thousand *navétanes* were despatched in 1943 'with the cooperation of the governments of these two territories'.[34] The battle for groundnuts was not so successful as had been anticipated; Governor Desschamps' target had been four hundred thousand tons; actually only 275,000 tons were produced, partly because of sales over the border in the Gambia where a higher price for them was paid.[35]

Such were the administrative pressures that chiefs were ordered to produce so much of a particular crop on pain of house arrest. Cercles were required to produce crops they did not grow. There is the famous story of the cable received by one French administrator instructing him to produce honey in his district. He replied. 'Agreed honey. Send Bees.'[36]

Richard-Molard wrote of the Free French exactions thus:

> . . . one cercle is required to produce so many tons of liana rubber, even though no liana grows in the territory. The native is therefore forced to travel on foot, sometimes over long distances, to buy rubber elsewhere, regardless of cost. He must sell this to the Commandant at the official price which is several times lower than the purchasing price to escape the hands of justice.[37]

While the peasant once again had access to cash income, the strict rationing of the war and the activities of German submarines meant that few imports reached West Africa. Furthermore the peasant had to devote time he would normally have spent on subsistence crops to export crops. Ivory Coast in particular suffered under the Free French regime. According to Amon d'Aby, farmers there were forced to change from one crop to another and since they were *sujets* they could not refuse. Plantations were destroyed on the grounds of disease, though often this was motivated by the fact that they were in competition with European plantations, which anyway benefited from the premiums paid on all plantations of twenty-five or more contiguous hectares—there being few African plantations of this

33 Discours prononcé par le Gouverneur-General de l'AOF, Monsieur Pierre Cournarie, à l'ouverture de la session du Conseil du Gouvernement, December 1943.
34 Deschamps, 'Discours du 23 aôut'.
35 Discours du M. Charles Dagain, Gouverneur du Sénégal, à l'ouverture de la session ordinaire du Conseil Colonial, 3 October 1944.
36 Jacques Richard-Molard, *Afrique Occidentale Française*, Paris, 1956, p. 167.
37 *Ibid.*, p. 167.

size. Europeans were also given priority in requisitioning forced labour.[38]

While the peasant suffered under the Free French regime with its determination to produce what the Allies needed regardless of the wants of the peasants, major developments took place in the cities where port and aviation facilities were improved. Furthermore, with half France occupied by the Germans and the Vichy portion opposed to de Gaulle, groundnuts could only be processed on the spot, which led to an expansion of industrial activity. There was also a revival of small industries which supplied other goods impossible to obtain from abroad. Perhaps the most significant development under the Free French was the intervention of the government in the economy on a scale it had never undertaken before. This was even more characteristic of the development of the economy in British West Africa during the war.

British West Africa

While plans for cooperation between France and Britain in West Africa were short-lived, a revolutionary new departure in the administration of the four British West African colonies resulted from the war. To ensure maximum exploitation of resources for the war effort, a Minister Resident in West Africa with Cabinet rank was appointed to coordinate economic policy and in particular the activities in West Africa of the British ministries of War, Transport and Food Supply. The Resident Minister could settle any immediate problems without reference to Whitehall. While this did not involve any change in the constitutions of the territories, it did mean for the first time that there was overall coordination of the economic activities of the four British West African administrations. In December 1943 Mr Noel Hall, the former Director of the British National Institute of Social and Economic Research, was appointed Development Adviser for West Africa.

This move towards coordination of economic policy in West Africa was accompanied by a related change as the government changed its previous attitude of laissez-faire to one of increasing control of, and intervention in, the economy of the four West African colonies.[39] Where the large commercial companies had had practically free rein in the pre-war years, the new systems of quotas, price-controls, and regulation of imports and exports meant their increasing subjection to government demands. The staple export of the big companies, the cash crops, were now purchased at fixed prices by the government, with the companies acting only as their agents. Marketing was supervised by the West African Produce Marketing Board. This is not to say that the large companies suffered as a result of government intervention, since as agents for purchase in a market which was expanding because of war-time demands, their level of activity

38 Amon d'Aby, p. 74.

39 For a more detailed discussion see my *West Africa under Colonial Rule*, London, 1968.

increased after the lean years of depression. Indeed government policy favoured the large firms, to whom it gave licences, rather than to more recently created firms, or to ones that had not in the past proved reliable. These firms organised into the Association of West African Merchants worked hand in hand with the administration. What *was* important was the fact that even though the government consolidated the stranglehold of these firms over the import–export business, it also introduced a major new element into the economy: that of direct participation in it by the government. From this point it was only a short step to the introduction after the war of permanent marketing boards to control prices paid to the peasant producers. Thus while African and Levantine firms suffered under the new government regime, the long-term implications of having prices regulated by government rather than the free play of the market, or worse still oligopolistic collusion, was immensely significant.

The major economic contribution to the war made by the West African peasant was in the form of indirect subsidies to the British Government through the fact that the peasant was paid a lower price for his produce than he could have commanded at prevailing world market prices. The substantial sterling reserves of the West African Governments, built up through direct taxation from duties levied on their export produce or the imported goods they purchased with their earnings from cash-crops, were put at the disposal of the British Government for its war effort. Over and above such indirect subsidies, the West African Governments were responsible for various contributions towards the war effort: in 1940 the Legislative Council of the Gold Coast voted £100,000 for the purchase of a bomber or fighter plane and made an interest-free loan of £500,000 for the purchase of bombers or Spitfires. In 1940, too, the Sierra Leone Legislative Council granted the Imperial War Fund £100,000 'in grateful recognition of the great benefits that Sierra Leone has received during the last 153 years under the British flag . . .'.[40] Similar contributions were made by Nigeria as well as the tiny Gambia.

In addition private subscriptions were made to 'The Win the War Fund'. Sokoto collected £6,216 11*s*. 0*d*. for the fund and the Sultan of Sokoto publicly denied German propaganda that 'the slave-driven natives of the English Colonies are compelled by force to contribute to the Win the War Fund'.[41] In Nigeria in 1941 an intensive campaign for Nigerian Saving Certificates had among its leaders the veteran nationalist and frequent thorn-in-the-flesh of the colonial government, Herbert Macaulay, who declared that 'Victory for Democracy and the freedom of Mankind depends on our contributions, our Determination, and our Loyalty.'[42]

40 *The Times*, 4 July 1940.
41 *The Times*, 7 October 1940.
42 *Lagos Calling for Nigerian Savings Certificates*, Lagos, 1941.

It was estimated officially that gifts of money made by colonial governments, native rulers, various bodies and individuals during the war were:

Gambia	11,478
Gold Coast	361,496
Sierra Leone	148,698
Nigeria	409,255
	£931,127[43]

While the economies of the West African countries had, as we have seen, stagnated in the inter-war years because of the prolonged depressed state of the world market, the war stimulated demand for West Africa's basic exports: groundnuts, palm-oil, cocoa, etc. In particular, after the Japanese occupation of Britain's Far Eastern colonies, which supplied her with tin and rubber, as well as palm-oil, demand for these increased steeply. In 1940 the whole cocoa crop of the Gold Coast was purchased by the British Government, and the only problem affecting the export of palm produce and iron ore from Sierra Leone was shortage of shipping space. Iron ore from Sierra Leone, in fact, replaced European sources lost to Britain after 1940. Such was the demand for tin after the fall of Malaya to the Japanese that Britain introduced forced labour on the mines in Jos, which invoked heavy criticism in the House of Commons. Not until April 1944, however, was this use of forced labour on the tin mines abandoned. In the Gold Coast, the urgent need for aluminium resulted in the building of a branch line from Dunkwa to Awaso, forty-six miles in length, to open up the bauxite reserves of the Western Province. As the Minister Resident for West Africa, Viscount Swinton, declared, there was an enormous increase in production in West Africa, which had become of vast importance since the fall of Malaya and Burma.[44] Most significant of all for the future of the economy, shortage of shipping space for imports led to the development of a wide range of secondary industries in West Africa, thus weakening the traditional economic pattern by which Africa exported unprocessed goods in return for manufactured goods. A tentative step was taken towards industrialisation.

Beside the rapid expansion of the economy, closely controlled by the government, was a rapid increase in infrastructural development. Some of this was a by-product of the very expansion of the economy, as in the case of the bauxite railway in Gold Coast. Some of it—airports, improved roads and expanded port facilities—resulted from military exigencies. But, most important of all was the introduction of government-guided

43 The *Colonial Empire* (1939–1947) Cmd. 7167.
44 *The Times*, 2 April 1943.

development for its own sake. This was initiated in 1940 as a result of the Colonial Development and Welfare Acts which resulted from the concern aroused by the riots in the West Indies, a cure for which was seen in economic and social development, a principle applied to other colonies as a prophylactic against the repetition of such a situation elsewhere in the Empire. Though the amounts made available at first were quite inadequate, as Julian Huxley was to complain in 1944[45]—£5 million a year for development projects and £500,000 for research—the implication of the new policy was emphasised in 1940 by Sir Bernard Bourdillon, Governor of Nigeria: 'The outstanding fact is the abandonment of the principle of self-sufficiency for the Colonies. Nigeria cannot balance her budget and make all the progress required.'[46] Thus in 1944 the Colonial Office announced a scheme for building 48,275 miles of road in Nigeria for which a free grant of only £1,810,000 would be made by C.D. and W.

The war saw not only an expansion in the economy in West Africa but other developments which were to be of major importance in the period of decolonisation. These took place in both French and British West Africa and can be summarised as: (1) increasing state intervention in the economy and a move towards control of the activities of the large companies, (2) the abandonment of the principle of colonial self-sufficiency: while C.D. and W. was brought into operation during the war, though not as a result of it, the French equivalent, F.I.D.E.S. (Fonds pour l'Investissement pour le Développement Economique and Social), was a direct result of the war and given in recognition of French Africa's services to Free France (see below p. 620) and (3) the creation of local industries which after the war would substitute for metropolitan industries, even when there were no physical barriers to freedom of trade between metropolis and colonies.

SOCIAL CHANGE IN THE WAR

For those Africans who were at all conversant with the language of their colonial masters, the war brought into question the whole fabric of the latter's authority. Not only did they hear denunciations of the methods of their colonial rulers from German propaganda and attacks on the British by the Vichy French and vice versa, but through American and to a lesser extent Soviet sources they became increasingly familiar with arguments about the basic 'immorality' of imperialism. In Dakar, for instance, the French, through the major newspaper *Paris-Dakar*, delighted in pointing out that everything was not well in the neighbouring British colonies[47] while at the same time they forbade their subjects publicly to listen to

45 *The Times*, 6 July 1944.
46 *West Africa*, 25 May 1940. Speech to the West African Students' Union.
47 See in particular *Paris–Dakar*, 1 July 1942.

British radio broadcasts in general and to any stations broadcasting 'anti-national' propaganda.[48]

For the prestige of the white man, on which colonial rule was so intimately based, the spectacle of Frenchmen attacking fellow Frenchmen and of France under occupation by another European nation was particularly damaging. White supremacy, coupled with the hitherto monolithic structure of colonial rule, was for ever shattered as far as educated French-speaking Africans were concerned. In Dakar they witnessed the Free French savagely denounce officials of the Vichy regime, sentencing them to imprisonment and national degradation.[49] They learnt that Pierre Boisson, former High Commissioner for Africa, had been disgraced and jailed. The benign Marshal Pétain, the hero of Verdun, whom countless Vichy posters had proclaimed the protector and saviour of France, was now discredited.

In British West Africa, though the administration was secure, Africans were well aware of the plight of the mother country. For the first time the administration *appealed* for the loyalty of their subjects rather than *assumed* it as of right. Even the Vichy administration in French West Africa, which abolished all consultative assemblies and reduced African *citoyens* to the effective status of *sujets,* appealed to the loyalty of the Africans. Pierre Boisson wrote in *Dakar Jeunes* that 'Africa is the white and the black associated. Let these two African groups of young people learn to know each other, learn to appreciate each other, learn to love each other.'[50] African intellectuals like Ouezzin-Coulibaly and Mamadou Dia were encouraged to, and did, contribute to this pro-Vichy journal. The pro-Vichy Légion Française de Combattants, formed in August 1940, grouped together both French and African veterans. The Légion's brochure reported a march past of six thousand members—'Whites and Blacks united in the same desire . . . to serve the flag and not to betray the memory of heroes'.[51]

The myth of British imperial invincibility was shattered in the eyes of educated Africans with the conquest of Britain's Far Eastern colonies by the Japanese, a coloured race. Nevertheless Africans remained overtly loyal. There were no revolts against British rule, nor indeed against French rule. In British West Africa, it seems clear that the educated elements were well aware of what the nature of any régime imposed on them by victorious Germans would be.[52] In Senegal, where elements of German-instilled

48 *Journal Officiel de l'AOF*, 21 December 1940.

49 See for instance the Gaullist paper, *Clarté*, which started publication in Dakar in January 1944.

50 *Dakar–Jeunes*, no. 1, 1941.

51 *Légion française des combattants de l'Afrique noire* (Brochure), Dakar, 1941.

52 *The Times*, 25 March 1940: 'The Gold Coast's enthusiasm for the war is something more than a natural expression of loyalty to the British connexion. It is a measure of the people's detestation of Hitler and the savage racial doctrines for which he is known to

(*continued opposite*)

racism had manifested themselves under the Vichy regime with the destitution of Jewish officials, in particular Governor Geismar,[53] who subsequently became Secretary-General of the Government under the Free French, a number of Africans were sufficiently motivated against the Vichy regime to cross into the Gambia to join the Free French Forces. Some, too, joined the Resistance and paid for this with their lives.[54]

Another important source of new ideas about the morality of colonial rule came from the huge numbers of white troops who passed through West Africa. African civilians and soldiers met, in the vast majority of cases for the first time ever, white men who were not of the ruling class. Some of them had radical views and communicated their ideas to the Africans, others demonstrated that there was little essential difference between the subject classes of Britain and those of Africa.[55] In French West Africa French Communists encouraged the organisation of *Groupes d'Etudes Communistes* (G.E.C.s) after the establishment of the Free French regime. These anti-imperialist G.E.C.s were to become the basis of French West Africa's mass party, the Rassemblement Démocratique Africain, after the war. Even more important was the impact of new ideas on African troops serving abroad: during the war West Africans served in Europe, North Africa, East Africa, the Middle East, India and Burma. West African troops saw societies in Egypt, India and Burma 'apparently even more illiterate and poverty-stricken than theirs, who seemed farther along the road to self-government'.[56]

In India they came into contact with a virulent nationalism which had extracted from the British the promise of independence in exchange for cooperation in the war effort. Over and above this all illiterate soldiers were taught to read and write and thus became more susceptible to the new

(52 *continued*)
stand. The Gold Coast has a large and fast-growing literate community which keeps in touch with world events, and long ago acquainted itself with "Mein Kampf". In that brazenly outspoken book, which is a more potent instrument of anti-Nazi propaganda than any Ministry of Information could have devised, Hitler refers contemptuously to the African as a "semi-ape" upon whom it is a "sin against reason" to lavish Western Education.' Also, *Nigerian Eastern Mail* (cited in *West Africa*, 10 February 1940), 'It is not merely martial ardour that prompts us to long for a more direct share in the struggle, but a certain knowledge of what our fate would be if the Nazi horror were victorious.' An interesting sidelight on the impact of German propaganda is that a number of Nigerians born during the war were christened Hitler. Among the Urhobo, where it is quite common to name children after important men, my gardener, a post-war child, was christened Governor. His elder brother, born during the war, was named Hitler.

53 See Arrêté published in *Journal Officiel de l'AOF*, 19 April 1941.

54 See Abdel Kader Diagne, *La résistance française au Sénégal et en A.O.F. pendant la Guerre 1939–1945* (Mimeographed), Dakar, n.d., but preface dated 16 August 1949.

55 *The Times*, 19 August 1949, reported that about 100,000 British troops passed through West Africa during the war. 'The standard of conduct of the troops has been excellent on the whole, but in such a large army there have been sufficient cases of human frailty, of one sort and another, for the alert African to realise that here was a creature with his own limitations.'

56 John Wilson, *Education and Changing West African Culture*, New York, 1963, p. 8.

horizons shown to them. A large number of soldiers received technical training so that at the end of the war many of them came home having asked: 'What improvements may we expect to find in the Divisions and in our villages when we return?'[57] In Senegal African veterans rioted in their camp at Tiaroye against the low wages they were paid by comparison with their European counterparts. The mutiny was brutally suppressed with many soldiers killed.[58]

Some returning soldiers were to play a vital role in the formation of the political parties that gained independence in the fifteen years that followed the war.[59] Many were no longer content with the colonial situation as they had left it: they wanted to improve their own lot and that of their people. In particular they saw that education was vital to the progress they sought.

The economic expansion brought about by the war was as important an instrument of rapid social change as the war itself. As *The Times* put it at the end of the war, 'a new occupational pattern has been woven into the social structure, but there is as yet no economic system to match it'.[60] The growing urban wage-earning class—some cities doubled in population during the war—the rapid growth of trade unions, the hardships caused by inflation despite price controls, inadequate housing that led to the sprawl of slums in Lagos and Accra and the notorious *bidonvilles* of Dakar, produced a large group of discontents ready to be mobilised by the political leaders who emerged after the war. The General Strike in Nigeria in 1945 and the skill with which Dr Nnamdi Azikiwe harnessed it to his newly founded National Council of Nigeria and the Cameroons was but a presage of things to come. Indeed the war proved, as Meyer Fortes correctly predicted in 1945, to be 'the outstanding instrument for social progress in West Africa for fifty years'.[61] By making a large group of Africans aware of the inequalities and injustices of the colonial system and by bringing the colonial powers into the debt of their African subjects, the war created a climate in which European master was prepared to concede reform and African subject was ready to profit from it.

REFORMS: PRELUDE TO DECOLONISATION

The brief, heady years of the Popular Front Administration in French West

57 *Nigeria Annual Reports for the Northern, Western and Eastern Provinces 1944*, cited by G. O. Olusanya, 'The Resettlement of the Nigerian ex-soldiers after World War II: A Guide for the Present', unpublished paper read at the Xth Annual Congress of the Historical Society of Nigeria, December 1969.

58 Ruth Schachter-Morgenthau, *Political Parties in French-speaking West Africa*, Oxford, 1964, p. 16.

59 G. O. Olusanya in 'The Role of Ex-Servicemen in Nigerian Politics', Journal of Modern African Studies, 6, 2, 1968, pp. 221–32 argues that in Nigeria returning soldiers had much less influence on post-war political developments than usually supposed.

60 *The Times*, 19 August 1946.

61 Meyer Fortes, 'The Impact of the War on British West Africa', *International Affairs*, xxi, 2, 1945, pp. 206–19.

Africa, during which there was considerable liberalisation of the regime under which the *sujets* lived, were succeeded by a tightening up of the indigénat under the Vichy regime.[62] Not until the Free French took over the administration was there hope of renewed liberalisation. And this did not come about until after the war. In the short run the exigencies of war in the form of need for increased crop production and large numbers of recruits for the army entailed a continuation of the hated indigénat. If reforms did not accompany the progress of the war as they did in British West Africa, they were at least promised as a result of African assistance to the Free French in the famous 1944 Conference of Brazzaville where the whole imperial future of France was discussed by the Free French Government. There was no equivalent conference concerning Britain's imperial future and reform took place on an *ad hoc* basis following the reaction to the changed climate that had developed in the Allied camp as a result of the war and a desire to avoid a repetition of the riots that had erupted in the West Indies. As far as British West Africa was concerned, the nearest thing to the Brazzaville declarations was the speech of Mr Clement Attlee, the Labour Deputy Prime Minister, on 15 August 1941, when he made it clear to the West African Students' Association that the principles of the Atlantic Charter applied to all races of the world, coloured as well as white: 'You will not find in the declarations which have been made on behalf of the Government of this country any suggestion that the freedom and social security for which we fight should be denied to any of the races of mankind.'[63] The first positive moves towards increasing participation of Africans in their own government came in September 1942 when Lord Cranborne approved the appointment of unofficial African members to the Executive Councils of Nigeria and the Gold Coast. Nana Sir Ofori Atta and Mr K. A. Korsah were appointed in the Gold Coast, while Mr A. A. Alakija and Mr S. B. Rhodes were appointed in Nigeria. These were of course 'safe' appointments in that none of them represented the radical strain of politics that was fast developing in British West Africa and in that they were nominees of the British Administration and were not elected by the people. But they did establish an important precedent: for the first time under British colonial rule West Africans were to be involved in the executive as distinct from legislative process of their own countries. This concession at the national level was followed by local government reforms whereby in 1943 Accra's new Town Council was given an elected majority. These two principles of an elected African majority in the legislative process and inclusion of Africans in the executive process were not extended to the post-war constitution for Nigeria, and only partially conceded for that of the Gold Coast. The so-called Richards

62 *Journal Officiel de l'AOF.*
63 *The Times*, 16 August 1941.

Constitution, while it broke important new ground in establishing a central legislative council for *all* Nigeria, did not extend the elective principle beyond Lagos and Calabar which had long since had elected members in the old Legislative Council. While there was an unofficial majority, nationalists could point out that this was from the Government's point of view a safe one since this majority contained government nominees and conservative emirs and chiefs. The Executive Council remained unchanged and while there were still nominated African members in it the actual responsibility for formulating policy remained with the British heads of departments.

The Richards Constitution, introduced without discussion with the politicians, who had greatly increased their constituencies during the war years, evoked widespread opposition since it did not seem in any substantial way to advance Nigeria on the road to self-government as implicitly promised by the Atlantic Charter.

The new Constitution of the Gold Coast was introduced after a measure of consultation with African political opinion. It was also much in advance of the Nigerian Constitution in that it provided for an elected majority. On the other hand the African members of the Executive Council were not chosen from the Legislative Council and did not head departments. In both the Nigerian and Gold Coast Constitutions drafted during the war, Africans, whether forming an elected or unofficial majority were limited to criticising policy not formulating it. However, the provision for an elected majority in the Gold Coast Constitution was sufficient to ensure it an enthusiastic reception where the Nigerian one had met with open hostility.

In Sierra Leone, while Paramount Chief A. C. Caulker had expressed the hope in a 1943 Legislative Council debate that 'the Sierra Leone Government has already drawn up plans to make this Dependency what it should be after the war in the spirit of the Atlantic Charter',[64] such was not the case. Sierra Leone did not have a new constitution until 1951. The only concession made was the appointment of two unofficial African members of the Executive Council in May 1943: Mr Sowell Boston and Paramount Chief A. C. Caulker. The Gambia, whose main newspaper, *The Gambia Echo* demanded in 1940 that the colony should have the same franchise as some British colonies had had for twenty years, waited until 1946 to get its first elected member on the Legislative Council.[65]

If the wartime constitutional reforms in Nigeria and Gold Coast seemed very tentative they did at least embark on the path to self-government in that the principle of an African legislative majority and African participation in the executive were conceded.

64 Cited in Martin Kilson, *Political Change in a West African State: a Study of the Modernisation process in Sierra Leone*, Cambridge, Mass., 1966, p.
65 *Gambia Echo*, cited in *West Africa*, 6 April 1940.

This principle of Africanisation was also hesitantly conceded in the Civil Service. Partly this was the result of difficulties in recruiting expatriates for senior administrative and technical posts because of the war, partly it was in the genuine spirit of reform. In 1941 the prospects for Africanisation in the Gold Coast were examined. In 1942 A. L. Adu and K. A. Busia were appointed Administrative Officers. In 1943 the Governor of the Gold Coast, Sir Alan Burns, declared that it was Government policy 'to appoint Africans to senior appointments wherever suitable Africans could be found, and further to take all practical measures by way of scholarships, training schemes and the like to augment the number of Africans suitable for such appointments'.[66] No such far-sighted policy was promulgated for the other British West African territories.

Since a policy of Africanisation, however tentative, presupposed a supply of educated Africans, one of the most significant advances made during the war in British West Africa was in the field of education. In 1939 the Conference of West African Governors recommended the establishment of an African University and emphasised that such a University 'must be West African in spirit and reality as well as in name, and not a colourless imitation of a British University'.[67] No doubt they had in mind here Fourah Bay College, then the only degree-granting institution in Tropical Africa, which was closely tied to Durham University. This pious hope was to be followed up during the war by the Elliott Commission which included African members and whose recommendations sired the University Colleges of Ghana and Ibadan. The Colonial Office, however, did not neglect mass education. In May 1941 the Advisory Committee on Education in the Colonies appointed a sub-committee to 'consider the best approach to the problem of mass literacy and adult education other than literacy in the more backward dependencies, taking into account the emphasis which the Advisory Committee laid in past years upon Community education'.[68] Without general education, the report concluded, 'true democracy cannot function and the rising hope of self-government will suffer inevitable frustration'.

In 1942 the Nigerian Government gave concrete recognition to the report in its Ten Year Educational Plan whereby it was proposed to meet 'the increasing demand for schools of all types throughout the country, especially in the Southern Provinces . . .'.[69]

Perhaps the most significant reform of all was in the attitude of the Colonial Government to development. The Colonial Development and

66 Cited in *The Report of the Select Committee of the Legislative Council on Africanisation*, Accra, 1950.
67 *The Times*, 1 September 1939.
68 Advisory Committee on Education in the Colonies, *Mass Education in African Society*, Colonial No. 186, 1944, p. 4.
69 *Nigeria Ten Year Educational Plan (1942)*, Lagos, 1944.

Welfare Fund introduced a new era in the relationship between metropolis and colony: the old laissez-faire attitude whereby the colonies were meant to pay for their own development was replaced by a philosophy whereby the metropolitan government recognised its obligation to assist its dependencies in both economic and social development. Where the old Colonial Development Fund had spent a total of some £5 million over a period of eleven years up to 1940 in the British colonies, it was, as we have seen, now proposed to spend £5 million a year for a period of ten years together with £500,000 a year for research. *The Gold Coast Spectator* welcomed this decision to increase development funds: 'when Britain grows old she shall not die, for we, the growing sons, will keep the flag flying'.[70] *West Africa* reported the enthusiastic reception by the West African Students' Union for Sir Bernard Bourdillon on 21 May 1940 where the news was welcomed by Dr K. Abayomi and Mr O. Coker.[71] Britain, despite the exigencies of war, showed good faith in her new determination to secure rational development for her West African colonies. In December 1943, as we have seen, Mr Noel Hall was appointed Development Adviser for West Africa and Mr Maxwell Fry, the architect, was appointed Town Planning Adviser, which in the context of the rapid growth of the major cities in the war was a vital move. Earlier, trade union advisers had been appointed to establish arbitration boards, conciliation boards and wage boards 'to ensure freedom of expression of opinion and representation, and to bring labour standards to a level comparable, so far as local conditions will allow, to those attained in Great Britain'.[72] By 1945 Nigeria had produced a Ten Year Development Plan for £40,000,000, the only precedent for this being the famous Guggisberg plan for the Gold Coast. Part of this sum was to be provided by the Colonial Development and Welfare Fund. In retrospect, the £4 million a year proposed for development may seem hardly appropriate for a country of Nigeria's vast size. But in the context of the almost total absence of planned development in the pre-war years, and the prior doctrine of colonial self-sufficiency, it marked a radical new departure in colonial rule in West Africa. These steps towards rational social and economic development, coupled with the new governmental restraints on the activities of the expatriate oligopolies, were to have a determining effect on the pattern of the period of decolonisation in British West Africa that followed the war.

While the Free French regime made heavier demands on its West African *sujets* than the Vichy regime had, it promised reform in return for sacrifice. Whereas the promise of reforms at the end of the First World War was never kept, de Gaulle, at his famous Brazzaville Conference of

70 Cited in *West Africa*, 2 May 1940.
71 *West Africa*, 25 May 1940.
72 *The Times*, 29 October 1941.

De Gaulle addressing the Brazzaville Conference

1944, was to initiate reforms which, with the advantage of hindsight, we can now see had as their logical outcome the independence of French-speaking Africa.

Though for the most part Africans had been passively conscripted into support of the Free French regime in French Black Africa, de Gaulle in recognition of the contribution Africa had made to the liberation of France offered them political, social and economic reforms. Without Equatorial Africa as an initial base for his Free France, without Black African troops and the food supplied by African peasants, it is doubtful whether de Gaulle could ever have achieved his goal of the rehabilitation of a defeated France. As de Gaulle put it himself at the Brazzaville Conference: France found in Africa 'her refuge and the starting point for her liberation'.[73]

For Africans, the conference, which consisted of the three Governors-General of West Africa, Equatorial Africa and Madagascar, together with the sixteen governors of the constituent colonies, and nine delegates from the Provisional Consultative Assemblies and a Bishop, all of them non-Africans, proposed timid reforms or concessions in the political spheres. While there was to be no question of self-government, Africans were to

73 *La Conférence Africaine Française*, Commissariat aux Colonies, Algiers, 1944. On the Brazzaville Conference see: Hubert Deschamps, *Méthodes et Doctrines Coloniales Françaises*, Paris, 1953 and Réné Viard, *La Fin de l'Empire Colonial Français*, Paris, 1963.

play a greater part in the political process. The colonies should be represented in the constituent assembly of the Fourth Republic; each colony should, like Senegal, have a representative assembly; and a federal assembly should be established for each of the Federations. The franchise for Africans, though not universal, should be as wide as possible. However while Africans were to be represented in the Constituent Assembly, the conference was against their continued representation in the National Assembly of France when it was established. The conference thus was essentially anti-assimilationist and advocated decentralisation with a coordinating Colonial Parliament.

In the social sphere the hated indigénat was to be abolished and forced labour was to be phased out over a period of five years. Labour conditions for African workers were to be improved. Education was to be increased down to the village level, though it was to be exclusively in French. Health services were to be improved, and a School of Medicine was to be established. More openings in the administration should be made available to Africans, though the higher posts would still be reserved for Frenchmen.

Finally in the economic spheres F.I.D.E.S. was established. Between 1945 and 1960 over £400 million was spent from this fund on roads, bridges, ports, agricultural and urban improvement. In the preceding years of colonial rule almost no aid had been given by France to her colonies, the notable exception being the dam across the Niger at Sansanding.

While French West Africans were largely passive participants in the events of the Second World War, and with the exception of two memoranda submitted by Fily Dabo Sissoko, a Cantonal Chief from French Soudan and a former school-teacher, had no voice in the Brazzaville Conference, the developments resulting from the replacement of the Vichy Administration by the Free French were of vital importance. Most commentators agree that the Brazzaville recommendations, many of them enshrined in the constitution of the Fourth Republic, paved the way for self-government and eventual independence for French Black Africa. Here it is interesting to note that the Brazzaville Conference was not even noted in the British press at the time.[74] And so, though we have argued that French-speaking Africans had been largely indifferent to which of the regimes ruled them during the war, and that indeed the Free French demanded more of them, the success of the Free French regime opened a new chapter in the social, economic and political history of French West Africa. For what Brazzaville so clearly laid down was that Africans, whilst they had no right to self-government, had the right to representative government on the model of the old Senegalese Conseil-Général; and that France had an obligation to invest in the economic development and social welfare of her African subjects.

74 Edward Mortimer, *France and the Africans 1944–1960*, London, 1969, p. 27.

For both British and French West Africa the Second World War marked a major turning point. In retrospect we can see that colonial rule as it had been practised since the occupation of West Africa had come to an end. France and Britain were now embarking, whether they were conscious of it or not, on a period of decolonisation which was to prove much shorter in duration than either power could envisage at the time.

CHAPTER 18

Politics of decolonisation in British West Africa, 1945–1960

OLAJIDE ALUKO

Within two decades of 1945, the whole of the former British West Africa had become independent. Indeed with the attainment of independence by Nigeria on 1 October 1960, one could say that the transfer of power in British West Africa was virtually complete; the Gold Coast which became independent in March 1957 and Nigeria together contained over 94 per cent of the total population of British West Africa. Sierra Leone achieved independence in April 1961. Tiny Gambia, the first British Colony in West Africa, could have been independent much earlier than February 1965 but for initial doubts about its political future.

This rapid rate of decolonisation was not envisaged by the British government in 1945, as we have seen in the previous chapter. Although the British Government had long accepted the principle of ultimate self-government for its 'Empire on which the sun never set', it was still thought in 1945 that this would occur in the dim, distant future. In September 1941 Winston Churchill, then the British Prime Minister, told the Commons that the reference in the Atlantic Charter of August 1941 to the freedom of people to choose the form of government under which they would live would be applied to territories liberated from the Axis Powers and that it could not be used to prise open the British Empire.[1] In June 1945 at San Francisco where the representatives of the United Nations had gathered to write the Charter of the new organisation, Lord Cranborne, then the leader of the British delegation, said it would be 'unrealistic and prejudicial to peace and security' to make independence the universal goal of colonial policy.[2] Recalling how essential for the Allied Powers, particularly Britain and the United States, had been the resources of the colonies and the bases at Freetown, Takoradi and Mombasa during the

1 *House of Commons Debates*, ccclxxiv, col. 69, 9 September 1941.
2 *UN Conference on International Organization*, Documents, viii, Commission 11 General Assembly 20 June 1945, p. 159.

war, and how these colonies had been 'welded into one vast machine for the defence of liberty', Lord Cranborne declared that it would be absurd for anybody to contemplate 'the destruction of this machine or its separation into its component parts'.[3] What, then, accounted for the rapid pace of decolonisation that did in fact take place in the succeeding fifteen years?

THE PACE OF DECOLONISATION

A number of complex factors were responsible for the accelerated rate of decolonisation of British West Africa. Some of these were interrelated. Most important were the emergence of the U.S. and the U.S.S.R., two anti-colonial powers, as super-powers at the end of the war; the principle of international accountability written into the U.N. Charter; the influence of the Labour Government that came into power in Britain in the July 1945 elections; the absence of white settler communities in British West Africa; and the agitation of the nationalists.

The influence of the United States on the British Government's decolonisation programme was profound. Historically, in principle and by self-interest, the U.S. had been against colonialism. During the war, the U.S. Government was almost as concerned with the abolition of the British Empire as with the defeat of the Axis Powers. At every possible point, the U.S. Government exerted pressure on the British Government to liquidate its empire. At the Yalta Conference in February 1945, Franklin Roosevelt backed by Marshal Stalin demanded that the British Empire be converted into international trusteeships: to which Churchill's reply was 'never, never, never'.[4] At the end of the war it seemed as though the disappearance of the British Empire was the first objective of American policy.

Although these American assaults on its empire were resented by the British Government, they nonetheless went some way towards modifying the thinking of the British leaders about decolonisation. For, after all, it was the entry into the war of the U.S. on the side of Britain that ensured the defeat of the Axis Powers. Furthermore, the reduced economic and military strength of Britain after the war, together with the emergence of the U.S. as the nuclear giant of the Western Powers, had made the security of Britain ultimately dependent on U.S. arms. In such circumstances, the voice of the U.S. Government on decolonisation could not but have some weight with the British leaders.

Still, the importance of this should not be exaggerated. For with the Communist *coup d'état* in Czechoslovakia in 1948, and the Berlin Blockade of July 1948–May 1949, the Soviet threat was from the viewpoint of the U.S. Government more menacing than British colonialism. Henceforth, with the formation of N.A.T.O. in April 1949, U.S. policy towards

3 *Ibid.*

4 Edward R. Stettinius, *Roosevelt and the Russians*, London, 1950, p. 212.

colonial issues became, as did all other U.S. policies, subject to the exigencies of defence, for which the continued strength of the Western colonial powers was vital.

In the case of decolonisation in British West Africa, there were no defence problems of any significance to complicate U.S. policy. The area was physically distant from the Communist world, and Sino–Soviet political and economic influence there was insignificant. U.S. policy was directed towards accelerating the process of decolonisation, which the British Government had decisively begun in 1951 with the introduction of representative governments in all its colonial territories in West Africa except the Gambia.

The Soviet Union, the other anti-colonial colossus which emerged at the end of the war, also had some influence on the liquidation of the British Empire. The U.S.S.R. of course had a highly developed doctrine concerning the nature of imperialism. Between 1941 and 1945, the U.S.S.R. and the U.S., as we have seen, worked together in an attempt to abolish British imperialism. At San Francisco in May 1945, V. M. Molotov, then the Soviet Foreign Minister, issued a press statement saying: 'We must first of all see to it that dependent territories are enabled as soon as possible to take the path of national independence.'[5] At the U.N. General Assembly, and at the meetings of the various U.N. organs dealing with colonial issues, the Soviet Union continued its searing attacks on colonialism.[6]

It is difficult to assess the impact of Soviet anti-colonial diatribes on Britain's programme of decolonisation. On the one hand, incessant Soviet criticism of British colonialism at the U.N. was counterproductive. It made Britain more intransigent, and less willing to submit its colonial affairs to international scrutiny.[7] On the other hand, it also placed the U.S.S.R. at the head of the radical anti-colonial Afro-Asian group of states at the U.N., a situation which neither the United Kingdom nor the United States could view with equanimity. So the British Government made some concessions to the anti-colonial forces in order not to alienate the Afro-Asian states.

Another factor contributing to the rapid transfer of power in British West Africa was the principle of international accountability which was to be applied to colonial administration under chapters XI–XIII of the U.N. Charter. It was chapter XI (Articles 73 and 74) titled 'Declarations Regarding Non-Self-Governing Territories' that applied to the former British West Africa. Under these Articles, the colonial powers undertook certain obligations. They were required 'to ensure, with due respect for the culture

5 *New York Times*, 8 May 1945.
6 As the cold war intensified, the U.S.S.R. began to associate the U.S. with Western colonialism.
7 Harold K. Jacobson, 'The UN and Colonialism: A Tentative Appraisal', *International Organisation*, xvi, 1, 1962, p. 40.

of the peoples concerned, their political, economic, social, and educational advancement, their just treatment, and their protection against abuses'; 'to develop self-government, to take due account of the political aspirations of the peoples, and to assist them in the progressive development of their free political institutions. . . .' The administering Powers were also required 'to transmit regularly to the Secretary-General for information purposes . . . statistical and other information of a technical nature relating to economic, social and educational conditions' in their non-self-governing territories.

The provisions concerning the development of self-government and the economic, social and political advancement of colonial peoples may have appeared imaginative. Since the information to be supplied to the Secretary-General was limited to that of a purely statistical and technical nature, the effects were not far-reaching. For what interested the anti-colonial forces within the United Nations was *political* information. Nor was there any provision under the 'Declarations Regarding Non-Self-Governing Territories' for the establishment of any committee or body to consider even the statistical and technical information supplied by the administering powers. However, through the effort of the Secretary-General and his staff, an *ad hoc* committee called the 'Committee on Information from Non-Self-Governing Territories' (usually abbreviated as the Committee on Information) was set up to consider all the information supplied by the Colonial powers.

Although the British Government was not enthusiastic about the work of the committee, it gave its cooperation as long as discussion was limited to statistical and technical information. However, it refused to provide the Secretary-General with political information on its colonial territories until September 1961.[8] By then all the territories that made up British West Africa, except Gambia, had attained independence. So the role played by the U.N. in the decolonisation of British West Africa was minimal. Nevertheless, by insisting that the colonial powers should periodically provide statistical and technical information on the economic, social, and educational conditions in the colonial territories, the U.N. showed an interest in the building up of national forces that were ultimately to undermine the basis of colonialism. In this sense, it can be said that the U.N. provided an environment which accelerated the process of decolonisation which was already under way.

The victory of the Labour Party in the July 1945 general elections in Britain was also an important factor in the decolonisation of British West Africa. The Labour Party came to office with the old Radical tradition of Little England and with emphasis on better living conditions at home. Ideologically the Labour Party was opposed to territorial acquisition or the retention of colonies abroad. This did not mean that the Labour Party

8 *Ibid.*

wanted to liquidate the British Empire. The question of decolonisation was not a party or an ideological question, but a national one. In the Commons, the Labour Government had to confront Winston Churchill, now Leader of the Opposition, who in 1942 had declared that he had not been appointed the first Minister of the Crown in order to preside over the liquidation of the empire. And, as Sir Charles Jeffries, formerly Deputy Under-Secretary of State for the Colonies, has said, 'Mr Attlee certainly did not propose to go down to history as a man who had done just that.'[9] So once in office, the Labour Party became cautious about decolonisation.

Even so, it was the Labour Party Government that led the way in making the programme of speedy decolonisation acceptable to a large section of the British public. It was only a Labour Colonial Secretary that could in 1948 have made the far-reaching policy statement of Arthur Creech-Jones: '... the central purpose of British colonial policy is simple. It is to guide the colonial territories to responsible government within the Commonwealth in conditions that ensure to the people concerned both a fair standard of living and freedom from oppression from any quarter.'[10] This liberal attitude of the Labour Government towards decolonisation after the war was strongly bolstered by the existence of substantial liberal elements within Britain.[11] It also received strength from the mood of the country at that time. Britain emerged from the war so militarily, economically and financially weak that there seemed to be lacking in the country not only the will but also the physical and financial power to resist nationalist demands for self-government.[12]

The absence of any white settler community in British West Africa meant that the struggle for independence was carried on directly between the indigenous peoples and the British Government. The transfer of power could be rapid since there was no white settler population to be conciliated before independence. West Africa was thus able to escape the difficulties that arose in Kenya and Zambia.

However, by far the most important single factor that made the attainment of independence complete in British West Africa within twenty years of the war was the nationalist agitation. As James Coleman has rightly pointed out, 'When a country is granted independence by the British government is determined largely by the time the nationalists were able to mobilise and display sufficient power and national unity to compel a

9 Sir Charles Jeffries, *Transfer of Power*, London, 1960, p. 59.
10 *Ibid.*, p. 15.
11 As far back as 1923, Lord Devonshire, then the Colonial Secretary, had said that the interests of the indigenous people should be paramount in the administration of a colonial territory. Although this principle was not totally followed after that year, it was an evidence of some liberal opinion on colonialism in Britain.
12 With the independence of Ceylon, India and Pakistan by 1948, the British Government had taken a decisive step towards decolonisation, which was soon to occur in its West African colonies.

commitment.'[13] The rest of this chapter will therefore be devoted to examining briefly for each territory such questions as who spearheaded the attack against colonial rule, how successful were they in achieving the cohesion needed to compel Britain to grant independence and to run the independent government, and to what extent was effective decolonisation achieved with the formal granting of independence.

THE STRUGGLE FOR INDEPENDENCE

While there were some variations from territory to territory, the struggle for independence was led largely by the educated, urban minority in British West Africa. Likewise, while there were some minor differences in the nature of the struggle from one territory to another, it was on the whole rather peaceful. There was some terrorism in the Gold Coast, but in none of these territories was the struggle for independence marked with large scale violence or guerrilla warfare. The timely concessions made by the British Government to the nationalists played a major part in minimising violence.

Gold Coast to Ghana

In the Gold Coast between 1945 and 1949, when the Convention People's Party (C.P.P.) of Dr Nkrumah was formed, the nationalist struggle was led principally by the intelligentsia, who in August 1947 founded the United Gold Coast Convention (U.G.C.C.) partly to protest against the 1946 Constitution, which provided for the first time for an unofficial majority in the Legislative Council, and partly to demand self-government 'in the shortest possible time'. The leaders of the U.G.C.C. were mostly professional men, well-to-do traders and teachers. The Chairman was A. G. Grant, a timber merchant. Its two Vice-Presidents, R. S. Blay and J. B. Danquah, were lawyers. Other leading members were E. A. Akufo Addo and J. W. de Graft Johnson, both lawyers, and W. E. Ofori Atta, a graduate teacher.[14] Their influence was not countrywide, but based on the support of the urban élite.

By late 1948 the U.G.C.C. leaders had virtually lost the leadership of the nationalist movement to younger and less educated men and women, who formed the C.P.P. in June 1949. The landslide victory of the C.P.P. at the 1951 general elections eliminated the U.G.C.C. as an effective force in the nationalist struggle; indeed, from June 1949 until independence in March 1957, the struggle for independence was organised and directed by the C.P.P.

There were several reasons why the U.G.C.C. lost the nationalist leadership to the C.P.P. Its programme of self-government 'in the shortest possible time' was not as attractive to the masses as the C.P.P.'s programme

13 J. S. Coleman, *Nigeria: Background to Nationalism*, 1963, p. 396.
14 Dennis Austin, *Politics in Ghana*, London, 1964, p. 52 n.

of 'self-government now'. Furthermore, the U.G.C.C. was miserably lacking in the energy, skill and organisational capability necessary to canalise the considerable discontent then existing in the country into an effective political weapon against the colonial administration. Since it was an élite party, its moderate appeals failed to attract any significant support among the working classes and the youth. The C.P.P.'s radical programme of a happy tomorrow for everybody with the elimination of colonialism 'Now' was in accord with the mood of the masses, especially the working classes and the youth that formed the core of the nationalist crusade. The C.P.P.'s programme also involved an attack against the existing indigenous power structure in the Gold Coast. All this made the C.P.P. more attractive to a wide range of commoners who had long been critical of the wide powers of the chiefs under the colonial administration. 'Its appeal to the "Common Man" and for Self-Government Now', wrote Dennis Austin, 'ran like a flame through the Colony and Ashanti chiefdoms and branches were formed in many instances without the knowledge of the national headquarters.'[15]

The militant C.P.P., which was a breakaway party from the United Gold Coast Convention, was a commoners' party. Unlike the U.G.C.C., the C.P.P. drew its support largely from elementary school leavers. The leading members of the party were primary school teachers, clerks in government and commercial offices, petty traders, storekeepers, local contractors and not very successful businessmen—the 'verandah boys', men without wealth, position or property to lose in any showdown with the Colonial Government. Unlike the executive committee of the U.G.C.C., made up almost entirely of university graduates, the first committee of the C.P.P. in 1949 consisted of only two graduates, while there were three with secondary school certificates, and four elementary school leavers.[16] The C.P.P. included in its hierarchy members of the Committee on Youth Organisation, which had branches all over the country. It also included the Trade Union Congress of the Gold Coast. Indeed, membership of the C.P.P. was indistinguishable from that of the T.U.C. The C.P.P. also drew support from the unemployed, of whom there were many after the war. Of great significance was the role of the market women within the C.P.P.

Apart from the mass following enjoyed by the C.P.P., there was the factor of Marxist–Leninist ideology which pervaded the thinking of the C.P.P. leaders, particularly that of Nkrumah himself who openly proclaimed having been influenced by Hegel, Marx, Engels, Lenin, Mazzini and Gandhi. Several C.P.P. leaders, especially those of the T.U.C. such as Anthony Woode and Turkson Ocran, held Marxist–Leninist views about colonialism. Some C.P.P. activists were said even to have gone to the

15 *Ibid.*, p. 114.
16 *Ibid.*, p. 116 n.

extent of maintaining relations with the Communist-dominated World Federation of Trade Unions (W.F.T.U.).[17] All these factors went a long way in determining the nature of the nationalist struggle.

From 1945 until January 1948, when Kwame Nkrumah became the general secretary of the U.G.C.C., the struggle for independence was conducted peacefully. In fact, it was not until the formation of the U.G.C.C. in August 1947 that one could speak of a truly organised nationalist movement seeking self-government for the Gold Coast. Even then the intelligentsia leadership of the U.G.C.C. was content to secure the transfer of power in a constitutional way.

From early 1948, however, until the general elections of February 1951 the nationalist struggle was marked by violent agitation, demonstrations, strikes, rioting and virulent newspaper agitation resulting in the imprisonment of many nationalist leaders. This period opened with the famous demonstrations by ex-servicemen. These demonstrations were brought to an end with the shootings at Christiansborg Castle Crossroads on 28 February 1948, in which two men were killed by the police and four wounded. This incident led to a wave of violence that spread like bushfire throughout most of the country. There was rioting and looting on an unprecedented scale. In Accra the offices of the United Africa Company and other European-owned stores were burned, and the gates of Ussher Fort prison were forced open by the mob and its prisoners released. By the time the wave of violence was subdued on 9 March 1948, twenty-nine people had been killed and 237 wounded.[18] Although the demonstrations were not organised by the U.G.C.C. leaders they attempted to exploit the situation for their own political advantage with the result that the Colonial Government quickly clamped down on them by arresting and detaining the 'Big Six' of the U.G.C.C., namely J. B. Danquah, Akuffo Addo, Ako Adjei, Obetsebi Lamptey, Ofori Atta and Kwame Nkrumah.[19]

The Christiansborg Castle Crossroads demonstration, and the scale of the violence that followed it, shocked the Colonial Government out of its complacent view of the Gold Coast as the 'model colony' and for some time it lost control of the situation. A state of emergency was declared all over the country. With the restoration of law and order, the British Government immediately instituted a Commission of Enquiry under Aiken Watson to investigate the causes of the disturbances. The report of this Commission led in turn to the setting up of the Coussey Committee to review the 1946 constitution.

Another wave of violence soon followed the publication of the report of the Coussey Committee which *inter alia* provided for a semi-responsible

17 Colin Legum, 'Pan-Africanism and Communism', *The Soviet Bloc, China, and Africa*, Uppsala, 1964, p. 21.
18 Kwame Nkrumah, *I Speak of Freedom*, New York, 1961, p. 6.
19 *Ibid.*, p. 7.

government: an executive council of three ex-officio and eight representative ministers and a nationally elected assembly. Owing to the dissatisfaction of the C.P.P., the T.U.C. and other nationalist elements with the report of the Coussey Committee, Kwame Nkrumah on 8 January 1950 declared 'Positive Action'—a civil disobedience campaign of agitation and propaganda, and 'as a last resort, the constitutional application of strikes, boycotts and non-cooperation based on the principles of absolute non-violence'[20]—to force the colonial government to establish through a general election a constituent assembly which would determine a full self-government constitution for the country. Earlier, on 6 January 1950, the T.U.C. had declared a general strike against the colonial government. So effective were Nkrumah's 'Positive Action' campaign and the general strike that all government services were paralysed. Transport and communications were dislocated. The whole economic life of the country came to a standstill.

Three days after the declaration of 'Positive Action', on 11 January 1950, the crowds took to the streets to demand 'Self-government Now'. This led to further violence, looting and rioting in many parts of the country. As there was no truce between the nationalists and the government, the ex-servicemen staged another demonstration which led to a clash with the police in which two African policemen were killed.[21]

The government dealt with the situation with a mailed fist. On 11 January 1950, the whole country was again placed under a state of emergency and on 17 January all the main towns in the country were placed under strict curfew. The T.U.C. and C.P.P. leaders, including Nkrumah, were swiftly rounded up and charged with various offences such as promoting an illegal strike, attempting to coerce the government, or even sedition. Nkrumah was found guilty on three charges and sentenced to a total of three years imprisonment. All others brought to court were sentenced to various terms of imprisonment.

The nationalist struggle during this period was also carried on through violent newspaper agitation. Through his chain of newspapers, namely the Accra *Evening News*, the Sekondi *Morning Telegraph* and the Cape Coast *Daily Mail*, Nkrumah and his comrades-in-arms waged continual verbal warfare against the colonial government. A columnist in the *Evening News* writing under the pseudonym 'The Rambler' was unrelenting in his denunciation of the imperialists.[22] The offices of these newspapers were frequently subjected to police search between 1948 and early 1951. During the period of 'Positive Action' in January 1950, all newspapers were banned. Many C.P.P. leaders, in particular Gbedemah, were imprisoned

20 Kwame Nkrumah, *Ghana: The Autobiography of Kwame Nkrumah*, London, 1959, p. 92.
21 *Ibid.*, p. 99.
22 *Ibid.*, pp. 76–7.

for seditious publications. Many were fined heavily for libellous or seditious articles and publications. Nkrumah said he had paid, by early 1951, about £4,750 as fines for libel suits largely brought by government officials.[23]

During the twenty months between the formation of the C.P.P. in June 1949 and the general election of February 1951 nearly all of the major leaders of the C.P.P. had been imprisoned by the Colonial Government for sedition, libel or promoting 'Positive Action'.[24] Apart from Nkrumah, the

23 Nkrumah, *I Speak of Freedom*, p. 17.
24 Austin, pp. 114–15.

Kwame Nkrumah

Chairman of the C.P.P., and Gbedemah, its Vice-Chairman, those imprisoned included Kojo Botsio, its general secretary, Krobo Edusei, Saki Scheck and G. K. Amegbe. The 'Prison Graduate Caps' became the symbol of Ghanaian patriotism and nationalism. This proved a great asset to the C.P.P. in the general elections that preceded the attainment of independence in March 1957. 'The cult of martyrdom' that grew around the C.P.P. leaders went a long way towards uniting the majority of the people behind them in the crusade for independence.

With the victory of the C.P.P. in the general elections of February 1951, and its acceptance of office under the Coussey Constitution which Nkrumah had earlier described as a 'bogus and fraudulent' constitution,[25] the character of the Gold Coast nationalist movement was completely changed.[26] The C.P.P. now moved from 'Positive' to 'Tactical Action'. From the time the C.P.P. took office in 1951 until the attainment of independence in 1957, the struggle for independence was carried out through a continual process of peaceful political and constitutional negotiation.

Pressure was henceforth exerted by the nationalists not through boycotts and mass demonstrations, but through speeches in the Legislative Assembly and articles in the newspapers. Owing to the continued criticism of the 1951 Constitution by the C.P.P. in the Legislative Assembly between early 1951 and early 1952, Nkrumah's title of Leader of Government Business was changed to that of Prime Minister on 5 March 1952; and on 21 March 1952, Nkrumah was allowed to form a cabinet. The C.P.P. resented the continued presence of three ex-officio expatriate members in the cabinet and the fact that the title of Prime Minister did not give Nkrumah any more powers than those he had previously. So the C.P.P. continued to press for further constitutional change. In June 1952, Nkrumah discussed the question of revising the Coussey Constitution with Oliver Lyttelton, the British Colonial Secretary, who was vising the Gold Coast at the time. Lyttelton promised that the British Government would consider the proposals of the Gold Coast Government after consultation with the chiefs and people of the Gold Coast. Again, in October 1952, Nkrumah raised the question in the Legislative Assembly. By March 1953, more than a hundred and thirty different organisations had submitted their views about the proposed constitutional review. The subsequent discussion of these views and the consultations that accompanied them formed the basis of the Government's White Paper on Constitutional Reform of July 1953. On 10 July 1953, Nkrumah moved his 'Motion of Destiny' in the Legislative Assembly in which he called on the British Government to grant independence to the country.[27] The motion was unanimously carried.

25 Nkrumah, *Autobiography*, p. 140.
26 Bob Fitch and Mary Oppenheimer, *Ghana: End of an Illusion*, 1966, p. 35.
27 Austin, p. 28.

The 1953 White Paper on Constitutional Reform, after months of negotiation with the Colonial Secretary, became the basis of the 1954 Constitution which granted full internal self-government to the Gold Coast. During the negotiations that preceded the inauguration of this Constitution there was a remarkable degree of accord and understanding between the C.P.P. government and the Colonial Governor, Sir Charles Arden-Clarke. It was in recognition of this that Nkrumah said the Constitution should be called the 'Nkrumah–Arden-Clarke Constitution'.[28]

In spite of a few extremists within the C.P.P. who wanted to take the nationalist struggle out of the Legislative Assembly into the streets to force the British Government to concede a more rapid rate of decolonisation, the constitutional reforms that took place between 1951 and 1954 were on the whole achieved peacefully. The achievement of full internal self-government within three and a half years after the C.P.P. came to office was remarkable. It was the fastest pace of decolonisation in British West Africa. The reasons for this are not far to seek. By then the Gold Coast, the 'model colony', appeared stable; the economy was substantially healthy, the country had great economic potentialities; there was a fundamentally homogeneous population which up till 1954 was free from acute ethnic or religious conflicts.[29]

But in 1954 there developed a serious threat to national unity in the form of ethnic, sectional and religious parties. The most important of these was the Ashanti-based National Liberation Movement (N.L.M.) formed in September 1954. Others were the Northern People's Party (N.P.P.), the Togoland Congress and the Muslim Association Party (M.A.P.). All these parties challenged not the colonial government but the C.P.P. government, for sectional rather than national reasons. The threat of the N.L.M. was the most serious for the C.P.P. government since it jeopardised the national unity which the British Government required before it would concede independence. The N.L.M. appealed basically to ethnic sentiment. But it also stood for the principle of federation before independence so that the Ashanti could have a substantial measure of local autonomy. To achieve this end, the N.L.M. resorted among other means to violence. There were riots, looting and general lawlessness in Ashanti, especially in Kumasi, for the greater part of the period from late 1954 until 1956.

Meanwhile, the C.P.P. Government began to press the British Government to fix a date for independence. Although Nkrumah said the C.P.P. had expected the country to be independent by 1956, the British Government would not set any firm date because of the wave of lawlessness and rioting in Ashanti. The new Colonial Secretary, Mr T. A. Lennox-Boyd,

28 Nkrumah, *Autobiography*, p. 147.

29 Over half of the population was Akan-speaking. Indeed Akan, which is a group of closely related dialects, is widely spoken all over Ghana.

wrote to Nkrumah that the British Government would be unable to fix a date until a substantial majority of the people had shown that they wanted independence in the very near future and had agreed upon a workable constitution for the country. He added that if the Constitutional Adviser, Sir Frederick Bourne, whom the Gold Coast Government had invited in September 1955 to advise on the devolution of powers to the regions, did not succeed in recommending proposals acceptable to the majority, the British would have no alternative but to call for a new general election to seek the views of the people.

All attempts of the C.P.P. to meet the N.L.M. and their allies half-way were rebuffed by them. The N.L.M. and the Asanteman Council, the traditional ruling council of Ashanti, refused to participate in the series of discussions with Sir Frederick Bourne on the form the independence constitution for the country should take unless the principle of federation was conceded. By December 1955 Sir Frederick had submitted his report to the Gold Coast Government. While not supporting a federal form of constitution as demanded by the N.L.M., the report recommended five regional assemblies, each to be entrusted with a wide range of powers delegated from the central government and guaranteed under the constitution once established.[30] The N.L.M. rejected the Bourne Report. They boycotted the Achimota Conference of February/March 1956, convened by Nkrumah to discuss not only the Bourne Report but also all other matters relating to the form of the independence constitution for the country. The Achimota Conference, however, agreed to almost all the recommendations of Sir Frederick Bourne. In addition, it recommended that there should be a House of Chiefs in each Region to discuss social and cultural legislation. In April 1956, the government formally accepted these recommendations.

In the meantime, rioting, arson and looting continued unabated in Ashanti. In view of this, and in view of the continued intransigence of the N.L.M. and its allies, the Colonial Secretary felt he had no alternative to asking Nkrumah to hold another general election in July 1956 to determine the people's support for the proposals on the form of an independence constitution. He intimated to Nkrumah in May 1956 that after the general election the British Government would be prepared to accept a motion calling for independence within the Commonwealth if passed by a reasonable majority in the newly elected Legislature, and then to declare a firm date for independence. The general elections of July 1956 returned the C.P.P. to office with a healthy majority, seventy-two out of the one hundred and four seats in the Legislature. On 23 August 1956, a motion was passed by a large majority in the new Legislative Assembly calling on the

30 For the powers to be delegated to the regional assembly, see section 8 of the *Report of the Constitutional Adviser*, Accra, 1955.

British Government to grant independence to the Gold Coast under the name of Ghana in March 1957. On 17 September 1956, the Governor conveyed the Colonial Secretary's approval to Nkrumah, agreeing to the country's independence on 6 March 1957.[31]

Although Nkrumah was critical of the Colonial Secretary for not publicly condemning the lawlessness and the violence being fomented by the N.L.M. and its allies in the country,[32] and although he did not take kindly to the demand for a general election before fixing a date for independence, there was hardly any bitterness in the negotiations that took place in the period 1954–7 between the C.P.P. leaders and the British Government. Indeed, 'the unique relationship and understanding' that had developed between Nkrumah and the Governor, Sir Charles Arden-Clarke, ensured cooperation between them throughout the turbulent months of 1954–6. Nkrumah himself acknowledged that without the help and cooperation of Sir Charles Arden-Clarke, he might well not have succeeded by September 1956 in getting the British Government to agree to the country's independence on 6 March 1957.[33]

Nigeria

Even though there were some businessmen and a few radical elements, the nationalist movement in Nigeria during the postwar period was led primarily by the intelligentsia; those who had studied abroad and had returned as lawyers, teachers, doctors and journalists. Most of these leaders, partly because of their wealth and partly because of their position within society, would not contemplate any head-on violent collision with the Colonial Government. For instance, Nnamdi Azikiwe, who was the leader of the nationalist movement in the 1940s, had investments of about £5,000 in Zik's Press Ltd.[34] Although in the 1940s he often spoke of 'violence' to evict the British from Nigeria, and of 'the blood of tyrants being used to water the tree of liberty', he was neither a revolutionary nor a man of violence. Late in the 1940s, when the Zikists were mounting in his name a campaign of violence against the Colonial Government, he criticised what he called 'the youthful impetuosity of the Zikists'.[35] Chief Obafemi Awolowo, who later emerged as a major nationalist leader, also opposed the use of violence to coerce the British Government into constitutional concessions.[36] There was never any nationalist leader in Nigeria with a self-government programme as radical as that of Nkrumah. One has only to

31 Nkrumah, *Autobiography*, p. 233.
32 *Ibid.*, p. 206.
33 *Ibid.*, p. 234.
34 Walter Schwarz, *Nigeria*, London, 1968, p. 96.
35 Anthony Enahoro, *Fugitive Offender*, London, 1965, p. 97.
36 Obafemi Awolowo, *Awo*, London, 1960, p. 251.

compare Dr Azikiwe's aim in 1948 of 'self-government within our lifetime' with Nkrumah's slogan in 1949 of 'self-government now'.[37]

Azikiwe and Awolowo had some support in the urban areas, especially among the workers. Between 1946 and 1951, the National Convention of Nigeria and the Cameroons (N.C.N.C.) gained support from the trade unions, especially that section led by Nduka Eze. The trade union movement was, however, hopelessly divided and generally opposed to joining in the formation of a left-wing party.[38] Not only did the trade union movement in Nigeria lack radicalism and internal unity, it also lacked that unity of purpose which existed between the Gold Coast T.U.C. and the C.P.P. during the struggle for independence. To this should be added the fact that the nationalist struggle to all intents and purposes was confined to the south and the urban centres there. The bulk of the population was still politically unconscious and so did not participate in the struggle. Furthermore, in the south, the nationalist parties were divided to such an extent that they expended more of their energy in challenging each other than in challenging the colonial government.

These factors go a long way to explain why the Nigerian struggle for independence was, on the whole, non-violent and unheroic. Chief Awolowo later lamented that 'our struggles for independence have produced no martyr—no single national hero who is held in reverence and affection by the vast majority of people in Nigeria'.[39] As Walter Schwarz has written, 'compared with the Gold Coast whose colonial experience was similar, the fight for freedom was sedate'.[40]

The campaign for independence in Nigeria can be divided into two distinct periods: from 1945 to 1951, when semi-representative government was established, and from 1951 to 1960, when responsible government devolved upon Nigerians.

During the period 1945 to 1951, the struggle for independence was largely a verbal one. Newspaper agitation, petitions, violent verbal threats in political rallies and meetings were the weapons the nationalists employed to put pressure on the Colonial Government. The only riots during this period were the sporadic ones that occurred in Port Harcourt, Onitsha, Calabar and Aba following the Enugu colliery shooting, in which twenty-nine persons were killed and fifty-one wounded in November 1949.

It was through his newspapers, such as the *West African Pilot*, the *Daily Comet* and the *Guardian*, that Azikiwe and his lieutenants waged war against the colonial administration. These newspapers were quick to point

37 Nnamdi Azikiwe, *Zik: A Selection from the Speeches of Nnamdi Azikiwe*, London, 1961, p. 157.
38 Enahoro, p. 98.
39 Obafemi Awolowo, p. 299.
40 Schwarz, p. 82.
41 For further details see Coleman, pp. 299–301.

out the iniquities of the colonial system of government. They also strove to mobilise public support for the nationalist cause. More often than not, the brimstone and fire which these newspapers rained on the Colonial Government landed them and their editors in court. In 1945 two of the papers, the *Pilot* and the *Daily Comet*, were banned by the government on the grounds that they misrepresented the nature of the General Strike of June 1945 organised by seventeen unions with about thirty thousand workers in government services, the railways and the postal system.[42] The editors were frequently fined or imprisoned for libel or sedition. Anthony Enahoro, for example, who was for some time the editor of the *Daily Comet*, was imprisoned twice on charges of sedition between 1945 and 1947. The situation was unlike that in the Gold Coast, in that it was a few young and radical elements within the nationalist parties, not their leaders, who were imprisoned by the Colonial Government. Neither Azikiwe nor Awolowo ever went to jail on the independence issue.

Petitions also were used as a means of demanding concessions from the Colonial Government. In April 1947, following an incident of racial discrimination at the Bristol Hotel in Lagos against a colonial officer of African descent, Mr Ivor Cummings, a United Front Committee—a coalition of nationalist parties formed specifically to protest against the incident—presented a petition to the Governor demanding *inter alia* an end to all forms of racial discrimination in the country. The United Front Committee succeeded in obtaining a circular[43] from the Governor banning racial discrimination in public places including those hospitals hitherto reserved exclusively for Europeans. Its petition also led eventually to an end to discrimination against Nigerians in appointments to the academic staff of the University College, Ibadan.[44] Then there was the petition which the N.C.N.C. delegation took to the Colonial Secretary in London in 1947, after making a nationwide tour seeking both financial and moral support. The petition sought among other things the revision of the Richards Constitution of 1946, which the nationalists condemned partly for the arbitrary way in which it had been introduced and partly for its content. For, apart from providing for an unofficial majority in the Legislature, the Richards Constitution did not advance the trend towards self-government. The N.C.N.C. delegation to London also submitted another memorandum listing thirty-three grievances for which they demanded redress. The delegation was brusquely rebuffed by the Colonial Secretary, Arthur Creech Jones, who told them to 'go home and cooperate with the administration'. Nevertheless, the petitions were not in vain for the British Government's decision in 1948 to revise the Richards

42 The N.C.N.C. did not organise the strike but sought without avail to exploit it for the nationalist cause.
43 Circular No. 25, 21 March 1947, Lagos, 1947.
44 Coleman, p. 299.

Constitution, which was originally envisaged to last nine years, was largely a concession to nationalist views expressed so frequently in these petitions.

In March 1948, after this concession, Dr Azikiwe announced that he was prepared to cooperate with the Colonial Secretary,[45] but his younger radical followers, the Zikists, remained alienated. Between 1948 and early 1950, they conducted a campaign of verbal assault and intimidation against the Colonial administration. In October 1948 a leading Zikist, Osita Agwuna, called for revolutionary struggle against British imperialism and its allies in Nigeria.[46] In February 1949 Ogedengbe Macaulay, son of Herbert Macaulay, another Zikist leader, talked of dragging the government down, and of seizing power by force. In the same month H. R. Abdallah, one of the first northerners prominent in the nationalist movement, then President of the Zikist Movement, echoed Nkrumah by calling for positive action to end British rule. The government quickly clamped down on the movement by arresting ten of its leaders, charging them with sedition and sentencing them to imprisonment. This did not completely eliminate the movement since those of its leaders still at large tried to make capital out of the Enugu colliery shootings on November 1949 by organising the sporadic risings in the four towns in the eastern part of the country. They failed, however, to spark off a revolution. On 18 February 1950 a Zikist leader tried unsuccessfully to assassinate Hugh Foot, the Chief Secretary to the government. This so shocked the government that it ordered the arrest of all Zikist leaders still at large. They were tried for sedition and imprisoned. In April 1950, the government imposed a ban on the Zikist Movement which not only led to the collapse of the youth front, but also decisively marked the end of militant nationalism in Nigeria.

Throughout the period 1945–51 the nationalist movement lacked unity except for the short-lived truce following the Bristol Hotel incident when all the nationalist forces formed the United Front Committee, and again after the Enugu colliery shooting in 1949 when they formed the National Emergency Committee. In fact, ever since the split in the Nigerian Youth Movement in 1941 over the contest between Akinsanya, a Yoruba, and Ernest Ikoli, an Ijaw, for a vacant seat on the Legislative Council—which Azikiwe attributed to tribalism—chronic disunity, mainly of an ethnic nature, had characterised the struggle for independence in Nigeria. After Azikiwe's assertion that an attempt to assassinate him in July 1945[47] had been ridiculed as cowardly by the *Daily Service*, organ of the Nigerian Youth Movement (N.Y.M.), there ensued a press war between the *Daily Service* and Azikiwe's *Pilot.* So bitter did the campaign become, and such was the polarisation between the Ibo who dominated the N.C.N.C. and

45 Azikiwe, p. 104.
46 Coleman, p. 298.
47 For details see Nnamdi Azikiwe, *Assassination Story: True or False*, Lagos, 1945.

the Yoruba who dominated the N.Y.M., that there were serious fears of violent conflict between the two ethnic groups in Lagos in 1948.

The proposed revision of the Richards Constitution through consultation at all levels between 1948 and 1950 did little to reduce tribal bitterness within the nationalist movement. Since the new constitution was clearly going to establish the power structure within Nigeria for the future, the struggle between leaders of different ethnic groups over such issues as representation in the Central Legislature, regional powers, and revenue allocation became crucial. Between 1948 and 1950, as Coleman has remarked, less was heard about '. . . British "autocracy" and "rapacious colonial exploitation", and more about Fulani threats either to continue their interrupted march to the sea or to withdraw to the Western Sudan, about Yoruba allegations of threatened Ibo domination . . .'.[48]

In anticipation of the enactment of the 1951 Constitution which was to establish representative assemblies in each of the three Regions, and a Central Legislature to which these regional assemblies would send delegates, two new parties, both regional in outlook, were formed—the Northern People's Congress (N.P.C.) for the North, and the Action Group (A.G.) for the West. When elections were held under the 1951 constitution, it was found that each of the three different parties had won one of the three Regions—the N.P.C. in the North, the N.C.N.C. (which still tried to present a national image) in the East, and the A.G. in the West. Indeed, as Ken Post has stated, by 1951 the divisions within the nationalist ranks had become 'final and complete'.[49] So bitter and irreconcilable had the rivalries and the antagonism within the ranks of the nationalist forces become that Azikiwe even thought of withdrawing from politics for the next five years.[50]

Under the new constitution, known as the Macpherson Constitution after the Governor who introduced it in 1951, the struggle for independence was conducted mainly through constructive, and constitutional, agitation. This was to characterise the succeeding years until independence was gained in 1960.

At party rallies and on the pages of newspapers, both the N.C.N.C. and the A.G. campaigned for revision of the Macpherson Constitution, which they considered unsatisfactory; apart from obvious defects such as the difficulty of dissolving any of the Regional Houses of Assembly without endangering the lives of other Legislatures in the country, the nationalists felt that the constitution made no substantial advance towards their goal of self-government. They resented the fact that the Executive Council of each Region was presided over by the Lieutenant-Governor, and included five ex-officio expatriate members as against nine elected members.

48 Coleman, p. 312; Enahoro, p. 98.
49 Ken Post, *The Nigerian Federal Elections of 1959*, London, 1963, p. 7.
50 Azikiwe, *Zik*, p. 168.

Furthermore, the powers given to the Nigerian Ministers were insubstantial. They were given charge of certain 'matters', but did not head departments and there was no provision for a Premier for each Regional government. Similarly, in the Central Legislature, the Ministers were not given any ministerial responsibility and the Central Executive Council which still included six European senior ex-officio members was presided over by the Governor. This was all fuel for nationalist agitation over the question of full self-government.

In 1951 both the N.C.N.C. and the A.G. called for full self-government for the country by 1956. At the annual conference of the A.G. in Benin late in 1952, a resolution calling for independence not later than December 1956 was unanimously carried.[51] The temperature of nationalist agitation was raised by Anthony Enahoro's motion in the central House of Representatives in March 1953 that 'this House accepts as a primary objective the attainment of self-government by Nigeria in 1956'. While the resolution was warmly welcomed by both the N.C.N.C. and the A.G., the leader of the Northern delegation, the Sardauna of Sokoto, proposed that the phrase 'self-government as soon as practicable' should be substituted for 'self-government in 1956'. The N.P.C. rejected the resolution, in the words of the Sardauna, because 'the Northern Region does not intend to accept the invitation to commit suicide'.[52] This echoed their suspicion and deep fear of domination of the educationally and economically backward North by the much more advanced South.

While Enahoro's 'self-government in 1956' motion led to a temporary rapprochement between the A.G. and the N.C.N.C., the N.P.C. leaders who opposed it earned for themselves such abuse from the Lagos masses that they found the idea of secession 'tempting'.[53] Still determined to push for self-government in 1956, the A.G. led a delegation to the North in May 1953 to the masses there, as it were, over the heads of the Northern leaders. The campaign led to ghastly riots that continued for three days—the Kano riots—in which at least thirty-six persons were killed and two hundred and forty wounded.[54]

Alarmed by the scale of the riots resulting from disagreement over the date for self-government, the Colonial Secretary, Oliver Lyttelton, convened a constitutional conference in London in July 1953 to revise the 1951 Constitution 'to provide for greater regional autonomy, and for the removal of powers of intervention by the Centre'.[55] Although the N.C.N.C. and the A.G. had some reservations about the Colonial Secretary's declaration it was finally agreed at the conference that a federal form of

51 Enahoro, p. 121.
52 Alhaji, Sir Ahmadu Bello, *My Life*, London, 1962, p. 119.
53 *Ibid.*, p. 135.
54 *Report on the Kano Disturbances*, Kaduna, 1953, p. 21.
55 *Report of the Conference on the Nigerian Constitution*, Cmd 8934, London, 1953, p. 3.

constitution would have to be established to allow each region to develop at its own pace. While the two Southern parties demanded self-government for Nigeria by 1956, the Northern delegation again opposed it. In the circumstances, the Colonial Secretary said the British could not grant self-government to the whole country by 1956. But, by way of compromise, he told the conference that the British Government would be willing to grant full internal self-government to any region that wanted it in 1956. The Southern leaders were mollified by this declaration. Azikiwe declared that this was 'the first time in British colonial history that Britain had offered self-government to a colonial people on a platter of gold'.

The work of the constitutional conference of 30 July–22 August 1953 in London was completed by a further conference in Lagos in January 1954. The result of these two conferences was the Lyttelton Constitution, which came into effect on 1 October 1954. Besides establishing a federal system of government, the constitution made further advances in the direction of decolonisation. Each region was to have a Premier who could preside over the regional executive council when the Governor was absent. In the executive councils of the Western and the Eastern regions there were to be no more ex-officio members, but the Northern executive council still retained three ex-officio members. The Nigerian Ministers were now given ministerial responsibility for individual government departments. The Central Executive Council in Lagos still retained three ex-officio members, and the Governor-General was to preside over it. No provision for the post of a Federal Prime Minister was made under the 1954 Constitution.

So impressive were the concessions made by this constitution that there was little over which the nationalists could agitate. The only major doubt in 1954 was when the whole country would become independent. But this depended not so much on the British Government as on the Northern leaders. Once the Northern leaders showed their readiness for full self-government and once this was accompanied by a substantial degree of national accord, the British Government indicated it would be prepared to consider a date for the country's independence.

The struggle was now largely internalised: with rival nationalists competing for power and trying to force the Northern leaders to change their attitude about the time for independence. As Chief Awolowo, the A.G. leader, later said, 'After the 1953 Constitutional Conference, excepting the occasional periods of our barren friendship with the N.P.C., we directed our campaign against that party; in this campaign the Action Group was condemned at every turn by the N.C.N.C.'[56] The N.C.N.C., which formed a coalition government with the N.P.C. in Lagos, wanted the N.P.C. leaders to be given time to decide on the date for independence and were anxious that no step should be taken on the issue which might push them towards secession.

56 Awolowo, p. 253.

Encouraged partly by the independence of Ghana on 6 March 1957 and partly by increased confidence in themselves, the Northern leaders announced early in 1957 that the North would be ready for full internal self-government by 1959. It was against this background that the House of Representatives unanimously passed a resolution on 26 March 1957 'to demand independence for Nigeria within the British Commonwealth in 1959'.[57]

At the beginning of the Constitutional Conference,[58] the leaders of the Nigerian delegations presented a joint memorandum requesting the British Government to grant the country independence in 1959. But the Colonial Secretary, Mr Lennox-Boyd, replied that this was a request for the British Government to draw 'a blank cheque' in favour of independence for Nigeria in 1959; and that he could not in good conscience go to his colleagues in the cabinet without knowing 'how the cheque will be filled in'.[59] He said that there was no question about full self-government being the objective but that the problem was the timing. He argued that Ghana was a different proposition. Unlike Ghana, Nigeria was a federal state with many problems, such as the balance between the Centre and the Regions, minority problems and fiscal allocation, all of which had to be resolved before independence. However, he indicated that if, after the federal election at the end of 1959, all had gone well and the people of Nigeria remained broadly united then the British Government would reasonably be able to consider a date for the granting of independence. He suggested that after the federal election there might be a further conference.

The Nigerian delegates were not entirely satisfied with these arguments. They wanted the Colonial Secretary to commit the British Government to granting independence to the country before 2 April 1960. The Colonial Secretary, however, refused to be tied down to any specific date in 1960. The Nigerian delegates were so disappointed that they said they reserved the right to pursue the issue further with a view to impressing upon the British Government the necessity for granting independence not later than 2 April 1960. But as Chief Anthony Enahoro wrote later, 'nobody really intended to do anything about it'.[60]

On all other points, there was general agreement. The East and the West were granted full internal self-government in 1957.[61] It was agreed

57 *Nigeria Debates Independence*—The Debate on Self Government in the Federal House of Representatives on 26 March 1957, Lagos, 1957.

58 This conference was scheduled for 1956, but was postponed till May 1957 as a result of the Foster–Sutton Commission's inquiry into Dr Azikiwe's connection with the African Continental Bank.

59 *Report of the Nigeria Constitutional Conference*, held in London, May and June 1957, Cmd 207, London, 1957.

60 Enahoro, p. 153.

61 For reasons of space we cannot examine here the Foster–Sutton Commission of Enquiry, which delayed this internal self-government by a year.

to appoint a Prime Minister and to have a cabinet at the federal level. The conference adjourned to await the reports of the Fiscal, the Minorities and other Commissions. After the conference, the A.G. was brought into the government coalition to give the appearance of national unity and thereby strengthen the hand of the leaders in negotiations with the British over independence. This undoubtedly made it easier for the British Government to be more specific about the date of independence during the resumed Constitutional Conference in 1958.

The conference met in London from 29 September to 27 October 1958. It agreed to regional self-government for the North in 1959. It also reached broad agreements on the reports of the various Commissions except that on the problem of minorities—the Willink Commission—which did not favour, as some had hoped, the creation of new states out of the existing three Regions before independence. Instead, it merely suggested ways to allay the fears of the minorities, and left the country with an intractable problem.

In retrospect one can say that the conclusions of the Willink Commission were a grave error of judgement. Why did it arrive at such conclusions? Three main reasons can be given. First, the British Government was opposed to the creation of new states before independence. This was clear from the terms of reference given to the Commission. Its primary task was 'to ascertain the facts about the fears of minorities . . . and to propose means of allaying those fears . . .'; and 'to advise what safeguards should be included for this purpose in the Constitution of Nigeria'.[62] It was only to recommend the creation of one or more new states as a last resort, if and 'only if no other solution' seemed to the Commission to meet the case. The British Government's attitude appeared to have been motivated by a desire to maintain the hegemony of the North in the post-independence politics of the country so as to have a moderating influence on its policies. Secondly, for political and other reasons, the two major parties—the N.P.C. and the N.C.N.C.—who had controlled the federal government since 1954 opposed the creation of new states out of the Regions under their control. The N.P.C. reiterated its well-known opposition to the creation of new states from the North, which it claimed was 'united by history and tradition'.[63] The N.C.N.C. Government told the Commission that while it was not opposed to the creation of states in the East, it could not agree to the creation of the Calabar and Ogoja Rivers (C.O.R.) state which, it claimed, was 'inspired only by a negative dislike for the Ibo tribe'.[64] Instead, it suggested the creation of three new states: namely, Ogoja, Rivers and Cross River, out of the Eastern Region. This was being disingenuous, as the Commission observed, for the N.C.N.C. leaders knew very

62 *Report of the Minorities Commission*, London, 1958, p. iii.
63 *Ibid.*, p. 72.
64 *Ibid.*, p. 47.

well the Colonial Secretary's position that under no circumstances could he allow the creation of more than one new state out of each existing Region.[65] Thirdly, the Commission based its conclusions on the assumption that Nigeria would continue to follow the path of liberal democracy and parliamentary government after independence,[66] and that under such circumstances each of the major parties would be compelled to woo the minorities both in and outside its own Regional base in order to win seats in the Federal House of Representatives. The basis for this assumption is difficult to establish. It was a mistaken and false assumption. No sooner had Nigeria attained independence than it deviated almost completely from the path of liberal democracy.[67] Once the basis of the Commission's report was detroyed its conclusions became totally irrelevant.

Meanwhile, when the report was being discussed at the Constitutional Conference in September/October 1958, only the A.G. and the delegates representing the minority groups put up a fight for the creation of new states before independence. But they were outmanoeuvred. The Colonial Secretary, Lennox-Boyd, presented them with a crucial choice. If new states were to be created out of the existing ones, he said, then the date for independence would have to be put off. To all the delegates, however, independence appeared to be the primary objective which must be achieved without further delay. It would have been politically suicidal for the A.G. to have been labelled as the main obstacle to independence; and so it gave in.

With the minority problem swept under the carpet, the Colonial Secretary told the conference that he did not think it would be physically possible to make adequate preparations for independence by 2 April 1960. Therefore, he stated, he was authorised to say that if a resolution was passed by the new Federal Parliament early in 1960 asking for independence, Her Majesty's Government would agree to that request and would introduce a Bill in Parliament to enable Nigeria to become a fully independent country on 1 October 1960. This was warmly received by the Nigerian delegates.

Sierra Leone

In Sierra Leone, as in Nigeria, it was the educated minority who led the struggle for freedom. Up till the late 1940s the struggle was directed and dominated by the educated Creoles of the colony such as Dr Bankole-Bright, a medical practitioner, Mr O. O. During, a lawyer, Mr I. T. A. Wallace-Johnson, a journalist, and Dr G. C. E. Refell, also a medical practitioner. These educated men formed the Sierra Leone National Congress which had a moderate programme aiming at eventual decolonisation.

65 *Report of the Nigeria Constitutional Conference*, para. 24.
66 *Report of the Minorities Commission*, p. 89.
67 All local and parliamentary elections between October 1960 and the coup of January 1966 were marked by arson, looting, fraud, rigging and suppression of opposition groups.

While advocating greater freedom for its people, it nonetheless adopted a moderate approach to achieve this end.

But from the late 1940s the leadership of the nationalist struggle passed to another educated minority group, predominantly based in the Protectorate, most of whom were products of the Bo government school in the Protectorate. Among these were men like Dr Milton Margai, a medical practitioner, his brother Albert Margai, a lawyer, and Siaka Stevens.[68] It was these men, nearly all from the Mende tribe, who formed the Sierra Leone Organisation Society in 1946. Their professional status and social standing meant that they were basically opposed to the use of violence to coerce the Colonial Government into making constitutional concessions. They could not always restrain their followers and there were sporadic riots between 1946 and 1951. But on the whole the struggle for independence was, as in Nigeria, conducted through constitutional agitation and a continual process of constitutional revision.

The main features of the 1947 Constitution and the composition of the Protectorate Assembly must be the starting point for any discussion of the nationalist struggle for independence in Sierra Leone. For it was against these that the nationalists directed their attack. The 1947 Constitution established a new Legislative Council with an unofficial majority. But the composition of the Council was weighted in favour of the traditional élite. The whole of the Protectorate, which contained about 90 per cent of the population and from which the bulk of the country's wealth came, was represented only by paramount chiefs to the exclusion of the Western-educated élite. There were only four members elected by the Creoles of the colony, usually from the educated élite. There was no change in the composition of the Executive Council, which had included two nominated Africans since 1943. Likewise, the Protectorate Assembly, a quasi-legislative body set up in 1945–6, was dominated by the traditional élite. Of its forty-two members, twenty-six were paramount chiefs, and only two were from the educated élite.[69]

It was the domination of the Protectorate Assembly by traditional leaders that led directly to the formation of the Sierra Leone Organisation Society (S.L.O.S.) in 1946 by the educated élite from the Protectorate to agitate for the democratisation of the Assembly. When the new Constitution was introduced in August 1947, the S.L.O.S. joined leaders of the Creoles in campaigning against it through petitions, memoranda, etc., which went unheeded. Charging the British Government with discriminating against the Protectorate's educated élite, the S.L.O.S. in 1948 called for a revision of the 1947 Constitution to ensure that the new progressive

68 *West Africa*, 20 September 1952, p. 873.

69 M. Kilson, *Political Change in a West African State: A Study of the Modernization Process in Sierra Leone*, Cambridge, Mass., 1966, p. 155.

and literate elements be better represented in the new Legislative Council. Again the British Government ignored their petition.

The Creole-dominated political groups, such as the Sierra Leone National Congress and the Sierra Leone Socialist Party, were also opposed to the composition of the new Legislative Council set up as a result of the 1947 Constitution, not because it did not go far enough in the direction of self-government but because it robbed them of the dominant position they had enjoyed in the Legislative Council since 1924. So instead of advocating the extension of the franchise in line with democratic principles, they sought, contrary to the provisions of the Constitution, to restrict it further by asking that only literate persons be allowed to sit in the Legislative Council because they feared the powers of conservative chiefs and British officials over the illiterate masses.

Because of the dissatisfaction of the educated élite of the Protectorate with the reforms of 1945–6, the Constitution of 1947, and the refusal of the British Government to take any notice of their petitions, three serious riots took place in the Protectorate and in the Southern Mende areas, between 1946 and 1951.[70] Whether it was leaders of the S.L.O.S. who engineered the riots could not be categorically established. What was certain was that they were not averse to the riots—which were probably organised by their younger, more radical followers in the Protectorate—since they hoped that such disturbances would pave the way for 'basic political change' in the country.

It was these outbreaks of violence that led the Governor, Sir George Beresford-Stooke, to send a dispatch to the Colonial Secretary in January 1949 to question the preference given to the traditional leaders over the educated élite in the Protectorate. These riots thus appear to have contributed to the revision of the 1947 Constitution in 1950 and to the inauguration of a new one in 1951 which provided for representative government as in the Gold Coast.

No account of the nature of the struggle for freedom in Sierra Leone between 1945 and 1951 can be complete without an understanding of the sharp cleavage between the Creoles of the Colony and the peoples of the Protectorate. The Creoles had joined the educated élite in opposing the composition of the Legislative Council set up in the 1947 Constitution because it was weighted in favour of the traditional élite. But the complaint of the Creole élite was not that the Protectorate was represented by the paramount chiefs but that the Protectorate was given a preponderant voice thus depriving them of their hegemony in the politics of the country. Without avail they employed all sorts of legal and other arguments between 1947 and 1951 to try and reverse the trend towards the domination of the country's politics by the Protectorate. In September 1948 they sent a petition to the Colonial Secretary demanding that any member of the

70 *Ibid.*, p. 159.

Legislative Council be required to pass a literacy test. In December 1948 the Legislative Council had to defer the introduction of the new constitution for six months owing to Creole opposition.[71] In 1950, a common front was formed by the Creoles in the colony against Protectorate domination. That same year saw the formation of a countering common front among all the Protectorate leaders. The Protectorate's educated élite, which in 1948 had criticised the franchise provisions of the 1947 Constitution, turned around in 1950 to defend them and to advocate their implementation so as to safeguard the dominant position of the Protectorate.[72] In 1950, Dr Milton Margai referred to the thirty thousand Creoles with disdain as 'foreigners' whose interests should be disregarded by the Protectorate people, 'the owners of the country'.[73] So sharp had the division between the Creoles and the Protectorate people become in 1951 that Dennis Austin has observed that 'by 1951 it began to look as if Sierra Leone was to be the only West African Colony to shape its constitution on the basis of communal differences'.[74]

The Protectorate chiefs joined the educated élite of the S.L.O.S. in launching a new political party, the Sierra Leone People's Party (S.L.P.P.), under the leadership of Dr Milton Margai. It was the S.L.P.P. that won the general election held under the 1951 Constitution and formed the first representative government in Sierra Leone. The subsequent road to independence was on the whole surprisingly smooth. The struggle for independence, if it could be called so, was carried on through continual constitutional revision, one stage being regarded as a jumping-off ground for another until the attainment of independence. Except for some feeble demonstrations by the Creoles during the visit of Lennox-Boyd, then the Colonial Secretary, to Freetown in May–June 1959, the agitation for independence was not marked by violence or even harsh words.

As a result of further pressure from the S.L.P.P. leaders, and in line with constitutional developments in the Gold Coast and Nigeria, the 1951 Constitution was revised in 1953. The new constitution gave African Ministers responsibility for a limited range of executive functions. Dr Margai was given the title of Chief Minister. But the Colonial Government still kept expatriate heads of departments outside the control of African Ministers. Several expatriate officials, such as the Directors of Medical Services, Education and Agriculture, continued to sit in the Central Legislature.

In 1956 a further constitutional advance was made. Dr Margai was given the title of Premier. In 1958 an African Minister was appointed as

71 *West Africa*, 20 September 1952, p. 873.

72 *Ibid.*

73 M. Kilson, pp. 169–70.

74 D. Austin, 'People and Constitution in Sierra Leone: 2. Three Years of Negotiation', *West Africa*, 20 September 1952, p. 873.

Minister of Finance for the first time although it was maintained that he was not directly responsible for policy on finance until 1959 when the expatriate Financial Secretary hitherto responsible for financial policies was reduced to being the administrative head of the Ministry.

Following the visit of the Colonial Secretary to Freetown in May–June 1959, the Sierra Leone political leaders and the Colonial Secretary agreed that there should be constitutional talks early in 1960 in London on the country's advance towards independence.[75] The conference met in London from 20 April to 4 May 1960. It was agreed by all the delegates, except Siaka Stevens—who insisted that there should be an election before independence as was the case in the Gold Coast and Nigeria—that Sierra Leone should become independent on 27 April 1961.[76]

As in Nigeria and the Gold Coast, the main difficulties in the way of independence came not from disagreement between the nationalists and the British but from the contest for power among the nationalists themselves. The S.L.P.P. had other political groups with which it had to contend for power at home and it needed to demonstrate that it had the majority of the people behind it in the struggle for independence. The objection of Siaka Stevens, during the constitutional conference in London in 1960 to independence in April 1961, was to presage a turbulent time for the S.L.P.P. before the attainment of independence.

On the eve of the 1960 constitutional conference, Dr Margai's S.L.P.P. brought together such rival parties as the United Progress Party (U.P.P.) and the People's National Party (P.N.P.) in a coalition called the United National Front (U.N.F.). After the conference, Dr Margai formed a national government from the members of the U.N.F.[77] But this effort at a united front did not prevent the emergence of a militant opposition to challenge the S.L.P.P. and its allies. Within a month of the formation of the coalition government there was an outbreak of disturbances in Kono district sparked off by a struggle for power between the supporters of the S.L.P.P. and another party, the Progressive Independent Movement.[78] But it was the threat of Siaka Stevens's movement—the Election Before Independence Movement (E.B.I.M.) which Dr Margai said was dominated by 'a few mischievous people'[79]—that appeared to S.L.P.P. leaders as the most menacing.

By September 1960 the E.B.I.M. had transformed itself into a political party—the All People's Congress (A.P.C.)—which drew most of its supporters from the workers in Freetown and other urban centres and could be expected to employ a general strike to achieve its objectives. The A.P.C.

75 *West Africa*, 13 June 1959, p. 569.
76 *West Africa*, 7 May 1960, p. 529.
77 *West Africa*, 23 April 1960, p. 464 and 4 June 1960, p. 635.
78 *West Africa*, 18 June 1960, p. 681.
79 *West Africa*, 17 September 1960, p. 1053.

mounted a militant campaign against independence before a general election; the government feared the radical nature of the A.P.C. and tended to overreact to any criticism from it. Late in February 1961 there was a violent clash between supporters of the A.P.C. and the United National Front and this resulted in the arrest of fifteen A.P.C. leaders. In March, just a little more than a month before independence, eighteen other A.P.C. leaders, including their chairman Siaka Stevens, were arrested and placed in detention for incitement.[80] So fearful of the A.P.C. was the national coalition government that on the day of independence a state of emergency was declared, and thirty other A.P.C. leaders were placed in detention.

In spite of these clashes the coalition government did not collapse and, though with some difficulty, was able to control the situation. In view of this the British Government felt that it could reasonably fulfil its promise to grant independence to Sierra Leone on 27 April 1961 and the celebrations took place without incident.

The Gambia

The movement for independence was initially dominated by a group of highly educated men in Bathurst and neighbouring Kombo. They included men like the Reverend John C. Faye, leader of the first Gambian political party, the Democratic Party; I. M. Garba-Jahumpa, a teacher and leader of the Muslim Congress Party; a well-known Bathurst lawyer, Mr P. S. N'Jie, leader of the United Party (U.P.) and his brother, E. D. N'Jie, also a lawyer. By 1960, as in Sierra Leone, leadership of the movement had passed into the hands of the educated élite from the Protectorate. There, the dominant figure was David Jawara, an Edinburgh-trained veterinarian who became leader of the Progressive People's Party (P.P.P.).

It is not surprising that neither organised pressure groups nor party politics emerged in the Gambia until 1951 when the first political party, the Democratic Party, was formed. The constitutional reforms that occurred between 1946 and 1951 stemmed almost entirely from unilateral decisions of the British Government to bring the Gambia into line with the constitutional advance taking place in other West African colonies. As other colonies advanced towards self-government, the Gambia appeared more and more as a liability to the British. It had a population of only three hundred thousand, in a narrow strip of territory seven to twelve miles on both sides of the Gambia River; carved, as it were, out of the Senegal. The economy was poor, being dependent on a single cash crop of groundnuts and the annual budget had to be subsidised by the British treasury. The Colonial Office was therefore not unwilling to involve the people more and more in the discussion of their affairs and the long-term future of the territory.

80 *West Africa*, 22 April 1961, p. 427.

It was on British initiative that the elective principle introduced in other British West African colonies since the 1920s was established in 1946 when Bathurst and Kombo were each allowed to elect one representative to the Legislative Council. The new constitution also introduced an unofficial majority of one in a thirteen-member Legislative Council as well as three nominated unofficial members in the Executive Council. This constitution was revised in 1951 to increase the unofficial members in the Legislative Council to four. The nominated unofficial members in the Executive Council were increased to four. What is more, the elective representation from the colony area was increased to two.[81] It was largely to take advantage of this limited reform that political parties—the Democratic Party, the Muslim Congress Party and the United Party— began to spring up in Bathurst during 1951 and 1952. They became a major factor affecting the pace of change. As is often the case, change produced its own momentum. The various political parties criticised the 1951 Constitution for not going far enough in the direction of giving more powers to Africans in the government of the country.

As a result of pressure from the colony-based parties, Governor Wyn-Harris in May 1953 invited thirty-four members and former members of the Legislative and Executive Councils to serve as a consultative committee to formulate proposals for a new constitution. The proposals of this body, with some modifications, became the basis of the 1954 Constitution.[82] Under the new constitution, the representatives from the Colony were increased to seven, four elected directly, the other three indirectly. The Protectorate was to be represented also by seven members selected by an electoral body consisting largely of the traditional élite. There was provision for expanding the Executive Council to include three of these representatives to be called Ministers, but without responsibility for determining policy in government departments.

The political leaders soon found many provisions of the 1954 Constitution unsatisfactory. Leaders of the Protectorate criticised the equal number of seats given to the Protectorate and the Colony when the Protectorate had five times the population. The educated élite, both in the Colony and in the Protectorate, criticised the indirect voting system prescribed, particularly in the Protectorate since it favoured the continued hegemony of the traditional élite who were regarded as amenable to government pressure. Others criticised the inadequate powers given to the Ministers. Late in 1954 the Democratic Party leader, the Reverend John Faye, called the constitution a kind of fraud.[83] A year later P. S. N'Jie, whose United Party had won the largest number of seats in the election to the Legislative Council and who

81 H. A. Gailey, *A History of the Gambia*, London, 1964, pp. 186–90; *West Africa*, 24 November 1951, p. 1095.
82 Gailey, p. 188.
83 *West Africa*, 17 October 1954, p. 971.

was appointed a Minister, resigned from the government. This precipitated a crisis in which the United Party mounted a relatively militant verbal attack on the colonial administration and the 1954 Constitution. They called for an extension of universal adult suffrage to the Protectorate and demanded that the elections scheduled in 1959 should usher in full internal self-government for the country, with a responsible cabinet on the Westminster model.[84]

Late in 1958 at the Brikama Conference of Protectorate chiefs and representatives of all Colony-based political parties, which was attended by Sir Edward Wimbley, the new Governor, all the Colony-based parties except the Muslim Congress demanded extensive reforms of the 1954 Constitution: that the principle of universal adult suffrage should operate throughout the country; that the number of members of the Legislative Council be increased to twenty-seven; and that there should be a responsible cabinet government of nine Ministers.[85] The Protectorate chiefs joined the Muslim Congress in opposing these demands, but the new Governor could not totally ignore them. He therefore convened a conference of all the political groups from Bathurst and the Protectorate to discuss the situation. Because of the emergence of the educated élite from the Protectorate, the conference was able to agree on some far-reaching proposals such as the establishment of a cabinet form of government under a Chief Minister and a Legislative Council elected on the principle of universal adult suffrage throughout the country.

The conference reported to the Governor shortly before the visit of the Colonial Secretary, Lennox-Boyd, to Bathurst in June 1959. Lennox-Boyd met representatives of all the political parties. But although he was unable to give them a firm promise about self-government and some misunderstanding led to a clash between the police and the crowd in front of Government House,[86] a change in British attitude to Gambia soon became noticeable. With the failure of the egg project intended to diversify the Gambian economy the British Government came to favour devolution of responsible government to the Gambia with every encouragement, if not pressure, to negotiate union with Senegal. With this end in view, a new constitution was introduced in 1960 which extended universal adult suffrage to the Protectorate. Membership of the Legislative Council was enlarged to thirty-four, consisting of nineteen elected representatives, seven from the colony and twelve from the Protectorate; eight to be appointed by the Proectorate chiefs, four ex-officio and three others nominated by the Governor. There was also to be ministerial government

84 Michael Crowder, 'The Gambian Political Scene', *West Africa*, 2 November 1957, p. 1035.
85 *West Africa*, 18 October 1958, p. 987.
86 *West Africa*, 20 June 1959, p. 580.

with a cabinet consisting of the Governor, the four ex-officio members and not more than six ministers.

It was the extension of the franchise to the Protectorate that led directly to the formation of a new Protectorate-based political party, the Progressive People's Party (P.P.P.), led by David Jawara. The P.P.P., which was henceforth to play a leading role in the struggle for self-government,[87] won eight of the twelve elective seats in the Protectorate. No party won a majority in the house and while Jawara and others were appointed Ministers, none was designated Chief Minister. In March 1961, the Protectorate chiefs threw their eight members behind P. S. N'Jie who had chosen to remain outside the government and got him appointed Chief Minister. Jawara resigned from the government and mounted a violent attack on the 1960 Constitution, particularly the power of some thirty-five chiefs to select eight members. Coming so soon after the strike of the Gambia Workers' Union and the most serious labour disturbances in Gambia's history earlier in the year, the crisis precipitated by the P.P.P. marked a definite advance towards independence. The P.P.P. emerged as the most militant advocates of rapid decolonisation.

The government responded by calling for conferences to review the constitution and to discuss the future of the country. Meetings were held in Bathurst in May and in London in July–August 1961. There was agreement on drastically reducing the seats reserved for the chiefs and on granting internal self-government. But as to the P.P.P. demand for independence in 1962,[88] the new Colonial Secretary, Mr Ian Macleod, was cautious. While not making union with Senegal a condition, the British Government was anxious that the new Gambian Government should negotiate with the Dakar Government before talking of independence.[89]

The general elections of May 1962 in which the P.P.P. won seventeen of the twenty-five elected seats in the Protectorate brought David Jawara to power as Chief Minister. It devolved on him to negotiate the final stages of decolonisation with Britain. Faced with the stark realities of the Gambian situation, the P.P.P. in power were less insistent that independence must come at once.

Ministers were sent to negotiate closer union with Dakar and U.N. legal and financial experts were brought in to advise but little progress was made. There was agreement on coordinated economic development of the Gambia basin, a defence pact and the sharing of diplomatic missions. While some Senegalese, especially from the Sine-Saloum region, who feared the competition of the port of Bathurst on the economic life of Kaolack and Zinguichor, were opposed, the Government of Senegal welcomed a merger of both countries. It was the political leaders of small

87 *West Africa*, 21 January 1961, p. 61.
88 *West Africa*, 19 August 1961, p. 907.
89 *West Africa*, 5 August 1961, p. 851.

Gambia who feared that they would be economically, culturally and politically submerged and become second-class citizens in a French-speaking Sene-gambia no matter what constitutional guarantees were provided. They feared that the little foreign investment and external aid they might expect on independence was likely, in a merger with Senegal, to be diverted to their more powerful neighbours.[90]

The political leaders were united in the belief that the interest of the Gambia in self-government and economic development, not the British anxiety to rid itself of a liability, should determine the future of the country. They opted for independence. The British could hold out no longer. A conference in London in July 1964 agreed that the Gambia, while continuing to cooperate with Senegal, should become an independent country within the Commonwealth on 18 February 1965.

EXTENT OF DECOLONISATION

By 1965, all the British colonies in West Africa had attained independence. But this does not mean that complete decolonisation was achieved by that year. The nationalist agitation had been directed at securing not only political independence, but also control of their countries' public services, armed forces, and economic and monetary institutions. Upon the introduction of representative government in each of the four British colonies in West Africa in 1951 the process of Africanisation was accelerated at varying rates according to the circumstance of each territory. To what extent had decolonisation been carried out in the public services, armed forces, the economy and monetary institutions in these countries by the time of independence? What major factors affected the pace of decolonisation in each country?

Ghana

In Ghana, where the process of decolonisation may be said to have gone furthest by the time of independence, it cannot be said to have progressed very far, outside the political and constitutional area, by March 1957. Ghanaianisation of the public service was the most rapid. In 1946 there were eighty-nine Ghanaians in top public service posts as against nine hundred and sixty expatriates.[91] In 1957 the number of Ghanaians holding senior posts in the public service had risen to 1,581 as against 1,204 expatriates.[92] Yet, three years after independence, the percentage of expatriates in the public service was not much below 40 per cent. Apart from the numerical strength of these expatriates, some held such strategic positions that they continued to exert undue influence on Ghana's

90 R. W. Howe, *The African Revolution*, London, 1968, pp. 159–61; *West Africa*, 25 April 1964, p. 473.
91 Austin, p. 8 n.
92 *Ibid.*

external as well as her domestic policies.[93] It was not until July 1960 with the adoption of a republican form of constitution that, according to E. N. Omaboe, the Ghana Government took 'some radical decisions which Ghanaianised the top posts of the civil service'.[94]

As for the armed forces, the process of decolonisation was slower. In 1957 the Ghana national army, which was originally part of the Royal West African Frontier Force, was in the main officered by the British. Up till September 1961, the G.O.C. of the Ghanaian armed forces was a Briton; and the majority of senior technical and administrative posts in the army were held by the British. Even after the dismissal of all British officers in the army in September 1961, the Ghana Government entered into an agreement in May 1962 with the British Government under which a Training Team Mission of over two hundred men was established in Ghana to train the Navy, the Air Force and the Army. When the Nkrumah regime was overturned in February 1966 there were still more than two hundred British naval and air force 'experts' in Ghana.

Decolonisation of monetary institutions and the economy was even slower. Upon independence Ghana chose to continue its membership of the sterling area. In March 1957 the country's foreign reserves were still, in the main, invested in London. Although the Bank of Ghana was established in 1957 and although an Ordinance of 1957 empowered the Bank to create a fiduciary issue up to a maximum value of £12 million, it was not until April 1961 that such a fiduciary element was actually created in the currency supply. Until that date all the Bank's currency liabilities remained backed 100 per cent by sterling.[95] Similarly, at the time of independence, two British-owned commercial banks—the Barclays Bank D.C.O. and the Standard Bank of West Africa—were responsible for handling the bulk of banking business in Ghana. It was not until 1961 that the government established a state-owned commercial bank, the Ghana Commercial Bank, to break the virtual monopoly of these two foreign banks.

The colonial economy, whereby the Gold Coast used to export most of its primary commodities to Britain and receive in return the bulk of its imports of manufactured goods from Britain, was not drastically altered on independence. Nor had it been so radically transformed thirteen years after independence, although Britain's share of Ghana's trade has since 1957 continued to decline. In 1957 about 72 per cent of Ghana's imports came from Britain, while about 52 per cent of its exports went there.[96] In 1960 the figures were 37 per cent of Ghana's imports and 31 per cent of its

93 Up till 1960, Nkrumah's private secretary and his closest advisers on law and order, economic policies and foreign affairs were British.

94 Walter Birmingham, I. Neustadt and E. N. Omaboe, eds., *A Study of Contemporary Ghana*, London, 1966, i, p. 452.

95 Birmingham, Neustadt and Omaboe, p. 312.

96 *Annual Report on External Trade of Ghana, 1957*, i, pp. 24–5.

exports.[97] In 1967 the figures were 30 per cent of imports and 28.5 per cent of exports.[98]

Although cocoa, which remains the backbone of the country's economy, has continued to be grown by Ghanaian private farmers, those who were licensed to buy cocoa in 1957 were still largely expatriates. In 1957–8, the expatriate buying agents were responsible for purchasing 72.5 per cent of all the country's cocoa. Even as late as 1960–1, 37.3 per cent of all the cocoa was purchased by them.[99]

Wholesale and retail trade was at the time of independence dominated by a few large foreign firms, such as the United Africa Company (U.A.C.), a Unilever affiliate, and the Swiss United Trading Company, as well as smaller Cypriot, Lebanese and Indian firms. The government tried to break their domination of the country's trade by establishing in 1961 the Ghana Trading Corporation.

Mining, which in 1957 contributed 5.7 per cent of the G.N.P., was in that year almost totally in the hands of foreign-owned or foreign-controlled firms. Indeed, until 1961 all seven of the firms engaged in gold mining—namely, the Amalgamated Banket Areas, Ariston Gold Mines, Ashanti Goldfields, Bibiani, Bremang Gold Dredging, Ghana Main Reef, and Konongo Gold Mines—were foreign-owned. The manganese industry has for years continued to be worked by the African Manganese Company, a wholly British-owned subsidiary of the American Union Carbide Corporation. Manufacturing and construction industry, which was very small at independence, was largely dominated by a few foreign firms. Sawmilling, the largest single manufacturing industry, was largely owned and operated by the U.A.C.

The tobacco and brewery industries were similarly dominated by foreign-owned firms in 1957. Indeed, independence has been followed by an increase in the establishment of foreign-owned enterprises with heavy investments, such as the soap factory built at Tema in 1961 by the U.A.C. But it is the aluminium smelter at Tema, built with an initial capital investment of about £45 million, which was envisaged as increasing to some £100 million by the 1980s, that is the most notable example of the introduction of large wholly foreign-owned and foreign-controlled enterprises which has been one of the features of industrial development in post-independence Africa. This aluminium smelter is owned and controlled by two American giant aluminium companies—the Kaiser Aluminum and Chemicals Corporation and the Reynolds Metal Company. Under the agreement signed between these companies and the Ghana Government in February 1962, the Volta Aluminium Company (VALCO)

97 *Economic Survey 1963*, Accra, 1964, p. 48.
98 *Economic Survey 1967*, Accra, 1968, p. 41.
99 *Report of the Committee of Enquiry on the Local Purchasing of Cocoa*, Accra, 1966, p. 159.
100 Birmingham, Neustadt and Omaboe, pp. 24–5.

—the name of the joint company formed by Kaiser and Reynolds to operate the aluminium smelter—is protected against nationalisation for the next thirty years. Furthermore, it was exempted from foreign exchange control, which has been in force in Ghana since late 1961.[101]

It is difficult to be precise as to the degree of Ghanaianisation of senior posts in these foreign-owned concerns. No doubt some Ghanaianisation has been taking place, especially in the commercial firms in their wholesale and retail trade. But in the manufacturing and construction industries, in mining, and in the smelting of aluminium at Tema, where the level of skills required is extremely high, the rate of Ghanaianisation has not been rapid.

Nigeria

In Nigeria at the time of independence in October 1960 decolonisation had not progressed even as far as it had in Ghana. But, as in Ghana, the process of decolonisation went furthest in the public service, except in that of the former Northern region. In March 1955, there were 464 Nigerians in top senior (superscale) posts of the federal civil service, representing about 25 per cent of all the top senior posts in the federal service.[102] By December 1960 the number of Nigerians in such posts had risen to 2,562, representing about 64 per cent of the total. The Nigerianisation of the civil service at the regional levels, except in the North, had been impressive by 1960. In the West, Nigerians held 83 per cent of all superscale posts in 1960; the corresponding percentages in the East and North were 80 and 28 per cent respectively.[103]

Although the percentage of expatriates in the Nigerian federal public service had been reduced by the time of independence, some continued to hold strategic posts. For instance, Peter Stallard, formerly a Northern Nigerian administrative officer, was retained until about eight months after independence as the Secretary to the Prime Minister and head of the federal civil service. Even after the replacement of Peter Stallard by a Nigerian, Stanley Wey, in 1961 there continued to exist some close consultation and a cordial working relationship between the Nigerian leaders and the British officials in the country. It was such cordiality that made it possible for the British High Commissioner, Sir J. Cumming-Bruce, to attend the last cabinet meetings of the outgoing civilian regime after the *coup d'état* of 15 January 1966.[104]

In the armed forces, the process of decolonisation was slower still. At independence the majority of the officers in the army were British. Indeed,

101 Arrangements were provided under the agreement for the training of Ghanaians who would ultimately take up the operation of the smelter.
102 Omorogbe Nwanwene, *The Civil Service of Nigeria: Problems and Progress of Its Nigerianisation, 1940 to the Present Day*, unpublished Ph.D. thesis, London, 1967, p. 358.
103 *Ibid.*, p. 385.
104 Schwarz, p. 119.

it was not until mid-1962 that the Federal Government arranged for about half of the three hundred senior posts in the army to be held by Nigerians.[105] Until 1965 the General Officer Commanding the Nigerian Army was a Briton, Major-General Welby-Everard. Even after the Anglo-Nigerian Defence Agreement was abrogated in 1962,[106] the training of the Nigerian military and naval officers continued to be entrusted to the British, though Indians have staffed the Nigerian Defence Academy.

The process of decolonisation has made the least progress in the financial and economic spheres. Like Ghana, Nigeria chose at independence to continue its membership of the sterling area. By the time of independence, and even during the immediate post-independence years, the currency liabilities of Nigeria were backed 100 per cent by sterling. The banking business of the country was, and even in 1971 still is, dominated by expatriate commercial banks, principally Barclays Bank D.C.O. and the Standard Bank of West Africa. The situation differed from that in Ghana, however, since there were a few indigenous banks such as the African Continental Bank, the National Bank and the Cooperative Bank to compete with the expatriate banks. And in post-independence years these have gained increasing strength.

The extent to which the foreign banks have Nigerianised their top posts is difficult to ascertain. What can be said is that since the late 1950s Nigerianisation has been a declared objective of these banks. Irrespective, however, of the number of Nigerians in senior posts of foreign banks, the effective control of their operations still rests with the expatriate officials. In December 1961, the then Federal Minister of Economic Development condemned the foreign-owned banks for not giving loans and overdrafts to Nigerian businessmen.[107] Early in 1970, this same censure was repeated against the expatriate banks by the chairman of the Warri Chamber of Commerce.[108]

The colonial economy has remained largely unchanged. In 1954, 72 per cent of Nigeria's exports went to Britain, while about 60 per cent of its imports came from there. In 1960 the corresponding figures were about 49 per cent of exports and 45 per cent of imports.[109] In 1967, the figures were 29.5 per cent of exports and 28.9 per cent of imports.[110] Although her share in Nigeria's trade has been steadily falling since independence, Britain remains Nigeria's largest customer for both exports and imports. The formulation of economic policies for Nigeria was at the time of independence, and for at least two years later, controlled in the main by the British.

105 *West Africa*, 6 January 1962, p. 5.
106 For details see *The Draft Agreement between the Governments of the U.K. and of the Federation of Nigeria*, Lagos, 1960.
107 *West Africa*, 9 December 1961, p. 1371.
108 *Daily Sketch* (Ibadan), 7 February 1970.
109 *Abstract of Statistics*, December 1961.
110 *Nigeria Year Book*, 1969, p. 148.

During the constitutional conference of 1958, on the advice of the Colonial Secretary, all the Nigerian leaders, except those of the West, agreed to abandon the Five-Year Plan 1955–60, in favour of a new Three-Year Plan 1959–62.[111] Thus the governments of the federation, except that of the West, continued till 1962 to develop the areas within their jurisdiction under a plan drawn up largely by the colonial administrators.[112]

Wholesale and retail trade in the country was in 1960 largely dominated by foreign merchant firms. The most important were U.A.C., John Holt, Paterson Zochonis, Compagnie Française de l'Afrique Occidentale (C.F.A.O.), and Société Commerciale de L'Ouest Africain (S.C.O.A.). Most of these expatriate firms were engaged not only in the extensive import/distribution trade, but also in the profitable produce-buying business. Some, especially the European firms, began in the 1950s to Nigerianise their management and to appoint some Nigerians as directors. But at the time of independence the managers, directors and top senior officials of these firms were still mostly expatriates. Although the process of Nigerianisation has since independence been intensified in some of these firms, it is difficult to say precisely what percentage of the senior posts in each of the firms had been Nigerianised. However, there were more indigenous people in the import/distribution business and in the produce-buying trade than in Ghana.

What has, in fact, happened since independence is that most of these foreign merchant firms have moved into technical trading and industry. A few established supermarkets for consumer retailing. In 1962 the Kingsway Stores, with an annual turnover of about £3 million, established a department store in Ibadan and in 1963 another in Port Harcourt.[113]

The manufacturing and construction industry was at the time of independence largely dominated by expatriate firms. Since independence there has been a steady increase in the number, as well as in the variety, of foreign firms engaged in manufacturing and construction.[114] At independence, Richard Costains & Co. and other expatriate firms dominated the construction industry, though there were a few indigenous construction companies at that time. In 1960 the manufacturing of cigarettes (from which Nigeria derived a revenue of £6 million in that year) was entirely in the hands of the Nigerian Tobacco Company, a subsidiary of the British American Tobacco Company.[115]

111 Awolowo, p. 266.

112 Indeed, a senior federal official recently confirmed this in a public lecture, and added that it could in fact be said that after 1962 the economic policies of the country were strongly influenced by the British.

113 Peter Kilby, *Industrialisation in an Open Economy: Nigeria 1945–1966*, London, 1969, p. 68.

114 *Ibid.*, pp. 68–84, for a list of the foreign firms that have entered the industrial and construction sector of the economy since 1960.

115 *West Africa*, 5 August 1961, p. 845.

The oil industry, which since 1958 has become very important in the country's exports—crude oil accounted in 1966 for 33 per cent of the total value of the country's exports, or £92 million of the total value of £278.7 million—has been entirely dominated by expatriate oil companies.[116] By 1960, the Shell-BP had invested over £60 million in the oil industry and maintained in entirety its ownership, control, and operation. Since then other foreign oil companies have obtained oil-prospecting licences from the Federal Government.[117] There is nothing to be said about the Nigerianisation of the oil companies in 1960 for the industry was then just getting on its feet. While Nigerianisation of the senior posts of various oil companies has not been totally ignored the rate has been painfully slow, mainly because of the difficulty in finding a sufficient number of Nigerians with skills appropriate to oil technology and industry.

Sierra Leone

The Sierra Leone story is little different. Here too, the most rapid rate of decolonising was in the public service, though this was still slower than that of the Gold Coast and Nigeria. In 1953, 24 per cent of the senior posts in the civil service were held by Sierra Leoneans. In 1959, the proportion had increased to 56 per cent. But if the number of expatriates on contract is taken into account, as it should be, Sierra Leoneans held only 45 per cent of the senior posts.[118] Until 1960 the chairman of the Public Service Commission was British; and the Commission itself was not fully 'Sierra Leoneanised' until that year.[119] At the time of independence the proportion of Sierra Leoneans in senior posts of the public service was not much above 50 per cent.

As of April 1961, decolonisation had made little headway in the armed forces. In 1959–60, 70 per cent of the country's military establishment, including internal security forces, was supported by a subvention from Britain.[120] At independence the armed forces remained under the command of British officers. Since then Britain has continued to provide men and facilities for the training of officers of the Sierra Leone armed forces and to supply all the necessary military hardware.

In the economic and financial area, decolonisation had not gone as far as it had at the time of Ghana's or Nigeria's independence. For one thing, Sierra Leone was much poorer than either of these two countries. Indeed the British government had to provide the sum of £7.5 million to tide the country over the immediate post-independence years.[121] At the time of

116 L. H. Schartzl, *Petroleum in Nigeria*, Ibadan, 1969, pp. 152–3.
117 *Ibid.*
118 *West Africa*, 13 February 1969, p. 179.
119 M. Kilson, p. 174.
120 *Ibid.*, p. 175.
121 *West Africa*, 7 May 1960, p. 527.

independence and for several years after, the mining industry, which formed the backbone of Sierra Leone's economy, was almost entirely owned and controlled by foreign mining companies. Early in 1969 the government began negotiating with these foreign mining companies—especially the rutile, the bauxite and the diamond mining companies—to increase Sierra Leone's share of the proceeds realised from exploitation of its natural resources.[122] Diamond exports, which traditionally have accounted for over 60 per cent of the total value of the country's exports, continued to come largely from mines owned and operated by a British mining company, a subsidiary of the Consolidated African Selection Trust, later called the Sierra Leone Selection Trust Ltd.

The commerce of the country was at independence dominated by a few foreign firms such as the U.A.C. and some French and in particular Lebanese firms. This pattern has not changed substantially. Likewise, in the banking business, it is the few foreign commercial banks such as Barclays Bank D.C.O. and the Standard Bank of West Africa that have continued to predominate. After nearly a decade of independence, these foreign-owned banks have taken only a few hesitant steps towards appointing Sierra Leoneans to senior posts in their business. The manufacturing and construction industry, which was much more meagre than in Ghana and Nigeria at the time of independence, was largely dominated by foreign firms such as the U.A.C.

The Gambia

Of all the four former colonies in West Africa, it was in the Gambia that decolonisation in areas other than the political had made the least headway by the time of independence. The reasons are obvious. The Gambia was not only the smallest of them but also the poorest and the least developed.

In the public service of the Gambia at the time of its independence in February 1965 the expatriates still held about 70 per cent of the total number of senior posts. Most held strategic positions, including, for example, the Attorney-General, who continued to be a member of the cabinet, the Secretary to the Prime Minister and the cabinet, the Chief Justice, the Registrar-General, the Director of Cooperatives, and the Commissioner of Police.[123]

At independence the few large firms in the country were owned and run by non-Gambians. There were a few British firms such as U.A.C., five French trading firms and important Lebanese and Indian firms.[124] These firms handled most of the export–import trade as well as local commerce. The picture has not changed much since independence.

122 *Annual Economic Review of Sierra Leone and the Gambia by the Standard Bank of West Africa*, 1969, p. 4.
123 B. Rice, *Enter Gambia: The Birth of an Improbable Nation*, 1968, p. 137.
124 *Ibid.*, p. 28.

At the time of independence industrialisation of the country was in its infancy. It was restricted to two small groundnut oil mills, a small fisheries plant, a small lime-processing plant, and a small factory turning out shirts, singlets, umbrellas and plastic shoes. Almost all these were owned and operated by foreign firms; and all their senior officials were expatriates. This was equally true of the only commercial bank in the country in 1965, the Standard Bank of West Africa, as of the International Bank for Commerce and Industry established in 1968.

Conclusion

What emerges from the foregoing analysis is that although all the British colonies in West Africa had by 1965 attained independence within the Commonwealth, the process of decolonisation had largely been confined to the political and constitutional areas. Nearly a decade after the independence of Nigeria, this situation has tended to persist, though in varying degrees, in the four ex-British West African colonies. Indeed, in some cases, as we have seen from the analysis above, political independence has been followed by an increase in the control of the vital sectors of the economy by foreign concerns—giving rise to the phenomenon which has been loosely described as 'neo-colonialism'.[125] Why was this so?

Broadly speaking, there were three reasons for this state of affairs. First, there was a lack of domestic capital resources and technical know-how in all these ex-British colonies, and yet their leaders wanted to modernise the countries. Second, there still remained with independence a substantial amount of goodwill towards Britain, especially among the new African leaders. Finally, there was the difficulty of reorienting established trade patterns and restructuring the existing economic arrangements.

At the time of independence, all these countries lacked adequate domestic capital resources, although Ghana and Nigeria were a little better off than the rest. Without exception, all were desperately short of men with the technical, managerial and entrepreneurial skills necessary for the task of economic development. The income per capita of Ghana, the wealthiest of them at independence, was £50, while that of Nigeria, the next highest, was about £25 in 1960. The small populations of Sierra Leone and the Gambia further compounded their poverty. Yet their leaders sought to develop these countries as rapidly as possible. This they could not undertake without foreign private investment, as well as capital aid and technical assistance from the developed countries. For all of them, at least during the immediate post-independence years, Britain was the main source of capital resources and the skills required for industrialising their countries.

125 The term 'neocolonialism' is used in the sense employed by Berkeley Rice to describe the situation in the post-independence Gambia (p. 137). It is not used here in the way Nkrumah used it in his book, *Neocolonialism: The Last Stage of Imperialism*, 1965.

Because of the timely way in which Britain granted independence to her colonies in West Africa, she was able to retain a substantial amount of goodwill and respect among the leaders of these newly independent states. As a result these leaders were somewhat unwilling, both before and for some time after independence, to embark on a radical programme of decolonising, not only in the public service but in particular in the economic and commercial life of their countries.

The policy of nationalisation, which could have increased the control of the local governments over the economy, was in principle opposed by all the leaders of the four ex-British colonies in West Africa, including Dr Nkrumah until 1961. The opposition of the Nigerian leaders to nationalisation was vigorous, particularly during the immediate post-independence years. Condemning those calling for nationalisation, Dr Okpara, then the Premier of the former Eastern Region, described them as Communists.[126] Furthermore, there are enormous difficulties in reorienting a country's established commercial relationships and in revolutionising overnight a well-established economic arrangement. Such difficulties, combined with common factors shared by Britain and her former colonies of West Africa—such as language, and the legal, administrative, educational and political system—made it extremely difficult for the leaders of these newly independent states to restructure their economy in a radical way.[127]

As we can see from the above, the process of decolonisation in the non-political areas of the former British West Africa, which has been going on since the end of the Second World War, has not gone very far. Recently, the tempo for decolonisation has received new momentum in all the four Commonwealth West African countries. Since 1968, when the Gambia ceased to rely on budgetary support from Britain, it has been putting greater emphasis on Gambianisation of its economy and finances and this is also true of Sierra Leone since Siaka Stevens came to power in April 1968. One of the main reasons behind the Business Promotion Act of Ghana of July 1970 is to increase the Ghanaian ownership, and control, of commercial and industrial enterprises in the country. But the most far-reaching policy statement on decolonising the economic–monetary areas can be found in the Nigerian Second National Development Plan, 1970–4. There it was stated that the policy of indigenisation of ownership, and control, of expatriate-owned enterprises would be vigorously pursued in order to attain 'economic independence for the nation and the defeat of neo-colonialist forces in Africa'.[128]

It is not easy to foresee just when complete decolonisation will occur. We can, however, say that full decolonisation in these non-political areas

126 *West Africa*, 12 November 1960, p. 1287.
127 It was the combination of these factors that frustrated Nkrumah's effort between 1961 and February 1966 to revolutionise the economic organisation of his country.
128 *Second National Development Plan 1970–74*, Lagos, 1970, p. 32.

will not come until the four English-speaking West African countries are able to mobilise sufficient capital resources and technical know-how for their own development without resort to large foreign investments; until their leaders have the political will to push for decolonisation, convinced at last that complete decolonisation in these areas will not result in unacceptable sacrifice for their countrymen.

CHAPTER 19

French West Africa, 1945–1960

MICHAEL CROWDER AND DONAL CRUISE O'BRIEN

While African political leaders in British West Africa were demanding at least some measure of self-government by the end of the war, and some like Nkrumah, Azikiwe and Wallace-Johnson had their eyes set firmly on the long-term goal of independence, this was a word that scarcely passed the public lips of any French West African leader of consequence until the late 1950s.[1]

Léopold Sédar Senghor, it is true, gave a newspaper interview in 1946 in which he talked in terms of independence,[2] but this was belied by his political record in the Constituent Assembly and throughout the period 1946–58.[3] Independence appears to have been discussed in African political circles in the immediate aftermath of the war, but only a few isolated individuals or splinter groups were ready to talk openly in terms of a struggle to achieve such an objective. The publication of the manifesto of the Senegalese *Parti Africain de l'Indépendance* (P.A.I.) in 1957, which made the claim to independence in the most unequivocal language, opened a new period in this respect; and the Constitutional Referendum of 1958 made African independence for the first time an acknowledged possibility within the framework of French official policy objectives.[4]

The main aim of the West African leaders in the immediate post-war years nonetheless was to extract from France the promises of reform made at Brazzaville: abolition of forced labour, abolition of the *indigénat*, abolition of the humiliating status of *sujet* (subject), greater participation in the political process of the French Empire and social and economic improvements for their people. The secessionist ideas of Ferhat Abbas who told the

1 See Michael Crowder, 'Independence as a Goal in French West African Politics: 1944–60', *French-speaking Africa: The Search for Identity*, ed. William H. Lewis, New York, 1965.
2 *Gavroche*, 8 August 1946. See Irving Leonard Markovitz, *Senghor and the Politics of Negritude*, London, 1970, p. 103.
3 *Journal Officiel: Assemblée Constituante*, 1946.
4 Refers to documents in the Senegalese National Archives at Dakar (ARSD), dossier 13G 17/36.

Second Constituent Assembly that he wanted 'an Algerian state, an Algerian parliament'[5] found little public response from the West African deputies. Perhaps the nearest any West African came to Ferhat Abbas's position was Sourou Migan Apithy, representing Dahomey–Togo, when he told the Second Constituent Assembly that 'our ideal is not to sit on the banks of the Seine nor impose ourselves in what are essentially metropolitan affairs, but to regulate the affairs of our own country on the banks of the Congo or of the Niger'. However, in the next breath he recognised that Africans would continue 'to discuss with the people of France matters interesting the *ensemble* which we form with them'.[6]

French West African politicians in the post-war years differed fundamentally from their English-speaking neighbours in that, though both were interested in reform, the French-speaking African leaders formally rejected independence in favour of greater participation in the political process of a French Union or Community of which Africa would be a constituent part.[7]

THE CONSTITUENT ASSEMBLIES OF 1945–6

Apart from the Deputies elected by the citizens of the *Quatre Communes* of Senegal, no French Black Africans had any previous practical experience of the parliamentary process. In 1945, following on the reformist spirit of Brazzaville, Africans from all the constituent colonies of French West and French Equatorial Africa were to be represented in the Constituent Assembly that was to draft a new constitution for France as well as her overseas territories. In each case elections were to be made on two separate rolls: one for citizens, mainly metropolitan French residing in the colonies, who alone legally could vote in the referendum which would eventually accept or reject the new constitution, and one for the subjects, who would have no such right. Effectively this at least meant that the subjects could make their views heard in the deliberations of the Constituent Assembly. The representatives of the subjects had little direct experience of the parliamentary process though, since the majority of them had been educated in Dakar, they were familiar with the elections to the Deputyship, the *Conseil Colonial* and the Municipal Councils for the Quatre Communes. They were also influenced by the relatively free political

5 *Journal Officiel*, Debates of the Second Constituent Assembly, 1st Session, 18 September 1946, p. 3791. See his speech of 18 September 1946 to the Second Constituent Assembly.

6 *Journal Officiel*, Debates of the Second Constituent Assembly, 2nd Session, 18 September 1946, p. 3802.

7 There are three major studies of the period which forms the subject of this chapter: Franz Ansprenger, *Politik im Schwarzen Afrika*, Köln and Opladen, 1961; Ruth Schachter-Morgenthau, *Political Parties in French-speaking West Africa*, Oxford, 1964; Edward Mortimer, *France and the Africans 1944–60: A Political History*, London, 1969, which is particularly useful for explaining the interrelationship of Metropolitan and 'Colonial' politics in the Constituent and National Assemblies.

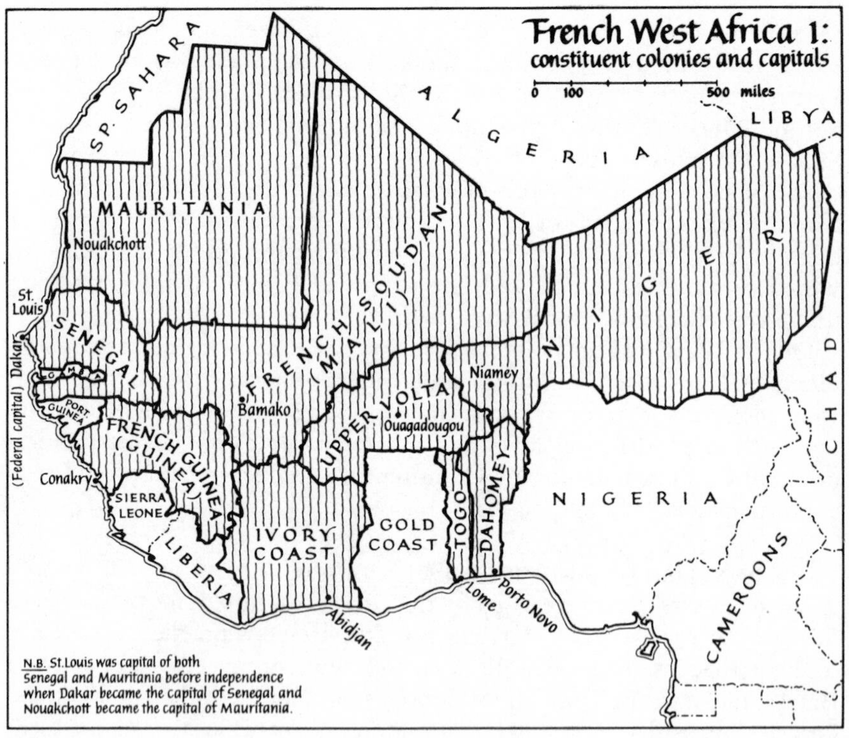

19:1

discussions conducted in the Senegalese press; in their own territories newspapers were subject to censorship or closure if political views inimical to the administration were vented.

Outside Senegal only two territories can be said to have been open to movements of opposition politics. In the Ivory Coast, of which Upper Volta formed an administrative unit, Free French policy favouring European planters as against African planters had built up great resentment against the administration (see Chapter 17, p. 607) and gave Félix Houphouët-Boigny, a *médecin africain*, chief and wealthy planter, a basis on which to campaign. In Togo and Dahomey, which for electoral purposes were linked together, despite the fact that Togo was not legally a French colony but had been administered as a League of Nations Mandate (and had now become a U.N. Trust Territory), the Togolese resented the inroads into their special status, especially the fact that they had been administered jointly with Dahomey for a period and had come under the general administrative aegis of Dakar. They also wished to maintain the special status acquired by them under the U.N. Charter. A remarkable conference was called on 11–12 May 1945 by the Commissioner for Togo to discuss the implications of the Brazzaville recommendations for Togo; forty-five Togolese including Sylvanus Olympio, Savi de Tové and a number of chiefs, as against seventeen Frenchmen, participated. Only one member, the Director of Public Works, voted against an autonomous status for Togo. The Togolese participants made it clear that they were primarily interested in being Togolese citizens and only wanted the freedoms associated with French citizenship. Sylvanus Olympio notably declared: '. . . we want a representative [in the Constituent Assembly] uniquely to deal with questions relating to the Togolese native, and not metropolitan questions. French affairs debated in the Parliament do not interest us.'[8] Dahomey itself, despite the existence of the indigénat, had a long tradition of African élite politics[9] and had a large Catholic educated group who, with the Togolese *administrés*, voted for Sourou Migan Apithy, a lawyer and accountant who had studied in Paris.

In Guinea the contest for the *sujet* seat was conducted on largely tribal lines between the two major ethnic groups of the country, the Fula and the Malinké, with Yacine Diallo, the nominee of the Fula chiefs, winning the seat. Similarly in Soudan Fily Dabo Sissoko, much better educated than Yacine Diallo, secured election largely through the support of the chiefs to whose class he belonged.

In Senegal, the dual role did not operate to the advantage of the white French citizens because of the large bloc of African citizens in the Quatre Communes. These citizens secured the election of Lamine Guèye, leader of the local branch of the S.F.I.O. who had unsuccessfully campaigned for the deputyship on several occasions before the war. His nominee for the subject roll, Léopold Sédar Senghor, who had been absent from Senegal in France since 1928 but was himself a subject, was elected largely through Lamine Guèye's support. In Senegal, as in the Ivory Coast, the campaign was based on hotly contested political issues. A number of racial incidents involving locally stationed French troops and the massacre at Tiaroye (see Chapter 17, p. 614), had inflamed many voters against the French. A youth movement centred on the town of Thiès had succeeded in attracting a substantial political following on the basis of the militantly anti-colonial declarations of its leaders.[10] Furthermore, while French women were allowed to vote, the wives of the *originaires* of the Quatre Communes, who legally enjoyed the status of French citizens, were initially excluded from voting. Only direct intercession by Lamine Guèye in Paris secured them the right.

The seven elected African deputies from French West Africa, together with three from Equatorial Africa and Cameroun, Félix Tchicaya, Prince Douala Manga Bell and the *métis*, Gabriel D'Arboussier, formed a tiny

8 *République Française: Territoire du Togo*, 'Procès-verbaux des séances des 11 mai et 12 mai 1945 de la Commission chargée d'exprimer son avis sur les modalités d'application au Togo des recommandations de la Conférence de Brazzaville', Lomé, n.d., p. 20.
9 See ch. 15.
10 ARSD Documents 13G 17/36.

minority in the French National Assembly, determined to secure the abolition of the worst abuses of the French colonial regime as envisioned at Brazzaville. They were aware that there was no hope of even limited self-government, let alone independence since all the metropolitan parties, including the Communists, were opposed to such a development. Brazzaville had made it clear that France did not envisage self-government for her colonies even in the distant future. They therefore focused their main attention on securing the same rights for subjects as for citizens. They had already had a taste of how difficult this would be. The Commission established to make recommendations on the basis of representation in the first Constituent Assembly was headed by Gaston Monerville, a French West Indian. Senghor and Apithy served on this Commission, which had its liberal proposals overruled. While it had recommended a single roll of citizens and subjects in the African colonies, double rolls were in fact established so that in all the colonies except Senegal, where African citizens formed a majority, metropolitan French citizens were able to secure a seat. Instead of the ninety-five colonial deputies recommended for the colonial empire only thirty-five seats out of 586 were eventually allocated to France's colonies and many of those elected were Metropolitan French because of the double roll. Although the African members realised that it would be difficult to extend universal suffrage as practised in France to Black Africa, they had not anticipated a franchise so restricted that Senegal, with a hundred years of representative institutions, had only 2.9 per cent of its population eligible to vote. Finally they resented the categorisation of the colonies as 'countries over which France must continue to exercise its domination'.

In the Constituent Assembly itself the African deputies allied themselves with either the S.F.I.O. or the Communist Party through its associate party the *Mouvement Unifié de la Résistance* (M.U.R.). The Communists proved particularly solicitous of the African deputies, helping them with housing and instructing them in Parliamentary procedure.

For the African deputies the most urgent questions were the abolition of forced labour and the *indigénat*. Both of these were realistic goals since, though the Assembly's main objective was the drafting of a constitution, it did have limited legislative powers. By laws of 22 December and 20 February the *indigénat* was abolished. Forced labour was formally abolished on 11 April by the *Loi Houphouët-Boigny*, which secured Houphouët-Boigny's political hegemony over the Ivory Coast (which included Upper Volta, the traditional French 'man-power' reserve) where forced labour was more of a burden on the population than in any of the other West African territories.

The status of 'colony' was formally abolished in January 1946 and colonies now became known as *Territoires d'Outre-Mer*. The economic and social reforms recommended by Brazzaville were incorporated in a formal programme of aid to the overseas territories by the establishment of F.I.D.E.S. on 12 April. On 7 May the humiliating status of subject was abolished and all Africans became *citoyens*, though not full *French* citizens so that this did not immediately give them all the right to vote. This law came into operation two days after the referendum on the constitution in which legally only French citizens were allowed to vote so that only those who had elected first roll deputies were entitled to vote. Thus the electors of the African deputies, with the exception of Lamine Guèye (representing the first roll in Senegal) had no say in the referendum.

The constitution put to referendum on 28 September was defeated for metropolitan rather than colonial reasons. Essentially the metropolitan electors rejected it because a small majority preferred a bicameral chamber. The liberal proposals for the 'Empire' were suppressed incidentally. Since the Second Constituent Assembly was more reactionary in its approach, it is important to see what Africans lost as a result. The provisions of the constitution that affected the Black African territories were as follows: (a) African deputies were to have seats in the National Assembly; (b) local assemblies were to be set up for each of the Territoires d'Outre-Mer, with larger powers than the *Conseils-Généraux* of the French *Départements*; (c) these assemblies together with the Conseils-Généraux would elect a *Conseil de l'Union Française* which would oversee the affairs of the French Union, although basic power with regard to colonial policy would reside with the National Assembly rather than the Conseil de l'Union. (d) While the Territorial Councils were limited in their legislative powers and certainly did not make any provision for responsible government, they were to be elected by universal suffrage. (e) Perhaps most important of all were the provisions that over one-fifth of the seats in the National Assembly were to be allocated to the Overseas Territories. (f) Finally French West Africa was to be headed not by a Governor-General but an Under-Secretary of State who would be responsible to the National Assembly, where his administration could be criticised by the French West African deputies.

The African deputies returned to the Second Constituent Assembly with massive majorities from their electors, but basically disillusioned; although the concessions they had gained from the First Constituent Assembly had secured their political position *vis-à-vis* their people, they now faced a group of deputies among whom colonial interests were strongly represented. Except in Senegal, with its majority of African citizens, the newly elected French first-roll deputies represented colonial business interests. Yacine Diallo seems to have been justified when he told the Second Constituent Assembly that as far as Africans were concerned the rejected constitution 'gave complete satisfaction'.[11] Though the sub-

11 *Journal Officiel*, Debates of the Second Constituent Assembly, 2nd Session, 18 September 1946, p. 3813.

jects *en masse* had not been allowed to vote on the constitution, they showed their approval indirectly in the massive support those of them with the vote gave to their deputies. Senghor was re-elected without any votes against him; Apithy gained 8,096 with fifteen against. These results were undoubtedly representative of broad popular feeling, although one should emphasise that the suffrage (outside the four communes of Senegal) was at this time still restricted to Africans in élite positions (chiefs, veterans, those literate in French).

Whereas in the First Constituent Assembly the African deputies had worked with the parties to which they were affiliated, they now formed a special group to face up to the pro-colonial voices which loudly echoed the interests of the colonial commercial houses and the *colons*, sponsors of the newly-created (September 1945) States-General for French Colonisation. The constitutional recommendations of the so-called Inter-Group were the nearest the African deputies came to advocating a political programme which paralleled that accepted as natural in British West Africa. By a majority of one, the Commission for Overseas Territories accepted the following principal recommendations: (a) that France renounce all ideas of 'unilateral sovereignty over her colonised people'; (b) that they have freedom to govern themselves and to run their own affairs democratically; (c)—most important of all—the postulate that the French Union was composed of nations and peoples who had freely consented to come together; (d) there would be local assemblies empowered to establish their own constitutions while representation in the proposed Constituent Assembly of the French Union would be in proportion to the population of the member states.[12] This was tantamount to demanding eventual independence, even if it were within the framework of the French Union. Not surprisingly, the proposals were rejected by the more conservative Second Constituent Assembly. The West African deputies themselves in subsequent debates, in which they were remarkably vocal given their small number and their lack of parliamentary experience, reassured the French that they were not at heart secessionists. Houphouët-Boigny, alluding to the position of his more radical North African colleagues, principally Ferhat Abbas, declared, 'Certainly some of my colleagues ask for independence within the framework of the French Union for the countries which they represent. But that is not contrary to the principles that you, yourselves, have accepted and often affirmed.'[13] However, the West African deputies were increasingly to assure the Assembly that they were loyal Frenchmen, not secessionists. They concentrated their main efforts on gaining for their countrymen the same rights and privileges as

12 Daniel Boisdon, *Les Institutions de l'Union Française*, Paris, 1949, pp. 41–2.
13 *Journal Officiel*, Debates of the Second Constituent Assembly, 2nd Session, 18 September 1946, p. 3802.

Metropolitan Frenchmen while retaining their *statut personnel* in the same manner as the Senegalese *originaires*. Assimilation became their avowed goal.[14]

As far as the colonial territories were concerned the constitution that emerged from the Second Constituent Assembly (narrowly accepted by the French citizens at the Referendum of 13 October 1946) was much less liberal than that of the First Constituent Assembly. Perhaps surprisingly, all the West African deputies voted for it in the Constituent Assembly.

The hated double roll was retained to ensure representation of the *colons* in the National Assembly. For the second roll only certain 'Union' citizens could vote. The majority of Union citizens remained without the vote. The Union citizens had very limited representation in the National Assembly. Nevertheless it became the main focus for African political leaders, since the local territorial assemblies and the Federal Assembly in Dakar had such restricted powers. Furthermore the Assembly of the French Union, which it had been hoped at Brazzaville would be the controlling body for 'colonial' affairs, was little more than a rubber stamp. The decentralising spirit of Brazzaville, to which the West African Deputies had subscribed, was now lost in a formally assimilationist, but non-egalitarian, political structure.

The new constitution had two major consequences for the development of a new style of political activity in French West Africa in the post-war years: politics centred on Paris rather than West Africa; and because of this African political parties developed strong ties with metropolitan French parties. Finally, in insisting that the Federation of West Africa and French Equatorial Africa be headed by civil servants rather than by Under-Secretaries of State answerable to the National Assembly, the administration retained much more power than it would have done under the May Constitution. The significance of this was that the French administration, to counteract the development of radical political parties, was able to use its control over the chiefs, the agents of local administration, to encourage the growth of pro-Administration parties and in many cases to ensure the election of their representatives to the National, Federal and Territorial assemblies.

THE BEGINNINGS OF FRENCH WEST AFRICAN POLITICS 1946–50

Although the *indigénat* had made formal political organisation outside the Communes of Senegal impossible except for the period immediately prior to the elections to the Constituent Assemblies, there had been a surprising amount of political activity both in West Africa itself and

14 *Ibid.*, pp. 3790–2, 3797–803, 3813–14, 3819–20, 3840, 3849–50, speeches by Senghor, Houphouët, Lamine Guèye, Yacine Diallo. Also Apithy, who flirted with secessionist ideas (see above p. 665).

among Africans studying and working in Paris (see Chapter 15). Thus while the Constitution of 1946 provided the legal framework within which political parties could operate, Boahen would hold that it was not the principal stimulus to such activity.[15] The parties that 'mushroomed' in French West Africa, he writes, 'certainly did not emerge from the blue'.[16] This is only partly true. Indeed, the majority of parties that 'mushroomed' in French West Africa after the war were directly sponsored by the French administration who recognised that, under the new constitution, if they were to continue to control West Africa, they would have to control the parties electing representatives to the National Assembly and other legislative bodies. The outstanding exceptions were the Togolese *Comité d'Unité Togolaise* (C.U.T.) and the *Parti Démocratique de la Côte d'Ivoire* (P.D.C.I.). In the Ivory Coast, which at that time included Upper Volta, the contrast between parties with genuine grievances and those sponsored by the administration was very clear. On the one hand African planters had formed in September 1944 the *Syndicat Agricole Africain*, based principally on resentment of the discriminatory measures practised hitherto by the administration in favour of European planters. On the other hand, in the Upper Volta, which was administratively integrated with the Ivory Coast, the Mogho Naba, with the assistance of an administration hostile to Houphouët-Boigny, 'mobilised' Voltaics, who wanted the re-creation of the Upper Volta as a separate colony, into the *Union pour la Défense des Intérêts de la Haute Volta* (U.D.I.H.V.). The Mogho Naba fielded one of his provincial ministers, the Baloum Naba, unsuccessfully against Houphouët-Biogny in the elections to the National Assembly.

While it can be shown that initially the single most important stimulus to the formation of political parties was the new constitution, which galvanised the administration into 'sponsoring' them, in the long run there were other considerations that explain the rise of politicians and the development of their parties during the post-war period. These may have been inter-ethnic hostility in Guinea and ethnic nationalism among the Mossi of Upper Volta, or more general grievances against the colonial regime. There was solid material on which the aspiring politicians could build. While constitutionally the worst abuses such as the *indigénat* and forced labour had been abolished, real power in West Africa still lay largely with the local French administration and their agents, the chiefs. Protest could be made in the Territorial Assemblies and the Federal *Grand Conseil* but major legislation could still only be initiated from Paris, where now, under the Constitution of the Fourth Republic, the colonies were at least

15 On this see A. Adu Boahen's critique of the chapters by Robert L. Delavignette and Crawford Young in *Colonialism in Africa 1870–1960. II: The History and Politics of Colonialism 1914–1960*, eds. L. H. Gann and Peter Duignan, Cambridge, 1970, pp. 516–17.
16 *Ibid.*, p. 517.

no longer ruled by decree but by laws passed after debate. Only 803,000 Africans had the vote in the 1946 elections; the majority of adults remained disenfranchised 'citizens'. Few French West Africans had education sufficient to call the administration to task, so that in the bush there often seemed to have been little change. The continuing strength of the French administration and their agents, the chiefs, was clearly demonstrated in 1951 when it brought Houphouët-Boigny and his radical Parti Démocratique de la Côte d'Ivoire to their knees (see below, p. 680) and in 1958, when it defeated the 'secessionist' Prime Minister of Niger, Djibo Bakary, in the referendum on the 1958 Constitution (see below, p. 694). It must be noted that although the political resources of the administration were most clearly demonstrated in crisis situations such as those of 1951 and 1958, they were wielded very effectively on very many occasions in a less conspicuous manner right up to the time of independence.

Outside Senegal the major problem of the politicians was an organisational one. With little practical political experience and a largely illiterate electorate scattered over a vast, often sparsely populated, countryside with poor communications, it is remarkable how quickly they developed their parties. In two respects they had an advantage over their British West African counterparts, despite their comparative lack of experience. In the first place civil servants were eligible for election, which gave them a ready-made French-speaking élite, familiar with the workings of the colonial administration, to stand as candidates. In the second place they could enlist the active support of metropolitan Frenchmen working in the colonies who were sympathetic to their cause. Here Communists, more particularly before the fall of the Socialist–Communist alliance, were able to assist aspiring African politicians. *Groupes d'Etudes Communistes* (G.E.C.s) were organised in the capital cities of French West Africa after 1943 by Free French Communists, who often held official posts. The G.E.C.s were to provide the *Rassemblement Démocratique Africain* (R.D.A., later the dominant party of French West Africa) with Communist models for the organisation of a political party that reached to the grassroots level. Similarly French Socialists helped Lamine Guèye and Senghor in the organisation of the S.F.I.O. as a pan-Senegalese party. Here we can see that the radical parties relied just as much on the support of radical administrators as the Mogho Naba did on conservative ones.

A distinctive feature of political activity in French West Africa was that it crossed the frontiers of individual territories. Politicians may have had their base in a particular territory, but initially most of them were '*AOFienne*' in outlook. A number of factors account for this. In the first place for French West Africa the whole was greater than the parts. As we have seen in Chapter 14, effective power in French West Africa lay with the Governor-General in Dakar, who had overriding authority over the

The Grand Conseil in Dakar

Territorial Governors.[17] In the second place a Federal Grand Conseil (Great Council) was established at Dakar to which the territorial assemblies sent General Councillors, and of which all the West African deputies were by law members. In the third place, and perhaps this was most significant of all, the leading West African politicians were federal in outlook. Where they were civil servants, and a vast majority were, they had worked not only in their territory of origin but also in many cases in a majority of the French West African and even Equatorial African territories. The same system of *rouage*, or frequent rotation from post to post, and colony to colony (see Chapter 14), operated for them as for white administrators. They were thus less territorially and ethnically oriented than if they had served only in their territory of birth. In addition their leadership had been largely trained at the Ecole Normale William Ponty in Dakar, the 'Eton' of French West Africa, and the old school tie proved a strong federalist agent.[18] The move in October 1946 to establish at Bamako a pan-French-African party, the Rassemblement Démocratique Africain, by *all* the leading African politicians, to form a common front to fight the elections to the National Assembly can be understood only in the

17 This is contrary to the view expressed by W. A. E. Skurnik in 'France and Fragmentation in West Africa', *Journal of African History*, *(J. A. H.)* viii, 2, 1967, as we shall argue in greater detail below when we tackle the question of the Loi Cadre and Balkanisation.

18 A good parallel is the role of Katsina College graduates in welding together a strong pan-Northern-Nigerian political party, the Northern People's Congress.

light of the above. That eventually the R.D.A. did not succeed in welding together a party including all the African deputies was the result of pressure by Moutet the Socialist Minister of Colonies on the S.F.I.O.-affiliated deputies, Lamine Guèye, Senghor and Yacine Diallo, who never went to Bamako, despite their original intention to do so. Moutet, fearing the strong influence that the Communists, now no longer in the French Government, would exercise on such a federal party, persuaded them not to go. Indeed, by showing disapproval of the conference Moutet in effect strengthened the influence of the Communists over the R.D.A. since they were the only metropolitan party which sent representatives to Bamako.

The R.D.A., born from the conference, was a radical anti-imperial party, but this attitude should not be confused with a 'secessionist' intent. The French Communist Party itself was assimilationist in spirit and opposed to local nationalism; it wanted to mobilise Africans in the *international* struggle against capitalist imperialism. Thus the R.D.A., and its territorial sections were 'agin' the government of the day not because they demanded independence but because they demanded further rapid social, economic and political reform to improve the lot of their people. They also harked back to the first, unsuccessful constitution, in demanding that Governors of Territories be responsible to the Territorial Assemblies, though that constitution had only provided for 'Governors- General' responsible to the National Assembly.

With the later withdrawal of Apithy and Fily Dabo Sissoko from the R.D.A., it had only one really solid base: Ivory Coast which of course also included Upper Volta. But it had ardent lieutenants in all the other territories and ten years later, purged of some of its most militant elements, it was able to gain sweeping victories in the majority of the territories of the federation.

In the long view, then, while the R.D.A. was put on the defensive, the true loss was Senegal's. Where before it had been the focus of leadership in West African politics, it ultimately relinquished this by identifying itself too closely with the metropolitan government of the day.[19] But it was even more French West Africa's loss. For Senegal, with its political experience and sophisticated leadership, could have given an immense fillip to the new pan-West African party. As Houphouët-Boigny remarked in 1952, 'If Lamine and Senghor had been at Bamako, we would have written another page of history.'[20]

Pari passu with these political developments, which were so radically to change the lives of West Africans, was the almost equally important establishment of F.I.D.E.S.—the French equivalent of the Colonial Development and Welfare Act. As with British West Africa, French

19 A point well-made by Thomas Hodgkin and Ruth Schachter (Morgenthau) in 'French West Africa in Transition', *International Conciliation*, no. 528, May 1960.
20 *L'Afrique noire*, 24 July 1952.

colonial practice before the war had been that the African colonies should finance their own development from their own resources. Between 1946 and 1958 it is estimated that over 70 per cent of total public investment and over 30 per cent of civil and military recurrent expenditure was financed by France.[21] Probably as much as £400 million was spent in West Africa on new roads, bridges, railways, ports, aerodromes, schools, hospitals and agriculture. While both economically and socially this tended to improve the lot of the African subjects and to increase the number of educated constituents for the African parties, it also made French West Africa almost inextricably dependent on the metropolis, a factor that was to have a profound influence on the subsequent character of politics in French West Africa.

In the ten years after the 1946 election the main focus for debate was to be not independence but the nature of the constitutional relationship between the African territories and France. The independence movements in Morocco and Algeria, the bloody wars in Madagascar and Indo-China seemed to pass French West Africans by. Except in the Ivory Coast, and to a lesser extent in Senegal, the French Administration remained the dominant force. In the Ivory Coast, at first, the R.D.A. had had administrative support from Governor Latrille, who had already been recalled to Paris for his pro-African views at the insistence of the French planters, only to be reappointed. Once again the planters succeeded in having him recalled and the R.D.A. was put on the defensive, identified by both settlers and the government of the day as a dangerous, Communist-backed group potentially disloyal to 'France'. Latrille's successor, Orselli, later said that he had been sent out 'to suppress the R.D.A.'.[22] The French had already made inroads into the R.D.A.'s power base in the Ivory Coast by detaching the enormous and populous hinterland of Upper Volta which was once again made into a separate colony.[23] Though ostensibly this move was said to reflect the wishes of the Mossi people, it in fact suited the French Government's purpose of weakening the R.D.A. as well as their plan to extend the Abidjan railway from non-Mossi Bobo Dioulasso to Mossi Ouagadougou. The Mogho Naba promised electoral support and labour to France in return for the re-creation of Upper Volta.[24] The R.D.A. was defeated in the elections for the new Territorial Council of Upper Volta as a result of the open interference of Governor Mouragues, who was to

21 Theresa Hayter, 'French Aid to Africa—Its Scope and Achievement', *International Affairs*, xli, 2, April 1965, pp. 239–40. See also *AOF 1957* published by the Haut Commissariat de la République en Afrique Occidentale Française, Dakar, 1957.

22 Mortimer, *France and the Africans*, p. 123.

23 Upper Volta was first created a colony in 1920 from parts of Niger, Soudan and Ivory Coast, only to be dismembered in 1932 (see ch. 14).

24 Elliott Skinner, 'The Changing Status of the "Emperor of the Mossi" under Colonial Rule and Independence', *West African Chiefs: Their Changing Status under Colonial Rule and Independence*, eds. Michael Crowder and Obaro Ikime, Ife and New York, 1970, p. 111.

demonstrate his skill as a manipulator of 'free and democratic' elections yet again in Mauretania a decade later.

However, Governor Orselli himself, instead of trying to suppress the R.D.A. in the Ivory Coast, pursued a policy of winning them over: he succeeded in neither objective, and under pressure from the French settlers was recalled. The R.D.A. became increasingly hostile to the Administration and its alliance with the Communists became more firm under the influence of Gabriel d'Arboussier, though there did remain some doubts as to the wisdom of this course among some of the R.D.A.'s leaders, who felt that the main source of their troubles with the Administration was their alliance with the Communist Party. At the Second R.D.A. Congress in Treichville, a suburb of Abidjan,[25] this pro-Communist line was confirmed. The Congress expressed its 'faith in the alliance with the great people of France, who with the working class and the Communist party at their head, struggle with courage and confidence against American imperialism for their national independence'.[26]

The R.D.A. had declared its stand, particularly with the appointment of Gabriel d'Arboussier as its Political Secretary-General. The administration responded with an active policy of repression under Governor Péchoux in which R.D.A. leaders were arrested, chiefs and civil servants favourable to the R.D.A. dismissed, and maximum support was given to rival political parties. The R.D.A. retaliated with a determination and violence that compares with the struggle for independence in Ghana. For fifteen days its eight jailed leaders went on hunger strike while its militants boycotted European goods. A mass demonstration of women, led by Mme Ouezzin Coulibaly, wife of Houphouët-Boigny's lieutenant in Upper Volta, was held outside the prison and was dispersed by force. Even house-boys went on strike.[27] The vehemence of popular reaction was such that the administration was clearly losing control of the situation. Desperately it tried to arrest Houphouët-Boigny himself at Yamoussoukro, his home village, but the troops withdrew in the face of Houphouët-Boigny's insistence on his Parliamentary immunity and because of the ugly mood of the large crowd of his supporters. Finally all meetings of the R.D.A. in the Ivory Coast and elsewhere in West Africa were banned after the shooting of thirteen Africans by the troops at Dimbokro.

On the surface, this struggle was as near as any French West African party came to a war of independence but this possibility was not entertained

25 Initially it had been planned to hold the Congress in Bobo-Dioulasso, second city of Upper Volta, which was strongly sympathetic to the R.D.A. and hostile to Mossi domination, but this was baulked by Governor Mouragues.

26 For a discussion of the attitude of the P.C.F. towards colonial independence see Michael Crowder, 'Independence as a Goal in French West African Politics', pp. 25–6.

27 Though to anyone who has read Ferdinand Oyono's *Une vie de Boy,* Paris, 1956, this is not surprising.

by Houphouët-Boigny or his lieutenant Ouezzin Coulibaly who continued to protest throughout their loyalty to the French Union.[28] In this context we can understand how Houphouët-Boigny and the majority of the party were able to do a *volte face* and not only cooperate with the administration but eventually play a leading role in the French governments that decided the future of Africa. The Communists had not ultimately proved useful to them as allies in securing the fruits of assimilation. There was a visible demonstration of this in the passage of the *Deuxième Loi Lamine Guèye* by which African Civil Servants were in future to be given the same conditions of service as Europeans (a fact that was to prove a terrible burden to the budgets of independent French-speaking states where in some cases the cost of the Civil Service represented as much as 60 per cent of recurrent expenditure). Not only had the Socialist Lamine Guèye secured this triumph, but the *Indépendants d'Outre-Mer* (I.O.M.), a group of African deputies belonging neither to the S.F.I.O. nor the R.D.A., were gaining increasing influence in the French Parliament. With new elections forthcoming in 1951, and the vivid example in Upper Volta of what a hostile administration could do in securing election of candidates hostile to the R.D.A., Houphouët-Boigny decided to negotiate peace.

The context of this peace, discussed below, must be seen in the advantages gained by those deputies who cooperated with the government of the day, namely the I.O.M., among whom Senghor was to play a leading role.

Senghor himself, in 1948, had broken with his sponsor, Lamine Guèye, and formed a new party, the *Bloc Démocratique Sénégalais* (B.D.S.). Senghor's move was directed against the urban African élite of the Four Communes and against the cantonal chiefs. He relied heavily on the *marabouts* in his campaign against the chiefs. The *marabouts* were on the whole an élite with effective popular support in contrast to the chiefs, and Senghor thus showed himself to be more sensitive to the political preferences of the peasants than Lamine Guèye. However, as Bakary Traoré has pointed out, the S.F.I.O. had on the surface cleverly united modern and traditional élites into a mass party.[29] By securing the appointment of Governor Wiltord, a Guadeloupian favourable to the S.F.I.O., and democratising the chieftaincy so that at the level of the village, canton, and province chiefs were elected and not appointed by the administration

28 This point is made strongly by Majhmout Diop in *Contribution à l'étude des problèmes politiques en Afrique Noire*, Paris, 1959. Diop, who founded a true Independence party, le Parti Africain de l'Indépendance (P.A.I.) emphasised that while the R.D.A. was 'the only anti-imperialist mass movement of consequence that has seen the light of day in this country' its main objective was the 'struggle for the political, economic and social emancipation of the Africans within the framework of the French Union founded on the equality of rights and obligations [duties]'.

29 Bakary Traoré, 'L'évolution des partis politiques au Sénégal depuis 1946', *Forces Politiques en Afrique Noire*, Paris, 1966, pp. 25–9.

according to their loyalty, or traditional claims to the post, the S.F.I.O. apparently extended its control to the bush since a pro-S.F.I.O. Administration controlled chiefs now elected by the people.[30] However, while the central administration was pro-S.F.I.O., many of the local administrators were Gaullist or supporters of the M.R.P. And when an alternative party was presented at the next elections, they did not hesitate to support it or remain neutral. Furthermore, while nominally the party had an increased constituency in the former Protectorate, as far as political spoils were concerned 'les vieilles cités continuent à être, par le truchement de leur bourgeoisie, le vecteur du parti, qui demeure très centralisé'.[31]

The domination of the élite of the Quatre Communes over the new élite from the 'Protectorate' was shown clearly in the way Lamine Guèye nominated élite friends from the Communes to the major political posts, including Metropolitan and West Indian Frenchmen, to the exclusion of those from the Protectorate. Senghor remained the latter's lone voice. Furthermore, the S.F.I.O. became increasingly reactionary in its political attitudes, losing ground among the intellectuals to the Senegalese section of the R.D.A. and declaring its true colours by siding with the government against the six-month-long railway strike of 1947, the subject of Sembène Ousmane's remarkable and bitter novel, *God's Little Bits of Wood*,[32] in which he has a savage portrait of Lamine Guèye. Senghor himself, while openly toeing the party line, counselled negotiation with the strikers.[33]

Having over a period of a year made clear his distaste for the clique politics of the S.F.I.O. and its disregard for the peasants and the poor, Senghor resigned on 27 September 1948 from the S.F.I.O. and created his own party, the Bloc Démocratique Sénégalais. In the National Assembly he joined the Indépendants d'Outre-Mer.

While in the rest of French West Africa, including Ivory Coast, no major leader advocated independence, the special status of Togo as a U.N. Trust Territory was to make this not only a goal of political leaders there, but a realistic one. Though as an 'Associated Territory of the French Union', it sent a Deputy to the National Assembly, two Senators to the Council of the Republic and one Councillor to the French Union, it had an important alternative outlet for its grievances: the United Nations. Apart from Visiting Missions to which grievances could be voiced, Togolese could send petitions to the United Nations Trusteeship Council.

30 Decree of 12 February 1947 on Indigenous Rule in Senegal: 'Article 2: Village chiefs... are elected by direct universal suffrage by the electors male and female.... Cantonal chiefs are elected by a restricted electoral college.' [This included such non-traditional electors as Civil Servants, Retired Civil Servants, Holders of the Certificat des Etudes Primaires Supérieures, Ex-Servicemen with at least Sergeant's rank, holders of the *Légion d'Honneur* or *Médaille Militaire*.]

31 Traoré, p. 29.

32 *Les Bouts de Bois de Dieu*, Paris, 1960 (English translation, New York, 1962).

33 Traoré, p. 40.

France, too, had to report annually on its conduct of affairs in Togo. France's aim was to keep Togo within the French Union, since attainment of independence would set an unhealthy example to her other French West African territories. While under the Trusteeship agreement of 1946 France had agreed 'to proceed, when the time comes, to such consultations as may be appropriate in order to allow these populations to pronounce freely on their political regime . . .' it was not until the Trusteeship Agreement of 1950 that she was committed to the obligation to promote 'progressive development towards self-government or independence'.

The background to the elections to the National Assembly of 1951, mandatorily falling five years after the 1946 elections, were coloured by three major developments: first the French administration had shown it could break a party that it considered inimical to it; secondly Senghor's B.D.S. had eclipsed the S.F.I.O. in Senegal, its principal base and therefore in French West Africa; and finally France had had to acknowledge the possibility of independence for one of its West African territories. The five years that followed the 1951 election and the passing of the *Loi Cadre* of 1956 were preoccupied both from the African and the French point of view with the question of the future structure of the French Union and within this context the rehabilitation of the R.D.A. which the administration would permit to participate in this dialogue. A shadow was cast during this period, from the French point of view, not only by the prospect of self-government or independence for Togo, but by the attainment of self-government and eventually independence by the British Colony of Gold Coast.

THE ELECTIONS OF 1951 AND 1952 AND THE DEBATE ON THE FUTURE OF FRANCE AND AFRICA

In February 1951, a few months before the elections to the French metropolitan assemblies, Kwame Nkrumah's nationalist Convention People's Party won the Gold Coast general election and the Governor invited him from prison to become Leader of Government Business. For French West Africans this highlighted the difference in approach to political activity on the part of the French and British administrations. In the elections of 1951 the French Administration made it quite clear how dangerous it was for a political party like the R.D.A. to oppose it. Even though Houphouët-Boigny had made an ostensible peace with the French Administration by breaking with the Communists, and despite the support given him by François Mitterand, the Minister for Overseas France, the administrators on the spot savaged the R.D.A. at the polls. While the new electoral law increased the franchise from 172,500 to 3,095,500 the administration ensured that known R.D.A. sympathisers did not get on the roll, and indeed some were struck off it. The result was the eclipse of the R.D.A. both in the 1951 election, and in the election to the territorial assemblies in 1952.

Léopold Senghor in Senegal, whose party (the B.D.S.) benefited at least from administrative neutrality, was allowed to win crushing victories in these elections over Lamine Guèye and the S.F.I.O.

Just how powerful was the local administration is shown by the fact that it was able to secure two seats, one in Mauritania and one in Ivory Coast, for the right-wing Gaullist *Rassemblement du Peuple Français* (R.P.F.), which it supported, though the French Government considered its loyalty to the Republic as dubious as that of the Communists. Furthermore the number of registered voters under the new law would have been even larger had the law not been introduced at the last minute and had the colonial administration not set out to keep R.D.A. and C.U.T. voters off the roll. Even though Mitterand gave strict instructions to ensure the proper implementation of the new law, as Edward Mortimer has pointed out, 'In the Ivory Coast especially it would have been necessary to sack virtually every administrator and destool every chief in order to secure a genuinely free election.'[34]

Though as far as representation was concerned the R.D.A. was to be formally eclipsed for the next five years, at the grass roots level it remained very powerful in every colony except Senegal where it was primarily supported by the intellectuals. Elsewhere, as a party persecuted by the colonial administration and the chiefs, it increased its real support among the masses. In Paris in the National Assembly, despite its reduced numbers—only three deputies[35]—the R.D.A. had influence as an African party because of its formal alliance with François Mitterand's *Union Démocratique et Sociale de la Résistance* (U.D.S.R.). Thus in the debate on the future relations between France and Africa, which dominated the decade, in particular the five years preceding the next elections, the R.D.A. was able to make its voice heard in Paris. In West Africa, despite the administration, it became the only truly Federal party.

The majority group of African deputies in Paris, the Indépendants d'Outre-Mer, were only very loosely united around a common denominator of anti-Communism and hostility to the R.D.A. The I.O.M. never took on the characteristics of a parliamentary party in Paris, but worked in informal alliance with the M.R.P. (Christian Democrats). Its members were predominantly Catholic, supported by the missions and the administration. Founded in 1948 by Apithy who was 'disillusioned' by the R.D.A.'s alliance with the Communists, it assumed an importance in the National Assembly beyond its size because of the slender majorities on which the rapid succession of French governments were built. In terms of real achievement the I.O.M. had little to show before the 1951 election, as

34 Mortimer, *France and the Africans*, p. 166.

35 These were Houphouët-Boigny (Ivory Coast); Mamadou Konaté (Soudan); and Tchicaya (Moyen-Congo).

compared with the isolated Socialist Lamine Guèye. Before his defeat at the 1951 election, Lamine Guèye had achieved the prestige of the Second *Loi Lamine Guèye*. See above, p. 678.

The 1951 election results, however, gave the I.O.M. enough deputies to form a parliamentary group independent of any metropolitan party, whilst the R.D.A. had to hang on to the coat-tails of the U.D.S.R. The R.D.A. remained important, however, in that it swelled the parliamentary ranks of the tiny U.D.S.R., any further reduction in which might have excluded Mitterand from consideration for future cabinet posts.

In Paris, after the election, the I.O.M. was able to claim a major role in securing the passage of the Labour Code for Africa, which had met with considerable right-wing opposition. Apart from confirming the earlier abolition of forced labour, it gave African workers and their trade unions roughly the same code as that for Metropolitan workers. Shortly after this Pinay was replaced as Premier by the rightist René Mayer. Aujoulat, the I.O.M. deputy from Cameroun, who had been Secretary of State for Overseas France in the seven preceding Cabinets (and had been largely responsible for the passage of the Overseas Labour Code), then refused Mayer's offer of the same post since it was now to be shared with a right-wing deputy, Cavaillet, who had vehemently opposed the passage of the Overseas Labour Code.

The I.O.M. was now opposed to the government in Paris, and had to its credit the passage of the Overseas Labour Code. It was, however, unable to exploit this position and to form itself into a coherent African political party, despite its attempt to achieve this goal at Bobo-Dioulasso in 1953. This was primarily the result of the fact that the majority of I.O.M. deputies had no real base in the territories they represented other than support from the administration, nor did they share a common political philosophy. Thus it was that in 1956, when elections were held without administrative sanctions against the R.D.A., the party, with its grass-roots organisation, and its record of anti-administration, anti-chief politics, was able to capture the majority of seats held by the I.O.M. deputies.

Perhaps the most important feature of the Bobo-Dioulasso Congress was its discussion of the future relationship between France and her African territories. Ghana was clearly on the road to self-government and independence. In Togo, while the French had secured the victory of a pro-administration candidate, Grunitzky, over Sylvanus Olympio's C.U.T., they had also started preparing in 1953 for limited autonomy for the territory to avert any possibility of Togo uniting with Ghana and leaving the French Union. In Cameroun, the R.D.A. affiliate, U.P.C. *(Union des Populations du Cameroun)*, had broken with Houphouët-Boigny and supported d'Arboussier and now, led by Reuben Um Nyobé, demanded not only reunification with the British Cameroons, but independence. In the larger world of the French Union, the Indo-Chinese were fighting a war of independence, and French North African demands for independence were becoming increasingly vocal.

Against this background, French West African political leaders had to consider their own stand on the future of their territories. Foremost was their relationship with France itself. Would they pursue a goal of independence or seek some constitutional relationship with France in which they could have political equality with the metropolis, exercise control over their local affairs, and yet continue to receive the massive aid which France was injecting into West Africa? Second to that was the problem of whether their political future should be as a French West African Federation or as individual territories. Here we must remember that while for administrative purposes French West Africa had been divided into eight 'colonies', it had been governed since 1900 as a highly centralised federation where all major services were under the control of the Governor-General, not the Lieutenant-Governors (later Governors) of the constituent colonies. The Governor-General alone had the right to promulgate decrees and any inclination to independence of action on the part of Lieutenant-Governors was subject to the fact that the Governor-General controlled the budget. He alone could raise loans, impose customs duties on exports and redistribute the proceeds to the colonies. Agriculture, Public Works, Posts and Telegraphs and Sanitary Services were all federal services. It has been argued by Skurnik[36] that the colonies were always regarded as the fundamental political units and that therefore the Federation was only a temporary structure. This certainly does not accord with the above description of the role of the Governor-General of the Federation *vis-à-vis* his Lieutenant-Governors, nor do the archives in Dakar bear out Skurnik's view. The Lieutenant-Governors reported to the Governor-General in much greater detail than the Lieutenant-Governors of Northern and Southern Nigeria did to the Governor, for example, while the Governor-General of French West Africa intervened in much greater detail in the day-to-day administration of the constituent colonies than did his counterpart in Nigeria.[37] We would therefore argue that by 1944, at the beginning of electoral politics in French West Africa, the Federation was more important than the constituent territories. And, as noted earlier, the French African political élite was much more Federal than Territorial in outlook. An appreciation of this is essential to the drama that unfolded over the next seven years.

At Bobo-Dioulasso the I.O.M. members advocated a federal future for

36 W. A. E. Skurnik, 'France and Fragmentation in West Africa'.

37 This is brought out clearly by the respective archival sources for the 1916 revolt against the French in Dahomean Borgu and the 1915 Bussa (Borgu) rebellion in Northern Nigeria. Correspondence between Governor-General William Ponty (1908–15) and the Territorial Governors, preserved in the Dakar Archives, consistently states Ponty's detailed supervision of the territories.

France and Africa in which there would be increasing devolution of powers on the territorial assemblies. These would have responsible governments. Senghor, who eventually became the chief advocate of a federal republic, insisted that only through a federation of French West and Equatorial Africa with France could Africans have any real equality with the mother country. This, as a corollary, emphasised the importance of the Grand Conseil at Dakar in relation to the Territorial Assemblies.

Senghor's particular solution to the problem of the future of France and Africa found sympathetic ears in the R.D.A., which was after all a federal party. But, apparently paradoxically, the leader of the R.D.A., Houphouët-Boigny, came out in favour of direct federation between the individual territories and France. But the paradox was more apparent than real; the Ivory Coast was by far the richest territory in West Africa, and Houphouët-Boigny feared that it would end up by subsidising the federation to its own detriment, and what is more, a federation with its capital, and concomitant benefits, in Dakar, not Abidjan.

In one sense Senghor and the I.O.M. were echoing the proposals of Brazzaville for a Federal Assembly for the French Union. But they differed in that they demanded self-government at the local level, which the Brazzaville Conference had categorically rejected.

In advocating the idea of a federal republic a major obstacle was the French conception of a *république une et indivisible.* For Senghor what was important was not the fact that at one level the republic would be divided into component parts but that, their local autonomy notwithstanding, the African Federal Republics and the Metropolis would be indissolubly linked.

Against this background of discussion, and in particular the demand for greater autonomy, Mendès-France in 1954 began to initiate reforms. The first was a new statute for Togo, by which it was to be given a Council of Government, five of whose members were to be elected by the Territorial Assembly, safely dominated by the pro-French Grunitzky. While the Council of Government had no more real executive powers than the Territorial Assembly had legislative powers, the new statute marked a significant step forward towards the idea of granting a measure of local autonomy to French African territories. The motive force for this move had clearly been to counter developments in neighbouring Ghana, which with its common Ewe population had always been a pole of attraction for Togolese, in particular Olympio and his C.U.T. Similarly, with U.N. surveillance, the dangerous word 'independence' could be mentioned openly in Togo whereas it could not in French West Africa.

Mendès-France's reforms for French West Africa did not come into being until 1956 by which time several more French governments had ruled. The reforms were embodied in an outline law known as the *Loi Cadre*, which was presented after the 1956 elections to the National Assembly, in which the I.O.M. lost ground to Houphouët-Boigny's

R.D.A.[38] and consequently the character of the new law reflected his views rather than those of Senghor. Indeed, Houphouët-Boigny became *Ministre-délégué* in the government formed by Guy Mollet after the election.

THE LOI CADRE

The Loi Cadre, which was implemented in the elections of March 1957 to the Territorial Assemblies and which gave a measure of responsible government to the constituent territories of French Black Africa, was intended as a palliative to Africans in a world where independence was being rapidly achieved by colonial peoples. Gaston Deferre, the Minister of Overseas France who was responsible for introducing the Loi Cadre, made this quite clear in his speech to the National Assembly on 21 March 1956 when he cited the fact that the British had transformed the political and administrative regimes of their territories and declared that this 'has certainly contributed to the growth of impatience among the peoples of French West and Equatorial Africa'.[39] But Britain alone could not be held responsible for this development. France herself had been humiliated in Indo-China, forced to grant independence to Morocco and Tunisia, and had already conceded the principle of autonomy to one of her West African territories, Togo. Yet Deferre insisted that the Loi Cadre reforms were designed 'to maintain and reinforce the necessary Union between Metropolitan France and the peoples of the overseas territories'. In the debate which followed the introduction of the Loi Cadre, the Black African leaders did not protest against the idea of this union, but merely the form it would take. They were concerned with the relative merits of direct association between the individual erritories and France on the one hand and the federation of France with the existing federations of West and Equatorial Africa on the other. Independence as such was not considered by either the African deputies or the major parties. It was a word that was certainly discussed in private by teachers, students and trade unionists, but it did not pass the lips of the elected representatives in the debate on the structure of the Franco-African Community. Indeed it first entered public political vocabulary in June 1958 when Sékou Touré addressed the Fourth Congress of the *Parti Démocratique de Guinée* (P.D.G.) in Conakry[40] when he declared that Guinea would not renounce her independence even though she would link her destiny with that of France. As Foltz put it, 'independence'

38 The R.D.A. increased its seats from three to nine—seven of whom were from West Africa: Houphouët-Boigny (Ivory Coast); Mamadou Konaté (Soudan); Hamani Diori (Niger); Ouezzin-Coulibaly (Upper Volta); Modibo Keita (Soudan); Sékou Touré (Guinea); Diallo Saifoulaye (Guinea).

The I.O.M. deputies were reduced in number from fourteen to seven, of whom only four were from French West Africa as such: Senghor (Senegal); Mamadou Dia (Senegal); Hubert Maga (Dahomey); Nazi Boni (Upper Volta); Grunitzky was elected for Togo.

39 *Journal Officiel*, Debates of the National Assembly, 1st Session, 21 March 1956, pp. 1108–12.

40 Sékou Touré, *Expérience Guinéenne et unité africaine*, Paris, 1961, p. 50.

thus first entered common political discussion in a qualified form.[41] Otherwise it had been espoused only by the tiny Marxist-oriented Parti Africain de l'Indépendance.

Before examining the nature of the debate on the Loi Cadre and the far-reaching consequences it had for French Black Africa, we must ask ourselves why French-speaking African leaders were so reluctant to pursue independence as a political goal. Clearly a major reason was that France continued to dominate the economic, political and administrative structures of their territories. The economic limitations on freedom of political choice have been discussed by Elliott Berg.[42] We would argue that the economic explanation should be considered along with the political, administrative and sociological situation of West African leaders.[43] An analysis of the economic dependence of the French African territories, both in terms of trade and aid, makes it clear that any sudden rupture with France would have had disastrous consequences for them. Berg has estimated that French public investment in West Africa between 1947 and 1956 may have been as large as a thousand million dollars. Primary exports were supported by subsidies.

Dependence on France through membership of the franc zone, bilateral agreements and lack of direct trading precedents outside the franc zone would have made it difficult, as Guinea found in 1958, to relocate trade at least in a transitional period. Over and above this the French subsidised the administrative services, which, because of the Second *Loi Lamine Guèye* introducing equal terms of service for Africans and Europeans, consumed up to 60 per cent of the recurrent budgets of some territories. And often the main support for the political parties came, as we have seen, from the *fonctionnaire* class.

The economy was not only dependent on France but it was run by Frenchmen, at both managerial and subaltern levels. In Dakar in this period such jobs as those of shop assistants, barmaids, taxi-drivers, chambermaids, and even cleaners and cobblers were held by Europeans.[44]

European domination of the economy was reflected in African politics. In the Loi Cadre governments 20 per cent of the ministers in French Africa as a whole were of European origin and they were chosen mainly because, as Guillemin points out, they represented 'the economic and social power in their territories'.[45] The administration, too, was still largely dominated

41 William J. Foltz, *From French West Africa to Mali Federation*, New Haven and London, 1965, p. 50.
42 Elliott Berg, 'The Economic Basis of Political Choice in West Africa', *American Political Science Review*, liv, 2, 1960.
43 See Donal Cruise O'Brien, 'The Limits of Political Choice in French West Africa 1956–60', *Civilisations*, xv, 2, 1965, pp. 1–15.
44 See Michael Crowder, *Pagans and Politicians*, London, 1959.
45 P. Guillemin, 'La Structure des Premiers Gouvernements Locaux en Afrique Noire', *Revue Française de Science Politique (R.F.S.P.)*, ix, 3, 1959, p. 683.

by Europeans both in the 'bush' and in the Secretariats. There had been no policy of Africanisation such as had been introduced in British West Africa and Frenchmen held posts of a much more junior scale than had ever been permitted in British West Africa even in the hey-day of colonial rule. Thus any break with France implied potential dislocation of services. Furthermore the French Administration operated the electoral machine and any attempt to achieve a breakthrough using this machine required an organisation that could effectively combat the stranglehold of the French Administration held over it. Moreover, while the 1955 and 1957 elections had been free of administrative interference, experience did not suggest that such freedom would be permitted to parties pursuing policies openly hostile to the French Government.

The African 'political' class was furthermore itself committed to France. The chiefs, whether traditional or administrative in character, were the agents of the French Administration and could safely be relied on to bring in favourable votes. The *Marabouts*, or Muslim leaders, had become a conservative force under colonial rule and in Senegal, especially, where they were deeply enmeshed in the groundnut economy they were able to deliver France both the produce and the votes of their faithful. For the African radical they had become 'féodaux enturbannés, couverts de chapelets et médailles, pures créatures et instruments dociles de l'Administration coloniale qui les a couverts de tous ses soins pour mieux endormir la masse musulmanè'.[46] The peasantry, Muslim or pagan, were not 'the revolutionary class' portrayed by Fanon.[47] Indeed as Ahmadou Dicko points out, it was not until the Referendum of 1958 that the vast majority of the peasantry first heard the word independence.[48]

This vast illiterate mass—only a very small minority of the population of French West Africa had been to school by 1957—was barren ground for the urban highly educated revolutionary. Only in the towns, particularly among the workers, was there radical material, but here too, the fact that so many of the 'educated workers' were government employees made them hesitate openly to join radical movements.

The political leadership itself was, as noted above, largely drawn from the administration,[49] or from positions controlled by the administration, and was thus limited in its potential for open opposition to French policy.

Within the context of these limitations, there is no doubt that the French West African leadership wanted as great an autonomy at the local level as

46 Ahmadou A. Dicko, *Journal d'une défaite autour du Référendum du 28 septembre 1958 en Afrique Noire*, Limoges, 1959.
47 F. Fanon, *Les Damnés de la Terre*, Paris, 1961, p. 46.
48 Dicko, p. 57.
49 Sixty-one out of seventy deputies elected to the Territorial Assembly of Upper Volta, and forty out of sixty of those elected to the Territorial Assembly of Niger in March 1957 were *fonctionnaires*; see André Blanchet, *L'itinéraire des partis Africains depuis Bamako*, Paris, 1958, p. 22.

possible. However, while the Loi Cadre granted them a considerable degree of autonomy, many of them felt that the way in which it was granted effectively weakened their political position *vis-à-vis* France. The primary unit of government under the Loi Cadre was the Territorial Assembly, not the Federal Grand Conseil. Territorial Assemblies now had legislative powers in many of the realms which had once been the preserve of the Federal Government, and, in addition, the majority party in each Territorial Assembly could choose a cabinet, over which the Governor would preside, with the leader of the majority designated as *Vice-Président du Conseil*, or effective Prime Minister. The Grand Conseil on the other hand was given no executive and its inter-territorial services were made *services d'état*, administered by the High Commissioner. The essential unit in the French Union was now to be the Territory, not the Federation.

In the debate over the structure of the relationship between France and Africa, Senghor stood as champion of maintaining the Federations, and Houphouët-Boigny as champion of direct association. Senghor had initially proposed that French West Africa be divided into two groups, i.e. one based on Dakar (Mauritania, Senegal, Guinea and Soudan) and one on Abidjan (Ivory Coast, Upper Volta, Niger and Dahomey) but on the passage of the Loi Cadre he supported the maintenance of the French West African Federation as a whole since he was convinced that only autonomy based on such federal groupings could give Africans any sense of political equality with France.

Senghor and his followers took up as their battle cry that France was deliberately trying to 'balkanise' Black Africa through the Loi Cadre. Until the French Archives are open, we shall not know the full truth of this but it does seem clear from the debates in the National Assembly and from the views of French administrators and some businessmen that the French considered that there was a much greater likelihood of safeguarding their African Empire from the fever of independence through the device of the Loi Cadre and individual association of territories with the Metropolis than through a 'Federal France' in which West and Equatorial Africa could have some parity with the Metropolis.[50]

Political events after the introduction of the Loi Cadre showed how strong the French West African federation was in the face of its impending break-up. If the French Administration were prepared to pull down an edifice which after a half a century they considered 'only as a temporary structure',[51] African leaders for the most part were not.

50 See Foltz, *From French West Africa to Mali Federation*, pp. 75–6 and especially n. 32. French business interests in West Africa were in fact divided on this issue: those based in Dakar tended to support the Federation which gave them wide markets, while those based in Abidjan were firmly territorialists. See R. Cruise O'Brien, *White Society in Black Africa*, Faber, London, 1972.

51 Skurnik, 'France and Fragmentation in West Africa'.

The first concrete steps to rescue the Federation from impending disintegration were taken by the I.O.M. and the S.F.I.O. Meeting simultaneously but separately in January 1957 both the I.O.M. and the S.F.I.O. formed themselves into inter-territorial African parties, independent of any French Metropolitan party. The I.O.M. changed its name to the *Convention Africaine* while the S.F.I.O. became the *Mouvement Socialiste Africain* (M.S.A.). Another major move towards inter-territorial unity was made at Cotonou, where representatives of the large majority of trade unionists met to form the *Union Générale des Travailleurs d'Afrique Noire* (U.G.T.A.N.) under the leadership of Sékou Touré, who within the R.D.A. was spearheading the attack on the Loi Cadre. The Loi Cadre was denounced by the U.G.T.A.N. Congress as having as 'its sole aim ... to divide us, and to disguise and perpetuate the colonial regime'.[52]

In March 1957 the R.D.A. and the two newly formed inter-territorial parties went to the polls to contest the control of the new territorial assemblies, however much their leaders may have disapproved of the law that introduced them. The R.D.A., freed from administrative restraints at the polls, gained control of four of the eight territorial assemblies: the Ivory Coast, Guinea, Upper Volta and Soudan. The Convention Africaine gained control of Senegal. Mauritania was won by the *Union des Peuples Mauritaniens* (U.P.M.), while in Niger, Djibo Bakary led the dissident faction of the R.D.A. (which had broken with Houphouët in 1950) into a coalition government under the banner of the M.S.A.: the official R.D.A. joined the opposition. In Dahomey the R.D.A. formed a small minority opposition to Apithy's *Parti Républicain du Dahomey*.

Despite the sweeping success of the R.D.A. in the Territorial Assemblies, and their consequent control of the emasculated Grand Conseil, it was a party divided within itself. Its leader, Houphouët-Boigny, had, as Minister of State, been largely responsible for introducing the Loi Cadre in its present form. Yet in the meeting of the Grand Conseil, where the R.D.A. majority secured all the offices, a unanimous resolution was passed in Houphouët-Boigny's temporary absence in Paris demanding the creation of a Federal Executive. Senghor's idea had triumphed in the hands of his political opponents.

The division between Houphouët-Boigny and Sékou Touré over the future of the West and Equatorial African federations came into the open at the R.D.A. Congress in Bamako in September 1957. There, Houphouët-Boigny, supported only by his own Ivory Coast delegation and the Gabonese, found himself isolated. A compromise resolution was adopted in which the federalist views of the Sékou Touré wing were espoused, and Houphouët-Boigny, whilst he had clearly lost face to the younger members of the party, remained as President. However, he promised to

52 *Présence Africaine*, 17–18, February–May 1958, p. 124.

introduce a law in the National Assembly to 'democratise' the existing Federal Executives but his bill was lost in the morass of parliamentary procedure, apparently with his connivance and certainly with no regrets on his part.

Though the R.D.A. Congress was preoccupied with the future structure of the relationship between France and Africa, the question of independence was briefly forced on it by M. Papiebo, representing the African students in Paris: 'The slogans of Franco-African Community and federalism have been made to deceive the real faithful of Africa. We the students of Black Africa, leaders of tomorrow, who represent five thousand in France, demand independence.'[53] While the R.D.A. leadership would not entertain such demands, it welcomed the thesis that in joining a Franco-African Community France should recognise the *right* of African peoples to independence. Thus the Congress resolved that 'the independence of peoples is an inalienable right . . . but it considers that interdependence is the golden rule of the life of peoples and manifests itself in the twentieth century by the establishment of large political and economic groupings'.[54]

For the next nine months the question of independence was conveniently sidestepped with the formula that, by having their right to independence acknowledged, they were freely entering the Franco-African Community. The main focus of discussion and activity was therefore centred not on achieving independence but on ensuring that Africans entered the Community in a position of strength. The majority felt that such a position could only be attained through the establishment of Federal Executives and the unification of the existing inter-territorial parties. To this end negotiations took place between rival parties at both the territorial and inter-territorial levels. On 15 February 1958 the West African parties met in Paris to try and establish the basis on which a single political party could be formed. While no agreement was reached on the question of unification, the leading parties did agree on a common programme of federation, autonomy and the right to independence. The P.A.I., a Senegalese party with considerable support from younger educated townsmen, asked that the assembled parties adopt a programme of immediate and unconditional independence and was immediately expelled.[55] The unification of the parties foundered on the issue of the name for the proposed new monolithic party. The R.D.A. insisted that its name be adopted, unchanged, and not surprisingly the other parties rejected this. Nevertheless, though they remained divided on this issue, all were officially united on the issue of the creation of a Federal Executive for French West Africa.

53 Roneotyped transcript of part of the proceedings of the 1957 Congress.

54 The full text of the political resolutions of the R.D.A. Congress is printed in Blanchet, *L'Itinéraire des partis Africains*, pp. 187–9.

55 Foltz, p. 84. It is interesting to note that the expulsion of the P.A.I. was engineered by two young 'radicals', Abdoulaye Ly (C.A.) and Modibo Keita (R.D.A.).

(a) Lamine Guèye; (b) Léopold Sédar Senghor; (c) Gabriel d'Arboussier; (d) Félix Houphouët-Boigny.

In Dakar on 5 April 1958 the parties again met to discuss unity. While the M.S.A. and Convention Africaine were keen on unification, it was clear the R.D.A. was not. The Convention Africaine and the M.S.A. therefore took a step towards greater unity on their own by forming the *Parti du Regroupement Africain* (P.R.A.).

On 5 April 1958 the Grand Conseil again demanded the establishment of a Federal Executive and internal autonomy for the Federation as a whole. This was subsequently rejected by the Ivory Coast but on the eve of the fall of the Fourth Republic French West Africa was, as Foltz remarks, 'closer to a practical form of political unity than it had been since the days of the colonial regime of the Third Republic'.[56] Two large parties, the R.D.A. and P.R.A., may have confronted each other as rivals for power, but the vast majority of the leaders of both parties were agreed that the power they wanted was at the level not of the Territory but of the Federation. The advent of de Gaulle to power on 13 May was to give immediacy to this issue, and ultimately was to secure the triumph of the dissident Ivory Coast minority over the majority.

13 MAY AND INDEPENDENCE

It may at first seem incongruous to suggest that the 13th of May and independence had any direct connection. And yet the referendum held by General de Gaulle for a new constitution for France and what was left of her empire led to the very thing he most hoped to avoid: the secession immediately after the referendum of a member of the *république une et indivisible* and, within two years, the division of the Franco-African community into its constituent parts.

For the African leaders, de Gaulle remained an enigma. He was the man of Brazzaville but for more than a decade he had been in the political wilderness, and his supporters represented what for Africans were reactionary views about the French colonial empire. But he did promise an entirely new constitution in which the structure of relations between France and her overseas territories would be re-examined. Houphouët-Boigny was given a senior position in his cabinet, though Senghor turned down such a position due to pressure from radicals in his party who feared de Gaulle's rightist attitudes. To the issue of direct association versus Federal Executives was added a new one: independence. At the P.R.A. Congress at Cotonou on 25–27 July the younger leaders of the party like Djibo Bakary and Abdoulaye Ly demanded immediate and total independence which, after all, had been the successful battle cry of Sylvanus Olympio in the April elections to the Togo Territorial Assembly. In the heady atmosphere of that Congress even a cautious politician like Lamine Guèye supported the demands of the young Turks. Senghor, who had presented a political

56 Foltz, p. 86.

report to the Congress in which he talked in terms of a new constitution granting the right to independence, rather than offering independence itself, now declared that the French African territories 'must become independent before September'.

On the morrow of the Congress, faced with the starker realities of their tight bonds to France, the P.R.A. leaders disengaged themselves from this demand by elegant verbal acrobatics. Djibo Bakary was the only P.R.A. Président de Conseil who remained faithful to this demand. The others retreated, though Senghor's lieutenant, Abdoulaye Ly, broke with him on the issue and formed an independence party confusingly known as *P.R.A.-Sénégal.*

Perhaps Philibert Tsirinana (the moderate Malagasy leader) best summed up the dilemma most of his colleagues faced when he declared at a press conference in Tananarive on 21 August 1958: 'When I let my heart talk, I am a partisan of total and immediate independence; when I make my reason speak I realise that it is impossible.'[57]

In the two months that elapsed between the Cotonou Congress and the referendum, the issue of independence began to give way to the old tussle between the partisans of Houphouët's view of the relationship between France and Africa and Senghor's view. Both leaders were on the Consultative Committee on the Constitution, but de Gaulle's government favoured Houphouët's view. However, while forcing through the minority thesis, de Gaulle made two important concessions. First he inserted the 'right to independence' of the African members of the French African Community into the constitution. Secondly, to mollify the champions of the Federal Executive, he included a clause permitting member states of the Community to regroup themselves.

These concessions were made after de Gaulle's tour of certain African states (21–26 August) where he began to realise how rapidly the demand for full independence was gaining ground in the influential urban areas. In Brazzaville, confronted by banners demanding immediate independence, he remarked brusquely that anyone who wanted it could take it. Later he was to add the threat that they could take it with 'all its consequences'.[58] On 10 August he had already told the Constitutional Consultative Committee that: 'One cannot conceive of both an independent territory and a France which continues to aid it.'[59]

Whilst de Gaulle's tour brought home to him the growing strength of the demand for independence, the tour in Guinea, at least, exacerbated this demand rather than assuaged it. The Guinean branch of the R.D.A., the Parti Démocratique de Guinée, had adopted a militant political position

57 Cited in René Viard, *La Fin de l'Empire Colonial Français*, Paris, 1963.
58 *Ibid.*, pp. 84–5.
59 Cited in Marcel Merle, 'La Constitution et les Problèmes d'Outre-Mer', *Revue Française de Science Politique (R.F.S.P.)*, March 1959, p. 148.

after its decisive victory in the territorial election of 1957. De Gaulle himself wrote afterwards that he felt a watershed had been reached in his trip to Guinea where he found himself, in his own words, 'enveloped by the organisation of a totalitarian republic. There was nothing that was hostile or offensive towards myself. But from the aerodrome to the centre of the town, the crowd lined up on both sides of the road in well distributed battalions, and cried Independence with a single voice.'[60] De Gaulle left Conakry saying portentously 'Adieu la Guinée!'[61]

Because of misgivings about the consequences of Independence, both the R.D.A. and the P.R.A. gave their territorial sections the freedom to vote as they chose at the referendum which was held on 28 September. At the referendum only two Présidents de Conseil decided to vote 'No' to the Constitution 'with all its consequences': Djibo Bakary of Niger, who announced his decision to opt for independence on 11 September at the U.G.T.A.N. Conference at Bamako, with Sékou Touré of Guinea following suit the next day. For Sékou Touré, the decision was an easier one in that he actually could control the electoral machinery of his Territory: chieftaincy had been abolished the year before[62] and the field administration had been considerably Africanised. In Niger, where Djibo Bakary had only a slight majority over Hamani Diori's R.D.A., the risk was considerable. The French did not, and could not, intervene in the Guinea elections: they did so with a vengeance in Niger where the adept at electoral manipulation, Governor Jean Colombani, was appointed Governor specifically to handle the election, and brought out chiefs and conducted troop exercises to ensure a vote favourable to France. While Sékou Touré was able to deliver a French-style result—95 per cent in favour of 'No'—Djibo Bakary was able to get only 22 per cent in Niger. Elsewhere the results were overwhelmingly in favour of the de Gaulle Constitution. Though many of the leaders would have liked to opt for independence the limitations on their choice were such as outlined earlier and under French pressure, described in detail by Ahmadou Dicko in his journal of the referendum in Upper Volta,[63] they had little option but to say 'Yes'. Furthermore there was the consideration that if some members of the French West African Federation took independence, and others did not,

60 Général de Gaulle, *Mémoirs d'Espoir: Le Renouveau 1958–1962*, Paris, 1970.

61 The official French version of the events following Guinea's independence is that de Gaulle foresaw that the other states would want independence and made provision for it in the Constitution of the Fifth Republic. 'The Constitution of the Fifth Republic established the procedure for accession to independence on the basis of self-determination. It offered the Overseas Territories, by means of a referendum, the free [sic] choice between immediate independence or membership in the institutional Community', *France Aid and Cooperation*, New York: French Embassy Press and Information Service, 1962, p. 13.

62 'Conférence des Commandants de Cercle de la Guinée', S. Touré, *Prélude à l'Indépendance*, Paris, 1959.

63 Ahmadou Dicko, *Journal d'une défaite*.

the prospect of a Federal Executive seemed remote. Some leaders felt that Sékou Touré would have been of more service to the struggle for an authentic independence if he had stayed within the Federation, and then championed a Federal Executive which could have been the basis of a powerful independent West African nation. It has even been argued that if Ouezzin Coulibaly had not died on 7 September, he might have acted as a bridge between Houphouët-Boigny and Sékou Touré, with both of whom he was a moderating influence. He might have persuaded the latter, whose Federalist ideals were well-known, to stay in the Federation and vote 'Yes'. As it was, when they saw the French in Guinea withdraw all administrative services and personnel, destroy all important files, even tear out telephones, these leaders had little cause for regret at their decision to vote 'Yes'. Guinea looked as though she might fall apart, particularly given the long-standing hostility between the Fula and the Malinké tribes. However, Sékou Touré, with his well-regimented party and with a skilful rapprochement of the Fula leaders, was able to confound the predictions of most observers. Guinea was at least fortunate in having resources that could attract other countries than France to aid her. It is doubtful whether countries as dependent as Senegal and Niger on one cash-crop, heavily subsidised by France, or Upper Volta living on French Army pensions, could have survived such a rupture.

From the start the future of the Community appeared unsettled even though independence was not initially a major issue. There was strong support throughout the Community for some form of federation, and on 29 and 30 December there was a meeting at Bamako of federalists from Senegal, Soudan, Upper Volta and Dahomey with observers from Mauritania. After the Bamako meeting authority was obtained from the individual legislatures for participation in a proposed constituent assembly in Dakar on 14 January. A constitution for the proposed Federation of Mali was approved and it was agreed that each of the four countries would submit the constitution to a referendum.

Federation, however, had powerful enemies. Houphouët-Boigny was as opposed to it as ever and France saw the continued Balkanisation of Africa as her only hope of maintaining the Community. Upper Volta was subjected to a combination of economic threats from Houphouët-Boigny, who exploited its dependence on the Ivory Coast, and astute manipulation of regional differences by M. Masson, the new High Commissioner, who rivalled Governor Colombani as a political tactician. Dahomey was coerced by the threat that France might not finance the projected deep-water port at Cotonou and also had genuine fears of cartographic isolation in a federation based on Dakar.[64] Both Upper Volta and Dahomey rejected the

64 See Gil Dugué, *Vers les Etats Unis d'Afrique*, Dakar, 1960, for an account of the complex manoeuvres in Dahomey and Upper Volta.

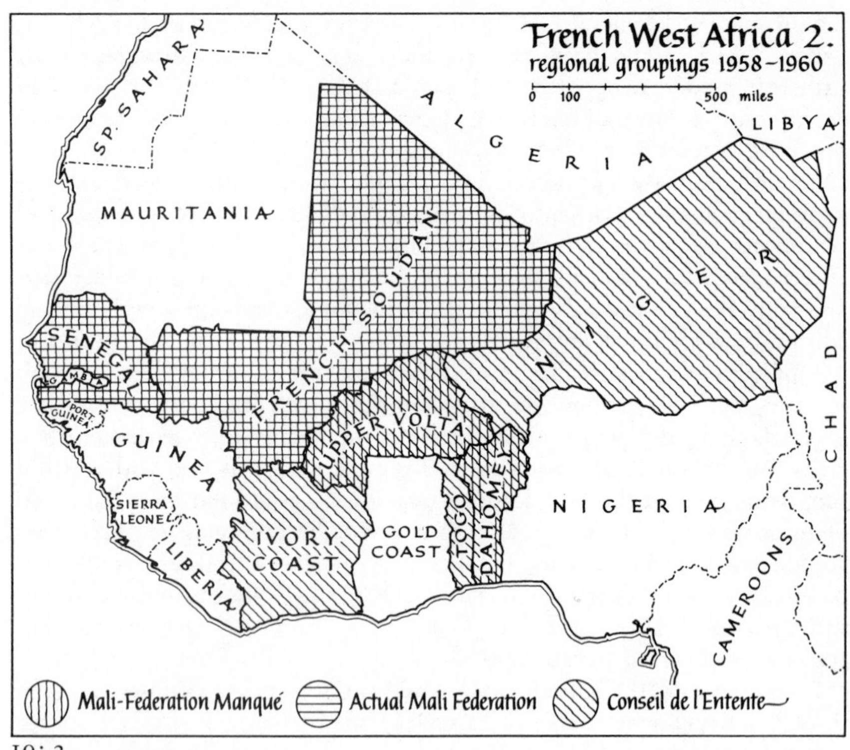

19:2

constitution and eventually the Federation of Mali comprised only Senegal and Soudan.

Houphouët-Boigny had succeeded in staving off the federation but was obliged to form a so-called *Conseil de l'Entente*, a loose economic grouping consisting initially of the Ivory Coast, its hinterland territories of Upper Volta, Niger and Dahomey and later joined by the independent Republic of Togo. The main reason for Houphouët-Boigny's action was the economic weakness of such desperately poor countries as Upper Volta and Dahomey, where Apithy had talked in terms of forming a Union of Benin States with Togo, but contributory factors may well have been the wish to snub the federalists and to maintain his leading role in West African politics.

On 1–3 July 1959 the inaugural meeting at Dakar of the *Parti de la Fédération Africaine*, formed from the governing parties of Mali with federalists from other West African states, made it clear that independence was now the paramount issue in French African politics. It was decided that Mali would seek independence and join with France in a confederation. The constitution of the Community allowed this but France's attitude, faced with Houphouët-Boigny's bitter hostility to the Commonwealth-type arrangements proposed, was uncertain. Houphouët-Boigny had wished to establish a tight community with federal institutions and when interviewed about the All-African People's Conference held in Accra in December 1958 he went so far as to describe it as destined to produce only idle talk and demands for illusory independence.[65]

After an initial refusal to recognise Mali's existence, France at last accepted the inevitability of African independence and at the sixth meeting of the Executive Council of the Community at Saint Louis on 11–12 December 1959 the right of an independent Mali to remain associated with the Community was recognised.[66]

The Mali Federation took its independence on 10 June 1960 after protracted negotiations over France's terms for continued aid. These included the right to retain bases in Dakar and Thiès and to station troops in the Soudan.

The transformation of the Community into a Commonwealth meant that Houphouët-Boigny had to reconsider his own position. On the eve of Mali's independence, Houphouët-Boigny declared that the four members of the Conseil de l'Entente would declare independence outside the Community and that any accord with France would be negotiated after international sovereignty had been assumed. Thus it was that the political leaders of Ivory Coast, Niger, Dahomey, Senegal and Upper Volta were able to attend Nigeria's independence celebrations as independent heads of state. The question was: how real was that independence?

CONCLUSION

To understand the apparent reluctance with which French-speaking African leaders pursued the goal of independence, and the dramatic contrast between their attitudes on the question and those of their English-speaking counterparts, one has to remember the poverty of French West Africa in resources and educated personnel. Heavily dependent on French aid and with the administrative services still largely manned by France, independence without France's goodwill seemed a bleak prospect, despite all its emotional attractions. Furthermore, the goal of independence was sidetracked by the quarrel over the Federal Executive. To many African leaders it was more important to maintain the federal structure than to rush into independence, since in the long run independence could only be meaningful at the level of a large grouping. The fear that a series of small states, without any obvious economic attraction for outside capital investors, would be tied to the apron-strings of France or else, in breaking with her, face a bleak economic future seems in the event to have been justified. None of the successor states of the former French West African Federation has had an easy passage, with the exception of the Ivory Coast,

65 John Marcum, 'French-speaking Africa at Accra', *Africa Report*, March 1961.
66 Jean Debay, *Evolutions en Afrique Noire*, Paris, 1962, has suggested that France agreed to the Federation of Mali's independence within the framework of the Community and did not cut her off for fear that she might join Guinea in a bloc hostile to France.

where rich resources have attracted a great deal of French investment and have produced an apparent boom situation. But here, as Samir Amin has shown, the boom has been largely to the advantage of the French investors, and while superficially there has been economic progress there has not been real development as far as the 'masses' who once rallied behind Houphouët-Boigny's P.D.C.I. are concerned.[67] Guinea, the only other territory with rich resources, has suffered an extended period of dislocation as a result of taking her independence and has had a great deal of difficulty in extricating herself from this situation. But here, at least, Guinea had something to attract alternative investors. Territories like Dahomey with a population of two million on independence and scant resources, Niger (population 2,666,000 in 1958) land-locked and largely desert, Upper Volta (population 4,090,000 in 1958) dependent for its income on remittances from migrant labour and French army pensions, could not afford to take independence without France's consent. The Republic of Mali, which tried under Modibo Keita to pursue policies which did not concur with those of France, is still not out of the economic wood. Thus it is that France is still able to exercise a control over so many of her former French Black African territories.

There is another, important, factor in the close relationship which the African leaders who acceded to independence had with France. Educated at 'assimilationist' schools, employed by and large in the French administration, finding their parliamentary feet in the National Assembly of France, devotees of French culture, there was an emotional bond between them and their former French masters which has little parallel in English-speaking West Africa. In 1957 Senghor, for instance, was quoted as telling the National Assembly:

> In Africa, when children have grown up, they leave the parents' hut, and build a hut of their own by its side. Believe me, we don't want to leave the French compound. We have grown up in it and it is good to be alive in it. We want simply to build our own huts.[68]

Furthermore, General de Gaulle, with all his international stature, was assiduous in his reception of often transient African Heads of States in Paris, enlisting them in his search for the construction of a *francophonie* that would be a force to reckon with in a 'free' world that was becoming increasingly Anglophone. But as a new generation of African leaders, many of them

67 S. Amin, *Le développement du capitalisme en Côte d'Ivoire*, Paris 1967.
68 *West Africa*, 7 September 1957, in a review of *Une Nouvelle Afrique* by Paul-Henri Siriex, Paris, 1957.

educated in the period of independence emerges, the emotional bonds to France recede. The problems of economic and administrative dependency of small states on *la Métropole* continue.[69]

69 For a detailed study of the extent of French aid to Africa see Theresa Hayter, 'French Aid to Africa—its Scope and Achievements', *International Affairs*, xli, 2, April 1965, pp. 236–51. L'AOF Haut Commissariat de la République en Afrique Occidentale Française. AOF, 1957 (Dakar, 1957); and Assemblée Nationale, Constitution du 4 octobre 1958, Première Législature, Première Session Ordinaire de 1961–62, Rapport, Annexe No. 10, Coopération, No. 1445; and Avis, Tome II, Coopération, No. 1459; Sénat, Première Session Ordinaire de 1961–62, Rapport Général, Tome III, Examen des Crédits et des Dispositions Spéciales, Annexe No. 8, Coopération. Most aid nowadays is channelled through the Ministry of Cooperation, but individual ministries can in fact make provision for aid to the independent French-speaking African states.

Index

Index

Index

- British-backed revolt of tributary states, 187–8
- British colonial administration, 450–2, 568, 571
- campaign, 493
- centralisation policy, 172
- 'Certificate of Apology', 232–4

Index